Project Management

Bernd-J. Madauss

Project Management

A comprehensive description of Theory and Practice

Springer

Bernd-J. Madauss
Bad Aibling, Germany

ISBN 978-3-662-69056-7 ISBN 978-3-662-69057-4 (eBook)
https://doi.org/10.1007/978-3-662-69057-4

This book is a translation of the original German edition "Projektmanagement," 8th edition, by Bernd-J. Madauss, published by Springer-Verlag GmbH, DE in 2020. The translation was done with the help of an artificial intelligence machine translation tool. A subsequent human revision was done primarily in terms of content, so that the book will read stylistically differently from a conventional translation. Springer Nature works continuously to further the development of tools for the production of books and on the related technologies to support the authors.

This Springer imprint is published by the registered company Springer-Verlag GmbH, DE, part of Springer Nature.
The registered company address is: Heidelberger Platz 3, 14197 Berlin, Germany

Prologue

This work is the English translation of the eighth edition of my book on "Project Management," which was published in 2020 by Springer Vieweg, Berlin in German language. I have reviewed the English translation for completeness and, where necessary, made relevant changes and additions. I would like to thank my friend and former colleague of MBB[1]-Space Munich, Ronald K. Larson (who later retired from NASA), for reviewing the English text and many figures and tables.

The business success of public and private organizations engaged in projects depends largely on the success of their projects. This results in the imperative need to introduce professional project management, which is the motto of the eighth edition of my book because experience shows that poor PM can lead to high risks. Mark A. Langley, former PMI president and CEO, confirms this: "*If an organization does not have a good PM, the organization is taking a very high risk in implementing its project goal.*"

Genesis of the book: In 1978, I was commissioned by the management of MBB Space to create internal training material on the topic of "Planning and Monitoring of R&D Projects" and subsequently conducted in-house training sessions. The training material (MBB UR-396/78), initially available to participants only, was later published as a seminar handbook due to widespread interest. The seminar handbook constituted a core component for my book on project management, which was first published in 1984 by C. E. Poeschel Verlag, Stuttgart. I continuously expanded the book by adding PM methods that had been tested in practice. Thus, in a sense, the individual chapters of the book reflect my professional life.

In 2017, the seventh edition was published by Springer Vieweg, Berlin, followed by the eighth edition in 2020. From the start with the first edition, it had been my goal to describe the entire spectrum of project management tasks as comprehensively as possible, to impart the full scope of all activities involved and their interrelations to each other. Thus, over the past 40 years, up to the eighth edition, there have been many additions, and the book has taken on the character of an encyclopedia.

As a target group, this book primarily addresses project managers and their teams in public authorities and companies worldwide. The project management tasks described here can be easily applied to small, medium, and large-scale

[1] Messerschmitt Bölkow Blohm GmbH, Munich.

projects of different industries and organizations, such as national and international government agencies, aerospace, automotive manufacturing, construction, municipal, chemical, and medical companies as well as social enterprises.

Bad Aibling, Germany Bernd-J. Madauss
November 2024

Foreword by Prof. Dr. Heinz Schelle, Honorary Chairman of the German Association for Project Management (GPM)

When the first edition of this book was published in 1984, I had the privilege of reviewing the work in the newsletter (forerunner of Projektmanagement aktuell) of the German Association for Project Management (*GPM*). My joy grew from page to page as I read the book. Since the IPMA[2] World Congress in Garmisch-Partenkirchen, Germany in 1979, it was realized that project management (*PM*) is more than a network planning technology but a management concept for first-time and one-time projects, which has already become more and more widespread. However, this important insight, which is taken for granted today, had hardly been reflected in the German-language teaching and handbook literature until then. Renowned scientific institutions more or less maintained this narrow view for long time. My colleagues in GPM and I became acquainted with a new world and view of project management, although we had already given up our fixation on the network planning technique.

There was talk, for example, of configuration management, a sub-discipline that was completely unknown to most project-oriented companies in Germany, of comprehensive quality management and of process models. We also got an idea of the US management literature, which was not accessible to most of us. As the high number of editions of this book shows, the work quickly became a bestseller and the bible of project management. Very soon, people spoke of "the Madauss" as they did of the Duden or the Brockhous. With the pioneering deed described above, Madauss became one of the fathers and pioneers of project management in Germany. Since he was also a founding member of GPM, our society has of course also benefited from him in an invaluable way. Madauss' publishing activities were not limited to the textbook. Many other publications were added to it. I was particularly impressed, to give just one example, by the standard work breakdown structure for ESA satellites, a very early fruitful approach which serves as a model for other industries.

What is the reason for the success of Madauss' work? The answer is easy: The author knows the problems of project management, especially for commercial spaceflight, not only from the literature, but also from an extremely extensive practical and responsible international management career. His curriculum vitae reads

[2] The original name INTERNET was renamed to International Project Management Association (IPMA).

like a Who's Who of European aerospace. Respected names such as MBB, ELDO, and SES Astra are included.

There is a somewhat cynical phrase in management circles: "If you can't do it, teach it. If you can't teach it, consult it." Madauss has shown time and again "that he can do it." It is all the more commendable that he has nevertheless also devoted himself to teaching and consulting for national and international institutions. For example, he passed on his knowledge and experience in 10 years of teaching at the faculties of computer science and aerospace engineering at the University of the Federal Armed Forces in Munich. Since he substituted for me in my research freedom, I know from my own experience how well his lectures were received by the students.

A detailed description of Madauss' activities would fill many pages. In summary, however, one could describe his long professional career by the following keywords: Space, systems engineering, planning, costs, organization, management, and business writing. In the course of his life, he has contributed a great deal to the advancement of systematic project management in Germany and around the world. For this he deserves the great thanks of the project management community.

Looking forward to the English edition which is based on the eighth German edition.

Garmisch-Partenkirchen, Germany
December 2016

Heinz Schelle

Foreword by Dr. Ludwig Bölkow, former Managing Director of MBB, Munich

The extraordinary technological advances of the past 100 years are largely related to the intensive deepening of knowledge by specialized scientists and expert engineers. Research and development applied in a planned manner have led to ever better methods in the use of the laws of nature in the service of mankind. The progress made has, in some cases, brought a paradisiacal standard of living to those who find themselves in the spaces of modern technologies, especially those of industrial societies.

In the last few years now, almost one would like to say before a climax of the growth of all these achievements, we have to increasingly recognize that the progress based on well-founded expertise confronts us with a magnitude that was not foreseen clearly enough so far, probably also not wanted to be perceived at the beginning, namely with the phenomenon of the limits to growth. Limits of feasibility are apparent in all areas. Raw materials are gradually becoming scarce. Habitats are threatened by destruction, research results reach the limits of ethical values and many a result achieved by technicians is no longer regarded in its entirety as a beneficial invention for mankind, even though it leads to progress in some areas. No one will deny that modern computer systems or the rapidly spreading robots are useful instruments for mankind. On the other hand, and this is also a fact, these developments give rise to considerable sociopolitical problems for our society.

For a long time, I have been of the opinion that we are at a turning point in the direction of thinking and that we must add a new dimension to specialism. A dimension through which a balance and more harmony between the various specialties can be achieved for the good of this earth. This can be expressed very understandably with the demand, repeated by me to my co-workers for over 25 years, to think more in terms of total systems, especially when searching for tasks. The cooperation of researchers from different disciplines, as demanded by the cyberneticist Nobert Wiener, is an urgent requirement of our time.

However, we find ourselves in a dilemma because our universities and companies still largely exclude the work of interdisciplinary teams. Students at the university as well as practitioners in the company are organized according to specialist disciplines and thus far too strictly separated from one another. Engineers, lawyers, and business economists are often hardly able to communicate with each other in a generally understandable way after receiving their diplomas; also, too close contacts are largely avoided in the future workplace. Even specialists in one field, for example engineers among themselves, often have communication difficulties.

On the other hand, the present time emphatically demands that we take to heart the calculation of reason demanded by Leibniz in all future developments, and that a certain priority therefore be given to interdisciplinary cooperation. After all, it has become clear to all of us, especially in recent times, how technologically advanced industrial products often meet with more rejection than approval. The development of modern nuclear power plants is a vivid example of this. The design of new industrial products therefore calls for better community solutions which, in addition to technological objectives, must also take into account general social objectives such as maximum possible safety, environmental protection, economy in the use of raw materials and public funds, and compliance with ethical values. In order to achieve this in the best possible way in the sense of a common goal, the following demand by Wiener with regard to interdisciplinary cooperation is especially indicative for our times: "*Everyone must be used to working together, be familiar with each other's habits of mind, and recognize the significance of a colleague's new proposal before it is fully formulated.*"

Interdisciplinary collaboration is a fundamental tenet of systems engineering that the aerospace industry embraced very early on. To a much more extreme degree than in other industries, aerospace products were dependent on systems engineering analyses because, for everyone to see, feasibility limits were being reached. Everything that worked on the ground without too many problems had to work reliably without maintenance in space, for example, under extreme weight and volume constraints. Interdisciplinary collaboration is most clearly demonstrated in the example of space travel. Building a manned spacecraft is tantamount to creating a miniature planet where living conditions must be similar to those on Earth. The scientists, engineers, biologists, physicians, astrophysicists, etc. involved in the development of such a system must join together in the interest of the common goal to form a well-functioning interdisciplinary team whose main task is to optimize the system. However, it was soon realized that the development of such complex systems required an expansion of the range of tasks. Experts for functional safety, scientists, lawyers, and controllers are now also part of such teams. The management of interdisciplinary working or project groups is increasingly being assigned to project engineers who, with the support of the team at their disposal, are responsible for system optimization and project implementation. It is my firm conviction that aerospace knowledge and know-how in the fields of modern project management, systems engineering, and the successful leadership of interdisciplinary working groups represent valuable potential for our country.

I have repeatedly and successfully challenged my staff to pass on their acquired knowledge at specialist conferences, seminars and at universities, and I very much hope that the interdisciplinary way of thinking will continue to prevail for the benefit of our people. It is a special pleasure for me that one of my former co-workers, Bernd Madauss, who among other things worked on the German-American solar probe project HELIOS, is now passing on his experience in the field of project management to the public. Particularly noteworthy is Mr. Madauss' organizational equalization of the individual project management disciplines, such as systems engineering, quality assurance, and project control, which makes it possible to achieve optimal project goals in the first place. In my opinion, the systematic treatment of the individual project phases and the associated consideration of life cycle costs are also of particular importance. The book created by Madauss treats the subject of project management as an overall theory and aims at the solution of tasks by interdisciplinary project groups. This publication is a valuable contribution that project management professionals, as well as newcomers to the various branches of study interested in the methods and procedures of project management, should definitely read.

I wish the author much success with this book and hope that it will contribute to the enrichment in science and practice in this new field.

Munich, Germany
August 1983

Ludwig Bölkow

Foreword by Romain Bausch, Chairman of the SES Board of Directors, Luxembourg

Satellites are used worldwide to transmit news and broadcast television signals and are now taken for granted. With the surprise launch of Sputnik in 1957, the satellite age began, and Arthur C. Clarke's vision (*use of communications satellites*) became a reality. In this context, the space visionaries Konstantin Tsiolkovsky and Hermann Oberth should also be mentioned. Tsiolkovsky formulated the basic mathematical principles of rocket propulsion in 1898, and Oberth described the basic equations of rocket technology as we know and use them today in his book "*The Rocket to Planetary Spaces*" in 1923.

As early as the 1970s, the Grand Duchy of Luxembourg was investigating ways of using the still young satellite technology to transmit radio and television programs throughout Europe. After examining various concepts, eleven shareholders under Luxembourg's leadership founded the satellite operating company Société Européenne des Satellites (*SES*) on March 1, 1985. The contract to build the first satellite, designated ASTRA 1A, was awarded to the US company RCA Astro Electronics. The RCA4000 satellite was launched into geostationary orbit at an altitude of 36,000 km on an Ariane IV launcher. With the launch of ASTRA 1A in 1988, SES became the first European satellite company to start broadcasting television programs. The success achieved quickly led to a demand for further satellite capacity. 2023 SES has a total of 70 satellites in two different orbits.

Already at the beginning of the first procurement activities of satellites, launchers, and ground facilities for satellite control, SES decided to assign the complex management tasks to an experienced project manager, thereby ensuring that all necessary project tasks would be carried out on the basis of proven management methods. In October 1985, SES was able to recruit space engineer Bernd Madauss, an experienced project manager with a proven track record of several years of practical industrial space experience, to take over the project control management of ASTRA 1A. He had previously been employed by a large German aerospace company, where he headed the main department of "Central Project Monitoring." In 1990, SES put him in charge of the Business Management Office (*BMO*), and in 1999 he was appointed Senior Program Manager of the Astra satellite fleet, a position which he held until his retirement in 2003.

During my time as President and CEO of SES—from 1995 to 2014—I have become familiar with the importance of project management over the past years and can confirm that SES' success is due, among other things, to the professionalism of its project work. This includes thorough project and system planning, as well as the preparation of technically and programmatically coordinated tender documentation (*functional specifications, specifications, plans, and model contract*) to provide the necessary prerequisite for vendors to prepare high-quality bids. In compliance with clear criteria and fair competitive conditions, bid evaluation and contract negotiation then take place and, after the contract has been awarded, project monitoring is carried out using established management standards.

I think it is worth mentioning that the management concepts for the procurement and handling of ASTRA projects implemented in the early nineties under the direction of BMO, which were partly prepared with reference to earlier editions of this book, are in principle still valid today. In individual cases, however, the procedures have been adapted to more recent requirements. The first edition of this book, which already covered all essential management functions for professional project management, was published in 1984, and the publication of the seventh edition is planned for spring 2017.

I am particularly pleased to say that since 1985, first as a consultant and then as a long-time SES employee, Bernd has passed on his extensive knowledge of project management, as described in detail in this book, to his colleagues. He has worked on many SES projects and, through his constructive contributions, has helpfully supported the company as well as his colleagues in overcoming manifold challenges. He also conducted in-house seminars on project management with success and great response.

I wish the author much success with the seventh edition of this book and am convinced that the work will contribute to the enrichment in science and practice in the field of project management.

Luxembourg, Luxembourg
March 2016

Romain Bausch

Foreword by Prof. Walter Peeters, President Emeritus and Professor in Space Business and Management of the International Space University (ISU), Strasbourg

It is very difficult to find a good book on project management for hi-tech projects. I believe there are two reasons for this: first, the subject of project management has only relatively recently become an essential part of engineering education, since traditionally engineers acquired their management experience by trial and error in practice. Then, in the seventies, management methods based on scientific principles ("Scientific Management") were increasingly used for project management at institutes, authorities, and companies. This process was accompanied by the creation of corresponding manuals and training courses based on them. For the aforementioned manuals, authors were needed who, on the one hand, had good scientific management skills and, on the other hand, could also look back on corresponding practical experience in project management. However, authors who could combine and harmonize both aspects were rather rare. A second reason was that project management methods had originally been applied in traditional sectors, for example in the design and construction of construction projects, and now had to be transferred to the so-called hi-tech projects.

However, hi-tech projects are generally subject to more difficult conditions, which are reflected in their technical execution (high complexity), schedules, and costing. Conditions are particularly extreme in the aerospace sector, due to the high project requirements, e.g. very high reliability. Communications satellites, for example, must be fully functional for at least 15 years under extreme environmental conditions without requiring maintenance. They therefore have a very high price, since they are often one-offs without a subsequent large series. Therefore, the result is often simulated with prototypes (so-called models) that are as true to life as possible. After the launch of a satellite, repairs are usually no longer possible in space. This means that there is no chance to try out the so-called beta variants in a realistic environment. In the space business, zero-defect tolerance is therefore a prerequisite, which means that project management requirements are correspondingly strict. All development and production processes are subject to thorough control in order to achieve the specified goal (technology, quality, deadlines, and costs). However, simulations can only partially determine the environmental conditions. For smaller components, microgravity can be simulated, for example in the Bremen drop tower or by parabolic flights, but long-term microgravity cannot be simulated on earth.

The previously mentioned prerequisites make it very difficult to find an author who has experience in project management of the various application areas and at the same time has a sound knowledge of the corresponding scientific management methods. The author of this book meets both conditions, both from the perspective of the client and from the perspective of the contractors. This unique combination makes his book particularly valuable and unique.

This book is an excellent handbook for engineers and managers involved in space projects. I have therefore highly recommended it to my students at the International Space University in Strasbourg. But I would like to go a step further and advocate that all those working in the hi-tech sector also use the book as a reference and apply the space examples it contains to their fields. The fact that it is now available in English will facilitate reaching out to a broader audience.

Many companies would like to play a role in the space sector, but from experience I know that the strict but necessary project management requirements are sometimes a deterrent. However, this book can be very helpful in overcoming that initial shyness. For other industries, such as vehicle manufacturing, it is advisable to adopt some or all of the lessons learned in the aerospace sector. As mentioned before, aerospace has been a pioneer in project management in the past. Very early on, theoretically established principles, such as systematic project structuring (e.g., work breakdown structure—WBS), proved their worth in the management of space projects, but also in the handling of non-space projects, such as in shipbuilding and the automotive industry. Therefore, this book is also highly valuable in the non-space industry.

The author of this book received a visiting professorship from the International Space University (ISU) Strasbourg in 2005 and is an ISU faculty member. This excellent book is a "must read" for all engineers and economists involved in hi-tech projects. I wish the author that this work will find a wide audience.

Strasbourg, France W. Peeters
July 2016

Preface of the Eighth Edition

The continuously advancing development and use of new products and services worldwide in areas such as *energy, nutrition, communications, biology, medicine, and transportation* are due to the innovative potential of scientists, engineers, physicians, and nutritionists. The focus is on meeting people's basic needs. Although novelties are usually gladly accepted (e.g., *fast cars, low-priced food, cell phones,* etc.), in the recent past skepticism is also increasingly noticeable and one realizes: "*not everything that is feasible is also useful for mankind.*" Innovations and their effects must be seen in context (i.e., *holistically*), and above all, their "sustainability" must be examined. Climate and environmental protection, for example, are outstanding aspects to consider.

Ludwig Bölkow, the former MBB CEO, wrote 1983 in the foreword to the first issue of this book: "*to think more in terms of overall systems.*" In this context, he pointed to the cyberneticist Nobert Wiener,[3] who called for collaboration between researchers from different disciplines as an urgent imperative. Interdisciplinary collaboration is an important principle of project management and systems engineering in the development of cutting-edge technologies. Therefore, important aspects such as environmental conditions feasibility, compatibility interaction with other systems, and the economics of the project must be examined in the early stages of a project to prevent problems from arising in subsequent phases. This is an absolute necessity for space projects such as satellites, which often have to function in orbit for up to 20 years without maintenance. Close cooperation between project management and systems engineering has proven its worth in practice, see Fig. 7.21. Both functions therefore have a permanent place in the organizational structures of space companies. The principle of interdisciplinary cooperation can be transferred to other industries by adapting it accordingly.

It was my intention from the outset to cover all aspects of project management comprehensively and in a practical manner in this book. For this reason, I have retained the structure and contents of the seventh edition, insofar as they are still valid, in the eighth edition and supplemented them with two new sections. In view of the book's subtitle "*Theory and Practice from a Single Source*," I have particularly worked to describe the theory, including its historical development and its implementation in practice by means of examples. The historical review of project

[3] Founder of *cybernetics (1884–1964).*

management is as old as the projects or project-like construction projects carried out in the past, such as the construction of the pyramids, the Great Wall of China and a variety of large architectural projects. In the past, too, the execution of construction projects, some of whose dimensions were gigantic, had to be managed and supervised but relatively little is known about their methods.

In modern times, the management methods we know today came about within the framework of aerospace research and development projects. In the sixties of the last century, the CPM and PERT network methods and management systems of the USAF and NASA were developed, which are mentioned in detail several times in the book. Many of today's common project management methods are based on these. It is important for their application that they are tested in practice, which is illustrated by means of practical examples. For the examples described in this book, I was able to draw on a large number of publications and field reports as well as my own management experience.

Over the past five decades, many management standards have been developed, which I have recorded in this book and clearly summarized for the reader in Sect. 27.2. The technological and economic success as well as the international competitiveness of a country depend to a large extent on the success of research and development of cutting-edge technologies and their utilization, and project management plays a key role in this context.

Bad Aibling, Germany Bernd-J. Madauss
June 2020

Preface of the First Edition

Whether it is the construction of a new hospital, a power plant, or the development of an airplane or a new transportation system, the projects, due to their particular tasks, represent complex systems in themselves and, since they are usually open systems, have a direct relationship (mutual influence) with their environment, that is, with other systems.

But it is only in the recent past that systems considerations have acquired a particularly great significance for society. Much more clearly than ever before, everyone is now aware of the extent to which the individual spheres of life are interrelated; they are, as it were, nested together to form a global system, and there is a tangible impression that the world has become smaller and narrower. The practical effects of this decision are well known to everyone when it comes to building a new road, a factory, or a municipal facility. New projects can now only be realized without difficulty if they can be integrated as smoothly as possible into the overall system surrounding them.

The term "*quality* of life," which corresponded to the "zeitgeist" of the era, referred to the achievement of optimal living conditions. The word standard of living, which had been used for a long time, had to give way to the new vocabulary because it was afflicted with considerable shortcomings. The term "standard of living" was aimed at the unconditional improvement of living conditions, but without taking into account the interactions with the environment. But everything has a price, and so does the constant improvement of living standards. Unpleasant "feedback" phenomena, for example, the problem of water pollution, make themselves clearly felt. It is therefore necessary to work toward achieving a better quality of life, if it is not to appear merely as an empty phrase or political slogan, by means of systems engineering methods and procedures in which control engineering processes *(closed loop processes)* are in the foreground.

Above all, the size and complexity of today's projects require early and thorough planning more than ever before. All-important influencing factors must be included in the planning process, and dynamic, i.e., control-oriented, project management must be introduced in order to better align the benefits of new development and construction projects with the needs of today's society than has been the case in the past. Today, more than ever, the demand for systems engineering solutions, which aims at interdisciplinary approaches to solutions during the planning stage, i.e., long before the production or construction of technical systems (large buildings, aircraft, etc.), must be met in order to prevent new systems from causing more harm than good. In this context, the time-oriented approach is also of utmost importance, since the benefit of a new system is often quite different in a long-term view than in a present-day view.

In the future, the realization of large and complex projects must therefore rely more than ever on methods and procedures of systems and project management, a demand to which the aircraft designer and co-founder of Germany's largest aerospace company, Messerschmitt-Bölkow-Blohm GmbH, Dr. Ludwig Bölkow, has repeatedly referred. Bölkow expressed this with the demand to *think as* a *whole* and decisively shaped the company's image as a leading systems company.

The intent of this book is to introduce the reader to the methods and procedures of project management of complex systems as well as interdisciplinary thinking in the project in all its breadth and diversity. This inevitably led to the fact that the topics could sometimes not be treated in the detail and depth required by the respective specialist for his subarea. However, this had to be consciously accepted in the interest of the intended objectives of this book, namely to present the field of project management as an overall doctrine. In this context, reference is made to the extensive bibliography of the individual chapters, which provides the interested reader with references to more in-depth literature.

In preparing the book material, the focus was on the applicability of the methods and procedures described in practice. At the same time, however, the theoretical background is also discussed in detail and the developmental history of project management is described. The broad scope of project management forces an equal treatment of all project-relevant topics from business administration and technology, which is another main concern of this book. The book is mainly aimed at two main groups, project staff in industrial and governmental practice who want to further complete their knowledge, and pupils and students who want to prepare themselves for management-oriented careers.

I am particularly indebted to my work colleague from Messerschmitt-Bölkow-Blohm GmbH, Professor Victor Billig, for the many valuable suggestions from theory and practice and the detailed review of the manuscript for this book. I would also like to thank Dr. Norbert Thom of the University of Cologne for his critical review of the individual chapters. For the stylistic revision of the book, I sincerely thank Birgit Neubacher, and for the typing of the extensive manuscript, Elisabeth Meinhardt deserves many thanks.

Bad Aibling, Germany Bernd-J. Madauss
August 1983

Introduction to the Book

The economic success of a company dealing with project tasks depends on the success or failure of the project work. This results in the imperative need to introduce professional and efficient project management. In principle, this is a worthwhile investment, because good project management (*PM*) can increase efficiency and reduce costs. It often entails organizational changes, because you have to give PM a place in the company, and more importantly, you have to give project managers authority commensurate with their responsibility.

The handling of projects in the industrial and governmental sector is an increasingly difficult task for those responsible due to the continuously increasing demands of technology, complexity as well as management under tougher international competitive conditions.

To achieve technologically and economically optimal project results in strict compliance with statutory environmental regulations, which in the industrial sector are subject to the laws of competition and in the regulatory sector are to be strived for in the interest of society, reinforce the call for efficient and system-oriented project management. The continuous improvement and widespread introduction of professional project management is of utmost importance for Germany as a business location and thus for the economy of our country.

Project management and systems engineering are closely linked, because systems-oriented approaches, which are discussed in detail in Chap. 7, lead to optimal solutions by taking into account all important influencing variables, for example technology, time schedules, costs as well as sociological factors. In close connection with systems engineering, project management has the potential to create optimal holistic solutions, which may also lead to the *sustainability* formulated by the United Nations in 1992.

The methods and procedures of project management are excellently suited to efficiently handle the processing of complex and/or large project plans of industry and public authorities with a clear project character. The strength of project management lies in the distinctive product- and goal-oriented way of working, which leads to a system-oriented way of thinking and to interdisciplinary cooperation.

In order to handle projects successfully, it is necessary for the upper management levels of authorities and companies to recognize project management (*PM*) as an important management task and therefore give PM a corresponding status. Project managers are the extended arm of upper management. They must therefore be given the appropriate authority. Above all, however, project management is also a creative task and not, as is often assumed, a purely administrative activity. The following statement by J.-J. Servan Schreiber (*The American Challenge, 1968*) is particularly true for project management: "Management is, after all, the most creative of all arts. It is the art of the arts; for it is the art of using talent properly."

The selection of chapters in this book reflects the versatility of the project management topic. The management of a project can best be compared with the management of a business, because the project manager assumes the responsibility of the complete spectrum of tasks of a project and should therefore also have the corresponding powers. The functions of planning, monitoring, and cost estimation are just as much a part of this as the tasks of systems engineering, quality assurance, contract management, and tender preparation. Of course, the head of a project, just like the head of a company, cannot in many cases perform these tasks alone and without appropriate professional support. For this reason, in the case of large projects, he must include appropriate specialists in his team, for example, systems engineers, planning experts, professional cost estimators, and experienced quality specialists. However, the overall responsibility lies with the project manager. In the case of small projects, however, the project manager is usually forced to carry out many tasks himself in personnel union *(see Chap. 5)*, but must rely on corresponding specialist departments for specialist activities, such as cost monitoring and quality assurance.

This book covers all aspects of project management as they may occur in practice. Some tasks, such as schedule and cost monitoring, are standard requirements that apply in principle to every project, while the topic of globalization, for example, does not apply in all cases. At several places of the book reference is made to project size which is defined in Tab. 1.1. A brief description of the contents of Chaps. 1–26 is given hereafter.

Chapter 1: This chapter describes the importance of project management for industry and public authorities as well as the history of the development of project management. It also discusses the importance of the systems engineering approach to project management and the decentralization processes required in large companies, which lead to *lean* management and can be accelerated by a project management approach. Project management as an intermediary between the company management and the development department is another important topic of this chapter. Reference is also made to the importance of aerospace as a pioneer of new management concepts and the need for project management among international authorities and companies. Project management should not be complicated, but as simple as possible, because especially small companies often shy away from the methods and procedures of project management, which seem difficult, because in the past they were mostly used only in connection with large projects; see subsection 1.9.2.

Chapter 2: Sometimes it is not clear whether a task is a project or a normal task without project character. However, a clear distinction is necessary for the use of appropriate management methods. In Sect. 2.1 character-features are compared to establish a clear feature for definition. The section also describes the terms program and project and how they are used in practice.

Chapter 3: The problem of management crises in the handling of projects is described in this chapter. The first section is called "Learning from mistakes," which is intended to express the fact that it is not about a sensational denunciation of mistakes made, but about the insight resulting from failure to rethink. In the past, there has been a change in the way tasks are performed in many areas of the economy, and this is particularly true for the handling of complex projects. The view into the abyss (*project crises*) should sharpen the senses for new and better ways, because it is necessary to prevent industrial and official decision makers renouncing the realization of necessary large-scale projects in the future due to failure experiences. Project crises express themselves, for example, through technical breakdowns, lack of adherence to deadlines and costs, high follow-up costs and organizational difficulties within the project or with other partners, and even international project plans are not spared from these.

Interestingly, the same mistakes are repeated again and again. It is easy to see that the mistakes made several decades ago are repeated today as a matter of principle, namely due to inadequate planning, organizational chaos, responsibility disasters, and so on. Mistakes made and their effects are not systematically processed and taken into account as *lessons learned* for new projects. The reasons for this are that in most cases there are no corresponding information databases and the know-how is therefore at best kept in the heads of the responsible persons.

Chapter 4: At the beginning of the concept phase, the main technical and commercial parameters of the project must be determined while maintaining the target costs. This is a difficult task for which the use of very experienced employees is recommended. It is also valuable if existing empirical values (key figures) from projects that have already been completed can be used. The use of cost estimation models also makes sense; see Sect. 24.5. Looking at Fig. 4.3, it is clear that the concept work has a large influence on the life cycle costs (LCC), while the costs for phase A (concept) are relatively low. Fig. 4.2 provides an overview of phases A to F including major milestones.

Planning a project is like projecting into the future, because at the beginning there is only the idea to realize a new project: hardware, software, service, etc. Everything is new and often there is little or no direct comparison to previous projects. In principle, a project plan is always an innovative task where success depends on many unknowns. This applies both to the actual task, e.g., the development of a new car model and to the programmatic aspects. In this chapter, the project flow from A to Z (*from cradle to grave*), which is to be understood as an iterative and a learning-oriented process, is discussed in detail. Chapter 4 discusses the different project phases and their varying importance to the project lifecycle, as well as the key project milestones and decision points, incorporating real-world examples from different industries.

The early phases of the *project life cycle* are of particular importance for the project success, because major decisions have to be made in them, which have a large influence on the further course of the project. In the early phases, feasibility studies and optimizations must be carried out in order to establish a basis (*baseline*) for the costly subsequent phases (development, production, and operation). At the end of this section, the individual phase flowcharts (*roadmap concept*) are described in detail.

Chapter 5: In this section, the principles of project organization and project management philosophy are dealt with in detail, because project organization is a central topic for companies and authorities. Often, however, project activities are hampered by the fact that the selected organizational concept has been poorly developed. Another problem is often that no balance has been created between the task, responsibility, and authority; i.e., the project manager bears full responsibility for the project, but lacks a clear definition of the task and often does not have the necessary authority. Subsequently, the topics of project management office (*PMO*), leadership tasks of multi-project management, concepts of matrix organization, organizational measures of industrial and governmental cooperation, and the importance of key personnel are discussed in detail. The subsequent chapters of this book are closely interwoven with Chap. 5, as reference is made repeatedly to the organizational principles described.

Chapter 6: Management tasks also need to be carefully planned; one should not leave this task to chance. This section describes in detail the steps and contents for creating project plans. In order to carry out projects efficiently, it is necessary that all processes required for project execution are defined in advance, clearly described and documented in appropriate plans. Here are some examples of typical project plans: the project management plan (*PMP*), development plans, cost plans, and test plans. Checklists for the most important project plans are given at the end of this chapter. For completeness, reference is made to the management concept of a plant project described in Chap. 23. The last subsection shows the outline and checklist of a management plan.

Chapter 7: The importance of systems *engineering in* the context of project management runs like a thread through this book. Chapter 7 describes typical systems engineering processes in the project in detail. Systems engineering is not only a general way of thinking, but also a task that is firmly anchored in project management. The systems engineering project tasks consist, for example, of performing analyses and calculations and creating system architectures and specifications. This chapter also covers topics such as the systems engineering control loop, system reviews, interface management, and monitoring the technical performance of the system (*task performance measurement*).

Chapter 8: Quality influences the economic efficiency of the system: this could be described as the slogan for the content of Chap. 8. After all, it is not just the mere function that is important for the operator of a system, but also the question of reliability and, derived from this, availability, MTBF (*mean time between failures*) and the safety of the system. A description of product liability and its significance for project management is another component of this section. Separate sections are devoted to documentation and configuration management (*see Chap. 12*). This chapter deals also with the importance of change control from the perspective of quality assurance (*change control board—CCB*).

Chapter 9: This chapter summarizes the methods and procedures of project planning and monitoring. A historical review of developments in the USA and Western Europe in this field as well as explanations of the significance of modern planning and monitoring methods for project management can be found in the first two sections. A particularly detailed section is devoted to the topic of planning instruments and their use. Building on this, the methods of project monitoring are then described in more detail with regard to an integrated approach (*tasks, times, costs*). This includes the creation of a work breakdown structure (*WBS*), the preparation of work package descriptions as well as schedule and cost plans. It is followed by project monitoring, consisting of status control and the projection of final results through trend analyses (*forecasting*), as well as the treatment of *earned value methodology* (*EVM*). Chapter 9 provides practical guidance for planning and monitoring.

Chapter 10: The cost framework, i.e., the available budget, is an important prerequisite for project implementation, because it is usually not possible without financial resources. This chapter is about resource and cost estimation. The first section points out the importance of cost analysis. Financial reference data is of great importance for future cost estimations. It is important to have a usable recalculation in which not only the costs but also technical reference data, schedules and all boundary conditions, such as false entries and side financing, are known. On the basis of an adjusted post-calculation in which all factors are known and have been taken into account, regression analyses can then be prepared as a basis for cost estimation relationships (*CERs*). This chapter discusses the use of various cost estimating methods (*top-down and bottom-up*) and their applicability. Finally, reference is made to the importance of project costs as design parameters and to cost optimization processes (*design-to-cost/life-cycle-cost—DTC/LCC*) and, in this context, to efficient cost reduction procedures.

Chapter 11: Information is the lifeblood of project work because many breakdowns occur due to information deficiencies. Project management and individual project members need to know quickly and in detail about events in the project that affect them. However, the passing on of information should be limited to factual aspects (*tasks, times, costs,* etc.) and should be carried out in a targeted manner. With exceptions, it is not appropriate to distribute information widely (*watering can principle*) and leave the recipients in uncertainty as to who should respond to the information. It has proven effective to structure information in such a way that each specific addressee is prompted to act and other employees receive the information for their review. This chapter describes the common information and reporting systems used in the project to inform management, customers, and partner companies at regular intervals about important aspects of project execution (e.g., *about the current schedule and cost status*).

Chapter 12: Projects often run through long periods of time until the planned product (*hardware, software,* etc.) is completed. An important aspect is the project documentation to be created during the development process, which systematically records all important construction documents of the future product in detail and without gaps. This includes, for example, specifications, plans, procedures, drawings, and manuals. Another important aspect is configuration management and the associated change control. This chapter describes in detail the documentation and configuration management procedures and methods that have been tried and tested in practice and are in common use today, as well as their implementation.

Chapter 13: Customer projects in which a customer commissions a supplier or manufacturer to carry out a project require a contractual arrangement between the project sponsor (*client*) and the supplier or manufacturer (*contractor*), e.g., by means of a contract for work and services, an assignment (*project order*) or a verbal order. The link between the partners is a written (*standard case*) or verbal contractual agreement. This section describes the basic principles of contract management (*types of contracts, pricing,* etc.) as well as the contract structure, i.e., the legal, the technical, and the programmatic part (*specifications, plans, functional specifications,* etc.).

Chapter 14: The award of a project contract by the customer is usually preceded by a competition in which the invited suppliers propose their products, delivery dates, prices, and contract terms. It is important that the customer formulates his requirements and wishes as precisely as possible, on the basis of which the suppliers can then respond without any gaps, while clearly highlighting their supposed competitive advantages, including management concepts. This section details the methods and procedures for the preparation of bidding documents by the customer and the preparation of promising bids by the supplier. Another aspect is bid evaluation and the contract negotiation based on it, which is also discussed in detail. Finally, questions on the subject of chances of winning are discussed.

Chapter 15: This chapter covers the aspects of project personnel. Motivation is a key to success. Project management is a difficult task and requires a high level of motivation. It is therefore important that government agencies and companies provide appropriate incentives for their project staff. For example, it is important to ask about development opportunities within the organizations. It is equally important that those responsible for the project are given appropriate authority in addition to the responsibility assigned, because key project personnel take on leadership roles. Motivation also includes success and recognition by management. The problem of lack of project management training is addressed in this chapter; see the model of an advanced training course for project management in Chap. 25. Leadership aspirations, leadership style, overcoming personnel bottlenecks, and the problems of staff reduction after the end of the project are further topics in this chapter.

Chapter 16: The management of international projects does not differ significantly from the management of national projects. However, international projects can involve a number of complications that require an increased level of management skill on the part of the project manager. This chapter therefore describes in detail the special features, such as cultural differences and behavioral patterns, which can lead to difficulties in the handling of international projects, using practical examples of international or multinational joint projects. A positive example of multinational cooperation is described using the example of the procurement and project management of the ASTRA satellite fleet by the company SES Luxembourg.

Chapter 17: The successful implementation of project management in a company often presents a particular difficulty, since considerable organizational hurdles usually have to be overcome. This topic is dealt with in detail in this chapter, including proposed solutions. The chapter focuses on the topics of organizational and personnel development, employee motivation, and the gradual introduction of project management in public authorities and companies. However, the introduction of project management should be made in such a way that existing and established structures are not destroyed without good reason. A balanced introduction is an important step toward achieving economic and social competitive advantages.

Chapter 18: This chapter deals with the topic of project management software. A recurring problem in practice is that although there are many excellent standalone software solutions available, e.g., for scheduling, cost planning, documentation, etc., there is a lack of integrated solutions. Another problem is often also the interconnection with the operating systems available in the company. These issues, including practical requirements from the project management perspective, are addressed in this chapter without making software recommendations. However, the problems of selecting and implementing PM software from the practitioner's point of view are pointed out and recommendations for solving the problem are given.

Chapter 19: The execution of project tasks always involves an entrepreneurial risk. In other words, there are no projects without any risk. This means that project managers and their teams must be prepared to deal with project risks, even if they are not fully aware of them. Project risks can be both technical and programmatic in nature and, if not counteracted in time, can lead to a deterioration of the planned outcome. In addition to technical risks that can lead to a catastrophe (e.g., *Challenger and ICE train disaster*), less spectacular risks should not be overlooked because, taken together, they can also lead to significant setbacks in the course of the project; i. e., the established project goals are not achieved, deadlines are not met, and cost limits are exceeded. This chapter describes practical and proven procedures of risk management in the project.

Chapter 20: The globalization process has become indispensable due to the worldwide networking of nations. This affects all areas of social life, including politics and the economy, and thus also influences cultural areas. Through modern communication technologies (*Internet as well as tele- and videoconferencing*) and world-spanning transport networks, globalization is advancing with great strides. Nationally oriented authorities and companies should therefore try to adapt to the globalization trend in the best possible way in the interest of their business objectives. Project management can play an important pioneering role in many cases, with authorities and/or companies from different regions initially initiating the globalization process as part of a joint project (*pilot project*) and then deciding on closer cooperation. The aerospace industry in the USA and Europe has been a pioneer of globalization using project management in recent decades.

Chapter 21: Space projects are frequently global projects, in part because satellite orbits cross countries and continents at high speed and also due to their complexity and high reliability requirements Therefore, cooperation between national agencies and companies is usually a given. The ISS space station is a good example of this; agencies and companies from Canada, Europe (ESA), Japan, Russia, and the USA are successfully pulling together here. You have to talk to each other if you want to place objects in space, because otherwise there will inevitably be technical and/or business collisions. Efficient management is needed for successful multinational cooperation.

Chapter 22: The author has complied with the frequently made request of project engineers from ESA and space industry, who are entrusted with the handling of international projects, to publish a standard Project Management Plan (*PMP*). In the past he has developed a PMP during the course of his international activities. The PMP is based on the author's experience and is recommended as a guideline for handling international projects. The title page, distribution list, and list of changes are not reproduced because the parties concerned usually use their own standards.

Chapter 23: The realization of a complex plant project requires the use of effective and proven management procedures. This chapter describes a management concept for a large-scale project that has been drawn up on the basis of management methods and procedures for large-scale projects that have been implemented and tested in practice. It describes particularly important project management focal points.

Chapter 24: From the beginning of a project, customers usually like to know what financial risk they are taking on. Is it possible to produce serious cost estimates at an early stage of a project? This is not an easy task and is difficult to solve without tools. Often those responsible rely on unreliable data and a lot is also omitted in an initial estimate. This is not surprising, because at the beginning of the project, no project structure or planning has been created. It could also be said that one groping in the dark, not only at the beginning of the project, but also for a long time during the course of the project. Therefore, the use of cost estimation models is recommended. An example of a parametric cost estimation of a military aircraft which can be performed within a short time is described.

Chapter 25: In accordance with Chap. 15, the author has developed a model of an advanced training plan for project management of which the structure is reproduced here in broad outline. The advanced training course for project management (*PM*), which is derived from this, consists of a theoretical and practical excursion part, the individual topics of which complement each other. One can select individual modules according to need.

Chapter 26: The efficiency of project management depends quite significantly on two aspects: the methods and tools available and the qualifications of the project personnel. The available project management (*PM*) methods and tools are largely fully developed and described in detail in the previous sections of this book. The required qualification can be acquired through training and/or practice. However, it is best if future project personnel (*project manager and his team*) are first thoroughly trained before they start their way into practice. The authors of this chapter taught at the International Space University (*ISU*) in Strasbourg. Based on their extensive PM experience, ISU requested them to develop a project management training program for ESA.

Contents

About the Author

Bernd-J. Madauss

Illustration: Leyla Sehar-Madauss

The author completed his studies of naval architecture at a technical College in Bremen in 1961. In 1966 the Hanseatic City Bremen granted him an engineering title. In 1980 the author finished an MBA study at Pacific States University (PSU) Los Angeles and 5 years later he received a PhD from the same University.

The author looks back on a successful professional life of more than five decades as a development engineer, project manager, management consultant, lecturer, and trainer.

Design and planning assignments at German aerospace industry

- 1960: detailed design work at the TRANSALL project office at VFW*, Bremen
- 1961: design work at the VJ 101 project office at EWR-Süd*, Munich
- 1962–65: ERNO Space*, Bremen
  - design of third stage feed system for Europa I/II Satellite Launch Vehicle (LV),
  - system work for high energy LV Europa III, precursor of Ariane LV family,
  - satellite proposal activities for ESRO satellites ESRO I/II and HEOS.
- 1966–67: Planning and control of ESRO satellite HEOS-A at JFM*, Munich.

Project control responsibilities at European Launch Vehicle Development Organization

- 1968–73: project control (PC) engineer at the ELDO** program office, Paris.

- work at the central planning and progressing control department,
- PC manager of the Europa III program office at ELDO, a precursor of the Ariane LV family.

**After several mergers the companies are part of Airbus Défense and Space.*
***since 1975 part of ESA.*

Assignment of engineering and management responsibilities at MBB space division

- 1974–1985: MBB space division, Munich
  - performing PC tasks for satellite projects, such as the HELIOS solar probe.
  - mid-seventies, assignment head of Project Operations,
    responsibility for planning and control of space and traffic projects,
    PC support tasks for magnetic levitation train development,
    acquisition support task for contractor selection of Metro Caracas.

Business management assignments at SES satellite operator

- 1985: project control manager of the first European commercial TV satellite Astra 1A.
- 1990: appointment head of the SES business management office (BMO).
- 1999: appointment senior program manager of the SES Astra satellite fleet.

Highlights of International Consulting and Training Assignments

- 2004–19: numerous assignments as consultant and trainer in Germany and abroad, focusing on project management.
- 2005/2006: involved in project planning of the YahSat satellite system at Abu Dhabi, UAE.
- 2007/08: employment as corporate advisor at YahSat and appointed acting CTO
- 2009: the author conducted in collaboration with some satellite experts a 25-day training program about Satellite Engineering and Management for the Army of the Sultanate of Oman.
- 2010: training of space management and engineering at CGWIC, Beijing
- 2012: the author performed at CGWIC and CAST, Beijing in collaboration with some satellite experts, a 9-month field office support task about production control of the Bolivian GEO communication satellite TKSAT-1.
- 2014: project management training at YUZHNOYE, Dnepropetrovsk, Ukraine
- 2018: project management support for moon rover project at Part Time Scientists (PTS), Berlin
- 2019: employment as corporate advisor at Space Communication Technology (SCT), Muscat, Sultanate of Oman.

Lecturing assignments

- 1980: until 2022 the author performed numerous lectures and seminars about project management. For 10 years he was a guest-lecturer at the University of the German Army in Neubiberg near Munich, and in the early nineties he taught at the TU Delft as part of Top Tech Studies.
- 1986: the author was appointed Full Professor by Pacific States University (PSU) Los Angeles.
- 2004: the International Space University (ISU) in Strasbourg approved the status of ISU faculty with the rank of Visiting Professor with the specialization of Space Project Management. In addition to annual lectures, in 2011 and 2012 the author led the 2-week project management module of ISU's Executive MBA program.

On August 1, 2024, ISU nominated the author to join the exclusive ranks of the university as a lifetime member of the the ISU Emeriti Faculty

Importance of Project Management for Industry and Authorities

1

Contents

B. -J. Madauss, *Project Management*,
https://doi.org/10.1007/978-3-662-69057-4_1

Abstract

By introducing professional project management, the business results of companies and contracting authorities will be significantly improved and secured. If a project gets out of hand due to technical or scheduling issues, it usually leads to cost increases, which in turn may reduce the expected business goal. This becomes particularly clear in an analysis of the Return on Investment (ROI). It is the responsibility of the assigned project manager to achieve the specified project results while adhering to schedule deadlines and cost targets.

Project management has become an indispensable part of today's business life in industry and government organizations. This is due to the fact that both industry and the public sector are increasingly involved in tasks that have a clear project character; see Chap. 2.

On the other hand, it can be assumed that the term "project management," similar to the term "systems engineering," for example, is often only used as a buzzword without having project management methodologies consistently introduced. For the term "project manager" or "project leader," this statement applies to the same extent. Although the execution of industrial and public projects with a clear project character requires economically optimal results and the implementation of modern and efficient management concepts, these necessary measures are often neglected. Instead, one finds project management that is correctly named, but not implemented in practice. Often enough, this results in project delays and/or cost overruns that are brought to public attention; see Chap. 3.

The development, testing, and implementation of modern project management is an absolute necessity for maintaining competitiveness in both national and international markets. Due to the ever-increasing complexity of modern products and the rapidly progressing internationalization of projects (see Chaps. 16 and 20), companies and public authorities are being forced more than ever to use more efficient management methods that have been tested and validated in practice. Of equal importance are the recruitment of qualified project personnel and the training of junior staff. The consistent introduction of project management requires teamwork and leads to a changed corporate culture. Lidke writes appropriately in his book "*Introducing project management in a company means initiating change processes. Project management changes the behavior and working methods of the employees concerned by introducing new methods and rules of conduct. Such changes generate resistance*" [1].

In this chapter, experience gained by the author of this book in various space projects has been described in order to provide the reader with appropriate practical examples that can also be applied to other industrial sectors.

1.1 Project Management—A Revolutionary Concept

1.1.1 Why Project Management?

Today, public authorities and companies are increasingly confronted with the problem of carrying out tasks that have a clear project character; see Chap. 2. However, the realization that the introduction of project management (PM) is advantageous for such activities is often only slowly gaining acceptance. There are two main reasons for this: First, the PM methods are often not well enough known in detail and/or not communicated effectively enough. Furthermore, the introduction of PM methods may lead to organizational changes, which often result in organizational and personnel conflicts and sometimes, even if temporarily, to chaos. Ultimately, however, the consequential introduction of PM in addition to the traditional staff/line organization leads to more optimal processes and thus to increased management efficiency; see Chap. 5. Aucoin comments: "*A project is the path that leads us from the imperfect to a better future. It is the means for embracing and capitalizing on an opportunity. It is the channel that leads from chaos to order.*" And further: "*A project is also a human endeavor*" [2].

The structural changes in our society that have occurred in recent decades have meant that university research projects, industrial development and production processes, and public construction projects have all become increasingly complex. This inevitably requires a greater convergence of disciplines towards each other and towards more interdisciplinary activity. The handling of project plans, therefore, increasingly requires the "*holistic designs*" demanded by aviation pioneer Ludwig Bölkow, which leads to optimal system solutions [3]. Bölkow was for long-time head of the German aerospace company Messerschmitt Bölkow Blohm (*MBB*). For this reason, the equal involvement of all the necessary specialist departments must be ensured from the first time the foundation stone is laid for a new project, i.e., as early as the preliminary phases; see Chap. 4. Project management is the best way of controlling activities aimed at achieving an "*optimum solution.*"

While the specialist departments of a company naturally represent the respective specialist point of view, the PM tasks essentially consist of integrating the individual solutions into an optimal overall system-level solution. From this point of view, project management (PM) in conjunction with Systems Engineering (SE) should not be regarded as an administrative task, but definitely as an essential creative overall task; see Chap. 7. Therefore, PM and SE should be given a correspondingly high priority in companies and public authorities. Kutscherer appropriately writes: "*Project management is a form of work that is geared towards adaptability, impact and innovation*" [4]. It should be pointed out that project managers are considered "integration specialists," for which on the one hand often the term "*generalist*" is correctly used, but on the other hand, neglects the previously mentioned specialist activity. However, his special knowledge is to be sought less in the traditional specialist area, but rather in the area of the integration of interdisciplinary subtasks to a higher level, the system level. In the following sections, the working methods of project management will be discussed in detail.

1.1.2 What Is a Project?—Definition of Terms

Projects are undertakings with a defined beginning and end that are characterized by the features of being time constrained, unique, complex, and new; in short, a project is an extraordinary undertaking. A detailed definition of the term is given in Chap. 2.

1.1.3 Project Management Can Cause Organizational Instability

Companies and public authorities have been dealing with the introduction of project management for several decades. Nevertheless, in many cases, serious conflicts still arise when trying to apply the topic in practice. How can this be explained? The main causes are:

- First, projects are undertakings with a defined start and end characterized by the features of time limitation, complexity, and relative novelty [5]; see Sect. 2.2. However, this often leads to conflicts with the processes of other projects and with the routine activities of functional departments
- Secondly, due to their interdisciplinary nature, projects require temporary organizational integration and associated with this, a division of responsibility and authority between the project and functional departments

Both can lead, at least temporarily, to organizational instability or chaos, in other words, the incorporation of a project is often a source of operational unrest. In principle, organizational changes are nothing new for a company. The assumption of additional tasks, company expansions, or mergers, for example, entail organizational adjustments. As a rule, however, management endeavors to re-establish organizational stability as quickly as possible. Because of their interdisciplinary nature, however, projects can often only be integrated into the traditional staff/line organization with some difficulty. In the line organization, it is difficult for them to be interdisciplinary, since the line staff are, and must be in principle subject-oriented, which makes it difficult for PM colleagues to push through interdisciplinary solutions.

A path must therefore be taken that is more in line with the project character, even if this means accepting temporary instability in the organization. Chapter 5 describes several organizational concepts involving the project management function. Former NASA Director James Webb expressed in his book, *Space Age Management*, that NASA has deliberately accepted organizational instability to some degree in order to gain sufficient maneuverability for project execution" [6].

1.1.4 Project Management as an Extended Arm of Company Management

Vollrath writes: "*Projects are always brought into being when a major undertaking goes beyond normal operational procedures. In such cases, several departments of a company may be involved, in order to work together in an interdisciplinary manner within a project organization*" [7]. If companies or authorities have to carry out project tasks, it must also be decided how and by whom the project management is to be carried out. In any case, this responsibility lies with the superordinate higher level organizational unit (*company and departmental management, etc.*), which has the necessary powers and bears the entrepreneurial risk associated with the task.

If a company or public authority has to implement several project tasks at the same time, it is necessary to delegate the tasks to a suitable organizational unit, e.g., the project management office (PMO); see Sect. 5.2.5. For the completion of interdisciplinary tasks, the implementation of a PMO is the most reasonable solution. However, full responsibility for project execution and the necessary degree of authority must be given to the appointed project manager by the higher-level organizational unit. This is the only way to ensure optimal project execution. The project manager's span of control must be broad enough to enable him to plan and monitor the interdisciplinary project work, in close cooperation with the functional departments. The project manager must carry out the project tasks on behalf of the company and in accordance with the specified objectives of the respective internal or external customers. In a sense, the Project Manager is what Bölkow calls the extended arm of the higher-level organizational unit [3]. This is illustrated in Fig. 1.1 using the principle of the matrix organization.

In many cases, the introduction of project teams leads to the organizational problems described above, as management is forced to add a new dimension, namely the horizontal structure of the project teams, in addition to the vertical structure of the functional departments. The resulting division of responsibility and authority between the functional and project areas often leads to management conflicts. It is

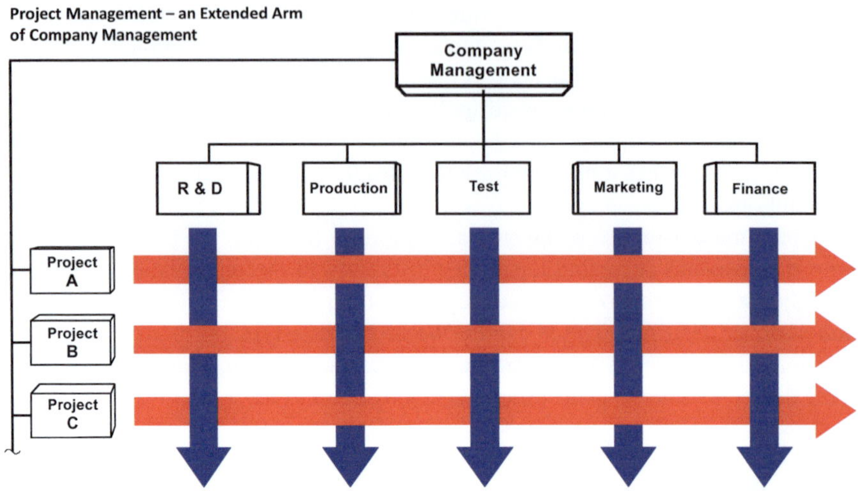

Fig. 1.1 The extended arm of management

therefore important that prior to the introduction of such organizational measures, an appropriate understanding and sufficient insight into the new circumstances is developed among all those involved; see Chap. 5. With regard to new forms of operational organization, such as those brought about by project management, Adamowsky mentioned as early as the 1960s: "*The organization of the specialist areas functions to a sufficient degree of satisfaction as long as the time factor does not play a significant or even determining role. The present, however, is characterized by a rapid change in the conditions of economic life*" [8].

1.1.5 Project Size

In the text of this book, frequent reference is made to project size. The question is difficult to answer because it looks different from the point of view of a small company than from that of a large company. On the other hand, any large project can be broken down into subsystems or assemblies down to the work package level using the project structure. A company handling a large project therefore also has to deal with small projects (*work packages*). In principle, the PM of a small project must act in exactly the same way as the PM of a large project. In summary, it is a matter of the well-known three factors: tasks, times, and costs, which must be planned and monitored; see Fig. 9.2. However, it must not be overlooked that large projects are often very difficult to plan and monitor because of their complexity. Three project sizes are suggested as a guideline, whereby a small company may have a different view of this than a large one. Guideline values which can be varied in individual cases are shown in Table 1.1 below.

Table 1.1 Definition of project sizes for small, medium, and large enterprises

Company/project size/ schedule	Small project (SP)	Medium size project (MP)	Large project (LP)
Large company			
• Project volume • Schedule	Max 2 million € Max 1 year	Max 10 million € Max 2 years	More than 10 million € More than 3 years
Medium size company			
• Project volume • Schedule	Max 1 million € Max 2 years	Max 5 million € Max 3 years	More than 5 million € More than 3 years
Small company			
• Project volume • Schedule	Max 100 thousand € Max 6 months	Max 500 thousand € Max 1 year	More than 500 thousand € More than 2 years

1.2 Historical Review

1.2.1 The Basic Idea of Project Management

The implementation of large projects is not only a phenomenon of modern times. We are already familiar with major projects from history, such as the construction of the Pyramids or the Great Wall of China. In more recent history, the Panama Canal and the Suez Canal are comparably gigantic projects. Construction of the Eiffel Tower for the World Exhibition in Paris can also be included in the series of large projects in world history. Very often, large projects were also realized within the framework of military objectives, such as the construction of large war fleets in the various historical epochs. The realization of all these projects undoubtedly required effective management.

In this context, Webb writes that the realization of very large projects during World War II posed particularly difficult problems for the government agencies of the USA: "*We had to initiate very large and complex special activities within the framework of the general rearmament* in order *to be able to complete complicated and urgent work, which, however, was not possible with the existing and established organizational methods*" [9]. In particular, the "*Manhattan Engineering District Project,*" the development of the first atomic bomb, which began in 1941, required entirely new organizational structures due to the enormous interdependence of scientists and engineers from universities, industry, and government. Webb further states that the execution of difficult tasks would not have been optimally possible using any of the then-existing government agencies. The ultimate solution was an independent specialized organization or an autonomous organization within the framework of an existing government agency.

The basic idea of today's project management is based on the major project plans in the USA. In this context, Webb speaks of "*Prototypes for Tomorrow*" and vividly describes how the basic idea of project management also played an important role within the framework of the Marshall Plan project, which was started in 1947, and continued with NASA projects initiated later. Significant milestones in the history of project management after 1945 were the realization of the Polaris program of the US Navy, the major programs of the US Air Force and the NASA's Apollo program. Closely related to these programs were the management systems *Phased Project Planning (PPP)* of NASA, the *Systems Program Management of* the USAF as well as the network planning systems PERT (*Program Appraisal & Review Technique*) of the US NAVY and CPM (*Critical Path Method*) of the Dupont company; see Sect. 9.3.4. The most comprehensive project management documentation, which was to become a guideline for all later project management developments, was the project management concept (*Systems Program Management*) of the US Air Force with the designation AFSCM 375, already mentioned and published under the direction of General Schriever in 1966 [10]. These developments were a guiding basis for project management as we know it today. NASA developed its own PM concepts incorporating the USAF experience and continuously evolved the project management concepts within the Apollo program. This led to the creation of extensive documentation consisting of manuals, procedures and regulations.

In close connection with the aforementioned PM developments of the US authorities, in which the aerospace industry was also involved, there were numerous publications on the subject of project management in the 1960s and 1970s, which contributed to the fact that the basic project management idea also spread to other branches of industry; for example, to the automotive industry [11].

The Project Management Institute (PMI), a professional association for project management based in Philadelphia, USA, was founded in 1969 and had 700,000 members worldwide in 2014 and published comprehensive documentation, the "*Project Management Body of Knowledge (PMBOK),*" as a training baseline for future project personnel [12]. PMI is the PM organization with the largest membership worldwide. The revised PMBOK edition became the official basis for diploma examinations and for reviewing formal graduate training programs in the USA on September 1, 1988, and is also the American ANSI standard and the basis for the ISO 21500 standard.

1.2.2 Project Management in Europe

In Europe, too, it was quickly recognized in the 60s that the management concepts newly developed in the USA for handling projects brought considerable advantages. The international organizations ELDO, ESRO, and NATO in particular adopted the management methods of the USA in whole or in part. However, the successor organization to ELDO and ESRO, the Paris-based ESA (*European Space Agency*), which was dependent on close cooperation with NASA in the Spacelab and ISS projects, adapted many of the project management methods originally developed at NASA to its peculiar needs. The transfer of methods has also had a positive impact on other branches of industry.

In parallel with the spread of the project management idea by the industry, project management associations and interest groups have contributed to its further dissemination. Particularly noteworthy in this context is the International Project Management Association (*IPMA*) based in Zurich. National project management associations have been founded in many countries, with IPMA as the umbrella organization. In Germany, the Society for Project Management (*GPM*) was founded in Garmisch-Partenkirchen in 1979 and had approximately 8000 members in 2016. GPM, headquartered in Nuremberg, Germany, aims to promote project management thinking and conducts conferences and various training programs; see Sect. 15.2. GPM has issued numerous publications on current project management topics, in 1989, for example, the "*Project Management Handbook*" [13].

1.3 Project Management and Systems Engineering as a Way of Thinking—The Integral View

1.3.1 New Perspectives

What is new in project management is the integral thought process. Project management and systems engineering are closely linked. Systems engineering (see Fig. 7.21) is to a certain extent at the heart of project management, since it too presupposes holistic thinking. Expressed in the words of Bölkow, systems engineering means "*thinking as a whole*" [14]. However, project management represents a further level of integration in which, for example, the goals of the engineer (systems engineering), the business economist (control), and the lawyer (contract execution) are combined into a unified objective in the interest of a common goal, namely the project result. The thoughts of von Weizsäcker, who emphasizes in his book "*The Garden of the Human*" the importance of the integral view in connection with the religious, scientific, and the social view, applies analogously to project management: "*Initial training of the integral view can consist in saying in each case in the language of the other (both) ways of thinking, what the phenomena seen from there, mean*" [15]. Project management inherently leads to an integrated technical-economic way of working and thus provides completely new perspectives in optimal project handling.

1.3.2 Thinking More Broadly Through Project Management

The adoption of modern project management methods, especially in connection with the knowledge of systems engineering, is a compelling necessity in all areas of our society where, for example, public construction projects, complex industrial projects or tasks in social areas are concerned. The efficiency of project management results from the goal-oriented way of thinking in connection with the idea of integration. In other words, all team members (project managers, technical specialists, system engineers, quality experts, lawyers, controllers, etc.) must have the same goal in mind and pull in the same direction. Interdisciplinary groups can also think more broadly than a unilaterally oriented group and, for example, better assess possible influences on neighboring areas (other communities, the environment, etc.). But at the same time, they also have greater abilities to combine individual results in the interest of an optimal overall result through an integration and optimization process. Effective teamwork is an important prerequisite in this context.

The respective technical specialists, namely engineers, chemists, sociologists, etc., are thus in the same boat as the project personnel. It goes without saying that the choice of specialists depends on the nature of the project. Here is an example:

Projects/Projects	Specialists
– Technical products	Engineers, physicists
– Terrestrial research	Geologists, meteorologists
– Economic aid	Economists, sociologists
– Medical projects	Physicians, paramedics

1.4 Project Management—A Significant Future Task for Industry and Public Authorities

1.4.1 Through Decentralization at the Scene of the Action

Decentralization often allows problems to be solved faster and more efficiently because you are directly at the scene of the action. Higher-level organizational units in a company often do not have the necessary access to detail because they lack the relevant background knowledge in the current situation. It is therefore advisable to have decentralized tasks solved by project groups. This is an effective principle. However, a prerequisite for successful problem solving by a project group is that the person in charge of the project is clearly given three things in verbal and/or written form: (1) clear terms of reference, (2) assignment of responsibility and (3) the granting of appropriate authority. It goes without saying that the three specifications must be balanced with each other. In many cases, however, the delegation principle leaves much to be desired. Often, supervisors delegate full responsibility for success but have difficulty defining tasks and granting the necessary authority.

At a time when company mergers and amalgamations are the order of the day, the consistent introduction of project management is an ideal way of counterbalancing the centralization efforts of continuously expanding organizations by means of partial decentralization; see Sect. 5.2. Responsibility and decision-making in the company must increasingly be decentralized for reasons of competence, and the delegation principle must therefore be applied. Although system tasks, such as the development of aircraft, power plants and traffic systems, can on the one hand only be handled by companies with a corresponding scale and the necessary infrastructure because of their complexity, there is, on the other hand, the necessity to handle many tasks, such as the management of a project, in a decentralized manner, since the managing director of a large company, who in addition to the realization of many medium and small projects, also has several large projects being handled at the same time, is themself hardly in a position to make project-relevant factual decisions. The project managers act as representatives on behalf of management.

Project management is not a passing fad, but a necessity. Campana writes: "*(Further) evidence that project management is not a passing fad can be derived from the fact that it is precisely the currently dynamic industries that are project-oriented: Information and Media Technology, Telecommunications, and Biotechnology*" [16]. An important reason for decentralizing project managers and giving them the appropriate authority is that they are close to the customer. The project manager acts on behalf of management, which presupposes good communication between the project and management; see Chap. 5. A quotation from Servan-Schreiber *is* appropriate here: "*Decentralization moves the decision-making power as close as possible to the scene of the action*" [17]. And Grau adds, "*Companies and managers are increasingly confronted with customers who demand 'quantum leaps' in product quality, development innovation, speed of delivery, and customer service, while they, in turn, have to get the job done with fewer and fewer resources*" [18].

1.4.2 Centralization Versus Decentralization

In the past, many companies have expanded through mergers or acquisitions, as this was the only way they were able to implement complex systems. After all, there must be competent and sufficiently strong functional departments with the corresponding know-how potential to be able to solve system tasks. For example, the development, production and marketing of aircraft, such as Airbus A380, Airbus A400M, or Boeing 747, requires an appropriate company size in order to have the necessary potential of scientists, engineers, technicians and managers, etc. available for development, production, and sales. Bölkow once put it this way: "*If only enough intelligent people are allowed to work together, their intelligence will reach a critical mass that will automatically lead to a chain reaction of creative achievements*" [19]. In order to be able to cope with the entire spectrum of tasks, qualified cooperation partners or subcontractors are involved in many cases. In this context, access to sales and capital markets also play a special role. Similar merger processes are also taking place in other sectors, such as the automotive and electronic industries. Another example is the expansion of the original small space company OHB based in Bremen with several German and European locations. In April 2016 the new OHB space center "Optics and Science," one of the most modern centers of the space industry for optical systems and scientific missions, was opened in Oberpfaffenhofen near Munich; OHB press release from April 18, 2016.

However, the continuous enlargement of business enterprises also brings with it, as a side effect, a constantly increasing centralization and bureaucracy, because "*the larger the business is, the more the interaction must be rationally organized*" [20]. On the other hand, too much centralization and the associated bureaucratization also leads to considerable problems. Increasing complexity of modern products requires system engineering decisions that can only be made sensibly in a decentralized manner, i.e., at the place of action. Over-centralization also leads to excessive paternalism and individual constriction at the work level, and thus to the alienation of many employees from the company. "*Alienation as the essence of industrial work reaches its highest form in the modern large-scale enterprise*" [20].

The managing director of the Bavarian Working Group of Engineers and Scientists in Industry, Anton Hauser, criticizes a German electrical corporation as follows: "*Too large an administration, too many managers and too many hierarchical levels lead, among other things, to too many people with different interests being involved in important decisions*" [21]. However, a company's employees need to recognize their influence on what happens in the company, which leads to more cooperation between employees and management, and thus to better performance. That is, each company employee must be able to recognize that his or her personal performance has a direct impact on the company's performance and that apathy and resignation add up to weakening his or her company.

Companies must make efficient use of the intellectual potential of their employees in order to increase performance. As a consequence, large companies, but also medium-sized and small enterprises, are therefore more than ever forced to set up decentralized managed work and project teams alongside central management, which have the necessary expertise and must therefore be granted a certain degree of autonomy. "*Decentralization requires understanding that the totality of good decisions is more valuable for a company and for the whole world than decisions made and controlled by a central body*" [22]. Decentralization, combined with autonomy, leads to a greater sense of responsibility and commitment on the part of employees, as this is the most likely way to appeal to the entrepreneurial consciousness of each individual, comparable to the circumstances of a self-employed person.

How can the relationship between centralization and decentralization be stabilized? There is no doubt that most companies have already initiated decentralization measures in the course of expanding their operations and diversifying their products. Large companies are usually structured according to product areas and/or locations, which are often managed according to profit center aspects. But is this sufficient to involve the execution level sufficiently in the decision-making process? Or is further decentralization necessary? A closer look at many large companies reveals that divisions and locations have long since reached the scale of a company, often with several thousand employees. Only one further headquarters, namely the corporate division management, has been installed, but it has not been decentralized deeply enough, which is only possible through subdivision into individual project areas; see Sect. 5.2.

1.4.3 Balance of Forces

"*Decentralization can only be successful if the leading forces realize that they cannot keep for themselves the powers they have delegated to their subordinates*" [22]. The development and production of a modern product, such as a modern machine tool, can no longer be handled along the lines of a single department. Today, this requires an interdisciplinary team of mechanical and electrical engineers, software specialists, quality assurance experts, as well as marketing and finance specialists, whose individual contributions are integrated by experienced systems engineers and project managers in the interest of an optimal overall product.

Since companies do not usually sell individual results but end products, such as a machine tool, an airplane, or a power plant, the area representing the interests of the optimally designed end product is of particular importance. Project management, in conjunction with systems engineers, has thus become an important and direct representative of business or division management. If one assumes that industrial companies usually produce not only one but many products at the same time, it is easy to see that the project management functions and system engineering tasks cannot be performed simultaneously by the management or divisional managers, but that competent project managers equipped with appropriate authority must be appointed for this purpose.

But it is precisely this step, namely the introduction of autonomous, decentralized project groups endowed with powers, that repeatedly leads to considerable problems, as many companies deeply distrust this redistribution of power. One reason for this is that the project managers appointed in each case often do not have sufficient experience to be able to represent the company in both technical and business terms. Therefore, high priority should be given to the selection and training of future project managers; see Sect. 15.2. In fact, a project manager with sufficient authority may act like a small businessman. Accordingly, this concept is often followed only reluctantly. The project manager often receives only a far too small portion of the required authority. The centrally controlled specialist departments are still decisive, but not the product-oriented project managers. This inevitably leads to failures and frustrations in the project business. As a result, project tasks are often not officially assigned to a competent project management team but are instead broken down into individual activities and competencies.

A true balance of power between the centralized and decentralized functions of a company or organization can only be achieved if technologically and administratively competent and empowered working or project groups are designated at the working level in addition to having a strong strategy-oriented company management. Bernhard quotes Pascale and Athos (The Art of Japanese Management) and recommends: "*Organizational structures tend to be broadly decentralized, with few central functions and with few hierarchical levels, in order to enable the most direct possible flow of information in both directions. Matrix relationships are used to integrate the independently operating units*" [23]. Bölkow urged the corporate leaders of Messerschmitt-Bölkow-Blohm GmbH, Munich, in his address on the occasion of his 70th birthday, in which he referred to the importance of holistic designs: "*Give your employees the necessary long leash. The higher up you are in management yourself, the less time and substance there is for detailed work*" [3].

In conclusion, both efforts, (centralization, which converges with the management and decentralization, which ends at its lowest level with working and project groups) must be reconciled; see Sect. 5.2. The advantages of centralization and decentralization are summarized in the following table [24]:

A) Advantages of Centralization
   - All legal powers lie with the management
   - Uniform policies, procedures, and decisions
   - All information goes to the management
   - Minimizing the duplication of functions
   - Reduction of the risk of company activities becoming independent
   - Detailed/extensive control procedures not required
   - Strong top management team
B) Advantages of Decentralization
   - Delegation of decision-making
   - Relief for top management
   - Generalist development
   - Promotion of cooperation (team spirit)
   - Product specialization
   - Increased management efficiency due to greater proximity to the scene

1.4.4 Project Managers on a Long Leash

Giving employees a long leash wherever possible is more important than ever today. This mindset results in an increase in employee creativity and responsibility and is an urgent necessity for modern companies. Project managers, whose basic attitude is that of a business manager, must be given a long leash. Only in this way can they and their teams deliver the required services in the interests of the company. Problems that impact several departments at the same time can be solved more quickly by decentralized project groups than would otherwise be possible. They operate much like a small business and therefore have more flexibility. "*The advantage of small companies is their flexibility*" [25]. However, the long leash should also be limited in its absolute length. In other words, management must obviously have appropriate methods for monitoring project leads.

Cooperation with the specialist departments is not always unproblematic, but it is certainly possible to accomplish to the satisfaction of all concerned. One method often cited in the literature is matrix organization (MO). As Grochla, the now deceased "*organization pope,*" put it: "*The main characteristic of the MO is the two-line management system on which it* is *based*" [26]. Although the MO is often criticized in comparison to other project organization forms and its implementation often fails because of the stubborn resistance of tradition-bound departments, it nevertheless represents an optimal solution for the execution of complex projects, because on the one hand, the enterprises must receive and develop their know-how stored with the functional departments, at the same time however the project departments must have access to this knowledge, in order to draw the largest use from it in the interest of the respective project. A balance must therefore be struck between the specialist functional departments and the project departments; see Sect. 5.2.

Grochla and Thom state the following in this regard: "... *the MO consequently offers a starting point for solving the fundamental organizational dilemma, namely to ensure at the same time continuity and purposefulness in the fulfillment of permanent tasks* and *to achieve responsiveness to changed or new tasks*" [26]. A well-known measure of organizational development is the introduction of matrix organization; see Sect. 5.2.4.

In order to handle projects efficiently, companies, government agencies, and offices can no longer do without the consistent introduction of project management. This must not remain merely lip service, but rather deliberate organizational changes must be accepted in order to give future project managers the necessary mandate to realize complex system tasks with all their consequences. Martin mentioned: "... *a project manager who makes good decisions 51 percent of the time is already successful*" [27]. If this is taken as a benchmark for future project managers, it should not be too difficult to find project managers in one's own ranks.

1.5 Cost Control Through Project Management

Project management is usually caught between the conflicting priorities of technical solution options, schedule constraints, project costs, international competition and government requirements. Lientz lists the following factors that have an influence on project management [28]:

- Technology
- Competition
- State requirements
- Politics and
- Several factors at the same time

It becomes clear here that project management is to be regarded as an important integration task. Many threads from different disciplines come together in PM and must be systematically processed. The focus here is on early consideration of the probable life cycle costs and the associated necessary decision-making processes; see Chap. 4.

1.5.1 Early Determination of Life Cycle Costs

Today, costs play a decisive role in the development of new technical systems. Particularly in large-scale projects in plant engineering and aerospace, it is clear that not everything that is technically feasible can for cost reasons be implemented. Rather, it is usually necessary to assess how a desired goal can be optimally realized within the given cost framework. This is the purpose of the early assessment and cost trade-offs of *life cycle costs (LCC)*. The term LCC covers the development costs and the construction and operating costs as well as the costs for system disposal at the end of the life cycle. This is particularly evident in the case of nuclear power plant decommissioning. Without an assessment of the probable LCC during the early project phases, catastrophic financial mistakes can occur, of which some large projects provide very clear examples; see Chaps. 4 (Phase model) and 10 (Cost management). Professional project management helps to avoid mistakes. Companies benefit from PM know-how [29].

However, the LCCs must also be examined at an early stage because, after all, not only the purchase price but also the subsequent operating costs play an important role for the later user of a product. Every project goes through various phases during its life cycle, which is structured as follows: A. concept selection, B. system definition, C. design and development, D. manufacturing, E. operation and maintenance, and F. phase-out; see Sect. 4.2. It is important that possible concept or system errors are identified early and eliminated by appropriate measures and are not dragged into the much more expensive development or even construction phase, since changes then inevitably lead to considerable additional costs.

The concept sets the course for all subsequent project phases. If, for example, the project goal is to improve communications between two cities or countries, several alternatives are available, e.g., cable, satellite, or microwave links. Each of these alternatives must already be examined conceptually with regard to its transmission capacity and quality, its expandability and its cost-effectiveness in order to be able to select a cost-optimal system in the future.

The concept selection is followed by the system definition, in which the following tasks, among others, should be solved:

- Creation of final and realistic specifications
- Definition of interfaces and responsibilities in the system
- Identification of risk areas
- Presentation of flow alternatives
- Selection of the best overall solution
- Formulation of realistic manufacturing concepts
- Description of preliminary concepts of operation
- Creation of a Work Breakdown Structure (WBS)
- Preparation of realistic schedules and cost plans
- Preliminary contract documents (sample contract & Statement of Work (SOW)

This list clearly shows how great the danger is, in the case of inadequate system definition, of dragging cost drivers into the development and production phase or unintentionally creating new ones. Over-specification and the tendency to achieve technical perfection are the next pitfalls in the project process. In this context, it is quite understandable that the future user of a project succumbs to the tendency to get only the best (nice to have). It is only with a view to the LCC that the view is often somewhat clouded because now the question arises as to what can be afforded (*affordability*).

For economic reasons, any over-specification must be avoided—according to the principle that "*good enough*" is enough. This applies to aerospace as well as plant engineering and other industries in order to avoid exceeding the cost framework. In this context, the Design to Cost (DTC) process has also proven its worth. In this procedure, the developer is given DTC targets in order to develop optimum technical and economic project solutions in the early project phases. In the process, regulatory constraints, such as quality standards and safety regulations, must be complied with; see Sects. 10.5 and 10.6.

1.5.2 Saving Costs Without Project Risk

Although aerospace equipment is very expensive and, for example, the premature failure of a satellite represents a high financial loss, in recent years the number of expensive test models has been reduced and replaced by simulations in order to reduce project costs. The next critical cost item is project documentation. Sufficient project documentation is absolutely essential, especially for large projects, which often involve several companies. This includes the full range of requirements, specifications, plans, procedures, reports, acceptance documents, manuals, contract documents, and drawings and lists. The commercial success of a project depends heavily on the quality of the documents. And the safety aspect *(safety)* plays a significant role in many industries and must therefore be thoroughly documented. The entire development effort, including safety and quality control, is part of the documentation. However, customer requirements should not be exaggerated in order to avoid increasing documentation costs.

Can the optimum documentation effort be quantified? Depending on the effort involved, one needs approximately 1–4% of the project costs for the writing and duplication work (*author's empirical value*). Realistically, however, one must consider the non-transparent documentation costs for project management, systems engineering, quality assurance, etc. The true costs are then likely to be between 15 and 25% of the project costs because initially only documents, specifications, plans, procedures, etc. are produced. To keep documentation costs down, it is recommended that companies and government agencies develop standards and that clients (customers) rely, as much as possible, on existing industry reporting procedures to limit the documentation effort to what is necessary. Electronic data processing and microfilming, as well as shipping on demand, are helpful measures to reduce costs; see Sect. 10.6 and Chap. 12.

1.5.3 Change Costs

Late project decisions can also be named among the "*cost drivers.*" The task of management is to react intelligently to changes, as the former US Secretary of Defense and later World Bank President Robert McNamara once put it [30]. Changes occur repeatedly during project implementation and must be responded to quickly and correctly to avoid jeopardizing the goal; see Sect. 12.2. Delays in the decision-making process quickly led to unnecessary additional costs. It is therefore important to implement a functioning control procedure to manage the necessary decision-making processes. In order to accelerate project decisions, the chaining of the decision-making process (customer, company, project) must be agreed upon first and foremost. Decision rules must be introduced as part of this agreement to allow, for example, limited ad hoc release of funds by the PM.

1.6 Project Management—Mediator Between Company Management and Development Department

1.6.1 Developer as Creative Problem Solver

High demands of creativity are placed on the developer. Therefore, he should already be involved in the bidding process in order to achieve good cooperation between the developer, the PM and the customer. Good cooperation can significantly increase the potential for innovation. Problems always arise when management passes on the demands of the market without involving the developer in the definition of goals and the feasibility check. To avoid this, the developer should be involved in the planning process as early as possible in close cooperation with the PM; see Sect. 14.4. This makes the best use of the developer's creativity to solve the technical-economic problem [31].

More than ever before, the developer must be a creative problem solver, since market demand is growing for newer and usually also better and, in some cases, lower-priced products. At the same time, the market power game has shifted very strongly from the national to the international market; see Chaps. 16 and 20.

1.6.2 Management Expectations: Creativity in a Given Time Frame

Industrial companies that are still successfully selling their products on the market today must at the same time ensure that they will be successful on the market tomorrow. This means that the future must exist on the drawing board today. The developer is at the center of the action here. He is expected to come up with new ideas and put them into practice creatively. Company management takes this for granted and selects personnel accordingly. In view of the international market situation, however, management expects more, namely creativity within planned time frames and a rational approach to material selection.

The management of a company, which is responsible for the technical and economic management, expects a maximum of creative activity from the developer in order to gain advantages in competition with other, equally ranked but competing companies. As a rule, there is full agreement on this between the company management and the chief designer. It is not uncommon for the designers of the competing companies to be in personal competition with each other. In the aviation industry, for example, it is the pride of every developer to work (*or have worked*) on the aircraft project that will then be completed and selected by the future operator (*airline/ air force*) as the best aircraft. It can often be observed that entire armies of development engineers and designers give up their previous homes and living quarters to follow the project after the project has been awarded to the selected company. This happens not only for financial reasons but also for career reasons. There is no doubt about the motivation of these engineers. They put up with inconveniences such as a change of residence and settling in at a new workplace just to be part of it. In short, the enthusiasm of development teams is usually very high and so is their willingness to create, provided the company is able to give them an interesting and challenging goal and the necessary recognition.

Time constraints for certain development activities are often perceived as an obstacle to creativity. Probably almost all developers are familiar with the evil of time constraints on the performance of creative activities. The management expects a maximum of creative activity within time limits, i.e., a creative activity superior to the competition is expected within a very short time, so that the new creation (new better product) comes to the market before the competitor's product.

Development engineers are constantly faced with the problem of material selection and, in the interest of the quality of the product they are developing, would be all too happy to use the best material available. However, for economic reasons, company management demands that less expensive material be used to keep manufacturing costs as low as possible. This very often represents a blatant contradiction to the creative thinking of the developer and often the quality of the product suffers as a result. As much as the demands of management for cheaper material may be justified, the developer is affected by this very personally, since he identifies with the newly developed project and also knows that the judgment of his peers about his creative abilities is directly and/or indirectly related to the quality of the products he designs.

1.6.3 Expectation of the Developer: Clear Goals and Creative Freedom

Developers usually expect clear objectives and the necessary freedom to create. The demands of the developer are not infrequently in stark contrast to the demands of management, from which conflicts frequently arise. Setting clear and achievable development goals is of particular importance, as the developer identifies with his creative work and accordingly reacts negatively to changes that may render weeks of creative activity superfluous. There is historical evidence that originally successful development teams were no longer creatively usable after months of work without a clear goal. This is probably related to the fact that the principally sensitive developers temporarily lost their self-esteem in the absence of a measurable sense of achievement.

Developers claim a certain degree of creative freedom. This claim is justified in principle since it can be objectively justified. However, this first requires a definition of the creative freedom of the developer: (a) with regard to the time in which he carries out the development activities (i.e., without strict deadlines, since ideas cannot be forced at any time), (b) with regard to the selection of materials and production methods, and (c) with regard to the procurement of information.

The claim of the creative freedom of the developer in the sense described above is basically justified, since creativity does not exist by wish or command, but arises quite suddenly, often even by chance, based on the talent, knowledge, and personal freedom of the developer. The creative developer's demand for creative freedom also aims at giving him the widest possible scope for the development of his personality.

1.6.4 Is There a Controversy Between Management and Developer?

As a rule, this controversy exists. However, the management/developer controversy is fought out more or less violently depending on the quality of the working climate, the company organization and the company product. It is probably also not possible to harmonize the expectations of both parties to such an extent that there is no longer any opposition. On the other hand, a certain degree of conflicting expectations is even helpful in keeping the optimization process going between what management expects, i.e., maximum creativity with time constraints and what the developer expects, i.e., clear goal setting and creative freedom. A conflicting understanding often arises on the following points:

- Maximum creativity under time constraints
- Use of inexpensive/cheap materials and manufacturing processes
- Demands for clarity of purpose
- Demands for creative freedom

This list can certainly be extended but is quite sufficient for the following discussions.

1.6.5 Problem Solving Through Project Management

Can the problems between the management and the developer be resolved? The opposing expectations between management and the developer are not based in principle on an often-assumed opposing mindset between the technicians on the one hand and the managers on the other, but rather result from the hard market demand which is brought to management from outside and which is then often passed on to the developer without explanation. This demand initially meets with resistance from the developer, since it does not correspond to the principle of creative freedom. Here the suspicion arises that it is basically a frequently occurring communication problem. An attempt should therefore be made by management to involve the developer in the management process as early as possible in order to achieve a stronger commitment to administrative problems (target, time, and costs) through the developer's shared responsibility.

The methods and procedures of modern project management offer a particularly good opportunity for conflict resolution between management and the developer. The demands made on the industry by the market are usually not directly translatable into the language of the developer but require precise analysis by a project team in which all relevant functions, such as product development, product assurance (quality control) and production control, are combined. In this process, the developer is at the center of the action. At the same time, he can contribute to the solution of the overall problem together with his colleagues from the parallel disciplines and is involved in the definition of detailed goals. He assumes joint responsibility and is in a better position to understand the demands that lead to deadline and cost pressure, i.e., to the restriction of creative freedom. Creativity under time and financial constraints is now declared a common goal and generates motivation for the developer to solve the task. In other words, in this constellation, the developer has a far greater task and responsibility than was previously the norm, which means that he is motivated to focus his creative capacity on the overall objective and not just on the sub-area of development.

In the aerospace industry, it has long been routine for the developer to participate in the planning process at an early stage as part of project management, and thus to be involved in setting detailed objectives. Aerospace projects are normally handled according to the following phases: Concept Selection, System Definition, Design and Development, Manufacturing, Operations, and Maintenance; see Sect. 4.2. During the first two phases in particular, there is constant iteration between market requirements (management), product idea (management/developer), and development proposal (developer).

In the detailed planning of the above-mentioned phases, special attention must also be paid to the special concerns of the developer. The planner must know that new developments are often affected by setbacks, especially when completely new territory is involved. "*The inventive process does not proceed schematically in a straight line, but involves aberrations, setbacks, leaps, and perplexing stalemate*" [32].

1.7 Aerospace as a Pioneer of New Management Concepts

1.7.1 Going Completely New Ways

Developments in modern project management point the way to solving complex system tasks and at the same time pave the way for new management and organizational concepts in government and industry. Aerospace often takes the role of a pioneer in this process, for example in the development of the often-cited concept of matrix organization [33]. Aerospace has traditionally been an industry of high innovation. Extreme project and system goals repeatedly challenged pioneering technological and administrative achievements. Technological in particular because the feat had to be accomplished to produce equipment that had to function flawlessly for long periods of time without maintenance or replacement during those periods, e.g., satellites in very high orbit. Modern communications satellites, for example, are expected to remain fully functional under space conditions for a period of more than 15 years, during which time, of course, they cannot easily be serviced or repaired. On the other hand, such top performance has to be achieved under very strict weight and space limitations, since the available rockets, e.g., Ariane 5 or SpaceX, have only a limited capacity and the transport, including insurance, is carried out under strictly economic aspects.

Achieving extreme technological goals require efficient management concepts; for example, in the areas of organization (*Chap. 5*), systems engineering (*Chap. 7*), product assurance (*Chap. 8*), schedule and cost management (*Chaps. 9 and 10*), and documentation and configuration control (*Chap. 12*). It was also an absolute necessity to leave the well-worn tracks in the area of communication and to break completely new ground. Mistakes were also made in the process. Some management procedures turned out to be too detailed and cumbersome. Difficulties also arose, above all, in transferring tried and tested methods to smaller and non-space projects. In most cases, however, it has been possible to successfully transfer the acquired know-how in the field of project management to other areas. For example, between 1977 and 1980, MBB successfully transferred its acquired know-how to the following major projects through management support in which the author was involved [34]:

- METRO Caracas
- MAGLEV Train
- Nuclear Power Plant
- Solar Tower Power Plant
- Ship of the Future

"*... space projects have advanced technology at a breathtaking pace. Hardly any other technical or scientific discipline can boast anything comparable. A number of fields benefit directly from it, and the economic impact is visible everywhere*" [35].

1.7.2 Management and Organizational Improvements

The development work of European spaceflight, financed by the national space agencies and ESA, has paid off. Just as in aircraft manufacturing (Airbus), European spaceflight is now commercially oriented. Examples include the global satellite operators INTELSAT, SES, and EUTELSAT; see Chaps. 16 and 20. "*One billion dollars is the amount spent by European governments on international space programs between 1966 and 1977*" [36]. The investment was raised by ESA member countries. In 1978, a study was carried out by Louis Pasteur University, Strasbourg, on behalf of ESA to examine the question of the profitability of space exploration. Four categories were considered in detail:

(1) Technological benefits
(2) Economic benefit
(3) Improvements in the area of management and organization
(4) Productivity improvements

The study group performed a cost-effectiveness analysis for each category to quantify the benefits. The benefits identified, relative to the total sum of all expenditures, were as follows:

- Benefit/expenditure = 2260/1000 = 2.26
- which means that for every euro invested, you got back more than twice as much.

For the third category, management, and organization, which is of particular interest here and primarily concerns the management functions of industrial firms, economic benefits were found to be about 17% of the total benefits mentioned above. "*One can conclude that space developments have strong positive impacts on industry. Not only do they improve their competitiveness, but they also improve their technical competence—and these benefits are reflected in every aspect of the firms; in their organization and in the quality of their personnel*" [36]. The transfer of space project management know-how to non-space industries are of great benefit for the European economy.

1.7.3 NASA Study: Transferring Aerospace Management Techniques to Other Fields

A similar study with the theme "*Application of Aerospace Management Techniques to Business Enterprises and Government* Agencies" was also conducted for NASA [33]. A total of 25 management techniques were reviewed for their applicability to other fields, including techniques such as systems analysis, systems engineering, value analysis, matrix organization, configuration management, information systems, performance evaluation methods, etc.

In their findings, the authors explicitly point out that the transfer of proven management methods from the aerospace industry to other industries only works if they are carefully adapted to the respective circumstances. The emphasis is on intelligent transfer, otherwise more harm than good can result; tact is absolutely necessary. It was found by 1971 that of the 25 management methods studied, only 56% could be transferred to other industries. The authors attributed this to the relative novelty of the methods and the typical time lag of technology transfer processes. In this context, they cite NASA Deputy Director George E. Mueller, who expressed that the constant migration of aerospace employees to other industries automatically leads to know-how transfer. It was also noted that some companies were already using methods similar to those used in the aerospace industry. However, these methods were not as advanced as the comparable aerospace methods. In the summary, the authors point out two issues important for method transfer:

(1) The method must be recognizably better than previous methods and
(2) a manager must be found who is willing to try out the new technique; for example, a man who comes from the aerospace industry and already knows the method or who is willing to take a calculable risk.

The transfer of management know-how from aerospace to other industries has accelerated since the study was performed through personnel transfers to other industries and training measures. In Germany, for example, the German Society for Project Management (*GPM*) has contributed significantly to the spread of PM know-how through various training measures, congresses, etc. based on aerospace findings.

1.7.4 HELIOS Solar Probe: Management Progress of German Space Industry

The HELIOS solar surface exploration project, launched in 1969, was a bilateral project between Germany and the USA and had two main objectives [37]:

- The successful fulfillment of the scientific mission, i.e., exploration of the Sun by approaching up to 40% of the mean Earth–Sun distance and
- Gain of management and technology know-how by US experts for the German space industry.

The first probe was launched in 1974 and the second in 1976. Both instruments met the mission objective 100%. The probes' contracted lifetime of 18 months was significantly exceeded. HELIOS A survived a full solar cycle of 11 years. The fulfillment of the scientific goal of the probes was thus proven. The third model, still flightworthy, was not launched for cost reasons and is now in the Deutsches Museum in Munich.

The fulfillment of the second main objective now also had to be demonstrated. HELIOS indeed became an important milestone for the German aerospace industry. The professional management methods and procedures used in the period 1969–1976, and more importantly, the mindset learned during that time, shaped to a very significant degree the understanding of modern management methods of MBB (*now Airbus Defense and Space*) and, moreover, of the subcontractors involved.

In addition to the development of improved procedures and regulations, the HELIOS project also invested in individual employees, who will continue to pass on their knowledge far beyond the borders of MBB and Germany into the new century; see Sect. 21.4.5.

Five former HELIOS employees later held management positions at the European Space Agency and effectively passed on their knowledge in this way. A large number of the former HELIOS employees (*including the author of this book*) were subsequently entrusted with management tasks of follow-on projects. This is evidence that the second goal of the HELIOS project outcome was achieved. The HELIOS school—*learning by doing*—was an important milestone for all those involved, who then passed on their knowledge with great motivation to carry out further projects.

1.7.5 Transfer of Project Management Know-How to Other Areas

In the context of an event held at the Batelle Institute in Frankfurt, the author tried to prove in 1981 that the project management know-how of aerospace can be made usable for the development of technical serial products [34]. On the basis of seven theses, which have not lost their validity, a way to transfer know-how was shown. However, it is important not to transfer the space management methods directly, i.e., one-to-one, but to adapt them to the respective situation. This is explained in detail using some typical examples. Particularly noteworthy in this context is the Handbook for Municipal Project Management published in 1977 by the Federal Minister of Research and Technology, whose authors are three German space companies, namely Dornier System, ERNO-Space and Messerschmitt-Bölkow-Blohm [38]. (Note: *the companies mentioned above became a part of Airbus Defense and Space after a merger.*)

The thesis paper, which has not lost any of its topicality, is reproduced below, including detailed explanations.

First Thesis: Space development projects, e.g., the development of launch vehicles, satellites, ground facilities, etc., are in principle comparable with large and complex industrial development projects in plant engineering, marine engineering, etc. The transfer of know-how has taken place successfully in practice.

Second Thesis: When handling European space development projects, project-specific constraints have to be taken into account, some of which also apply to other industries:

1. High degree of technical innovation, i.e., extremely high technical performance specifications and quality and reliability requirements.
2. Tight deadlines, as there are usually fixed launch dates, so-called launch windows, which cannot be postponed for scientific reasons.
3. Participation of many national and international organizations, companies, and institutions in the project; this results in the demand for an optimal division of labor (*work sharing*).
4. Established national and international client organizations that often prescribe different management methods or requirements.
5. Most companies have entered into firm industrial consortium ties, which can make contracting more difficult.

Third Thesis: In order to be able to take account of the project-specific boundary conditions, an efficient and dynamic approach is required. This results in the use of efficient management concepts such as:

1. Project-oriented organizational structures with matrix relationships in the company
2. Granting technical/administrative powers to the project manager who is responsible for the selection of the technical concept, the overall design of the system, and the selection of subcontractors and suppliers
3. An experienced project team, i.e., a practiced project team and experienced key personnel, are important for success
4. The application of efficient and proven management methods, such as phasing, work breakdown structures, planning techniques, effective cost estimating methods, make-or-buy decision concepts, configuration control methods, review techniques, functional safety analyses, trade-off techniques, etc., is a prerequisite for project success

Fourth Thesis: The willingness and insight to apply efficient management techniques is relatively high in space projects. Practical experience has shown that the success of a mission, e.g., an intelligence satellite designed for several years of operation without repair options or a manned space station, depends to a considerable extent on the quality of project management, i.e., on the leadership and control behavior of the project management. This can be transferred very well to other industries because although system failures can be repaired, unlike in space systems, the economic loss is often not insignificant.

Fifth Thesis: The experience gained in the field of project management in the aerospace industry can definitely be transferred to other industries. Here are some examples of methods [39]:

- Decision Analysis
- Technology Forecast
- Systems engineering
- Reliability analyses
- Value analyses
- Project management principles
- Matrix organization
- Industrial contracts with contract bonuses/penalties
- Company valuation methods
- Management information systems
- Network techniques
- Configuration management
- Product assurance methods

Fabian writes in this context: "*However, it would certainly be worth considering whether, as an industrialized state, we should not also move toward new forms of organization for government work that are oriented toward the proven management techniques of large-scale enterprises. Most of the higher-level work of the government enters into almost all departments to such an extent that an organizational structure corresponding to modern project management is appropriate*" [40].

Sixth Thesis: The efficient transfer of management know-how from the aerospace industry to other areas precludes a simple adoption of the methods. Rather, a tailor-made adaptation of the methods is required. Typical problems of know-how transfer are:

1. The use of new methods is usually associated with additional effort. Attention must therefore be paid to the proportionality of the means
2. When applying new techniques and methods, an implementation and adaptation phase must be provided to ensure tailor-made use, i.e., exact adjustment to the requirements and timely training of the user personnel
3. The compatibility with already existing systems, e.g., systems from the business management area, must be checked; if necessary, already existing subsystems can be included
4. It is often overlooked that the introduction of a new system using a new term, especially when using the English language, is unnecessary because the procedure already exists in a different form and under a different name and may have been used successfully for years.
5. The successful transfer of space management methods and know-how is only possible if comparable conditions are created, e.g., tight management discipline, consistent adherence to the planning logic once provided, systematic status control, etc.

Seventh Thesis: The transfer of management know-how from aerospace to other industries began as early as the 60s. Exemplary approaches to the transfer/dissemination of management know-how in Germany are:

a) German Management Foundation (*AKM*) in June 1969. On September 14, 1967, the Federal Ministry of Research and Technology (*BMFT*) organized a project management symposium. At the final discussion of this symposium, the head of the Department of Space Research at the BMFT, suggested the foundation of the AKM with the aim of drawing up recommendations for the project management of major development projects for the client side and the contractor side. The AKM endeavored to develop a project management system adapted to the European and, in particular, the conditions in the Federal Republic of Germany and not simply to adopt American management ideas, practices and principles. 1977, a brochure was *published* entitled "*Recommendations for the Management of Major Projects*" [41]
b) Publication of the Handbook of Municipal Project Management. This handbook was jointly prepared by the three companies referenced in Sect. 1.7.5 was published in 1977 by the Federal Minister of Research and Technology (*BMFT*) [38]. The project management procedures proposed in the book have been successfully applied in the aerospace industry. The handbook was intended to show the knowledge gained and offers suggestions for efficient planning and streamlined execution of projects in local governments. The suggestions made are based on experience in the aerospace industry

c) Industrial consulting by space companies. As already mentioned, MBB Space Division has been involved several times in know-how transfer. The companies Dornier System and ERNO Space were also involved in management know-how transfer tasks
d) Seminar Manual "*Planning and Monitoring of Research and Development Projects—With Contributions from Space Technology*" [42]. The seminar manual, published by the author in 1978, summarizes in a clear form modern methods and procedures of planning and monitoring R&D projects with the addition of many practical examples from the space industry. The methods presented have been tested in practice and, in addition, have been used in the management of non-space projects. The seminar manual, reissued in 1981, served as the basis of several successfully conducted industry seminars and was also the basis of this book.

In the past, space projects, such as the HELIOS solar probe, were mostly one-offs. But that changed, as in 1985 the US company Motorola began planning the Iridium project, based on a constellation of 77 low-flying satellites for worldwide voice and data transmission via satellite phones. The system went into operation in September 1998.

Space companies are increasingly moving toward mass-producing satellites of identical design. Mostly, however, these are small satellites with a still small number of units. In the meantime, there are considerations for the construction and operation of several hundred identical satellites, resulting in series production for cost reasons. Therefore, it should not go unmentioned at this point that space companies have to deal more intensively with series production of small satellites. There are therefore many considerations to transfer the knowledge of series production from non-space companies, e.g., the automotive industry, to space.

1.8 International Authorities Cannot Do Without Project Management

1.8.1 Sand in the International Gears

The implementation of international and bilateral projects, in particular, requires particularly efficient and effective project management in order to prevent mishaps like those that occurred in the ELDO projects (See Sect. 3.8), because national interests, cultural differences, etc. can quickly introduce a great deal of sand in the gears. Especially in the case of international projects, where language barriers are yet another obstacle, project management must be implemented that handles all project-relevant processes as objectively and non-politically as possible. The second Secretary General of ELDO, M. Aubiniere, in his inaugural speech to the ELDO staff in 1970, expressed himself as follows: " ... *in the past, many staff members here in the House kept bringing up politics in addition to technical arguments. I would now like to point out to you that in future you may only act and argue in a purely technical manner. All political questions are to be solved or answered by me alone*" [43]. This statement had a great signal effect for the ELDO staff because the international team consisting of six nations could now concentrate more appropriately on the actual technical-administrative problems.

But even so, international projects often have problems that are not always easy to solve; see Sect. 16.2. Problems arise, for example, in the following areas:

- Funding and return of funds
- Compensation transactions (*bartering*)
- Currency problems—exchange rates
- Cooperation of international authorities using the example of ESA
- Different levels of technology among member countries
- Use of national and/or international facilities
- Project decisions

Lessons had been learned from past problems, as can be seen perfectly in the example of the Ariane launch vehicle, which was developed under the ESA program and operated commercially by ARIANESPACE. After initial failures, the managers responsible for European space projects have learned that professional project management is a prerequisite for international competitiveness and the associated commercial success.

In this context, European participation in the development and operation of the ISS (*International Space Station*) is worth mentioning. The ISS is a successful global joint project of NASA, the Russian space agency ROSKOSMOS, ESA, and the Canadian and Japanese space agencies CSA and JAXA. It is an excellent example of global collaboration. It can be seen as a model of peaceful cooperation because the development of the ISS involved overcoming political, legal, programmatic, and technical hurdles, e.g., interfaces and differing standards; see Chaps. 16 and 20.

1.8.2 Funding and Return of Funds

An important factor for project implementation is funding. In the case of international and bilateral public-sector projects, this aspect is further extended by the question of the return of funds. If, for example, member countries of ESA finance a project according to a very specific country key, experience shows that the participating countries are expected to receive a financial return in proportion to the financial contribution. This means that the ESA project office must ensure that project tasks are assigned to domestic industry in proportion to the financial contributions of the country concerned.

This makes the awarding of contracts more difficult, since in addition to technical and economic decision-making criteria, the financial geographical distribution, i.e., return on investment, must also be taken into account. In addition, experience shows that member countries expect challenging task packages to be awarded, which further complicates the goal of finding cost-optimal solutions. Therefore, the factors of best technology, lowest cost, and best geographic task assignment play a significant role in the number of eligible firms. Vandenherkhove wrote as early as 1970, and this is still true 40 years later, "*Geographic distribution is undoubtedly a problem, but under certain conditions this problem can be solved satisfactorily*" [44]. The problem can be solved as follows:

- By specifying a range according to which the financial return must be made
- Waiver of 100% flowback solution per project
- Compensation under several projects

Vandenherkhove further wrote, "*In fact, our goal is only that each country should receive no less than 70 percent of the total amount due to it on the basis of proportional representation of all project contracts awarded by ESA over several years. This gives ESA the flexibility to select the best firms in each case*" [44]. The return coefficient of ESA member countries, i.e., the ratio of return to payment, has settled around the value of 1.0 over a longer period of time for all participating countries [45].

1.8.3 Compensation Transactions

The manager of an international project is confronted with a variety of special features, e.g., compensation transactions in which the financial contribution of a country or a company is made in the form of provisions, products, or services. Section 13.2.3 discusses the special features of compensation transactions.

1.8.4 Currency Problems

When handling international or bilateral projects involving industrial contract awards in countries with different currencies, it is advisable to define a lead currency so that the various contract packages can be compared with each other. However, shifting exchange rates pose a particular problem. The exchange rate risk can often not be assumed by the companies when contracts are concluded in a foreign currency. It is advisable to conclude industrial contracts in a lead currency. The problem remains that the companies concerned must bear the exchange rate risk. If necessary, the exchange rate must be hedged by appropriate insurance policies.

1.8.5 Cooperation of European Authorities Using the Example of ESA

When handling European space projects, industrial companies from the individual countries have to work closely together. The European Space Agency (*ESA*) was founded in May 1975 in its current form, merging ELDO and ESRO for the development, construction, and operation of joint European space projects. The Paris-based ESA currently has 22 European member states and maintains several sub-organizations, such as:

- ESA Headquarters in Paris, France
- European Space Research and Technology Centre (*ESTEC*) in Noordwijk, Holland
- European Space Operations Centre (*ESOC*) in Darmstadt, Germany
- European Space Research and Information Centre (*ESRIN*) in Frascati, Italy

In addition, most European countries still maintain their national space agencies, e.g., DLR (*Deutsches Zentrum für Luft- und Raumfahrt*) in Germany and CNES (*Centre National d'Etudes Spatiales*) in France. When ESA and national agencies collaborate on a joint project, a clear management approach must be developed to clearly define the roles and responsibilities of all involved. Abate, referring to the negative experience in this regard during the development of the European launch vehicle Europa I/II, wrote: "*... the member countries of ELDO acted in two ways in the implementation of the initial program, first as superior council members of ELDO, and then as national clients of industry acting on behalf of ELDO*" [46]; see Sect. 3.8.

For European projects, the organizational linking of the respective project partners plays a particularly important role. The tasks, responsibilities, and powers of all participants, the client, contractor, and subcontractor, must be clearly and unambiguously defined. The main contact person between the participating institutions and companies is the respective project manager. In order to improve communication between the international and national project offices, it is recommended to use the mirror image described in Sect. 5.3, according to which the project manager has a contact person with the same competencies; see Fig. 5.19. Furthermore, when handling European projects, care must be taken to ensure that uniform guidelines and procedures are drawn up according to which all project participants must proceed.

1.9 Project Management—An Integral Part of Mastering Complex Project Tasks

1.9.1 Project Management Does Not Have to Be Complicated and Should Be Simple

In the past, the impression has often arisen that project management is costly and complicated and therefore not applicable for many projects in the free economy and, here in particular, not for small projects. However, this is not true because the methods are in principle not complicated and mostly easy to understand for scientists and engineers. Therefore, it is important to follow simple procedures. The slogan "*KISS—Keep it simple stupid*" fits this paradigm. However, managers often find it difficult to transfer modern and efficient methods, which were mostly developed in connection with large-scale projects, to small and medium-sized projects or to projects within other industries. As already mentioned, in 1977 the German Federal Minister of Research and Technology published a handbook on project planning and implementation entitled "*Municipal Project Management,*" which offered simplified project management methods of the aerospace industry to municipal institutions [38]. In 1983, the German society for management training (*Wuppertaler Kreis*) published a book sponsored by the German Federal Ministry of Economics entitled "*Project Management in medium size companies*" [47]. The authors Groth; Erbslöh, Hugolshofer and Strombach did an excellent job of presenting the common project management methods in a simple and understandable form.

The principle that should apply to many disciplines applies also to project management; i.e., to use simple and pragmatic management methods! In other words, even complicated contexts should not be unnecessarily complicated but presented and described as simply as possible. Especially for such a versatile discipline as project management, simplicity is an absolute must!

1.9.2 Approaches to Simplification

Many new project management methods were developed in the USA and then introduced to Europe as imported goods. Often through use of the English language the impression arose that it must concern a complicated procedure, particularly since some management systems with an English name had no appropriate counterpart within Europe, for example configuration management. This important management instrument led to confusion, although some important functions of the configuration management are quite well-known for example in Germany, however, under another designation. Thus, drawing control, i.e., the collection and change control of technical drawings, well-known in Germany, is an important component of configuration management. The American-coined configuration management refers however to the entire technical documentation, i.e., specifications, building regulations etc., also to the collection and the change process. Particularly in the case of very complex systems, such as the development of a modern fighter aircraft or a power plant, the change process is particularly important since changes are known to cost money. The change process and the associated monitoring procedures are then also the main subject of configuration management; see Sect. 12.2.

However, in order to disseminate contemporary project management methods as quickly as possible in the interest of greater management efficiency, it is urgently necessary to clarify keyword-like management terms to all project staff in their original form and basic simplicity. Among other things, this is also a main concern of this book. In the following chapters, therefore, an attempt is made to present the authoritative project management methods in a simple manner.

1.9.3 Good Project Results Ensure the Success of the Performing Organizations

Projects, whether they are R&D projects or construction projects, are subject to economic constraints. In short, they must be completed within the framework of predefined budgets and the technical and schedule targets set have to be met. But this is a problem because the targets are usually estimates, and the range of accuracy can vary greatly. Of course, the degree of innovation plays a special role here. However, the determination of realistic specifications also depends on the thoroughness of the planning, the time available and the experience of the people involved. It is always disastrous if technical, scheduling, and financial planning is inadequate. As a rule, a lack of planning depth always leads to failure.

This leads to the conclusion that thorough project planning is indispensable for ensuring the success of project plans, but also of the companies concerned. However, this is just as true for projects of public authorities and institutions. The management crises described in Chap. 3 clearly show that newer problems of project management hardly differ from the problems of past times. To begin with, it is always the well-known planning deficits, which must also include organizational and management planning aspects; see Chap. 6.

Planning requires time and the knowledge of all parties involved. The deadline pressure imposed by the customer (*internally as well as externally*) is often the cause of highly compressed and therefore inadequate planning. What can be done to improve the situation? The catalog of what could be done is certainly very long. Often, it would be good if the customers would also be reasonable and allow the companies a little more preparation time for planning tasks because they themselves could also benefit from this. In addition, however, competition also plays a role because competitors often outbid each other.

The question now is, how much time do you need for planning? Are there time guidelines that should be adhered to? This is discussed in detail in Chap. 9. It is important that authorities, companies, or institutions effectively implement the *lessons learned* during project planning and project execution. Unfortunately, it happens repeatedly that the planning tasks to be performed are assigned to new teams that do not have the corresponding experience. In addition, there are usually no experience reports or they are not easily accessible. It is easy to see that it is of great, sometimes even existential importance for companies that planning tasks for complex project proposals are carried out by appropriate specialists and on the basis of planning standards; see Chaps. 9, 10, and 14.

1.9.4 Planning Effort Pays Off

Experienced project managers know that thorough planning in the early phases is important. The question is, however, how much planning effort and time is required, because planning also costs money, and one needs the necessary time. Mature planning, as described above, can certainly lead to the acceleration of the subsequent phases or, in any case, to strict adherence to all deadlines. However, it is quite understandable that the responsible managers find it very difficult to enforce a necessary planning time frame under the usually tough competitive conditions.

The reader might conclude that this is a theoretical and impractical claim and it is indeed difficult to prove with published hard figures that an increased time planning effort leads to improved results. The author has had the opportunity to observe the progress of various projects over the past five decades as a project staff member, and lack of planning has almost always been a major reason for failure. Chapter 3 describes three crisis projects that lacked consistent goal-oriented planning. The author's experience is that thorough planning, including alternatives, is essential for technical and economic project success. If you want to build a house or an airport, you must make a realistic plan beforehand, and you have to make a lot of assumptions. However, planning does not exclude errors and one must be prepared to change goals and/or plans if necessary. This is illustrated in Fig. 9.3 as a feedback loop. The project management processes developed in the 1960s were based on dynamics and flexibility. This was particularly evident in the Apollo program. This also underlines the importance of Phase A, in which trade-offs and compatibility analyses are used to simulate the future course of the project. As a result of the early phases, binding planning documents must be available, because otherwise you cannot successfully manage a program with a duration of several years. What counts at the end of a project is the success or sometimes the failure. The success story of SES-ASTRA is based on thorough preliminary planning work, consisting of binding specifications, the functional specification and detailed plans [48]. Further explanations on this subject are described in Chaps. 4 (*life cycle*) and 14 (*competition and bid preparation*).

Literature

1. Litke, Hans-D.: Projektmanagement, Carl Hanser Verlag, Munich, Vienna, 1991
2. Aucoin, B.: Michael, from Engineer to manager mastering the transition, Artech House, Inc., Boston - London, 2002, p. 155ff.
3. Ludwig Bölkow 70 years, in: MBB Aktuell 7/8-82, p. 3
4. Kutscherer, A.: Is Project Management and Organizational Development Economical and Efficient Even for Medium-Sized Companies? Battelle Conference Proceedings, Management of Development Projects, Frankfurt 26/27 March 1981, p. 91
5. Grochla, Erwin (ed.): Handwörterbuch der Organisation, C. E. Poeschel Verlag Stuttgart, 1980, p.1960
6. Webb, James, F.: Space Age Management, McGraw-Hill Book Company, New York, N. Y., 1969, p. 3.
7. Vollrath, Klaus: Die Chance es besser zu machen: Projektmanagement, in: Management + Seminar, 3/81
8. Adamowsky, Siegmar: The Second Organizational Revolution, in: Rationalization 19. yr. 1968 - 5, p.100
9. Source 6, p. 35
10. USAF: System Program Management Procedures, in: AFSCM 375, May 1966.
11. Hab, Gerhard, Wagner, Reinhard: Projektmanagement in der Automobilindustrie, Springer Verlag, 2004

12. PMI: Project Management Body of Knowledge (PMBOK), Drexel H., Pennsylvania, 1987.
13. Reschke, Hasso, et al: Handbuch Projektmanagement, Verlag TÜV Rheinland GmbH, Cologne, 1989.
14. Bölkow, Ludwig: Systems engineering means foreseeing behavior, in: VDI-Nachrichten No. 26, 25. 06. 82, p. 28.
15. Von Weizsäcker, Carl Friedrich: The Garden of the Human, Carl Hanser Verlag Munich, 1977, p. 19.
16. Campana, Christophe: Why Project Management is a Critical Factor for Every Company, in: Schott Eric and Campana Christophe, Strategic Project Management, Springer Berlin Heidelberg New York, 2005; p. 9.
17. Servan-Schreiber; J.-J.: Die amerikanische Herausforderung, Hoffmann und Gampe Verlag, Hamburg, 1968, p. 266.
18. Grau, Nino: The Importance of Project Management in *Business Process Engineering*, in: Lange Dietmar (Ed.) von Management von Projekten, Schäffer Poeschel Stuttgart, 1995, p. 170.
19. MBB: the nation's multi-purpose forge, in: GEO, April 1982, p. 85 and 99
20. Mellerowicz, Konrad: Betriebswirtschaftslehre der Industrie, Rudolf Haufe Verlag Freiburg 1981, vol. 1, 7th edition, p. 26/27
21. Siemens engineers oppose: The crisis of tomorrow is pre-programmed, in: Handelsblatt v. 16. 03.1982
22. Source 17, p. 267
23. Bernhard, Alfred: Japanisches Management - eine Kurzdiagnose, in: Management-Zeitschrift 20, 51(1982) No. 3, S. 106.
24. Terry, George R.: Principles of Management, Richard D. Irwin, Inc. Homewood, Illinois, 7th edition, 1977, p. 279.
25. Günthe, Joachim: Small is beautiful, in: Blick durch die Wirtschaft v. 15. 06. 1982
26. Grochla, Erwin and Thorn, Norbert: Die Matrix-Organization, Chancen und Risiken einer anspruchsvollen Strukturierungskonzeption, in: ZFBF-Kontaktstudium 29 (1977), S. 192
27. Martin, Charles C.: Project Management - How to Make it Work, Amacom, 1976, p. 49.
28. Lientz, Bennet P., Rea, Kathryn P., Project Management for the 21st Century, 2nd Ed., Academic Press, San Diego, 1995, p. 13.
29. Madauss, Bernd-Joachim: Nach Plan an der Katastrophe vorbei, in: Industriemagazin, March 1982.
30. Source 17, p. 94
31. Madauss, Bernd-Joachim: Creativity under Cost and Time Constraints, Construction & Design, February 1982.
32. Schubert, Johannes: Kreativität in der Großindustrie, lecture manuscript from the conference of the Protestant Academy in Tutzing, 17. 11. 1979.
33. Milliken, J. Gordon and Morrison, Edward J.: Aerospace Management Techniques: Commercial and Governmental Applications, Denver Research Institute - University of Denver Colorado, 1971.
34. Madauss, Bernd-Joachim: Utilization of Project Management from Aerospace for the Development of Technical Serial Products, Conference Proceedings on "Management of Development Projects", Battelle-Institut e.V., Frankfurt am Main, March 26/27, 1981.
35. Horn, Peter: Luft- und Raumfahrt in Europa, in: Süddeutsche Zeitung, May 18, 1982.
36. Economic Benefits of ESA Contracts, ESA Brochure BR-02, October 1979.
37. MDB: HELIOS solar probe final report, research and development contract QC-90-00 VP6/70, 1976.
38. BMFT: "Kommunales Projektmanagement", Deutscher Gemeindeverlag, Bonn, 1977
39. Source 33
40. Fabian, H.: Thoughts on the Policy of an Industrial State, in: MBB Report (not published), 1972.
41. AKM: Recommendations for the Management of Major Projects, 1977
42. Madauss, Bernd-J.: Planung und Überwachung von Forschungs- und Entwicklungsprojekten - Mit Beiträgen aus der Raumfahrttechnik, AIB-Fachliteratur; Gerberstr. 3b, 8202 Bad Aibling, 1978/81.

43. Madauss, Bernd-J., participated in the presentation. ELDO Paris 1972
44. Vandenkerkhove, J. A.: The Experience of ESRO in Managing International Space Projects, 3rd Orbital International Laboratory Symposium, Konstanz, 5-6 October; 1970 (Proceedings).
45. Palacios J.: Quelques aspects de la Politique Industrielle de I'ESA, in: ESA Bulletin, No. 12.
46. Abate, T.: Problems of Management of ELDOs Initial Programme, Sciences et Industries Spatiales 1/2, 1966.
47. Groth, Rainer; et al: Projektmanagement in Mittelbetrieben, Deutscher Instituts-Verlag, Cologne, 1983.
48. Ahrens, Wilfrid: ASTRA - Fernsehen ohne Grenzen, ECON Verlag, Düsseldorf, 1993

Definition of Project and Program

2

Contents

Abstract

Projects are undertakings with a defined Beginning and End, which are characterized by the features of time limitation, uniqueness, complexity, and novelty. Due to their interdisciplinary cross-sectional character, projects can bring about a temporary organizational change and, associated with this, also a redefinition of the areas of responsibility in the company (Madauss, Bernd-J., Handbook of Project Management, Schäffer-Poeschel Verlag, 1993, 5th edition, p. 9); In short: ***the term project refers to extraordinary undertakings.***

B. -J. Madauss, *Project Management*,
https://doi.org/10.1007/978-3-662-69057-4_2

The terms program and project are often used alternately. However, the term program is sometimes used for very large undertakings as a summary of related or specific projects [1]; see subsection 2.2.4.

2.1 A Project—What Is It?

2.1.1 Need to Define the Term Project

The questions "*What actually is a project*" and "*How can a project best be defined*" are of great interest because both questions are closely related to the application of project management methods and procedures. If one expects effective help from project management in the execution of difficult governmental and industrial tasks, corresponding success can only be achieved if these are actually projects because what is effective in the execution of projects may be useless or even disruptive in the handling of non-project tasks, such as routine activities. In order to avoid such unpleasant effects in advance, it is worthwhile first to check whether there are any project tasks at all before introducing project management methods. This is to be seen in particular also in connection with the generally observable trend of the introduction of fashionable terms. This development must be vigorously counteracted in order to avoid disappointment. The first important step is, therefore, to determine whether project tasks exist in the specific case.

The definition of the project term seems simple at first, but on closer inspection, it is then easy to see that many of the definitions used are also applicable to non-project tasks. Dülfer writes in this context [2]: "*The project term belongs to those terms which everyone understands and believes to know exactly, but whose precise characteristic definition reveals unexpected difficulties.*"

2.1.2 Special Features of a Project

When defining the term "*project,*" there is no getting around the need to determine with all clarity the special features that characterize a project, in order to draw a clear distinction from non-project tasks. Rüsberg already described the typical project characteristics in 1977 [3]. "*The term project should be understood to mean unusual undertakings which are characterized by*

- *one-time (acyclic) processes*
- *definable start and end times*
- *Task and objective*
- *the participation of several or numerous people, working groups, companies or institutions and*
- *Complexity*"

A characterization of projects used by all authors is the time limited by a beginning and an end. If one examines the life cycle of a project, one finds that every project, no matter what type of project it may be (*research, organization, plant construction, etc.*), runs through a very specific cycle that is always similar. The life cycle of a project begins with the planning and preparation process (*planning phase*) in order to be able to carry out the actual project task (*realization phase*) in a planned manner, and the project results then represent the project conclusion; see Chap. 4.

Projects are undertakings with a well-defined beginning and end; see Sect. 4.2. Although this definition is trivial and absolutely correct, in practice the beginning and end are not always clearly recognizable. Many times, a project begins diffusely and ends in the same way. However, the importance of defining a project as thoroughly as possible from A to Z cannot be emphasized clearly enough. It is of great importance to clearly grasp all elements of a project in order to be able to make a correct investment assessment that includes all phases. Often, baseline studies are carried out for a future project at a very early stage, which is sometimes almost forgotten by the time the project officially starts. However, financially, and more importantly, from the point of view of forming opinions, they clearly belong to the future project. The end of the project has a much more important meaning for the evaluation of the project events. If the end of the project is not clearly ascertainable according to the motto "*open end,*" then the project costs of all individual tasks are often also not clearly ascertainable in a post-calculation.

Only with the clear end of the project can you finally give an answer to the question "*was the project worthwhile*" because now you are sure that the project is not burdened by further costs and nothing stands in the way of a profit and loss statement.

2.1.3 Historical Review

See subsection 1.2 "*Space Age Management*" of Webb [4].

2.1.4 Systems Engineering and Project Management Are Required for Implementation of Projects

The realization of industrial and public projects of today is predominantly aimed at systems engineering solutions. Even in projects where the terms system or systems engineering are not explicitly used, it is often easy to recognize from the objective that in principle it is about the realization of systems after all. Systems engineering and project management are closely linked; see Fig. 7.21. Today, the construction of a factory building, a power plant, or a public facility, such as a modern swimming pool, requires systems engineering approaches more than ever before. In this regard, Kappel and Schwarz stated [5]: "*It is hard to overlook the fact that systems thinking has not only changed science, but beyond that, it has also changed daily life in a certain sense.*"

Today, the realization of projects depends on external influencing factors to a far greater extent than was the case a few years ago. Technological projects, which can generally be regarded as open systems, must be carefully networked with their environment, i.e., with external systems, as early as the planning stage so as not to appear as a sudden disruptive factor. This means that in the early stages of a project (*for example, in the concept phase*), compatibility studies must already be carried out in order to analyse in detail the interactions (*input/output variables*) of the system to be created with the environment or other technical systems; see Chap. 4. In this context, not only technical and economic factors play a role, but increasingly also political, ecological, and sociological factors. The development and implementation of microprocessors is a vivid example of this. Henning writes in this context [6]: "*The systems engineering work on the project includes also economic, political, sociological, and ecological dimensions in addition to technical ones.*" The unintended input–output factors, for example, the occupational changes that occur when microprocessors are introduced in industry and government, must be studied with the same meticulousness as the intended interactions.

The realization of a project begins with the project objective which is or should be manifested by a requirements catalog. At this point, the future system is not yet defined in its configuration but only the requirements that will be placed on the future system are known. During the concept phase, the future system is formulated taking into account the specified boundary conditions. The result of the concept phase is to be recorded in the concept documentation, consisting of preliminary planning documents and a preliminary system specification. The issuance of the system specification (first/preliminary edition) is an important milestone in the project's history. It anchors the future system and provides an important reference point for all subsequent activities.

The developments of the management of projects are trend-setting in the solution of complex system tasks and at the same time pave the way for new management and organization concepts at the authorities and the industry. Aerospace often takes the role of pioneer in this process, for example in the development of the often-cited concept of matrix organization [7]. Aerospace has always been an industry with a high degree of innovation. Extreme project and system goals repeatedly challenge pioneering technological and administrative achievements. Technological in particular because there is the challenge of producing equipment that must function flawlessly for long periods of time without the ability to be maintained or replaced during those periods. Modern communications satellites, for example, are expected to remain fully functional for a period of more than 15 years under space conditions without maintenance. On the other hand, such peak performance must be achieved under very strict weight and space constraints, as the available transport equipment has only a limited payload capacity.

The use of modern project management methods in combination with the knowledge of systems engineering is recommended in all areas of our society, where for example public construction projects, industrial projects, or tasks in social areas are concerned.

2.2 Literature Analysis

2.2.1 Literature Compilations

A look at the project management literature shows that several approaches have already been taken to define the term project. The typical project characteristics highlighted are largely compatible. However, characteristics have also been used which may be typical of projects, but which are also characteristic of non-project tasks.

In the following section, existing literature is first used to examine how the term project has been defined in current works. For this purpose, the author has drawn on what he considers to be the most authoritative publications. Let us now turn to the various definitions of the term from different sources:

(1) Duden 1980 [8]:

Although the Duden (a German dictionary) is not to be regarded as a technically competent work in this context, it is included because of its wide distribution. There, the twin terms program and project are defined as follows:

Program:

a) planned procedure
b) the details of a project precisely defined according to a plan

Project:

In addition to the term "*project.*" the Duden also describes the twin term "*program.*" After all, both terms are often used side by side and completely equivalent to each other. Sometimes even in mixed form in one and the same document:

a) [large-scale] investigation planned or already underway;
b) [large-scale] project
   - construction project
   - an interesting, bold, fantastic project
   - the highway bridge is a gigantic project
   - a project for the development of solar energy
   - prepare, realize, discard a project
   - deal with a project involving space travel

(2) NASA, 1963 [9]:

Now here is NASA's definition of terms, taken from the Apollo Terminology Manual:

- Project—A scheduled undertaking, within a program, which may involve the research and development, design, construction, and operation of a system and associated hardware, or hardware only, to accomplish a scientific or technical objective

- Program—A related series of undertakings designated to accomplish a broad scientific or technical goal. Attainment of such long-range goals may be accomplished by the implementation of specific projects.
- Space Flight Project—a task within the real program of space flight that leads to the accomplishment of mission objectives, e.g., collection of scientific data, transportation of cargo and personnel between two terminals, and satisfaction of commercial requirements.

(3) NASA, 1965 [10]:
NASA's Program Planning Manual specifies program project levels: "*Manned space flight schedules will be segregated into appropriate levels based on a program work breakdown structure and in accordance with the existing organization.*" Example: Overall Manned Space Flight Program

- Level 1: Individual Flight Programs
- Level 2: Projects
- Level 3: Prime Systems
- Level 4: Subordinate Systems

(4) NASA, 1967 [11]:
NASA's Apollo Program Management Manual describes the program levels according to their management importance:

- Level 0: Policy Level
- Level 1: Program-wide management in the Apollo Program Office
- Level 2: Management at the NASA Field Center level
- Level 3: Management at the Prime Contractor level

(5) Rüsberg, 1971 [3]:
"The term project shall be understood to mean unusual undertakings which are characterized by:

- one-time (acyclic) processes
- definable start and end times
- Task and objective
- the participation of several or numerous people, working groups, companies or institutions, and
- complexity is characterized"

(6) Dreger, 1975 [12]:
A project is a complex of measures, which is characterized by the fact that it has a finite extent, is carried out only once, and has a defined starting point and a defined

goal. A project is further characterized by the fact that it must have a definite predetermined goal, which can also be controlled in its fulfillment. It is carried out "*one-way-through.*"

A project is always an independent or at least localizable task but can also be a part of a superordinate task. This seems to be the definitional interface between project and program.

(7) Archibald, 1976 [13]:
Archibald defines the terms program, project, and task as follows:

- Program: A long-term undertaking that is usually made up of more than one project. Sometimes used synonymously with project
- Project: A complex effort, usually less than 3 years in duration, made up of interrelated tasks performed by various organizations, with a well-defined objective, schedule, and budget
- Task: A short-term effort (*usually 3–6 months*) performed by one organization, which may combine with other tasks from a project

(8) Dülfer, 1982 [2]:
The following project characteristics are named and critically examined by Dülfer with the inclusion of various sources; the applicable characteristics are marked with a plus sign (+):

- Target: the target is considered a constitutive characteristic of the project concept (+)
- Time determination; an undisputed project feature (+)
- Uniqueness: although relatable, this characteristic can be considered true (+)
- Novelty: not constitutive for the project concept
- Complexity; has no constitutive significance as a project characteristic, but can be used as a criterion for differentiation (+)
- Task-related budget: can hardly be used for project definition
- Legal-organizational assignment: this characteristic obviously plays a relevant role in the differentiation of various project types (+)

(9) Groth, Erbslöh, Hugelshofer, and Strombach, 1983 [14]:
The above cited authors use the following short form: "*project is a temporary (time-limited) project.*"

(10) Reschke and Svoboda, 1983 [15]:
"*A project is a task that is particularly characterized by the following features:*

- *Existence of a set of project objectives (regardless of the sharpness or vagueness of their formulation)*

- *Time limit, with a relatively long interval between defined start and end time (task with relatively long repetition cycle)*
- *Interdisciplinary character of the task, which requires a special coordination that goes beyond the existing departmental structure (interdepartmental combination of specialists)*
- *Relative novelty that makes a project distinguishable from other tasks (not a routine task) and carries a relative risk of goal achievement*
- *Financial framework for the costs and investments incurred in connection with the project processing (project budget)"*

(11) Rosenau, 1984 [16]:
Rosenau first defines the term project:

- Projects are one-of-a-kind undertakings that originate when something has to be done. Projects are temporary undertakings with a specific objective that are accomplished with appropriate resources
- As a general rule, if, for example, an amplifier is different from one another, then building each is a project. If each amplifier is virtually identical, you have a production line and are not engaged in a project per se
- There are many ways to characterize projects. "*Hardware project*" and "*software project*" are common terms, depending on whether the final result is a tangible product (hardware) or a computer program accompanied by a report or some other form of documentation (software)
- Thus, the product or result of a project is a second characteristic.
- A project is not an ongoing activity, but rather an undertaking that ends with a specific accomplishment.

The second step is a definition of the term program:

- "*Program*" is sometimes used synonymously with "*project.*" Thus, the expression "*program management*" may be used interchangeably with "*project management.*" However, programs are not normally considered to be a collection of interrelated projects. Some organizations use task management as weIl
- Program management, project management, and task management as well
- Program management, project management, and task management are generally identical. But programs are usually larger than projects and projects are usually larger than tasks

(12) Platz and Schmelzer, 1986 [17]:
Here is the definition according to Platz and Schmelzer: "*R&D projects are characterized by certain features:*

- *Target agreement*
- *Time limit*

- *Limited resources*
- *Uniqueness*
- *Complexity*
- *Size*
- *Novelty*
- *Uncertainty and risk*
- *Dynamics*
- *Interdisciplinary collaboration*"

According to the authors, the first three are concept-defining project characteristics.

(13) Burghardt 1988 [18]:
Let us now consider the main criteria of a project as defined by Burghardt:

- Unambiguity of the task
- Defined duration with a fixed end date
- Coordinated cost constraints and
- Clear responsibilities

Burghardt further speaks of the trinity, "*product-project-process*" and states "*the product is understood as the product or output of an activity project; it is the result of the development and projecting efforts and thus the 'output' of the development or a projecting.*"

He then continues: "*A project, on the other hand, is the goal-oriented plan for the manufacture of this product in the aforementioned sense. A project is necessarily always clearly delimited in its temporal sequence, i.e. it has a start and end date. The uniqueness in the general conditions of a project is probably the decisive characteristic of a project.*"

(14) Schelle, 1989 [19]:
Schelle takes up the project characteristics described by Dülfer [2] and essentially confirms the concepts of

- Target
- Time determination
- Uniqueness
- Complexity and
- Legal-organizational assignment

Accordingly, these are constitutive features, while the features have no relevance for term differentiation.

- Novelty and
- Task-based budget

Table 2.1 Definition of project characteristics

	B) Definition notes (see Sect. 2.2)												
A) Project characteristics	1	2–4	5	6	7	8+14	9	10	11	12	13	15	16
a. One of a kind—uniqueness	X	X	X	X		X	X		X	X	X		
b. Time constrained with clear start and end times		X	X	X	X	X	X	X	X	X	X	X	X
c. Well-defined assignment, responsibility, and finished product		X	X	X	X	X		X	X	X	X	X	X
d. Complexity			X	X	X					X			
e. Participation of many individuals, groups, companies, etc.			X										
f. Interdisciplinary character of the assigned project								X		X			
g. Relative innovation								X		X			
h. Financial constraints and limited resources					X			X		X	X	X	X
i. Size										X			
k. Risk										X			
l. Dynamics										X			
m. Separation from other efforts													X
n. Project-specific organization						X							X

(15) PMI, 1989 [20]:

PMI, the American professional organization for project management, defines a project as follows: "*Any undertaking with a defined starting point and defined objectives by which completion is identified. In practice, most projects depend on finite or limited resources by which the objectives are to be accomplished.*"

(16) DIN 69901 [21]:

The DIN Committee for Project Management has defined the term project as follows:

"*A project that is characterized essentially by uniqueness of conditions in its entirety, such as.*

- *Target*
- *Time, financial, personnel or other limitations*
- *Differentiation from other projects and*
- *Project-specific organizations*"

Table 2.1 below summarizes the characteristics used to define project features.

The literature excerpt reproduced here is a representative cross-section of all previous definitions. In the following Sect. 2.2.2, all matching project characteristics are identified. This is followed by a critical review of the statements made in Sect. 2.2.3 and finally by a consideration of the terms "*project*" and "*program*" in Sect. 2.2.4.

2.2.2 Matching Project Characteristics

Out of 16 definition sources, only 12 could be used; see Table 2.1. Definition sources (1), (3) and (4) do not contain universally valid and comparable project characteristics. Definition sources (8) and (14) are in principle congruent, which is why only definition source (8) was used. The determined congruence of the definitions used for a project leads to the following ranking of the project characteristics (see values in brackets):

a) One-time (acyclic) sequence/uniqueness (8/12)
b) Time limit/clear start and end date (12/12)
c) Clear task, responsibility, and goal setting (11/12)
d) Complexity (4/12)
e) Participation of many people, working groups, companies, etc. (1/12)
f) Interdisciplinary character of the task (2/12)
g) Relative novelty (2/12)
h) Financial framework and limited resources (6/12)
i) Size and mass (1/12)*
k) Uncertainty and risk (1/12)
l) Dynamics (1/12)
m) Demarcation from other projects (1/12)
n) Project-specific organization (2/12)

**Letter "j" was intentionally omitted.*

It is clear from this list that only the first three project characteristics show a sufficiently high level of agreement, i.e., they are used by at least two-thirds (8/12) of the 12 usable definition notes. The remaining project characteristics are in the range of up to 50% (6/12).

Assuming that the sources used are competent statements, the quantitative cross-sectional statement made here carries appropriate weight.

2.2.3 Critical Examination of the Statements

A critical examination of the statements reveals that only features (a) and (c) are unambiguous project features because feature (b) is in principle also applicable to non-project tasks. The time limit is an essential project characteristic for clearly distinguishing projects from routine tasks. The completion of an R&D task is the best example of this. But the uniqueness of the task is also a typical project characteristic. The task did not exist before and will not exist again afterward in exactly the same form, because otherwise, it would be a routine task and not a project. In addition, however, other characteristics, such as complexity, are typical of projects, since they can be increasingly determined in connection with project tasks. At this point, the author will now attempt to justify in detail the definition given at the beginning of this article:

- Projects are undertakings with a defined beginning and conclusion
- Up to this point, the statement is indisputable and is confirmed by all sources.
- ...which are characterized by the features of temporal limitation, uniqueness, complexity, and novelty

In the case of the additional characteristics, only the time limit is still undisputed, while the uniqueness, complexity, and novelty are only partially recognized as typical characteristics.

However, the author believes that the characteristics of uniqueness, complexity, and novelty, while not unique to projects, can be considered typical for projects; especially for R&D.

Uniqueness is distinctive for projects because it describes the prototype-like project character very well. Projects are made-to-measure suits and thus definitely unique. Even plant projects in which standard components are used are unique projects in many respects, e.g., because of their location.

Of course, complexity does not only occur in the context of projects. Nevertheless, complexity is a typical project phenomenon, both in technical and managerial terms. Technical complexity often results from complex tasks with a high degree of innovation, which can often only be solved with the help of systems engineering; see Sect. 1.4. The example of the German-American solar probe HELIOS is a good illustration of technical complexity [22]:

- Extreme mission objective: approaching the Sun to within 0.3 astronomical units (AU), resulting in severe thermodynamic stresses
- Ten (10) physically very different experiments, whose mutual compatibility (mutual non-interference) had to be ensured
- Large magnetic purity requirement
- Fault-free interaction of several technically different systems (e.g., mechanical structures, position control components, data transmission equipment, energy supply systems, thermal control units)
- Maximum communication distance 2 AU

Closely related to technical complexity is managerial project complexity, for example, when a large number of technically complex interfaces need to be defined, specified, and monitored. In this context, the involvement of a large number of companies and/or authorities leads to further management complexity. At times, the Apollo program involved up to 275,000 people and more than 4000 industrial contractors and suppliers [11].

The characteristic of novelty is also not only applicable to projects but at the same time it is also typical for projects. An R&D project is by definition a novel task. New ground is often broken, especially when very complex tasks are involved. In short: If we break new ground, which is usually the case with projects, we can speak of the novelty of the project, and this is then also a typical project characteristic.

- …and, because of their interdisciplinary cross-sectional character, can bring about a temporary organizational change and, associated with this, a redefinition of the areas of responsibility in the company.

In the vast majority of cases, projects have an interdisciplinary cross-sectional character. Vollrath [23]: "*Projects are always initiated when a larger project goes beyond the normal operational process. In such cases, several departments of a company are almost always affected, and they work together in an interdisciplinary manner within a project organization.*"

If a company or an authority intends to realize a project task, it has to be decided who is to be made responsible for the realization. Assuming that the company or authority in question realizes several project tasks at the same time, it is necessary to delegate the task to a suitable organization. In the case of interdisciplinary project tasks, the most sensible solution is to involve a project group. The project manager must carry out his tasks in accordance with the specified objectives and, to a certain extent, represents the extended arm of the company management.

2.2.4 Project or Program

As already mentioned, the terms project and program are often used alternately. The definition in the Duden [8] does not distinguish clearly enough to be able to use the terms unambiguously. Here, however, the use of the term program for intended processes is already pointed out. In fact, the term "*program*" is widely used for processes, especially in the field of software, program languages, etc.

However, let us turn again to the concept of program in the sense of project management. NASA defined the term "*program*" in 1968 in the context of the Apollo project [9], as a summary of related undertakings or specific projects. This relationship is described even more precisely in NASA's Program Planning Manual [10]. Here a clear hierarchy was defined, according to which the Apollo manned space flight program was structured as follows (*excerpt*):

– Level 1:	Individual flight programs
– Level 2:	Individual projects (*examples*)
	– Saturn V rocket
	– Lunar vehicle (*LEM*)
	– Ground facilities
	– etc.
– Level 3:	Main systems (*examples*)
	– Rocket stages
	– Communication center
	– Test facilities
	– etc.

– Level 4:	Subsystems (*examples*)
	– Engines
	– Tanks
	– Onboard computer
	– etc.

The term program was clearly used here as a collective term for individual projects; certainly, a sensible approach. Rosenau [16] also points out the frequent mixing of the terms program and project and finally says "*However, programs are normally considered to be a collection of interrelated projects.*" Archibald [13] defines a program as follows: "*A long term undertaking which is usually made up of more than one project.*" To avoid misunderstandings, it is therefore suggested at this point not to use the term program alternately with the term project, but only as a collective term for several projects.

2.3 Final Consideration

The definition of a project given by the author at the beginning of this Chapter is, as can be seen from the literature selection reproduced in Sect. 2.2.1, only congruent with the definitions of other authors in some points. The first two project characteristics, namely "*time limitation*" and "*uniqueness,*" can be clearly confirmed by the sources examined. These two characteristics are indisputable for projects.

The additionally used characteristics "*complexity*" and "*novelty,*" on the other hand, are not considered by many authors to be concept-defining characteristics; see Sect. 2.2.2. Detailed reasons were therefore given as to why the use of these project characteristics was included by the author as typical for projects. This also refers to the observation that projects are characterized by an interdisciplinary cross-sectional character, which then often leads to organizational changes.

The final statement: " ... *a project is an extraordinary undertaking*" summarizes the project concept in a short and concise form.

Literature

1. Madauss, Bernd J.: Handbuch Projektmanagement, Schäffer-Poeschel Verlag, 5th edition Stuttgart 1993, p. 9
2. Dülfer, Eberhard: Projektmanagement INTERNATIONAL, C.E. Poeschel Verlag, Stuttgart 1982, p. 4 and 7
3. Rüsberg, Karl-Heinz: Die Praxis des Projekt-Management, Verlag Modeme Industrie, Munich 1971, p. 20.
4. Webb, James F.: Space Age Management, McGraw-HiIl, New York, NY, 1969, p. 35.
5. KappeI, Rolf and Schwarz, Ingo A: Systems Research 1970-1980, Vandenhoeck & Ruprecht, Göttingen, 1981, p. 27.
6. Henning, Th.: Can Power Plant Planning be Improved with the Help of Systems Engineering? VGB Kraftwerkstechnik 60, March 1980, p. 164

7. Milliken, J. Gordon and Morrison, Edward J.: Aerospace Management Techniques: Commercial and Governmental Applications, Denver Institute - University of Denver Colorado, 1971.
8. Duden, Vol. 5, Bibliographisches Institut Mannheim/Vienna/Zurich, Dudenverlag, 1980 edition; p. 2050.
9. NASA: APOLLO Terminology, NASA SP-6001, 1963, p. 75f, p. 91.
10. NASA Handbook NHB 2330.1: Program Scheduling and Review Handbook, 1965, p. 9.
11. NASA Handbook: NASA-APOLLO Program Management, 1967, pp. 1 to 7.
12. Dreger, Wolfgang: Projekt-Management, Bauverlag GmbH, Wiesbaden and Berlin, 1975, p. 2.
13. Archibald, Russel D.: Managing, High-Technology Programs and Projects, John Wiley & Sons, Inc, New York, London, Sydney, Toronto, 1976, p. 18 and p. 25.
14. Groth, Rainer, Erbslöh, Fritz Dieter, Hugeslhofer, Hans-Jacob, Strombach, Manfred E.: Projektmanagement in Mittelbetrieben, Deutscher Instituts-Verlag, Cologne, 1983, S. 10
15. Reschke, Hasso and Svoboda, Michael: Projektmanagement - konzeptionelle Grundlagen, Beiträge der Artikelreihe in Frankfurter Zeitung Blick durch die Wirtschaft (GPM-Verlag, 1983), p. 6.
16. Rosenau, Jr. Milton: Project Management for Engineers, Lifetime Learning Publications, Belmont, Cal. 1984, p. 2ff.
17. Platz, Jochen and Schmelzer, Hermann J.: Projektmanagement in der industriellen Forschung und Entwicklung, Springer-Verlag, Berlin, Heidelberg, New York, London, Paris, Tokyo, 1986, p. 2.
18. Burghardt, Manfred: Projektmanagement (Leitfaden für die Planung, Überwachung und Steuerung von Entwicklungsprojekten), Siemens AG, 1988, p. 16ff.
19. Schelle, Heinz: Zur Lehre vom Projektmanagement, from: Reschke, Schelle, Schnopp, Handbuch Projektmanagement, Vol. 1, Verlag TÜV Rheinland, Cologne 1989, p. 5ff.
20. Project Management Institute (PMI): Project Management - Body of Knowledge (PMBOK), January 1989, p. 4-3.
21. Project definition according to DIN 69901
22. Madauss, Bernd-J: Planung und Überwachung von Forschungs- und Entwicklungsprojekten, AIB Fachliteratur, Bad Aibling, 1978/82, p. II-2.
23. Vollrath, Klaus: Die Chance es besser zu machen. Project Management, in: Management + Seminar, 3/81

Crises During Project Execution

3

Contents

B. -J. Madauss, *Project Management*,
https://doi.org/10.1007/978-3-662-69057-4_3

Abstract

We should learn from mistakes and, if possible, not repeat them. This presupposes that mistakes made are recognized as such, documented, evaluated, and ways found to make the findings available to future project staff, and above all to the decision-makers. However, this is often neglected. The failure is usually justified by "short time" and "a lack of budget." A lot of know-how is lost as a result and the cycle begins again—mistakes made are repeated. It is important that companies make it their duty to review what went well and what did not go so well after the completion of a project. As a further measure, project managers should report on the project status, problems, and lessons learned at regular intervals during project implementation.

Although public authorities and companies are generally aware that the successful completion of projects requires the consistent implementation and application of proven professional project management, it happens time and again that those responsible fail to observe this principle for a variety of reasons. This is true for large construction sites, complex industrial plants, public sector, and IT projects. But such mishaps also occur in smaller industrial projects, which can lead to financial losses or damage to the company's image. Sometimes those affected are faced with the question "*abort or continue.*" On the one hand, aborting can be financially sensible to avoid greater damage, but on the other hand, it can mean the loss of a customer and be understood as an admission that mistakes have been made. Often one can see that management mistakes of the past are repeated. For example, one should have conducted a detailed feasibility study first, and/or reserved more time for the planning phase, and/or clarified the resource issue to identify project risks early on and include them in the planning process.

Project crises exist in both hardware and software projects, but corresponding empirical values are rarely available. Therefore, the initiative of the "Standish Group," founded in 1985, which has set itself the goal of improving the management of projects in the software sector, is gratifying, and these findings also apply analogously to hardware projects. The group publishes annually an appropriate report, "*Standish chaos report,*" in which the successes and failures of projects are reflected. Approximately 60,000 software projects were evaluated, 50% from the USA, 25% from Europe, and another 25% from other parts of the world. Long- and short-term IT projects from all industries in the areas of development, production, and service were affected. The following are the results of the 2012 survey [1].

Class	Description	Proportionate (%)
1	Projects successfully completed on request, on time, and without cost overruns	29
2	Partially successful projects where not all targets (performance, deadlines, costs) were achieved as planned	52
3	Projects that were not successful and were canceled	19

However, it should not go unmentioned in this context that the professional handling of projects is usually an enormous challenge for those involved, since the product and thus also the project complexity is increasing to a higher degree, which leads to demands for new technologies and difficult interfaces and entails industrial cooperation. In the following, seven typical project requirements are described, which can occur both individually and simultaneously:

1. Project size: system projects, e.g., transportation systems, IT networks, aerospace projects, and construction projects including infrastructure
2. Quality requirements: e.g., high quality and system reliability standards and improvement of availability, maintenance, and technical safety
3. Innovation requirements: Development of new technologies and processes, e.g., in plant projects, IT projects, and aerospace tasks
4. Project complexity: Projects with a high degree of networking and technical as well as programmatic interfaces, e.g., integrated solutions (*mechanical, electronics, and software*)
5. Schedule targets: deadline limits (*targets*) set by the competition
6. Cost targets and financing: benchmarking, competitive pricing, and financing strategies
7. International project management: international collaboration and teams, cultural barriers, and language rules

Section 3.1 describes the problem of repeating past mistakes because the transmission of management experience, including mistakes made, from previous projects is poorly controlled, especially when project teams have not been adequately prepared and trained before starting their task.

The fact that projects often get out of hand is a well-known phenomenon and is discussed in Sect. 3.2. But what are the reasons for this? Is it a failure of management, or are there special circumstances, such as overriding political influence, increased deadline and cost pressure, or other external and/or internal influences? In this chapter, the documented problems of the major projects Capital Airport Berlin (*BER*) and Hamburg Elbe Philharmonic Hall (*ELB*) were examined and contrasted with the project crises of the Development of the European Rocket described in Sect. 3.8, as they occurred about 50 years ago. The crisis progression of the three projects is similar in principle. In conclusion, the main causes identified are presented in 10 problem points to provide future project managers with practical and promising recommendations for professional project management. Of course, the problems identified did not only occur in the major projects mentioned here, because the list of projects with misalignment could be extended. But project mishaps of medium-sized and small projects, which are usually not in the public spotlight, also cause concern for investors.

Section 3.3 is devoted to the topic of "*learning from mistakes.*" The preparation of experience reports is a focus of this section regarding the involvement of the taxpayer in public projects. Furthermore, attention is drawn to the special nature of the task (*projects are unique*) and the lack of interdisciplinary project training evident in this context, although there are professional associations, such as the Society for Project Management (*GPM*) in Germany, which has been conducting certified PM training for several years; see Sect. 15.2.3.

Section 3.4 deals with the problem of inadequate scheduling, cost planning, and adherence to planned values, which can be observed time and again during project execution. This deficiency can be found in many projects, which in consequence usually also considerably worsens the economic result of the company.

The problem of consequential costing is addressed in Sect. 3.5. The focus is on the importance of life cycle costs (*LCC*) and their influence in the early project phases. In many projects, life cycle costing is an important factor for investment planning and, in conjunction with the calculated revenues and operating costs, the prerequisite for determining the *return on investment* (*ROI*).

Section 3.6 analyses organizational problems because another serious problem that can often be found in project management is organizational deficiencies. These are often internal company organizational problems, for example when the task, responsibility, and authority of the project manager and the cooperation between the project and specialist departments are not clearly defined. The creation of a project management office (PMO) can help to improve the situation within the framework of a matrix organization; see Sect. 5.2.5.

Section 3.7 describes cooperation with partner companies within the framework of joint ventures or consortium solutions, which often leads to considerable organizational problems. In the handling of international projects, this is often further complicated by political boundary conditions. Crises in project management are often caused by inadequate preparation and a lack of management training for project personnel.

The topic of the ELDO crisis and its management is presented in Sect. 3.8 is a project of the past but provides a typical lesson-learned example of how to master the future from a project nightmare of European spaceflight by specifically initiating corrective measures and learning from the mistakes made.

Europe has taken an initiative for improvement. On May 30, 1975, the European Space Agency (*ESA*) was founded by 22 member states. The predecessor organizations ELDO (*launcher*) and ESRO (*satellites*) were merged as part of ESA. At this time the development of a new launch system started under the name "Ariane." It was understood that better management under the control of ESA would lead to success. In this respect, it is worth mentioning that the performance of the Ariane launch vehicles has been very successful, and that Ariane 5 is the world's most reliable launch vehicle; i.e., with a reliability of close to 100%.

On December 25, 2021, NASAs James Webb Space Telescope (*JWST*) was launched by Europe's Ariane 5 to its final destination; i.e., the Lagrange-Point (*L2*), at which the gravitational forces of the Sun and Earth are approximately equal, creating a stable orbital location.

3.1 Repetition of Past Mistakes

Why is it difficult to transfer management know-how already gained in overcoming crises to follow-up projects as "*lessons learned*"?

Problems in project execution, such as technical, scheduling, and financial miscalculations, are not a new phenomenon but a familiar one from the past. It has a lot to do with the difficulty of predicting the future with certainty. In projects with high innovation content, it is particularly difficult to determine schedules and costs. C.F. v. Weizsäcker states: "*Economic forecasting, like every forecast, is an eminently difficult art*" [2]. To a certain extent, this is an excuse for planning errors. On the other hand, we now (*in 2023*) have widely tested management methods of professional project planning and monitoring, and computer-aided cost estimation models to minimize the degree of misestimation. This is discussed in detail in Chaps. 9 and 10.

In most cases, it is not just a matter of misjudgment on the part of those responsible, but of serious disruptions caused by external influences. In national and/or international public projects, there are often overlapping political squabbles to the detriment of the project. This results in major delays, as important decisions are not made clearly and unambiguously or are made too late. It is of utmost importance that the project executing agency (*e.g., one or more authorities or companies*) has already clearly resolved all technical, administrative, and so-called political problems prior to or at the beginning of the project, so that the appointed project management can start project implementation (*planning, execution, and monitoring*) on the basis of fully coordinated project requirements. This is by no means an exaggerated theoretical demand, but a common practice that is obvious to every project expert. In order to avoid chaotic conditions, this process should be carried out thoroughly in the early phases (*see Chap. 4*), and the future project manager should, if possible, be involved in the preparatory processes from the very beginning.

Organizational deficiencies and unclear responsibilities often lead to delays and associated cost overruns. For this reason, the responsibility and authority of the respective project management should be clearly defined at the start of the project or phases. This helps all parties involved, i.e., the project sponsor and the project management at all levels; see Chap. 5.

3.2 Projects That Got Out of Hand

The project failures of two major German construction sites described in this Section are based on serious technical and management shortcomings. It is not the intention to expose the responsible parties by enumerating serious management failures, but rather to contribute to raising awareness of management problems. Learning from mistakes (*lessons learned*) is the goal of following subsections and in no way an improper criticism. It is important to understand that project management is a complex cross-cutting task that should not be complicated by obstructive interventions, such as improper political influences.

At the time of writing the eighth edition of the German version of this book (*2021/2022*), two major projects, Berlin's new Capital Airport (*BER*) and Hamburg's Elbe Philharmonic Hall (*ELB*), were in the public eye with devastating negative headlines regarding completion dates and cost overruns. Media reports such as "*the project is coming apart at the seams*" or "*it is running haphazardly toward disaster*" were the order of the day. We have refrained from discussing additional projects, as the management errors of the previously mentioned BER and ELB projects are exemplary for many project mishaps.

Project problems usually arise from an accumulation of errors that disrupt the process. The following problem descriptions of the Elbe Philharmonic Hall Hamburg and Capital Airport Berlin projects represent a snapshot at the time of writing the eighth edition of the German version of the book. The backgrounds that led to considerable problems in project execution were analysed. The unbalanced nature of these projects was denounced in numerous media reports and portrayed as a blatant management crisis. The problems were also hotly debated among the general public; however, due to a lack of information, objective assessments are difficult for many people.

Finally, at this point, the causes of the problems encountered in the BER (*Sect. 3.2.1*) and ELB (*Sect. 3.2.2*) projects were analysed in the context of the causes of the mistakes made 40 years ago in the development of the Europa III launcher (*Sect. 3.8*). The failure scenario is strikingly consistent for the three projects studied (*BER, ELB, and ELDO*). In conclusion, the main causes identified have been clearly highlighted in Table 3.1 (*Sect. 3.2.3*) in order to provide future project managers with practical and promising recommendations for professional project management.

3.2.1 Berlin Capital Airport

For orientation, Fig. 3.1 shows the terminal of the new capital city airport Berlin (*BER*). When can BER finally be put into operation was a question often asked in recent years, and how much more will the airport cost? In the meantime (*September 2017*), it looks like completion will not be possible until 2018 or 2019. Around the world, airport projects of similar size and with comparable infrastructure have been successfully completed in recent years. It can therefore be assumed that the handling of complex projects is basically not new territory for architects and engineers.

Table 3.1 Comparison of identified management deficiencies of the projects BER, ELB, and ELDO

Problem areas	Problem description	Reference[a]
1. Schedule planning—An important PM task	BER—The established schedule plans were not realistic ELB—Integrated schedule planning was missing for a long period of time ELDO—Schedule plans were frequently not coordinated	Chap. 9
2. RFP based on completed planning	BER—RFP premature—detailed plans missing ELB—RFP premature—detailed plans missing ELDO—Contracts released by National agents with planning errors	Chaps. 4 and 14
3. Project control—An important PM task	BER—Generally ineffective ELB—Generally ineffective ELDO—Only coordination without central control	Chap. 9
4. Realistic cost estimates	BER—Deficient cost planning—Missing planning documentation ELB—Deficient cost planning—Missing detailed plans ELDO—Missing cost competition, cost reimbursement pricing	Chap. 10
5. Management control and competence	BER—Unclear management control—political interference ELB—Unclear authority between architect and general contractor ELDO—National authorities are contract partners. ELDO is coordinator only	Chap. 5
6. Change management	BER—Professional change management absent ELB—Professional change management absent ELDO—Professional change management absent, delayed implementation	Chap. 13
7. Reporting	BER—Consistent report management not observed ELB—Consistent report management not observed ELDO—Report management deficient	Chap. 11
8. Project management	BER—Weak appointments and lack of qualified personnel ELB—Weak appointments and lack of qualified personnel ELDO—Weak National PM, ELDO was coordinator only	Chap. 5
9. Weak customer oversight	BER—Missing effective oversight ELB—Missing effective oversight ELDO—Missing effective oversight	Chap. 5
10. Risk management	BER—Not used consistently as control instrument ELB—Not used consistently as control instrument ELDO—Not explicitly required at that time	Chap. 19

[a]Chapters in which the correct method is described

Fig. 3.1 Terminal of new Berlin Capital Airport (BER). (Source: Creative Commons)

It is therefore all the more surprising that the schedule for the completion of the capital city airport had to be postponed again and again, even though the airlines were already selling tickets in 2010. The following chronology of the most important events was reproduced in Focus magazine November 2014, p. 48:

- 1996—Berlin, Brandenburg and the federal government decided in June to build a major airport for the capital
- 1999—The planning approval is submitted by the Schönefeld airport company
- 2004—The green light is given for the construction of BBI—Berlin Brandenburg International, which was the first abbreviation
- 2008—construction began on a never-ending series of mishaps and obstacles
- 2009—The first setback hit the airport due to the flight restriction: no takeoffs from midnight to 5:00 am
- 2010—Fixed flight routes are not correct, new ones are needed
- 2012—Construction mishaps and planning errors, especially in fire protection, lead to another postponement of the opening
- 2013—After another postponement, Hartmut Mehdorn becomes the new BER boss in March
- 2014—The costs have more than doubled. Experts now expect 6 billion Euros in the end. There is no test operation in the north pier and the north runway cannot be renovated in the summer

Finally, Focus (*a German Journal*) asks the question: will Mehdorn still see the BER opening in office, since his contract ends in February 2016? In mid-2016, the Chairman of the Airport Investigation Committee (*Delius*) saw the targeted opening date in the second half of 2017 as "*highly endangered,*" and completion of the government section is not expected before 2021.

The issue of meeting deadlines is a general management problem in the execution of projects, as unforeseen technical and/or programmatic problems can always arise and lead to delays. However, it should not happen to the extent that it did with the Berlin Capital Airport, because the analysis of the causes clearly shows, after all, that the delays were politically motivated. In 2013, Rainer W. During wrote in his book "*Laughing Matter BER*" that the opening date had been postponed no less than nine times [3].

A prerequisite for the successful management of projects is that the customer (*client*) sets clear goals and creates the conditions for professional project management. For the handling of large and complex project plans such as the BER, the customer must first create the conditions for a systematic approach. It is therefore necessary that the client, for his part, prepares a very careful planning of the project and this includes all conditions and necessary approvals. All so-called political boundary conditions should be clarified by the customer beforehand so as not to overload the project process with them. This is particularly important in the case of multiple owners, as in the BER case. Mehdorn mentions in Spiegel (*36/2014*) that many controversies are related to the complicated ownership structure, with the federal government, Berlin and Brandenburg as shareholders, each with their own interests and limited financial resources, and refers to the airport in the SZ of June 28/29, 2014, p. 3, as a "*political construction site.*"

Project managers and their teams are responsible for fulfilling the project tasks. The tasks specified at the beginning of the project must be completed in a quality manner and in compliance with the deadlines and costs. This is often a very difficult goal to achieve. Since the task specification (*functional specification*) cannot usually be changed, this leads to missed deadlines and closely related to this, to cost increases. During again: "*The postponement threatens to derail the cost plan for BER*" [4]. In addition to the previously mentioned schedule overruns at BER, the costs rose from the originally planned almost 2 billion Euros at the start of construction in 2006 to more than 5 billion Euros in 2014, and "*one lacks, it says in a letter* to *the airport company, an overall cost plan for the completion and opening of the airport ... A long-term financing and business plan is also missing to this day*" [Christoph Pauly, Andreas Wassermann, Spiegel 25/2014, p. 45].

So, what are the causes of the problem? They are certainly manifold. Project problems usually have to do with the beginnings. The implementation of projects requires clear objectives and planning at the beginning, as well as the use of effective project management. This is particularly true for complex large-scale projects such as the new capital city airport in Berlin. In this context, it is also important that the project manager and team members have many years of practical experience. During a review, European auditors identified significant weaknesses in project preparation and drew particular attention to the fact that, when the contract was awarded, the execution planning was not yet ready. [Christoph Pauly, Andreas Wassermann, Spiegel 25/2014, p. 45] When it comes to carrying out very demanding project tasks, public authorities are usually overwhelmed because they do not usually have qualified project staff. A common way out is to engage an experienced consulting firm and award the project to a general contractor. In both cases, appropriate funds must be made available. A consortium offer from Hochtief for general contracting of DM 50 million (*approx. 25 million Euros*) was rejected as unacceptable by the Red/Green coalition government [5], and "*on February 7, 2013, the three shareholders declared the negotiations with the Hochtief/IVG consortium to have failed*" [6]. Compared to the total budget, the previously mentioned amount for the general contractor was about 1%, based on the original budget of 2 billion euros.

"*The world Mehdorn found consisted of a construction site without a general contractor, with 65 engineering firms coordinating the work*" [SZ 146, June 28/29, 2014, p. 3]. In other words, there was no overarching project office where the threads came together. However, this is a basic requirement in project management because it is necessary to look at things from the perspective of the overall system and to keep control of the interface problems. "*BER is increasingly becoming a legal perennial and has long been feeding legions of lawyers and experts*" [7]. To end the chaos, the use of efficient project management was urgently needed. "*... as of May 1 (2013), the Project Management Office (PMO) implemented by Mehdorn began operations. It was to ensure optimal control, short reaction times, quick decisions, hierarchy-free working, and transparency in the work still to be done*" [8]. Mehdorn mentioned a number of problems:

- "*A head of the supervisory board is not a senior construction manager. He controls and advises the management, which is solely responsible for everything operational*" [9]
- "*At BER, politics and factual issues are constantly mixed. On the Supervisory Board, politicians and ministerial staff are confronted with issues for which they are not trained*" [9]
- "*Many on the construction site are overly cautious, we hear. They decide nothing without hedging their bets -...*" [SZ 146, June 28/29, 2014, p. 3]

It was admittedly very late that the owners turned the rudder around to finally introduce efficient management which Mehdorn accomplished by introducing a PMO organization in the hope that the situation would then improve. "*Don't make it too bad with BER*" Mehdorn wrote in an Email to Rainer During—"this will be a super airport [10]. In December 2014, Mehdorn announced his resignation from the chief post of BER for health reasons and retired in April 2015.

In 2013, During wrote "*... the opening date was postponed no less than nine times, and the capital city airport was demoted from European hub to major to single airport*" [3]. During also points out that the all-too-willing acceptance of change requests from politicians and airlines led to the cost explosion. The originally planned 1.7 billion Euros now amounts to more than 5 billion Euros; 5.4 billion Euros were already mentioned in 2015. "*While people around the globe are* *amused by the capital region and ask about the legendary precision of German engineers, the commentary of Berlin's governing mayor Klaus Wowereit on each new bankruptcy ends with the statement that the capital airport, which has been bumbling along for years, is a success story and that the constantly rising additional expenditures ... would be well spent*" [3].

According to During, the top management of the airport company and its supervisory body, led by the state rulers, increasingly lost track of the situation and overlooked the fact that chaos had long reigned on the major construction site due to a lack of sufficient coordination. To avoid chaotic situations that lead to time delays, additional costs, and plenty of frustration, the handling of large construction sites like Berlin's new Capital Airport should be performed by professional management, and "*political influence on the project processes should be avoided if possible*" said Mehdorn after taking office. During also mentions in his book, using the terminal as an example, the problem of contract award practices, which led to schedule delays. He also says that BER is increasingly becoming an ongoing legal issue [8]. Problems of communication between top managers, as mentioned by Roth (*an insider*), [9] *are* also described by During, "*They didn't talk to each other, only communicated through their anterooms*" [10]. Another problem arose from the termination without notice of the general planner responsible for site supervision who was accused by the head of engineering of inadequate coordination of construction supervision according to a capital mistake [7]. Wowereit reiterated to the Berlin House of Representatives that BER remains a *success story but* admits that the dismissal of the general planner threw the schedule out of joint, which was painful but unavoidable [7]. The original plan was for BER to open in 2007 [3]. Meanwhile, the completion date is 2018 or 2019.

Roth, an insider at the BER airport construction site for 18 months from July 2010 to December 2011, observes and reports on the conditions in the company's central controlling department [9]. His experiences, described like a novel, are amusing to read. He describes the lax work and management style, the fact that he was not assigned a clear area of responsibility by his boss despite repeated requests, and that he mainly had nothing to do. Furthermore, although the boss made coffee before departmental meetings there was no agenda or minutes at meetings, so that meetings sometimes had to be repeated because people had forgotten what had been decided at the previous meeting. His question "*shouldn't we keep better minutes of the meetings?*" led to irritated faces. Another problem was that the invited employees showed up late or not at all for the Great Controlling Round. There was no need to cancel, it was enough to simply not show up. Roth also complained about the lack of communication and described the "*grapevine*" as the most reliable source of information.

Regarding organization, Roth quotes a colleague as saying, "*we are part of an unprofessional, disorganized, unstructured, and indifferent bunch, that upper management is at odds and that everything has to go through the bosses' desks and no responsibility is delegated.*" He also points to the complicated contractual entanglements of the airport energy supplier, a subsidiary of the airport company that does not physically exist but is present in the mind and imagination as a zombie.

According to Roth, the issue of scheduling is also problematic. External consultants were called in for this. A colleague at the airport received the following answer to his request for inspection: the plan is secret and cannot be inspected. Only the designated contact persons (*chosen ones*) are entitled to inspect it. This statement led to a lack of understanding on the part of the employees, and the management team made a confused impression.

After 5 months, Roth then jumps onto the vacant high-priority topic of "*improving risk management*" in order to finally be able to assume independent responsibility. Now, however, the management intervenes with a request for a consulting firm, because there is an unused consulting budget that they still want to use anyway. The consulting firm sends five consultants to first determine how risk management works. Roth notes that most airport employees perceive it as a problem and consider it to be exposing a weak point. As a result, discussions with managers are not very productive because there are (*supposedly*) no risks, while employees mention some, but with reservations. Several risk workshops produced risk lists with a few risks in the "*force majeure*" category as a result. However, this did not lead to the desired change in the risk culture.

Gerkan calls the "*seven plagues*" of construction and primarily means the following inadequacies which, based on his experience, occur in many projects right at the beginning [11]:

1. Wrongly defined construction tasks or unspecific and politically controversial wish lists
2. Unrealistic construction costs
3. Unrealistic time frame
4. Lack of communication from all parties involved
5. Non-architectural trades and committees that question everything
6. Permanent changes
7. Vanity of the builders and architects

He also mentions the generally applicable "*plague of wishing*" and, in this context, the Oxford professor of urban planning Bent Flyvbjerg, who attests that most managers of major projects have a fatal tendency to lie and deceive, and adds "*In order to win construction contracts, the costs of their proposal are usually glossed over and set far too low*" [12]. "It was *against the backdrop of an increasingly ambiguous wishful thinking that the endless drama surrounding the Capital Airport played out*" [13].

Gerkan denounces the arbitrariness of politically imposed scheduling, which has resulted in serious cost consequences, and points out that it is imperative to use expert planners and construction managers to draw up realistic schedules and progress plans [14]. He then summarizes his view of the biggest mistakes in construction projects of our time and describes BER as exemplary for the failure of many large construction projects [15].

1. Joint and several liability—architect's liability with insufficient possibility of control
2. Running behind the demand—constant changes of the builder
3. Lump sums—lead to cost explosion in case of late changes
4. Diversity of use—expensive rescheduling without the benefit of functionality
5. Black build—the deadline-driven start of construction without a permit
6. Reporting—embellished reports to the Supervisory Board
7. Controlling—unrealistic reports and figures are ignored
8. Commissioning before manufacturing—lack of pragmatism
9. Delegating responsibility away—delegation of responsibility
10. Cheating with built-in overconfidence—scheduling problems often lead to frenzied actions and the search for easy and quick solutions (*however, instead professional planning would be necessary*)

In Sect. 3.2.3, the 10 most serious management errors are summarized in comparison with the projects BER (Sect. 3.2.1), ELB (Sect. 3.2.2), and the Europa rocket (Sect. 3.8). Cross-references lead to the relevant sections of this book in which the recommended PM methods (lessons learned) are described.

Finally, the question remains will BER still open in mid-2017, 2018 or not until 2019, as Mehdorn predicted, and will the cost of BER rise from the original estimate of about 2–6.5 billion Euros (*or more*)?

In November 2019, the BER Airport Supervisory Board announced the new opening date and roughly advised October 2020. The seventh date is 9 years after the original start date. Will it work this time? "*We'll see, we'll see*"*. Finally, the BER airport was opened on October 31, 2020, and given the name "Willy Brandt."

**A famous quote from football player Franz Beckenbauer*

3.2.2 Elbe Philharmonic Hall

Figure 3.2 shows the side view of the Elbe Philharmonic Hall, and in this context, it is referred to as a new landmark of Hamburg.

Almost 8 years after the foundation stone was laid, the "Hamburg Elbphilharmonie" was opened by Hamburg's mayor Olaf Scholz* on January 11, 2017 [16].

*Since December 8, 2021, Scholz is Chancellor of Germany.

When the foundation stone was laid in April 2007, the opening date was set for the summer of 2010, and the general contractor Hochtief wanted to hand over the completed building to the city in October 2016. The plaza, which offers a spectacular panoramic view of the city and the Elbe River with the harbor at a height of 37 m, was to be open to visitors in November 2016.

Architect Mergenthaler had words of praise for Mayor Scholz. "*Without his courage and vision, we wouldn't be standing here today,*" he said, referring to the Hanseatic city in Spring 2014. At that time, after years of dispute and standstill at the construction site, the city, architects and a construction group had found a

Fig. 3.2 Side view of the Elbe Philharmonic Hall in Hamburg (Creative Commons)

solution for further cooperation. Now all sides again praised the new spirit on the construction site [16].

There is a 6½-year delay between the originally planned opening date in mid-2010 and January 2017, the date now officially announced. At the same time, total costs rose from an original 186 million Euros (*2005*) to 865 million Euros (2013) within 8 years [17].

– 2005 Feasibility study	186 M€
– 2006 Conclusion of contract	348 M€
– 2008 Supplement 4	572 M€
– 2013 Supplement 5	865 M€

The authors Dey, Haarmeyer, and Mischke have described the (*unbelievable*) development history of the project in great detail, and their report lists many omissions and errors that led to the misalignment (*delay and cost increases*) of the construction project. The most serious errors can be summarized as follows [17]:

- Former Hamburg mayor Ole von Beust said, "*...organizationally, some things could have been done differently, that's obvious.*"
- Hamburg Mayor Olaf Scholz said *in* 2013, "*You should have had the courage at the beginning of the project to put many tens of millions of Euros into planning and preparation and only then decide whether to build the concert hall.*"
- Marcelino Fernandez Verdes, the new CEO of Hochtief AG, made the following comments in 2013: "*You don't get anywhere with mutual recriminations. I ignored the previous history and asked myself the question: What must happen for this project to be successfully completed? How can we find a solution together and rebuild lost trust? In the meantime, the city, the realization company (ReGe),*

the planners and the builders are taking the same approach; we're pulling in the same direction."

- Alexander Gerard, the architect and project developer, the spiritual father of the Elbphilharmonie, received the Semper Medal of the Architecture-Centrum on September 1, 2005. Because of internal tension, he left the project in 2013 after several years. "*That those then responsible made so many serious mistakes and that the project is now gobbling up so much money unnecessarily is madness. Nevertheless, the Elbphilharmonie will be a great success for the city.*"

"*Around the turn of the millennium, plans for HafenCity, which Mayor Henning Voscherau had initiated in 1997, began to take shape in Hamburg*" [17]. In 2000, a master plan for the cultural use of HafenCity was approved by the Senate. On December 16, 2003, the Hamburg Senate decided (*printed matter 1773924*) to pursue the Elbe Philharmonic Hall project on Kaispeicher A, and on April 2, 2004, the foundation stone was laid by Hamburg Mayor Ole von Beust together with Henner Mahlstedt (*Hochtief*), project coordinator Hartmut Wegener, Culture Senator Karin von Welch, and architect Pierre de Meuron.

The Elbphilharmonie is to be a unique building in the world. On January 19, 2005, the city took over the direction of the project. The services of architects and engineers are regulated in principle, as is customary in Germany, according to HOAI (*Honorarordnung für Architekten- und Ingenieurleistungen*) (*Compensation Regulation for Architectural and Engineering Services*):

1. Basic determination
2. Preliminary planning and cost estimation
3. Design planning and cost calculation
4. Approval planning
5. Implementation planning
6. Preparation of contract award (*including determination of quantity estimates and preparation of bills of quantities*)
7. Participation in the awarding of contracts including cost estimate
8. Construction supervision and documentation
9. Object supervision

In this context, it should be mentioned that problems arose during the work on Service Phase 5 (*implementation planning*) because the services required by the architects, such as rounded and curved components, led to massive conflicts with the general contractor, who was anxious to apply a cost-effective solution with standard components. This is a classic case of what you would like to have (*nice to have*) and what you can afford (*good enough*) because special features have their price. This phenomenon also exists in other industries, such as high-tech projects and aerospace.

There was no cost cap at the time, but there were efforts to introduce a bonus-malus scheme. However, this did not come about, and in November 2005 it was agreed to drop the idea. A realistic determination of target costs was also a problem.

In a feasibility study that was later presented to the parliamentarians, construction costs of 2886 Euros per square meter were assumed. Costs of between 3000 and 7800 Euros were known for comparable concert halls. After completion of the preliminary design, in April 2005 the architects determined a total cost of 196.7 million Euros, which Hamburg's mayor von Beust found high but thought it could be done.

For project control, the realization company (*ReGe*) engaged the Braunschweig engineering company Assmann, whose main task was to ensure that deadlines were met. ReGe trusts the random checks made and records the total costs at 186 million Euros although the architects stick to their 10 million higher figures. In an amendment to the minutes of May 25, 2005, they wrote, "*The general planners have submitted a correct cost estimate.*" In July 2005, Ole von Beust mentioned that the cost estimates are pessimistic, but they are quite confident that budget overruns, if any, will be minor. However, reference is made to the tight schedule and that any delays could then lead to additional costs.

Hamburg's renowned architect and expert Volkwin Marg says, "*No one has ever applied petty standards to a cathedral of the spirit,*" but at the same time warns against planning developments and special requests, which can influence costs. He goes on to say that one can be proud if one achieves an accuracy of 5–10% under hard premises. However, if the premises are soft, one has to leave room for growth. From this, it could be concluded that the calculated figure of around 200 million Euros is of little significance. ReGe points out that the cost estimate is serious because it is based on a great depth of planning [18].

Architect Marg explains that the biggest cost surprises always occur when planning changes are made during the construction period. If the Elbe Philharmonic Hall becomes a *wish-fulfillment project* in which the city continually introduces change requests into the ongoing process, there may be considerable additional costs.

Project planning for the Elbe Philharmonic Hall is proving to be an ongoing problem. In June 2006, the architects of the Swiss architectural firm Herzog & de Meuron warn in writing that the planning had not been completed and point out the enormous complexity of the building. Against this background, they also note that a call for tenders for the general contractor (*GC*) currently entails numerous additions and changes. Coupled with State Councilor Volkmar Schön's comment that it's hard to say no in the *wish-what phase* because no one wants to look like a killjoy, the architects' following statement fits: "*The cost target, the targeted quality and the completion date are in the highest danger.*" The question of whether the Hamburg State Chancellery has received the architects' cry for help remains open. Ole von Beust is also said to have heard nothing [17].

Although it became increasingly clear in the summer of 2006 that more preparation time would be needed to clarify other important open points and improve the depth of planning, on August 10 ReGe called for binding bids by September 15. In addition, technical problems are looming, e.g., a static calculation of the supporting structure that has not yet been completed, which still has aftermath, because the Technical University of Hamburg later determines that the load-bearing capacity of the foundation piles is not sufficient. Other problems include the lack of design documents for heating, ventilation, electrical, and lighting, as well as a sprinkler system in the large hall that is required according to the fire protection report. Worth mentioning in this context is the problem of the execution planning being divided between the architects and the general contractor (*Hochtief*), which unnecessarily led to increased conflictual interface problems. In the same period, ReGe receives a wish list from the cultural authority: Cafeteria, third hall, etc., which does not contribute to cost stability. It is already easy to see here that the processes described had led to scheduling and cost problems.

In addition to the cost issues, risk management should always remain in focus. A bidder proposed a non-recourse financing model for the hotel construction, which was positively accepted by ReGe because of the calculated cost advantage of 14 million Euros, although the city of Hamburg thus assumed the full investment risk of the hotel construction as investor and owner. During a meeting between the mayor, the Senator for Culture, ReGe, and others, the proposal was approved. However, there are no minutes of this meeting [17].

This was followed by lengthy negotiations between ReGe, the general contractor (*Hochtief*), and the architects. The main issues were the cost level and the staking out of the services, as the stage technology was not included in this. Further cost increases are on the horizon because of redesign requirements from the cultural authorities. By the end of 2006, the total costs had already reached almost 300 million Euros.

Only a few days before the planned signing of the contract, the architects point out contradictions in the dates and warn against signing, because if the plans are not coherent, the architects and the general contractor will work at cross purposes. Hochtief is not able to sign the contract finally because of the lack of fixed schedules and a fierce dispute arises. But the contracts with the general contractor were signed in December 2006, despite internal protests, with a cost volume of 241 million Euros and a construction period of 36 months. At the same time, ReGe demanded that the schedules be synchronized by March but this was not achieved.

In January 2007, ReGe (*Wegener*) confirmed that the construction costs for the public sector are low compared to high-profile concert halls and that there is therefore no unpredictable risk. Only subsequent planning changes by the city can lead to cost increases. And already in March 2007, ReGe received the first project change notice (PÄM #001) and more will follow; in June 2007, ReGe already reached PÄM #100. ReGe is understaffed and therefore unable to cope with the flood of changes. Hochtief's well-organized claim management overwhelms ReGe. It is obvious that the city completely underestimated the complexity of the project and the required management effort. A well-organized Project Management Office *(PMO)* would have been the solution. A frustrated ReGe employee left the team and was replaced by a new employee who takes the precaution of pointing out that schedules and costs are not her specialty. The city informs that it has no suitable specialist staff in the area of cost control, and anti-claim management to stand up to the opposition does not exist.

One fatal error is the mayor's lax management of ReGe, as von Beust himself later admits. Reporting is poor, and it becomes abundantly clear that the contents of the quarterly reports are often glossed over and aspects such as cost developments are not included. Although the architects clearly pointed out in writing the danger of schedule and cost problems, the Senate was not informed of this. In addition, tensions between ReGe and the architects increased. The contract concluded in 2006 proved to be completely inadequate at the end of 2007. ReGe was in danger of becoming a pawn between the general contractor and the architects [17].

In 2008, responsibility for the Elbe Philharmonic Hall project was transferred from the building authority to the cultural authority following a Senate resolution. The desolate situation on the construction site and the additional demands by the general contractor led to a crisis summit at the mayor's office on July 1, 2008, to discuss the status of the project, the delays, and the expected risks. By now, at the latest, those responsible had an inkling that the city would be faced with additional demands in the hundreds of millions. The mayor was quite taken aback by the situation because his hope was that they would go home from this conversation "*cheerfully purged,*" but the opposite was the case.

In the meantime, the relationship between the cultural authority and the project coordinator of ReGe had visibly deteriorated due to insufficient information about the project status and risks. There was also still no synchronized schedule. The mayor had lost confidence in the ReGe head and suggested that he resign. His successor is the former ReGe employee Leutner. In November 2008, David Koch again demanded a clear definition from the architect's office of exactly what was to be built, because otherwise there was a risk that Hochtief would subsequently claim additional costs. At that time, it was already clear that the total costs were well over half a billion euros. The Senator for Culture, Karin von Welck, commented: "*We all underestimated the complexity of the project.*" Addendum 4, with its gigantic cost increases, is once again only an interim solution.

In January 2009, the Senator for Culture described Addendum 4 as a great success. Wegener, the previous head of ReGe, assesses the situation correctly in that he considers Addendum 4 to be a disaster for the city, which is also confirmed very soon, because less than a year after the conclusion of Addendum 4, there are already 40 construction obstruction notices from Hochtief. It is also clear that the completion scheduled for the end of 2011 is not feasible. Möller, head of Hochtief, says "*we can't build without plans.*" Plans must be provided by the client or his architect, but the latter is constantly changing the plans. In this context, he cites an example concerning ventilation. Instead of a cheap variant, a luxury variant is chosen, and he adds, that the one who has arranged it must pay; city, ReGe, civil engineering, etc. [17].

At the turn of the year 2010/2011, the situation is sobering. Delayed schedules, public disputes, and farewells of those previously in charge; Mayor Ole von Beust, Culture Senator Karin von Welck, the head of the Senate Chancellery, and the Economics Senator all take their leave. A new political era began in February 2011. Olaf Scholz becomes Hamburg's new mayor and Barbara Kissler becomes Senator for Culture. At the same time, the situation comes to a head, and the starting position for 2012 is more disastrous than ever before. Kissler takes an energetic approach, threatening the construction company with the words "*no more games*"; at the same time, Hochtief threatens to halt construction and since fall 2011 there has been a de facto construction standstill in large areas. The architectural firm from Switzerland is also seriously considering suspending all activities due to overlapping competencies with the general contractor. In the media, the Elbe Philharmonic Hall is referred to as a chaotic construction site along with the Berlin Metropolitan Airport and the Stuttgart train station.

In the spring of 2012, Marcelino Fernandez becomes Chief Operating Officer (*COO*) at Hochtief. His appeal in May 2012: we can't get anywhere by blaming each other—how can we rebuild trust? At a one-on-one meeting in July 2012 in the office of the Senator for Culture, a good understanding quickly emerged between Kissler and Fernandez. The Elbe Philharmonic Hall is very important for Hochtief's reputation and must therefore be completed properly.

After the construction group had met almost all the city's demands in December 2012, the newly elected mayor of the Hanseatic City of Hamburg, Olaf Scholz, together with the Senator for Culture, Barbara Kissler, announced on March 1, 2013, "*We have reached agreement with Hochtief on a new version of the contract (Addendum 5)*" [17]. In September 2013, it became apparent that the new arrangement seemed to work. Hochtief has not submitted any further addenda.

Following up on the detailed situation description "*Elbphilharmonie*," Haarmeyer and Mischke ask Hamburg architect Volkwin Marg, "*How could so much go wrong with the Elbphilharmonie project? Who failed, politics, common sense, or both?*" [18]. Marg mentions the simple triangular relationship of responsibility between client, architect, and contractor that is common in construction and criticizes the dissolution of responsibilities in a multitude of competencies. Risks are due to increasing complexity.

Regarding cost forecasting, Marg points out, based on his experience, that it would have been correct to include an appropriate costing tolerance of 75% at the time of commissioning:

- 15%: Surcharge for remodeling
- 45%: Approach for mixed use (*Elbphilharmonie, hotel, and apartments*).
- 15%: Allowance for early-stage cost determinations (*missing and unreconciled detailed plans*)

He remarks: "*We architects know all too well the concert of desires, because they lead to costly changes. The client bears the responsibility for additional orders. Whoever orders must also pay; this is the minimum qualification required of a builder. In this context, it must also be understood that the construction company must charge additional costs and/or postpone deadlines in order not to have to declare insolvency.*"

The Taxpayers' Association wrote on 5 Jul 2014 on the subject: "*A major cause of the enormous cost increase was that the project was put out to tender too early, i.e. before the design planning was completed. Taxpayer money could have been saved if the city had completed the design planning first before putting the project out to bid*" [19].

The opening of the Elbphilharmonie on January 11, 2017 was a striking event.

3.2.3 Comparison of Management Failures at the New Capital Airport Berlin (*BER*), the Elbphilharmonie Hamburg (*ELB*) and the Euro-Rocket (*ELDO*)

This subsection summarizes the 10 most serious management mistakes identified for projects BER (*Sect. 3.2.1*), ELB (*Sect. 3.2.2*), and the Euro-rocket (*Sect. 3.8*). References lead to the relevant parts of this book where the recommended PM methods (*lessons learned*) are described.

The mistakes described could certainly have been partially or completely avoided by professional management, because, compared to the lack of management experience in the development of the Euro-rocket more than 40 years ago, there was sufficient theoretical and practical management experience for the handling of complex large-scale projects at the beginning of the BER and ELB projects. However, one cannot expect complex projects to be successfully managed by management laymen. At project start (*Phase A*) the responsible authorities in Berlin and Hamburg should have assigned a project management team with demonstrated practical experience. Most problems could have been avoided by a competent project management office *(PMO)* for BER and ELB, as suggested by Mehdorn. It is fundamentally wrong to hand over important management tasks such as project planning to a planning office without tight control of the project. The appointed project manager is in principle also the top planner! He needs a planning team and can then, of course, assign tasks to a planning office without losing control.

As can be seen from the explanations in Sects. 3.2.1, 3.2.2, and 3.8, there were several causes that led to serious schedule overruns and cost increases. In Table 3.1, 10 examples of striking management problems identified at the BER and ELB projects, which are quite similar to the management problems identified about four decades ago at the European rocket development project ELDO. Management mistakes made 40 years ago have been repeated. The comparison shows a similar root cause in all three cases. The problem was not due to a lack of project management methods, which have indeed been improved and tested over the past 40 years, but in all three cases, it was due to the amateurish approach and political interference in factual issues. Mehdorn again commented on this: "*At BER, politics and factual issues are constantly being mixed. On the Supervisory Board, politicians and ministerial staff are confronted with issues for which they are not trained*" [9].

Unfortunately, however, experience gained is often not passed on to others. The transfer of know-how often works poorly. There are many reasons for this. On the one hand, there is often the opinion that one does not need any special knowledge to manage a project, but only "*common sense*," and therefore one is resistant to experiences that others have had. A second reason is that the experiences made are not passed on by the know-how bearers for various reasons. Managers often do not have enough time to write down the experiences they have had or deliberately withhold their knowledge according to the motto "*knowledge is power.*" Authorities and companies should therefore create a project culture in their own interest with the following steps:

- Step 1: Prepare a brief field report upon completion of the project
- Step 2: Archiving the project documentation—know-how database
- Step 3: Implementation of training measures (*lectures, seminars, workshops, etc.*)
- Step 4: Project coaching by experienced project managers
- Step 5: Creation of a project manual as a binding guideline

In today's fast-paced world, experience reports should be designed and disseminated in such a way that they are easily accessible through publications, internal reports, and databases, and written in an understandable form. Regular training of junior staff and the holding of workshops are important measures to pass on current experience. Project personnel is usually a scarce resource and therefore one is often forced to use junior staff. However, in order not to endanger the project, the principle of project coaching should be applied in such cases by having an experienced manager available as a mentor.

Managing directors of authorities and companies as well as line managers must recognize the importance of project management and support the necessary measures in the interest of the respective company. This subject is dealt with in detail in Chaps. 5 (*project organization*) and 15 (*project personnel*).

3.3 Learning from Mistakes

It is necessary to record experiences made in writing, including all mistakes made, according to the motto "learn from mistakes" because due to the frequent change of project personnel, valuable management knowledge is often lost or destroyed, and newcomers must reinvent the wheel. To prevent this, it is an important prerequisite that those responsible summarize the most important findings in the areas of engineering, quality, risk, time, costs, contract, etc. as part of a project review. The result should be a *lessons-learned report* that clearly summarizes the most important findings. This measure does not require an unjustifiably large amount of effort, but it has a great benefit for new projects and should be mandatory for future project managers. Valuable management experience should be systematically recorded to ensure that it is not lost. This is especially true for projects with durations of several years since one must expect personnel changes and employees often do not pass on their knowledge.

3.3.1 Field Reports

The field reports should not include extensive details but should include a summary of lessons learned for each of the previously mentioned areas (*planning, costs, etc.*). The information is best recorded in tabular form as follows:

a) Problem statement (*short description*)
b) Background information on how the problem arose

c) Abstract of a recommendation to avoid similar mistakes

It is also recommended that the lessons learned be incorporated into internally conducted PM seminars. The outline of the experience report could be divided according to the following categories to facilitate the search for important recommendations:

a) Project Management
   - Customer relationship
   - Project organization
   - Industrial organization
   - Subcontractor Management
   - Project Personnel
   - Matrix organization
b) Engineering
   - System processes
   - Specification creation
   - Technical reviews (*reviews*)
   - Tests (at system and subsystem level)
   - Assembly, Integration and Test (*AIT*)
   - Operational operation
c) Product assurance
   - Reliability analyses
   - Quality control
   - FMECA
   - Audits
   - Material review board (*MRB*)
d) Programmatic aspects
   - Statement of Work (*SOW*)
   - Work Breakdown Structure (*WBS*)
   - Work Package Descriptions (*WPD*'s)
   - Documentation and configuration control
   - Changes (*Engineering Change Proposal—ECP, Contract Change Notice—CCN*)
   - Change Control Board (*CCB*)
   - Construction deviations (*Request for Deviation—RFD and Request for Waiver—RFW*)
e) Scheduling
   - Base schedule as well as its trustworthiness
   - Identified delays, their cause, and actions to meet the overall schedule.
   - Critical path (*range*)
f) Financial Control
   - Basic cost plan
   - Cost overruns, their cause, and actions to comply with the total cost.
   - Cost projections (*Cost to and Cost at Completion—CTC and CAC*)

g) Contract management
h) Other

3.3.2 The Taxpayer as Controller

There are an increasing number of public discussions about inadequacies in the handling of large projects. Citizens who in the past were hardly interested in what was going on in public projects because they lacked information as well as the necessary expertise and were of the opinion that everything was in good hands, are suddenly taking an interest. How can this be explained? Without excluding at this point that the popular knowledge, as well as the understanding of democracy of the average citizen, has been advanced, the increase of public discussions results to a substantial part certainly from the fact that many large-scale projects and smaller local projects derailed recently in plain view for everyone. The project goals set were not achieved, or only after years of delays, and the planned costs were exceeded many times over. The absolute magnitude of the project costs, which often represent a significant percentage of the available budget of the federal government, state, or municipality, literally forces the taxpayer to take on the role of controller.

By their very nature, people are less interested in detailed technical explanations of the specific project problem, as this requires very specialized knowledge, and more interested in "*top management information*" about how to proceed. Although the problems encountered in the project which led to considerable schedule and cost overruns may be explainable in detail, there remains a considerable degree of mistrust in the project management, and one wonders whether the technology was not yet mature, the goals were set too high, the schedule and cost estimates were completely off, or simply that management failed. What remains is considerable doubt about the leadership qualities of the authorities and industrial companies involved.

However, project failures are not only limited to the public sector but are also well known in the private sector of industry, although this has no direct significance for the general public since it is an entirely private risk of the company. Certainly, however, the additional costs are later passed on to the product costs and thus do hit the "*little man in the street.*" In some cases, the company must declare bankruptcy to the detriment of the company's workforce which can, of course, also lead to a burden on a broader section of the population, if one thinks of the unemployment problem in this context.

3.3.3 Change in the Task Definition

Are management crises in the handling of projects symptomatic of our times? And how can this be changed? As can be seen from the following remarks in this chapter, the explanation for the large number of management crises that occur is rooted in a lack of political, organizational, managerial, and systems management. What superficially appears to be a failure on the part of those responsible turns out, on closer examination, to be part of a lack of ability to cope with complex system tasks, in which it is essential that the various *players* (politicians, managers, engineers, IT specialists, economists, lawyers, etc.) see themselves as team members and work closely together under the guidance of professional project management. The technological changes of our time and increasing international competition (see Chap. 14) are forcing public authorities and companies to become more professional.

Many offices, authorities, and industrial companies have had to undergo a considerable change in their tasks in the last decades without the improvement of management methods and corresponding personnel training being provided at the same pace. This is particularly true in the area of project management. Federal agencies that were traditionally responsible for national issues are now confronted with the management of international projects. The tasks of municipal offices have also become visibly more complex and therefore require better project management methods. In addition, the initiation of modern industrial projects requires a particularly high level of knowledge and experience of systems engineering and management to be able to handle them within economically justifiable limits. For various reasons, many industrial projects are also carried out in industrial consortia without adequate organizational forms having been developed for this purpose.

3.3.4 Lack of Training

A main reason for management crises in the project area can also be seen in insufficient preparation and training of project personnel. Good preparation and training are indispensable for mastering project tasks. Lack of insight into training results in most cases from the fact that companies and authorities do not see project management as a task area requiring special knowledge of methods, tools, and management skills. It is often wrongly assumed that every experienced specialist can automatically lead a project.

Servan-Schreiber quoted former US Secretary of Defense Robert McNamara over 40 years ago as saying, "*Europe is ..., particularly weak in operations and management training*" [20]. This was probably true in the past but has since changed, as there are many public and in-house training courses on project management available in Europe and especially in Germany. In this context, reference is made to publications and seminar events of the German Society for Project Management (*GPM*) and the Project Management Institute (*PMI*); see Sect. 15.2.2.

Many German universities have now included the subject of "*project management*" in their curricula. Project management is an interdisciplinary activity in which it is important that engineers, business economists, and lawyers, for example, work closely together. It is therefore necessary for institutes and universities to consider PM training as an integral cross-cutting topic. Offers for PM training should address both university graduates and employees with professional experience who want to qualify for new tasks in project management through additional or postgraduate studies.

3.4 Lack of Cost and Schedule Adherence—A Serious Problem for Business

3.4.1 Cost Explosion Threatens the National Economy

When cost overruns in the billions become known, this usually has a shocking effect, although some people may already have become accustomed to such news. But cost overruns in the billions, as they have become known in Germany in the past in connection with major construction sites, power plants, and aircraft construction projects, lead to reflection, since their magnitude in the national budget can no longer be dismissed as a digit behind the decimal point, and the question of how much the national economy can still cope with without suffering serious damage arises of its own accord. However, cost and deadline overruns are not only a phenomenon of today, but an unpleasant consequence, which, as Tauber writes, is already known to us from antiquity: "*The most classic example in the truest sense of the word is probably the Acropolis, which could only be completed when the Athenians unceremoniously plundered the war chest of the Attic League in Delos to finance the remaining construction costs*" [21]. Grün mentions that the Nürburgring, completed in 1927 and estimated to cost 2.5 million marks, ultimately cost 14 million marks [22]. Does that seem reassuring to us? Is it perhaps even a phenomenon that one simply must live with?

As mentioned at the outset, this is not just about the theoretical abstract consideration of budget overruns, but about the resulting threatening effects on the national economy. It is irrelevant whether the sum of overruns results from many individual projects or a few major projects. In this regard, Grün says again: "*The public discussion of cost target overruns has drawn attention to a considerable dilemma. On the one hand, especially in the case of new types of projects, the target costs that were decisive for starting the project were often exceeded many times over. On the other hand, many of the facilities now accepted as necessary would not have been created if the extent of the cost increases had been known before project approval.*" Grün quotes the Viennese mayor Gratz in this context, who comments on the gigantic cost explosion at the Vienna General Hospital (*7.5-fold increase in costs, Grün 1977*) with the following statement: "But without the almost suicidal willingness to act in this way, not half of what has actually been realized would have been realized since 1945." But will this procedure still work in the future? It can be assumed as certain that a general cost increase of modern products caused by a continuous increase in complexity, which is then overlaid by cost increases resulting from deficient project management, is not sustainable for the economy in the long run.

The striking cost increases for the Berlin Capital Airport (*Sect. 3.2.1*) and the Elbe Philharmonic Hall (*Sect. 3.2.2*) projects described above should be included hereto.

3.4.2 Collapse in 2054

In this context, a consideration by Augustine is interesting [23]. According to his research, from the Wright Brothers (*1911*) until the introduction of high-performance fighter aircraft, the unit price per aircraft in the USA increased by a factor of four in each 10-year period. If one compares this trend with the extrapolated development curve of the US defense budget, according to Augustine's exaggerated statement, the defense budget in 2054 is only sufficient for the purchase of a single aircraft, which the Air Force and the Navy would also have to share. While the validity of this statement for the year 2054 may be questionable, the trend toward cost increases due to greater product complexity, which is evident everywhere and does not consider cost increases due to inflation, is a serious problem for the economy. Cost overruns due to inaccurate cost forecasts and poor management are even more significant because they represent a further, unforeseeable increase in the cost of the product, especially when the originally planned values are exceeded many times over, for example in the construction of power plants and some aircraft construction projects; *see Chap. 10* about professional cost estimation.

3.4.3 Only the Tip of the Iceberg

Now, however, it can be assumed that the cost problems discussed in public are only the tip of the iceberg and that these are only the most spectacular cases. In fact, schedule and cost overruns are nothing unknown to municipalities, companies, and even private households. One encounters this problem almost everywhere in our society. However, the question arises as to how the deviations (*meaning budget overruns*) relate to the baseline budget value. If, for example, it was a matter of a regularly occurring overrun, then this could already be considered in the budget planning. However, it is not as simple as that, because the schedule and cost deviations between the planned and actual values can vary considerably. However, there is no discernible pattern here. There are probably three main reasons for this: firstly, the deviation ratio depends on the quality of the estimate; secondly, it depends on the quality of management; and thirdly, external factors that cannot be influenced, such as strikes, construction stoppages, inflation, etc., are often responsible for cost and schedule overruns. The phase in which the respective project is currently in must also be considered because a more precise cost estimate is naturally possible for a manufacturing project than for a research and development (*R&D*) project.

If, due to ignorance or tactics, too few costs have been included from the outset, the calculation may ultimately not add up. Furthermore, the implemented monitoring mechanisms and management behavior are of decisive importance for cost and schedule compliance. For example, it is essential for management to ensure that no project changes are made for which there is no financial cover. In this regard, former Federal Research Minister von Bülow *said, "The high cost of cutting-edge technical equipment, on which modern basic research depends, necessitates not only scientific risk-taking but also excellent management skills to ensure that costs are effectively controlled*" [24].

The most unpredictable and least subject to regularity are all external factors that cannot be influenced by the project or can only be influenced to a minor extent, such as strikes, unexpected cost increases, and, in the case of foreign transactions, shifts in currency parities. Although the delays resulting from approval procedures can be planned in principle and influenced by the project to a certain extent (*well-prepared project documentation is certainly of great benefit here*), they are nevertheless subject to the time rhythm of an organization external to the project.

3.4.4 Improving Cost and Schedule Adherence

Improving cost and schedule adherence is of paramount importance in the interests of our economy. Financial and schedule-related project disasters, which unfortunately occurred all too frequently in the past, must be prevented at all costs. To this end, the following measures would be urgent:

(1) Improvement of estimation methods (see Sects. 9.3.4 and 10.1.1)

(2) Conducting project analysis:

Just as preliminary studies are part of every project, it is urgently necessary to analyse completed projects in the interest of future projects. The financial and scheduled development of completed projects must be seen in connection with the project history and the detailed technical events. However, since final reports cannot be prepared free of charge, there is usually not the insight on the part of the respective project sponsor to make this necessary investment in the future. Nevertheless, government and industrial project offices should pay more attention to this task in the future. The company Messerschmitt-Bölkow-Blohm GmbH (*MBB*) in Munich, was, for example, commissioned by its client, the German Research and Testing Institute for Aeronautics and Astronautics (*DFVLR*), after completion of the HELIOS solar probe project in 1976, to prepare a final report in order to record the experience gained from this project (*lessons learned*) for the future [25]. The report, which can be regarded as a novelty, clearly summarizes the experience gained in the areas of management and technology and contains the most essential project data in the chapter "*Reference data and statistics*"

(3) Development of contemporary estimation procedures:

Estimating deadlines and costs cannot be left to chance, because with every estimate, be it costs or deadlines, the important question arises "*how good is the estimate?*" Therefore, when submitting a bid, some clients expect that not only a final date as well as the calculated total price are given, but that in addition appropriate details as well as statements about the credibility or quality of the estimated unit costs (*estimating credibility*) are available. Of course, the quality of the estimate for the individual items can vary. It depends essentially on the following factors:

- the knowledge of the product
- the experience of the estimator
- the available data material
- the time available for estimation

For each estimation process (*times and costs*), specific rules must be followed, in order to get better estimation quality; see Sects. 9.3.4 and 10.1.2

(4) Improvement of management quality

Management quality is directly related to schedule and cost compliance, and a lack of management quality easily leads to unwanted overruns. This connection is not always visible, but it can be easily explained with a few examples. Most problems start at the very beginning. Every project has its own life cycle (Chap. 4), and in the first phase, the concept phase, the absolutely necessary measures, namely the selection of a clear concept, execution of *feasibility analysis,* and formulation of a clear objective, are often not perceived thoroughly enough, because, after completion of the concept phase, a transparent *baseline concept* must be available, as a basis for all following steps. If this important measure is neglected, fundamental problems, such as the location question for a construction project or the power supply for a satellite project (*solar cells or isotope battery*), are transferred to the subsequent definition phase, which can then lead to schedule and cost overruns in the subsequent phases. Consistent project execution according to predefined phases (*Chap. 4*) is an essential step to improve management quality

(5) Another management weakness frequently identified in the execution of projects is in project organization

In many projects, either not enough attention is paid to this aspect, or compromises are tolerated which make it impossible for the appointed project managers to carry out the necessary tasks efficiently. The problems described in the following sections are symptomatic of this. Ambiguities in the assignment of responsibility and authority as well as the organizational integration of project management inevitably lead to schedule delays and cost overruns. By contrast, observance of the principles of project organization described in Chap. 5 leads to improved management quality

3.5 Follow-Up Costs—Who Pays Them?

3.5.1 Follow-Up Costs 15-Fold

The question of follow-up costs has become increasingly urgent in recent years. While the development and procurement costs for new projects may often already seem shocking, the issue of follow-up costs is many times more serious. What is meant by this? Follow-up costs or consequential costs are the expenses incurred by the operator of a newly acquired plant during its use, for example, the personnel costs for operating the plant, material, and fuel costs, and the costs for maintenance and servicing. With the increasing complexity of modern products, the consequential costs also increase quite considerably.

The evaluation of 10-year statistics of the procurement office on the entirety of the defense material of the German army revealed a cost ratio of 1:3:6 for development, procurement, and utilization [26]. This means that if the enormous sum of 1 billion Euros has to be spent on the development of a new weapon system, additional production costs of 3 billion Euros and utilization costs of 6 billion Euros have to be expected, i.e., the program will then cost a total of 10 billion Euros. The above-mentioned ratio of 1:3:6 depends, of course, on the size of the series and represents here only an average value for the totality of the defense material. However, the order of magnitude mentioned is almost congruent with the budget figures of the US Department of Defense for 1979: development 1, procurement 2.6, and utilization 6.2 [27]. However, the ratio can reach even higher values, related to a specific project, since procurement and utilization costs are quantity dependent. For example, the calculated values for a German main battle tank project are 1:5:15, reportedly a common value for highly complex military systems [26].

However, the problem of considering follow-up costs is not limited to areas such as defense technology or aircraft construction but is also increasingly affecting public institutions. "*Indeed, the ongoing and, in the case of individual projects, considerable follow-up expenditures of public investments can narrow the scope for investment activities*" [28].

3.5.2 Significance of the Consequential Cost Approach for the Economy

Life cycle costing is of far-reaching importance for future projects. In the early 1960s, the US Department of Defense introduced the concept of *life cycle costing* [29]. The objective was to view a project from the perspective of the owner (*ownership consideration*), who must give equal weight to development, procurement, and follow-on costs when ordering a new product. This consideration is comparable with the way of thinking of a passenger car buyer, who considers not only the car purchase price but also fuel, maintenance, garage costs, etc. In fact, it turned out that procurement and follow-on costs exceeded the original development costs many times over and placed a considerable burden on the defense budget. If one examines the US defense budget of 1979, the picture already mentioned emerges [27]:

– Research and development (1.0)	9.92%
– Procurement (2.6)	25.56%
– Subsequent costs (6.2)	61.11%
– Other costs	3.41%

The consideration of life cycle costs takes on a special significance *when the following consideration is made: the life cycle for modern technology projects (for example, for combat aircraft)* is approximately 30–50 years: 10 years of development, 5–10 years of procurement and, overlapping this, a utilization phase of up to 30 years. The parameters influencing procurement and follow-up costs are largely defined in the early phases of the project: After completion of the definition phase, 85% of all parameters have already been determined. See Fig. 10.20. The production companies are therefore required at an early stage to prepare an estimate of procurement and follow-on costs in addition to the cost estimate for research and development. Special procedures were developed to highlight the *cost drivers* so that targeted cost-reduction measures could be introduced as early as possible. The reason was to do everything possible in the development phase to keep procurement and follow-on costs as low as possible so as not to clog up the budgets for future necessary developments. Taking into account the ratios of 1:3:6 quoted above it can be seen that the services of the US Department of Defense were quite willing to agree to an increase in development costs in favor of considerably lower follow-up costs; see Sect. 10.5.2.

3.5.3 Transparency of Life Cycle Costs

The consideration of life cycle costs is of serious importance for future projects. In Germany, too, there are already clear signals to this effect. However, determining the total life cycle costs is not always a simple undertaking, as the subsequent costs may not even be recognized as costs associated with the project. Therefore, the starting point for any life cycle cost analysis is the definition of the life cycle (Chap. 4). It is important that the individual project phases are precisely defined and described in terms of content. Only on this basis it is possible to make cost predictions for all phases at an early stage of a project. Equally important, however, is the identification of cost influencing factors in order to provide developers with an instrument for cost reduction; see Sect. 10.6.

3.6 Where To Go With the Project Organization—Help from the Matrix?

3.6.1 Project Management—An Outsider Role

"*The project manager is still in an outsider role in companies and administrations; he does not fit into the classic staff-line organization*" [30]. In fact, many project managers have not yet found a permanent position with their employers. They often hang in the air. At best, they work in staff departments or are integrated at different places in the development department, for example. The cases in which a company or authority has introduced project departments as an independent organizational branch above a certain size are still relatively rare. Even in companies that live predominantly from projects, the consistent introduction of an organizational area called Project Execution or "Project Management Office *(PMO),*" which is also equipped with the necessary powers, responsibilities, and personnel, often meets with little approval. There are reasons for this, as it involves the redistribution of power within the company. Traditionally, power in the company has been structured according to the principle of line management, i.e., according to specialist areas such as development, production, sales, and finance. Specialist decisions are made in the respective department and higher-level decisions are made by the company management. But in today's world, this model no longer works well, because there are many cross-cutting tasks and therefore the silo mentality must be replaced by new organizational structures, such as the matrix organization; see Sect. 5.2.4.

3.6.2 Compulsion for Reorientation

As the tasks of industry and government have changed, so must the management. Traditionally, most industrial companies were engaged in the production of goods in their sector, construction companies in the construction of private and public buildings, and public authorities in the provision of services in their field. The degree of innovation used to happen in relatively slow steps and, most importantly, only within the scope of the particular industry. Of course, a company that manufactured machine tools, for example, carried out development work to improve the production line in parallel with the current series. However, in principle, the innovation usually represented only a relatively simple improvement within the branch area. But in other areas, such as civil engineering or government agencies, a very strong industry linkage was a matter of course. There was little or no overlap. The few existing interfaces, for example between the mechanical engineers and the electrical engineers in the installation of a pumping plant, were relatively simple and straightforward. That is, it was possible to perform industrial tasks with relatively little management overhead. The construction workers built the foundation, the mechanical engineers supplied and installed the pump, and the electricians supplied the electric motor and associated wiring and, if necessary, a control panel. The assigned departments were the design, mechanical and electrical departments, and the construction crew. The situation here is certainly highly simplified, but it reflects the traditional working methods of many industrial companies. Analogous was the way of working in other industrial sectors and at the authorities. Company or authority management could fall back on specialist departments which were relatively easy to supervise due to their clear and unambiguous task structure. Furthermore, competent service departments, such as the finance and purchasing departments, were available to the management.

It is clearly recognizable that in such a company constellation, the managing director(s) simultaneously perform tasks or project management. If the tasks became too extensive, corresponding staff departments were set up or the assistants were entrusted with coordination tasks. The practice described here for managing a business represented a functioning management concept for a very long time.

However, the structural change that has increasingly occurred in industrialized countries in recent decades is forcing a reorientation. At this point, a brief analysis of this structural change is important. In almost all areas of industry, the building trades and the authorities, a new understanding of tasks is discernible. Construction authorities can no longer make subject-specific decisions as in the past but are forced more than ever to consult with other authorities, for example, the environmental authority, before making a decision. Industrial companies fall hopelessly behind if they do not apply the idea of systems engineering, which, for example, unites the mechanical, electrical, and electronic aspects. But new construction projects also show how necessary it is to coordinate structural engineering aspects with those of other disciplines before the first foundation walls are erected. But what is the essence of this easily verifiable brief analysis? The disciplines have moved closer together, interfaced more with each other, and become more interdependent. The degree of linkage has increased considerably. A modern machine tool is no longer essentially a mechanical engineering task as it used to be, but simultaneously an electronics and software product. To put it better, the modern machine tool is an electromechanical system consisting of mechanical, electrical, and electronic subsystems and the associated software. In the development of such a system, it is therefore necessary that the individual subsystems are integrated into an optimal system, which requires that the project manager has knowledge of systems engineering; see Sect. 7.1. It is clear that this management task in all its complexity cannot be assigned to any of the above-mentioned specialist areas. Traditionally, there is no regular position for this. In practice, however, there are further points of connection. The above-mentioned machine tool product must be produced on time, to quality standards, and at low cost if one wants to hold one's own on the market; see Sects. 8.1 and 10.2.

This results in further skin-tight interfaces, for example marketing, production management, quality assurance, and the cost and finance department. However, these departments must not only fulfill passive control tasks but must be integrated into the project team as active partners during the early project phases. Marketing, quality, production, and cost analyses must also be carried out in the early stages. Only then is the project team complete! However, it must be led.

At this point, at the latest, a decision must be made as to who should be responsible for project management and thus for the difficult task of integrating the technical-administrative work described above. Ultimately, this task lies with the entrepreneur or managing director, because, as already mentioned at the beginning, he has traditionally always been the top project manager. However, since he cannot manage several projects at the same time, he must use empowered managers to manage the projects; see Fig. 17.5. After weighing all the advantages and disadvantages, the matrix organization represents the best compromise for integrating project teams into a company; see Sect. 5.2.4.

3.7 The Working Group (*ARGE*) Without a Helmsman—The Dilemma in the Management of Complex Large-Scale Projects

3.7.1 Who Controls the ARGE?

On the outskirts of major German construction sites, one often finds signs indicating that the contract is being carried out by a consortium, or ARGE for short. The usually quite respectable company names belonging to the consortium are also mentioned, as well as their specific area of responsibility for the project. The companies involved are mostly experienced in civil engineering, building construction, electrification, etc. Similar rules of the game have developed around industrial projects, for example, many federal research and development contracts are handled in joint ventures.

The development of a new traffic or messaging system, for example, is assigned to eligible companies from the electrical or traffic industry after a bidding phase. So, what is the main characteristic of such a consortium? The essential feature is already expressed by the designation ARGE Companies that are working on a project at the same time join together to form consortiums for the purpose of better work coordination. To ensure smooth cooperation, it is necessary for the consortium partners to define their areas of responsibility and interface to each other as precisely as possible to avoid discrepancies. It is equally important that the partners agree on deadlines. But who controls the consortium? Who is responsible for the overall system?

The handling of technologically simple and conventional joint projects certainly requires only relatively minor control mechanisms, which can be performed either by the client of the project itself or by an architectural or engineering firm brought in. But who takes over the control of a consortium in technologically complex projects, where the problems that arise can usually lead to considerable financial consequences? The subject of project cooperation is discussed in detail in Sect. 5.3.

3.7.2 Assistance from the Leader

At this point, it is important to analyse the management process referred to a little more closely. Projects carried out by a consortium do not automatically have joint management, unlike projects carried out by a single company. This is in fact a management gap, quite unlike the prime contractor solution where one company takes over the overall contract (*turnkey*) and subcontracts where necessary. So, a way must be found to solve the management gap.

Without going into details, the possibility of joint ARGE management suggests itself. At the same time, however, it must be clear that a functioning ARGE management, i.e., one with powers of attorney, can impair the individual interests of the ARGE companies. This is the decisive point why the above-mentioned solution is usually not realized. The problems often start with the staffing of the ARGE management office. Questions such as: which company provides the ARGE project manager, and what powers may the project manager have, etc. are often difficult to solve. In this context, it should not be forgotten that the interests of the individual companies, which may even be competitors in another project, are considerably influenced by a strong ARGE management and that, under certain circumstances, the competition could exert an unfavorable influence on their own business. These reasons are quite understandable.

In the interest of the project to be carried out, however, the solution of a strong ARGE management is definitely to be favored, since only in this way can an optimum system solution be achieved; see Sect. 5.3. A clear conflict arises from the conflicting interests between the project and company viewpoints. If one assumes that the unsatisfactory handling of a major project is ultimately detrimental to the image of the companies involved, however, a satisfactory solution aimed at increased system or project management must be sought in the long term. In many joint projects, therefore, the decision is made to adopt the so-called lead management concept. This means that, with the agreement of all the companies involved, one of the participating ARGE companies could also take on the role of lead manager (*Federführer*). This company then has the difficult task of coordinating the work to be carried out at the individual companies, although the coordination task requires further definition with regard to the powers and responsibilities granted to it. This is because in many cases the coordination task is limited only to the compilation of information provided voluntarily but does not include any control rights. This means that the right to control and make decisions at the project level still lies with the individual ARGE companies. If the companies disagree on individual points, compromises that are detrimental to the project repeatedly occur without the lead manager being able to intervene in a controlling manner. This situation can be compared quite well to a ship without a helmsman.

3.7.3 Compulsion to Design Wholeness

So, what are the problems that repeatedly lead to such behavior? Traditionally, industrial and construction companies were usually able to carry out large-scale projects on their own, such as the construction of a power plant, the development of an aircraft, or the planning and construction of a public facility or a hospital. The companies carried out such projects on their own initiative (*aircraft construction*), on behalf of a utility company (*power plant facility*) or a public authority (*hospital*). The competence of the companies was undisputed, since, to stay with the above examples, the tasks were mainly sector-related, namely: plant construction, aircraft construction, and building construction. Wherever necessary, subcontractors were brought in, but the lead was clearly with a single company.

Two factors have significantly changed the situation in recent decades: project size and the complexity of modern projects. Many major projects have taken on such a large financial volume that a single commercial enterprise is hardly in a position today to carry out the tasks alone for reasons of capacity, capital, and risk. In the case of state projects, moreover, it is often stipulated that, for reasons of proportionality, several national and often also international companies are to be equally involved in the project. The aim is to ensure that the return on investment is as widely distributed as possible. This is often referred to as the "watering can principle."

The sharp increase in project complexity is another reason why companies are increasingly moving toward handling major projects as part of a joint venture. Using the example of development work for a modern fighter aircraft, the situation will be examined here in a little more detail. In contrast to earlier aircraft, where the aerodynamic aspects, i.e., flight characteristics, were paramount, today two other factors play an equally important role, namely the engine design and the electronic equipment. In other words, the optimum performance design of a modern combat aircraft is equally dependent on the basic aerodynamic, engine, and electronic concepts and on the coordination of these three factors. While it used to be the case that the actual aircraft manufacturers designed an aircraft essentially according to the aerodynamic aspects and then purchased the engines and the hydraulic and electrical equipment to go with them, it is now essential to strive for the design according to systems engineering aspects, in which a holistic solution is paramount, right from the early stages of an aircraft project. "*The advances in technology of the last twenty years repeatedly call for (such) holistic designs,*" says aircraft designer Bölkow [31]. If one now compares other projects, such as power plant construction or the construction of modern hospitals, with the situation in aircraft construction, a similar tendency can be seen. These projects have also greatly increased in complexity over the last 20 years due to a greatly refined infrastructure. Here, too, it is advisable to manage projects from a system engineering point of view. Conclusion: In the interest of a system-optimal project solution, the lead engineer must also be given a mandate for project and system management, otherwise the success of the project will be severely jeopardized.

Whatever the interests of the companies involved in a large-scale project, the problem of managing large-scale projects according to technologically and economically optimal aspects must be solved by the companies involved. This task cannot be undertaken by governmental or semi-governmental organizations or empowered engineering firms, as it involves entrepreneurial considerations such as industrial responsibility and risk-taking. It is only necessary to realize that the relationship between project size and complexity and the average company size has shifted considerably in recent years and to draw the necessary conclusions about industrial cooperation from this. In the future, joint ventures must be managed like commercial enterprises to a much greater extent than before. Several tried and tested options are available for this purpose:

- A working group (*ARGE*) with a strong leader concept
- Establishment of an industrial consortium with a strong leadership or
- Establishment of a management company

Further details are described in Sect. 5.3.

3.8 International Projects Influenced by Politics—The Tragedy of ELDO

3.8.1 Problems, Crises, Permanent Crises, Project Disasters

Increasingly, government agencies and industry are involved in conducting international business. In the area of government projects, this results from the need to share the enormous costs associated with implementing major projects, such as various aerospace projects, the Spacelab space laboratory, the International Space Station (*ISS*), and NATO's Tornado and Eurofighter Typhoon aircraft projects. "*Nationally, research becomes virtually unaffordable for modern defense*" [32].

In addition to the generally known problems that arise in international business, such as differences in legislation, tax law, language, and general customs, multinational but also bilateral government projects are often subject to the political influences of the participating member states. The political-technological tragedy of ELDO (*European Space Vehicle Launcher Development Organization*) shows most blatantly where immature management solutions mixed with political self-interest can lead, namely from problems to temporary and then to permanent crises, ultimately ending in technological and financial project disasters. But the successor organization ESA (*European Space Agency*) is not unaffected by the political currents of the 22 European member countries. However, it can assert itself far more convincingly than ELDO vis-à-vis the member countries because of its considerably improved management conditions.

The example of NATO's Tornado aircraft project also demonstrates how political disagreement reduces management efficiency [33]. Even in bilateral projects, jealous political behavior often impedes objectively reasonable project management. Sometimes even the clarification of the location issue for the joint project office represents an insurmountable hurdle or at least leads to project delays. The example of the ELDO tragedy is intended to illustrate the problems of international project handling, and it can only be hoped that similar mistakes will not be repeated.

3.8.2 Program Discontinued After 11 Years of Development

"*The delusion is short, the regret is long,*" this quote of Schiller was used by the Association of International and European Civil Servants to describe the situation on the occasion of the liquidation of the supporting program EUROPA II, decided at the 65th Council of May 25, 1973, and issued a special trilingual letter to commemorate this piece of European history.

The EUROPA II launcher should have carried a 200 kg payload into a 36,000 km high geostationary orbit. ELDO and the development work for the European satellite launch vehicle carried out within the framework of this organization were just 11 years old, and already everything was over again. Press reports such as "*EUROPA II goes to the German Museum*" or "*Europe's rocket development a total hullabaloo*" accompanied the end of EUROPA II and the Paris-based ELDO. The first stage of the EUROPA II launcher (*flight F 12*) was not launched but donated to the Deutsches Museum in Munich. How could this happen? Were the Europeans not capable of developing such a complicated device as a launch vehicle? On the other hand, hadn't the English, French, and Italians developed excellent airplanes? Even in the Federal Republic of Germany it was possible to build airplanes, and didn't the missile development come largely from Germany anyway? Questions like these were asked again and again to the staff involved in ELDO's missile development. How can the development errors, which cost the ELDO member states millions of Euros, be explained? Was it ultimately a management problem?

3.8.3 On the History of ELDO

As a result of a proposal made in early 1962 by England and France to establish a European organization for the development of a launch vehicle for the launch of satellites serving commercial purposes, ELDO was created. The founding members were Australia, Belgium, the Federal Republic of Germany, France, Italy, the Netherlands, and the United Kingdom. Even before ELDO was founded, work had begun on the ELDO initial program, i.e., the development of the launch vehicle to be launched from the test site in Womera (*Australia*), designated ELDO A (*later named EUROPA I*), with which a payload of about 1 ton was to be placed in a polar circular orbit of 300 km altitude. By 1970, several test launches were made to test the three-stage launch vehicle, but the interaction of the three stages did not work out [34]. The first stage (*Blue Streak*) was developed and provided by England, the second stage (*Coralie*) by France, and the third stage (*Astris*) by Germany. Contracting was on a national basis and ELDO had the role of overall coordination [35].

Later, it was decided to abandon the Australian launch base in favor of the more favorable equatorial launch site in French Guiana, and it was decided at the same time to develop the four-stage EUROPA II launcher, which was intended to launch payloads into a geostationary orbit of 36,000 km altitude to operate Earth-synchronous communications satellites. The equatorial launch (*Guyana is almost at the equator*) allowed payload capacity to be significantly increased for easterly launches due to Earth's rotation gain.

The first three liquid stages were taken from the EUROPA I launch vehicle, and the fourth stage, a solid perigee engine was additionally developed by Italy. The first test firing, designated F 11, was not a total failure from a technological point of view, but it did not achieve the desired orbit.

Due to a fault in the control electronics, the rocket had to be destroyed prematurely. The designation F 11 was derived from flight designations F 1 to F 10 of EUROPA I and gave the public the impression that the 11th unit was now also malfunctioning. However, it was EUROPA II, F 1, an important PR consideration that was not made clear to the public. The F 12 follow-on launch planned for 1973 was not carried out because of the decision to discontinue the ELDO launch vehicle program.

In 1969, building on several previously conducted industrial studies in parallel with EUROPA I/II development, ELDO began preliminary work on a new and more advanced launch vehicle program, designated EUROPA III, which was to be used to launch a 750 kg communications satellite into a GEO orbit. For the first time, high-energy propellants of the type previously used only in the USA as part of the Apollo program were used. An upper stage with cryogenic propellants was planned, i.e., the use of liquid hydrogen and oxygen, the handling of which placed particularly high demands on the planners. Development was to be completed in mid-1980 after five test flights, and the first operational firing was scheduled for 1985. A development budget of 470 million accounting units (*in 1972, this was equivalent to $470 million U.S.*) and a 20% planning reserve were provided for this purpose. With the discontinuation of ELDO activities, the EUROPA III program was also canceled. A short time later, based on a French initiative and building on the existing preliminary development results of EUROPA III, the European launch vehicle Ariane was brought into the program of the newly founded European Space Agency ESA under the leadership of the French national space agency CNES (*Centre National d'Etudes Spatiales*). Thereafter, ARIANE launchers were marketed with great success by ARIANESPACE, an international company based near Paris. The Ariane 4 and 5 launchers became a hot seller worldwide.

It is important to be mentioned that Ariane 5 has in 2023 become the most successful Satellite launch vehicle worldwide.

3.8.4 Europe's Missile Development—A Total Tohubohu

Scolding for ELDO was just one of the many press reports following the July 1972 failure of the first EUROPA II (*F11 rocket*) launched in French Guiana. The Stuttgarter Nachrichten of July 18, 1972, continued: "*The last Europa rocket burst because an English company in South America had installed electronics in the German stage without taking enough care of the grounding. After the failure, a technician of the European rocket organization ELDO said: Watch out, now we are to blame again. I don't even dare to go home anymore. He was right. A little later, ELDO's head of administration, Renzo di Carrobi, said that the whole thing should be thrown out. These days, the German Federal Minister of Science, von Dohnanyi, has also spoken out in favor of throwing it out.*" The same report further states, "*In truth, Europe's missile development is a total hullabaloo, with some not knowing what they want and others not knowing what they are allowed to do.*" This press release very aptly summarized the management problems of European launch vehicle development. The following analysis gives some indication of this.

3.8.5 Analysis of the ELDO Problem

How could the above-mentioned misguided development come about? Were the Europeans incapable of developing their own launch vehicle? Or was it rather a management and leadership problem whose roots were deeply rooted in European politics? The previously quoted Stuttgarter Nachrichten newspaper of July 18, 1972, writes very aptly *on* this subject: "*To rail at ELDO in this situation is a stop-the-thief method designed to divert attention from political culpability.*" And elsewhere it says: "*This state of affairs, however, is the responsibility of the ministers and diplomats who today advocate letting the EUROPA missile die. They have neither endowed the ELDO with rights enabling it to work meaningfully, nor set it a fixed task. Missile construction, one connoisseur once said, has always been a kind of alibi for Europe's governments, a charitable employment program for their own industry that made them feel noble and technologically advanced at the same time.*"

A look at the ELDO Convention ratified in March 1962 by the seven countries Australia, Belgium, the Federal Republic of Germany, France, Italy, the Netherlands, and the United Kingdom clearly shows the birth defect of this organization. It states that "*the Organization shall, within the framework of its activities, promote the joint development of relevant techniques in the Member States and shall assist Member States 'at their request' to bring to bear the techniques developed within the framework of its activities*" [35]. This already sealed the ELDO's impotence, as its role was clearly that of a coordinator without powers. The member states had secured themselves against unwelcome interference by the ELDO by the restriction "*at its request.*" Article 6 (1) of the Convention then explicitly stated that industrial contracts for the implementation of the Initial Program were to be concluded by the individual states in accordance with their usual procedures. In 1966, the long-time program director of the ELDO Initial Program (*EUROPA I*), Brigadier Abate, wrote in this regard, "*The overall project control of the ELDO Initial Program shall therefore be operated without the approval of the group awarding the contracts*" [36].

Abate also addresses the management problem. The seven ELDO member states were involved in the implementation of the initial program in two ways: super ordinately as members of the ELDO Council and simultaneously as the national contracting agencies. Under this condition, ELDO could not bring the missile program, which was in any case poorly prepared, under its complete control because it could not escape national power influences. Servan-Schreiber described the situation as follows: "*The example of ELDO has shown what confederative cooperation looks like. The main characteristic of this type of organization is that decisions can only be made by unanimous decision of the members. Therefore, the joint institutions are limited to a secretariat whose powers extend only to certain methods of implementation, but not to any of the major decisions. Such institutions decide little, slowly, and poorly*" [37].

Further problems resulted from the different development statuses of the partner countries. At the start of the development work, England and France brought their far more advanced rocket stages into the program. This led to considerable problems, which can be summarized as follows [38]:

- Different deadline situations: that is, increased deadline pressure on the German and Italian companies, which led to (a) planning problems and (b) team utilization difficulties.

- Inadequate interface definition: the two lower stages were already in the hardware phase, and coordination of interfaces with the upper stages and electronic components yet to be developed was therefore mostly carried out to their detriment, i.e., not always in a system-optimal manner.
- Non-optimal system design: the Blue-Streak and Coralie stages, most of which had already been completed, made it difficult to achieve an optimal design for the overall system.

In the end, poor interface definition and control was probably the biggest management problem in the development of the EUROPA I and II launchers, because the development of the individual stages was firmly in national hands, and the powerless ELDO had the difficult task of system management and interface control for the overall missile as part of its coordination mission and that was a very difficult management task.

However, interfaces can only be coordinated sensibly if one also has the necessary powers of attorney. How else could it happen that, as mentioned above, an English company installed electronic equipment on the German stage without satisfactorily solving the problem of grounding? Whose responsibility was this now? Was it the responsibility of the English or the German company? This was a typical interface problem. It was not until the follow-up developments of Europa III and Ariane that an effective management procedure called "*Interface Control Document—ICD*" was introduced; see Sect. 7.2.3.

In successfully completed projects, the task of interface control has therefore also been elevated to a very important project management task. The term interface management exists for this. In NASA's Apollo program, the US company General Electric was responsible for *interface management*. A basic rule of modern project management states, for example, that in the case of integration processes of subsystems into a system, the guarantee for the proper functioning of the same should reside in the area of responsibility of the system representative.

So, did the buck stop with Germany? The German companies involved naturally denied any responsibility, and the situation was also far more complicated. The previously mentioned basic rule can only be implemented if the responsibilities and powers of attorney are clearly regulated via a contract so that the system owner can convince himself in advance of the quality of the subsystems and the problem-free nature of the interfaces to each other. In contrast, the two companies involved from Germany and England, which were contractually bound to their national clients, had no clear mandate for interface control; in fact, they were left hanging in the air in this respect. A classic case of how not to do it.

Of course, on the other hand, ELDO nevertheless organized regular meetings to coordinate interfaces, but the extremely complicated organizational structures between the customer and *the* contractor often led, as already mentioned, to situations that the Stuttgarter Nachrichten of July 18, 1972, described with the phrase "*some don't know what they want, and others don't know what they're allowed to do.*" For example, it could happen that at an interface meeting between the third stage for EUROPA II to be developed by Germany and the launch facility in French

Guiana, the following parties sat at the same table without the overriding powers and responsibilities being clearly clarified: ELDO as Europe's rocket coordinator without authority; SETIS (*Société Européenne d'Etude et d'Intégration des Systèmes Spatiaux*), a company set up by several European aerospace companies to assist ELDO in its system coordination role, but which could only make recommendations; and the French national space agency CNES, which was responsible for the launch facilities in French Guiana but was not under ELDO but under a responsible French ministry. The German space agency GfW (*Gesellschaft für Weltraumforschung mbH—today DLR*) was also at the table but was also not under ELDO, but under the responsible German research ministry. GfW was the client for the German industry, i.e., the companies Bölkow KG in Ottobrunn and ERNO Raumfahrttechnik GmbH in Bremen, which in turn were represented by ASAT (*Arbeitsgemeinschaft Satellitenträger*). Under these circumstances, it was very difficult to develop such a complicated device as a launch vehicle. One must compliment the responsible technicians for having made it this far at all, which was also clearly confirmed by an investigative commission staffed with European and American experts.

3.8.6 Learning from Negative Experiences

The abandonment of European launch vehicle development, decided in December 1972 and executed in May 1973, set Europe back almost 10 years. By 1973, European rocket builders had spent nearly 2 billion Euros without a working launch vehicle to show for it. The disunity of the Europeans manifested itself at this time by the Brussels Council of Ministers decisions in favor of three *special programs*, namely: SPACELAB, the special program favored by the Germans (*a manned space laboratory first launched into orbit by the American Space Shuttle in 1983*); MAROTS, the special program desired by the British (*a generation of intelligence satellites*); and ARIANE, the special program sought by the French (*a multi-stage launch vehicle*), which today, more than 50 years after the ELDO collapse, restores to the Europeans a sense of self-confidence in rocketry with Ariane V. The *European Space Agency* (ESA), newly founded in 1974 and headquartered in Paris, is responsible not only for the European basic program (*development of scientific satellite projects*) but also for the three special programs mentioned earlier.

There is no question that Western Europeans have learned from the bad experiences of the ELDO dilemma, and this must also be clearly confirmed. ESA's projects have been successful so far, and it is no exaggeration to say that ESA also has the necessary experience to manage complex international projects. These include various satellite programs, contributions to the International Space Station (*ISS*), as well as the developments of the European launch vehicle Ariane 5 and the continuation with Ariane 6.

Fig. 3.3 Ariane 5 at the launch facility (Creative Commons)

3.9 All Well That Ends Well

Ariane 5, a European success story—what takes a long time finally comes out good! After years of rocket chaos, Europe managed to climb into the league of the most successful launch service providers in international competition with the Ariane rocket family. After initially two flawed launches, Ariane 5 has developed into the world's most reliable rocket in international competition with a success rate of 98% [39]. Figure 3.3 shows Ariane 5 at the launch facility.

Since 1989, the Ariane 5 launch vehicle has launched 117 times into space from the European Spaceport Kourou in French Guiana, of which 115 are successful in a row. On December 25, 2021, Ariane 5 brought NASA's highly complex James Webb Space Telescope (JWST) very precisely into the intended orbit position and thereby saved fuel which may be used for lifetime extension. The JWST orbit at

Lagrange-Point (*L2*), is a position at which the gravitational forces of the Sun and Earth are approximately equal, creating a stable orbital location. The 117th and final launch of Ariane 5 carried the German communications satellite Heinrich Hertz, a small GEO, into the intended orbit.

The success of Ariane 5 is due to an improved management structure between ESA (client) and industry (contractor) as well as rigid quality management. Ariane 5 was developed by the space companies of ESA member states on behalf of ESA. ESA commissioned the French space agency CNES to carry out the project. CNES had technical management and was responsible for financial management and the distribution of tasks to the individual companies in the member countries. Both ESA and CNES have demonstrated that good project management leads to technical and economic success.

The maiden launch of the subsequent and larger carrier rocket Ariane 6 took place on 10 July 2024.

Literature

1. The Standish Group 2015 Chaos Report,www.standishgroup.com
2. Weizsäcker, Carl Friedrich von: Wege in die Gefahr; Carl Hanser Verlag, Munich 1976, p. 47
3. During, Rainer W: Lachnummer BER, 1st ed., Rotbuch Verlag, Berlin, 2013, p. 7f
4. Q.3, p. 205ff
5. Q. 3, P. 59
6. Q 3, P. 64
7. Q 3, P. 196
8. Q.3, P.277
9. Roth, Mathias: Der Hauptstadtflughafen, Klampen Verlag, Springe, 2013
10. Q. 3, P. 199
11. Gerkan, Meinhard von: Black Box BER, Quadriga Verlag (www.quadrigaverlag.de), 2013, p. 49
12. Q. 11, P. 55
13. Q. 11, P. 60
14. Q. 11, P. 66
15. Q. 11, P. 82
16. Rehrmann, Marc-Oliver, NDR, 12.01.2015
17. Dey, Andreas, Haarmeyer, Jan, Mischke Joachim: "Wunschkonzert". A documentary on how a brilliant idea became the biggest construction scandal in Hamburg's history. The incredible story of the Elbphilharmonie. Hamburger Abendblatt, December 13, 2013
18. Haarmeyer, Jan, Mischke, Joachim: Elbphilharmonie - eine Kathedrale, Hamburger Abendblatt, December 4/15, 2013
19. Bund der Steuerzahler, Elbphilharmonie wird immer teurer, www.schwarzbuch.de/print/286, dated 5.7.2014
20. Servan-Schreiber, J.-J.: Die amerikanische Herausforderung, Hoffmann und Campe Verlag Hamburg, 1968, p. 96.
21. Tauber, Christian H.: Kostengünstigeres Bauen: Planning ahead can get a better grip on construction costs, in: Handelsblatt, 9/10. 07. 1982.
22. Grün, Oskar: Beiträge zur Projektorganisation; Heft 3: Kosten und Kosten-rechnung in der Projektorganisation, Vienna, 1977, DBW-Depot 78-3-3, p. 3, 2 (C. E. Poeschel Verlag Stuttgart).

23. Augustine, Norman R.: Augustine's Laws and Major System Development Programs, in: Astronautics & Aeronautics, 4/80, p. 36.
24. BMF-Mitteilungen 4/82, p. 74
25. MBB: Helios solar probe final report, research and development contract QC-90-00 VP 6/70, 1976.
26. Bethke Rodin: Modern Material Preservation for Missile Systems, Wehrtechnik 3/81.
27. Seldon, M. Robert: Life Cycle Costing: A Better Method of Government Procurement, Westview Press, Boulder Colorado, 1979, p. 2.
28. Lenk, Reinhard and Lang, Eva: Herstellungskosten und Folgekosten öffentlicher Investitionen, IFO Institut für Wirtschaftsforschung e. V., Munich, 1981.
29. Life Cycle Procurement Guide, Joint AFSC/AFLC Commanders Working Group on Life Cycle Cost, ASD/ACL, PAFB, Ohio 45 433, July 1976.
30. Hegi, Othmar: Project Management, a Foreign Body in the Staff-Line Organization, from: Project Management
31. MBB - the nation's multipurpose forge, in: GEO, April 1982, p. 85.
32. Tiede, Heinz: Waffenbrüder, in: Capital 7/82, p. 136.
33. Köppl, J. Bruno: Probleme des multinationalen Rüstungsmanagements und deren Auswirkungen auf die Verteidigungsfähigkeit der NATO-Staaten unter dem Aspekt der wachsenden sowjetischen Bedrohung, Donau-Verlag, Straubing-München, 1979.
34. Ruppe, Harry O.: Die grenzenlose Dimension Raumfahrt, Econ Verlag, Düsseldorf-Wien, 1980, vol. 2p. 128.
35. Convention for the Establishment of a European Organisation for the Development and Construction of Space Vehicle Launchers, London 29 March 1962.
36. Abate, Arthur T.: Problems of Management of ELDOs Initial Programs, Sciences et Industries Spatiales 1/2, 1966.
37. Cf. Q. 20, p. 185
38. Madauss, Bernd-J.: Projektmanagement bei internationalen Vorhaben, from Projektmanagement (*Saynisch et al.*), Oldenbourg-Verlag, Munich-Vienna, 1977, p. 412.
39. de.wikipedia.org/wiki/Ariane-5

Life Cycle of a Project

4

Contents

Abstract

Projects run in individual phases that must be logically coordinated. At the beginning, the project objective must be clearly defined and the end of the life cycle is determined by the specified end result. To prevent problems, it is important that the rules of a logical process are followed. Especially the early phases play an important role, because in them the plans for the realization are created. Mistakes made in the early phases often have a major impact on the expensive subsequent phases. The life cycle, which is divided into phases, is the basis for determining the life cycle costs (LCC).

B. -J. Madauss, *Project Management*,
https://doi.org/10.1007/978-3-662-69057-4_4

The efficient execution of a project requires that a clear and unambiguous project objective is set at the beginning of the project. In this way, a balance can be achieved between the objectives set at the start of the project and the available budget (investment); see Fig. 4.1. Adherence to this balance during project execution is necessary to complete the project successfully. Any instability of the balance that becomes apparent requires either the correction of the project requirements (*change*), the adjustment of the budget, or a change of both factors.

At the beginning of a project, it is relatively difficult to assess the financial balance between the above-mentioned two factors, since the translation of the project requirements into a clear financial statement is usually relatively imprecise. This is mainly due to the fact that projects are, in principle, innovative undertakings and inherently contain a greater or lesser number of unknowns, because projects are one-offs (*never done before*). Steinberg lists five success factors that play a decisive role in innovation competition and points to a strong fixation of companies on past successes and criticizes the lack of systematic "*thinking ahead*" in planning for the future [1]:

- Market orientation and differentiation
- Speed of innovation
- Strategic timing
- Mastering complexity
- Development efficiency

Therefore, it is important to start with a concept phase in which the conceptual solutions as well as alternatives are developed. The calculated costs and deadlines are usually still relatively uncertain at the beginning of the project, but after completion of the concept phase, the degree of accuracy has usually already improved

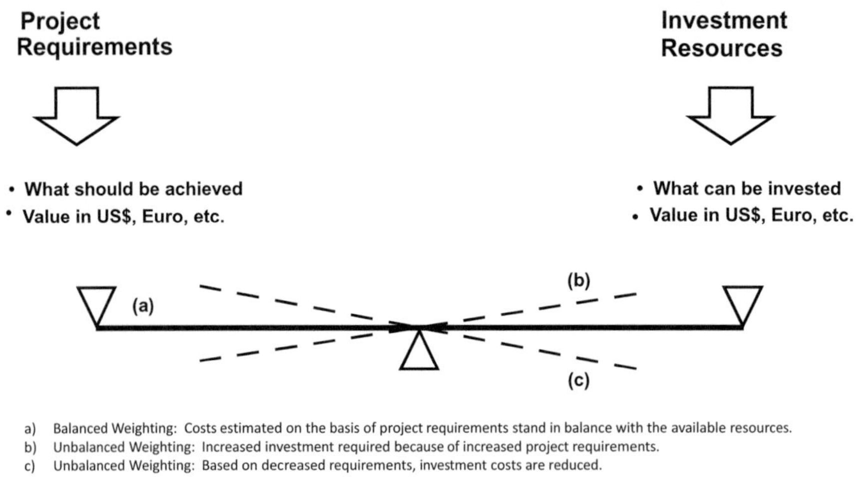

Fig. 4.1 Balance between project requirements and budget

considerably. The estimation accuracy can be further increased by means of software-supported cost estimation models; see Sect. 10.3. Based on the initial project plans, an initial version of the life cycle costs (LCC) and ROI (*return on investment*) can then be calculated; see subsect. 10.5.4.

Thorough planning and monitoring are important for controlling risks because as project complexity increases, so does the entrepreneurial project risk; see Chap. 19. It is equally important that the most diverse management methods are also applied at the right time because only then do they have an optimal effect; see Chap. 6. Methods that are important at the beginning of the project often no longer have any effect at a later point in time; the same applies to the reverse situation. In order to be able to take the right path, the course of a project, and its life cycle, must be precisely known. Every project goes through a very specific path that can be divided into individual phases. Saynisch wrote: "*The project process with its life cycle phases is understood as a sequential decision procedure oriented towards learning*" [2]. Despite planning, there are almost always unforeseen problems in the project process that need to be resolved. However, in some cases, there are also unsolvable problems. As Augustine puts it, "*In project work, you are dealing with known unknowns and with unknown unknowns*" [3]. To safeguard against the known unknowns, one should provide for appropriate planning reserves (*margins*), while with the unknown unknowns, the so-called lightning strikes out of the blue, one is often faced with a problem for which there are no standard solutions available.

4.1 The Life Cycle—An Iterative Process

At the beginning of a project, there is a goal to be set. However, it often happens that the goal is not defined enough, is unrealistic, or is set too high—nice to have. This is the problem of over specification. "*The wish is the father of the thought*" says a German proverb. Some decades ago, the author was involved in a metro project in a South American country and experienced how increased riding comfort led to significant cost increases. For space projects, the reliability requirements are often a strong cost driver. There are many more parameters to be considered when setting a goal. For example, political and environmental requirements or restraints, as well as time and cost limitations.

However, it is not uncommon for such influences to only become apparent at a later point in time. This is not just the case in the private sector. Authorities and companies are also aware of the problem. Projects are often carried out without sophisticated plans and alternatives and ignorance of imponderables, such as hidden costs and supply chain problems, must also be taken into account. A project has been started and suddenly some technical, organizational, and financial problems for which you were not prepared to come across which leads to additional costs. On the other hand, cost estimations in the early phases of a project are extremely difficult. In order to reach at the start of a project a realistic balance between the defined goal, schedule, and costs (see Figs. 4.1 and 9.2) appropriate preparatory work must be carried out.

4.1.1 Project Requirements

Depending on the task, size and complexity of a project, its duration can range from a few weeks to a decade or more. Particularly large and complex projects, such as the development, construction and commissioning of a new type of aircraft, may even require a period of up to 20 years; e.g., the Concorde supersonic aircraft. In the beginning, there is the project idea, which is usually preceded by more or less detailed considerations and studies before being translated into a concrete objective. The conversion of the project idea into a concrete project requirement (*specification*) is a first important project step, which should be carried out as thoroughly as possible and recorded in a requirement specification. Here is the definition in DIN 69901-5: "*The requirements specification, also known as requirements description, requirements catalog, product sketch, data sheet or customer specification defines the totality of the tasks, deliverables and services required by the customer*" [4]. It can be an external or an internal customer. For smaller projects, it is often sufficient to create a data sheet in which all *requirements* are recorded. For scientific undertakings, sometimes the term "*Mission Requirement*" or "*Mission Requirements Document—MRD*" is applied.

The requirements specification is the reference document for all subsequent project work and also serves as the basic document for project tenders; see Fig. 9.4. The procedure described above is logical and basically self-evident, but in practice a systematic procedure is often dispensed with for various reasons, which then could lead to problems later on. There are two main reasons for this: (1) people are of the opinion that it is not worth the effort because the people concerned already know what is involved, or (2) people are of the opinion that it is too early to commit oneself because there are still many open points to be resolved. However, both views are wrong and do not lead to success. They are a fatal fallacy because they only postpone the problems to a later point in time.

Experience teaches that it is correct to proceed step-by-step when drawing up a specification by first defining the objectives of the system level and then completing them successively in the top–down principle to the next level. In cases where it is not yet possible to define precise requirements, assumptions can be made which are verified at a later stage and thereafter converted into specifications. At this point, one should not make the mistake of prematurely dealing with solutions that are reserved for the subsequent phases but rather focus strictly on the definition of *requirements*. The creation of a requirements specification is an important management task and its importance should not be underestimated.

4.1.2 Birdseye View

As already mentioned, the project task can only be optimally solved by a systematic step-by-step approach. When looking at the project from a birds' eye point of view, (*the top-down approach*), it can be seen that the course of a project runs according to a very specific rhythm and in phases. During each of the project phases, the required tasks and the personnel involvement may change. For a better understanding it is therefore necessary to consider a project in its full length (*life cycle*), i.e., from the beginning of the conception to the commissioning. At the start of the project, conceptual project studies and compatibility analyses should be carried out, followed by detailed definition and specification work. Development work, which ends in the creation of reproducible prototypes, should not begin until the specification work has been completed. The next step is serial production, followed by commissioning and retirement. The project steps described, the phases, are the basic building blocks of the life cycle of a project and, because of the very different range of activities, also may require a completely different employee structure and the differentiated use of methods, procedures and software support appropriate to the respective phase.

The division of a project into individual project phases is of great importance, especially for large and complex projects, because at the beginning of the project it is not necessary to decide on the project as a whole, but only on the first phase in the first step, and then on the second phase, and so on. The cost of the preliminary stages is relatively low compared to the more expensive subsequent stages (development, production and operation). The overall project is broken down into stages, and problems are approached cautiously. Disciplined adherence to this system prevents the expensive phases from being started without the relatively cheaper preliminary phases having been properly completed first. Some may counter that it takes too long for the development and/or procurement of a new system to complete the preliminary phases (concept and definition phases). On the other hand, however, the results of the preliminary phases represent important information for the release of the expensive subsequent phases (*development and/or procurement*). A way out is often a phase overlapping as described in Sect. 4.3.5. The meaning of the systematic procedure becomes clearly visible from the cost relationship of the individual phases shown in Fig. 4.3.

In the early phases, the foundation for the project is laid. It starts with the requirements of the system (see Sect. 4.1.1), on the basis of which concepts and alternatives are developed. Looking at the cost comparison in Fig. 4.3, it is quite reasonable to perform detailed system analyses, including parameter and cost comparisons (*trade-offs*) on the basis of which the starting signal for the subsequent phases is given. Often, from the point of view that conceptual changes in the later phases lead to high additional costs, it makes sense to carry out several system analyses in parallel in the low-cost early phases, including life cycle costs (*LCC*); see Sect. 10.5.4.

4.1.3 Approaching the Situation

The question posed earlier about the balance between the project requirements and the planned investment funds remains in principle, but the absolute values must be verified. In the case of a small and/or simple project costing a few hundred thousand Euros, a comparison with the intended budget (*investment costs*) can be established relatively quickly by means of a thorough description of the task and the cost determination based on this. For large and complex projects, however, a step-by-step approach must be taken. Figure 4.2 shows the standard phase model for project plans developed by Hopf and Madauss in 1981 and revised by the author in 2015 [5]. The development of this model was originally based on existing concepts of ESA and NASA, as well as the authors' own experience in participating in projects from space travel, plant engineering, and traffic engineering. The standard phase model shows the entire *life cycle of* a project, i.e., from the project idea to the end of the project (*from cradle to grave*), including all main milestones of the top project level.

4.1.4 Major Milestones of the System Level

Milestones are important measurement points that allow the project manager to see the status at the planned time; see Sect. 9.3.4. Below is a list of typical project milestones including their meaning; see Fig. 4.2.

(1) MRR—Mission Requirements Review (*Project Requirements Review*): Project requirements are defined in conjunction with the preparation of a business plan. In the MRR review, the project requirements (*specifications*) are to be checked for completeness and feasibility.
(2) SCR—System Concept Review: Based on the MRR initial technical concepts, schedules and financial plans, including cost projections, impact assessments, and the investigation of alternatives shall be performed. The SCR meeting is used to review the proposed baseline concept and its alternatives.
(3) SSR—System Specification Review: At this stage, the system-level specifications are to be reviewed. They specify the required technical requirements for the system, including all functional and environmental tests and quality standards. For larger projects, the lower-level subsystem and component specifications are also to be reviewed, since the functioning of a system understandably depends on the functionality and quality of the subsystems and components.
(4) PDR—Preliminary Design Review: The SSR outputs are the released specifications of the system level and, if necessary, also of the lower levels and the inputs for the development process. Design reviews shall be conducted to assess the development process. The PDR is, in a sense, the test run for the CDR; see V-Model in Fig. 7.15. During the reviews, all development documents (*parts lists, drawings, procedures, etc.*) must be checked for completeness and content. Based on the PDR results, functional tests are performed at the subsystem and component levels.
(5) CDR—Critical Design Review: The development work is often supported by functional prototype tests running in parallel and the results obtained are incorporated into the CDR—development documentation (*parts lists, drawings, procedures, etc.*). During the CDR, all documentation comes under scrutiny and is checked for completeness and content. Successful CDR completion (*design*

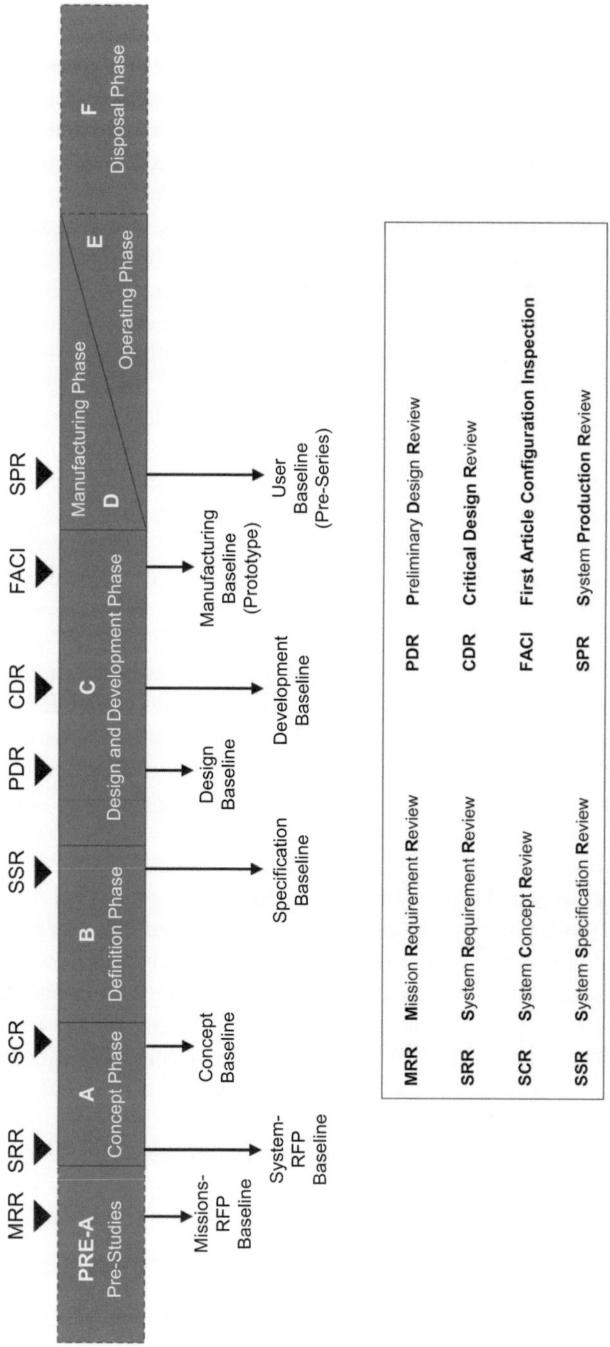

Fig. 4.2 Standard phase model for R&D projects

freeze) is the prerequisite for all subsequent activities, e.g., building and testing prototypes.

(6) FACI—First Article Configuration Inspection (*configuration inspection of the product created according to the construction documents*): To determine if the first item was built according to the construction documents or if there are construction deviations (*deviations*). Large deviations in the agreement between the construction documents (*as designed*) and the created product (*as built)* lead to rework, otherwise an acceptance by the customer cannot take place.

(7) AT—Acceptance Test: The recorded acceptance test is associated with the product acceptance and often also the official handover or delivery of the product (*hardware, software or services*) to the customer. This can involve the handover of a building, a factory facility, a satellite in orbit or an IT solution. With the acceptance and handover, the product with all rights and obligations becomes the property of the customer.

(8) SPR—System Production Reviews: During the production process of series products, e.g., vehicles, machines or systems regular functional and quality checks as well as acceptance tests are required. Corresponding milestones are defined on an operational basis, e.g., qualification reviews.

(9) OR—Operation Reviews: Periodic functional reviews are required during the operation of systems (*e.g., vehicles, equipment, etc*.). Corresponding milestones are defined on an operational basis.

The major milestones described here are at the top level of a system. In larger projects in which a system is divided into smaller units, e.g., modules or subsystems, the previously mentioned reviews in connection with the well-known systems engineering V-Model (*see Fig. 7.15*) must also be performed at the levels below.

Project *reviews* should be carried out by independent individuals or teams (*review teams*) who were not themselves involved in the preparation of the respective project work. The topic of reviews is discussed in detail in Sect. 7.3.3.

As already mentioned, the status of the milestones at a specific point in time is a yardstick for the project manager to determine the status. The start of a further phase requires the total completion (*close-out*) of the preceding phase. In a multi-phase project, the main milestones are supplemented by more detailed milestones of the lower levels; see Sect. 9.3.4. In the practical implementation of the phase model shown in Fig. 4.2 for a very specific project, there may or will be, deviations in terminology and milestones due to adaptation. However, it is important that the basic phase logic is adhered to.

In the case of larger project plans, a distinction must also be made between the customer and supplier viewpoints. While the customer (client) must consider the entire life cycle, i.e., phases A to F, the supplier (contractor) is often only responsible for a sub-area, e.g., phase C and/or D.

At this point, it should not go unmentioned that the term "gates" is also used synonymously instead of "milestone." The term "gates" is intended to emphasize in particular that a decision to continue must be made at certain points in the project planning. This can of course also be achieved by defining major milestones, provided that the project team strictly adheres to them.

4.1.5 Relative Cost Comparison of the Project Phases

In line with the illustration in Fig. 4.2, Fig. 4.3 shows the relative cost comparison of the individual project phases. At the beginning of the project, i.e., at milestone 0, an initial relatively rough cost estimation must be prepared for the entire project on the basis of the before mentioned project requirements (*specifications*); see Sect. 10.2). These preliminary total project costs must be reconciled with the planned investment funds; see Fig. 4.1. In the course of the continuous phase progress, detailed cost determinations are then made on the basis of further advanced studies and these are then also reconciled with the planned budgets.

As shown in Fig. 4.3, the life cycle costs (LCC) of a project are the sum of the costs determined for phases A to F. The cost ratios are based on empirical values and are only approximate values. As can be easily seen, the costs of the first two phases A and B are low compared to the subsequent phases. On the other hand, however, the early phases are of great importance because important decisions about the project work are to be made during the early phases. The ratio between a relatively small amount of costs spent for the early project phases and the benefits at the more expensive follow-on phases gained from smart technological and financial solutions will be further discussed in detail in Sect. 10.5.

4.2 Project Sequence from A to Z

4.2.1 Project Start and End Often Unclear

Projects are undertakings with a defined beginning and end; see Chap. 2. Although this definition is trivial and absolutely correct, in practice the beginning and end are not always clearly recognizable. Many projects begin unclearly and end accordingly. However, the importance of covering a project as thoroughly as possible from A to Z cannot be emphasized clearly enough. It is important to clearly identify all costs of a project life cycle (LCC) in order to be able to make a total investment calculation. Often, important preliminary work for a new project is done at a very early stage, i.e., prior to the official project go-ahead, and sometimes forgotten by the time the project officially begins. However, financially any pre-work clearly belongs to the project in the context of an overall view. The end of the project is even more important for the final assessment of the project result and the total project costs incurred. If the end of the project cannot be clearly determined according to the motto "open end," the project costs of all individual tasks cannot be clearly determined either. Only when the end of the project is clear can the question be asked "was the project technically successful, on schedule and also economically a success"?

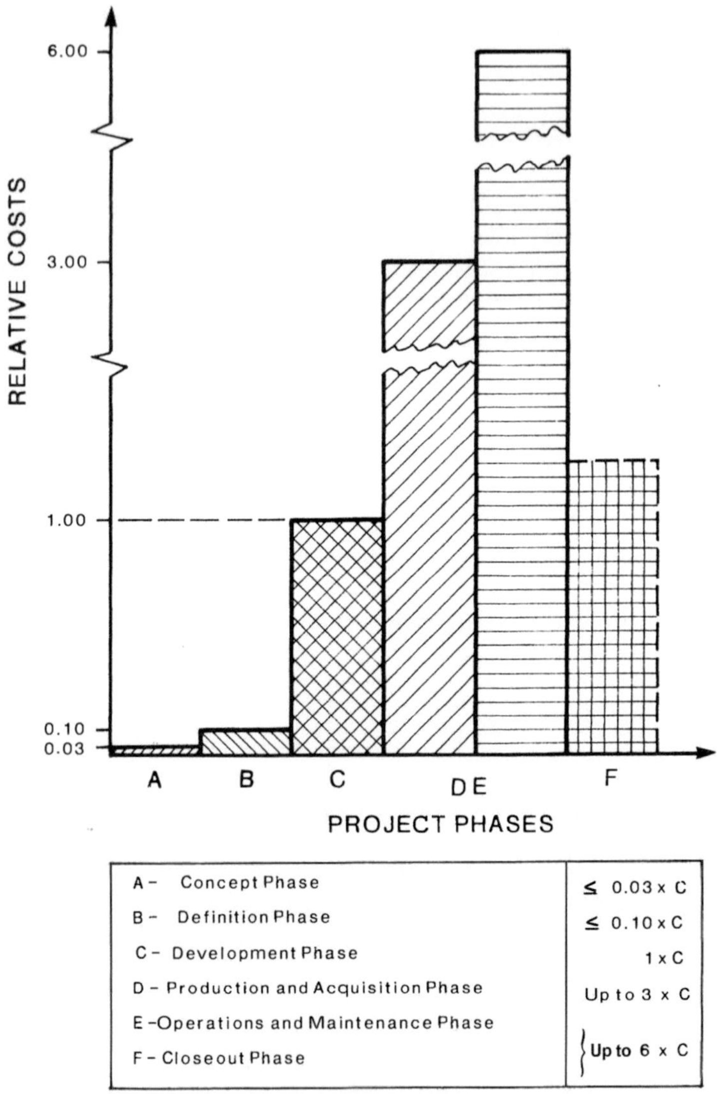

Fig. 4.3 Relative cost comparison of the individual project phases

What does this mean in business management terms? What is the best way to assess whether a project has been economically worthwhile? In other words, is there a positive financial return (*return on investment—ROI*) that is greater than the investment? There are two different ways of looking at this. The manufacturer or supplier assumes the development and production costs including warranty by apportionment and a profit rate as part of the sales price or the costs are assumed by the customer or purchaser (*e.g., the public sector, system manufacturer, plant operator, etc.*); see Sect. 10.1.

4.2.2 Phased Approach Eliminates Uncertainty

The creation of clear starting and ending points for the individual project phases is a prerequisite for well-functioning project management. The life cycle is divided into the following phases; see Fig. 4.2:

A Concept formulation*
B Project/system definition
C Research and development (R&D)
D Production and procurement
E Operation and maintenance
F Disposal

**In certain situations where special clarification is still required, it is common practice to add another phase "0" (pre-phase A) upstream.*

The course of the project phases is made controllable by establishing milestones. By considering all project phases as a common whole, an ROI consideration can be carried out from the perspective of the project sponsor or owner; see Sect. 10.5.4 (*Life Cycle Costs—LCC*).

The division of a project into individual project phases goes hand in hand with different levels of responsibility. Responsibility for the overall project can, for example, be subdivided according to the project phases, but overall responsibility should be left in the hands of a higher-level overall project management. The reason for this lies in the necessary specialization, e.g., preliminary phases, development, and production, and in the need for higher-level overall control. The need for overall control is slowly gaining acceptance in large and complex projects, which often extend over several decades because the importance and influence of the early phases on the subsequent phases have been recognized; see Sect. 10.5.4. The organizational integration of overall project management into the company organization is dealt with in Sect. 5.1.3.

Companies are often not responsible for all project phases, e.g., only for the development and/or production of a product, while the customer or operator is responsible for all phases, including the preliminary phases and the operation, maintenance, and disposal phases. In the automotive industry, the car manufacturer handles all phases from A to D and after the sale of the produced vehicle, the buyer takes over the operation and maintenance responsibility.

For the construction of a factory plant for example, the future operator or owner makes a corresponding investment in the hope that the profits generated will cover the acquisition and operating costs (*contribution margin*) and that additional profits can be generated. From the operator's point of view, investment decisions depend

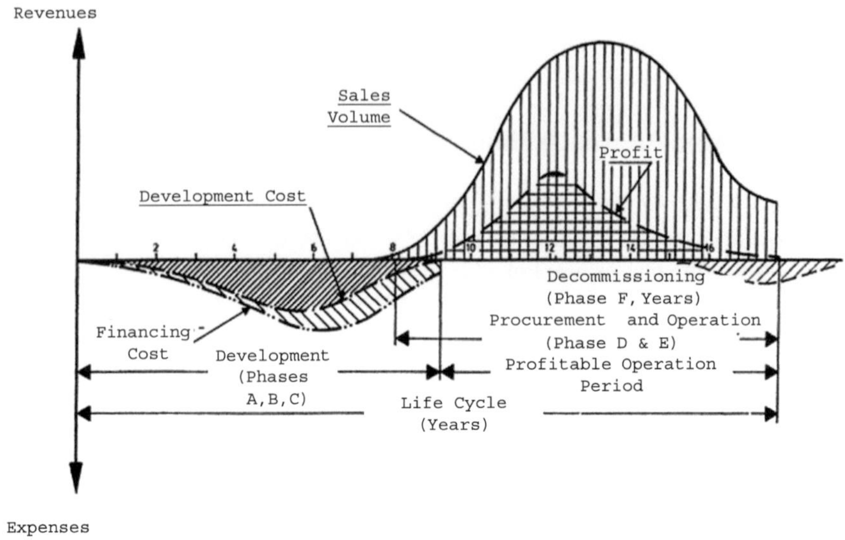

Fig. 4.4 ROI consideration of a project

on investment considerations taking into account the sales opportunities, the LCC and operating costs, the return on capital and the calculated profit rate.

The complexity of large system projects is often difficult to manage in its technological and economic entirety due to the use of new technologies and multiple interfaces. Efficient management in the early phases increases the chance of reducing the LCC due to optimization and, consequently, of achieving an improvement in the economic result, which is illustrated schematically by the return on investment (*ROI*) shown in Fig. 4.4; Sect. 10.5.4 provides further details on the LCC and ROI considerations.

4.3 Planning the Project Phases

4.3.1 Comparison of Different Phase Concepts

In Sect. 4.2, reference was made to the precise definition of the project life cycle to be able to combine the integral responsibility for all project phases (*end-to-end* responsibility) in one point. In the second step, the individual project phases and the main milestones of a project phases must be determined; see Fig. 4.2. Finally, the sequence of each individual phase must be planned in detail in advance.

The definition of project phases is not uniform for the various industries, but in principle builds on the phase breakdown shown in Fig. 4.2. For R&D projects which may only involve the manufacturing of prototypes but no series production, a slightly modified (*shortened*) phase model is often used by combining phases C (*development*) and D (*production*) to form a phase C/D for example. This is also often the case for the development of new one-off products. This is also the case in the development of unique products, e.g., research satellites or power plants. Another example is to combine the production and operation phases for a small series undertaking.

If, on the other hand, commercial or military aircrafts are procured, for example Airbus A380 or A400M, of which a larger series is required, the life cycle includes very distinct production and operational phases. In this respect, for example, the phase concepts of NASA and the US Department of Defense (*DOD*) also differ, as do the phase concepts of ESA and the German Ministry of Defense. ESA and NASA usually only conduct R&D projects and small series at most, while the German and US DODs mostly handle projects with large series of production. Saynisch compared different phase concepts and found that while there are industry-specific differences in phase design and terminology, the phase content, e.g., concept, definition, execution, is very similar [6]. The use of phase models is an important prerequisite for professional project management, and it does not matter that industry-specific phase models are used.

4.3.2 Phase Models of Different Industries

The phase models presented hereafter are in principle compatible with the standard phase model shown in Fig. 4.2, and recognizable deviations are justified on an industry-specific basis. In this context, reference is also made to the comparable phase model of the PMI (*Project Management Institute*), which is described in Chap. 2 (*Project Life Cycle and Organization*) of the project management guideline PMBOK (*Body of Knowledge)* [7].

For projects without subsequent series production, the phase concepts of ESA and NASA, which have been tested in practice, are to be preferred for the implementation of R&D projects. Figure 4.5 shows the phase concept of the ESA [8] and an abbreviated representation of the phase concept of NASA is shown in Fig. 4.6 [9].

After a series of mishaps with the Atlas launch vehicle in the early 1960s, the US Air Force (*USAF*) developed a new program management procedure (*AFSCM 375*) in 1966 under the direction of General Schriever, which reproduced a detailed phase model on which many of today's phase models are based [10]. NASA's Apollo program also benefited from the USAF's experience and based its own phase model (Fig. 4.6) on the USAF phase model.

In 1977, the US Department of Defense *(DOD*) introduced the phase model outlined below with four major milestones (0 through III) as a new regulation with the expectation that the release of each subsequent phase can only occur when the preceding milestone is 100% complete [11]:

Milestone 0: Project initiation

- Determination of the project requirements (*requirements*)
- Budget release (*concept phase*)
- Decision to carry out the concept phase

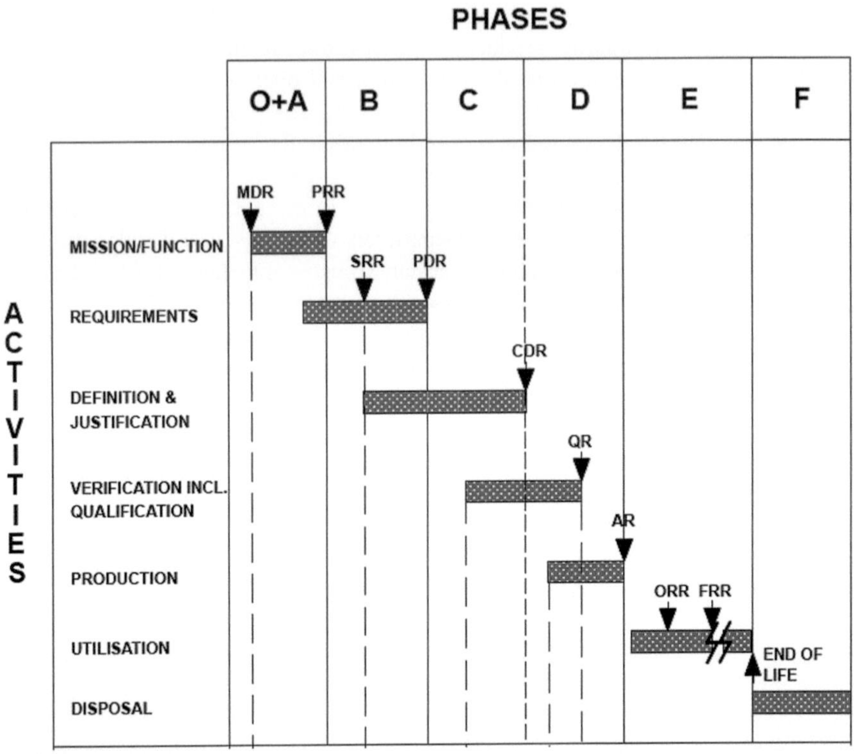

Fig. 4.5 ESA Project Phase Model

Milestone I: Release of the definition phase

- Demonstration of the examined concepts (*concept selection*)
- Budget release (*definition phase*)
- Decision to carry out the definition phase

Milestone II: Release of the development phase

- Completion of the definition phase (*detailed evaluation: technology, costs, deadlines*)
- Budget release (*development phase*)
- Decision to carry out the development phase

Milestone III: Release of the subsequent phases

- Completion of the development phase (*test evaluation*)
- Budget release (*subsequent phases*)
- Decision to carry out the subsequent phases.

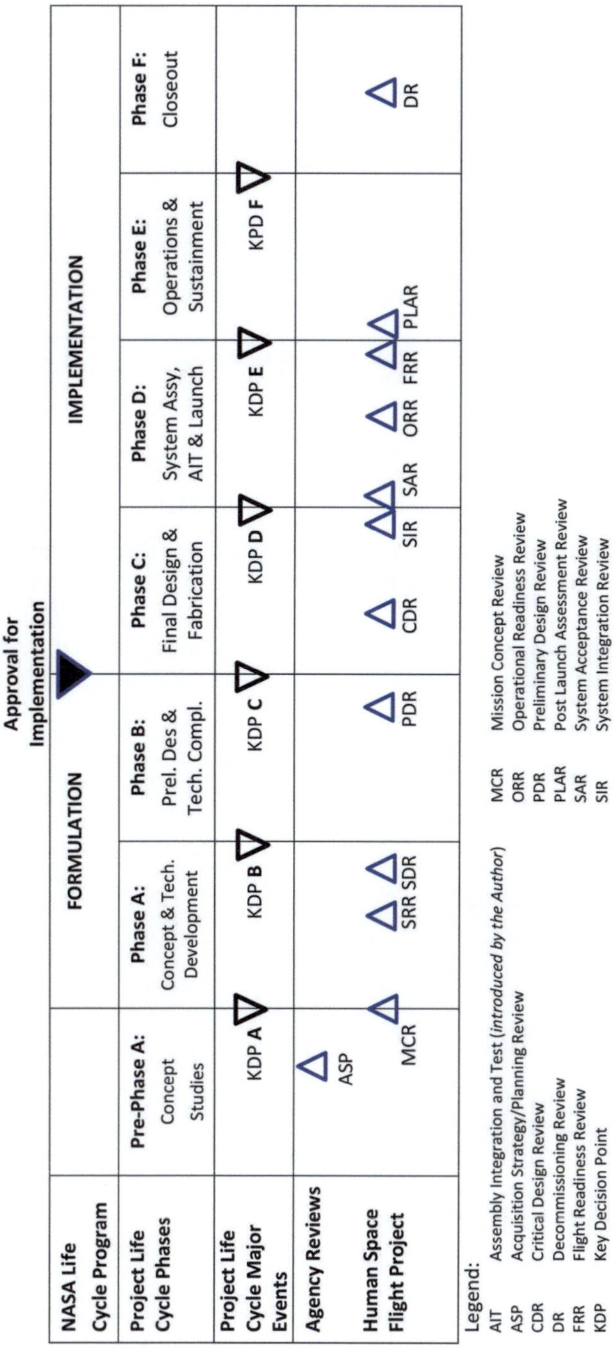

Fig. 4.6 Abbreviated NASA Phase Model

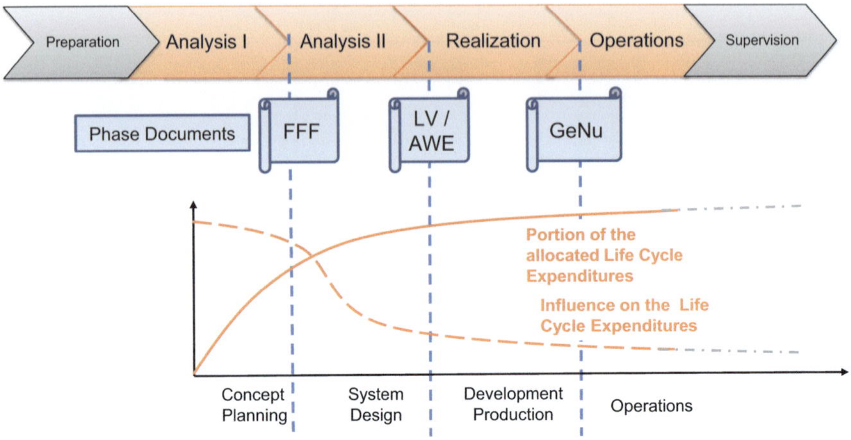

Fig. 4.7 Phase Model of the German DOD (BMVg). FFF = capability gap and continuation; LV = proposed solutions; AWE = selection decision; GeNu= permission to use

The results of Milestones I to III are to be submitted to the Procurement Review Committee in order to advise on how to proceed. The decision-making process described here on the basis of measurable results (*Milestones I to III*) is important not only for the military sector, but also for any new civilian product to be developed.

The phasing concept of the construction industry requires engineers and architects apply to perform the following services [12]:

1. Basic determination
2. Preliminary planning
3. Design planning
4. Approval planning
5. Implementation planning
6. Preparation of the award
7. Participation in the awarding
8. Object monitoring
9. Object support and documentation

Lennartz presented on September 28, 2016 a lecture the life-cycle cost management of the German DOD; see Fig. 4.7 [13].

4.3.3 Application of Phase Models

The examples of different phase models described above (*Figs. 4.5–4.7*) are in principle comparable with the standard model shown in Fig. 4.2 because, despite the use of different designations, they all start with detailed preparatory phases before the actual execution phase can start. This is an important aspect because a hasty step without thorough planning (*early phases*) easily leads to major problems and often also to considerable cost increases.

Even in the private sector, e.g., in the construction planning of a home of one's own, it pays off to first prepare a thorough plan in close cooperation with the architect before starting the expensive construction measures. After all, it is a truism that one must proceed step-by-step, and, for example, the installation of the interior fittings should not take place before the roof is completed. This principle can be applied *mutatis mutandis* to all project planning, even in the construction of a television satellite. Early phases should be fully completed before starting development and/or construction work.

If a medium-sized company wants to implement a project, it is also possible to refer to the phase sequence described in Fig. 4.2. It is not necessary to use the terms literally. One should only make sure that the principles remain the same. In this context, it should also be mentioned that reviews must take place, for example, as part of a project meeting.

Conclusion: the standard phase model shown in Fig. 4.2 is applicable regardless of size and industry. In July 2015, the 5th workshop for the passenger transport system "SpaceLiner" of DLR (*SART*) was held in Bremen, Germany. The SpaceLiner is a visionary rocket-powered flying object that can transport passengers in a suborbital trajectory at several times the speed of sound between continents in record time. Following Fig. 4.2, the phase model shown in Fig. 4.8 was presented at the workshop. In 2015, the SpaceLiner project was still in Phase "0"; i.e., at Pre-Phase A [14]. As shown in Fig. 4.8 the Mission Requirement Document (*MRD*) was adopted during a Mission Requirement Review (*MRR*) in July 2016 [15]. The MRD is a prerequisite for the initiation of Phase A.

4.3.4 Life Cycle and Project Decisions

The introduction of a new system, for example, a new aircraft or a new IT system, often requires enormous resources to finance the project. In this context, Oesterer writes: "*The execution of large projects represents a considerable risk for the general contractor, both in terms of their financial scope and their time frame ... The decision must therefore be made at risk, i.e., accepting the possibility that the project will fail*" [16]. It is therefore natural that, in order to estimate the risk, the ordering party (*public or private clients*) prefers a staggered, i.e., phased, project sequence with the possibility of making a go/no-go decision at the end of each phase.

A look at the cost comparison of the individual phases (Fig. 4.3) makes it clear how relatively low the risk is for releasing the early phases (*concept and definition*) compared with the cost-intensive subsequent phases (*development, production, and operation*). For public projects, the decision-making process for project continuation must also be in line with the budget release process.

The phased approach is primarily intended to help identify and prevent emerging project disasters in advance. In the early phases, and especially in Phase A, the aim is to investigate whether the envisaged project is technically and/or economically

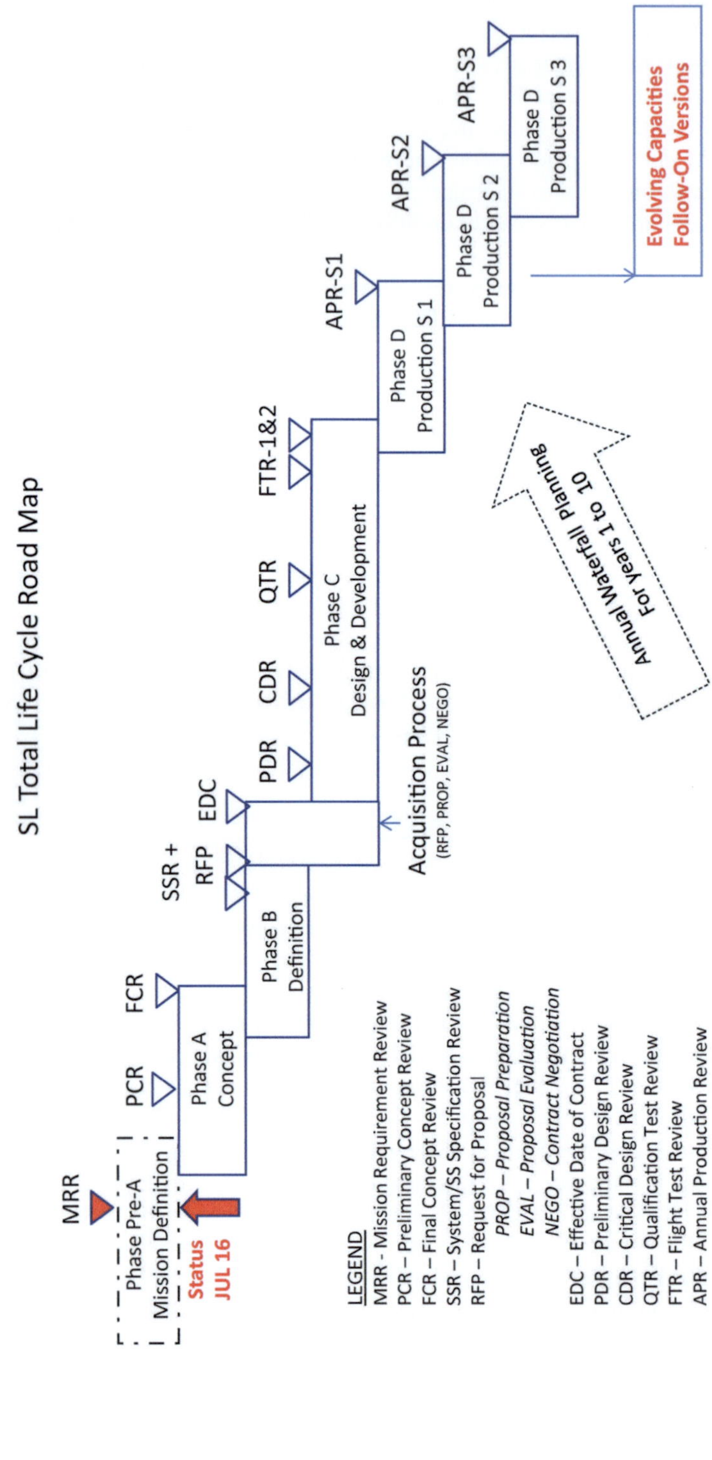

Fig. 4.8 Phase model of the DLR SpaceLiner project

feasible. For this purpose, a "feasibility study" as part of Phase A is also used to check feasibility at an early stage of the project. A problem that is often encountered is that customers (*clients*) tend to over-specify (*nice to have*) in the first approach. A feasibility study is helpful to define realistic (*good enough*) project requirements (*specifications*).

The milestones defined in the early phases also serve as a decision-making aid for the course of the project. Pre-defined termination criteria can help project managers make go-no-go decisions. In the case of problems identified early on that are serious and technically and/or economically unfeasible to resolve, project managers should be encouraged to use the facts (*termination criteria*) to initiate a termination (*show stopper*) or modify project requirements in the early phases, which may not be popular most of the time but may be necessary from a technical/economic perspective of view. The HERMES space shuttle project, which was terminated in 1992, is a good example.

Figure 4.3 shows that the costs of Phase A (*concept*) in relation to Phase C (*development*) are only about 3%, and the costs for Phase B are up to about 10%. On the other hand, as already mentioned, important decisions for the future course of the project have to be made in Phases A and B. Figure 4.9 illustrates the relationships between

a) the course of life cycle costs (*curve on the right of the figure*),
b) of influence in the early phases (*curve on the left in the figure*) and
c) of the percentage cost determination during the early stages (*curve in the middle of the figure*) [14].

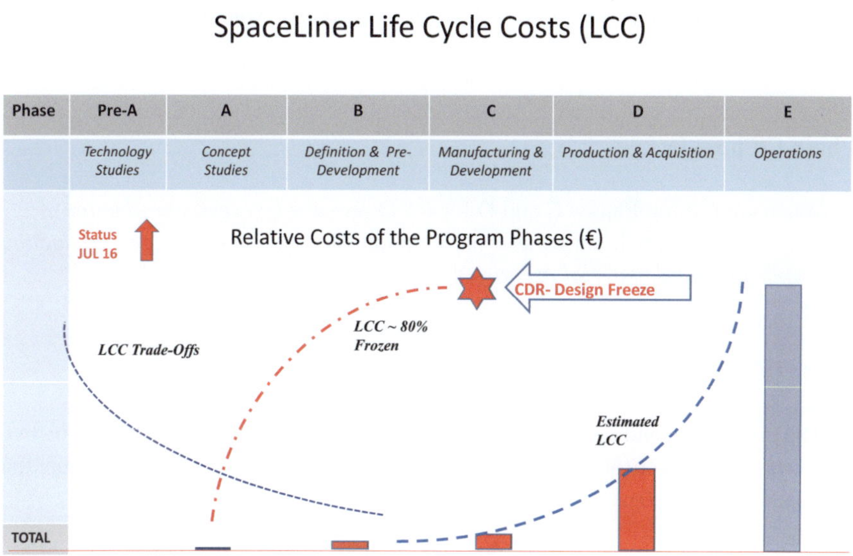

Fig. 4.9 LCC decision parameters

It is clearly evident that the influence on project events is greatest in the early phases and that, according to experience, up to approximately 80% of the specifications are fixed (*frozen*) after the completion of Phase B [17]. These relationships underscore the importance of the early phases. Project managers should therefore pay corresponding attention to the early phases; see Sect. 10.5.4.

4.3.5 Phase Overlaps

All too often, the individual phases overlap in practice, which easily leads to considerable additional costs and the very greatest problems in decision-making. As the member of the German Parliament (*Bundestag*) Brigitte Traupe said in this regard back in the early 80s: "*The procurement of the weapon system often begins with a provisional production status. At the same time, massive further development takes place. The weapons built first are then converted to the final development and design status after two or three years. Even a layman understands that this costs additional money*" [18].

In principle, phase overlaps should be avoided as far as possible. However, there are situations that justify a controlled phase overlap in order to defuse the often precarious schedule situation somewhat. Phase overlaps are possible in the following areas and are often necessary to avoid additional costs:

Phase B/Phase C:

- Early material procurement (*for example: components with extremely long delivery times—so-called long lead items—LLIs*)
- Pre-development of critical project areas (*for example: tapping into new unproven technologies*).

Phase C/Phase D:

- When combining Phases C and D for R&D projects (*no series production*).
- Initiation of a small series after completion of the first prototype (*production preparation and material ordering*).

Phase D/Phase E:

- The production and operating Phases (D/E) run visually in parallel, but if the two phases are broken down according to the individual devices/lots, a sequential arrangement is possible.

Phase overlaps must be planned with particular care in any case. The main milestones (Fig. 4.2) must not be influenced by anything. If there are overlaps between

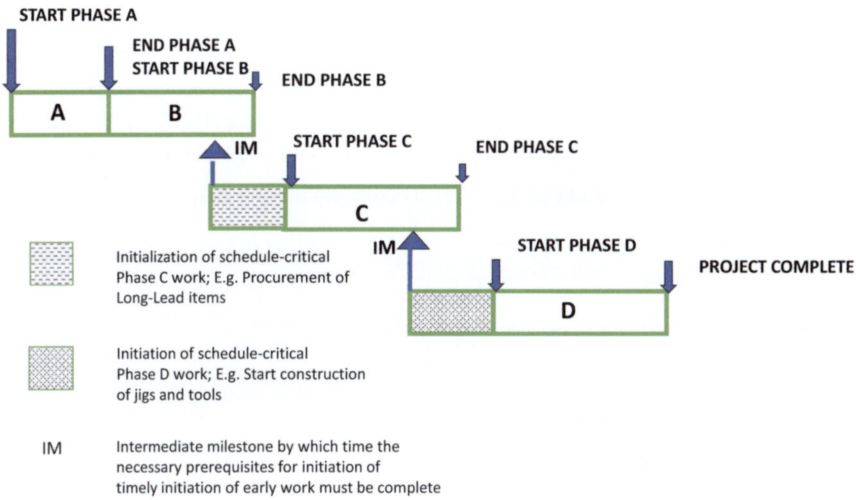

Fig. 4.10 Possible phase overlap

the individual phases, the overlap must be precisely identified. Figure 4.10 illustrates this with an example. In principle, the early ordering of materials is a clearly identifiable measure that usually does not have too much financial impact. If, for example, the project is canceled early, only relatively minor commitments have been made. The situation is similar for the early development of particularly critical components. It is important to make such measures (*minor phase overlaps*) exceptions and to plan and limit the overlap precisely in technical, financial, and scheduling terms.

Retaining the principles of the standard phase model shown in Fig. 4.2, it can be meaningful to summarize individual phases, to optimize the procurement process. Under certain conditions one can combine Phases A and B, C and D or even the Phases B, C and D, if one pays strict attention to the fact that the definition of the milestones remains. Figure 4.11 shows a variant of merging project phases typical for ESA projects [19].

4.3.6 The Road Map Concept

Before starting a journey by car or bicycle to a specific destination, but without knowing the road connection, it is prudent to consult a road map. If you want to reach the destination as quickly as possible and without long detours, you need to study the road map carefully in order to identify the optimal route. The situation is similar when trying to reach a project goal as quickly and cost-effectively as possible. This also requires a road map or a detailed project schedule (*phase schedule*). The division of a project into individual phases is only the first step in this process.

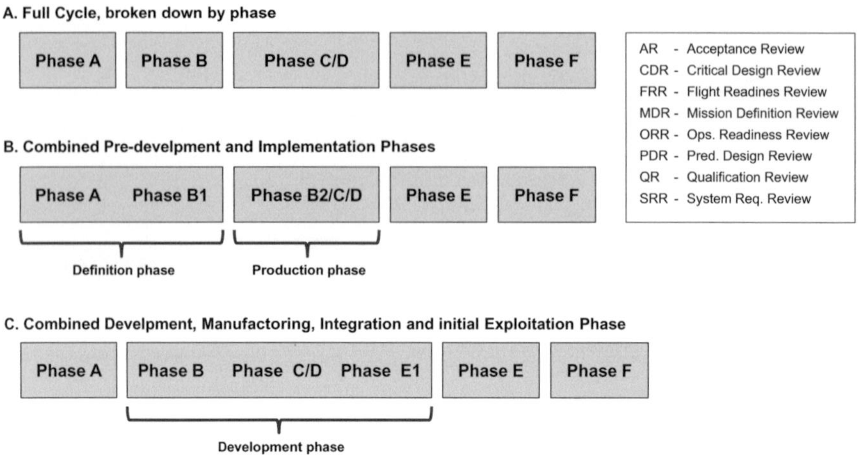

Fig. 4.11 Combining of project phases for ESA projects

The second step is to define the main stages of each phase individually. Even experienced project team members are often not able to provide information right away about which main tasks are to be performed in the individual phases. This is especially true for the early phases of a project (*conception, definition, and development*). The processes of the later phases (*production and operation*), which, as far as IT projects are not concerned, are often also referred to as hardware phases, are usually better known to most engineers, since they involve more generally comprehensible tasks such as manufacturing, assembly, testing, etc. Nevertheless, here as well as for the early phases, a detailed phase sequence plan should be created at the beginning of the project, which precisely defines the phase sequence. This phase flow chart, which is not to be confused with the detailed project flow chart to be drawn up later, describes in logical sequence the necessary processes of the phase concerned and creates the necessary orientation framework for all future measures. The phase flow chart is a further refinement of the respective phase and is best presented in the form of a flow chart. It describes the procedural strategy of each phase, and it must clearly indicate the logical sequence of individual project actions. For example, it expresses that a detailed cost estimate can only be made after the work breakdown structure (*WBS*) and the project specification have been adopted.

The USAF and NASA recognized the importance of a planned phase sequence early on and created detailed phase sequence plans for the execution of their projects. In both cases, standardized sequence plans were developed based on the realization that the phase sequences for technologically similar projects are largely identical.

It is important that phase flow charts, which are an important orientation aid for all project members, are used when handling large project projects. Phase sequence

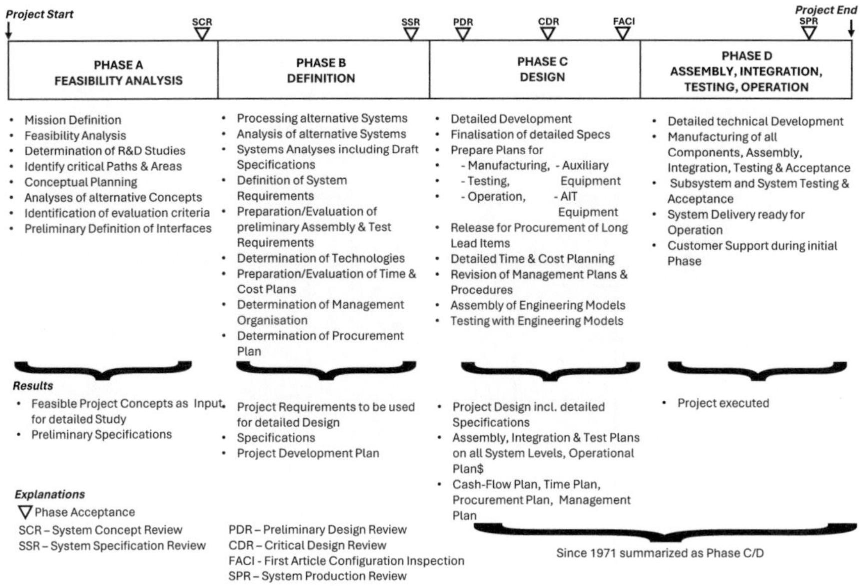

Fig. 4.12 Abbreviated description of phases A to D (source NASA)

plans are a prerequisite for understanding the individual project measures and should therefore not be missing from any project. They indicate which main task is to be performed, when, and in what detail, and also represent an ideal checklist for project management. With reference to NASA [9], a brief description of phases A to D is given in Fig. 4.12.

The following descriptions of the individual phases are intended as a suggestion for the development of industry-specific phase flowcharts and, within the scope of the objective of this book, can only be of a guiding nature.

Phase A: Feasibility and concept studies (see Fig. 4.13)

- *Task and significance:* In this phase, system concepts are studied, the possibilities for feasibility and fulfillment of customer requirements are studied, and the compatibility of the system with other system environments is analysed. At the same time, preliminary schedules, cost plans, and management concepts are developed during this phase. Phase A is concluded with a final report, consisting of the system specification and the preliminary schedule and cost plans, and is used to prepare and decide on the initiation of the subsequent phase (*Phase B*).
- *Required qualifications of Phase A staff:* Predominantly system engineering and business management staff. Phase A staff should have a great deal of all-round knowledge and must, of course, be aware of the state of the art in technology as

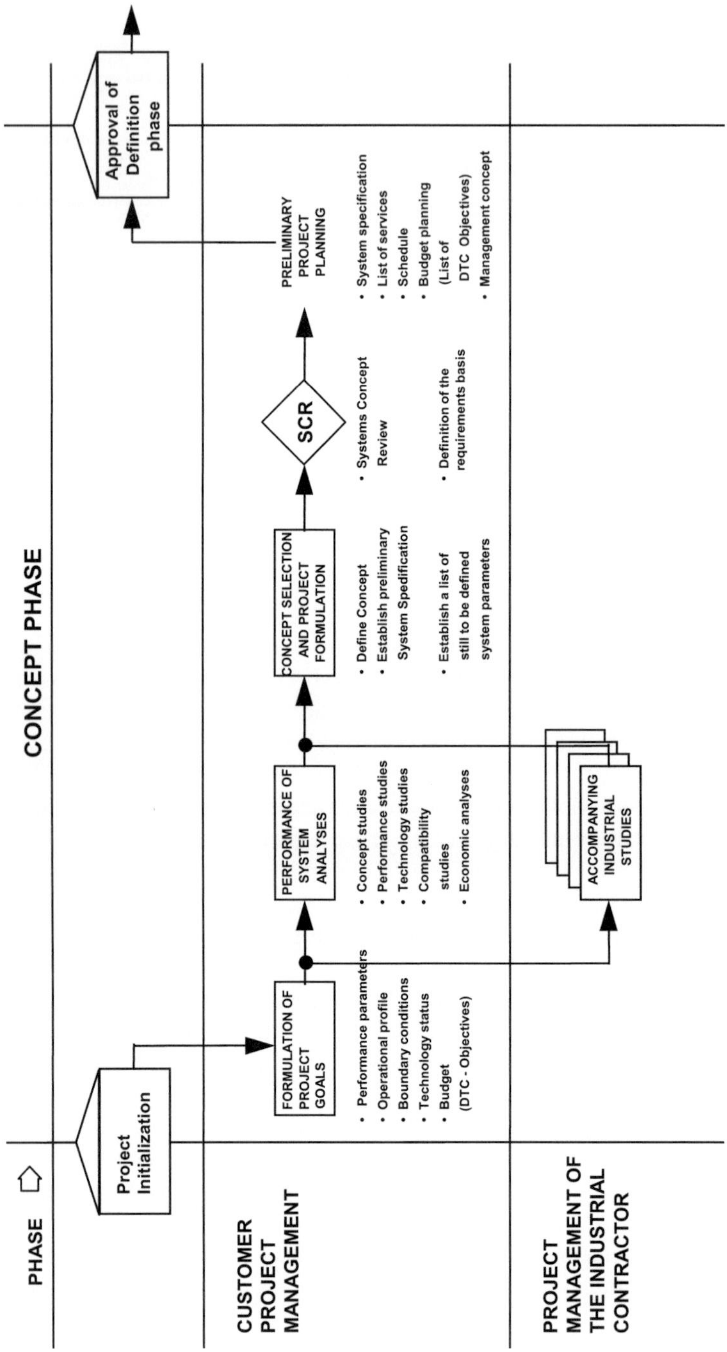

Fig. 4.13 Concept Phase

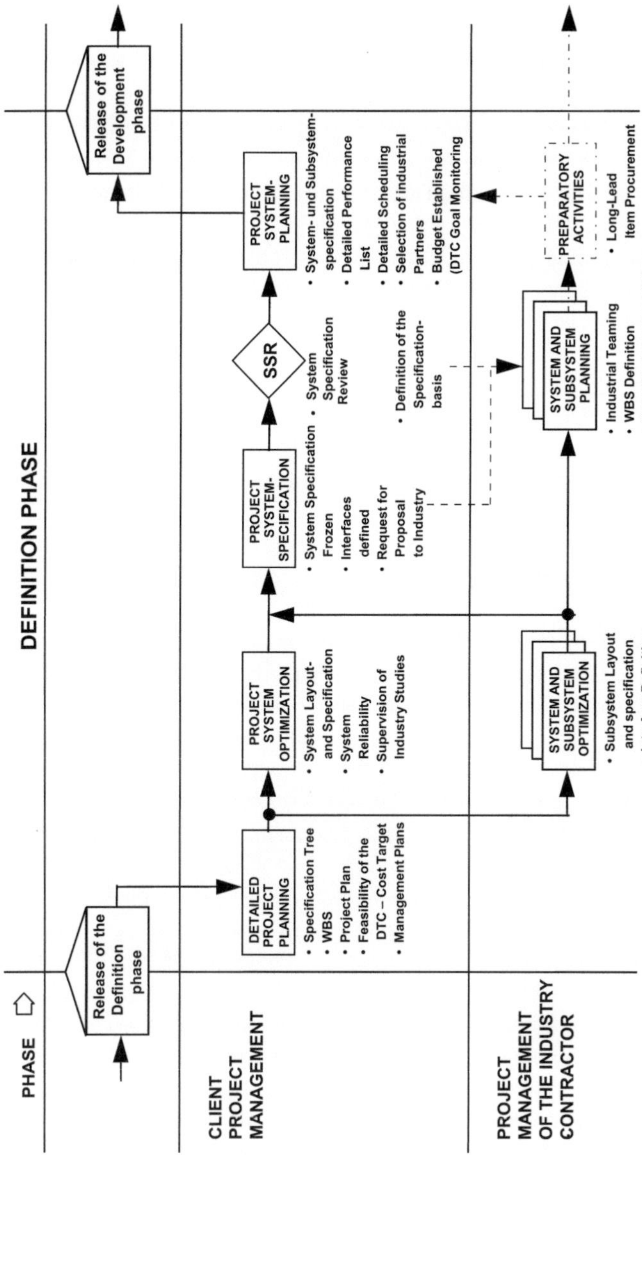

Fig. 4.14 Definition Phase

well as the innovations under development. They provide the direct link to universities and research institutes and to future users. The influence on authoritative field offices should not be underestimated. The inventiveness and competence of Phase A employees shape the company's image and allow influence to be exerted on future projects. The employees in this phase must therefore also be good salespeople in the sense of new but usable and realistic ideas. In summary, Phase A teams are usually staffed with top systems engineering professionals (*physicists and engineers*) and project-specific subject matter experts (*engineers, chemists, mathematicians, etc.*). Employees from Project Control and Product Assurance are also brought in to help, but are not yet at the forefront of the action. However, an important trend can already be observed in the concept phase: The future user also wants to have credible costs as early as possible to assess the initial technical designs. However, these cost analyses cannot be carried out by a businessman alone. Rather, this is an interdisciplinary task that requires both commercial and technical knowledge/experience; see Sect. 10.1.

Phase B: Definition work (see Fig. 4.14)

- *Task and significance:* On the basis of the results from Phase A (*concept selection*), in Phase B the project/system (*for example satellite, ground stations, power plant, etc.*) is technically structured, defined, and planned in detail. In this phase, the sub-elements of the system are described in detail and the development measures are pre-planned. In addition, this phase also investigates the extent to which components already available on the market (*off-the-shelf equipment*) can be used. Other important tasks of Phase B are the identification of competent entities for the development and supply of equipment and the preparation of binding schedules, costs, organization, and management plans for the product development to be carried out in the subsequent phase. Phase B ends with completed planning documents (*individual specifications, binding schedules and cost plans, and mature management concepts*) for the preparation and decision to initiate the follow-up phase (*Phases C or C/D*).
- *Required qualification of Phase B employees*: The required qualification of employees for Phase B is basically the same as for Phase A. Thus, it is often the same team members who previously performed Phase A. On the other hand, however, a more practice-oriented attitude is already expected for Phase B activities, and for this reason, Phase B is very often led by engineers who have previously performed development work. In summary, the recruitment of personnel for Phase B is largely still to be filled by experienced project staff from Phase A, but the leadership function is often already in the hands of managers and engineers who have experience in the planning, specification, and execution of the subsequent phases (*development, production, and operation*). The project team also includes experienced project controllers, QA experts, project business peo-

ple as well as contract lawyers. For large projects, they are team members of the PMO core team (Sect. 5.1), and for medium and small projects, they are deployed on a temporary basis according to the project plan.

Phase C/D: Development and Production Phase (see Figs. 4.15 and 4.16)
In projects without series production, Phases C (*development and prototype production*) and D (*production including trial operation*) are often combined into a single phase, Phase C/D, due to the already high degree of overlap and for cost-saving reasons (Fig. 4.2).

- *Task and significance:* In contrast to Phases A and B, this is a development-oriented phase. The focus is on the creation of a product in accordance with specifications within the framework of specified deadlines and cost limits. In this phase, design and laboratory work (*including IT tasks*) must be carried out and, building on these results, manufacturing, integration and test work must be completed, which is then followed by the tasks for trial operation. This phase ends with the delivery and functional handover of the agreed end product (*or end products*) on the agreed date.
- *Required qualifications of Phase C/D employees:* Phase C/D employees (*physicists, engineers, economists, and lawyers*) differ from the study-oriented employees of Phases A and B essentially in their strongly pronounced orientation towards the realization of software and/or hardware products. Aspects of production, integration, and testing, as well as deadlines and costs, are naturally in the foreground in this phase. In general, the following can be said about the staffing for Phase C/D: engineers, physicists, economists, and lawyers with basic knowledge of systems engineering, quality assurance, IT, design (*laboratory*), manufacturing, integration and testing, project control, contract and financial management, logistics, and organization are sought. A clear demarcation between Phase B and C/D employees is not possible and also not meaningful; i.e., employees often move from Phase B to Phase C/D.

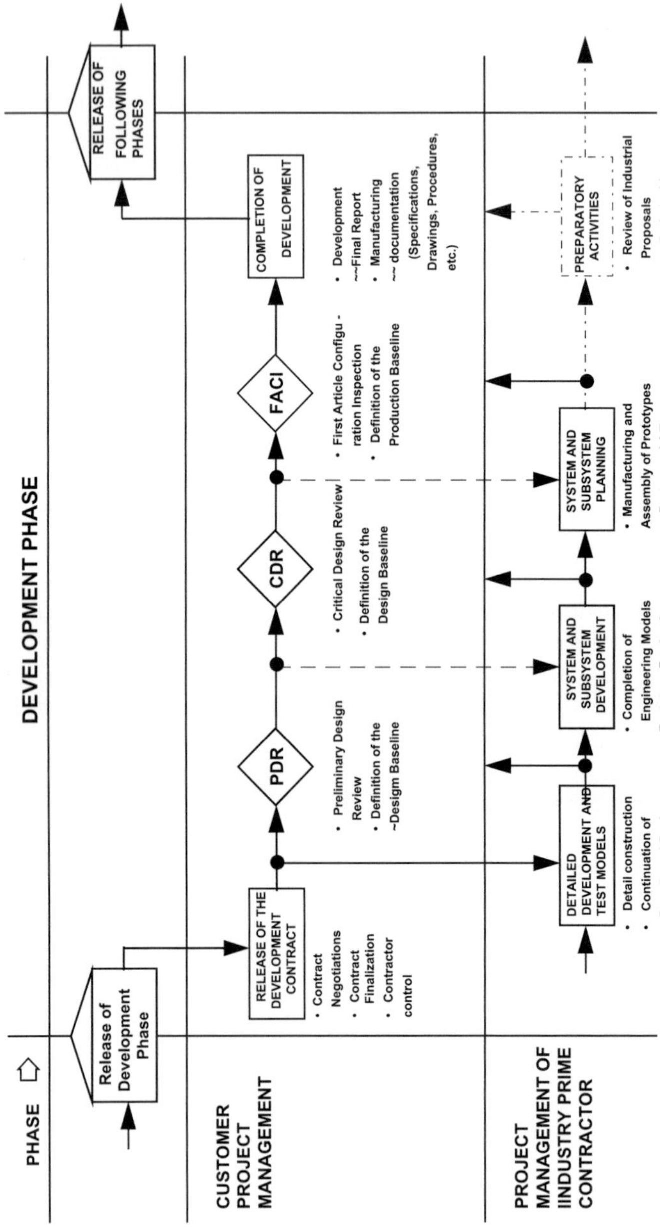

Fig. 4.15 Development Phase

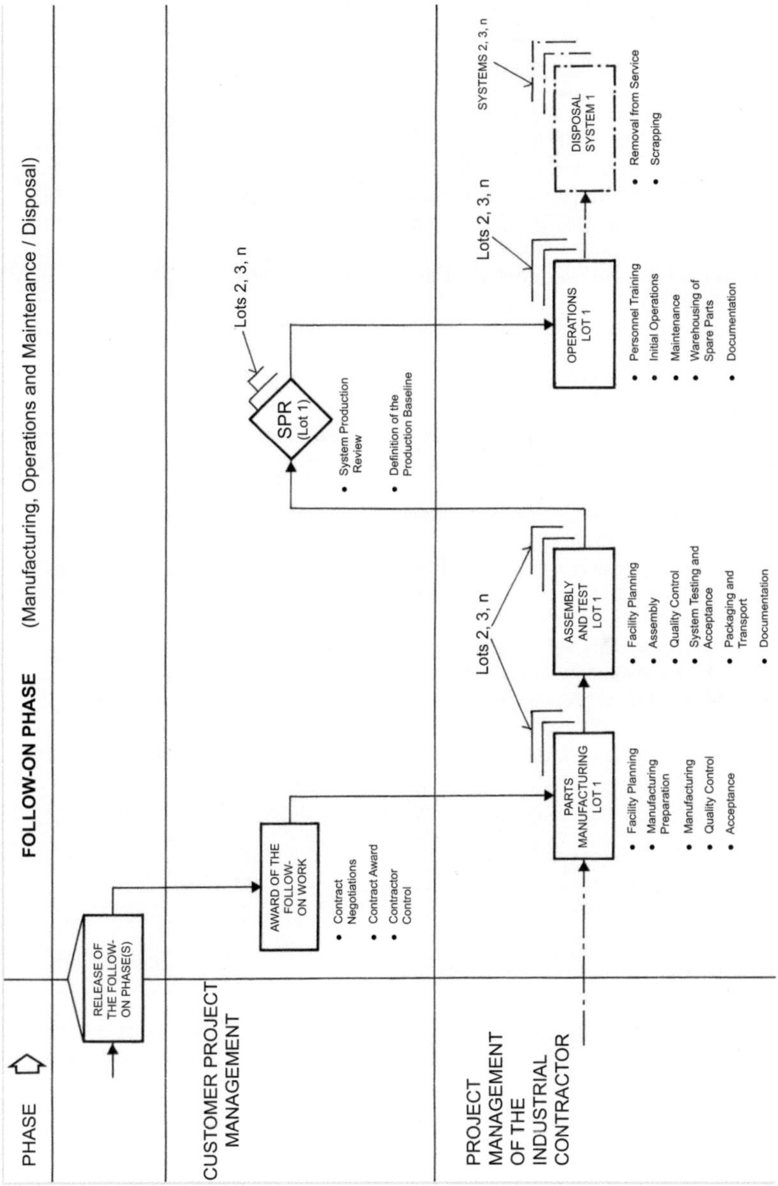

Fig. 4.16 Follow-ON Phase

Literature

1. Steinberg, Claus: Projektmanagement in der Praxis, Schäffer Verlag, Stuttgart, 1990, p. 3
2. Saynisch, Manfred: Grundlagen des phasenweisen Projektablaufes, from Saynisch, Schelle, Schub: Projektmanagement, R. Oldenbourg Verlag, Munich-Vienna, 1979, p. 33.
3. Augustine, Norman: Augustine's Laws, 6th Edition, American Institute of Aeronautics and Astronautics, Inc. Reston, Virginia.
4. DIN 69901, Project management - Project management systems - Part 5: Terms, 2009
5. Hopf, Hans-Dieter and Madauss, Bernd-J.: Projektmanagement in Forschung und Entwicklung, GPM-Industrie-Seminar (MBB Report UR-471-81 0), 11-13. 2. 1981.
6. Saynisch, Manfred: Phasenweiser Projektablauf bei Entwicklungsvorhaben, from Saynisch, Schelle, Schub, Projektmanagement, R. Oldenbourg Verlag, Munich-Vienna, 1979, p. 103.
7. PMI, Body of Knowledge (PMBOOK), 5th edition, ch. 3, May 2014.
8. ESA, ECSS-M-ST-10C Rev. 1, 2009
9. NASA-HDBK-0008; NASA Product Data and Life Cycle Management (PDLM) Handbook, 2012.
10. USAF System Program Management Procedures, USAF Document AFSCM 375-4, May 31, 1966.
11. DOD: "Major System Acquisitions≫, DoD Document DoDD-5000.1, 18 January 1977.
12. HAOI, 2013
13. Lennartz, Ludwig, Captain at Sea: Life Cycle Cost Management (LCCM) as a Method in the Business Unit of the BMVg and the Supporting Tools, 4Cost Conference on September 28 and 29, 2016.
14. Madauss, Bernd-J., Trivailo, Olga, SpaceLiner - Engineering and Management Approach, DLR-SART Workshop, Bremen July 2015.
15. DLR-SART, Bremen, SpaceLiner Mission Requirements Document, SL-MR-SART-00001 ½. Issue 1, Rev. 2, 08.07.2016
16. Oesterer, Dieter: Die Bewertung von Großprojekten nach ihrem technischen und wirtschaftlichen Risiko, Wiss. Bericht. AEG-Telefunken 50 (1977) 4/5, p. 146.
17. Gaffney, R., L.: "A Parametric Analyst Faces Reality," Astronautical Systems Division.
18. Traupe, Brigitte: Plenary Minutes of the German Bundestag 9/138 of December 14, 1982, p. 8643.
19. Madauss, Bernd-J. and Brand, Günther, Project Management Seminar Document, June 2015.

Principles of Project Organization

5

Contents

Abstract

The success of a project depends to a large extent on the organizational measures undertaken. These include the organizational structure, staffing, and the mandate of the project management. The following principle must be observed: The project manager is responsible for project implementation and execution

B. -J. Madauss, *Project Management*,
https://doi.org/10.1007/978-3-662-69057-4_5

but must be given appropriate authority. This also applies to cooperative projects in the homeland and abroad.

The execution of projects for which several persons with specialized knowledge are required, presupposes in the interest of a smooth process, an organizational division of the overall task (*top level*) into several subtasks *(lower level*). However, this entails the problem that the subtasks performed by specialists must ultimately be combined again into an overall task. This is particularly true for large and/or complex projects (*systems*), such as for industrial plants, vehicle construction, IT software, and aerospace developments. It may, however, also apply occasionally to medium-sized and small projects, for example for the supply of critical components and complex software. In some cases, it may become necessary that the execution of projects requires cooperation with international partners and companies (see Chap. 16), which could result in an additional problem in dealing with a different language(s), culture(s), and currency(ies). Steinmann and Schreyögg mentioned: "*Division of labor creates complexity: the creation of specialized jobs and departments each represents interruptions in the overall flow of services. Parts of tasks are done by different people, in different places, at different times, and this inevitably raises the problem of bringing all these separately performed parts back together so that a cohesive performance unit can emerge*" [1].

The implementation of a tightly structured project organization, which clearly and unambiguously shows the responsibilities, accountabilities and powers of the participating organizations (*companies, institutes, etc.*) and/or employees, cannot be overestimated in its importance. In this regard, Dale states "*When two or more people are working together for the same goal, each person involved needs to know what part of the overall work he or she should take on*" [2].

Many projects fail not because of a lack of technical competence on the part of the employees involved in the project, but because of organizational confusion. The timely implementation of an effective organizational concept, combined with the nomination of appropriate key personnel and the clear definition of responsibilities, accountabilities, and powers, is therefore an important prerequisite for successful project execution. The implementation of projects by several companies or organizations also requires a clear organizational arrangement of inter-industry relations.

In Germany, a distinction is usually made in organizational theory between organizational structure and process organization. "*While in the Anglo-American literature under organization largely structural-organizational aspects are dealt with, the German organization theory in the treatment of organizational phenomena assumes the separation into an organizational structure and process organization*" [3]. This chapter deals with the previously mentioned organizational structure in the context of project management, because according to the German definition, the project organization is an organizational structure in contrast to the process organization, which is dealt with in Chap. 4 (*life cycle*) and Sect. 9.2 (*project planning and control*).

Every organizational structure measure should be preceded by an analysis of the functions and tasks to be performed to ensure that the organizational structure meets the necessities. Care should be taken to ensure that the organizational interfaces are meaningful and clearly defined. Embedding the project organization in the company organization and ensuring that the project and line organizations interact as harmoniously as possible is of particular importance for successful cooperation.

In this chapter, reference is often made to large-scale projects. However, this does not mean that the methods and procedures described are not applicable to small projects. It is true that they are very applicable because responsibility should also be clearly defined for small projects. In large projects, the management rules are only more visible and therefore they have a pilot character for smaller projects. For small projects, the project manager often has to perform several management tasks simultaneously, for which dedicated employees are available in large projects.

5.1 Project Management and Project Office

For further consideration, it is important to distinguish between the different types of projects, as they may occur in government agencies, companies, universities, etc. Many a student says "*I have a project*" and means, for example, his diploma thesis. It is undoubtedly a project for which a structure (*outline*) is necessary, for which deadlines must be met, and there are interfaces with the professor or mentor. However, it is usually a project in which there is a project manager who is also his own collaborator: a one-man band. A book with several authors, on the other hand, is a real project. A very similar situation is the creation of an offer with contributions from several colleagues to a fixed deadline. This requires a proposal manager or coordinator; see Chap. 14. The other extreme is, for example, the development of an aircraft such as the Airbus A380, in which many hundreds of people were involved throughout Europe. This requires a well-functioning management organization. The following is a selection of various project types:

- Study projects
- Research projects
- Development projects
- Production projects
- IT projects
- Logistic projects
- Manufacturing projects
- Procurement projects
- Plant projects

The list can certainly be extended. If the project is processed together with domestic and/or foreign partners, this leads to special difficulties with impacts on the project complexity and also on the organizational structure.

For all projects, regardless of the industry, there are general rules of planning and monitoring that should be followed to avoid project mishaps. This applies equally to large and small projects. Often problems only become apparent at a later stage, and savings are made at the wrong end, namely at the beginning, which results in insufficient project planning. The consequences are then often very painful.

Development projects are usually very demanding undertakings where, if you want it to be good, you usually have to pull out all the stops of project management. In other words, development projects usually require the use of all known PM tools. For a better understanding and orientation of the common organizational aspects, this chapter refers to the experiences of handling development projects. It is assumed that adaptation to other project types is possible.

5.1.1 Task, Responsibility, and Authority of the Project Manager: "*Like a Company Boss*"

The task and responsibility of the project manager can be defined as follows: "*Achieving the pre-specified project goal within the specified cost and schedule*" [4]. The project manager's task and responsibility encompass the entire spectrum of project execution. In other words, the Project Manager (*PM*) is fully responsible for both the technical and administrative execution of the project. The responsibility of the project manager at the project level is, in principle, comparable to the responsibility of a managing director at the company level. This comparison is often clearly visible, especially in large projects. Any isolated approach, e.g., into a separate technical and economic mode of operation, does not make sense, since technical decisions in most cases have economic consequences, and economic decisions just as often have technical consequences. The project tasks and all technical and economic decision-making processes of a project should therefore be performed holistically and in an integrated manner by the project management. It goes without saying that the PM should enlist the support of the company's service departments and, if necessary, also technical specialists in carrying out his tasks.

If one compares the management principles with those of managing a company, the project manager must be given overall responsibility. As a rule, only the integrated technical and economic view leads to optimal success. Mellerowicz comments: "*The industrial company, more than any other company, is a synthesis of technology and business. The two cannot be separated from each other; in their mutual dependence they determine the design of the industrial operation, both following their laws ...*" [5]. For the project manager's area of responsibility, the combination of both parts is a basic prerequisite for success. In practice, however, this is still often overlooked or even prevented. McKinley points out the necessity to merge business and technical knowledge to achieve more innovation and states: "*A new relationship between technicians and business people is necessary, because so far they still form two clearly separated groups with different professional interests*" [6]. However, this requires that the respective company consciously delegates full responsibility to the PM. The principle of fully responsible delegation is often not very strong because of fears that the appointed PM might not do it well enough. Experience in the aerospace industry has shown that organizational forms that give the project manager and his team both parameters, technology and business in their hands, makes at the same time efficient project management possible. In this regard, McKinley again states, "*It is the melding of technical and business expertise that will prevent the idling of an excellently trained and highly paid scientific workforce and drive innovation in the marketplace*" [6].

In order to master this difficult task, the project manager must be equipped with sufficient power and competence, which can be summarized as follows [4]:

- Planning, management, and control of the technical task
- Selection of subcontractors and suppliers
- Planning, approval, and control of project costs
- Deadline scheduling and control
- Implementation of a functional project organization and selection of key personnel.

It is an organizational principle that the tasks, responsibilities, and powers assigned to the project manager must be in balance with each other. It is not appropriate if the project manager is given full responsibility for the execution of a project but does not have the powers necessary to accomplish the task. However, it must be said with reservation that the project manager must necessarily operate within the framework of the rules and guidelines of the company in which he works. This means that the project manager must recognize and comply with constraints such as commercial requirements, general organizational regulations, and basic requirements of the technical departments; see Sect. 5.2.

5.1.2 Management Functions in the Project

Defining the necessary management functions is an important prerequisite for setting up the organizational structure. As already described in the previous section, the project management function must be viewed in its entirety as an integrating activity. It follows from this that the project manager should be first and foremost an integrator or generalist with appropriate leadership qualities and not so much a full-blown technical or administrative specialist [7]. However, it should be pointed out in this context that the function of project management is also a specialty, namely a specialist in management activities, which consists of optimally linking the various subtasks at a higher level, the project, and system level. If we look at the career history of successful project managers, we can see that they usually come from a very specific field and then become qualified as systems engineers and later as project managers; see Chap. 15. In this regard, Heinz Hermann Koelle mentioned: "*Project management is primarily the art of getting the desired work done by people within the promised time and the available resources with success*" [8].

Although the project manager must be given full responsibility for technical and commercial execution in the interests of maximum efficiency, he cannot be expected to master all the fields of knowledge pertaining to the project. He must therefore draw on experts in the relevant fields and delegate tasks. "*Being able to delegate is a critical skill for success or failure for any manager*" [9]. However, important decisions must be made solely by the project manager after weighing all technical and economic aspects of the project. In large and complex projects, the project manager may have to call on appropriate specialists from a wide range of disciplines to solve his tasks and, if necessary, temporarily include them in the project team; see Fig. 1.1 (Mtrixorganization).

It is expedient to break down the project task into individual functions, which in turn represent a self-contained task and for which a corresponding project employee can also assume responsibility. This requires a thorough analysis of the required functions in each individual case. The result of the analysis will be different for different projects because different functions need to be performed for a research and development project than for a production or service project. For example, in the execution of large and complex aerospace projects in Europe and the USA, the structure of the project task according to the following functions has proven successful [7]:

- Planning and Control
- Product Assurance
- Systems Engineering
- Subcontractor Control
- Manufacturing Control
- Assembly, Integration, and Verification (Assembling, Integrating, and Testing)
- Operational

In contrast, the following project functions would be conceivable for a charitable aid program in a disaster area [10]:

- Headquarters coordination
- Press release
- Financial control and administration
- Auxiliary measures
- Logistics and transportation
- Construction measures
- Liaison Office

The structure according to project functions must be made according to logical and, above all, practicable aspects so that corresponding responsibilities can be assigned to the individual functions. For example, it would be wrong to place the tasks of product assurance and systems engineering on the same hand, as these are completely different tasks and responsibilities. Systems engineering is predominantly a creative development task, while product assurance is to a greater extent a monitoring function, e.g., quality, reliability, and safety; (see Chap. 8).

When defining the project functions or project tasks, it is usually not sufficient to define only the main terms. It is important that further detailing into subtasks is made in order to be able to distribute the responsibilities more precisely. The following is an example of typical PM functions for hi-tech projects [4]:

(1) Project administration (*Chaps. 9 and 13*):
  - Information Management
  - Contract Management
  - Statement of Work (*SOW*)

(2) Sequence planning and monitoring (*Chap. 9*):
  - Work breakdown structure
  - Work package descriptions
  - Schedules
  - Status control
  - Performance Evaluation
  - Reporting

(3) Cost planning and monitoring (*Chaps. 9 and 10*):
  - Personnel and cost plans
  - Payment plans
  - Work releases
  - Cost control
  - Reporting

(4) Configuration and documentation control (*Chap. 12*):
  - Documentation structure
  - Documentation control
  - Configuration monitoring
  - Filing and distribution
  - Reporting

(5) Risk management (*Chap. 19*):
  - Risk identification
  - Risk assessment and classification
  - Risk reduction (*mitigation*)
  - Reporting

The project management functions defined by Martin are mostly identical to the PM functions described by the author but are not structured in the same way. However, they can also be understood in part as a supplement and refinement [11]:

- Control of the project work
- Defines the project mission together with the project decision-maker
- Develops the project order (feasibility)

- Obtaining all information relevant to the project and information from team members
- Information on the line management
- Plan the project tasks
- Develop a milestone plan
- Determines resource and capacity needs in coordination with the line management
- Planning of resources, costs, and intermediate goals
- Suggests team members to the decision maker*
- Appoints deputy
- In coordination with the line management, sets team member work schedules, including vacations, classes, etc.
- Plans and leads team meetings
- Determines scope of work, processing sequences, procedures, and deadlines in coordination with team members
- Ensures the elaboration of proposed solutions and evaluates them
- Performs target/actual comparisons for dates/costs/intermediate results
- Prepares milestone meetings
- Reports serious deviations between milestone meetings
- Controls all project documentation
- Coordinates a project-specific competency framework with the decision maker*
- Ensures that the project is handled according to project management rules
- Monitors the risks and informs the decision-maker if necessary*

* higher-level management authority

As mentioned above, the Project Manager has to perform a wide range of technical and administrative tasks. To do this, he needs appropriate authority, see Sect. 5.1.1. In addition to the list of tasks described above, Kraus has defined the following authority for project managers [12]:

- Disciplinary authority to issue instructions to team members
- Functional authority to issue directives to team members
- Technical authority to issue instructions to the team members
- Access to and management of approved project resources
- Compliance with instructions of the decision maker*, but not to instructions of the line management
- Nomination of team members
- Decision on proposed solutions and their submission to the decision maker*.
- Decision on the type and scope of information to the line management
- Decide on the participation of team members and project staff in milestone meetings.
- Veto power over team decisions.
- Decisions authority in case of conflict within the team
- Access to all information relevant to the project

* higher-level management direction

5.1.3 Project Team and Organizational Structure

Project management is an interdisciplinary task that can for large and complex projects only be efficiently handled by a team. Martin states, "*A complex project requires a team of very different employees and departments. The project manager must have a leadership style that is acceptable to all*" [11]. Vollrath makes similar comments, stating, "*Teamwork is the adequate work organization for handling projects*" [13].

Team strength depends to a large extent on the project size and complexity. While it is basically correct to try to keep the project team as small as possible in order to minimize project management costs, it is extremely uneconomical if important management functions are neglected, because this can become expensive. Project management is a control function, and if a project, regardless of size and complexity, is poorly managed due to a lack of sufficient control power, the project losses are often much greater than the financial outlay required for more appropriate project management personnel. The cases of trying to manage a very large and complex project with a very small project group are unfortunately very common in practice. The opinion sometimes expressed in the literature that project teams should be very small, i.e., limited to a few persons only, is therefore wrong in principle. What is correct, is that the management team should be adequately staffed in order to be able to complete the defined tasks without gaps and in a professional manner. Even in the case of small projects with a limited financial volume, the project manager must have the appropriate resources (*time/resources*) to carry out his or her tasks. At the same time, of course, care must be taken to ensure that the project team does not become disproportionately large. There are many examples of this. It would be completely uneconomical if project management is similarly staffed to the departments entrusted with the technical work.

The previous explanations already make it clear that the appropriate ratio between project and specialist areas is important; see Fig. 5.1. However, the exact ratio cannot be determined across the board and depends on a wide variety of factors, such as the type of project, company structure, etc. Complex and innovative research and development projects (R&D projects) in which, for example, the systems

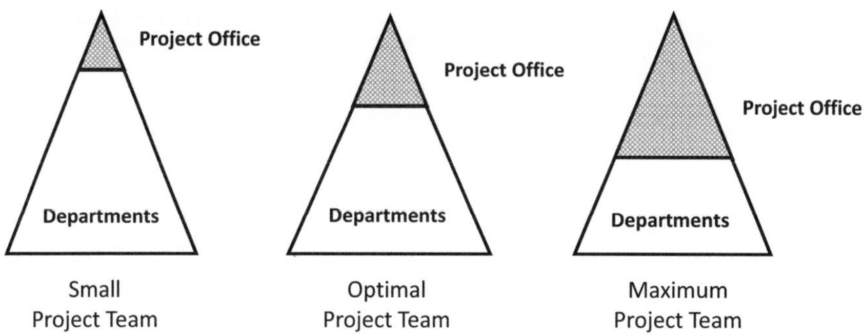

Fig. 5.1 Relative project team strength (project/subject area)

engineering area of responsibility is strongly emphasized, require a larger project team than a production project; see Sect. 5.1.5.

The division of tasks between the project and specialist department teams requires a clear definition and delimitation of the individual tasks. For example, the project team is not required to carry out detailed design and development work, which requires corresponding specialist knowledge. The task of the project team is rather the planning, control and integration of all project work with regard to the efficient achievement of the project goal under integration of all important project parameters, such as: system technology, quality, safety, deadlines, costs, environmental protection, etc. From this it can be derived that the project team for the fulfillment of the goals set for it cannot consist only of administrators. The areas of responsibility already mentioned must be the responsibility of and managed by appropriate experts who are firmly integrated in the project team. It is not sufficient for the company to have corresponding experts in other departments. Rather, the project manager must be assisted for the duration of the project by appropriate personnel who, just like him, represent the interests of the project in question. An optimal division of tasks between the project department and the specialist departments should be made based on function, as shown in Fig. 5.2. In Sect. 5.1.5 the topic of project team strength and PM effort is described in detail.

The project team (*project office*) must be equipped with adequate technical competence in order to be able to carry out technical and economic project management as required but no technical work should be carried out in the project office. The interface between the project office and the specialist departments must be clearly and unambiguously defined in every case. In practice, however, it is precisely here that problems arise time and again since the definition of this interface usually causes the greatest difficulties for project and department staff. Very often it is simply not recognized that the tasks of the project team, in contrast to the tasks of the departments, are not execution-oriented but planning and monitoring-oriented; see

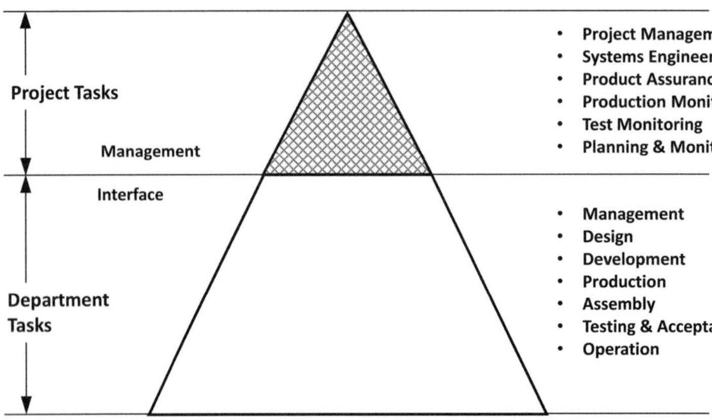

Fig. 5.2 Optimal division of tasks between the project and the departments

Fig. 5.2. The typical tasks of the project office and the departments are compared below.

Project management office	Department
– Goal setting – Creation of project plans – System analyses and design – Functional safety analyses (*overall system*) – Creation of system specifications – Creation of functional specifications – Scheduling and monitoring – Cost planning and monitoring – Risk management – Documentation and configuration management – etc.	– Design and construction – Mathematical models – Analyses (*structure, thermal control, EMC, etc.*) – Laboratory developments – Model production – Production planning – Manufacturing – Quality control – Test planning – Functional tests – Assembly/Integration – Fixture construction – etc.

The prime responsibilities of the system PMO, which are summarized below, are mandatory for successful project execution, regardless of the project size (*small, medium, or large*); see also Fig. 5.3.

- System engineering (*see Chap. 7*)
- Product assurance (*see Chap. 8*)
- Project Control (*see Chap. 9*)

For very small projects the assigned PM must perform the above-referenced responsibilities to the extent possible in personnel union and through short-term support provided by specialist organizations, whereas for very large projects the necessary specialists may be integrated into the PMO team through temporary assignment agreements.

Figure 5.4 shows an example of an extended project organization for Phase C of a large R&D project [7]:

- System engineering (*see Chap. 7*)
- Quality assurance (*see Chap. 8*)
- Project Control (*see Chap. 9*)
- Cost/Finance management (*Chap. 10*)
- Information control (*Chap. 11*)
- Documentation & configuration control (*Chap. 12*)
- Contract management (*Chap. 13*)
- Acquisition & proposal management (*Chap. 14*)
- Risk management (*Chap. 19*)

The development of the project organization must be based on a detailed study to determine the necessary project responsibilities. In addition, a detailed breakdown

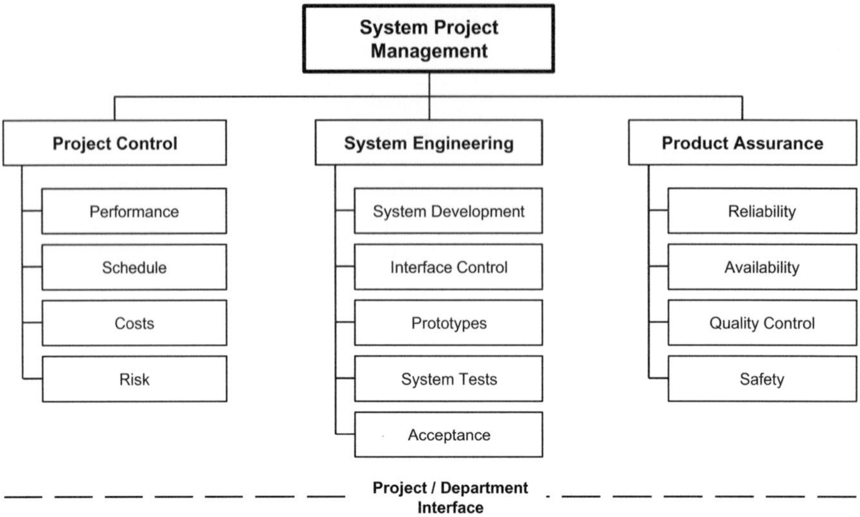

Fig. 5.3 Standard PM organization with three prime functions

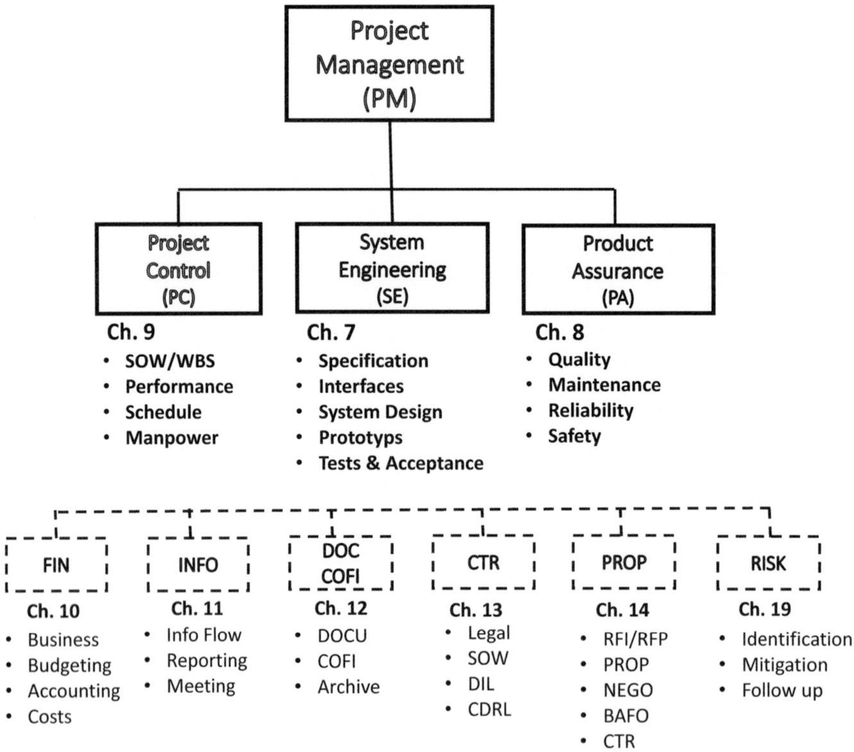

Fig. 5.4 Extended PM organization with multiple functions

of the project organization must be made in order to define the exact work content of the respective organizational units. The preparation of corresponding job descriptions is recommended in any case; see Sect. 5.1.4 (*key personnel*). In the case of small projects, the same procedure is to be followed, but the individual functions may be performed by the project manager with assistance provided by specialists. It is sometimes the case that the project manager is also responsible for specialist tasks in addition to his or her project tasks; in this case, he or she wears two hats, figuratively speaking. However, special topics, such as quality assurance or cost control, must be handled by appropriate specialists.

5.1.4 Key Personnel

Creating and implementing a functional project organization plan is an important step for efficient project execution. However, the appropriate staffing of the project team is just as important. Above all, the greatest care must be taken when selecting key personnel, i.e., when filling key management positions (*project management, planning and monitoring, product assurance, systems engineering, etc.*). Education, practical experience, and the ability to work in a team are the decisive factors in filling the top project management positions. Practice shows time and again that management tools alone are not enough to successfully complete a project. The right staffing and, in particular, the careful selection of key personnel is of the utmost importance for project success.

The nomination of the project manager often causes the very greatest difficulties, because he/she must be accepted equally by the technicians and economists in the project as well as by the specialist departments. His or her ability to integrate is a necessary prerequisite for managing a project, along with a balanced expertise in all project areas, such as planning and monitoring, systems engineering, etc., and good leadership skills. Conflicts arise repeatedly in project activities, which must be resolved objectively and in the interest of the project. The integration ability must be sufficient to balance conflicting goals, such as a deadline-cost problem, in the interest of an optimal project solution. His expertise must not be one-sided, e.g., only technically oriented, but must include the economic and administrative areas. This does not mean, of course, that he must be a top planning expert, for example. As with any manager, his leadership qualities must be motivating and, moreover, team-building. These premises should definitely be taken into account when selecting a project manager. "*A (good) project stands and falls with the people involved*" [14].

The view sometimes expressed in professional circles that the project manager should essentially only be an administrator and need not understand too much about the technology, i.e., the factual task of the project, is not correct. In practice, one can experience time and again how technically oriented disputes arise, the solution (*compromise*) of which primarily involves technical-economic aspects. Thus, in practice, project managers of engineering projects are indeed usually recruited from the technical areas. A typical development is that a technician with good technical knowledge of one area, e.g., electronics, and, in addition, good general technical knowledge of other areas, e.g., mechanics and testing, advances to become a systems engineer and, building on this is entrusted with the management of a small project, e.g., a work package. The recruitment of a PM for a chemical or biological project should follow the principle described above for a technical project. The successful completion of a smaller project is then the platform for larger project management tasks. Accompanying project manager training that provides future project managers with the necessary knowledge of project administration tasks, i.e., contract, financial management, project control, etc., and leadership behavior, in addition to on-the-job training, is highly recommended. Many companies run courses to achieve this result [15]. This is explicitly discussed in Chaps. 15 and 25.

Staffing the lower management level with appropriate key personnel, as is necessary for major projects, is also of critical importance for project success; see Fig. 5.4. The holders of these positions must also perform integration and management tasks and have the ability to work in teams. Their career path is very similar to that of the project manager. They are usually employees with a sound education in engineering, industrial engineering or business administration and with practical experience in a specialized field. For many technicians, being part of the project team is a prerequisite for later managing a project themselves. The project manager must have a very decisive say in the selection of second-level key personnel. It would be an indefensible practice if the team composition were to be made in an autocratic manner by another unit, e.g., the HR department, without consulting the appointed PM, because the project manager depends on good cooperation, which is not solely dependent on technical competence, but also on the team composition and team spirit. A successful project manager of a large German aerospace company once put it this way: "*A team of top people does not automatically guarantee the greatest success; rather, it depends on the right composition of the team*" [15]. Of course, in practice, circumstances, personnel availability, etc. must be considered. However, a prerequisite is a personnel discussion between all relevant departments with the participation of the project manager.

The project manager must draw up appropriate job descriptions which clearly specify the tasks, responsibilities and competencies of the future job holder. As a practical example, the job description of a PC manager for an R&D project is given below; see Sect. 9.5.1 [7].

(1) Duties and responsibilities of the PC Manager:

    The PC Manager who reports to the PM, is primarily responsible for planning and monitoring the project sequence, the deployment of resources and cost control. The planning and monitoring of the project sequence, the deployment of resources and the costs are all in one hand. Starting from the contract inception, the planning documents are to be prepared, on the basis of which the work releases are then made. The monitoring process includes the functions, status determination, deviation analysis and initiation of corrective measures; see Table 5.1.

(2) Competencies of the PC Manager:

Table 5.1 Typical PC tasks

Project control tasks	Detailed task description
PC—Primary Tasks • *Schedule Control* – WBS – WPD – Scheduling – Schedule Status Control – Performance Control – Report Preparation • *Cost Control* – Personnel Planning – Cost Planning & Control – Financial & Resource Planning – Work Authorization – Personnel/Cost Status Control – Report Preparation	The PC Tasks Listed describes the primary activities of the PC Team. The PC Manager must be qualified to process the listed activities. For large projects the PC Manager shall lead an expert team consisting of qualified personnel to CONTROL following major PC tasks: • Statement of work (SOW) • Work Breakdown Structure (WBS) • Work Package Descriptions (WPDs) • Schedule Status Control and • Cost Control and Monitoring • Earned Value Management (EVM) • Monitoring of project reviews • Action item control • Meeting coordination
Additional PC Tasks • *General Administration* – Contract Administration – Performance Description – Additional Tasks – Bonus System – Information Systems – *Data Management* – Control Rooms – Report Preparation – *Documentation and Configuration Control*	When implementing additional PC Tasks, the PC Team will normally be assigned responsibility based on organizational and/or financial grounds. The tasks can be performed by the PC Manager only when he possesses the appropriate experience. • It is advantageous to allow the PC team to perform the administrative tasks in close cooperation with the primary Tasks. • For the tasks of Documentation and Configuration Control, there are alternative organizational solutions: 1) As an element of Quality Assurance 2) As an element of PC 3) As an independent major function of Project Management • In exceptional cases the PC Team may be responsible for Subcontractor Control.

In order to be able to perform the described PC tasks and responsibilities effectively, the PC manager must be equipped with the necessary competencies. The PC manager is equivalent to the project functions of the second management level (*system engineering, product assurance, testing, etc.*); see Fig. 5.4. The competencies of the PC manager are:

– Selection of qualified PC personnel
– Participate in schedule, cost, and contract negotiations
– Authority to issue directives in the PC area

(3) Organizational PC interfaces:

When performing the above-mentioned PC tasks, the PC manager has to control project-internal and project-external interfaces. The project-internal interfaces mainly refer to the areas of planning and monitoring and their related coordination tasks. For example, in cooperation with the heads of systems engineering and quality assurance, the PC manager must prepare the required PC inputs (*planning documents and status reports*) in the prescribed form and on time. Project-external managerial interfaces are to be established to the following departments:

- Accounting/Finance Department
- IT/Software Department
- Other projects (*for example, plant occupancy*)
- Central planning (*standard methods, etc.*)

(4) PC Manager Qualification:

In order to be able to perform the PC task as an integral management function, this position is to be filled either with a technician who has correspondingly good commercial knowledge or with an administrator with a correspondingly good technical understanding. A combined or technical and commercial double degree (*for example an MBA degree*) is an advantage for this position. The incumbent should also have appropriate practical experience and have worked independently as a planner and/or project business person on a project for an extended period of time. In summary, the PC Manager must be able to demonstrate the following characteristics:

- Good technical and economic knowledge
- Creativity/ideas
- Organizational skills
- Sense of responsibility
- Empathy/team experience
- Flexibility/adaptability
- Critical faculties
- View for the overall situation.

The example of a job description for the PC Manager (*Head of Planning and Monitoring*) reproduced here is based on experience from the aerospace industry, which can however easily be adapted to other branches.

In addition to the PC Manager job description, the standard ESA job descriptions for the Planning Engineer and Project Clerk (*third level of management*) are reproduced below [16].

(1) *Scheduling Engineer* (*Schedule Control Engineer*). The qualifications of the scheduling engineer should be as follows:

a) Basic knowledge of general project operations; i.e., the project planner for an engineering project must have knowledge of general engineering operations on the project, for example, in order to establish and follow a project schedule and lead related planning meetings in collaboration with project staff responsible for engineering.

b) Thorough knowledge of relevant planning and monitoring methods:
   – Creation of work breakdown structures (*WBS*)
   – Drafting work package descriptions
   – Establishment of flow charts
      Planning network design
      Gantt chart creation
      Milestone definition
   – Conduct progress reviews.
   – Lead planning and project status meetings and reporting
   – Handling of common planning software
c) Basic knowledge of the project merchant's area of responsibility.

(2) *Cost/Business Administration* (*Cost Control Manager*). The qualifications of the cost/business control manager should be as follows:
   a) Detailed knowledge of commercial project management; that is:
      – Collaborate in the creation of work breakdown structures,
      – Preparation of detailed staffing schedules,
      – Preparation of detailed cost plans,
         Record quantity estimates in accordance with project schedules,
         Conversion of the quantity estimates into costs using the operationally determined cost rates (*hourly rate, etc.*)
      – Cash Flow Plan
      – Release for work, which must be in accordance with the prepared design documents
      – Perform personnel/cost status checks (target/actual comparison)
      – Final Costing (*Cost to Completion*)
      – Reporting (staffing and cost status).
   b) Knowledge of Accounting:
      – Invoicing and auditing
      – Inventory assessment and evaluation
      – Payment transactions
      – Revenue analysis
      – Cost category review
      – Transfers.
   c) Basic knowledge of the project planner's area of responsibility.

5.1.5 Team Strength and Costs of the Project Office

The estimation of personnel and costs for the project office (*project management team*) cannot be answered in a generalized way. As already indicated in Figs. 5.1 and 5.2, it depends on having the correct team strength in relation to the overall project and thus also to the functional departments. Ratios, however, are only helpful if one has a good knowledge of the project in question and the circumstances. It is easy to see that an R&D project is far more management-intensive than a production project, for example. It depends primarily on the definition of the management content, the complexity, the degree of novelty and also on the particular industry. Considering the tasks of an R&D project office (see Fig. 5.4), as a guideline, 10–15% of the total project costs is realistic for a direct account assignment for a R&D project management. On the other hand, the costs are well spent if they can be used to avoid incorrect planning and excessive cost increases. This can be easily understood by looking at negative examples such as those mentioned in Sects. 3.2 and 3.8.

In the development of space projects, the management costs of the industrial prime contractor are in the order of magnitude as mentioned above. It must be taken into account that the management team (*project office*), in addition to administrative activities, e.g., at the system level, also performs system engineering tasks (*Chap. 7*) and product assurance (*Chap. 8*), etc.; see Fig. 5.3. For very complex projects, the organizational structure may be supplemented by additional functions; see Fig. 5.4.

If, on the other hand, the system engineering tasks are carried out in the individual departments, as is sometimes the case in practice to the detriment of efficient project management, the related costs are naturally not incurred in connection with the centrally operating project management but are distributed among the individual departmental work packages. In the following discussion, however, the organizational structure shown in Fig. 5.3 will be used as a starting point for a better understanding, because at a minimum the management functions shown in Fig. 5.3 are to be performed in the same or a modified form in the execution of every project.

However, if the project is very small, e.g., up to a hundred thousand Euros, the functions of planning and monitoring, product assurance, systems engineering, etc. must also be performed, but the project manager can perform some of the tasks in personnel union. For special tasks, e.g., for professional cost monitoring and contract management, he must call on in-house services. This statement is important because project managers of small projects sometimes make the mistake of forgetting or not correctly estimating the cost of services and then perform important things themselves for which they lack the expertise.

The project management effort depends mainly on the general project tasks listed in Fig. 5.3 and the factors described below:

(1) Project-specific parameters:
   - Industry type (*construction industry, plant engineering, etc.*)?
   - Project type (*R&D, production, etc.*)?
   - Project size (*or volume*)?
   - Project complexity (*interdisciplinary or within a discipline*)?
(2) Order conditions:
   - Internal or external client?
   - Multi-level project organization (*subcontractors*)?
   - Management within the framework of a consortium of companies?
   - etc.
(3) Special Management Requirements:
   - Organizational structure (*range of functions*)?
   - Planning detailing?

- Monitoring cycle (*weekly, monthly*)?
- Reporting (*frequency, distribution list*)?
- etc.

Here is a practical example of a development project: In a management study for the European launch vehicle program Europa III (*later Ariane*), the costs for the project office were determined based on the following influencing variables [17]:

(1) Project-specific parameters:
   - Space project
   - R&D projects
   - Project volume; order of magnitude 1 billion Euros
   - Estimation was based on a fiscal year of 1970
   - Schedule approximately 10 years
   - Technologically complex project (*development of cryogenic systems*).

(2) Project Conditions:
   - Multi-level project organization (*client; prime contractor; subcontractors and suppliers*)
   - Cooperation with European national authorities.

(3) Special management requirements (*management functions*):
   - Project Management (*including Project Control, Documentation Control, etc.*)
   - Quality assurance (*overall system*)
   - Contract management (*companies from six nations*)
   - Systems Integration
   - System technology for steps, payload fairing, equipment section, ground test equipment, and launch equipment.

The financial outlay for the client's project management office (*PMO*) amounted to about 10–15% of the total program costs, a typical amount for complex development projects; see Sect. 5.2.3.

5.1.6 Organizational Changes

As desirable as a stable project organization may be, it is almost impossible to achieve It in practice. On the contrary, organizational changes are often firmly planned from the very beginning. To understand this better, a look at the project life cycle and the individual project phases is helpful; see Chap. 4. The completely different tasks on the long path from the concept phase to the production and/or operating phase force organizational changes. However, any organizational change should always be preceded by a thorough functional analysis. Figure 5.5 illustrates this with an example. The project organization for phase A (*concept phase*) corresponds to the objectives and main functions to be performed in this phase:

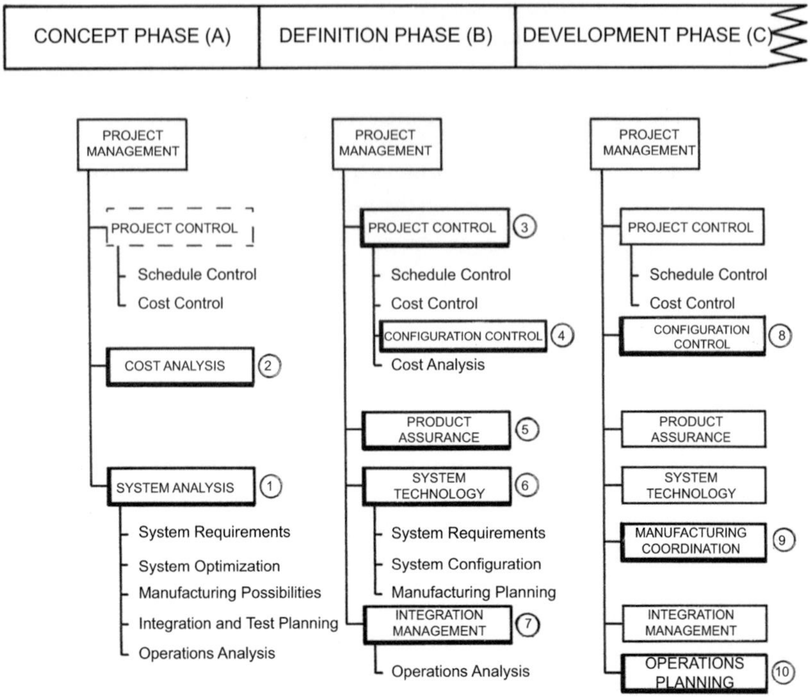

Fig. 5.5 Organizational changes during the project life cycle

(1) System analysis and
(2) Cost Analysis

PC management staffing is usually not required at this stage; the project manager performs this function, if necessary, without team assistance but with the support of partially seconded schedule and cost control experts.

During the definition phase (*Phase B*), the project organization must already be expanded by several functions due to the expanded range of tasks:

(3) Project Control Management
(4) Configuration control
(5) Product assurance
(6) Systems engineering (*previously systems analyses*)
(7) Integration Management.

Some of the additional functions mentioned here were already included as individual tasks in Phase A but did not yet have the significance of an independent project function in the organizational plan. Other functions. in this example cost analysis (2), are omitted completely in the subsequent phases or are carried out as part of project control management (3).

In Phase C, there is a transition from purely paper-based activities (*e.g., analyses, plans, specifications, etc.*) to development activities (*e.g., design work and the production and testing of functional models and prototypes).* This is also reflected in the project organization in that additional functions are added:

(8) Configuration control (*see Chap. 12*),
(9) Manufacturing coordination (*with a view to future manufacturing)*
(10) Operational planning (*in terms of future operational planning*).

5.2 Project Organization in the Company

5.2.1 Operational Organizational Structures

In every business, very specific functions must be performed in order to achieve the business goal, and in larger businesses these functions inevitably lead to specialization. Dale writes in this regard, "*The difference between large and small enterprises is that in the smaller one there is less specialization and one person can take on different functions*" [18]. Typical functions of an industrial enterprise are:

- Research and development (R&D)
- Project Management Office (PMO)
- Production
- Purchasing
- Marketing/Sales
- Financial Management
- Human Resources.

In larger companies, the creation of corresponding organizational units, through which the overall operational task is segmented into specialized areas of

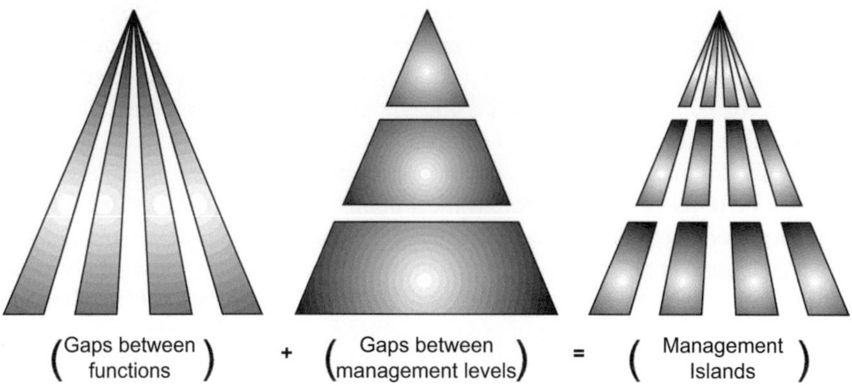

Fig. 5.6 Principle of company structuring according to functions and hierarchy levels

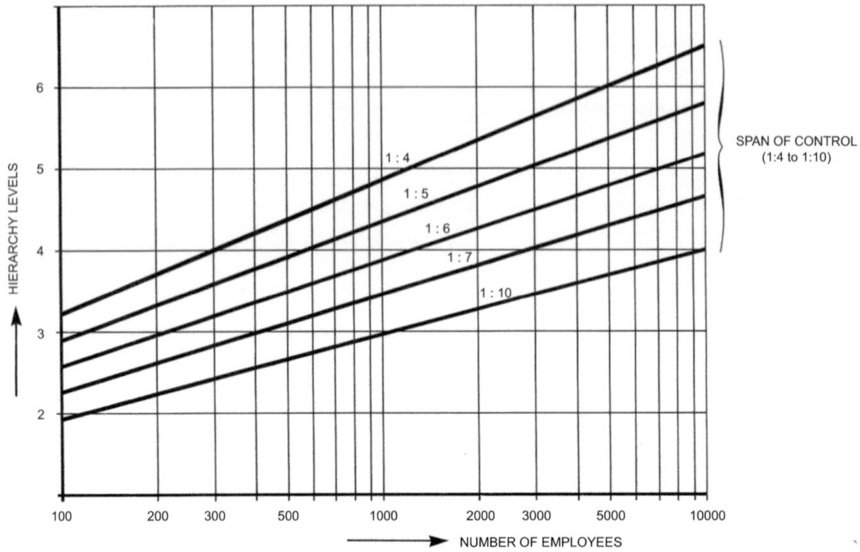

Fig. 5.7 Determining the hierarchy levels

responsibility, is obvious. This results in decentralization. The head of the company no longer has to take care of every detail himself, i.e., from a central location, but has the various operational tasks handled decentral by specialists. He or she is the only one with a pyramid structure. However, large companies with several hundred or several thousand employees are forced by the complexity of their tasks to structure the company not only according to functional aspects but also according to hierarchy and responsibilities. As a result, the entire range of tasks of a large company is broken down into manageable and controllable organizational units (*management islands*); see Fig. 5.6 [19].

The required hierarchy levels of a company depend in principle on the average ratio of the number of employees per organizational unit. If the average ratio between the head of an organizational unit and his or her employees (*span of control*) is known, the ratio between the number of employees and the hierarchy levels can be easily determined; see Fig. 5.7. If, for example, the average span of control has a ratio of 1:7 and the company has 400 employees, exactly three hierarchy levels are sufficient, whereas a four-level company hierarchy would be required for 2800 employees. In practice, however, it can be assumed that different hierarchy levels will also be required due to different focal points in the company. In an industrial company, there are naturally many more employees working in R&D and production than in marketing and financial management, for example. Consequently, there will be more hierarchy levels in some areas and fewer hierarchy levels in others.

However, the decentralization process in large companies cannot be limited to structuring according to functional and hierarchy-oriented aspects. Terry defines seven categories according to which a company can be sensibly structured (*decentralized*) [19]:

(1) Breakdown by function:
  a. The breakdown by function is usually common at the top of the organization as well as at the lowest organizational level (*Terry*)
  b. Examples (*author*):
    – Research and development (*R&D*)
    – Purchasing
    – Production
    – etc.
(2) Breakdown by product:
  a. This breakdown is very common among companies with different product groups due to the specialized industry knowledge required (*Terry*)
  b. Example (*author*):
    – Aircraft
    – Helicopter
    – Satellites
    – etc.
(3) Breakdown by location:
  a. The organization of the company by location can be done for various reasons: existing capacities (*equipment, know-how, etc.*), favorable labor costs, market segmentation, etc. (*Terry*)
  b. Examples (*author*):
    – Berlin
    – Munich
    – Paris
    – etc.
(4) Breakdown by client (*customer*):
  a. To improve competitive opportunities, organizational reflection (*specialization*) in this regard can be useful (*Terry*)
  b. Examples (*author*):
    – National authorities (*CNES France, DLR Germany, National Governments, etc.*)
    – International authorities (*ESA, NASA, NATO, etc.*)
    – Private clients
    – etc.
(5) Process-oriented outline:
  a. The process-oriented company structure is particularly suitable when work processes can be advantageously combined in independent organizational units for economic or technological reasons (*Terry*)
  b. Examples (*author*):
    – Production
    – Assembly
    – Testing
    – etc.

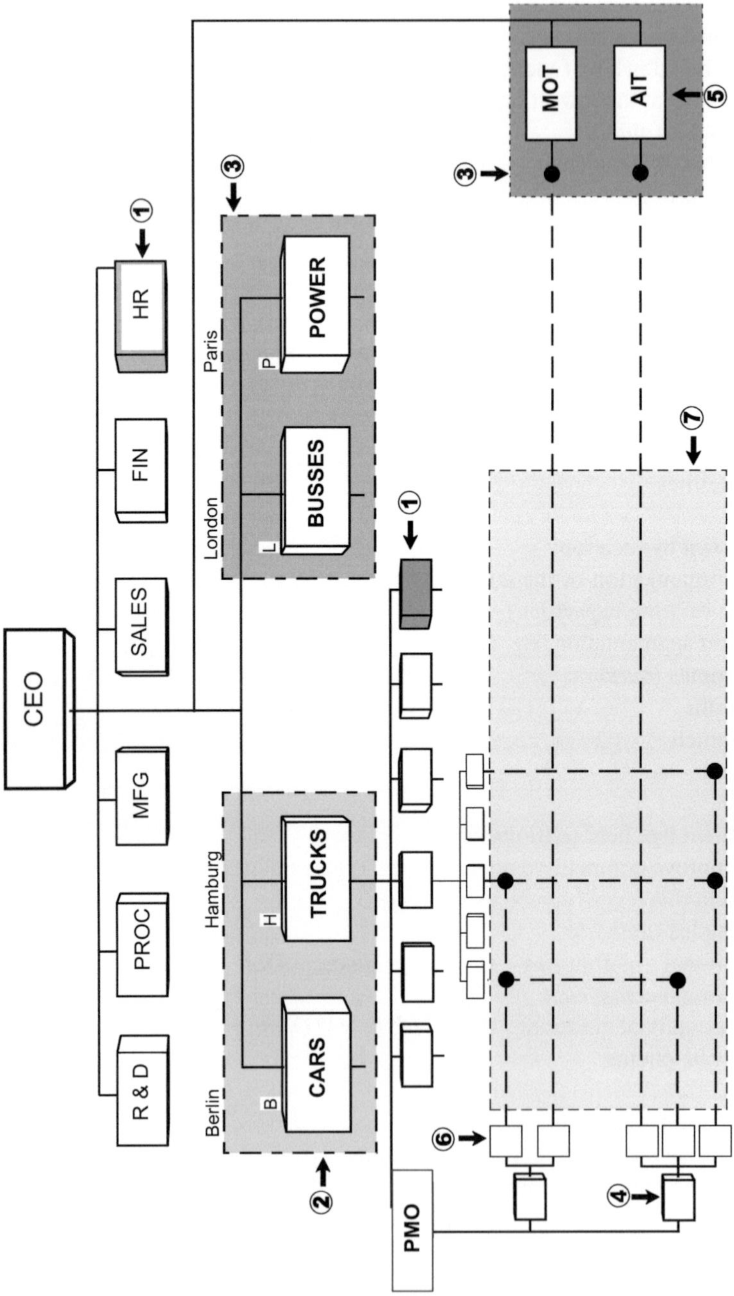

Fig. 5.8 Organization chart of a large company with decentralization levels

(6) Project-oriented outline (*task force*):
   a. The structure according to projects requires the use of autonomous project teams, which themselves have a small staff with experts who have the necessary knowledge. Project teams (*task force groups*) are usually relatively small and can be dissolved once the task has been completed (*Terry*)
   b. Examples (*author*):
      – Automobile development and production
      – Construction of a new production plant
      – Investigation of a new propulsion system
      – etc.
(7) Matrix-oriented outline:
   a. The matrix organization, also known as the lattice work organization, has proven to be a viable answer for solving constantly more complex and growing undertakings (*projects*), for the realization of which more flexible and, above all, systems-oriented organizational structures are imperative (*Terry*)
   b. Examples (*author*):
      – R&D project team (*example: helicopter development*) that draws on specialist areas (*aerodynamics, design, etc.*) to fulfill the project task
      – Plant project management (*example: power plant*), which handles the project with the support of the specialist departments (*turbine construction, heavy current area, etc.*)
      – Study management (*example: solar tower power plant*) conducting a system analysis for a modern solar power plant in cooperation with the specialist departments (*thermodynamics, hydromechanics, etc.*)
      – etc.

Of course, the organizational concepts of large companies defined by Terry are not necessarily all to be applied at the same time. Figure 5.8 shows an example of an organization of a large company in which all forms of decentralization defined by Terry are applied.

Many large European companies have already implemented several decentralization processes in the past while scaling up operations (*new markets, expanding the product range, increasing sales, etc.*). This is especially true for the decentralization measures defined by Terry (1) to (5). However, difficulties often arise when implementing project management within the matrix organization or a central project management office (*PMO*); see steps (6) and (7). All too often, project teams are hopelessly understaffed and lack the necessary authority. Yet it is precisely the creation of powerful project teams (*task force groups*) that results in success for a company. Project teams are best suited to open up new territory. The development of new and complex systems in particular requires project management; see Sect. 2.1.

In 1970, the German aerospace company Messerschmitt-Bölkow-Blohm (*MBB*), a company with about 20,000 employees, introduced a corporate structure with eight central divisions and eight business units each, which remained unchanged for many years [20]. It was an exemplary organization because it was lean and efficient. Projects, such as the development and construction of aircraft, helicopters, and satellites, were handled in the respective divisions.

Example of MBB Management:

- Central Services
  1. New development
  2. General development
  3. Technology
  4. Manufacturing
  5. Personnel (HR)
  6. Economics
  7. External Relations
  8. International Relations
- Business Units
  1. Military aircraft
  2. Civil aircraft
  3. Defense technology
  4. Traffic engineering
  5. Space travel
  6. Helicopter
  7. Cybernetics
  8. Operating

5.2.2 Organizational Integration of Project Management into the Company Organization

Although project business is booming in many companies, project management is often only half-heartedly introduced; see Chap. 17. "*The project manager is still in an outsider role in companies and administrations; he does not fit into the classic staff-line organization*" [21]. This statement, made by Hegi as early as 1972, is often still true in 2016. The organizational integration of project management into the organizational structure encounters considerable resistance in many companies and thus often leads to conflicts. In the process, two completely contradictory developments sometimes occur in practice. In some companies, project management is unable to free itself from its staff position, while in others, mammoth project structures are created that do not even stop at the incorporation of specialist departments. In both cases, serious conflicts between project management and the specialist departments occur time and again.

The project managers located in staff departments usually do not have the necessary authority and staff to be able to control a project effectively, and frustration and resignation very quickly arise among the project managers. The specialist departments, which are disproportionately strong compared to the project management (see Fig. 5.1, Case 1), can usually assert themselves successfully due to their specialist preponderance, but this is often not in the interest of optimal system design, because specialist departments are biased and represent specialist interests. Therefore, it is reasonable for the PM to be given the authority to design the system. This, of course, requires system engineering expertise. In the second case, where the PM is disproportionately strong and can act completely autonomously, the concerns of the specialist departments are completely subordinated to the project interests, which can lead in the long term to the dismantling of the specialist departments, which are oriented toward accumulated know-how; see Fig. 5.1, Case 3.

A major cause of conflict in the introduction of modern project management concepts results from the divergent interests of project and specialist departments. In contrast to project departments, specialist departments naturally pursue long-term goals and thus also secure the company's know-how stockpile. They are the bearers of the company's knowledge and the guarantors of continuity and progress in research and development, production and other areas. Dismantling their competence leads to a reduction in a company's competitiveness in the long term.

An essential project characteristic is the time limitation by having clear start and end points. Projects resemble a rising star that burns out again after a more or less long lifetime. This is the fundamental difference to the specialist areas which are designed for permanence. The objective of a project is the temporary accomplishment of a complex task (*system task*) involving the entire know-how of a company. Projects act like transformers that convert the entire know-how potential of a company to the project task. This analogy implies the importance of a partnership and balanced cooperation between the project and specialist departments.

5.2.3 Forms of Project Organization

Due to the rapid development of new products including IT solutions, the push for project solutions is without a doubt an imperative for public authorities and companies. There is no getting around it anymore because it is a logical response to the demand for project management to cope with complex system tasks.

But both the tentative approach of placing project management in staff departments and the path to all-dominant project management can only be seen as a short-term solution and a field of experimentation (*learning phase*). A long-term solution that is designed to last leads to the firm anchoring of project management with competencies in the company organization should be the goal. However, it must be ensured that the historically grown specialist areas are not neglected.

What is the best way to integrate "*the project organization*" into a company? First, one must examine the different forms of organization. Grochla defines three forms, which are reproduced in Fig. 5.9 [22]. Burghardt mentions five forms of project organizations, the first three being identical to those defined by Grochla [23]. There is only a slight difference between the matrix and contract project organization because the matrix organization can also be used for external tasks.

- Pure project organization (*Fig. 5.9—2*)
- Influence project organization (*Fig. 5.9—1*)

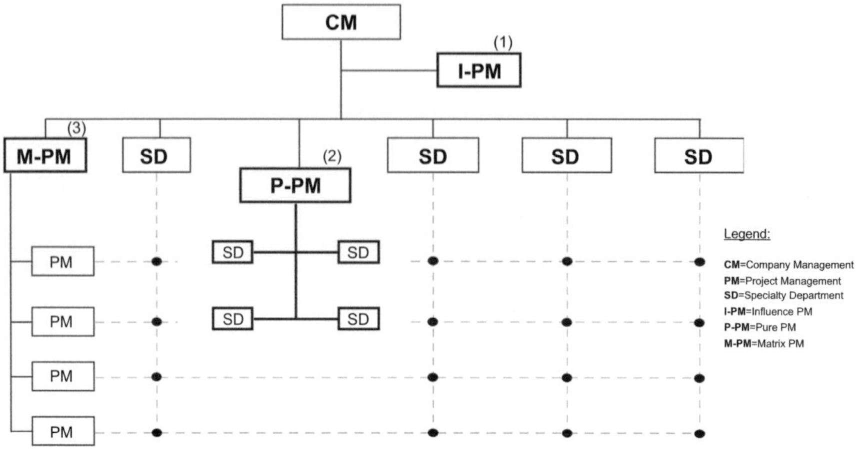

Fig. 5.9 Forms of project organization (Grochla)

- Matrix project organization (*Fig. 5.9—3*)
- Order project organization
- Project management in line

Table 5.2 below shows the advantages and disadvantages of the five project organization variants described by Burghardt [23].

In some companies, the different forms are used in parallel, which means a lot of duplication of work because there is often no central office for PM matters. In general, however, there should be a balance between the specialist and project areas. In this context, the matrix organization should be emphasized. The indicated solution shown is a fair distribution of the task increase of competences and influence between both areas. The increased demands on companies have brought with them new forms of management, from which the demand for further decentralization results due to the ever-increasing specialization: e.g., the matrix organization. In the effort to achieve the best possible balance of tasks and competencies between the traditional specialist areas and project management, the matrix principle is emerging as a viable solution.

Task and competence splitting can best be realized with matrix organization (*decentralization level 7; Terry*) [19]. The matrix form of organization is a logical step toward multi-project management with optimum utilization of departmental capacities for all projects. Of course, matrix management requires an increased willingness to cooperate on the part of all employees involved. Project priorities are often the cause of conflicts between the specialist and project areas, which must be resolved jointly, and only if this is not successful should the next higher level be called in. In this context, it should be noted that both project management and the specialist area serve the same corporate goal. Priority conflicts must therefore be resolved in a factual manner with the involvement of all parties concerned in the interests of achieving an optimum corporate objective.

Table 5.2 Advantages and disadvantages of the individual organizational forms (Source: Burghardt)

Type of project organization	Advantages	Disadvantages
Total project organization	▷ PM has full authority ▷ Shortest communication channels and lowest overhead ▷ Optimal alignment with the project objective	▷ Risk of establishing the project group after project end ▷ Transfer problems after project end ▷ Risk of parallel developments in project and neighboring line
Influence project organization	▷ Separately suspended developmental areas can be induced to cooperate in a managed way ▷ Minimal changes in the existing organization	▷ PM has little authority to issue directives ▷ No personalized responsibility ▷ High coordination effort
Matrix project organization	▷ Rapid grouping of interdisciplinary groups ▷ No dislocation problems at the start and end of the project ▷ Promotion of the synergy effect	▷ Project staff serve "two masters" ▷ High conflict potential between project and line
Order project organization	▷ Clear delimitation of competencies between project and line ▷ Easy integration of any subcontractors (also outside the own company) ▷ Great flexibility for multi-projects	▷ Need for a separate organizational pillar ▷ Competitive thinking of the organizational pillars ▷ Danger of "bureaucratization" of project management
Project management within a Functional Department	▷ All advantages of the pure project organization ▷ No need for staff transfer	▷ Only minor development tasks possible ▷ Not always the right technical and qualitative personnel available

5.2.4 Matrix Project Management in Practice

There must be a fair and balanced division of labor and competencies between project management and the specialist departments, which leads to the best possible fulfillment of the company's objective and motivation of the personnel. This means that, on the one hand, the company's performance must be optimally sold to customers by means of projects and, at the same time, the technological know-how of the specialist departments must be maintained over the long term. Both of these factors ensure the company's continued existence. In a management colloquium at the German aerospace company MBB, the different goals and management roles of the project and specialist departments were analyzed in detail with regard to a fair division of responsibilities. The result of the colloquium, which brought about a reorganization of project management, is summarized in Fig. 5.10 [24].

However, the introduction of matrix organization often encounters considerable difficulties in practice, as it presupposes a division of competencies between the traditional specialist departments and the project departments. The project

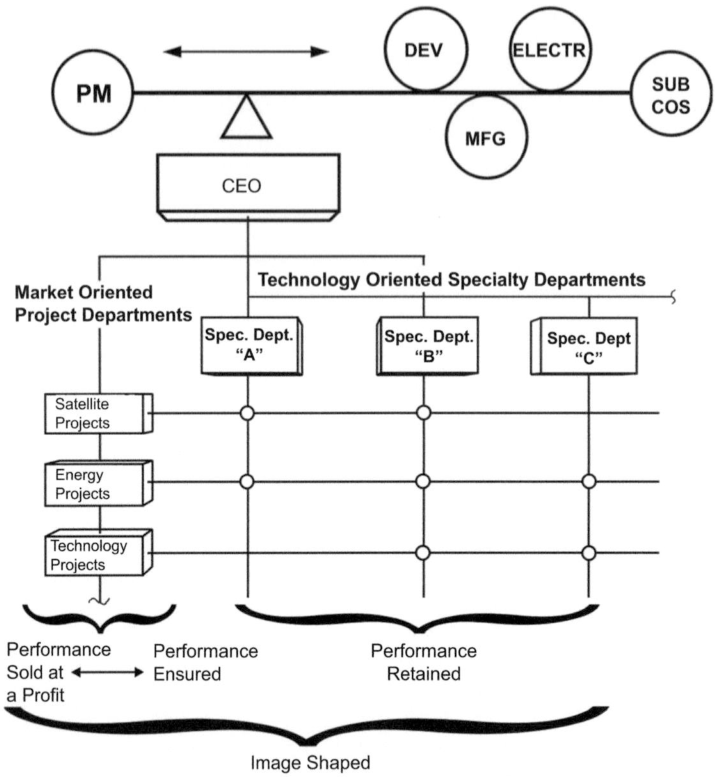

Fig. 5.10 Balance of management between project and specialist area

managers, who can also be regarded as the extended arm of the management (*see Fig. 1.1*), frequently encounter resistance from the specialist departments in their efforts to find system-optimal solutions. This conflict, which takes place at the nodes of the matrix, essentially results from different interests. The "*aerodynamics in aircraft construction*" department, for example, is in principle interested in an optimally designed aircraft in terms of flight dynamics, while the "*airframe designer*" tends toward a statically safe, weight-oversized aircraft fuselage. The design of an optimal system, however, requires compromises between the different disciplines. The engineering manager must succeed in achieving an acceptable compromise. With the increasing system complexity of modern projects, the importance of system optimization (*system engineering architecture*) has grown enormously. In addition to the influencing variables relevant to development, the concerns of production, testing, quality, logistics, operations, sales, financing, etc. must also be included in the optimization process.

The management of complex projects (*e.g., system projects*) requires the involvement of a very broad spectrum of specialist knowledge. This means that partners and subcontractors are also involved in the process. If, for example, only one major project

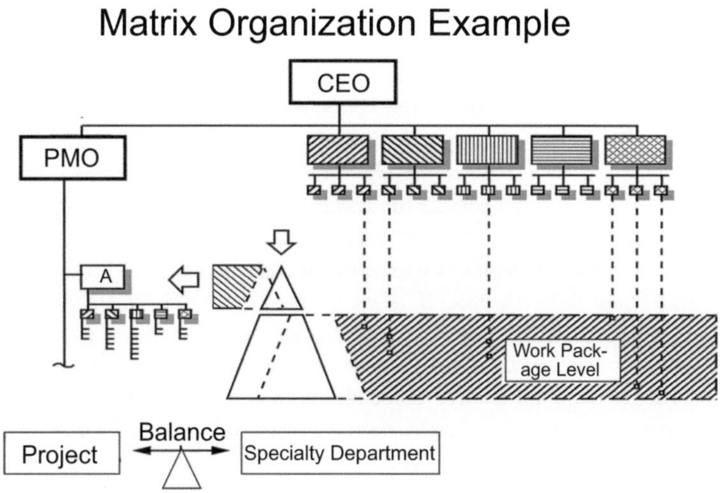

Fig. 5.11 Intersection points in the matrix organization

is being carried out in a company, an organizational solution as shown in Fig. 5.9, version 2 (*total project management*) is probably appropriate. If, however, several projects of equal priority are to be managed simultaneously in a company, the idea of a matrix organization is obvious. It is a logical step to transfer the idea with the extended arm of the management (*see Fig. 1.1*) from the staff unit milieu to a matrix concept.

The manifestations of matrix organization are not uniform in practice. In addition to projects that involve almost all specialist divisions of the company, many companies naturally also have specialist division-related projects, for example R&D and/or manufacturing projects only. In such cases, it can make sense to integrate project management into the departments. Since project management is not a need for itself, but a necessary measure for project execution, the objective should be a decisive criterion for the organizational structure in any organizational planning. Grochla and Thom have described in detail the possibilities for the introduction of different matrix organization forms in an essay from 1977 [25].

To ensure that the project runs smoothly within the matrix organization, the management interface between the technical and project departments must be clearly defined; see Fig. 5.2. In principle, the project department (*project office*) is responsible for the overall technical planning and monitoring (*control*) of the entire project. On the project side, specifications, plans, performance descriptions (*specifications*), guidelines, and procedures must therefore be drawn up and coordinated with one another. These documents must be prepared in as close cooperation as possible with the departments concerned, since this is where responsibility and competence are transferred to the departments concerned. In practice, the points of intersection in the matrix organization are defined by detailed work package documents (*specifications, service descriptions and schedules and cost plans*); see Fig. 5.11. An important prerequisite for the definition of the interface documents (*work package documents*) is mutual acceptance, since in the interest of efficient project execution, a binding internal agreement must be reached; see Sect. 9.4. In many companies, compliance is ensured by the internal order system. Both parties (project and department) agree on the scope of work, deadlines and costs in the project/work order.

The introduction of a matrix organization represents a compromise solution acceptable to all involved, because the basic project management forms of influence, project management and total project management, defined according to Grochla are nevertheless afflicted with considerable deficiencies in the long term. Grochla and Thom write to this: "*The project-oriented MO (matrix organization) appears as an effective solution, if the enterprise must assert itself in a relatively dynamic environment*" [25]. Since the matrix organization strives for a compromise. and this certainly does not always succeed, the matrix organization is also heavily criticized again and again. McCarthy therefore correctly writes: "*The introduction of the matrix organization is not a panacea, and conflicts between project and functional areas will continue to occur constantly*" [26]. In this context, another argument is of particular importance. Projects, by definition, have a definite beginning and an equally definite end. The problem of staffing up and down at the end of the project can also be best solved within the matrix organization.

Another point of criticism is the argument that management efficiency in the matrix organization is severely impaired by the willingness to compromise (*division of tasks and competencies between the specialist departments and the project management*). For example, the project manager may have to integrate employees into his team by temporary transfer who do not report directly to him. However, Thamhain and Wilemon, after surveying 68 project managers, concluded that management efficiency, i.e., the control options by the project manager, is almost the same in the matrix organization as in total project management; see Fig. 5.12. In this regard, they write: "..., *the evaluated data show that project managers in matrix organizations have almost the same control over departmental personnel as project managers in total project management*" [27]. They further conclude that the efficiency of project managers depends much more on leadership style and working conditions. Litke writes, "*The matrix organization may be the costliest, but it is also the most versatile organizational form for projects*" [28]. Kerzner confirms that by stating: "*The matrix organization is an attempt to combine the advantages of the pure line organization and those of the project organization*" [29]. In the automotive industry, due to the technical complexity and strategic importance of projects, the matrix organization is the most common and the most appropriate form of organization [30].

5.2.5 Project Management Office

By creating a central organizational unit for project management, the Project Management Office (PMO), as shown in the example of the matrix organization in Fig. 5.11, a management balance can be achieved with the specialist department.

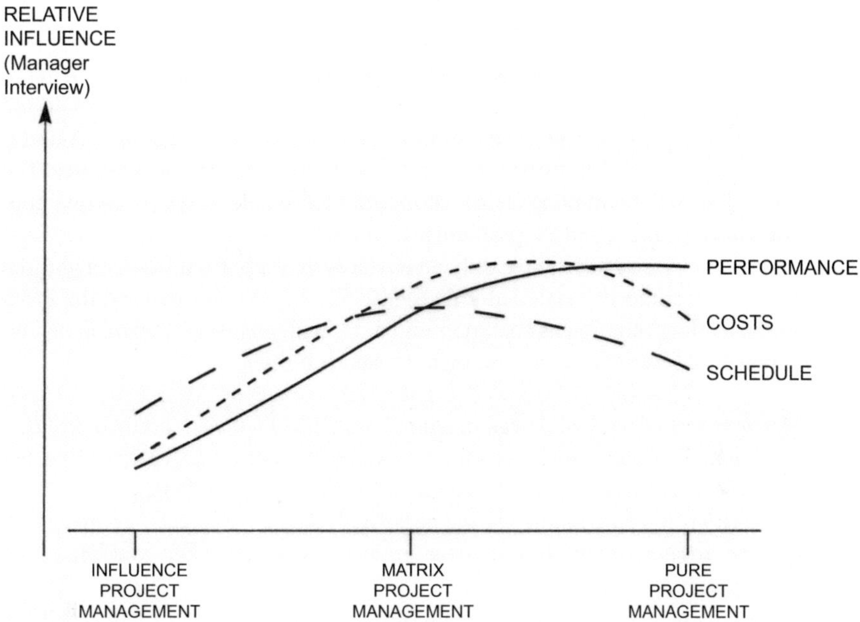

Fig. 5.12 Relative project management control span of different organizational forms

Due to the organizational equality with the specialist department, the PMO has the necessary authority to negotiate at eye level. The PMO should therefore not be active in an advisory capacity, as can often be read, but should have overriding full responsibility for the project activities of the company; proverbially: "*that is where the music plays*"; see Fig. 5.11.

In their book "*The Program Management Office Advantage,*" authors Tjahjaa, Dwyer and Habib describe the economic significance of PMO implementation. In this context, they refer to an MIT study according to which PMO implementation leads to improved control measures and, as a consequence, to lower cost overruns and higher profits. The authors define PMO as an organizational unit that not only monitors and supports project units but plays an important organizational role in improving project operations [31]. They then mention that much is published on the topic of PMO, although the topic is not new. In the past, other terms such as *Project Center of Excellence, Project Management Group,* or simply *Program Management* were used instead [32].

Since the term PMO is not standardized internationally, it is used very differently, on the one hand as the central project office of a company, but also as the project management office of a major project. Another term for a management office with almost identical tasks was the System Project Office (SPO) of the USAF, which was introduced in the 1960s. All USAF programs with systems responsibilities were placed under an SPO. "*The SPO is the only over-all field systems management organization during the Definition and Acquisition Phases of the system life cycle. Effective systems management is directly dependent upon positive day-to-day actions of personnel in the SPO and their authority to act for the organizations they represent. This principle requires senior representation and stable assignments of personnel*" [33].

The tasks of a PMO are very versatile. In order to be able to act as a counterweight to the specialist department, the PMO must have appropriate authority. The PMO's position in the company is crucial in order to be able to intervene in a controlling manner in the event of problems. In the opinion of the author, the PMO should therefore by no means have only an advisory or supporting function without decision-making authority (stated at DIN 69901-5) [34]. On the contrary, the PMO most have a leading function as shown in Fig. 5.11, with precisely defined decision-making powers. This includes the following responsibilities:

(1) Management of the PMO organizational unit. The PMO is a permanent organizational unit that ensures a balance between the traditionally strong line and the projects at the same hierarchical level as the specialist departments.
(2) The PMO is the professional home port for all project managers reporting to it (*mostly large projects*) but also for project managers in the specialist area (*mostly small projects*).
(3) Management and supervision of the project staff. The head of the PMO is responsible for managing the employees who report directly to him, as well as those temporarily seconded to special projects, and for providing them with professional and personnel support. He must be consulted in the event of conflicts.
(4) Coordination and standardization of all project processes according to ISO 9001.
(5) Integration of all top-level project plans. The *PMO is* responsible for creating a Master Project Plan (*MPP*) as the basis for the company's business and capacity planning and for updating it regularly; see Chap. 9.
(6) Budget Control. The PMO should coordinate overall project budgeting, financing, and cost control in conjunction with the finance department.
(7) Capacity planning and regulation of priorities. Appropriate resources are required for the execution of project tasks. Particularly scarce resources, such as key personnel, often cause headaches for the company. In the event of priority conflicts, the head of PMO is responsible for working out a compromise acceptable to all parties involved in the interests of the company. This must be done in close cooperation with the PM concerned and representative(s) of the specialist department, taking into account all aspects such as the Master Project Plan (*MPP*).
(8) Improving communication. The PMO must ensure a regular exchange of information within the PMO, but also horizontally between the PMO and the functional departments. This can be done verbally (*conversations and meetings*) and formally (*reports and written communications*); see Chap. 11 and Fig. 5.11. In addition, the PMO is responsible for the standardization of reporting.

(9) Use of uniform standards. The head of PMO is responsible for creating and implementing uniform PM standards and guidelines: e.g., manuals, procedures, formats, etc. A practical example (*Planning & Monitoring guideline*) is given in Sect. 6.1. The standards and guidelines should be valid for the entire company, including the PM for small projects in the functional departments.

(10) PM training. Taking into account the standards that have been created, the PMO must regularly conduct appropriate PM training for PM employees and also for department employees who are responsible for work packages.

The consistent introduction of a PMO organization parallel to the line results from the change in circumstances. New technologies and their networking, often in connection with complex IT requirements, require strong management (*PMO*) including system work. This brings changes and often leads to conflicts between the project and line management, because it is about the distribution of tasks, powers and the claim to leadership. However, a balanced compromise must be reached; see Fig. 5.10. Heintel and Kranz confirm "*Project management brings hardly any advantages if its relationship to the hierarchy is not clarified, is constantly redefined so that it constantly wears itself out or if too much energy is tied up in it. One simply has to take note that organizations resist change, and the introduction of project management is undoubtedly a change*" [35]. Another important aspect is the systems engineering approach mentioned earlier, which should be firmly embedded in the PMO. According to Meier, "*In the development of IT systems, in addition to the actual system development process (project management), systems thinking is an elementary and indispensable component on the path to successful project implementation*" [36].

5.2.6 Project Steering Committee

The Project Steering Committee is an important control body of the company for the overall supervision of project activities in the company. The steering committee is the superordinate project authority in the company. The committee must have strategic decision-making authority, e.g., to decide on project partnerships and the regulation of priorities for projects running in parallel. The staffing of the steering committee should consist of the following individuals: the head of the company or of a business unit as well as representatives of the specialist departments; e.g., Marketing, Development, Production, Controlling & Finance and Human Resources as well as the PMO manager.

The steering committee provides the link between the company organization and the project organization. To create a functioning permanent link between the project steering committee and the PMO, a regular exchange of information is required as described in Sect. 11.2. The information exchange is based on the preparation of regular status reports from the project managers to the head of PMO for forwarding to the project steering committee; see subsect. 11.2.4. The information loop is then closed in regular status meetings and/or reviews as described in Sect. 11.3; see Fig. 11.2.

5.3 Project Management in the Context of Industrial Cooperation

5.3.1 Need for Community Projects

More and more projects are being carried out within the framework of industrial cooperation. The reasons for this are certainly diverse, but can essentially be summarized as follows:

(1) Project Size:
The volume of orders exceeds the capacity of a single company (*personnel, machinery, equipment, etc.*).

(2) Project Complexity:
Due to their enormous complexity, modern projects often require interdisciplinary work teams (*construction specialists, mechanical engineers; electrical and electronic specialists, IT specialists, etc.*), which in many cases a single company may not have. Lack of practical experience may be another problem.

(3) Project Funding:
Many major projects, e.g., turnkey construction projects (*Turn-Key Contracts—TKC*), must be fully or partially financed by the manufacturer (*general contractor*). However, a single company is often unable to provide the necessary sureties and guarantees on its own.

(4) Project Risk:
Project execution is often associated with a high level of risk, which can only be borne by a few companies alone in the case of major projects.

(5) Conditions of the purchaser:
In certain cases, e.g., in international projects, it may happen that the customer requires the participation of certain companies from the customer's country (*local content*). Here are two examples:
- Foreign clients often stipulate the involvement of domestic companies for certain reasons, e.g., assembly work to be carried out in the country.
- In international government projects, proportional representation is very much in the foreground. The international customers then expect the main contractor to involve companies from other contributing countries as partners and/or subcontractors in the project in order to guarantee a corresponding return on investment.

The situations listed here are very often the reason for joint projects. That is, companies that want to participate in a project and cannot do so for the reasons listed above are forced to join forces with one or more appropriate partners for project collaboration.

5.3.2 Project Management in the Context of Cooperation

The management and organizational principles for handling a project within the framework of cooperation are basically the same as for an in-house project because cooperative projects must also be managed cost-effectively according to the usual business management principles. In most cases, competition also forces cooperative projects to find cost-effective solutions.

International governmental projects must also follow this principle, although it must be acknowledged that the implementation can lead to some difficulties and also to additional costs due to cultural barriers. The European Space Agency (*ESA*) with currently 23 member States is a good example of how development projects can be realized in a multinational environment at low cost and in many cases under fixed-price conditions. This proves that through good management and clear organizational structures, international joint undertakings can be carried out cost-effectively. In this context, reference can also be made to international projects. In international projects, the aspects of "*work-sharing,*" i.e., the division of work packages, and "*industrial organization*" are the focus of cooperation negotiations.

The agreed division of tasks is an important indicator for the creation of an effective industrial organization. In practice, however, the division of a project into subprojects and work packages (*task sharing*) often leads to disgruntlement and management crises because the project partners are initially unable to agree on the division of labor and also on a joint and company-neutral project office (*system project management*). The aspect of "*noble vsnon-noble-work*" can also play an important role in the division of tasks. This involves, for example, access to key technologies and leadership claims. Questions of joint project management and the location of the joint office are also often contentious. In principle, the solutions shown in Fig. 5.13 are possible structures for cooperation.

- In the first case, the partner companies establish a consortium with a joint *project management office (PMO)* to manage the project tasks. However, this does not clarify who is the customer's contractual partner, a joint venture (*JV*), consortium or management company; see Sect. 5.3.3.
- In the second case, a company with system experience takes over the general contractor role (*turn-key contractor—TKC*) vis-a-viz the customer and involves the partners as subcontractors.

Two organizational solutions are shown in Fig. 5.13. To avoid misunderstandings, it is pointed out that the project office for complex international projects is usually referred to as Project Management Office (*PMO*). In the first case, the participating companies must assign appropriate personnel for the joint system project management according to the proportional representation in the project, and the tasks and competences of the project management must be defined. These employees must look after the interests of the system project management for the time of their secondment. This is an important principle, otherwise the system goal is jeopardized. However, this often leads to personnel conflicts, since the seconded

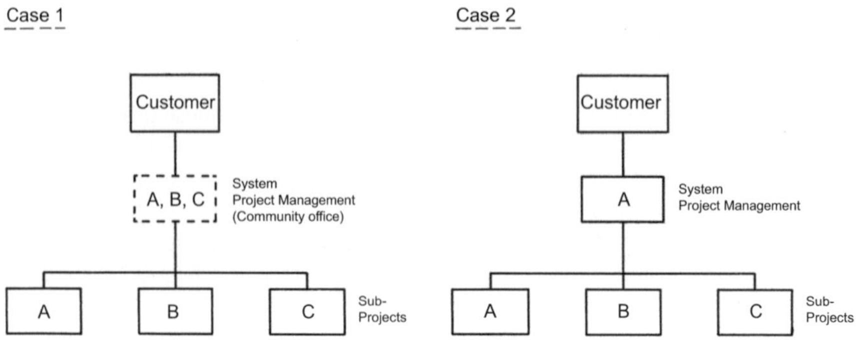

Case 1 — WORK DISTRIBUTION :

Company	Sub-Project Tasks	System Project Management
A	x	X (proportional)
B	x	X (proportional)
C	x	X (proportional)

Case 2 — WORK DISTRIBUTION :

Company	Sub-Project Tasks	System Project Management
A	x	x
B	x	
C	x	

Fig. 5.13 Forms of industrial project collaboration

employees return to the parent company after completing their task and may not have acted in the fullest interests of their own company with the decisions made at the system level. To avoid these problems, the partners should make an arrangement before work begins that gives the employees seconded to system project management a neutral status and protects them from later problems. The second case is simpler from an organizational point of view, since one company provides system project management for all partners. However, there is the problem of taking an advantage, because the lead company can gain competitive advantage through the leadership role. In both cases, it is advisable to set up a higher-level Board of Management *(BOM)* to represent the interests of all partners and monitor compliance. However, it is essential to avoid the BOM interfering in the day-to-day routine of system project management, e.g., conducting direct negotiations with the customer without the involvement of the project manager.

A common mistake should be avoided at all costs, namely the half-hearted establishment of a system level PMO. In many cases, the so-called system level PMO may exist only virtually. However, this is not acceptable for complex system projects, despite modern communication, because the members of the system project management remain at their workplace and only come together for formal meetings. As a result, team building suffers. The system-level PMO must look after the interests of the system (*overall project*) and should therefore, if possible, be housed in a separate office complex and separate from the main offices. Housing the system-level PMO in a closed office complex at one of the parent companies is the second-best solution. Section 5.3.3 below discusses this aspect in detail.

5.3.3 Proven Cooperation Models

In the past, public authorities and companies have tried out various cooperation models, from consortiums to General Contractor (GC) solutions, and have been able to gather relevant experience. What worked well in one case was a failure in another. Depending on the situation, all cooperation models can be used successfully, but the *pros* and *cons* must be examined carefully beforehand. Some well-known cooperation models are described below.

ARGE

The joint work cooperation model (*in Germany referred to as Arbeitsgemeinschaft—ARGE*) is a relatively loose association of companies for joint project implementation. In principle, the ARGE is a community of interests without legally binding character in the sense of civil law. This means that the ARGE partners enter into individual agreements (*contracts*) with the client (*customer*). This means that the joint project office (*system project management*) has no actual right of instruction vis-à-vis the partner companies, since the contractual powers do not lie with the system project management. The system project management is therefore largely dependent on voluntary information from the partner companies. The consortium lacks important control options, e.g., cost control. Therefore, regardless of the original motivation, the system project management is often degraded to an administrative enforcement body, since it has little influence on system technical aspects. Egotism is usually stronger than the jointly supported thought (*system design*). However, the system idea presupposes that an optimal system is created and that the companies involved are subordinate to this requirement. Many examples from practice testify to this.

Conclusion *The ARGE cooperation model is a viable option for handling projects that are relatively simple from a technological point of view,* e.g., *large construction sites without too many interface problems. What matters here is the motivation of the companies involved to cooperate. However, for complex high-tech projects where strong system project management and engineering is important, other forms of cooperation should be sought. The success of cooperation within the framework of a consortium depends not only on the agreed rules but also very much on the willingness of the partners to work together constructively.*

Joint Venture (JV)

When companies want to realize a project together, the term joint venture is frequently used. Similar to the ARGE or the open consortium, a joint venture (JV) is a relatively loose form of cooperation without a central system project management. A joint venture is one in which several legally and economically independent companies are involved, but which have joint management responsibility and jointly bear the financial risk. The partner companies must make internal arrangements regarding JV management and risk assumption, similar to a consortium.

Conclusion *If there is a common interest in the project and the intention to share the entrepreneurial risk, the formation of a JV can make sense. However, the mutual interests, goals and economic interests should be contractually regulated. The success of cooperation within the framework of a joint venture depends not only on the agreed rules but also very much on the willingness of the partners to work together constructively.*

Consortium In many cases, the handling of cooperative projects is aimed at system development, and this requires strong and efficient system project management and the joint liability of all project partners. Project handling within the framework of a consortium lends itself to this, since the consortium can be regarded as a civil law partnership and, almost without exception, can be described in the legal form of a company. According to Delorme and Hoessrich, "*The prerequisite for the emergence of a civil law partnership is the conclusion of a partnership agreement, also known as a consortium agreement or consortium agreements in the case of a consortium*" [37].

According to VDI 1991 (*Society of German Engineers*), a distinction can be made between an open and closed consortium [38]. Figure 5.14 illustrates both variants. In the open consortium, the General Contractor (*GC*) is in charge of the overall system project management and has a direct contract with the Customer and a direct contractual relationship with all subcontractors. Another solution is the closed consortium, in which the consortium companies appoint a prime contractor from among themselves to lead the consortium and represent it externally; i.e., to the customer, as a mouthpiece.

The decision as to which company takes on the leading role is not always easy, because the system leader, has competitive advantages. If the partners make a long-term commitment through a consortium agreement, there is the possibility of a rotating leader role. In addition, a Board of Management *(BOM*), in which the managing directors of the partner companies are equally represented, can support the

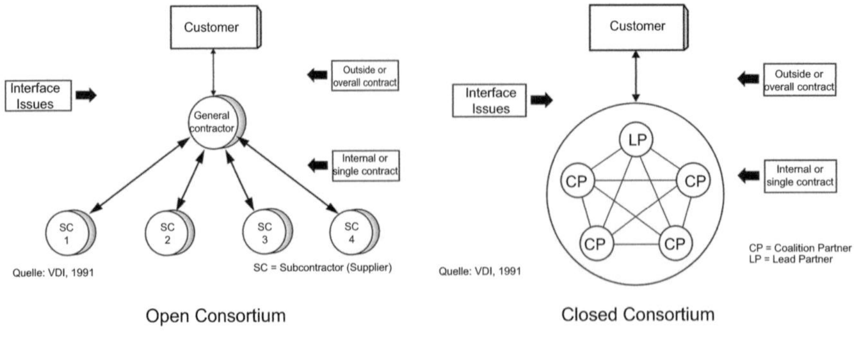

Fig. 5.14 Example of an open and closed consortium

chosen system leader in the fulfillment of its role and also monitor his performance at the same time. Figure 5.15 shows the consortium structure, including the customer relationship, in a practical example.

The chosen leading company "A" (*Prime Contractor*) acts on behalf of the consortium and is the contractual partner vis-a-vise the customer and at the same time also the contractual partner of the subcontractors (*"B to F"*). The prime contractor (*the "Prime"*) is entrusted with the overall system project management. He has the task of managing the day-to-day business with all rights and obligations. Depending on the urgency, monthly to quarterly BOM meetings (*consortium meetings*) are held, at which the manager of the Prime reports on the respective project status. It should be avoided at all costs that the supervisory body interferes in the day-to-day business of the Prime in the form of a bypass. In summary, there are three types of contracts:

I. Main contractor contract (Customer/Prime)
II. Subcontractor contracts (Prime/subcontractor)
III. Consortium agreement (all consortium partners)

Contracts I and II are normal contracts for work and services in which the supplies and services are anchored; see Chap. 13. Consortium transactions usually also require the conclusion of a consortium agreement; Steinrücke defines the content of the consortium agreement as follows [39]:

(1) Purpose of the consortium formation
(2) Names of the members
(3) Consortium shares (*quotas*) of the members

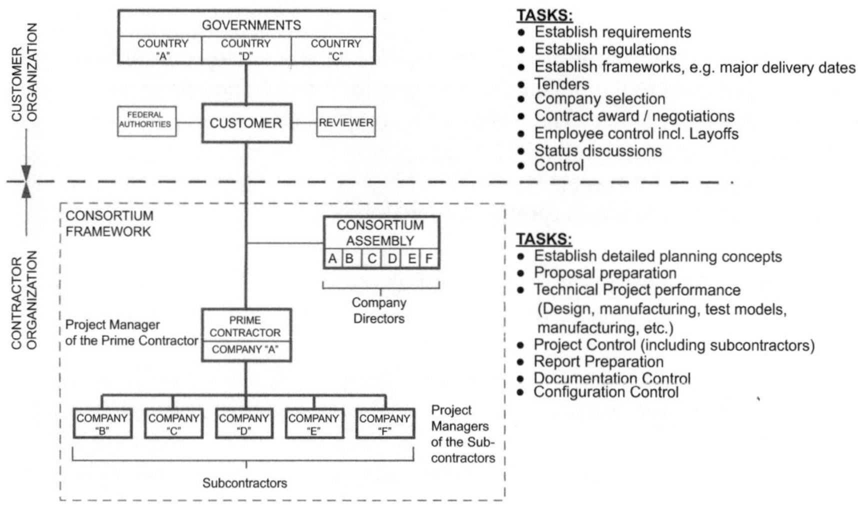

Fig. 5.15 Management structure of a consortium

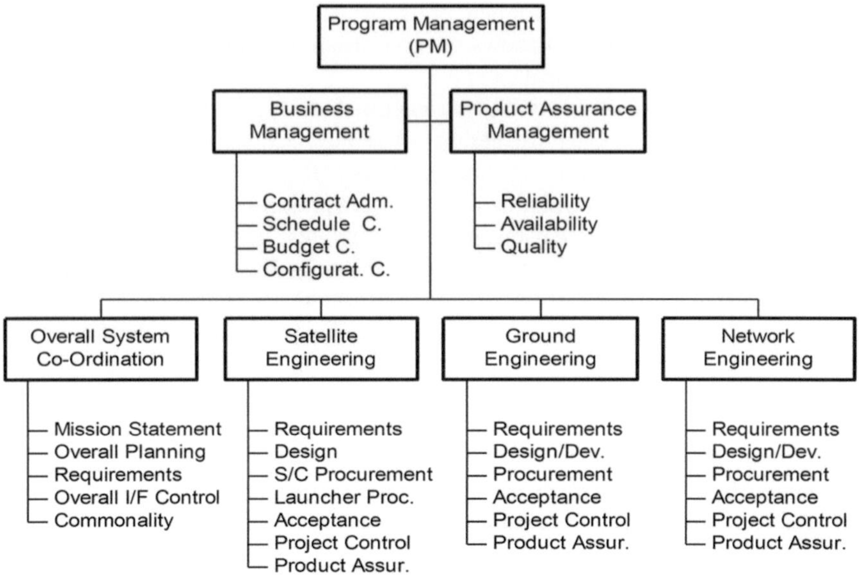

Fig. 5.16 Project organization of a system operating company

(4) Ownership
(5) Representation of the consortium members
(6) Management of the consortium*
(7) Duties and rights of the consortium members (*in particular the contingent liabilities and profit sharing*)
(8) Termination of the Consortium.

**The Manager of the consortium is not synonymous with the project manager.*

The project organization in the internal relationship of the consortium corresponds in principle to the organizational form for a technology project of which a model is shown in Fig. 5.4. The mixed team composition with employees from different partner companies, often also from different countries, can lead to problems. Figure 5.16 shows a practical example of a project organization for an operating company with system responsibility.

Conclusion *The closed consortium solution represents a viable compromise for implementing a joint project. In many cases, however, this compromise is the only way to carry out a complex technology project with major financial risks, which would often not be possible on a stand-alone basis. Sometimes it is also a requirement of the customer that companies from partner countries must participate in order to maintain the participation proportion. The success of cooperation within a consortium depends not only on the agreed rules but also very much on the willingness of the partners to work together constructively.*

Management Company For the implementation of complex projects (*system contracts*), another form of cooperation is possible, namely the establishment of a joint company, a so-called management company. The management company acts as the general contractor (*GC*) vis-à-vis the customer (*client*) and becomes its contractual partner. On the other hand, contracts must be concluded with subcontractors who are also owners of the management company. There is no question that this solution, quite similar to the consortium, is also a compromise. As a rule, the subdivision of the contract is carried out exactly as in the case of a closed consortium, according to the participation quotas. As shown in Fig. 16.1, a Board of Management *(BOM*), in which the managing directors of the parent companies are equally represented, takes over the supervision of the company. The project management (*representative vis-à-vis the customer*) does not have to be identical with the management of the management company.

Management companies are often set up for risk sharing, strategic reasons and have often only a thin capital base. Guarantees, as required by some customers, may therefore have to be provided by the founding companies. The project organization in the internal relationship of the management company corresponds in principle to the organizational form shown as a model in Fig. 5.4. The team composition with employees from different partner companies, often also from different countries, can lead to problems just as in a consortium.

Conclusion *The establishment of a management company represents a workable compromise for the implementation of a joint project. In many cases, however, this compromise is a good approach to jointly implementing a complex technology project with major financial risks. The reasons for a management company are basically the same as those for a closed consortium. The success of cooperation within a consortium depends not only on the agreed rules but also very much on the willingness of the partners to work together constructively.*

Integrated Project Team (IPT)

The integrated project team, as shown in Fig. 5.17, is being used more and more in the context of collaborations, such as a consortium or a management company, both nationally and internationally, and it provides a good opportunity for companies to work closely together. Of course, legal aspects and the problem of confidentiality must be handled delicately. Cultural aspects also need to be considered in advance, such as political, religious and language differences. It is important that the future IPT employees are prepared and trained for their area of responsibility. A checklist of important IPT aspects is shown below.

I. Need for Integrated Project Team (*IPT*)
   - Today many projects are joint ventures
   - This requires intelligent management concepts
   - Organization of partnerships and teams is important

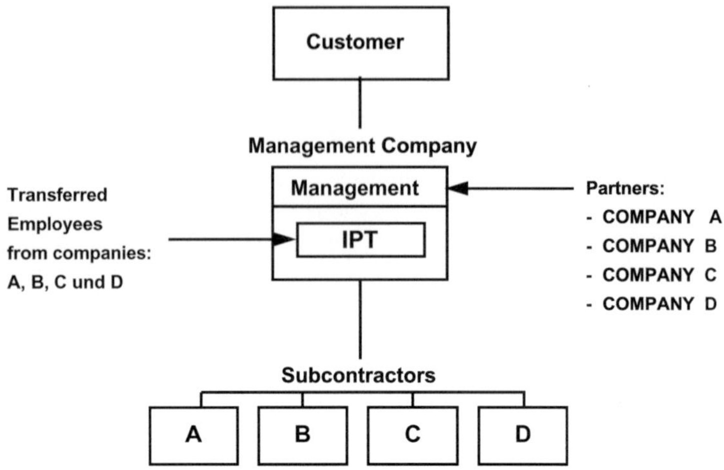

Fig. 5.17 IPT—embedding in a management company

- Project team members have to co-operate effectively
- Good communication between the teams is mandatory
- It is necessary that all partners form an Integrated Project Team (*IPT*)
- Partner companies must also implement a PMO
- IPT staffing with team members as shown in Fig. 5.3
- Required Leadership of IPT

II. IPT Requirements to be considered
- Definition of common project objectives
- Implementation of IPT organization
- Requirements of IPT culture
- IPT cooperation agreements
- Legal implications for IPT
- Recruitment of IPT staff
- IPT office location
- Co-located project offices
- Communication requirements
- Language to be applied

III. Problems with IPT implementation
- General Requirements
  Appointment of project manager
  Selection of team members
  Authority of PM & his team
  Identification of support teams
  Legal applications
- IPT location, respectively co-location
- Staffing requirements of IPT staff
- Delegation of PM from parent companies

- Language to be applied
- Law and rules to be applied

Figure 5.17 shows how the Integrated Project Team (*IPT*) is embedded in the management company. The company management is subordinate to the partner companies, which in turn assign employees to the IPT.

Conclusion *The use of integrated project teams (IPT) is constantly increasing due to an increasing number of collaborations. If the aforementioned aspects (Need, Requirements, Problems) are taken into account when collaborating with employees from other companies, countries and cultures, and if rules are created that apply equally to all participants, management tasks can be carried out efficiently by an IPT. Employees must approach each other in order to learn from each other. The quality of cooperation in an IPT depends on the willingness of employees to work together constructively. The IPT is supervised by a supervisory body (BOM). It is important that the BOM consists of experts and is not too large. "The optimum committee has no members" says Augustine* [40]*, it probably means it should be small.*

Prime/Turnkey Contractor (TKC)

Without question, the prime contractor concept, (Fig. 5.13 Case 2), offers the best solution for strong and competent project management; see Figs. 5.3 and 5.4. The prime contractor bears full responsibility and entrepreneurial risk for the project and is therefore dependent on competent and powerful project management. In contrast to project management under a cooperative arrangement, subcontractors can be selected according to competitive rules. As a result, subcontractors are selected to a much greater extent on the basis of free market criteria: best technology and/or lowest costs; see Chap. 14.

In some cases, customers may require a turnkey solution. In such case the prime contractor, respectively the TKC, hands over the fully completed and tested project (system), including all required hard- and software deliverables, training and services to the customer (*buyer*). Title will be handed over from TKC to Buyer.

Conclusion *The TKC solution is an attractive business model for customers without much management expertise: see Fig. 5.13, case 2. After handover of the project (title) to the customer (buyer) ownership, including all rights and obligations, are transferred from the TKC to the customer.*

The author of this book issued a TKC checklist as part of a bidder's guideline for a Metro project in South America, which is shown below.

a) Is the organization and cooperation between the TKC, subcontractors and suppliers clearly described?
b) Has there been a clear and unambiguous description of how the TKC will control and monitor all project areas?

c) Has there been a clear and unambiguous description of how the *JV will* control and monitor all project areas?
d) Have work authorization been clearly defined at all project levels?
e) Do the contractual arrangements between the TKC, subcontractors, and suppliers allow the TKC to have the right to monitor status at all levels without restriction?
f) Are the payment modalities to subcontractors and suppliers and the release by the TKC clearly regulated according to a milestone plan?
g) Has it been clearly specified that the overall development responsibility (*system design responsibility*) lies within the responsibility of the TKC?
h) Is there an overview list of all companies, TKC, subcontractors and suppliers involved in the project, including second source suppliers and consulting firms?
i) Has the TKC created and approved an internal project organization?
j) Have the individual management functions been clearly and unambiguously defined by the TKC?
k) Has the TKC established a central PMO and clearly defined the tasks, responsibilities, and authorities?
l) Has the TKC identified the key PMO personnel required?
m) Do the assigned PMO team members have an appropriate qualification?
n) Have the TKC and subcontractors assigned the required key personnel?
o) Can the proposed key personnel demonstrate the required qualifications?
p) Is full availability of key personnel ensured?
q) Have all necessary communications media between the parties involved been defined by the TKC?
r) Has the TKC made sure that all subcontractors also have an effective PM and contact person planned?
s) Can the TKC provide appropriate office space and communications facilities?
t) Has the TKC provided proven PM procedures?
u) Do the subcontractors also have proven PM procedures available, and are they compatible with the JV procedures?

5.3.4 Multi-Level Project Organization

If several institutions or companies are involved in the execution of a project, this leads to the introduction of a multi-level project organization. The realization of large-scale projects often requires the participation of many institutions and industrial companies, which are related to each other in a multi-level hierarchical relationship according to their capability, competence, workload, and responsibility. The structure of the project organization depends to a large extent on the division of work in connection with the work breakdown structure (*WBS*). The WBS, whose basic structure should largely follow the system breakdown, defines the individual hierarchy levels: Program, Project, System, Subsystem, etc. until the work package level is reached; see Sect. 9.3. The project organization of the companies involved should be as identical as possible, i.e., arranged as a mirror-image.

Because of its compelling logic, the Apollo program is a good example to explain this. Management responsibility for the upper levels of the program, e.g., launch vehicle, lunar lander, etc., was the responsibility of NASA (*the contracting agency*) and the responsibility for the individual systems, subsystems, etc., was with the space industry; the overall project organization for the Apollo program is illustrated in Fig. 5.18 [41]. All lines from the three project offices in Houston, Texas (*lunar flight vehicle and mission control*), Huntsville, Alabama (*Saturn V launch vehicle*) and the Kennedy Space Center, Florida (*launch and ground facilities*) ran centrally to the Apollo program management at NASA headquarters in Washington D. C. The internal project organization of program management in Washington, D. C., was staffed under the leadership of the Program Manager (*PM*) with the following key positions:

PC Program Control (*schedule, cost, configuration, etc.*)
PA Product Assurance (*quality, reliability, safety, etc.*)
SE System Engineering (*analyses, specification, design, etc.*)
AIV Assembly, Integration & Verification (*facilities, procedures, tests, etc.*)
OPS Operation (*space and ground*)

All organizational entities at NASA and the industry shown in Fig. 5.18 introduced a mirror-image project organization with identical management functions in order to improve the communication between counterparts.

The concept of mirror-image organizational structures created contact persons at all project levels who, in principle, had similar tasks, responsibilities, and authorities. The SE Manager of the NASA program management office, for example, had direct contacts at the SE managers of the three NASA centers and the industrial

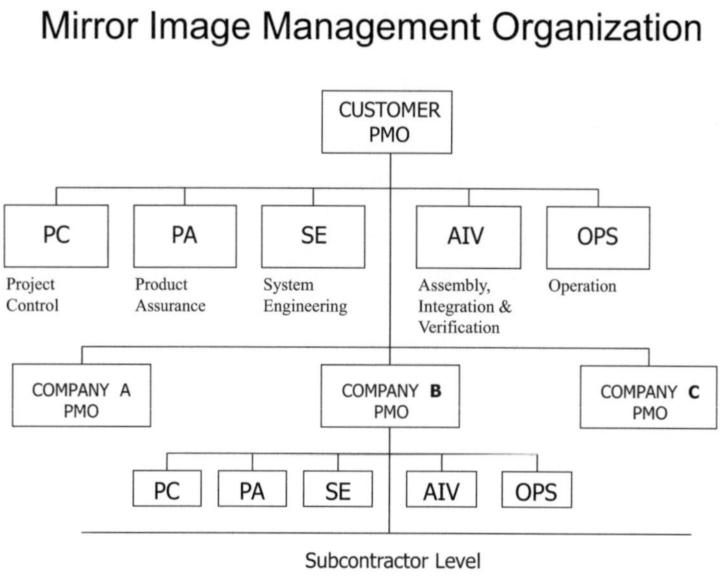

Fig. 5.18 Program/project organization of the Apollo program

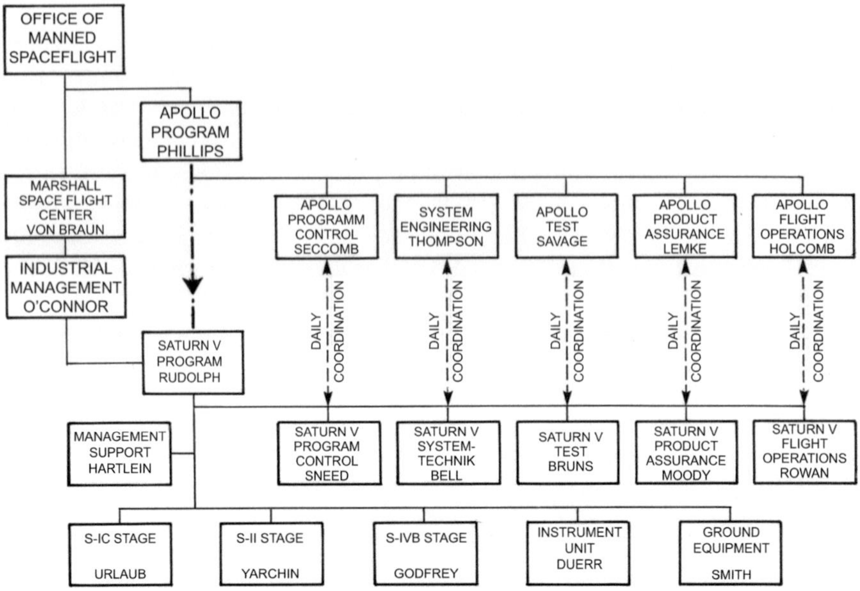

Fig. 5.19 Mirror-Image Management Organization of the Apollo Program

companies, who carried out their functions according to identical guidelines and regulations.

Parallel to the official contacts between the individual project offices, which were carried out by the respective project managers at the various project levels, there was a daily informal exchange of information between the functionaries (*Project Control, Systems Engineering, etc.)*; see Fig. 5.19 [42]. "*The Mirror Image Concept*" was introduced by NASA's Director of Human Space Flight George E. Mueller [42]. The concept of mirror-image organizational structures, where the customer and the contracted companies align their organizational structure accordingly. The principle of mirror-image organization is a logical approach and can serve as a model for multi-level project organizational structures. In his role as proposal manager at MBB space division, the author has in the past at several times successfully applied the Mirror-Image Concept for several European space projects. Subcontractors from different countries had been convinced to follow the prime contractor's guideline for a mirror-image project organization.

5.3.5 Multinational Cooperations

International cooperation projects are the order of the day at the current time and are constantly growing in importance. In many cases, it is an industrial cooperation in which contracts are concluded across national borders. The relationship of the external project organization (*industrial organization*) is the same as for any national cooperation. However, the respective national legal standards must be taken into account as a basis when drawing up contracts; see Chap. 16.

In principle, international collaborations are handled in exactly the same way as national collaborations. Well-proven collaboration models should be used; see Sect. 5.3.3. However, the principles described there should be applied even more rigorously, because two important factors are added: the language barrier and the cultural background. It is therefore essential that employees prepare themselves thoroughly before starting an international project. Companies are well advised to offer appropriate training courses that also teach behavioral patterns; see Chaps. 16 and 20.

HELIOS Solar Probe However, there are exceptions where, on the customer side, the establishment of a joint project office is for some reasons neither desired, possible nor necessary. In the implementation of the HELIOS solar probe program for example, the German and US governments agreed on a sensible division of tasks that made it possible to dispense with a joint system project office. The decisive factor was that

- the division of tasks (*Germany-USA*) had no system-critical interfaces and
- German and American engineers consulted each other regularly as part of a firmly established joint working group.

Under this premise, efficient customer cooperation was possible without creating a joint project office. The division of tasks basically provided for two independent systems: on the German side the development of a highly complex satellite system (*the probe*) and on the US side the provision of a launch vehicle, the launch facilities and the performance of the launch. The satellite/launcher interface, which involved the exchange of technical documentation as well as joint testing, was clearly defined and relatively easy to manage. The creation of the Helios Joint Working Group *(HJWG)* as a permanent management forum was a mainstay of German-US cooperation [43]. Note: Two HELIOS satellites, of which a flight-ready model is exhibited in the Deutsches Museum in Munich, orbited the Sun at a perihelion of 0.3 astronomical units (*AU*); 1 AU corresponds to the mean distance between the Earth and the Sun. Figure 5.20 shows the concept of the HELIOS collaboration.

SPACELAB Space Laboratory The description of this form of cooperation, which, by the way, was chosen for the cooperation between ESA and NASA for the Spacelab/Space Shuttle interface, shows a feasible way for cooperation between individual nations that perform tasks in parallel to each other that lead to a common goal. The prerequisite, however, as already mentioned, is a precise and controllable interface definition; see Fig. 5.21 [44].

MRCA-Tornado The MRCA-Tornado military aircraft project was realized between the three countries Germany, England, and Italy. Program management was entrusted to Panavia, a Munich-based management company, founded by the three partner countries. "... *Panavia was an interesting management organization:*

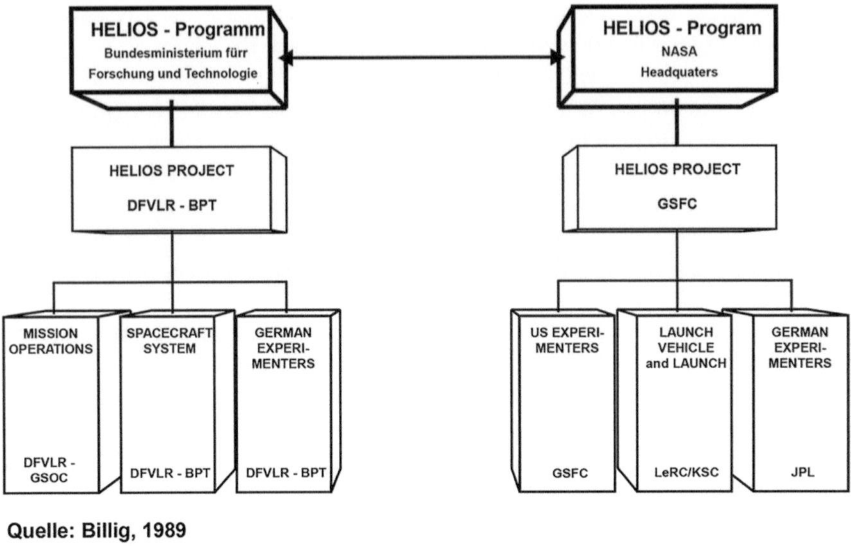

Fig. 5.20 HELIOS Program Cooperation between Germany (GFW) and USA (NASA)

a limited liability company under German law with the usual English organizational form of a board of directors consisting of two representatives of each of the three shareholders …" [45]. The MRCA program organization is shown in Fig. 5.22.

International Space Station (*ISS*) The aforementioned cooperation model was also successfully applied to the ISS project. "*The International Space Station (ISS) Program's greatest accomplishment is as much a human achievement as it is a technological one - how best to plan, coordinate, and monitor the varied activities of the Program's many organizations. An international partnership of space agencies provides and operates the elements of the ISS. The principals are the space agencies of the United States, Russia, Europe (ESA), Japan, and Canada. The ISS has been the most politically complex space exploration program ever undertaken*" [46]. The ISS program is not only a technical success story, but also a political one. The participating space agencies agreed on overarching system specifications and interface definitions between elements that remained within national jurisdictions. The elements shown in Fig. 5.23 were separately orbited and then integrated in space. The following five partner countries/organizations worked in parallel on the joint project:

- USA—NASA
- Russia—ROSCOSMOS
- Canada—CSA
- Japan—JAXA
- Europe—ESA—Member Countries

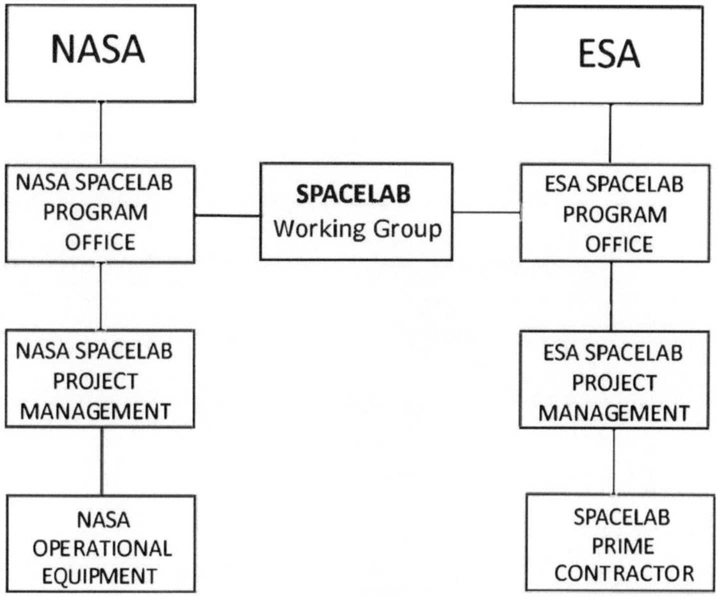

Fig. 5.21 SPACELAB Cooperation between ESA and NASA

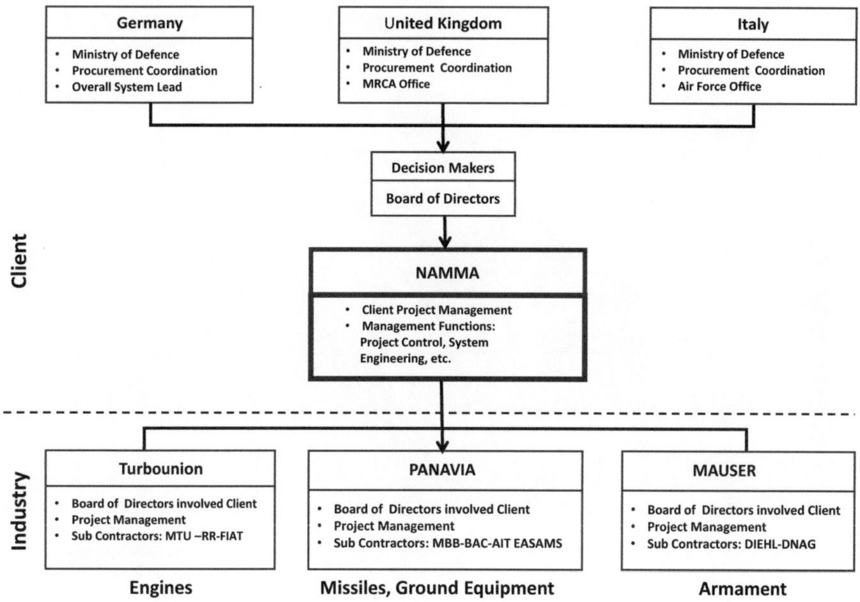

Fig. 5.22 Cooperation concept MRCA-Tornado

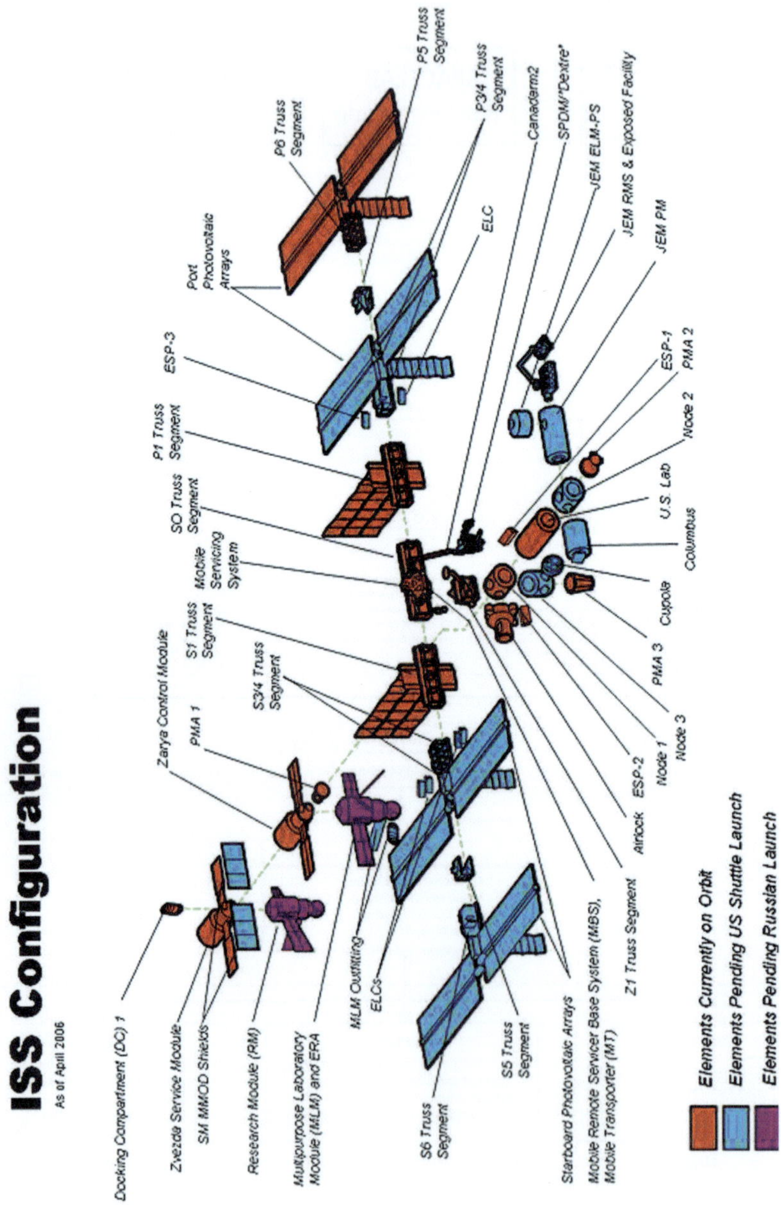

Fig. 5.23 Configuration of the International Space Station [Creative Commons]

- Belgium
- Denmark
- France
- Germany
- Italy
- Netherlands
- Norway
- Spain
- Sweden
- Switzerland
- United Kingdom

International Launch Service (*ILS*) Another form of international cooperation is the management company ILS; see Sect. 5.3.3. In the 90s, the companies Lockheed Martin (*USA*) together with the Russian companies Khrunichev and Energia founded the management company LKE (*changed to International Launch Service—ILS)* to market the Russian launch vehicle PROTON. As Business Manager of SES, Luxembourg, the author had 1993 a leading role in the contract negotiations between SES and LKE.

Public-Private-Partnership (*PPP*) Reutzel and Rullmann point out in a book that the topic of public-private partnerships, also known as "*Public-Private-Partnership*" (*PPP*), is in Germany a well-known and practiced organizational approach, whereby the state uses private third parties to fulfill its tasks. Reutzel writes in this regard "*as early as 1490, Franz von Taxis took over the organization of the postal service in the domain of the House of Habsburg*" [47]. PPP projects can be divided into two main categories:

- Institutionalized PPP and
- Project or contract PPP

Reutzel and Pullmann also describe the difference between the two main categories in such a way that the institutionalized PPP, unlike the contract PPP, cannot decide on exchange agreements of services. Part 2 of the book contains details of the project-based PPP [48]. A contract is to be concluded with a partner which has the task and financing as its object. Various contract models are described in the book.

Wolff adds that PPP concepts have currently become the popular way for public institutions to provide services worldwide and points out differences in PPP activities worldwide in this context. According to Wolff, the Anglo-Saxon countries rely most heavily on PPP models and are also following this trend in continental Europe. In Germany, mainly the building and civil engineering sectors apply PPP models. Under Chancellor Schröder, the goal was for at least 15% of all public investments to be PPP contracts [49].

In the multinational PM environment, the topic of public-private partnerships (PPP) has made inroads. This involves contractually regulated cooperation between the public sector, e.g., the space agencies DLR or ESA, and European space companies pursuing a common goal in a partnership of convenience. The PPP objective may, for example, be the implementation of a joint project involving a public authority representing the interests of the public and a private-sector industrial partner. The contribution of the public partner is, for example, the sovereign role as a governmental body to prepare and supervise a project proposal, while the industrial partner contributes its technological know-how to the joint project. Kong defines PPP as follows: "*PPP is a generic term for the relationships formed between the private sector and public bodies often with the aim of introducing private sector resources and/or expertise in order to help provide and deliver public sector assets and services*" [50].

The EU Commission first issued "*Guidelines for Successful PPP Projects*" in March 2003, which were intended to help the project manager reconcile public and private sector objectives. The European Commission distinguishes between contract-based PPP and institutionalized PPP. In contractual PPP projects, the cooperation relationship between the public and private partners is purely contractual, while in institutionalized PPP there is a business start-up financed by mixed capital from public and private investors [51].

Literature

1. Steinmann, Horst and Schreyögg, Georg, Management: Gabler Verlag, Wiesbaden, 4th ed. 1999, p. 407.
2. Dale, Ernest: Management, Econ-Verlag, 1972, 5. 147.
3. Kosiol, Frich: Aufbauorganisation, in: Handwörterbuch der Organisation. Ed. by E. Grochla, C. E. Poeschel Verlag, Stuttgart, 2nd edition, 1980, p. 180.
4. Madauss, Bernd-J.: Principles of Project Organization in Research and Development Projects in MBB Report UR-476/81 0,1981.
5. Mellerowicz, Konrad: Betriebswirtschaftslehre der Industrie, Rudolf Haufe Verlag, Freiburg, 1981, 5. 23.
6. McKinley, John K.: Durch Verschmelzung von kaufmännischem und technischen Wissen zu mehr Innovation, in: Handelsblatt, 21. 9. 1982.
7. Madauss, Bernd-J.: Planung und Überwachung von FuF-Projekten, in: AIB-Fachliteratur, Bad Aibling, 1978/81, 5. 11-8.
8. Koelle, Heinz Hermann: Projekt-Management, Lecture Manuscript, TU-Berlin.
9. Hobbs, Peter: Professionelles Projektmanagement (German edition) mvg-verlag im verlag moderne industrie, Landsberg am Lech, 2000, p. 70ff.
10. Madauss, Bernd-J.: State of the Art Project Management, lecture manuscript (not published), PSU/EAP Los Angeles/Markt Endorf, 1981.
11. Martin, Charles C.: Project Management, AMACOM, 1976, 5. 44.
12. Kraus, Georg: Projektmanagement mit System, Gabler Verlag, Wiesbaden, 1995, p. 35f
13. Vollrath, Klaus: Projektmanagement, in: Management + Seminar, 3/1981.
14. Lohoff, Günter and Lohoff, Petra: Von der Idee zur Implementierung, in Claus Steinle, Heike Bruch, Dieter Lawa, Projektmanagement, Edition Blickbuch Wirtschaft, FAZ, 1995, p.74.

15. Madauss, Bernd-J.: Result Report of the First Practical Seminar - Planning and Monitoring of R&D Projects, MBB Education Program (RW11), 12. 6. 1978.
16. Gehriger, Helmut: Planning and Control for European Space Satellite Projects, ESRO-Document SP-70, July 1972.
17. General Electric, BMBW Europe III Management Study, GE Contract 1P6, September 21, 1970, 5. 1-29.
18. Source 2, p. 279.
19. Terry, George R.: Principles of Management, Richard D. Irwin, Inc. Homewood, Illinois, 1977, pp. 268-274.
20. Bölkow, Ludwig: Der Zukunft verpflichtet, Herbig Verlagsbuchhandlung, Munich 2000, p. 234f.
21. Hegi, Othmar: Projekt-Management ein Fremdkörper in der Stab-Linien-Organisation, from: Projekt-Management, (systematische Grundlagen und Beispiele aus der Praxis), Vol. 1, Verlag Industrielle Organisation, Zurich, 1972, 5. 77.
22. Grochla, Erwin: Projektmanagement - Organisatorische Gestaltungsmöglichkeiten, ZIU Management Seminar; Zurich, November 1980, p. 4.
23. Burghardt, Manfred: Einführung in Projektmanagement, Siemens AG, Berlin und München, 2002, 4th edition, p. 55ff.
24. MBB, Minutes of Results of Executive Colloquium - Space Division, July 1977.
25. Grochla, Erwin and Thom, Norbert: Die Matrix-Organisation - Chancen und Risiken einer anspruchsvollen Strukturierungskonzeption, in: zfbf-Kontaktstudium 29 (1977), 5. 195.
26. McCarthy, Jr. John F.: Matrix Management for Aerospace 2000, NASA Technical Memorandum 81509, May 1980.
27. Thamhain, Hans J. and Wilemon, David L.: Leadership Effectiveness in Program Management, in: IEEE Transactions on Engineering Management, Vol. EM-24, No. 3, August 1977, 5. 102.
28. Litke, Hans-D.: Projektmanagement, Carl Hanser Verlag, 2004, 4th ed. S. 75.
29. Kerzner, Harold: Projekt Management (German translation), mitp-Verlag, Bonn, 2003, p. 87.
30. Hab, Gerhard, Wagner, Reinhard: Projektmanagement in der Automobilindustrie, Springer Gabler, 4th edition, 2013, p. 41.
31. Tjahjaa Lia, Dwyer, Paul and Habib, Mohsin: The Program Management Office Advantage, Amacom 2009, p. 4ff.
32. Source 30, P. 75.
33. USAF AFR 375-2, System Program Office (SPO), February 1962.
34. DIN 69901-5: Project Management Office (PMO), 2009.
35. Heintel, Peter, and Krainz, Ewald E.: Projektmanagement, 2. Aufl., Gabler, Wiesbaden, 1990, 29ff.
36. Meier, Markus: Projektmanagement, Schäffer Pöschel Verlag, Stuttgart, 2003, p. 3.
37. Delorme, Hermann and Hoessrich, Hans-Joachim, (von Möhring und Rittershausen): Konsortial- und Emissionsgeschäft, Fritz Knapp Verlag, Frankfurt a.M., 2nd edition, 1971, pp. 11, 12.
38. Madauss, Bernd: Organization and Staffing, ISU/ACT Training for Satellite Engineering and Management at Royal Army of Oman, 20010, p. 16 (not public).
39. Steinrücke, Bernhard (new edition of Scholze, Herbert: Das Konsortialgeschäft der deutschen Banken, Duncker & Humbolt, Berlin, 2nd edition, 1973, pp. 9-11.
40. Augustine, Norman R.: Augustine's Laws, American Institute of Aeronautics and Astronautics, Inc, Reston Virginia, 6th Edition, Law 31.
41. NASA: NASA Apollo Program Management, Volumes 1 through 5, Dec. 1967.
42. Bilstein, Roger E.: The Saturn Management Concept, NASA Report CR-129029, June 1, 1974.
43. Billig, Victor, Madauss, Bernd, Schneider, Kuno: Industrial Cooperation through Project Management, GPM Schriftenreihe, Tüv Verlag Reinland1989, p. 79.
44. Yardly, John F.: Role of Spacelab in the Space Transportation System, in: Journal of the British Interplanetary Society, Vol. 30, 5. 43 - 46, 1977.
45. Source 20, p. 265.

46. Wikipedia 2016 (ISS Kooperationsmodell)
47. Reutzel, Andre, Rullmann, Jörg: Public-Private Partnership, Kommunal- und Schul-Verlag Wiesbaden 2008, p. 16.
48. Source 47, p. 59.
49. Wollf, Sarah: Public-Private Partnerships in Germany, Springer Gabler, Wiesbaden 2014, p. 3ff.
50. Yong, Hee, Kong: Different Models of PPP, PPIAF, Asian Development Bank, p. 4.
51. EU guideline (Leitlinie) Wikipedia 2016

Project Management Planning

6

Contents

Abstract

It is a mistake to assume that good management comes about by itself or by a lucky coincidence. Management processes must be thoroughly planned and implemented.

The term *management planning* is rarely used and its meaning is often not understood. By contrast, word combinations, such as scheduling, cost planning, and personnel planning are familiar terms that everyone understands. In the Anglo-Saxon, and especially in the US-American management literature, however, the word *management planning* is an established term. There, the word *planning,* in keeping with its original meaning, is understood in a much more comprehensive and substantive way, e.g., also in the area of management and organization. It is quite naturally understood, that management processes must also be thoroughly prepared and planned.

B. -J. Madauss, *Project Management*,
https://doi.org/10.1007/978-3-662-69057-4_6

It is no exaggeration to claim that many management problems are causally related to a lack of preparation and planning. It is always a surprise to see how large projects are started without any clarity about the organization and management processes in detail at the beginning of the project. However, management is, in most cases, not a simple task, but a very complicated one, and this is particularly true for project management. There must be clarity right from the start about *who* has to do *what, why, when,* and *where.* Work plans, organizational plans, and flow charts provide the necessary clarity for all technical and administrative staff involved in the project. Carefully prepared management plans are the guiding principles of functioning project management and are of particular importance for large-scale and very complex projects.

6.1 Avoiding Ambiguity Through Management Planning

6.1.1 The Five W-Questions

Clarification of the previously mentioned W-questions is a prerequisite for successful project management. Each W-question must be clarified as early as possible for each area of the particular project. The management plans must therefore be drawn up at the same time as the planning of the phases; see Chap. 4. In this context, reference is also made to the synchronous specification process; see Chap. 7. For large projects, the planning of the project processes is of particular importance because it involves the hierarchical competence of all PM processes. It is important that all W-questions are answered in advance in order to largely avoid ambiguities and competence disputes from the outset. The necessity and importance of creating management plans usually become apparent very quickly, for example, when it comes to coordinating technical and planning interfaces.

6.1.2 Identification of the Required Management Plans

A systematic approach is important when planning PM tasks. It is therefore recommended that the PM team prepares an initial set of required management documents (*plans, procedures, specifications, etc.*) prior to the start of the project or at the beginning of a particular project phase. This process is also part of documentation management, which is described in detail in Chap. 12.

Figure 6.1 shows an example of the PM planning tree structure with various management plans for a large project starting with the *overall project plan,* in which the most important project goals and the summaries and tasks of the subordinate individual plans are described. At this point, it is clear that, depending on the size and complexity of the project, a decision must be made as to whether the various plans are to be prepared as individual documents or combined into one overall document with an appropriate chapter structure. For very small projects, a few pages are often sufficient for the entire management planning, while for large projects the

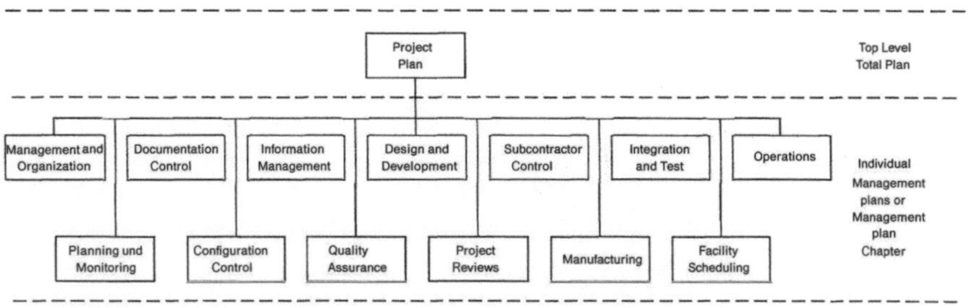

Note: The responsibility for the individual Management Plans is as follows:

Management Functions	Management Plans
• Project Manager	– Top Level Project Plan
	– Management & Organization
• Planning & Monitoring	– Planning & Monitoring
	– Documentation Control
	– Configuration Control
	– Information Management
• Quality Assurance	– Quality Assurance
• System Technology	– Design and Development
	– Project Reviews
• Subcontractor Control	– Subcontractor Control
• Manufacturing Control	– Manufacturing
• Integration & Test	– Integration & Test
	– Facility Scheduling
• Operations	– Operations

Fig. 6.1 A structural example of different management plans

preparation of individual plans is advantageous. The tree structure for project plans is an example of other trees required, such as the specification tree.

At the beginning of a project, it must be considered whether all or only some of the project plans mentioned in Fig. 6.1 are required for a specific project in order to adequately define the necessary project functions. The basic principle is that an independent management plan or a corresponding chapter must be drawn up for each clearly identifiable management function. It may well be the case that one organizational unit has simultaneous responsibility for the preparation of two or more project plans. Conversely, however, two organizational units should never be responsible for a common plan. If there is an overlap in responsibility, the corresponding plan should be divided between those two units. Here is an example: If the management functions of documentation and configuration control are in one hand, it may make sense to combine the documentation and configuration management plans into a single document; however, it is not mandatory. On the other hand, if documentation and configuration management are managed by two separate organizational units, two separate plans or chapters must also be prepared. Of course, the structure must be supplemented if necessary. All management plans must be recorded in a corresponding list; see Table 12.1.

6.1.3 Implementation of the Management Plans

The official implementation of the management plans is an important measure for the project because they prescribe the management process in a binding manner. The implementation must be coordinated with the main project schedule (see Chap. 9) and approved by the project manager. Once the management plans have been approved by the PM, the implementation process is initiated. Of course, the management plans, like all other project documents, are subject to change and must be adapted to the respective project situation.

6.2 Management Plan Contents and Evaluation Criteria

6.2.1 Management Plan Contents

When creating management plans, it is recommended that corresponding scenarios are created in advance in order to achieve a uniform structuring of the plans. Therefore, a procedure has been developed in which the contents of the documents to be created, in this case the management plans, are to be described precisely. This procedure is known as *Document/Data Requirements Description (DRD)*; see Sect. 12.1. In the context of the development work for the European launcher program *EUROPA-III* (*ARIANE precursor*), a management study was prepared by General Electric on behalf of the German Federal Ministry of Education and Science (*BMBW*), now BMFT, which contained, among other things, documentation requirements descriptions of the most important management plans [1]. The management study, prepared in 1970, can be considered a valuable document for designing a project management system. Based on this, the following structure of a typical overall project management plan is reproduced:

(1) Introduction
- Project goals
- Brief description of the entire project
- Brief description of the project phases
- Summary of the overall schedule (*main milestones*)
- Reference to the subordinate management plans (*planning tree*)
- Reference documents
- Introduction of the following sections, (2) to (9).

(2) Project management
- Company and project organization (*superordinate*)
- Project organization, responsibility, and functions
- Industrial holdings/subcontractors and suppliers
- Project directives and strategies
- Project reviews and reports

(3) Work plan
- Technical requirements

- Boundary conditions
- Required attachments
- System structure
- Subsystems
- Technical development work
- Work breakdown structure (*WBS*)
- Summary description of services
- Project flowchart or network diagram
- Description of the reporting levels

(4) Financing and procurement
- Financing concept
- Funding/Phase
- Procurement plan

(5) Development Plan
- Definition of required system/subsystem analyses
- Definition of development areas and development strategies (*e.g., test philosophy, simulations, studies, etc.*)
- Schedule concepts and constraints of the above-mentioned development steps and dependencies
- Drawing hierarchy, drawing release mode, and review cycle
- Specification approval and drawing release process, drawings, identification of critical areas
- Role of project management in system integration—technical directives
- Design documentation and release mode
- Description of the change monitoring

(6) Integrated test plan
- Summary of the entire test program
- Identification of test models (*laboratory samples and prototypes*)
- Timing of the test programs
- Definition of the test categories:

  Development testing: material, components, and subsystems before design work is completed

  Qualification testing: hardware testing at all system levels to demonstrate that the design/construction meets specification requirements

  Acceptance testing: hardware testing of all system levels to demonstrate that the hardware was created, manufactured, and assembled as designed

(7) Quality assurance (*QA*)
- Definition of the most important QA requirements
- List of applicable QA regulations
- If necessary, highlighting of special QA requirements
- Schedule of the QA program
- QA responsibility of the project management
- QA reporting procedures
- Component and material specifications

(8) Commissioning

- Operational planning and definition of the required facilities
- Description of the personal training plans and the corresponding time schedule
- Definition of the standard procedures and plans to be applied

(9) Documentation
- Identification of the required overall documentation
- Documentation structure
- Definition of responsibility for the creation and release of all documents
- Description of the documentation control process.

Seery has created a list of project plans typical for technology projects, in which he mentions technical project plans in addition to traditional management plans [2]:

- Technical, Schedule, and Cost Control Plan
- Safety and Mission Assurance Plan
- Risk Management Plan
- Acquisition Plan
- System Engineering Management Plan
- Technology Development Plan
- Software Management Plan
- Surveillance Plan
- Review plan
- Mission Operations Plan
- Environmental Management Plan
- Logistics Plan
- Data Management Plan
- Configuration Management Plan
- Security Plan
- Export Control Plan
- Waiver Log
- Change Log

6.2.2 Management Plan Checklists

In practice, it happens time and again that project management is able to enforce the creation of management plans, but the quality of the plans is often lacking. In order to ensure that the plans are drawn up without gaps according to a uniform scheme and contain all the necessary customer requirements, it is advisable to use appropriate checklists for verifying completeness. Based on the management study mentioned above, the following checklists are reproduced in accordance with Fig. 6.1 [3].

(1) Overall project management plan
- Does the plan include a summary description of the project/system: technical requirements, constraints, equipment needed, etc.?

  - Does the plan describe the individual subsystems, auxiliary systems, etc.?
  - Are the brief descriptions of the various subordinate individual management plans included?
  - Does the plan describe relationships with subordinate management plans?
  - Does the plan include references to management policies, methods, procedures, and reports?
  - Does the plan define the complete scope of deliverables?
  - Does the document show the higher-level work breakdown structure (*WBS*)?
  - Does the plan include the overall schedule, including key milestones?
  - Does the plan show the distribution of tasks of all companies/departments involved (especially important for joint ventures and international projects)?
  - Does the plan include flowcharts, networks, and other graphic representations to simplify and explain individual operations?
  - Does the plan show the content, structure, and responsibilities of key downstream planning documents?
  - Does the plan describe the reporting system for the technical, schedule, and financial progress report?
  - Does the plan describe the techniques for authorizing and approving budgets?
  - Have the key deliverables (end items) been thoroughly defined and listed?
  - Are the key delivery dates in line with the specifications?
  - Have all major work elements and activities been identified and listed?
  - Have all planning dependencies been defined and shown?
  - Are the estimates for staffing, materials, equipment, etc. reasonable and complete?
  - Have sufficient reserves been provided for any problems that may arise?

(2) Management and organization
  - Does the plan contain the organizational structures of the company, the project office, and, if applicable, other participating companies, as well as the organizational interfaces?
  - Does the plan specify organizational requirements for subcontractors?
  - Does the plan include detailed descriptions of organizational responsibilities?
  - Are key personnel identified in the organizational structure?
  - Does the plan describe the experience and qualifications of key personnel?
  - Does the plan define the areas of responsibility to be performed by the positions shown on the organizational chart?
  - Were all the project office functions to be performed covered?
  - Does the plan describe the role of monitoring bodies (*councils, boards, property leases*) to be specifically established?
  - Does the plan include key project strategies?

(3) Planning and Monitoring/Project Control
  - Does the plan describe the main principles of project monitoring?
  - Does the plan describe the basic planning concept?
  - Does the plan describe how project and work releases are to be made?

- Does the plan explain how status monitoring and assessments are to be conducted?
- Does the plan describe how plan deviations can be quickly identified?
- Does the plan describe how corrective actions will be initiated?
- Does the plan specify the requirements for a planning and monitoring center (major projects)?
- Does the plan describe the control procedures for schedule and costs?
- Are procedures for performance measurement described in the plan (*schedule, costs, technology*)?
- Does the plan include reporting procedures (*status, critical path/area, problems*)?
- Does the plan describe how status meetings are to be conducted (*agenda, dates, minutes of meeting, etc.*)?
- Does the plan include information on the review concept?
- Does the plan indicate the use of standard formats and symbols?

(4) Documentation control
- Does the plan describe how documentation requirements are defined?
- Does the plan specify the method of documentation storage and identification?
- Does the plan specify the use of formal documentation lists and documentation descriptions?
- Does the plan describe where to view the descriptions if they are not part of the plan?
- Does the plan describe the methods for reviewing documents to ensure that the documents delivered are consistent with the approved descriptions?
- Does the plan include a description of how the documents will be reproduced and distributed?
- Are the interfaces to other organizational units (*e.g., central services*) described in the plan?
- Does the plan define the methods by which documents will be identified, modified, controlled, and released?
- Is a documentation manager and the management procedures identified in the plan?

(5) Configuration control
- Are configuration control requirements included in the plan?
- Does the plan describe the procedure for establishing the baseline configuration and the associated change control?
- Does the plan describe what physical and functional characteristics are included in the baseline configuration?
- Does the plan include the procedures for creating the configuration for the system, design, and product requirements?
- Does the plan describe the associated review techniques and documentation packages to be reviewed?
- Does the plan describe the change control process for the baseline configuration?

- Does the plan include a procedure for controlling all software programs for hardware assemblies and operating programs?
- Does the plan describe how configuration deviations will be handled?
- Does the plan describe the procedures for regular configuration definition?
- Does the plan name the nominated configuration manager?

(6) Subcontractor control
- Does the plan describe the basic methods for subcontractor control?
- Does the plan include the procedures for make-or-buy *decisions*?
- Are subcontractor selection review procedures included in the plan?
- Does the plan identify alternatives in the event that a subcontractor(s) needs to be replaced?
- Does the plan specify how the subcontractors' work breakdown structures (*WBS*) will be integrated into the overall WBS?
- Does the plan describe the WBS numbering system to be used for all subcontractors?
- Does the plan describe how the prime contractor will manage and integrate subcontractor project control groups (*e.g., review and acceptance procedures, schedule and cost control, reporting, etc.*)?
- Does the plan discuss the procedure for monitoring subcontractors' quality control methods?
- Does the plan describe how the prime contractor will coordinate development efforts among subcontractors?
- Have arrangements been made in the plan for development and manufacturing coordination between the prime and subcontractor facilities?
- Are the integration and acceptance activities at the main contractor and subcontractors coordinated and described in the plan?
- What arrangements for continuous contact between prime and subcontractors have been made and are they described in the plan?
- Have rules for early termination of the contract relationship between prime and subcontractors been established and described in the plan?
- Has the prime contractor assigned an organizational unit and/or individual to subcontractor control and specified this in the plan?

(7) Manufacturing Coordination
- Does the plan describe all manufacturing activities?
- Does the plan describe the manufacturing organization, the functions to be performed, and responsibilities (on the project side and in the manufacturing area)?
- Are the documents (specifications and drawings) to be provided by the development department precisely listed and provided with delivery schedules?
- Are the engineering tasks that accompany production described in the plan (e.g., manufacturability studies, tool development, material and process control, and work preparation)?
- Does the plan describe an integrated hardware sequence (*all companies/departments involved*) and the associated schedule?

- Does the plan thoroughly describe shop procedures to ensure that the company in question has the necessary equipment, tools, and experienced staff?
- Are the material procurement and control procedures adequately described?
- Have provisions for rework and repair during the production process been made and described in the plan?
- Does the plan include a list of all items to be produced?
- Are the production considerations for the make-or-buy decisions described in the plan?
- Does the plan describe how the company in question has scheduled production operations and control costs?
- Is a manufacturing manager/coordinator nominated for the project?

(8) Facility coordination

- Does the plan describe existing and/or new equipment to be procured that can be provided for the project in question?
- Does the plan identify milestones for equipment deployment?
- Does the plan include guidelines for the companies/organizations involved, indicating which production, test, and operating facilities are required?
- Does the plan include, if required, a facility construction program, including possible modification activities of existing facilities?
- Are the asset procurement/provisioning activities adequately costed?
- Does the plan provide for facility occupancy procedures to avoid conflicts with other projects?
- Does the plan include alternative solutions?
- Does the plan specify who is responsible for facility coordination on the project side?

6.3 Management Concept for a Plant Project

The realization of a complex plant project requires the application of effective project management. A typical management concept for a plant project is described in Chap. 23. In addition, a standard Project Management Plan (PMP) is provided in Chap. 22.

Literature

1. BMBW Management Study for EUROPA **III** - Final Report, in GE Contract 1 P 6, September 21, 1970.
2. Seery, Bernhard D., Project Planning, in Larson Wiley J. (managing editor) Applied Project Management for Space Systems, McGraw Hill Companies, 2008, p. 78ff.
3. Madauss, Bernd-J.: Konzeptentwurf zur Planung und Steuerung des SNQ- Projekts, 1984 (not published)

Systems Engineering Management for Projects

7

Contents

B. -J. Madauss, *Project Management*,
https://doi.org/10.1007/978-3-662-69057-4_7

Abstract

Systems engineering and project management are closely linked. The selected system architecture is a decisive factor for the development and production of technically and economically optimal systems in order to compete successfully in the global market.

System engineering is a prime management function (SEM); see Figs. 5.3 and 5.4.

The successful realization of public and industrial project tasks under competitive conditions is an absolute necessity today. Therefore, system manufacturers must start thinking during the development process about how the specified project goal, including deadlines and optimal life cycle costs, can be achieved or, compared to the competition, even exceeded. To survive in competition over the long run means to win the battle, i.e., to be better compared to the competition in the following three areas:

a) Performance	better (*performance & quality*)
b) Time schedule	faster (*time to market*)
c) Costs/price	cheaper (*LCC*)

The goal of becoming and remaining a market leader requires having a competitive advantage in all three areas. Of course, this is not an easy task because improving performance should not be achieved by overburdening employees, but through innovation. Improving performance often requires the development of new technologies. Here is a fitting statement from Zahn: "*Technologies are essential mechanisms of economic and social change. Technological advances stimulate the ability of companies to innovate.*" [1] At the same time, however, the interaction between performance, schedule, and LCC must also be considered, because an improvement in performance should not lead to a worsening of the schedule-cost situation. Ideally, the innovation should also have a positive influence on these factors. In no case should quality suffer as a result; see Chap. 8. It is thus clear that the key to competition lies in the area of technology development and innovation.

Sometimes it takes a long time for an idea to lead to realization. Ludwig Bölkow, an aerospace pioneer and former chairman of MBB's management board, mentions in his book "*Der Zukunft verpflichtet*" *(Committed to the Future)* that he developed the "*tilt nose*" for high-speed aircraft as early as 1942 for which he received a patent [2]. In the 1960s, French and British aviation companies jointly developed the Concorde passenger airplane, which flew at 2.4 times the speed of sound, and also used the idea of a tilting nose to give pilots a better view during takeoff and landing; see Fig. 7.1.

Fig. 7.1 Concorde with tilt nose (*Source: Creative Commons*)

International competition is in many cases the driving force for innovative ideas for the development of challenging products; i.e., with better performance, faster time to market, and lower LCC. This in particular is the case for the worldwide commercial aerospace market. During recent years, two very interesting aerospace examples of innovation have led to a paradigm shift that is worth listening to, namely:

- The development of new communications satellites in which traditional chemical propulsion has been replaced by electrical propulsion (*fully* electrical). This leads to an increase in performance and a longer service life in orbit and thus also to an increase in economic efficiency
- The return of stages of the Falcon 9 launch vehicle (*LV*) to the launch site, which can then be reused after refurbishment, which will save costs, especially the cost of expensive engines; see Fig. 7.2. Falcon 9 is an innovative space product of SpaceX, a company owned by Elon Musk

Both cases are groundbreaking innovations in which systems engineering (*systems architecture*) played a decisive role. Most of the time, after all, it is initially just an idea, the implementation of which is often initially met with resistance. But there are also positive examples. The satellite operator SES ASTRA, Luxembourg welcomes the development of new technologies for in-orbit refueling of satellites and simultaneous replacement of obsolete payloads [3].

Fig. 7.2 Two Falcon 9 boosters return safely to Earth (*Source: Creative Commons*)

7.1 System Technical Processes in the Project

7.1.1 Overarching System View

It is often a long journey from the initial idea to the realization of a project; see Fig. 4.2. The feasibility of a project is checked in the early phase by means of systems engineering processes. Even in projects where the terms system or systems engineering are not explicitly used, it is often easy to recognize from the objective that in principle it is about the realization of a system. Systems engineering and project management belong together; see Fig. 7.21. Today, the construction of a factory building, a power plant, or a public facility, such as a modern swimming pool, requires systems engineering approaches more than ever before. In this respect, Kappel and Schwarz mentioned: "*It can hardly be overlooked that 'systems thinking' has changed not only science, but beyond that also 'daily life' in a certain sense.*" [4]

And Madzar writes that the systems architect is held in high esteem in medical technology, giving the following reasons: "*The complexity of products and systems continues to increase. Whereas in the past a few developers for hardware, software, and design were enough to develop a product, today rapid teams of more than 20 developers ....*" [5] Systems engineering research originated in the United States. Its beginning can be traced to the methods of cybernetics and systems theory developed in the 1940s. Blanchard and Fabrycky understand the 40s as the transition point from the machine age to the system age [6].

The rapidly increasing complexity of today's industrial products and government projects makes the use of systems engineering thinking more and more necessary. Systems engineering must be regarded as an integral part of industrial and public project management. Organizational solutions for integrating the systems engineering team into the project team are described in Sect. 5.1; see Figs. 5.3 and 5.4.

Systems engineering should, as far as necessary, take into account the concept of "*Industry 4.0*" or the so-called fourth industrial revolution when performing tasks. The aim is to combine industrial production with modern information and communication technologies. Industry 4.0 is a concept consisting of four organizational principles to help companies identify and implement:

- Networking: Machines, devices, sensors, and people can network with each other and then communicate via the Internet
- Information transparency: Sensor data enhances information systems of digital factory models to create a virtual image of the real world
- Technical assistance: Assistance systems to help people make informed decisions using aggregated, visualized, and understandable information
- Decentralized decisions: Cyber-physical systems are capable of making independent decisions and completing tasks as autonomously as possible. In exceptional cases, e.g., in the event of malfunctions or conflicting goals, it transfers the tasks to a higher authority

7.1.2 The Black Box Approach

In principle, systems engineering solves tasks by means of a global approach, in which a matter is viewed from an all-embracing bird's-eye view, so to speak, in order to then proceed further into the details. This approach is also known in the literature as a *top-down approach*. In the first step, the entire system, e.g., a new database to be set up or a power supply system, is viewed as a *black box* that has a specified task to perform and must cooperate with its environment. In the first step, it is not yet necessary to break the *black box* down into its component parts. Rather, the functions of the system must first be analyzed and specified. Therefore, ESA recommends first, i.e., in the early stages, to create a *function tree* to translate the project requirements into individual functions, such as power generation, transformation, power transportation, and distribution, etc. [7].

In the second step, derived from the function tree, a refined structuring of the system (*black box*) into subsystems and components takes place. The *product tree* may include e.g. solar generators, substations, power lines, control systems, software, etc. In the case of a database, this results in a breakdown into data acquisition, storage, data output, and database management. In a satellite project, as shown in Fig. 7.3, it is basically no different than a ground power system. At such an early

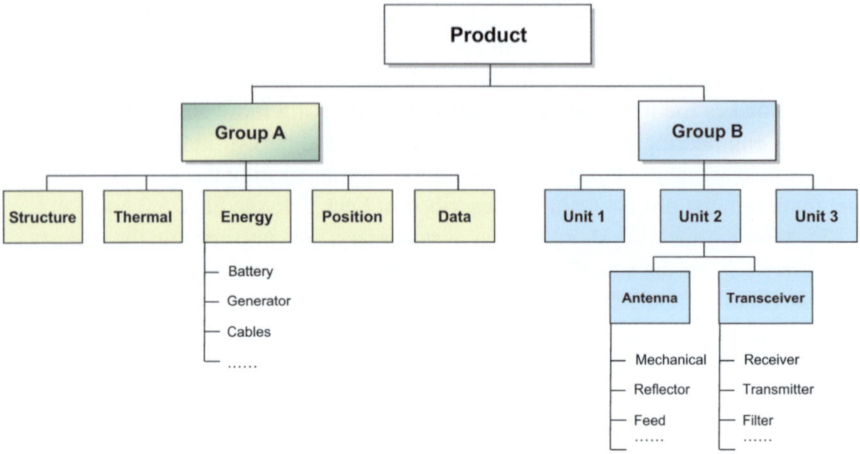

Fig. 7.3 Project tree for the space and ground segment of a satellite project [7]

stage of the project, an outline of the specifications (*specification tree*) should be made.

The example shown in Fig. 7.3 implies that the system (*product*) consists of four levels. That is, corresponding specifications must be created at all levels: System Specification (*level 1*), Segment Specifications (*level 2*), Subsystem Specifications (*level 3*), and Component Specifications (*level 4*). This is an example that can be varied in practice. At this stage (*early stage*), it is not yet necessary to provide further detail. That is reserved for later steps.

It is worth noting that representatives of industry and ESA formed a working group in the late 1980s to develop a standardized *product tree* for space projects. The product tree forms the basis for the creation of the Work Breakdown Structure (*WBS*). The *WBS* contains the elements of the product tree as well as the management, *system engineering,* and *quality/product assurance* functions; see Sect. 9.3.3.

7.1.3 Closed and Open Systems

In systems engineering, a distinction is made between closed and open systems. Closed systems are characterized by the fact that they have little or no significant interaction with their environment. A hermetically sealed diving sphere that remains motionless on the seabed is an example of this. The processes inside the diver's sphere have hardly any influence on the water surrounding it. Conversely, the environment (*water pressure, cold, etc.*) also has hardly any influence on the interior of the diver's sphere. Figure 7.4 shows the mode of action of a closed system using the example of a diving sphere. Due to the pressure vessel and sufficient thermal insulation, the diving sphere system is excluded from almost any communication with the environment.

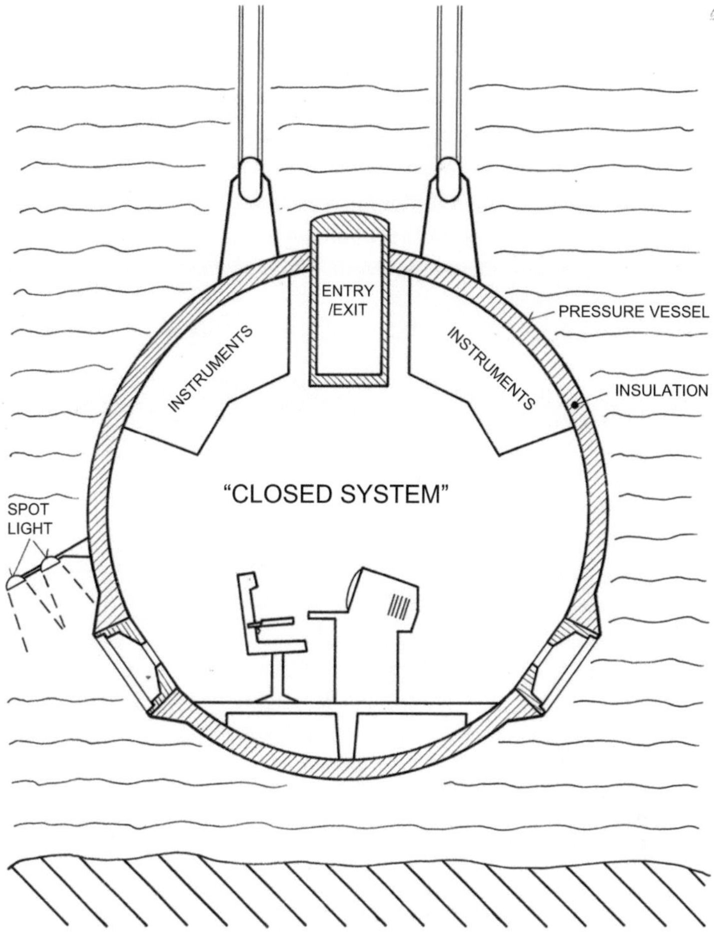

Fig. 7.4 The diving sphere as a closed system (Illustration: Christian Madauss)

Open systems, on the other hand, are in constant interaction with their environment. Information, energy, and/or matter enter or leave the system. For example, the living beings on Earth are open systems, which can be illustrated most clearly by the inhalation and exhalation processes. In fact, however, most systems that are considered closed are also systems that have a minor interaction with their environment. Ernst von Weizsäcker writes, for example: "*Strictly speaking, closed systems do not seem to exist at all ....*" [8] In the case of the diving sphere shown in Fig. 7.4, as already mentioned, there is also a minor interaction with the environment, but this, e.g., a minor heat flux or the radiant energy of the observation spotlight, is negligible for short-term consideration.

In most cases, technical systems, e.g., a power plant, a database or a spacecraft, are open systems. The intended interactions with the environment, i.e., the planned inputs and outputs, are important design variables: e.g., the solar energy supply in a satellite system (*input*) and the information transfer from the satellite to the ground station (*output*); see Fig. 7.5. In addition, however, there is usually an additional, unintended interaction with the environment, e.g., heating of the satellite on the side facing the sun and cooling on the side facing away from the sun. Therefore, when designing a system, not only the intended input/output quantities but additionally the unintended interactions should be included in the calculation. This principle should always be followed when designing a system.

In the past, the consideration of the unintended interaction of a system with the environment often did not seem significant. For this reason, many systems were

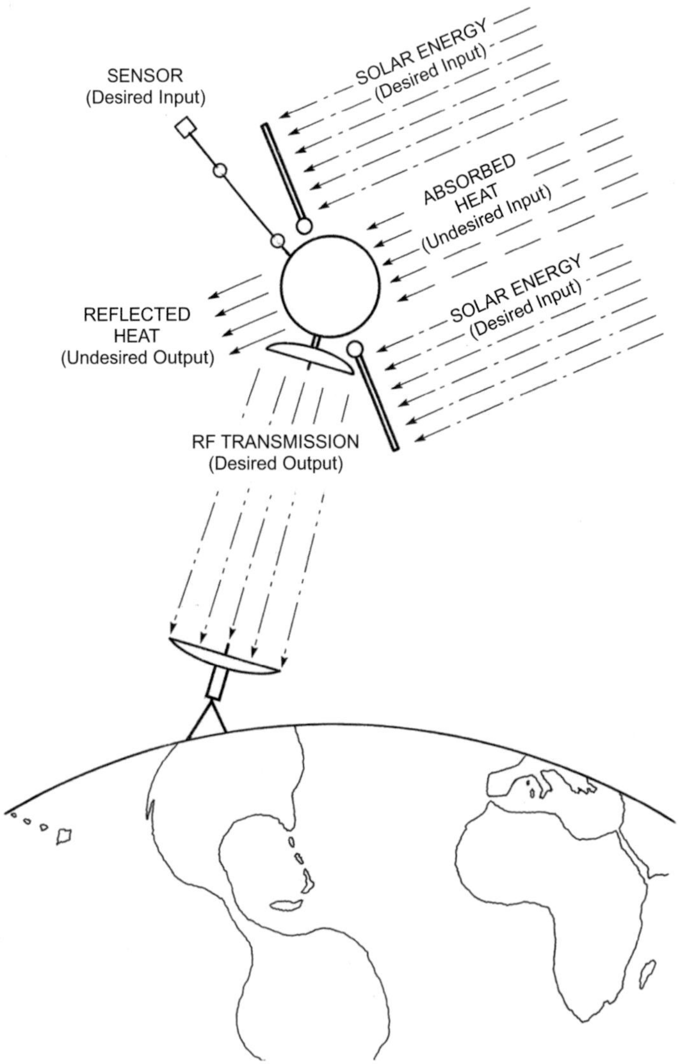

Fig. 7.5 Input/output variables of a satellite system (Illustration: Christian Madauss)

developed without sufficient input/output consideration. The history of the development of the internal combustion engine is a very illustrative example of this because the exhaust gases of the internal combustion engine represent an output variable that has only recently, namely in connection with the problem of increasing environmental pollution, attracted increased attention and led to the catalytic converter in conjunction with unleaded gasoline. The VW emissions scandal attracted worldwide attention in 2015, leading to investigations into diesel vehicles from all German manufacturers that were found to contain excessive levels of pollutants.

While in the case of the steam engine, in the interest of greater economy, the aim was deliberately to achieve a system that was as closed as possible—special insulation measures were taken to prevent heat radiation (unwanted output)—the system of the internal combustion engine was deliberately opened wide at the exhaust, neglecting the later consequences. The system-technical input/output consideration, which is vital for a satellite system, is a trend-setting thought model for the necessary system consideration of technical products of our time. This is as true for a power plant as it is for a database, a vehicle, or a commercial aircraft. Precise knowledge of the interactions of a system with the environment, which, as we know today, can be regarded as a global system, is a prerequisite for the design of technical systems.

The black box approach, a recognized method of systems engineering, which has been simplified here, may seem trivial, but it is the only feasible pathway for the realization of complex systems, which will be explained in more detail below. The design and construction of large and complex systems, such as power plants, airplanes, stadiums, hospitals, databases, etc., requires that at each step all interrelationships are first analyzed in detail before proceeding to the next stage. To stay with the database, in the first step it is only a matter of analyzing the main parameters, such as storage capacity, access possibilities, data backup, expandability of the system, sharing or connection to existing databases, data preparation (*tabular, graphical, etc.*), location of the database and operating personnel. Only when these factors have been satisfactorily matched by an initial system analysis should the second step, refinement of the analysis to the module or subsystem level, be initiated. Possible subsequent steps should be taken step by step according to the same scheme.

The main advantage of this approach can be seen in the unbiased questioning of possible solutions to a specific requirement or function. For example, the requirement to transport goods between two locations can be solved by different systems: trucks, train connections, airplanes or ships. The parameters to be investigated are:

(1) Transport requirements
   - Quantity
   - Type of transported goods
   - Required speed
(2) Infrastructure
   - Road condition
   - Available airfields
   - Existing shipping lanes
   - Condition of the railroad network

(3) Transport capacity available/to be expanded
   - Trucks
   - Aircraft
   - Ships
   - Trains
(4) Deadline and cost limits

In the first step, again a black-box consideration of the first level has to be carried out. Based on the transport requirements, the existing infrastructure and transport capacity as well as the given time and cost limits, the most favorable system has to be selected. It is easy to see that at this stage a detailed analysis, e.g., on the loading options by ship, would be too early, as it may still turn out during the higher-level analysis that the goods cannot be transported fast enough by ship. A very similar situation would occur if the requirement for a new weapon system were immediately answered by the development of an aircraft without first considering other possibilities, such as the use of a missile system.

However, system analysis has another outstanding advantage. The black box approach creates the conditions for easy communication between the various disciplines. The questions concerning the system are, as it were, brought to a common denominator in a generally understandable language. The experts from different disciplines involved in the project meet at the system level in a generally understandable language in order to agree on the system-level interfaces. The specialist disciplines bring their specific knowledge to the system in a pyramid-like condensed form. It is the task of the system architect (*system engineer*) to coordinate the individual disciplines in the interest of an optimal system.

7.1.4 The System Engineering Control Loop

The task of project management is to complete very specific objectives as efficiently as possible within a time and financially limited framework. The focus of project management is generally the project task, which varies depending on the industry. In industrial projects, it is usually a technical product that is to perform a very specific function and that exists only as an idea at the start of the project. In terms of systems engineering, the future product (*project goal*) can be viewed as a system, e.g., a transportation system, an intelligence system, or a manufacturing system. The intended system can be divided into modules, subsystems, components, etc. with corresponding sub-functions; see Fig. 7.3. The definition of system is not limited solely to physically identifiable systems or sub-systems but also includes imaginary systems, such as functions, which can also be part of a system. In the case of non-industrial projects, e.g., a charitable aid program, the systems engineering approach can also be applied, although such projects naturally involve hardware-oriented subsystems to a much lesser extent.

In principle, a project can be regarded as an open system, since there is almost always an interaction with the environment, which can be understood as an external control loop consideration. For a new power plant to be built or a rocket launch facility to be installed, this becomes particularly clear. In addition to the functional links, i.e., the integration of the new system with already existing systems, e.g., the connection of the power plant to the existing energy grid, there are links to global systems such as the environment; see Fig. 7.6. Similar processes are known from almost all areas of modern life, although they have only recently been subjected to systematic consideration.

The networking of open systems with their environment as well as the investigation of mutual influences and feedback between the systems is an important task of systems engineering. Vester writes: "*Reality is not an incoherent catalog of topics, but always a network of feedback and nested control loops.*" [9] In fact, all influencing parameters of a new system must be included in a control loop consideration from the beginning of the project. In practical terms, extensive system analysis must be performed in the early stages of the project (*Phase A*) to investigate the compatibility of the system (*black box*) with its environment. In other words, the planners of a new project must carefully analyze all the effects of the system on the environment and the resulting *feed-back on* the planned system.

This is particularly important in today's world, as the mutual influence of systems is not limited to technological parameters alone, but increasingly includes ecological, economic, and sociological factors. In aerospace, the importance of mutual influence, i.e., the feedback effect, had been recognized quite early, since the effects

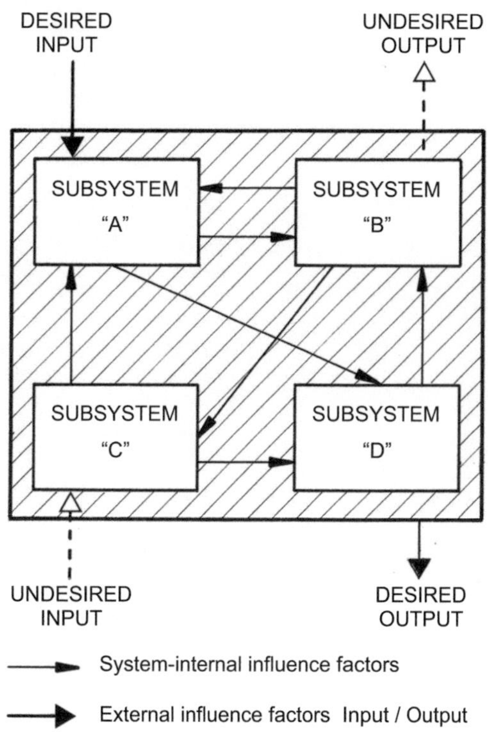

Fig. 7.6 Internal and external influences of a system

of external influences (*feed-back)* were felt immediately due to stringent weight and volume limitations. Manned spacecraft are literally mini-planets. For example, they have their own power supply system, their own environmental system (ventilation and temperature control), etc. They are therefore an ideal object of study for systems engineering processes.

In addition to the external control loop of a system to the environment, there are of course also control loop processes within the system. In the case of a nuclear power plant, for example, the system consists of the subsystems reactor, turbine, steam boiler, circuits, control systems, etc.; see Fig. 7.7. There is also a mutual influence between the subsystems with *feed-back reactions*, i.e., there is an internal control loop. All subsystems are interconnected. Actions taken in one subsystem can influence the function of other subsystems. There is no fundamental difference between the outer and inner control loops. Depending on the point of view, one could also speak of control loops in the macro- and microsystem areas.

Systems engineering processes require a high degree of interdisciplinary cooperation in order to be able to perceive the control loop functions, i.e., the objective treatment of all influencing parameters, as precisely as possible. Ropohl writes in this context: "*The eco-technical and socio-technical interdependencies, especially in large-scale projects, will in the future force the participation of chemists, biologists, urban planners, psychologists, sociologists, administrative experts, lawyers,*

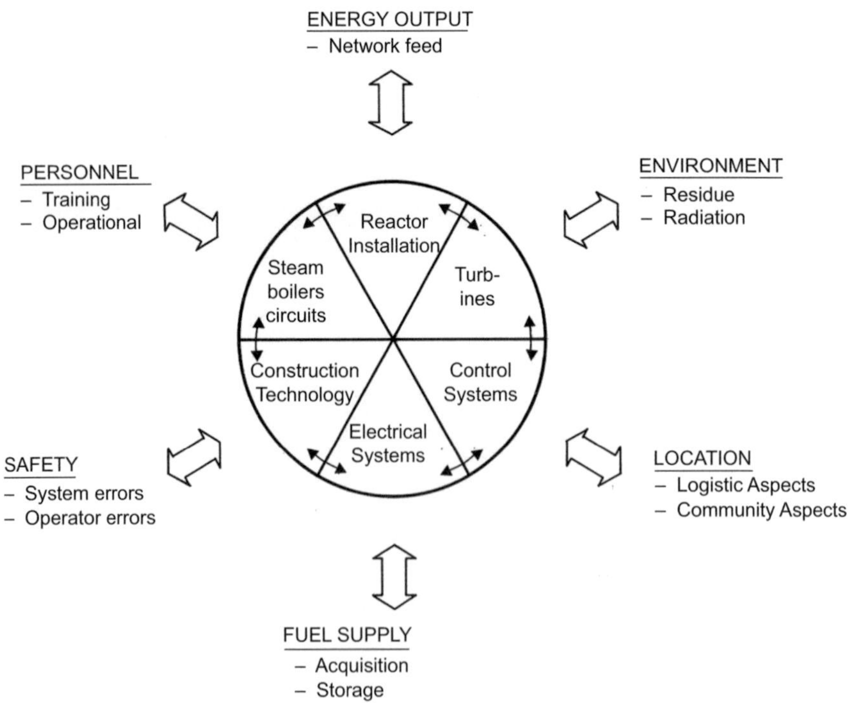

Fig. 7.7 Internal and external influence factors of a power plant

etc." [10] In practice, interdisciplinary cooperation is often associated with greater difficulties since companies are often unable to enforce close cooperation due to their rigid organizational structure. Ropohl appropriately states, "*The necessity of interdisciplinary cooperation is hardly questioned today; however, considerable difficulties usually arise when one wants to get serious about this principle in practice.*" [10]

One reason for this may be the lack of generally useful tools for thinking and understanding, which Ropohl criticizes. In addition, there is certainly an organizational problem, because genuine cooperation presupposes the will to jointly solve a problem, which can best be realized through teamwork. Here again, the aerospace industry played a pioneering role due to its special tasks, because the development of high-performance aircraft, but especially manned space flight, required the use of interdisciplinary teams at an early stage, e.g., the involvement of physicians and biologists. The Apollo project team, which was responsible for the development of the lunar module, naturally not only occasionally consulted an aerospace physician, but also firmly integrated the project function "*medical care*" as a key position in the project team. This was a crucial step in bringing about a community solution. The necessary control loop function was regulated by this measure analogously to the teamwork between engineers and business people. If this model is transferred to major projects in Germany and Europe, the disciplines mentioned by Ropohl will have to be integrated into the project team more than before.

One or the other will probably fear that this will degrade the project management to a non-decision-making instrument. Certainly, it is also difficult to unite several and often conflicting views under one banner, i.e., to find a common denominator after weighing all influencing parameters (*control loop simulations*). But it should not go unmentioned here that problems are not solved by suppressing important influencing parameters of the control loop, e.g., aircraft noise, radiation hazards, etc., in the planning phase, only to have them come to light as incontrovertible facts in the operating phase. Recent history has given us a very impressive illustration of the consequences of influencing factors that came to light too late in connection with major German projects; see Sect. 3.2. Later discussions are often not conducted in a factual manner but in a politicizing manner. For this reason, the project manager should have sufficient and, above all, balanced knowledge of all the disciplines in question in order to bring about a system-optimal compromise in the event of conflicting views among the experts.

The aerospace industry solves many problems by investigating system engineering solutions in the planning stage (*Phases A and B*) using simulations (*control loop considerations*). A good example of this is accident and reliability analysis, and also simulation programs for noise abatement (*throttling of engine noise*). The focus of all these simulations is the system behavior of the aircraft, i.e., the external and internal control mechanisms, i.e., aircraft and environment as well as aircraft and subsystems. Simulations are used to test the system's behavior. An attempt is made to examine the entire spectrum of all influencing factors in order to learn about the sensitivity of the system. The simulation results are important basic data for the design of the system to be developed.

Systems engineering analyses basically follow the control loop principles. Starting from the target specifications, the system inputs, the systems engineering team carries out the systems analysis process and compares the results obtained with the target specifications. Ropohl defines the control loop for project work based on a typical control loop from technology, which can also be applied to the functions of systems engineering; see Fig. 7.8 [11]. The systems engineering control loop processes are described in detail below:

(1) *Project management*: Target specifications are passed on from project management to systems engineering. In practice, these are the system requirements, e.g., the performance specifications (*mission requirements*), reliability and safety regulations, weight and volume limits, environmental and site regulations, etc.
(2) *Systems engineering management*: The systems engineering management, an important function of project management (see Figs. 5.3 and 5.4), is the main control body in the systems engineering control loop. From here, the target specifications are passed on to the individual teams and the results achieved are

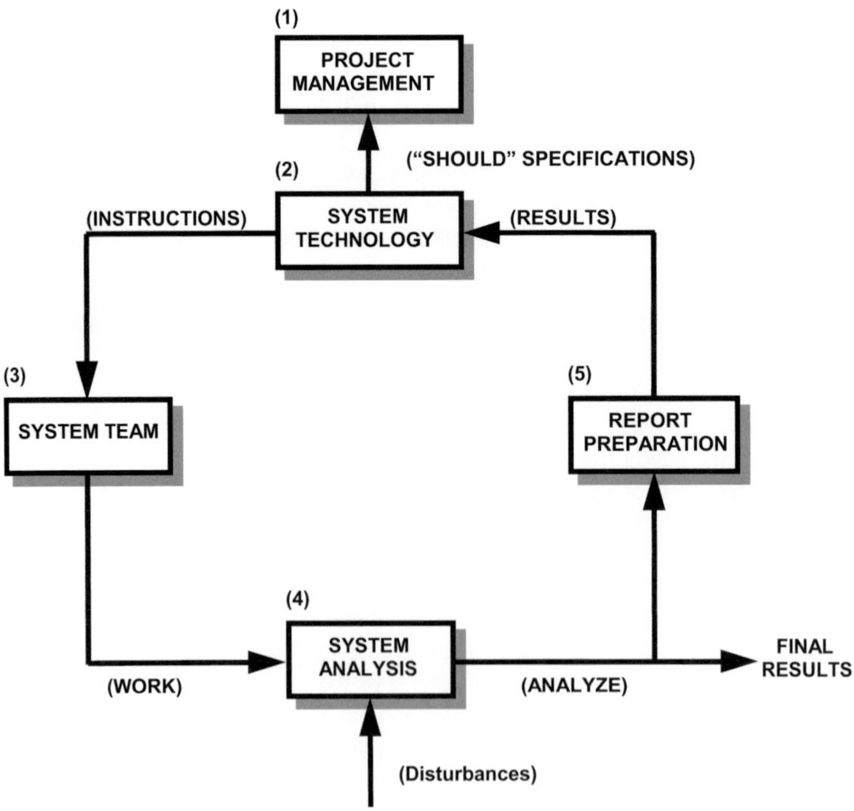

Fig. 7.8 System technical control loop

compared with the target specifications. Target/actual deviations must be checked for their significance and translated into decisions. Deviations can lead to changes in the target specifications, which, however, result in an intervention in the higher-level management control loop (*above point 1*), and new specifications for the system team.

(3) *System team*: An appropriate team that can fully represent the system engineering aspects must be provided to handle the system engineering task. Even at this level, the use of an interdisciplinary team may be necessary.

(4) *System Analyses*: Detailed system analyses shall be performed by the system team and shall include the following individual analyses:
   - Compatibility studies (*system/environment, system/feasibility, etc.*)
   - Performance analyses (*system configuration/performance, system/subsystems, etc.*)
   - Incident and reliability analyses (*system/incident variables, system/reliability, etc.*)
   - Manufacturing analysis (*system/manufacturing technology, system/material sourcing, etc.*)
   - Cost analyses (*system/development costs, system/life cycle costs, etc.*)
   - If necessary, the investigations are to be carried out in cooperation with other project teams and include extensive parametric analyses. External influencing variables (*disturbance variables*) are also to be included in the investigation. The often-raised objection that external influencing variables that are not yet known in absolute terms can be solved with parametric studies by simulating disturbance factors in a certain range by means of assumptions. This means that the effects depending on assumed influencing factors are known long before the absolute values become known, which can then be incorporated and adjusted later.

(5) *Reporting*: The results of the system analyses (*above point 4*) are to be summarized clearly and forwarded in report form to the management of systems engineering. In practice, these are mostly technical reports that are recorded via a documentation system; see Sect. 12.1.

In practice, the control loop shown in Fig. 7.8 is usually iterated several times until a balanced final result, e.g., the system specification is ready for signature.

7.1.5 Systems Engineering Applications in Project Management

Today, the realization of projects depends on external influencing factors to a far greater extent than was the case a few years ago. Technological projects, which can generally be regarded as open systems, must be carefully networked with their environment, i.e., with external systems, as early as the planning stage in order not to appear as a sudden disruptive factor. That is, in the early stages of a project (*Phase A*), compatibility studies must be carried out to analyze in detail the interactions (*input/output variables*) of the system to be created with the environment or other technical systems. Not only technical and economic factors play a role, but more often also political, ecological and sociological parameters. The development and implementation of microprocessors is a vivid example of this. Henning writes in this context: "*The systems engineering work on the project today knows besides the technical also the economic, political, sociological and ecological dimension.*" [12] The unintended input-output factors, e.g., the job changes that occur when microprocessors are introduced in industry and government, must be examined with the same meticulous care as the intended interactions.

Particular importance must be attached to the strategic dimension of systems engineering approaches in the context of project management, because only the consistent initiation of systems engineering processes at the concept stage, i.e., in the earliest phase of a project (*Phases A and B*), can ensure the greatest possible degree of harmonization between the *New System* and its environment. The task of Phase A (*concept phase*) is essentially to carry out system compatibility studies. In this phase, starting from the system requirement and taking into account the boundary conditions, system analyses are carried out with the aim of achieving the optimum design of the system. The requirement for an optimal design of the system refers to the system itself (*maximum efficiency of the system*) and to the interaction with the environment (*input/output relationships*). For example, when designing a new commercial aircraft (*aircraft system*), the system's internal parameters (*aerodynamics, propulsion, etc.*) must be matched with external factors (*flight frequency, fare policy, aircraft noise, etc.*). I.e., a system-optimal aircraft concept (*output*) must be developed based on the project requirements and boundary conditions (*inputs*) defined at the beginning of the concept phase; see Fig. 7.9.

Based on the defined project requirements and the given boundary conditions, the remaining free parameters must be coordinated with each other according to the

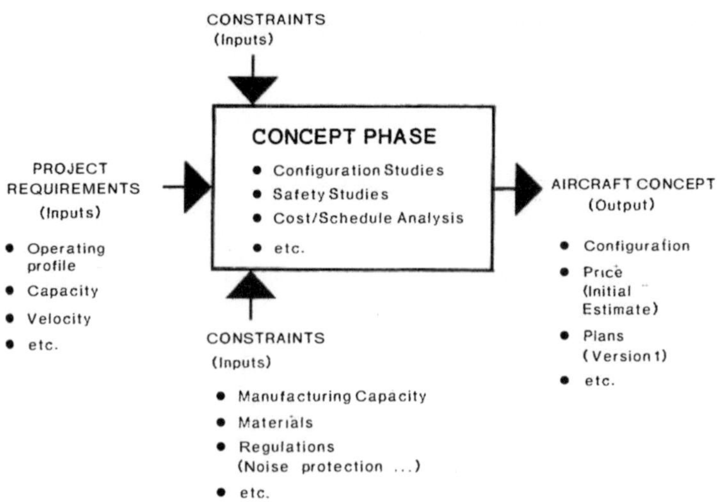

Fig. 7.9 System engineering processes during project design

principles of the control loop (see Fig. 7.8) in such a way that an optimum concept can be offered while complying with the specifications (*requirements and boundary conditions*). Of course, in practice there are often feedbacks which also influence the requirements and boundary conditions. For example, the originally specified aircraft capacity (*requirement*) or the available material (*boundary condition*) can be included in the analysis process (*optimization*). This is especially true for cutting-edge technological projects. In the recent past, for example, many major projects included the rapidly advancing microminiaturization in the development process. In other cases, changes in boundary conditions occurred even during the development process. This was the case, for example, with the construction of nuclear power plants, for which increased reactor safety requirements were introduced during the course of the project.

The advantages of the systems engineering way of thinking in the project can be seen in the simulation-like approach. That is, long before it comes to the creation of facts cast in concrete, the behaviors of new systems, including the effects on the environment, can still be examined in the development phase. There is still enough time to initiate necessary changes to the system.

7.2 Specification Processes in the Project

7.2.1 The System Specification

The realization of a project begins with the setting of the project objective, which is or should be manifested by a *mission requirements document*; see Fig. 4.2. At this point, the future system is not yet defined in its configuration, but only the requirements that will be placed on the future system are known. During the concept phase, the future system is formulated taking into account the given constraints. The result of the concept phase is to be recorded in the concept documentation, consisting of preliminary planning documents and a preliminary system specification; see Fig. 4.13. The creation of the system specification (*first/preliminary edition*) is an important milestone in the project history. It lays down the future system: It establishes an important reference point for all subsequent activities. The system specification is the overarching reference document for the system engineering design of the overall system and an important part of the development contracts; see Sect. 13.2.

The official publication of the system specification brings the new system into being, so to speak. In the so-called paper phase, i.e., at the transition from the concept to the definition phase (see Fig. 4.2), the system specification describes how the future system is to be structured. It also defines the interactions (*input/output variables*) with the environment. Typical contents of the system specification are (*here an example*) [13]:

- Main criteria of the technical objective
- Definition of system requirements
- Development/Production baseline

- Functional and performance requirements
- Subsystem interfaces

The suggested outline of a system specification based on the MIL-STD-490 is as follows [14]:

(1) Overview
(2) Reference documents
(3) Requirements
    (a) System definition
        – General description
        – System task
        – System diagram
        – Interface definition
        – Existing assemblies to be included
        – Operational and organizational concepts
    (b) Characteristics
        – Performance characteristics
        – Physical characteristics
        – Reliability
        – Maintenance requirements
        – System availability
        – System efficiency
        – Environmental conditions
        – Transportability, if necessary
    (c) Design and construction
        – Materials, machining, and parts
        – Electromagnetic radiation
        – Product identification
        – Work execution
        – Interchangeability of the components
        – Safety requirements
        – Human Engineering
    (d) Documentation (*specifications, drawings, etc.*)
    (e) Logistics
        – Maintenance
        – Warehousing
        – Plant and equipment
    (f) Personnel and training
        – Personnel (*qualification, number, etc.*)
        – Training (*time, place, etc.*)
(4) Quality assurance
    (a) General description
        – Test responsibility
        – Special tests and examinations

(b) Quality controls

(5) Delivery instructions (*packaging, transport, etc.*)

(6) Notes

(7) Attachments, if applicable.

The system specification can only be released after a critical *review*; see Sect. 7.4. In practice, however, timely adoption of the system specification is often made difficult by the fact that the individual specification areas have not yet been completely defined. Often these points are then omitted in order to define them at a later stage; a common stopgap is the so-called TBD (*to be defined*). In principle, however, this should be largely avoided for such an important document, because open points (*TBDs*) often have to be paid for dearly. In the interest of a speedy project process, however, a *draft of* the system specification can be circulated as an orientation framework. It is also possible for suppliers to use TBDs to propose cost-saving solutions under competitive conditions.

The system specification is an orientation framework for preliminary schedule and cost planning. Even in the early stages (*Phases A and B*), the respective project sponsor is interested in the most accurate schedule and cost forecast possible. Changes to the system specification have a direct impact on the schedule and cost forecasts. The project team must be aware of the direct relationship between the legs of the task performance, schedule and cost triangle; see Fig. 9.2 and Sect. 9.4. Special attention must be paid to all cost-driving factors such as quality and reliability requirements and critical functional features. Often, however, it is precisely the knowledge of these interrelationships that leads to a hesitant attitude. This is basically understandable, because one is then left sitting on the costs once they have been mentioned, even though this was only preliminary cost information. Nevertheless, it is necessary to point out the necessity of a courageous attitude when adopting the system specification.

In the future, the authorities and industry must realize more than before that the early phases of the project (*Phases A and B*) are, in principle, analysis phases designed to design an optimal system, taking into account and weighing all influencing factors. The technical goal, which is expressed by the system specification, is set against the schedule and cost factors. These three factors, along with the given boundary conditions (*fixed specifications*), form the main parameters of the iteration process of the project's early phases. The timely issuance of the system specification during this process initiates the reduction of change of all free parameters, at the end of which (*completion of Phase B*) the final adoption of the system specification and the schedule and cost plans is completed; see Fig. 7.10.

7.2.2 The Specification System

The system specification of a project represents the highest level and central technical document of technological projects. It summarizes the main data of the planned system. The system specification is the main reference document for all system

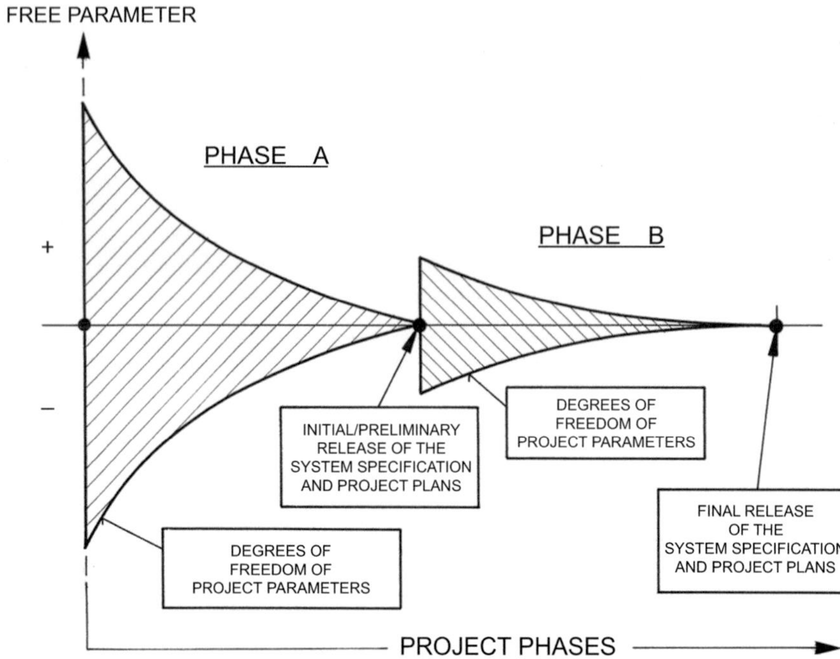

Fig. 7.10 Phase-dependent definition flexibility of a system

technology areas in the project and serves as a reference document for scheduling and cost planning. In addition, it is usually part of the contract; see Sect. 13.2. During project execution, the system specification, in conjunction with the planning documents (*schedules and costs*), is an important reference document (*baseline document*) for controlling the project processes.

In addition to specifying the overall system, larger projects also require that the subareas of the system be specified in detail. The system specification represents the top of the *specification tree*. The US Department of Defense has defined the different specification types as follows [15]:

Type A—System specification
Type B—Development specification
- B.1—Prime components
- B.2—Critical components
- B.3—Uncomplicated component
- B.4—Facilities Shipping
- B.5—EDP programs product specifications

Type C—Product Specification
- C.1a—Major Item Function
- C.1b—Major Item Fabrication
- C.2a—Critical Item Function

C.2b—Critical Item Fabrication
C3—Uncomplicated components (fabrication)
C4—Inventory Item
C5—EDP programs

Type D—Process specification
Type E—Material specification

This proposed outline can be used as a guide for major projects.

The application of specification types A to E during the various project phases can be seen in Fig. 7.11 [15]. The specifications are created in the respective preceding phase. As already mentioned, the system specification is created during the concept phase (*preliminary output*) in order to use it as a reference document in the definition phase.

The outline of the specification system prepared by the European Space Agency (*ESA*) is given below [16]:

SPECIFICATION TYPE	Application in Project Phases:		
	B) Definition	C) Development	D) Production
A. System Specification	X	X	
B. Development Specific			
BI. Major parts		X	
B2. Critical parts		X	
B3. Simple parts		X	
B4. Facilities		X	
B5. Software Programs		X	
C. **Production Specific**			
CI. Major Parts			
a. Function			X
b. Production			X
C2. Critical Parts			
a. Function			X
b. Production			X
C3. Simple Parts			X
C4. Facilities			X
C5. Software Programs			X
D. Manufacture specific			X
E. Material Specific			X

Fig. 7.11 Specification application matrix (MIL-S-83490)

(1) System specification
This specification contains the functional and performance requirements as well as the design criteria for the development, test and production of the complete system.
(2) System requirement specifications
Here, in addition to the system specification, the requirements of a specific discipline are described, for example, environmental conditions, safety requirements, etc.
(3) Sub (*partial*) system specifications
The subsystem specifications contain the functional and performance requirements and the design criteria for the development of the main components (*subsystems*) of the system.
(4) Device specifications
These specifications contain the requirements for the creation/procurement of individual devices.
(5) Interface specifications
The interface specifications define the requirements for the interfaces between the subsystems.
(6) Test specifications
The test specifications record the requirements for the test methods to be used. However, test specifications should only be created if special test requirements are involved; otherwise, the test requirements should be included as part of the system or subsystem specification.

The interaction of specifications (1)–(6) and the resulting hierarchy can best be represented in the form of a *specification tree*; see Fig. 7.12. The system specification for the Spacelab space laboratory, for example, is supplemented in the second level by five *interface control documents* (*ICDs*) and in the third level by ten *system requirement* specifications. Levels 4, 5, and 6 then reflect the subsystem and device specifications for the Spacelab project [17].

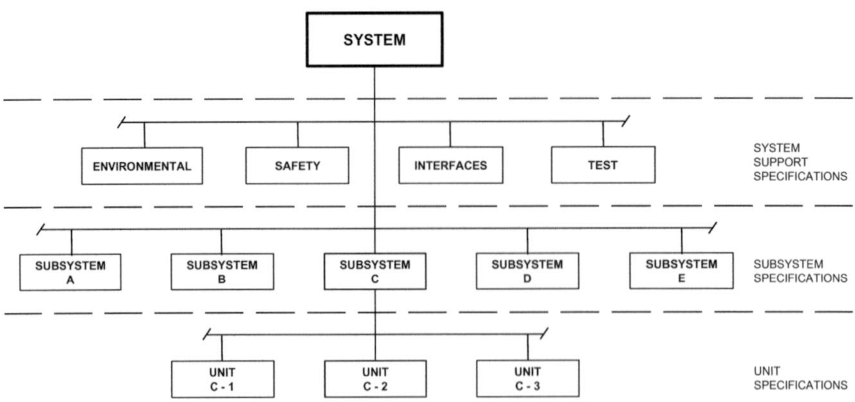

Fig. 7.12 Example of a specification tree

7.2.3 Significance of the Interface Specification

Monitoring interfaces is an important management task in large and complex projects. This is especially true for international projects. As can be seen from Fig. 7.6, the subsystems of a system are interconnected to a greater or lesser extent. The mutual influence of the subsystems of a system, which can be identified as input–output parameters at the interfaces between the respective subsystems, is often part of detailed system analyses performed at the beginning of the project (*Phases A and B*). This involves the precise determination of the boundary lines and definition of the interface. The end result of this process is the interface specification. Technical interfaces can be mechanical or electrical interfaces and also software links. Typical interfaces are for example [16]:

(1) Mechanical interfaces
   - Dimensions
   - Weights (*mass*)
   - Center of gravity locations
   - Fixing points
   - Clearances and tolerances
   - Heat transfers
   - etc.

(2) Electrical and electronic interfaces
   - Power
   - Voltage
   - Signal characteristic
   - Switch
   - Plug-in contacts
   - Wiring
   - Grounding
   - etc.

The interface specification ensures that the individual subsystems fit together exactly. For example, if a ship is being built, the shipbuilders must know the exact dimensions (*interfaces*) of the ship's engine when creating the engine foundation in order to ensure that it will fit when installed. This is particularly important on large projects where many companies are working simultaneously. For example, an electronics box for a satellite system manufactured in England must fit into the system mechanically as well as electrically or electronically 100% during final assembly in Germany. Examples of this kind can be continued at will. The larger and more complex the system, the more important interface management becomes, because situations in which the subsystems did not fit together have unfortunately often occurred in the past. The consequences are usually delays and additional costs.

Billig uses an example of the Europa I launch vehicle to describe how a simple interface problem can lead to disaster. For example, the missing "*J*" in the Italian alphabet led to the breakdown of the eighth launch of the Europa rocket in 1969

[18]. The plug contacts between the German-made third stage and the payload fairing coming from Italy were numbered alphabetically on both sides. The different alphabet, i.e., the missing "*J*," led to the fatal error; see Sect. 3.8.

In practice, the interfaces between two subsystems are often recorded in the respective subsystem specifications so that the connection points of the subsystems are completely congruent mechanically and/or electrically; illustrative examples are mechanical fastening and connection points and electrical plug contacts. For projects with a large number of subsystems, it is recommended to introduce a special interface specification; *the Interface Control Document—ICD.* All interfaces between the subsystems are summarized centrally in the ICD. In large projects, the use of special interface management is also recommended in order to better monitor the interface changes that occur during the development process. Interfaces exist at all project levels. For the systematic recording of all interfaces, the creation of an interface matrix is recommended; see Fig. 7.13.

Dreger defines the most important tasks of interface management as follows:

(1) Localization of the interfaces
(2) Ensuring the compatibility of the boundary elements involved in the interfaces (*the cardinal task of system integration*) and
(3) Interface monitoring, so that the interfaces are observed by all participants and secured against unauthorized manipulation [19]

During project implementation, system interfaces between customers and suppliers must be contractually defined. However, it must be taken into account that in

WBS	SUBS	STR	TC	COMS	AOCS	UPS	OBC	POW	HK
210	STR		X	X	X	X	X	X	X
220	TC			X	X	X	X	X	
230	COMS				X	X	X	X	X
240	AOCS					X	X	X	X
250	UPS						X	X	X
260	OBC							X	X
270	POW								X
280	HK								

Legend:

AOCS Attitude and Orbit Control System
COMS Communications
HK Housekeeping
OBC On-Board Computer
POW Power

STR Structure
SUBS Subsystems
TC Thermal Control
UPS Unified Propulsion System

X Mechanical, electrical and/or SW Interfaces

Fig. 7.13 Example of an interface matrix

order to avoid problems at the interface between two systems, subsystems, etc., appropriate clearances (*margins*) are provided. For example, in the early phase of the project, the energy output of the electricity supplier should be set at a higher value than the value for the consumer's energy requirement (*input*); see Fig. 9.39. This margin gives the project management the opportunity to compensate for any emerging bottlenecks early enough to allow for changes with minimum impact.

7.3 System Reviews

7.3.1 Concept and Procedure

System reviews are an important tool of modern project management. In the context of major aerospace projects, effective *review procedures* were developed in the 1960s to ensure successful project completion. The idea was to subject the technical development status to a complete review when very specific milestones had been reached. For this reason, *review milestones* are carried out along the individual project phases; see Fig. 4.2. These milestones are to be arranged in such a way that the completion of very specific development steps coincides with these milestones. Review milestones shall be timed to coincide with the completion of system specification, design work, etc. Ten typical system level review milestones applied for space projects are shown below [20, 21]:

- MRR Mission Requirements Review (*Mission Objective*)
- SRR System Requirements Review
- SCR System Concept Review
- SSR System Specification Review
- PDR Preliminary Design Review
- CDR Critical Design Review
- TRR Test Readiness Review
- TR Test Review
- AR Acceptance Review
- FACI First Article Configuration Inspection
- SPR Systems Production Review

Figure 7.14 shows an example of ESA review milestones of the respective project phases [21].

The milestones mentioned above and their assignment to the project phases are to be understood as examples and can also be modified according to the respective situation and supplemented by additional review milestones as required. In practice, it has proven useful to include the review milestones in the project planning for control purposes. Therefore, the final results of the reviews must be precisely defined. The consistent execution of reviews and the subsequent processing of the resulting improvement measures are of great importance for the economic success of the project. Project reviews must be carried out at all levels of the project, as

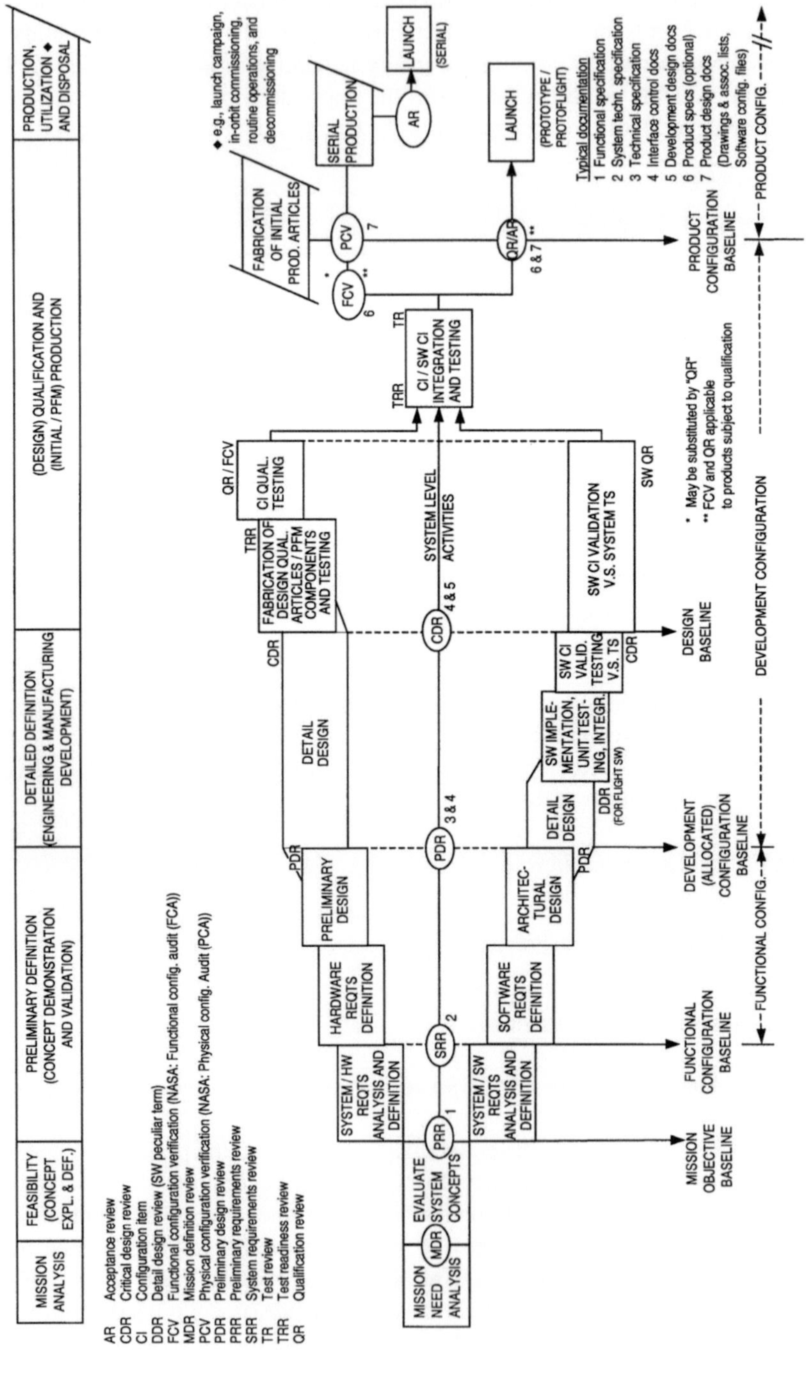

Fig. 7.14 Project phases and review milestones

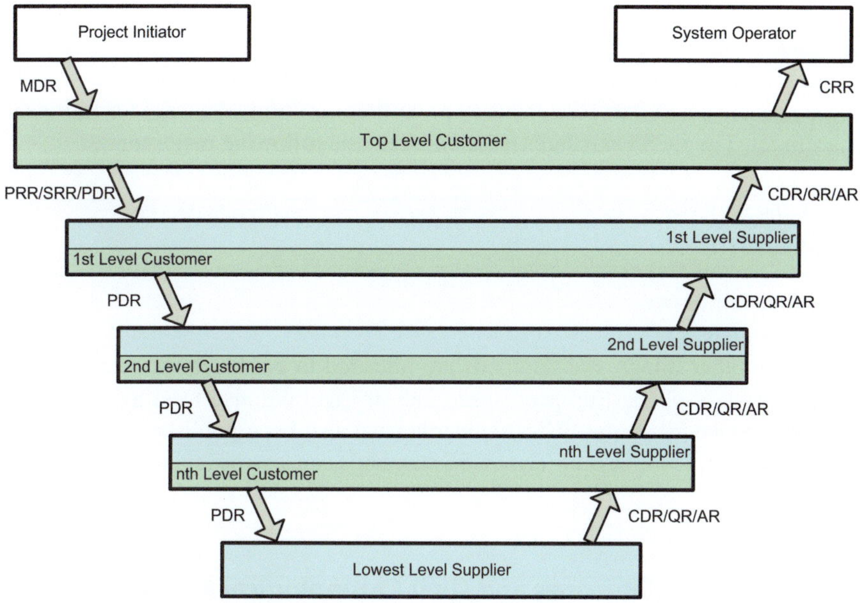

Fig. 7.15 V-Model—review milestones at all project levels (*ESA*)

demonstrated by the ESA example of the systems engineering V model; see Fig. 7.15.

The planning and consistent execution of project reviews is an important instrument of project management independent of the industry and should therefore not be missing from any government or commercial project.

The review methodology developed in the USA was adopted by the European aerospace agencies and industry. The idea of subjecting the respective development steps to a review at regular intervals is not new in principle. What is new, however, is the methodology and discipline associated with the process. Zeiss writes in this regard, "*However, this would not actually involve any new procedure in the development areas, because such meetings have of course always been common practice in all branches of industry.*" [22]

NASA Specification SP-6502 defines the purpose of reviews as, "*The purpose, wherever possible, is product improvement through formalization of the review process as well as thorough documentation that is a guarantee to contractor and client management that an optimum design proposal has been selected.*" [23] Arthur D. Little, concluded in a study for ESA about review techniques, that most knowledge of common review methods is found primarily in the aerospace industry, and that other industries have likely copied aerospace review methods [24]. This study suggests that ESA would need to create standard procedures and terminology for the review process to ensure that ESA and industry project managers speak a common language.

To standardize management methods, ESA has developed standard management procedures in close cooperation with European industry: ECSS (*European Cooperation for Space Standardization*). This is an initiative of ESA, the national space agencies, and the space industry to develop uniform standards for European spaceflight. The ECSS standards issued address the following major topics:

- Engineering (E)
- Quality (Q)
- Management (M) and
- Sustainability (S)

Among other things, the standards are intended to ensure the cooperation and competitiveness of the European space industry and are also used as a basis for contracts. Since 2014, the ECSS standards have also been published as European standards (*EN*); www.ecss.nl. The ECSS standards can also be applied to non-space projects.

7.3.2 Review Milestones and Their Chronological Assignment

Review milestones must be scheduled in accordance with the development flow of a project in terms of content and time. As shown in Fig. 4.2, the review milestones should be defined in such a way that the development steps coming to completion can still be subjected to a review before the final definition (*freezing*) of the basic documentation belonging to the respective development step. Following common management practice, the creation of phase-oriented basic documents, e.g., the requirements basis, specification, etc., is recommended. This ensures that the individual development steps are completed clearly and unambiguously prior to the start of the subsequent steps which can then be initiated based on a valid status; see Fig. 7.14.

The freezing of the respective *baseline documentation* requires the successful completion of the preceding review. However, a scheduled review does not always lead to the adoption of the baseline documentation. Missing or not yet fully completed documentation must be identified during the respective review and leads to ad-hoc scheduled special actions; e.g., the creation of Review Item Discrepancies (*RIDs*). Only after these special actions (*RID close out*) have been completed can the corresponding baseline documentation be released. The purpose of holding reviews is to uncover serious deficiencies, e.g., glaring interface problems. In extreme cases, the review may therefore have to be repeated. Tolerating grievances or improper completion of review milestones usually always leads to problems in the expensive subsequent phases.

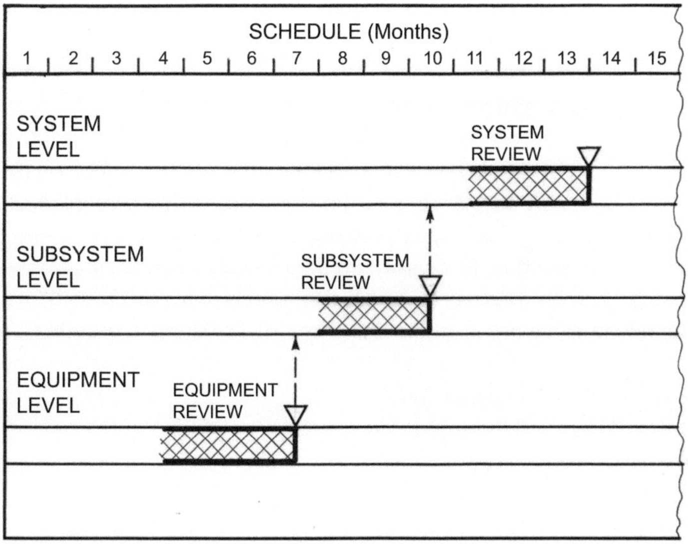

Fig. 7.16 Review integration process

7.3.3 Implementation of the Review Concept

The review concept of a project consists of a series of reviews that are performed at different project levels, i.e., design reviews (PDR and CDR) are to be performed, for example, at the system, subsystem, and equipment levels; see the V-model in Fig. 7.15. It must be ensured that the results of the lower-level reviews are incorporated into the review process of the next higher level in accordance with the V-model; see Fig. 7.16.

The main measures of the review process are as follows (*NASA*) [25]:

(1) Composition of the review team
(2) Definition of the responsibilities and powers of the review team and its chairperson
(3) Involvement of the client (*customer*)
(4) Determination of the review frequency
(5) Determination of the review input data
(6) Determination of the review output data
(7) Follow-up actions; e.g., RIDs (*Review Item Discrepancies*).

In order to successfully meet each review objective, the review team should have a collective technical competence possibly greater than that of the developer whose documents are being reviewed. NASA document SP-6502 states: "*If this collective advantage does not exist, the review program is unlikely to be successful.*" [25] In the same document, the following suggestions are made for staffing the review team:

(1) Assignment of permanent review specialists
(2) Recruitment of specialists
(3) Involvement of external experts (consultants)
(4) Combination of solutions (1)–(3)

The composition of the review team certainly depends very much on the circumstances of the company in question. In practice, one will be tempted to staff the review team as far as possible with permanent review specialists from the project area: depending on the type of review and the product, internal or external specialists will be called in for a certain period of time (*a few hours or days*). The following reflects the review team composition according to a NASA suggestion [20]:

(1) *Permanent review team members*
   - Project Manager/Committee Chairman
   - Systems Engineering Manager
   - Quality Assurance Manager
   - Production Manager
   - Test Manager
   - Reliability Manager
   - Head of Contract Management
   - Representative of the Client
(2) *Supporting/associated review team members*
   - Maintenance experts
   - Value analysis experts
   - Procurement experts
   - Other.

The project management is responsible for creating a review-friendly climate and must provide for each review appropriate personnel and the necessary preparation time, etc. For each review (*SCR, SSR, PDR, etc.*), an appropriate team shall be assembled in a timely manner and a chairman shall be appointed. The review chair, who is responsible for defining the input data, review schedule, etc., must be given the necessary responsibility and authority by the project management. For example, the review team must have the ability to initiate investigations to correct grievances in technically problematic areas or to present alternative proposals for discussion. In the case of smaller projects, the PM can take over the review management function as part of a project meeting.

In the case of reviews at the system level, the involvement of the client (*customer*) in the review process may generally be necessary in order to convince the client that the selected approach will lead to success and desirable in order to give him the opportunity to still intervene in a controlling manner. The participation of the client is particularly important in the early stages of the project, as he can still contribute his wishes and experience during this time. After the official system specification has been agreed at the end of Phase B (*SSR*), the customer's influence is only possible in the context of an official change; see Chap. 12.

Table 7.1 Example of a review plan (Source: Hermann)

Functional content	Responsible
Preparation:	
(1) Creation of the review plan including input requirements	Project Manager/Review Chairperson
(2) Template of Review Inputs (Data/Documents)	Project/Specialty Areas
(3) Pre-evaluation of Review Inputs	Review Team
Review Meeting:	
(4) Presentation of Review Inputs	Project/Specialty Areas
(5) Evaluation of technical documentation and preparation of a catalog of corrective measures	Review Team
(6) Release of the corrective measures catalog	Review Chairperson
Follow-on Tasks:	
(7) Publication of the corrective measures catalog	Project Manager
(8) Preparation of the Review Report	Review Chairperson
(9) Follow-on Work	Affected Project/Specialty Areas
(10) Final Review and Preparation of the Final Report	Review Team/Review Chairperson/Project Manager

The planning of reviews should be included in the overall project planning as early as possible. The review frequency, i.e., at which milestones and at which level the reviews are to be performed, is not always easy to determine. The importance of the project and the complexity of the system are critical factors in determining the reviews. It is easy to see that a complex system requires more review cycles than a less complex system, not only because of the greater number of system elements, but especially because of the greater interface issues. A detailed review schedule should be developed for each review process, showing the times for committee meeting preparations and follow-up work. A typical review procedure is presented by Hermann; see Table 7.1 [26]. It is important to mention in this context that the responsible engineers of the contracting parties at all levels (*top-down from the system to the components*) must record the agreements and confirm them by signature.

Explanations to Table 7.1:

1.: The review schedule should be known to all parties involved as early as possible. The main dates are usually part of the overall schedule. For a project lasting at least 2 years, a detailed review schedule, including all input requirements, should be distributed to all review participants several weeks before the scheduled review meeting.

2. and 3.: In the case of large projects, the review team should receive the inputs approx. 1–2 weeks before the review so that the review team can review documentation and carry out an appropriate preliminary evaluation. It can be assumed that the review team will also perform other tasks. Therefore, the aforementioned time frame is realistic for large projects.

4. to 6.: The review event can last from several days to a week for large development projects. However, for smaller projects, a few hours or 1–2 days are usually sufficient. The time to be provided depends on the subsystems to be reviewed and the quality of the review inputs.

7. to 10.: The follow-up work is to be seen chronologically in connection with the problems identified. However, the following benchmarks should be used here as well: The *action item and review report*, including RIDs (*review item discrepancies*), should be distributed within 1 week. Rework should take no more than 1 to a maximum of 2 months, so that the final report is available after 2 months at the latest. When setting these dates, it should be noted that schedule delays of the review process will have an impact on the overall project schedule.

The recommended durations are to be understood as a guideline for complex and large projects: For small and medium size projects, the suggested times should be shortened accordingly. For a small projects, a review can take place during a design meeting if necessary. It is important that the reviewer is independent and has sufficient time to review documentation.

The review inputs (technical documentation) to be provided to the review team differ in terms of their content for the respective review milestone, e.g., SCR, SSR, PDR, etc. For the System Concept Review (*SCR*), the main documentation to be submitted is the system requirements documents and the draft system specification, while for the Design Reviews (*PDR and CDR*), the main content is the submission of design documents (*detail specifications, block diagrams, drawings, etc.*). A typical documentation package (*input*) for the design review process (*PDR/CDR*) is reproduced below [20]:

Technology

(1) Equipment description
(2) Specifications, drawings, and sketches
(3) Logic diagrams (*for electrical and hydraulic equipment*)
(4) Operating concept of the device in the system
(5) Performance requirements
(6) Components list
(7) Detailed drawings for each custom design
(8) Test data for parts and components
(9) Tolerance analyses
(10) Design alternatives
(11) Development schedule
(12) Problem areas
(13) Known design flaws
(14) Operating descriptions
(15) Technical information for manuals and operating instructions

Project Management

(1) System technical aspects
(2) Overall project schedule
(3) Operating plan

(4) Maintenance plan
(5) Test requirements
(6) Storage and transport requirements
(7) Training program for maintenance personnel
(8) Training program for operating personnel

Reliability

(1) Reliability requirements
(2) Maintenance requirements
(3) Service requirements
(4) Safety requirements
(5) Reliability actual values during the equipment test
(6) Redundancy analysis
(7) Reliability analysis
(8) Security analysis
(9) Reliability predictions
(10) Comparison prediction/requirement
(11) Failure rate of components and parts
(12) Component selection
(13) Test requirements.

The availability of complete documents (*inputs*) is of great importance for a successful review. For this reason, appropriate checklists should be prepared for checking completeness. Figure 7.17 shows an example of a review checklist (*short version*), which is often much more comprehensive in practice [27]. The review outputs must essentially fulfill the following tasks:

- Establish a technical basis that gives the client's and contractor's management confidence that development is on the right track
- the requirements for taking corrective action, and
- a proposal for the release of the subsequent phase

System reviews are in principle important decision points for project management. The thorough analysis of the technical development status creates the basis for sound decisions. The output documentation must therefore contain the following information [20, 23]:

- Review organization (permanent review team)
- A complete list of all review participants
- Available input material
- Decisions during the review
- Action list (responsibility for action fulfillment and deadline)
- Schedule for follow-up actions

Question	Yes	No	N/A
1. Does the Specification contain all Customer requirements?			
2. Does the draft comply with all functional requirements?			
3. Does the draft reflect all environmental requirements (temperature, vibration, corrosion, etc.)?			
4. Was available information on similar developments reviewed and included (as applicable)?			
5. Were standard components used wherever possible?			
6. Were the required tolerances complied with during production?			
7. Does the design lead to optimal installation conditions?			
8. Does the design allow optimum maintenance conditions?			
9. Was a detailed value analysis performed?			
10. Were all safety requirements complied with?			

Fig. 7.17 Typical checklist for design reviews

7.4 System Engineering Management

7.4.1 Tackling Complex Problems

Systems engineering as a management function in the project is still a rarity. Even if the insight to the systems engineering mode of operation is present, the necessary organizational integration measures are often enough lacking. The implementation of the systems engineering management function in the project organization is an important prerequisite for solving systems engineering tasks satisfactorily. Haberfellner writes: "*SE (systems engineering) is a way of tackling complex problems and organizing their solution*" [28]. As shown in Fig. 5.3, the systems engineering task is an important management function of project management. Haberfellner answers the question "why is SE (*systems engineering*) necessary at all?" As follows: "*The constantly growing complexity in all areas makes it increasingly difficult or even impossible to gain and maintain an overview of a complex issue. Because of the close interconnection of problems, however, a holistic approach is absolutely necessary. Sub-problems that are interrelated may no longer be treated in isolation if the success of the whole, i.e., the smooth interaction of all elements, is not to be called into question*" [28]. The most important systems engineering tasks can be summarized as follows [29]:

- System analyses
- System specification
- System design
- Technical planning and control
  - System technical planning
  - Design monitoring
  - Review monitoring
  - Technical performance monitoring
  - Configuration control; (see Sect. 12.2)
  - Technical documentation
- System integration
  - Integration planning
  - Integration design
- System tests
  - Development tests (*simulations, functional tests, etc.—development models*)
  - Qualification tests (*functional tests, environmental tests, etc.—prototypes*)
  - Acceptance tests (*functional tests, environmental tests, etc.—series equipment*)
- Systems Engineering Reporting

7.4.2 Technical Performance Monitoring

Technical performance control *(TPC)* is a useful supplement to schedule and cost monitoring. Integrated project monitoring can only be carried out successfully with the help of functioning technical performance monitoring. The principle of technical performance control is described in detail in Sect. 9.4.7 in connection with integrated project monitoring.

7.4.3 Configuration Monitoring

Freezing of the baseline documentation after previously defined development steps (see Fig. 4.2) requires strict configuration monitoring. The developed *baseline documentation* represents an absolute technical reference value for the project. Changes or deviations from the baseline *(baseline changes)* can only be made if the resulting effects (*technical, scheduling, and financial*) are checked in advance and the change is approved by the Change Control Board (*CCB*). The basic principles of configuration management are described in detail in Sect. 12.2.

7.4.4 Industrial Systems Engineering in Germany

In 1981, the Association of German Engineers (*VDI*), together with the German Aerospace Society (*DGLR*) and the Society for Quality Management, held a working conference on the topic of *industrial systems engineering*. The Joint Committee on German Industrial Systems Engineering (*GIS*) assigned the overall topic to six working groups.

The VDI summarized the interim results as follows: "*They contain the clear statement that the application of systems engineering approaches is a necessity of modern industrial management in order to be able to hold one's own technologically in international competition in the future—quite particularly with regard to the competitive situation with Japan and the United States of America*" [30]. The individual results can be summarized as follows:

1. Public Relations
2. Requirements profile: Systems Engineer:
   - Training to become a systems engineer must build on a solid basic training as an engineer
   - Basic knowledge of various cooperation techniques
   - Basic knowledge of economics and business administration
   - Proficiency in at least one foreign language (*especially in the European environment*)
   - Basic knowledge of forecasting methods
   - Basic monitoring and control knowledge
   - Mastery of systems engineering planning
   - Management Methods
3. Systems engineering education and training for professional engineers:
   - Engineers of all disciplines need to be introduced to the thought process of systems engineering and informed about the possibilities and interrelationships of working methods
   - Elements of the systems engineering methods should be addressed prior to the pre-exam
   - The deepening of the acquired knowledge must take place towards the end of the study program through a compulsory course of at least one semester on the fundamentals of systems engineering for all engineers
   - It is necessary to point out the economic importance of the methods and procedures of systems engineering
   - Incorporate systems engineering into the corporate leadership development program
4. Systems engineering education and training in administration and non-technical management:
   - Often technical and commercial problems can only be solved holistically
   - All participants in a project (e.g. also the business administrators, financial people, lawyers, etc.) must have knowledge of system technology
   - Acquisition of basic systems engineering knowledge in the course of a general university/technical college education
   - In-service training
   - Working meetings between technicians and non-technicians
   - *On-the-job* training

5. System technical requirements for clients:
   - Systems thinking, a prerequisite for tackling major projects, requires interdisciplinary action
   - Systems engineering offers suitable tools for handling projects more efficiently
   - In the future, technical projects are to be tackled using systems engineering methodology
   - Emphasis must be placed on analytically sound planning
   - Decisions must be made on the basis of systematically prepared criteria
6. System technical requirements for contractors:
   - Systems engineering to make the required knowledge manageable and thus accessible for decision-making purposes
   - Consideration of systems engineering concerns especially where the limits of financial and technological capability are reached
   - System-technical reasonable precautions with regard to organization, motivation, training and methodology

7.5 Innovation Management

7.5.1 In the Beginning Is the Invention

The nucleus of all technological processes begins with the idea for something new—a new product, process or service—in short: a new business idea. The inventor or inventors can be private individuals or employees of a company. Inventors who are independent of any ties to a company often go their own way. For example, the two entrepreneurs "*Bill*" Gates and Elon Musk (*two of the best-known entrepreneurs of recent times*). Gates' ideas created the software company Microsoft with worldwide market penetration and Musk who created Tesla, a manufacturer for electrically driven motor cars and SpaceX which confronts the international launch service providers with highly competitive low-cost launch solutions.

However, ideas often arise in the workplace environment. The corporate climate plays a major role here. The German aerospace company MBB, headquartered in Munich, was considered "*Germany's think tank*" from its founding in 1968 until it was taken over by Deutsche Aerospace AG (*DASA*) in 1989. The innovative spirit fostered by Bölkow was a major key to the company's success. Bölkow mentions a selection of groundbreaking innovations in his book "*Committed to the Future*:

- The Bo 105 helicopter with a hinge-less rotor "System Bölkow" [31]
- A technical innovation in the transport sector "*... the developed magnetic levitation technology. The first ideas for this began in the mid-sixties*" [32]
- Development of a "*ten-Newton control engine*" for rockets and satellites [33]
- In 1971, the safety system >Airbag< was developed in close cooperation with the automotive industry as a kind of waste product of a military development [34]

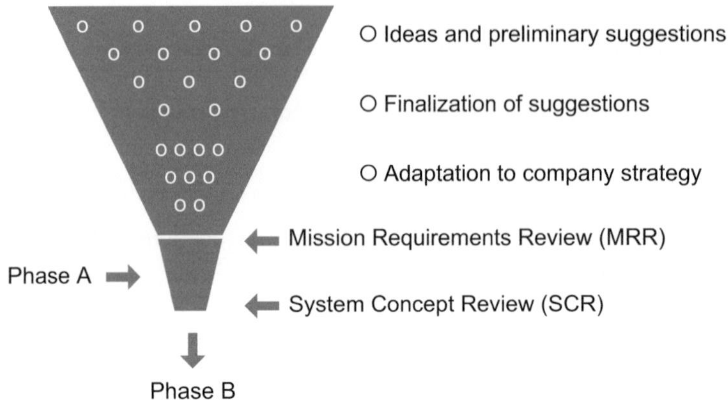

Fig. 7.18 Funnel for idea collection and presorting

Schubert, former head of MBB's space division, mentions on November 17, 1979, on the occasion of a conference of the Protestant Academy in Tutzing (*near Munich*), in his lecture "*Creativity in Big Industry,*" that in the past often very young scientists between the ages of 20 and 40 achieved analytical-mathematical top performances. He made reference to Planck, Newton, Galois, Gauss, Einstein, Heisenberg, v. Weizsäcker and Dirac [35]. In this context, the first German space pioneer Hermann Oberth, who at the age of 29 wrote the book "*The Rocket to Planetary Spaces,*" should also be included in the ranks of very young creative personalities [36].

7.5.2 Implementing Inventions Through Systems Engineering

How can inventors' ideas be successfully implemented? Are there tried-and-tested systems engineering methods? First, of course, you have to separate the wheat from the chaff, because not every idea has the potential to be a success story. Sometimes you have to *try things out,* because "the *proof of the pudding is in the eating*. But there is often a trivial problem, because trying costs money. The question is, how can the system engineer or the system architect (see Sect. 7.5.3), help and capture the ideas with a funnel (see Fig. 7.18) and pre-sort them accordingly? One way of capturing is the *capture plan,* in which e.g. the following questions have to be answered:

- Does the idea work in a practical test (*laboratory, simulation, etc.*)?
- Does the idea fit the company strategy?
- Is there an interested party or a market?
- Can you create a market interest?
- Are there internal/external customer contacts:
  - Possibly already existing customer connections?
  - Any existing partner connections?

- Are market strategies provided?
- Are there sponsors to market the idea?

In order to convert an invention that has successfully passed the initial filtering process after a review into a concrete project, the phase model described in Chap. 4 is applied, starting with Phase A or Pre-Phase A.

Thom points out that the economic earning power of companies can be significantly increased by the "*Betriebliches Vorschlagswesen*" (*BVW*)* and thus also contributes to the prosperity of a society [37]. Despite its great importance for companies, BVW has only been used to a limited extent in recent years, according to Thom.

Schubert points out that the creation of an innovation climate is an important prerequisite and lists the following aspects to be considered for long-term and successful R&D work [38]:

a) The volume of long-term R&D work must be proportionate to overall development. An appropriate ratio is 10–20%
b) The research area must not become the subject of controversial, short-term board discussions or decisions
c) The management of the research area should be assigned to an experienced scientist or development engineer who applies their management authority extremely cautiously and sees his task primarily in stimulating and advising the young team
d) The research teams should consist of extremely young employees with a high level of initiative
e) The (spatial) accommodation should promote team and group building
f) In general, the working method is oriented towards particularly creative groups of the present and the past
g) With regard to task finding, one deliberately ventures into very large problems
h) The inventive process does not proceed schematically in a straight line, but includes aberrations, setbacks, leaps and perplexing standstills

**BVW company internal suggestion scheme*

7.5.3 Systems Engineering Versus Systems Architecture

The terms systems engineering and systems architecture are twin terms and not opposites. In technical projects, the term "*systems engineering*" is mostly used, while the term "*systems architecture*" is preferably applied in the construction industry and in the IT and communications sector. However, the tasks are largely identical. In the previous sections, the systems engineering processes were described in detail with examples from the aerospace industry. The processes are transferable to technology projects in other industries. However, they can also serve as a guide for non-technology projects.

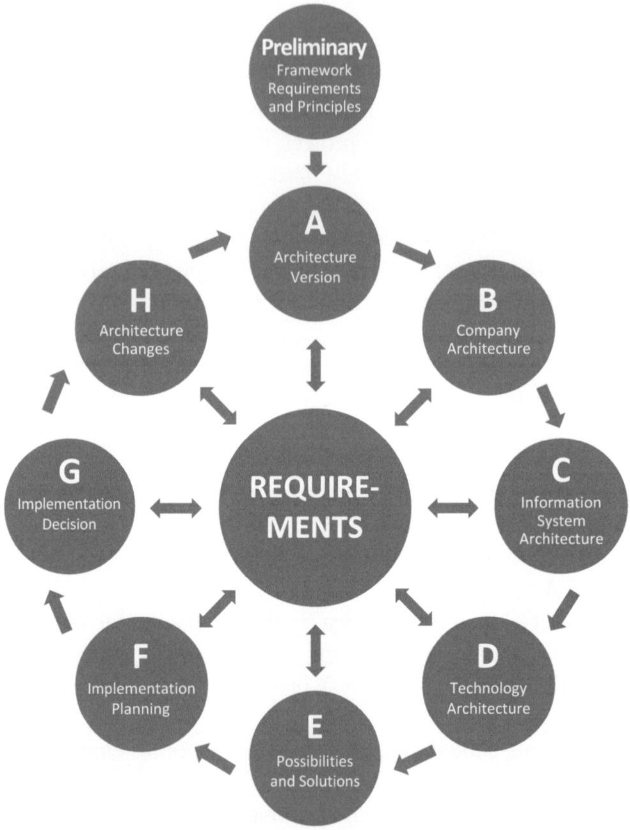

Fig. 7.19 Architecture Development Method (ADM)

For completeness, reference is made to TOGAF (*The Open Group Architecture Framework*) to define the development process (*Architecture Development Method—ADM*) with ten steps; see Fig. 7.19 [39].

1. Initial step: clarify integration of other models, define model adjustments and establish important principles
2. Step A (*Architecture Vision*): Defining the goals and stakeholders in updating the enterprise architecture
3. Step B (*Business Architecture*): Description of the current business architecture and the desired state
4. Step C (*Information and Systems Architectures*): Description of the current and desired application, information, and data architecture
5. Step D (*Technology Architecture*): Description of the current and desired technology architecture, including specific hardware systems
6. Step E (*Opportunities and Solutions*): Determination of the projects for the transformation from the current situation to the target state

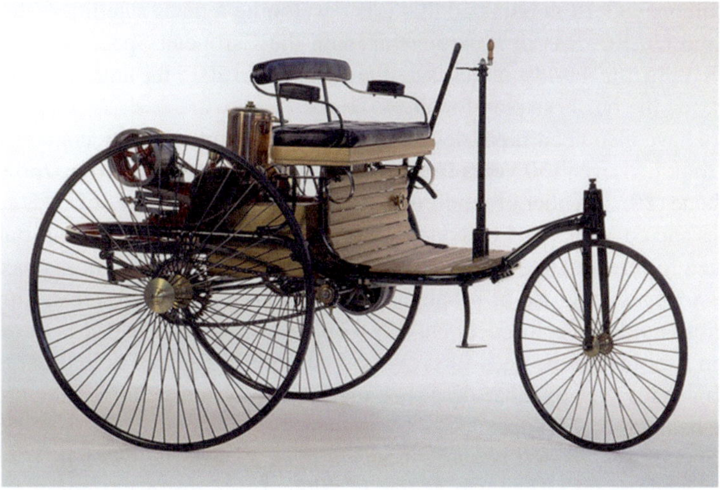

Fig. 7.20 Benz Patent Motor Car—The first automobile (1885–1886) (Source: Creative Commons)

7. Step F (*Migration Planning*): Planned transition from the current state to the target state
8. Step G (*Implementation Governance*): Monitoring of implementation to the target state
9. Step H (*Architecture Change Management*): Collection of requirements and external influences, which then serve as the basis for the next step of the process
10. Requirements Management: Requirements management continuously drives the ADM process and is therefore at the heart of the process

7.5.4 System Engineering Challenges of the Future

Technical Progress Leads to Prosperity

The technical progress evident today in many areas of life has led to prosperity and comfort in many countries. In the areas of transport (*land, air, and water*), energy supply and communications, major successes have been achieved in recent decades. From the first passenger car (*Benz Patent-Motorwagen, 1885/86*) to 2015, it took only 130 years for nearly 90 million vehicles to be produced worldwide, according to the World Automobile Association (*OICA*). Figure 7.20 shows the first automobile patented by Benz. Continuing, the automotive industry is currently focusing, among other things, on the development of electric vehicles and driverless driving with an autopilot, as has been practiced in aviation for some time.

There is an interesting parallel between electrically powered satellites and self-driving electric cars. These are systems engineering feats. The European space industry has also recently been developing satellites with electric rather than chemical propulsion systems. In a press release dated March 11, 2016, Bremen-based space company OHB announced that it is developing a news satellite with electric propulsion (*ELECTRA*) in a partnership with the European Space Agency (*ESA*) and SES Astra, the world's largest satellite operator. In 2015 the author was involved in preparing the business plan for ELECTRA.

There have also been rapid developments in communications technology in the past. It took just over 150 years from the first telephone (*Philipp Reis, 1861*) to the smartphone (*2017*). Other groundbreaking developments, such as railroads connecting cities and countries, radio and television, which make our lives comfortable, are now taken for granted. New ground is being broken in the field of unmanned aerial vehicles (*drones*), the use of robots and the networking of IT systems. The list of innovations could be continued, but this would be too extensive.* It is the author's intention to point out the need for a systems engineering approach in order to avoid project mishaps such as those described in Chap. 3. Systems engineering processes must be handled by SE specialists.

**Technically interested people are encouraged to visit the German Museum in Munich, founded by Oskar von Miller in 1906. There are a variety of technical exhibits to admire there, ranging from steam engines to rocket propulsion and satellites; see "Space Cowboys" published in Capital magazine, Oct. 20, 2016.*

The development of products and processes is a continuous process, and the tasks of project management and systems engineering are extremely important because in modern times it is no longer just a question of developing and producing individual assemblies, e.g., an engine or a pump, but of integrated overall solutions. Often, the customer also wants support in the operation of a system—in short, a complete solution. This leads to systems engineering processes in which the interaction of the various engineering disciplines is crucial. The system concept is also gaining ground in the networking of subsystems to form a system. The networking of robots in production is similar to the global networking of ground stations during the launch of a satellite.

The fourth industrial revolution is cited in various publications. The term "*Industry 4.0*" is associated with the goal of ushering in the so-called fourth industrial revolution. Modern information and communication technologies are to be used to integrate and digitally network subsystems (*robots*). The system is to enable largely self-organized production. The individual subsystems, i.e., machines, plants, etc. communicate and cooperate directly with each other. The realization of such a system is a classic SE task that must include a technical and economic optimization process with consideration of life cycle costs (*LCC*).

When talking about the fourth industrial revolution "*Industry 4.0,*" the question arises about the previous revolutions. Bauernhansl from the Fraunhofer Institute (*IPA*) Stuttgart, Germany mentions in an article [40]:

- 1750—Beginning of the first industrial revolution. Beginning of mechanization with water and steam power
- 1870—Start of the second industrial revolution. Beginning of mass production with the help of assembly lines and electrical energy

- 1960—Start of the third industrial revolution. Use of electronics and IT to automate production
- 2014*—The fourth industrial revolution. Intelligent and digitally networked systems designed to enable largely self-organized production
- **At CEBIT 2014, Prof. Scheer described in a keynote speech (You Tube) the impact of "Industry 4.0" with regard to the digital future of industry and rounded this off by referring to the "Real Time Smart Factory."*

Roland Berger's publication "*The digital transformation of industry*" summarizes the need for industrial digitization in *the* following three points [41]:

"1. The digital transformation of industry offers Europe enormous opportunities—and presents it with major challenges. The promising opportunities of networked, more efficient production and new business models are offset by dramatic risks. By 2025, Europe could see an increase of €1.25 trillion in industrial gross value added, but also suffer a loss of €605 billion in value added.

2. The digital transformation of industry is confronting Europe's economies with radical structural change. New data, networking, automation and the digital customer interface are disrupting existing value chains. Companies need to question their products and capabilities and increase their digital maturity in order to identify, develop, and quickly implement new opportunities.

3. The digital transformation of industry calls for joint action by Europe. A rebalanced, unified, regulatory framework needs to be put in place, as well as a comprehensive, high-performance information, and communications infrastructure to secure the conditions for the competitiveness of European industry. A concrete next step for Europe's businesses and policymakers is to aggressively address the standardization of the digital economy."

Now, of course, the question arises as to how the digitization of industry will affect the job market. In the development offices, project managers and systems engineers will continue to have plenty to do to handle the upcoming projects. But how will it affect production? Spath from the Fraunhofer Institute takes a positive view: "*In this regard, I am absolutely certain that working people will continue to be the focus, even in an end-to-end virtualized and computerized factory. Our qualified employees' close sensory gaps that will always exist. They have many years of experience in assessing and solving exceptional situations. And as a workforce, they bring their creativity and flexibility to the processes. At the same time, the new technologies and support options also offer previously unthinkable development opportunities for employees. They can contribute their skills in a meaningful way and are relieved of monotonous or physically demanding tasks*" [42].

7.5.5 Space Drives Engineers Towards Effective Management

7.5.5.1 Interaction of Systems Engineering and Project Management

Spaceflight has been an important driver for effective management methods in recent decades due to the successful implementation of international and national

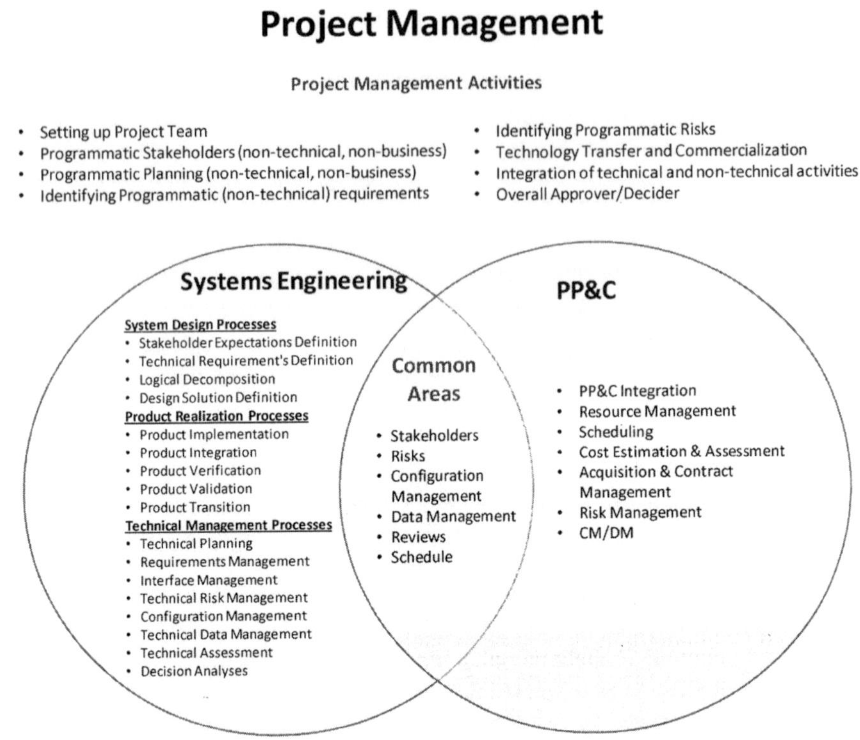

Fig. 7.21 Systems engineering as part of project management (Source: NASA)

projects. Chapter 21 describes in detail the historical course of space travel and the associated management development, starting with Sputnik. In this context, special mention should be made of the Apollo program. NASA fulfilled the goal set by President Kennedy on May 25, 1961, to land on the Moon on July 21, 1969, the day Neil Armstrong set foot on the Moon, within a little more than 8 years. This was an outstanding achievement by the technicians and managers in charge. To achieve the ambitious goal, NASA, in close cooperation with other agencies such as the USAF and the space industry, developed effective management procedures that continue to have an impact today. They were the basis for more advanced developments in systems engineering and project management as required for the efficient execution of technology projects. Former NASA Director James E. Webb emphasizes the importance of management in his book *Space Age Management* [43].

In the latest edition of NASA's "*System Engineering Handbook,*" [44] the close relationship between the responsibilities of systems engineering (*System Engineering*) and project planning and control (*Project Planning and Control*) is described under the generic term project management. This is illustrated in Fig. 7.21.

In addition, quality and reliability are naturally a particularly important issue for space travel. This is true for commercial and manned space flight. Commercially operated satellite systems, for example, must function maintenance-free in orbit for many years. Reliability is therefore an important economic factor. And astronauts should be able to return safely to Earth after their work in space is done. It is not surprising that the responsible space agencies and private operators have set correspondingly high standards for management, including system engineering and quality management.

The proven success rate (*heritage*) has always played a major role in the selection and procurement of systems for space missions; see Chap. 14. According to a scale developed by NASA for assessing product maturity (*Technology Readiness Level—TRL*), the highest TRL level (*TRL-9*) should be targeted in commercial or manned missions in order to avoid all risks, e.g., that of failure; see TRL definition in Sect. 9.4.7. Walter, science astronaut and full professor for space technology at the Technical University of Munich, considers it essential that the manned NASA Mars mission system planned for the coming years, including all subsystems and components, must have reached a high qualification level (*TRL level-8/9)* in order not to endanger the astronauts' lives [45].

The management methods developed and tested in recent decades by NASA, ESA, and other agencies have in principle proved their worth. However, there is also justified criticism that the requirements for industry are often excessive and too bureaucratic. Especially for commercial space projects, a simplification of the requirements is necessary. Streamlined methods have therefore been applied under the title of "*lean management*". However, this is not to be equated with the term "*agile project management*" because thorough planning and monitoring is still the basic prerequisite for efficient project management.

Rather, it is a matter of applying the delegation principle more efficiently. Public and private customers must delegate planning and monitoring authority to the contracted companies more than before and hold back on overly detailed execution regulations. This can best be done in connection with the WBS levels. After all, there is a serious difference between, for example, understanding the ECSS* standards developed by ESA in cooperation with the space industry as a mandatory requirement or as a guideline and, in the latter case, allowing companies more flexibility in execution.

**European Cooperation for Space Standardization*

7.5.5.2 Management of Future Space Projects

The international and national space agencies as well as the space industry can point to considerable management successes in the handling of complex projects. In this context, special mention should be made of the European Space Agency ESA, which currently has 22 member countries and has developed a balanced and competitive management concept over the past decades. Its methods for handling complex projects are also exemplary for non-space projects.

In addition to the standard programs of national and international aerospace agencies, which should not go unmentioned here, exotic-sounding manned and unmanned space missions are being considered worldwide in the long term. These include unmanned and manned flights to Moon and Mars, as well as unmanned missions to discover and mine ore deposits on asteroids using robots. NASA is

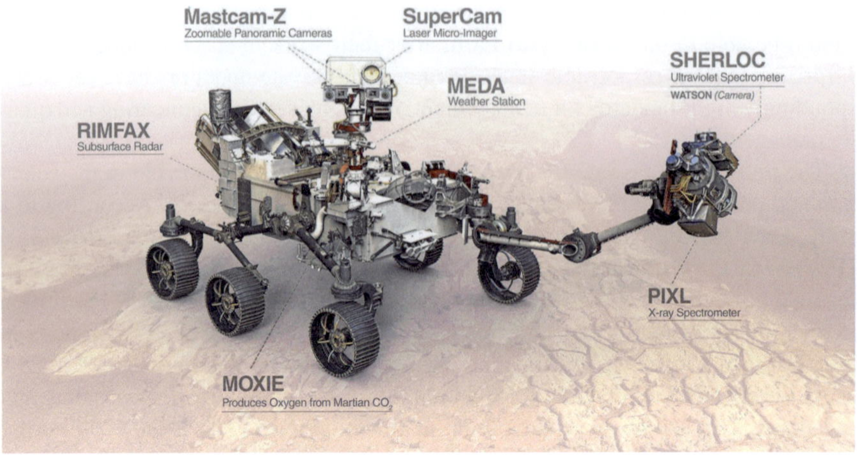

Fig. 7.22 NASA's Mars Rover Perseverance (*Source: Creative Commons*)

investigating advanced space systems and propulsion technologies for an "*asteroid mining*" project and the Luxembourg government developed a legal framework on the subject. Etienne Schneider, Deputy Prime Minister of Luxembourg, stated: "*The legal framework we put in place is perfectly in line with the Outer Space Treaty. Our law does not suggest either establishing or implying in any way sovereignty over a territory or over a celestial body. Only the appropriation of space resources is addressed in the legal framework*" [46]. In a press release dated April 4, 2017, OHB Bremen announced the establishment of the company "*Blue Horizon s.a.r.l*" based in Luxembourg. The company's objective is the extraction of space resources.

A manned flight to Mars is a goal of NASA, ESA and the Russian space agency ROSKOSMOS; Fig. 7.22 shows NASA'7s unmanned Mars Rover Perseverance. China and India are also interested in corresponding space projects. Mars is also a goal for newcomer and entrepreneur Elon Musk when he says, "*I'm pretty sure we (SpaceX) will get a contract from NASA to bring landers and rovers to Mars*" [47]. In addition, SpaceX is planning a manned lunar orbiter for four people [48].

Walter suspects, however, that a manned Mars mission will not be possible until 2048 in view of the technology work yet to be completed and an optimal Earth–Mars orbital configuration [45].

Systems engineering processes and project management play an important role in the planning and realization of the aforementioned projects. In the case of commercial space projects, the focus is not only on systems engineering solutions but also on economic viability, i.e., consideration of project costs and forecast profits. This applies to satellite operators, e.g., Eutelsat, Inmarsat, Intelsat, and SES, as well as to satellite and launch vehicle manufacturers. It also applies to the SpaceLiner (*SL*) project investigated by DLR, Germany, in which the author was involved with contributions, because the SL project should, according to DLR, be able to be operated economically. Project studies were carried out to investigate the required

Fig. 7.23 SpaceLiner flight configuration (*Source: DLR-SART*)

technologies, business areas and economic viability. System engineering processes and project management play an important role in this.

7.5.5.3 *"SpaceLiner"* Rocket Aircraft from DLR

In 2016, DLR's System Analysis department for Space Transportation (*SART*) completed a preliminary study (*Pre-Phase A*) for the "*SpaceLiner*" (*SL*) rocket plane. The SL is a suborbital hypersonic winged passenger transporter. During the multi-year preliminary study, technological and commercial system studies were performed and the binding mission requirements were defined at the end of 2016; [see *Mission Requirements Document (MRD)*] [49]. Figure 7.23 shows the SpaceLiner flight configuration 7. The release of the MRD set the stage for the start of a Phase A. At the completion of development (*Phases A through C*), including flight testing with two prototypes, a Technical Readiness Level* of TRL-9 must be achieved before volume production can begin.

**Maturity level TRL—see definition at Sect. 9.4.7.*

Future projects must pass through several go/no-go decision points before they can be realized. At certain milestones (*gates*), *feasibility* is examined and a decision is made as to whether the project in question can be approved. In addition to technical aspects, economic and political aspects are also decisive. It is important that phase A is carried out thoroughly to avoid surprises (*mishaps*) as described in Chap. 3.

The most essential technical aspects of the SpaceLiner project are described by DLR SART shown below:

(1) *Concept: The two-stage, vertical-launch configuration consisting of an unmanned booster and a manned passenger stage (orbiter) is designed for 50 passengers and has a total of eleven liquid rocket engines (booster 9, orbiter 2), which are powered by cryogenic oxygen (LOX) and hydrogen (LH2). Once the engines have burned out, the Orbiter stage can glide over large intercontinental distances in a very short time. Depending on the mission, flight altitudes of 80 kilometers and Mach numbers above 20 can be achieved. Flight times with the SpaceLiner (SL) are to be no more than 90 minutes on the Australia-Europe route or no more than 60 minutes on the Europe-California route. With maximum accelerations of 2.5 g occurring briefly during takeoff, the stresses for passengers on these missions remain below those of the Space Shuttle astronauts. In addition, the concept envisages the design of the passenger cabin in the form of a separate escape capsule, which can be separated from the vehicle in an emergency and should enable the passengers to return safely to Earth.*

According to DLR, the first operational system is conceivable between 2040 and 2050. The core aspect of the concept is the complete reusability of rocket stages in conjunction with series production comparable to aviation, which is expected to result in increased cost efficiency compared to current space transportation systems. A key challenge is to improve the safety and reliability of space components such as rocket engines so that they are suitable for daily use in passenger transport.

The development of the concept is supported by internal DLR funds as well as within the framework of EU research projects. In addition to DLR, various partners from the European aerospace sector are also involved.

(2) *Development to date: Currently, the SL is in the preliminary design stage. Based on the results of the respective previous studies, the design will be advanced in the course of development and with increasing consideration of subsystem development and integration. In parallel, studies are being conducted with regard to variants with modified requirements and specifications, the results of which can also be incorporated into the overall design process.*

The SL7 version is the current configuration, which is currently under investigation at DLR. As part of a numerical optimization, the double delta wing was replaced by a single delta wing, among other things, to improve the aerodynamic, thermal and structural mechanical properties in hypersonic flight. In addition to the 50-passenger reference design, the possibility of deriving a 100-passenger version of the SL suitable for routes with a larger market volume but shorter distance than the Australia Europe reference case was also investigated. To fulfill the different missions, a long or a short version of the booster stage will be used, depending on the required range, and will be combined with the 50- or 100-passenger version of the orbiter, respectively.

The mission requirements of the SL project were defined in 2016 as part of an MRR [49]. This includes the possibility of a special variant of the upper stage for transporting satellites into space, taking advantage of the cost benefits of joint series production with the passenger variant. Based on the requirements, a targeted development of the concept can be carried out as part of Phase A.

(3) *System data: There are two different SL versions*; see Fig. 20.3:
  a) *the point-to-point version for ultra-fast transport of passengers and*
  b) *the orbital version for transporting goods (payloads) Orbital positions.*
(4) *Most important technical parameters of the SL system:*

Identification size	Dimensions
o *Length:*	*84 m* *
o *Wingspan:*	*38 m* *
o *Height:*	*22 m* **
o *Cabin length:*	*15 m*
o *Take-off mass:*	*1840 t* *
o *Max. Flight altitude:*	*80 km* ***
o *Max. Speed*	*25,000 km/h* ***
o *Max. range:*	*18,000 km* ***

*Large version
**Total (Australia-Europe)
***Orbiter

(5) *Propulsion: The SL concept assumes that only a single version of reusable liquid rocket engine will be used, operating on the main power cycle with full pre-combustion. The expansion ratio of the nozzles is adapted to the different missions of the booster and orbiter. The propellant will be the high-energy and environmentally friendly combination of liquid hydrogen with liquid oxygen in a 6:1 mixing ratio.*

For SL development, including flight testing, a timeframe of approximately 20 years is envisaged [49] to ensure that the SL system is fully qualified before commercial entry into service. For the implementation of Phase C (*development including flight testing*) of the basic 50-passenger version, the investment volume was determined on the 2015 price basis [49]. The cost determination was performed as part of a dissertation effort and was based on a thorough analysis of preliminary technology approaches and the use of proven cost estimation methods [50]. Multiple sources and the results of mathematical cost estimation models were considered to increase the confidence of the cost determination. In subsequent phases, life cycle

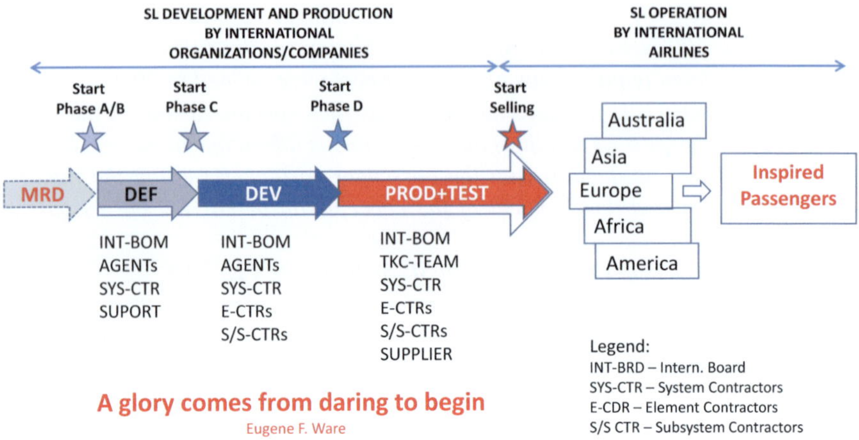

Fig. 7.24 SpaceLiner program phases

costs (*LCC*) should be optimized or reduced by appropriate measures, such as DTC methods (*see also Sect. 10.6*). Concrete starting points could be, for example, the inclusion of commercially available off-the-shelf components and the use of assemblies from other programs.

In July 2016, the author gave a presentation at the seventh SL workshop entitled "*SpaceLiner System—Business Plan Considerations*" in which he outlined the program phases; see Fig. 7.24. He pointed out the need for global cooperation in order to spread the high development costs over several countries. The use of efficient international management that has been tested in practice is an important prerequisite for the success of the project. In this context, reference is made to the successful cooperative model of the ISS space station, involving Europe (*ESA*), Japan, Canada, Russia and the USA *(NASA)* [51].

Literature

1. Zahn, Erich: Handbuch Technologiemanagement, Schäffer-Poeschel Verlag, Stuttgart 1995, p. 9
2. Bölkow, Ludwig: Der Zukunft verpflichtet, Herbig Verlagsbuchhandlung, Munich 2000, p. 57ff.
3. Selding, Peter B.: SES ready to invest in reusable rockets, in-orbit satellite services, Space News, 8th April 2016.
4. Kappel, Rolf and Schwarz, Ingo A.: Systems Research 1970-1980, Vandenhoeck & Ruprecht, Göttingen, 1981, p. 27.
5. Madzar, Goran: System architects - why we need them and what they should be able to do?, Medtech Engineer, 23.5.2015.
6. Blanchard, Benjamin S. and Fabrycky, Wolter J.: Systems Engineering and Analysis, Prentice-Hall, Inc, Englewood Cliffs, New Jersey, 1981, p. 11.
7. ESA, ECSS-M-ST-10C, Rev. 1, p.16
8. Von Weizsäcker, Ernst: Offene Systeme I, Ernst Klett Verlag, Stuttgart, 1974, p. 10.

9. Vester, Frederic: Überlebenschance für Wirtschaft und Gesellschaft, in: Management-Zeitschrift zfo 51 (1982) No. 2, p. 89.
10. Ropohl, Günter: Systemtechnik - Grundlagen und Anwendung, Carl Hanser Verlag, Munich-Vienna, 1975, p. 7
11. Source 10, p. 16
12. Henning, Th.: Can Power Plant Planning be Improved with the Aid of Systems Engineering? VGB Kraftwerkstechnik 60, March 1980, p. 164.
13. Miller, Dale Eugene: Managing a Contract (Project) Specification System for Satellite Projects, in: Project Management and Project Control, 10th ESRO Summer School, Frascati, Sept. 1972 (ESRO SP-90), p. 129.
14. DoD: Military Standard-Specification Practices, MIL-STD-490A, 4/6/1985, Enclosure 1.
15. DoD: Military Specification - Specifications, Types and Forms, MIL-S-83490, Aug. 31, 1995.
16. ESA (ESRO): Uniform Project Specification - Format, Content, and Control, QRA4, Issue I, 1972.
17. Sperling, F. B.: The Spacelab System Verification Programme - A Powerful Engineering-Management Tool, in: ESA Bulletin No. 26, p. 64.
18. Billig, Victor: Space Project Management, Politics and International Cooperation, in Industrial Cooperation, Verlag TÜV Rheinland, 1989.
19. Dreger Wolfgang: Systems Engineering - A New Discipline or a Systematization of Banalities?, in: Siegener Hochschulblätter, 3 (1980) Heft 1, S. 37-54, University GH Siegen.
20. NASA: Design Review Requirements, SR-QUAL-66-1, January 20, 1966.
21. ESA, ECSS-M-ST-40C, Rev. 1, 2009, p.32.
22. Zeiss, Chr. H.: Design Review bei Gemeinschaftsprojekten - Beispiel Nachrichtensatellit Symphonie, VDI-Berichte No. 307, 1978, p. 29.
23. NASA: Elements of Design Review for Space Systems, NASA Document SP-6502, 1967, p. 3.
24. Little, Arthur D.: A comparative Study on Project Review Techniques, Final Report, ESA Contract No. *4846/81fNLfPP* (SC), 1983, p. 3.
25. Source 23, p. 5
26. Hermann J. J.: Design reviews for satellites and space probes, VDI Reports No. 192,1973.
27. Jacobs, Richard M.: Implementing Formal Design Review, Proceedings of the 1967 Annual Symposium on Reliability, Washington D. C., January 1967, p. 124.
28. Haberfelner, Reinhard: Systems Engineering (SE) - A Method for Solving Complex Problems, in: ZO, 7/73, pp. 373-374.
29. Madauss, Bernd-J.: Planung und Überwachung von Forschungs- und Entwicklungsprojekten, AIBFachliteratur; Bad Aibling, *1978f81*, p. II-8.
30. VDI: The VDI joint connection "Industrial Systems Engineering (GIS)" introduces itself; Annex: Recommendations for education and training, procedure plan for public relations, Düsseldorf, April 11, 1983.
31. Source 2, p. 204
32. Source 2, p. 208
33. Source 2, p. 336f
34. Source 2, p. 253
35. Schubert, Johannes: MBB, lecture on the occasion of a conference of the Protestant Academy in Tutzing, 17.11.1979; p. 8.
36. Oberth, Hermann: The Rocket to Planetary Spaces, Verlag R. Oldenburg, Munich, 1023.
37. Thom, Norbert, Betriebliches Vorschlagswesen, 6th ed., 2003, Verlag Peter Lang, Bern, Switzerland
38. Source 36, p. 15ff
39. TOGAF (The Open Group Architecture Framework) https://de.wikipedia.org/wiki/TOGAF
40. Bauernhansl, Thomas: The Fourth Industrial Revolution - The Road to a Value-Creating Production Paradigm, Industry 4.0 in Production, Automation and Logistics, DOI 10.1007/978-3-658-04682-8_1, @ Springer Fachmedien Wiesbaden 2014.
41. Berger Roland, The digital transformation of industry, BDI

42. Spath, Dieter, Fraunhofer IAO, Study: Production Work of the Future - Industry 4.0 [ISDN 978-3-8396-05 70-7]
43. Webb, E. James, Space Age Management, McGraw-Hill Book Company, New York, 1968.
44. NASA, System Engineering Handbook (SE 20170001761), 2016.
45. Walter Ulrich. Im schwarzen Loch ist der Teufel los, Verlag Komplett-Media, 4th ed.
46. Space Mining gets a Boost through Luxembourg's new Law. Source not published.
47. Vance, Ashlee: Elon Musk, FinanzBuch Verlag, Munich, 2015, p. 228
48. SpaceX to Send Privately Crewed Dragon Spacecraft Beyond the Moon Next Year
49. DLR-SART, Bremen. SpaceLiner Mission Requirements Document, SL-MR-SART-00001-1/2, Issue 1, Rev. 2, 08.07.2016 (not published).
50. Trivailo, Olga: Innovative Cost Engineering Approaches, Analysis and methods Applied to SpaceLiner - an advanced, Hypersonic, Suborbital Spaceplane Case-Study, PhD thesis, March 2015, p. 33.
51. Madauss, Bernd-J.: SpaceLiner Business Considerations, presentation on the occasion of the 7th SpaceLiner Workshop of DLR SART in Julie 2016.

Quality Assurance Management for Projects

8

Contents

Abstract

***In addition to functionality, quality is another main feature of a product or system, because quality is also to be seen as an economic factor. "It's not enough that we do our best, sometimes we have to do what's required,"* Sir Winston Churchill (1874–1965).**

Product assurance is a prime management function (PAM); see Fig. 5.3.

The value of industrial products cannot be measured solely on the basis of performance and beauty features but also includes the characteristics generally described by the term quality. The buyer of a car usually compares, for example, the performance, purchase price, operating costs, and quality of the vehicle. In addition to

B. -J. Madauss, *Project Management*,
https://doi.org/10.1007/978-3-662-69057-4_8

rational decision-making criteria, subjective factors such as previous experience and design (*beauty*) also play a role in the purchase decision. In this regard, see the statement by Raymond Loewy "*Ugliness sells badly.*"

When comparing the performance characteristics of vehicles (*capacity, speed, acceleration, etc.*), we may be dealing with necessary (*minimum requirements*) performance characteristics on the one hand and additional, desirable (*extras*) performance characteristics on the other. In both cases, there is a direct relationship between performance and price. The situation is very similar with product beauty because it too has its price. Although the beauty attribute certainly does not play an immediate role in a number of industrial products, e.g., factory equipment, beauty, or a certain style or taste is very important for goods in the consumer industry. In addition, as a further important factor, the quality of a product is of very special significance for the buyer, because the smooth and safe use (*availability, safety, etc.*) largely depends on it. Product quality is directly related to acquisition costs (*purchase price*) and operating costs because quality also costs money. Quality can also be a decisive argument for some customers.

Quality is a high asset and a key issue, especially for high-tech projects, because other factors such as reliability and availability depend on quality. Quality is particularly important for space projects, because errors can easily lead to catastrophes. With the introduction of the term "*Product Assurance*" (*PA),* the tasks of *quality*, *material selection*, *reliability*, *availability*, *maintainability* and *safety* are summarized. See the definitions of NASA [1], ESA [2, 3], and DLR [4]. A comparable PA definition has been made by the Chinese aerospace company CAST [5]. Thomann defines product assurance as follows: "*A modern method for economically ensuring the quality, reliability, maintainability, and safety of complex technical systems.*" [6] Arianespace uses the term RAMS for the Ariane 5 and 6 launch vehicles to cover Reliability, Availability, Maintainability, and Safety [7].

In other industries, however, only the term quality assurance according to DIN EN ISO 8402 is usually applied. According to this definition, quality assurance is the sum of all measures to ensure constant product quality; see Sect. 8.2.1.

8.1 Product Assurance Is an Important PM Function

8.1.1 Detailed Task Description of Product Assurance

The overriding function of "*product assurance*" is to ensure the safe creation of the respective product. This includes, for example, quality, reliability, and safety. Product assurance is to be understood as an integral overall function and not merely a simple addition of the individual functions because quality, reliability, maintenance, etc. are closely linked. In order to achieve the specified goal, certain tasks must be performed, which are described in detail below:

1. Quality monitoring
   (a) Requirements

   (b) Inspections
   (c) Quality checks
   (d) Quality controls
   (e) Design backups
2. Material selection
   (a) Selection
   (b) Rating
   (c) Storage
3. Reliability
   (a) Analyses
   (b) Trade-offs
   (c) FMECA or FMEA
   (d) Tests
4. Availability
5. Maintainability (*serviceability*)
6. Safety
   (a) Safety aspects
   (b) Safety checks
   (c) Manufacturing procedures

Although quality, reliability, and safety of a product have always been an important banner for the manufacturer and the potential buyer certainly attached considerable importance to these characteristics for the purchase decision in the past, reliability research and with it the quantifiability and often also the understanding of these increasingly important factors is still in its infancy in many branches of industry. Thomann writes in this context: "*But with the growing complexity of technical systems*, the *importance of reliability, maintainability and availability came to the fore. In the American literature one speaks in this connection also of the* "ilities"! [6] In connection with product liability, which is becoming increasingly important, Thomann also points out the importance of safety aspects, which in turn are closely linked to reliability.

In the 1950s, the creation of mathematically based reliability concepts began in the USA as part of the development programs for modern weapon systems. Juran and Gryna wrote: "*Beginning in the 1950s, there was a movement to quantify reliability, availability, maintainability, and so on*" [8]. In this context, they pointed out the changing times; for example, if in the past it was about the production of a piston pump for garden tools or a bicycle, modern products are about complicated electronic devices, computers, or even nuclear reactors. Figure 8.1 compares the characteristics of traditional and modern products according to an elaboration by Juran and Gryna [8].

Relevance of Product Characteristics
and their changes with time

Product Characteristics:	Traditional Products	"Modern" Products*
Complexity:	Low (simple to complicated)	High (complicated to complex)
System behavior:	Mostly static	Mostly dynamic
Appearance:	Visible, tactile experience	Invisible services integrated (e.g. full maintenance contracts, Updates, database access, ...)
Precision:	Generally low (mm)	Often high (ym, nm)
Reliability:	Best possible consideration (through vague suggestions)	Quantified (often statutory) requirements
Modularity:	Low (primarily manufacturing)	High (primarily assembly)
Standardization:	Low (primarily functional)	High (primarily re-use)
Lifetime:	Often short-lived	Tending to long-life
	Easy to repair	Sustainability (incl. disposal)
Failure Causes:	Generally, production errors	Generally, design weaknesses

*Lit. [8] updated by Prof. Dr. Steffen Rietz, University Offenburg, Germany

Fig. 8.1 Characteristics of traditional and modern products (Juran and Gryna [8])

8.1.2 Initiate PA Measures in the Early Phases of the Project

The reliability of a system is an important economic factor; see Sect. 8.2. Juran and Gryna wrote that the costly failures that occurred in military electronics equipment in the 1950s were originally attributed to production and inspection errors, but were then proven wrong because the main problem was in the design [9]. They found that for medium to highly complex products, the cause of failure was as follows:

– Design	40%
– Manufacturing	30%
– Operation	30%

This means that developers must take product assurance goals into account as early as the design stage. Thomann puts it this way: "*Since quality, reliability, maintainability, and safety must be preemptively developed into the product, the most authoritative product assurance activities fall into the early project phases*" [6]. Typical reliability tasks in the early project phases include:

- Setting the system reliability target
- Breakdown of the reliability target among the subsystems, equipment, components, etc.
- Reliability analysis and *prediction* (*analysis and prediction*)
- Failure modes and their effects (*fault tree analyses*)
- Identification of critical components, materials, and procedures
- Pre-selection of components and suppliers.

8.1.3 Anchoring Product Assurance in the PM Team

Product assurance aspects must be preemptively developed into the product or system. For this reason, the project-specific product assurance (*PA*) team (*or person*) should be included as an integral part of the project management team of projects, because this promotes team building; see Figs. 5.3 and 5.4. In practice, however, there are often differences of opinion in this respect. For example, there are opinions that the product assurance team or quality assurance team should not be integrated into the project team in order to maintain a certain independence from the project management. However, this is quite obviously a misunderstanding, because anchoring the product assurance team as part of the project team does not mean that PA should be hindered in its specialist work and as shown in Figs. 5.3 and 5.4, PA also reports to the central PA department. On the other hand, the integration of the PA team into the project management team results in a much greater understanding of the project issues for those concerned, for example when it comes to creating project-optimal solutions.

It is important to understand that the temporary secondment of Product Assurance (*PA*) to the project management team does not mean that they lose their professional independence. The PA colleagues retain their technical supervisor and continue to be responsible for complying with the current requirements and regulations of authorities such as TÜV, VDE, etc. As team members, they have the task of looking over the developers' shoulders, so to speak, even at the design stage, and making sure that all the necessary PA regulations are complied with. They assume joint responsibility, but do not have the role of a controlling policeman who reviews a completed design and rejects it if necessary. Organizationally, the PA manager is firmly integrated into the review and change control boards. In the event of a conflict, i.e., if certain regulations are not adhered to and the PM does not respond to complaints, he must pull the ripcord and involve his specialist superior. Different rules apply to small projects, i.e., the PA responsible or quality expert is consulted by the PM.

The product assurance or quality management department is usually organizationally integrated into a corresponding specialist department in the company organization. The PA department is equally responsible for supervising the development and project teams as well as for quality control measures during production. When carrying out large projects, however, it is advisable, as already mentioned, to temporarily second the product assurance expert(s) to the project, as this is the best way to achieve interdisciplinary cooperation between the developers, economists, and product assurance specialists. Seghezzi underlines the need for interdisciplinary cooperation by saying, "*In order to put even the non-specialists and especially the managers in a more favorable position, business administration and quality specialists would have to work more closely together and jointly remove some difficult obstacles.*" [10]

At this point, reference is made to the International Standard ISO 9000. This standard specifies the verification requirements of quality assurance systems (*QA*), which must be complied with by the subcontractor/supplier during development and implementation.

"*ISO 9000 is a guide to the selection and application of the standards ISO 9001 to ISO 9004*" [11]. A precise designation of the ISO 9000 to ISO 9004 standards is reproduced below:

- ISO 9000—Guide to the selection and application of quality management standards, elements of a quality assurance system and quality assurance levels.
- ISO 9001—quality assurance systems; quality assurance verification level for development and design, production, assembly, and after-sales service.
- ISO 9002—quality assurance systems; quality assurance verification level for production, and assembly.
- ISO 9003—quality systems; quality assurance demonstration level for final inspections.
- ISO 9004 quality management and elements of a quality system; Guide.

8.1.4 Product Assurance Documentation Using ESA for Example

The European Space Agency (*ESA*), successor to the ELDO and ESRO organizations, has been involved in the development of complex space systems since the early 1960s, where the aspects of product assurance play a particularly serious role, because a fault in the system can mean the partial or total loss of the mission. A current example is the "ExoMars" mission of ESA and the Russian space agency Roskosmos, which partially failed in October 2016 and lost the signal to the "Schiaparelli" lander. Space projects are usually very expensive undertakings. Errors or non-functioning systems mean a great financial loss. In manned spaceflight, particular caution is required because an error entails the risk of personal injury, as in the Space Shuttle disasters of 1986 and 2003. However, this also applies to the transportation of people by public transport systems*.

*more than 100 people were killed in June 1998 resulting from the catastrophy of the high-speed ICE train at Eschede, Germany

Reliability and safety therefore have a very high priority in space projects. Space technology has largely overcome the difficulties associated with reliability through extensive test programs and simulations. Commercial news satellites function faultlessly in orbit for many years without any maintenance. Orderers of television satellites today assume a lifetime in orbit of more than 15 years in orbit. The first HELIOS solar probe (*a scientific mission*) developed by the German company MBB, which was launched into orbit around the sun in 1974, was still functioning perfectly after 10 years in the solar orbit.

The aerospace industry is undoubtedly leading the way because passengers assume when they buy a ticket that the aircraft used is safe. Satellite TV viewers are interested in error-free TV broadcasts. For other high-tech companies, the issue of system reliability is also of great importance. They usually have the ability to quickly fix system failures, but it is a cost factor. Aerospace's experience in product safety also benefits other industries.

The successes of the space industry in the field of reliability of highly complex systems are based to a large extent on the systematic approach. In all project phases. PA, along with systems engineering (*Chap. 7*), is a mainstay in all spaceflight projects; see Fig. 5.3. In the past, the European Space Agency (*ESA*) has developed a large number of guiding standard specifications, which Capella and von Högen have summarized as follows [12, 13]:

- PSS-01-00 Basic PA requirements
- PSS-01-10 PA Management
- PSS-01-20 Quality assurance (*QA*)
- PSS-01-30 Reliability
- PSS-01-40 Safety
- PSS-01-50 Maintainability and availability
- PSS-01-60 Component selection PSS-01-70 Material and machining selection
- PSS-01-80 80 Software quality assurance.

In total, over 50 ECSS Quality Procedures have been created by ESA, describing topics such as material selection, processes, quality checks, etc; *see ECSS Procedures: www.ecss.nl.*

8.2 Quality Influences the Economic Efficiency of the System

8.2.1 Quality Assurance and Its Importance for the System

More and more technical products are to be understood as complex systems for which a certain level of quality is a basic prerequisite for smooth, i.e., uninterrupted and safe use for a given period of time. The uninterrupted, i.e., the actually available, usage time of a system (*product*) is an important economic factor. A vehicle or a machine used in a household or industrial operation that cannot be used without interruption for a longer period of time cannot be used economically. Ensuring a minimum level of quality therefore takes on a high priority when creating a new system. Seghezzi writes in this context: "..., *because quality in a market-economic system means suitability of a project for the use or fulfillment of the customer's need.*" [10]

The quality of a system can be used *to draw* conclusions about its *reliability* and, from this, about its availability, maintainability, and safety. The popular term "quality," which Juran and Gryna refer to as fitness for use [14], represents an important characteristic of any system, from which the previously mentioned properties can then be derived. Juran describes the "Juran Trilogy," which consists of the following three management processes: quality planning, quality control, and quality improvement; see Fig. 8.2 [15].

8.2.2 Reliability of the System

The term reliability is used to express the probability with which the system or product will function faultlessly over a certain period of time. Highly complex transportation systems, such as the Airbus A80, the Intercity Express (*ICE*) or the Ariane 5 launch vehicle, are composed of a very large number of components. Their proper functioning is crucial for system reliability. In order to achieve a high reliability of close to 100% for passenger transportation, a sophisticated redundancy concept is required. However, very high reliability, as required for passenger transport, is also a cost issue. Since manned space flight naturally also strives for a very high, i.e., almost 100%, reliability value, the costs of manned space flight are very high.

Multi-tested commercial launch vehicles, such as Ariane 5 (*Europe*), Proton (*Russia*), Long March 3B (*China*), and Atlas (*USA*), achieve a reliability (R) of at least 95% ($R \rightarrow 0.95$) or more with the current state of the art. That is, the cases where the missiles fail to reach their target are equal to or less than 5% ($1-R \leq$

Quality Management		
Quality Planning	*Quality Control*	*Quality Improvement*
Establishment of Quality goals	Assessment of the current Quality level	Verification of the need
Identification of the customers	Comparison of current performance with the Quality goals	Establishment of the Infrastructure
Determining customer needs	Implementation of necessary actions in the event of deviations	Identification of improvement projects
Development of product characteristics to fulfill customer needs		Establishment of project teams
Development of processes to produce the product characteristics		Providing the teams with resources, training and motivation to identify causes and to stimulate corrective measures
Development of procedures for process regulation: Transfer the plans to manufacturing		Introduction of controls for preservation of the improved Quality standards

Fig. 8.2 The three universal processes of quality management (Source *Juran*)

0.05). To increase the reliability to almost 100%, i.e., "man-rated," as required for passenger transportation, one needs significant additional costs to further qualify the system and components, as mentioned earlier. Therefore, for commercial missions, one accepts a lower reliability value and accepts a calculated 5% failure. Possible losses can be covered by launch insurance; see Sect. 19.5.2. Reliability considerations also exist in other industries, of course, e.g., aircraft and vehicle manufacturing. On the one hand, high system reliability is an important economic factor, but on the other hand, it is also associated with high investment costs; see Sect. 10.5. This applies equally to all technical products, whether passenger cars, ICE trains, airplanes, or satellites. In other words, the weakest link in the chain poses the greatest problem. The failure of a small component can lead to the failure of the entire mission, unless remedial action has been taken through appropriate redundancies or on-board spare units. Any technical breakdown can lead to accidents and additional costs. That is why reliability analyses are important from the very beginning, i.e., in the early phases of the project; see Chap. 4. To demonstrate how this works, see the following example of a rocket. With a reliability of 95%, 105.263, i.e., rounded up, 106 rockets must be built. This results in a cost increase of the system by 6 from 100 to 106%.

As already mentioned, the weakest link in the chain poses a particularly serious problem. There have been some illustrative examples of this in the past. In a weather satellite, an important component (*resistor*) failed in the 1980s. German television stations had to make do with substitute images from an American satellite until Meteosat 2 was ready for use. The functional reliability (R) is of course also a function of time (t), but this will not be discussed in detail here. The aerospace example described here can be applied to other products, such as traffic engineering, medical technology, etc. One of the decisive factors for high system reliability is the quality of the components used. This requires that the development, production, and testing of the components (*resistors, capacitors, integrated circuits, etc.*) is carried out using correspondingly high-quality standards. The selection and procurement of components is a very responsible task that must be carried out in close cooperation between quality management and the developers; see Sect. 7.2.

In practice, however, it can be assumed that not every failure of a component immediately leads to a total failure of the system. Rather, it is the task of reliability engineers to ensure that appropriate redundancies are provided to take over the function in the event of failure of a subsystem, device, or component. In aerospace engineering, almost all components are redundant—sometimes having multiple redundancies. For example, the astronauts of Apollo still had a third fuel cell available after two fuel cells exploded. On the other hand, however, the redundancy concept cannot be designed to be arbitrarily generous, because redundancy also costs mass margins and money.

In the case of systems on the ground, lack of reliability does not automatically lead to disasters, as in the case of fast vehicles, aircraft, rockets, etc. When the system fails, appropriate repairs are initiated. In the meantime, the function of the system must be dispensed with or a back-up unit must be put into operation. However, the frequent failure of a system leads to increased costs and it may then no longer be economically feasible, firstly due to the failure (*reduction in operating time*) and secondly due to the repair costs incurred. In this context, reference is also made to the topic of FMECA (*failure modes, effects and criticality analysis*), which is discussed in connection with risk management in Chap. 19. "*It's not enough that we do our best, sometimes we have to do what's required,*" Sir Winston Churchill.

8.2.3 Availability of the System

The availability of the system is an important factor for the user because the economic efficiency can be directly related to the system availability. The availability expresses the percentage of the operating time the user can operate a machine. A system (*machine tool, IT system, vehicle, etc.*) that fails, for example, after 10 h of operation, only to be available again after another 10 h of repair (*downtime*), is uneconomical. In this case, availability is only 50%.

The operational availability (*OA*) of a system is an important metric for determining life cycle costs is defined as follows [16]:

$$\mathrm{OA} = \mathrm{UT} / \left(\mathrm{UT} + \mathrm{DT}\right)$$

UT = operational up time
DT = down time
Example: OA = 900 / (900 – 100) = 0.9 (90%)

The mean-failure distance, i.e., the mean usable time between failures (*mean-time-between-failure*) and the mean-downtime are important parameters from which the operational availability of the system can be derived [17, 18].

The total operational uptime represents a multiple of the mean time between failures ($z \times MTBF$) and the total downtime represents an equal multiple of the mean time between failures ($z \times MDT$); see Fig. 8.3.

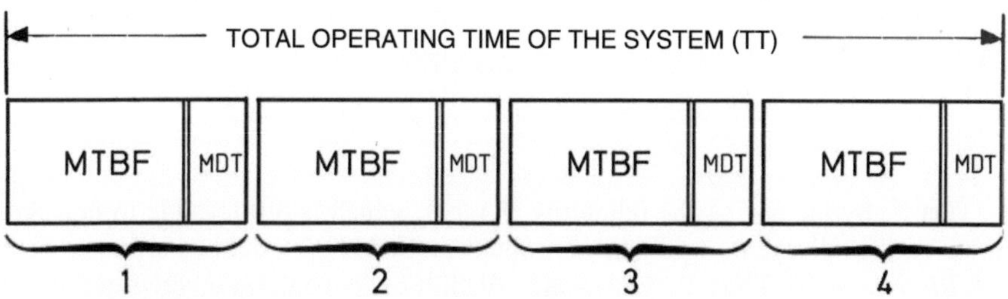

TT= OT + DT
MTBF OT (Total operating time) = n X MTBF
DT (Total downtime) = n X MDT
n = Number of downtimes I1, 2, 3,...)
MTBF = Mean Time Between Failures
MDT = Mean Down Time
OA = Operational Availability

$$\mathrm{OA} = \frac{\mathrm{OT}}{\mathrm{OT} + \mathrm{DT}} \triangleq \frac{\mathrm{MTBF}}{\mathrm{MTBF} + \mathrm{MDT}}$$

Fig. 8.3 Total operational uptime of a system

$$OA = MTBF / (MTBF - MDT)$$

MTBF = mean time between failure
MDT = mean downtime
z = number of cycles
t = time (hours)

It is clear that the time extension of the MTBF factor and/or the time reduction of the MDT factor are directly responsible for the operational availability (OA); see Sect. 10.6.3. An ideal state would be reached if there were no errors (*i.e., no system failures*). The factor MTBF is directly related to the reliability (R) of the system components [19]:

$$MTBF = 1 / \lambda$$
$$\lambda = \text{Fehlerrate}(\text{failure rate})$$
$$\lambda = (1 - R) / t$$

R = Reliability

When t = 1 h is specified, the following relationship between MTBF and R is obtained, namely [20]:

MTBF	R
5	0.82
10	0.90
20	0.95
100	0.99

That is, by increasing the reliability (R) to 1, which is not feasible in practice, however, the failure rate would be zero and thus the MTBF would be infinitely high (*MTBF = infinity*). Thus, by increasing reliability, the mean time between failures (*MTBF*) can be maximized. However, it must be remembered that increasing reliability cannot be done for free either. Aerospace developments clearly show the limits of feasibility, because the pursuit of high reliability, i.e., tending towards 1, as targeted for manned space flight, is expensive and, in extreme cases, unaffordable.

8.3 Producer Liability and Safety

The producer of a product placed on the market is liable for damages that are demonstrably related to the system's malfunction and are not due to operator error. Thoma writes: "*Anyone who puts a defective or damaging product on the market, causing injury or damage, is liable,*" and Brendl defines liability and producer liability as follows: "*Liability means 'having to answer for,' producer liability (PL) for the consequences of product and instructional defects for which one's domain is responsible*" [21, 22]. Illustrative examples of producer liability are traffic accidents caused by production errors in tire manufacturing, or aircraft accidents resulting from design errors; example: DC-10 crash caused by a defective cargo hatch.

Brendl writes on the importance of producer liability: "*The worldwide tightening of producer liability has reached a risk dimension that demands fundamental measures from a growing number of undertakings ... The PL tightening is an expression of a comprehensive change in social conditions*" [22]. That is, manufacturers of complex systems must be concerned, even at the development stage, with maximizing the reliability and safety of future systems.

In the aerospace sector, safety thinking has a particularly high priority because technical failure in this industry is often tantamount to disaster. In the past, the safety of aerospace equipment has been improved enormously. However, absolute protection against accidents is not possible for economic reasons, since an aircraft designed with such safety requirements would leave no weight and space reserves for passengers or other payloads. It is therefore necessary to accept a residual risk that is well balanced with the technical possibilities and the economic conditions. This principle can be applied to all areas of our lives and should not be forgotten in the discussion about reactor safety. Brendl states in this context: "*In a changing world of imperfect people, which is becoming more and more dependent on the division of labor, there are always residual risks.*" [22]

During the realization of the Apollo program, everything possible was done to bring the astronauts safely to the moon and then back to Earth. They were prepared to invest a great deal in the safety of the system, consisting of the lunar rocket, the lunar landing devices, and a large number of ground stations, but they were also aware of a residual risk.

Although failures of technical products (*vehicles, aircraft, machinery, power plants, etc.*) cannot be avoided in the future, everything must be done to maximize safety under economically justifiable aspects. This requires safety analyses both in terms of failure modes and their prevention, as well as for failure control. The aerospace industry has gained particularly impressive knowledge in this field in the past and recorded its most impressive successes with the Apollo program, Space Shuttle, Ariane 5 launch vehicle, ISS and modern aircraft construction. Analytical models and simulation techniques play a major role in this.

Producer liability should in principle be treated as an economic problem, in that safety risks are largely contained to avert financial damage to the producer. Brendl writes in this context: "*The way a venture deals with its PL risks is now already used by American venture capitalists as a criterion to judge its ability to shape its future at all promising.*" [22] The problem of system safety is intimately linked to system reliability, and this in turn is linked to the quality of the design (*redundancy, etc.*) and the components used. "*Preventive safety means thinking ahead and investing ahead,*" writes Brendl [23]. If one follows the chain of causality quality, reliability, safety, producer liability, it is easy to see that there is a direct connection between quality and producer liability and that corresponding *economic trade-offs* are worthwhile.

8.4 Product Assurance: Quality, Availability, Safety, Costs

The parameters of quality, availability, safety, and cost are closely linked because the availability and safety of a system depend on its reliability and thus primarily on the quality of the system. There is also a direct relationship between quality and costs. Thus, the costs of generating quality and the losses incurred by flawed quality must be contrasted, because flawed quality means limited availability (*repair costs and lower utilization and reduced revenue*) and/or technical risk (*producer liability*).

Also to be considered in this context are the costs of rework, scrap, and warranty obligations, which Crosby puts at 80% of quality costs for many companies [24]. Crosby assumes that these companies spend more than 10% of the sales volume on quality, but considers 4%, broken down as follows, to be the maximum justifiable:

– Rework	0.6% of sales
– Committee Works, Rejected work	0.4% of sales
– Warranty costs	0.5% of sales
– Quality Control Costs	2.5% of sales

The costs mentioned by Crosby, which certainly only apply to very specific industries of the bulk goods industry, do not give any information about the necessary costs for the entire product assurance process of a project, i.e., for quality assurance measures (*design and production*), reliability analyses, and controls, etc. of a system. The cost of product assurance for a system is closely related to the complexity of the system and must be based on the parameters of availability and safety. Burbridge, a very experienced systems engineer, puts the product assurance cost for complex large projects at about 7% of the total project cost [25].

8.5 Product Assurance Checklist of a Technology Project

The following checklist was created for a technologically complex transportation system [26]:

(1) Has the *turn key contractor (TKC)* prepared a product assurance (*PA*) plan?
(2) Has the TKC developed a reliability concept?
(3) Do the reliability results meet the specified targets?
(4) Is the reliability concept transferred to subcontractors and suppliers?
(5) Are TKC quality control procedures applied to the entire project, including subcontractors and suppliers?
(6) Has the TKC established *safety regulations* and can they be followed?
(7) Has the TKC created availability rules?

(8) Have all project stakeholders (*TKC, subcontractors, and suppliers*) planned for sufficient redundancies in development?
(9) Are quality and reliability reviews (*audits*) included in the PA plan for all project levels (TKC, subcontractors, and suppliers)?
(10) Has the TKC established a selection procedure for the procurement of critical components and materials?
(11) Are special procedures available for Hi-REL components?
(12) Are all procurement activities of components and materials listed in the schedules?
(13) Does the TKC have qualified PA staff with project experience?

Literature

1. NASA specification SSP 50191
2. ESA Procedure ECSS-P-00A, April 2000
3. ESA Space Product Assurance, ECSS/SEC (96)2, February 1996.
4. DLR, Product Assurance & Safety Requirements for DLR Space Projects, DLR-RF-PS-001, March 2009.
5. Liming, Ren, Changhong, Gu, Shousong, Qing: Space Product Assurance in China, CAST, Beijing.
6. Thomann, Adolf: Ist Ihr Wissen (über Produktsicherung) noch io?, in: Management-Zeitschrift io, 51 (1982), No. 5, p. 203ff.
7. CENELEC standard EN 50126
8. Juran, J. M. and Gryna, Frank M. (Jr.): Quality Planning and Analysis, McGraw-Hill Book Company, N. Y., 1980, 5. 168.
9. Source 8, p. 170
10. Seghezzi, Hans Dieter: Integrated Quality Assurance is a Component of Corporate Management, in: Management-Zeitschrift 51(1982) No. 4, 5. 158
11. DIN: ISO 9001, May 1987, 5. 1
12. Capella, A. J. and von Högen, M.: The ESA Product Assurance Specification System, Proceedings of 2nd ESA Product Assurance Symposium, ESTEG, 10-12 November 1981, 5. 14.
13. Von Högen, M.: The New Product-Assurance Documentation System, ESA Bulletin No. 31, 5. 101.
14. Source 8, p. 1
15. Juran J.M.: Der neue Juran - Qualität von Anfang an, Verlag mi, Landsberg/Lech, 1993, p. 28.
16. Hickey, Leo J.: Measuring Operational Availability, Logistics Spectrum, Summer 1973, p. 21.
17. Q. 8, P. 199
18. VDI Guideline 4004, Sheet 4 (draft), May 1982.
19. Dombrowski, Edgar: Introduction to the Reliability of Electronic Devices and Systems, AEG-Telefunken, 1970, p. 46.
20. Source 8; p. 208
21. Thoma, W: Product Liability: Present Status, Trends and Preventive Measures, Proceedings of 2nd ESA Product Assurance Symposium, ESTEG (Noordwijk, Holland) 11-12 November 1981, p. 32.
22. Brendl, Erich: Der Produzenten-Haftung gewachsen sein: Ein Maßnahmen-Programm, in: Management Zeitschrift io 51(1982) No. 5, S. 209.
23. Source 22, p. 212
24. Crosby, Philip B.: Quality costs less, 3rd edition, 1979, Alfred Holz, Großbottwar 3

25. Burbridge, Keith: Low Cost Program Practices for Future NASA Space Programs, Final Report, Volume II, Appendix, NASA Report CR-143657, Engineering Memorandum LCPP-6, 5. 30.
26. Madauss, Bernd-J.: Project Management Checklist, Management and Organizational Concept for the Metro Medellin Project, September 1983 (*unpublished*), "*Product Assurance*," p. A.11.

Project Control Management

9

Contents

B. -J. Madauss, *Project Management*,
https://doi.org/10.1007/978-3-662-69057-4_9

Abstract

Project Planning and Control is a central topic of project management and of great importance for technical and economic success. It involves the efficient control of the technical and financial aspects of a project, which must be planned and monitored in close cooperation from the outset, since the results are usually incomplete when considered on their own. The goal is to ensure that the specified technical performance has been achieved to the satisfaction of the client within the agreed time and cost target.

Project control is a prime management function (PCM); see Fig. 5.3.

Planning and status monitoring are two very important tasks of project management. Both functions are in principle separable from each other, but they are so closely linked that a combined exercise of the task complex "*planning and monitoring*" is absolutely necessary. After all, the planning of a project is only the very first and still unfinished step towards project control. Planning is a projection into the future and it is therefore burdened with all the shortcomings that are usually inherent in a forecast. It therefore requires constant iteration through the addition of new information brought to it through the execution process. The plan, once created, must be kept up to date through continual revision; this then leads to dynamic planning. Gehriger defines planning as follows: "*Planning means creating a mental model of how the future will be handled through foresight and coordinating the requirements and means through contemplation and analysis in such a way that the mental concept of the project can be decided upon and rationally realized as a feasible, written order.*" [1]

About 100 years ago, the French civil engineer Henri Fayol, who also became known as the founder of management theory, defined 14 management principles that are still valid [2]. The six management functions derived from these are shown in Fig. 9.1. This is a management process that is to be understood as a control loop. The individual stations are passed through at regular intervals and feedback effects may also occur.

The PM monitoring process is basically comparable to the process defined by Fayal. Deviations from the plan must be identified and corrected as quickly as possible so that the project does not get out of hand. The project monitoring process described in this book includes the eight processes defined below, which are described in detail in Sect. 9.4.1:

1. Project requirements
2. Baseline planning
3. Implementation and work release
4. Task execution
5. Actual achievements
6. Comparison of actuals vs. targets
7. Status analysis
8. Management decision

The planning and monitoring methods described herein, and in particular the *earned value* method presented in Sect. 9.4.6, are primarily intended to provide the PM with an "early warning system" to identify deviations from the plan and their

Fig. 9.1 Project monitoring cycle according to Fayal

technical and financial effects as quickly as possible so that appropriate corrective measures can be initiated without delay.

The planning and monitoring function plays a key role in project management. In international usage, the combined function of planning and monitoring is known as *Project Control (PC)* and includes both functions [3]. Control is understood in this context as the generic term for the planning and monitoring process in the sense of control and the planning part is assumed to be a necessary component.

9.1 Project Control in Historical Retrospect

The Project Control (*PC*) task is a key function of project management. Experienced project managers therefore rank it equally with, for example, systems engineering and quality assurance; see Figs. 5.3 and 5.4. The term *Project Control* summarizes the key factors performance (P), time (T), and cost (C). This is described in detail in Sect. 9.5. The goal of integrated project control is to align at Time Now (*TN*) the value of the three factors P, T, and C, so that the planned equilibrium is always maintained during the course of the project; see Fig. 9.2. Continuous monitoring of these three keys factors (*for example monthly*) should be seen as a primary PM task in order to achieve the established total project goals [4].

The literature on this subject is literally overflowing with theoretical approaches to the integration of these three factors, but only rarely are any real successes demonstrable. There are a number of reasons for this. The most important reasons are:

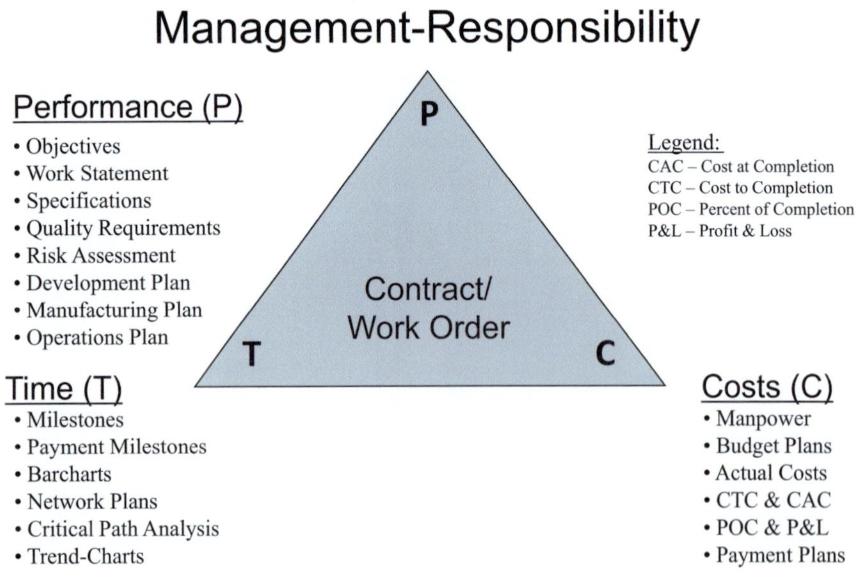

Fig. 9.2 Management triangle: Performance—Schedule—Costs

- Insufficient embedding of the PC team in the project organization
- Unqualified PC staffing
- Schedule and cost control not integrated but performed separately by a technical and a business person
- Integration of schedule and cost control on a very abstract and theoretical basis.

Project Control is a very important task of project management. Therefore, the responsible PC manager must have unrestricted and direct access to all technical and administrative information (*i.e., performance, time, and cost*) as required for project monitoring and control. The organizational embedding of PC must therefore be as shown in Figs. 5.3 and 5.4. For large projects the appointed PC manager must become a full team member (*i.e., 100%*) of the respective Project Management Office (*PMO*); see subsect. 5.2.5. For medium size and small projects the PC responsibility may be outsourced to a central PMO of the respective company. Another solution for small projects is that the PM performs the PC tasks in personnel union. The staffing of the PC team often causes problems, since the PC activity lies directly at the intersection of the technical and commercial project tasks. Thus, the question arises again and again whether the person responsible for the PC should be a technician (*engineer*) or a businessman or business economist. The PC task (*see Table 5.1*) requires competence in both areas from the job holder. The person responsible for the PC should therefore have sufficient technical and business experience. Because of the dual experience required, the future Project Controller should ideally have a double degree and practical experience in both areas of work. Alternatively, it should be at least a technically or commercially trained employee with sufficient practical experience in the second field of knowledge; see Sect. 15.2.

In the past, many attempts were made to integrate the factors of *performance, time, and costs with* the help of relatively formalistic, one might almost say fully automatic methods. However, this approach did not lead to the desired success. NASA's introduction of the PERT-cost method, a procedure for integrating time schedules and costs, was initially hailed as a future monitoring system but did not have any practically demonstrable success. A major reason for this was probably that the approach was too abstract and also too detailed. PERT plans were ideally suited for scheduling and critical path determination, but less well suited for cost planning and monitoring. Subsections 9.1.1–9.1.5 describe the PC development history on which the methodology used today is based.

9.1.1 The Planning and Monitoring System C/SCSC

In June 1966, the *US Air Force Systems Command (AFSC)* published a specification for integrated cost and schedule control called *Cost/Schedule Control System Criteria (C/SCSC).* The system was adopted by the services and US industry involved in major Air Force projects [5]. In December 1967, the US *Department of Defense (DOD)* adopted the specification for handling all DOD projects and introduced the specification as Directive 7000.2 [6].

As early as the 1950s, cost and schedule overruns on defense projects in the United States brought the question of better management methods very clearly to the attention of responsible DOD managers [6]. Particularly vexing to DOD managers was the fact that many of these overruns became known far too late, leaving little room to maneuver to make program changes except by adding costs to bring the programs to completion after all. This led to the development of new and better control and information systems. Cost overruns were of course, not unique to the USA but also existed in Europe, and are discussed elsewhere in this book.

During this time, among other things, the PERT planning system was introduced and became very widely used in the US and Europe. PERT, *Program Evaluation* and *Review Technique* was developed by the US Navy in 1958 as part of the Polaris Missile Project. It was in this context that the basic concepts for systematic project structuring were established. Methods emerged such as *Work Breakdown Structure (WBS)* and *Work Package (WP),* which later was also adopted in the German-speaking countries [7]. The definition of work packages (*WPs*) at the lowest WBS level provided the basis for efficient planning and monitoring of project tasks.

To improve cost and schedule awareness, former US Secretary of Defense Robert S. McNamara introduced a new weapons procurement strategy in 1961 that had two main objectives [5]:

(1) Direct coordination of the various plans and their weapon system requirements ... and require better comparative analyses (*trade-offs*) between competing weapon systems.
(2) Procure necessary weapon systems in accordance with expected schedules and technical performance requirements at the lowest possible cost.

It was clear that these goals required a particularly efficient management system. In response to the goals set by McNamara, therefore, the C/SCSC specification was developed, incorporating such well-known procedures as PERT, the WBS concept, and the earned value method. The C/SCSC specification became an important component of the *Resource Management System (RMS)* introduced by McNamara in August 1966 as DOD Directive 7000.1 for all US Department of Defense programs.

Five criteria are addressed in the C/SCSC specification [8]:

- *Organization:* Definition of contractual tasks and determination of task responsibility.
- *Planning and budgeting:* planning, scheduling, budgeting, and work release.
- *Data collection:* Recording of current schedule dates and costs for personnel and materials.
- *Analysis:* comparing planned and actual schedule dates and costs and performing variance analysis.
- *Revision and projection:* change control and final costing.

This specification (*C-Spec for short*) can be seen as the foundational document for what is today called "*modern*" project management. The creation of the C/SCSC specification aimed at watertight management. The first step was to work towards the creation of a systematically designed work organization. The creation of a work breakdown structure (*WBS*) and the definition of clear responsibilities for each work package in the organization chart created the basic conditions for efficient project management; see Sect. 9.3.

Derived from the WBS, the planning and budgeting process can now be initiated, the first step of which is the preparation of a work package description; see Sect. 9.3. In the next stage, the workflow and cost plans are to be prepared, which are to be jointly oriented to the WBS and/or work package. This will become the basis for work authorization. The carefully prepared planning documents, time schedules, and costs are the basis for a systematic collection of actual data, which are then compared with the plan data in the analysis process so that deviations from the plan can be identified at an early stage.

The last criterion of the C-Spec deals with the change process as well as the principles of final cost accounting. Project measures that involve a change in the schedule and/or cost budgets must be recorded and assigned to the relevant WBS elements and work packages. At the same time, any resulting rescheduling or planning changes must be initiated. The integration of cost and schedule data is of particular importance. An important method of project monitoring is, among other things, the regular cost extrapolation "*Estimated Cost at Completion,*" which is an important component of the C-Spec; see Sect. 9.4.

The introduction of the C-Spec can be considered an important milestone of modern project management. The procedures introduced with this specification have made a decisive contribution to improving the project planning and monitoring methodology. Some procedures implemented in this context, such as the mandatory introduction of work breakdown structures at the start of the project or the principles of integrated schedule and cost monitoring, have also led to further successes in project monitoring in Europe.

Several US studies have demonstrated that the introduction of C-Spec has led to qualitative improvement in project management among government agencies and industry in the United States [5, 6, 9]. Strach wrote: "*The most significant introduction was the earned value concept*" [9] and Wright particularly emphasizes that there was no attempt to proscribe in detail to industrial partners what systems they had to use, but only what information had to be provided [6]. Thus, companies were free to develop their own planning and monitoring systems and IT programs (*software*) or to improve existing systems and/or programs. It was only necessary to ensure that the information was prepared in accordance with the C-Spec.

9.1.2 Project Control—NASA Developments

Along with the US DOD, NASA was the pacemaker of modern project management for a long time and has contributed significantly to the development of standards. The project control methods developed by NASA are not only convincing due to their impressive logic and practicability but have also been excellently documented and, above all, repeatedly tested in practice. Particularly in the development of the Apollo program, especially simple but very efficient PC techniques were developed, which have also been applied in Europe, e.g., in the context of the ESA programs; see Sect. 9.1.3.

In 1962, NASA issued the *PERT and Companion Cost* System manual, which was released the same year by US Secretary of Defense Robert S. McNamara as joint planning guidance for the Army, Navy, Air Force, and DOD services [10]. NASA, building on PERT developments begun at the Navy in 1958, had achieved an important milestone in the creation of modern-day management systems. NASA's PERT manual became the starting point for many subsequent PC developments and related publications.

NASA's Apollo program, whose objective was to land men on the moon, is not only a pioneering technological achievement but also a masterpiece of efficient project management. The extreme demands of the *schedule-cost-performance triangle* forced those responsible at NASA to constantly perfect the PC methodology. In the mid-60s, therefore, NASA published a series of outstanding Apollo documents that created new dimensions in the field of project control.

The *Program Scheduling and Review Handbook,* published in 1965, can be seen as an important step in linking detailed PERT plans with the simplified bar chart diagrams necessary for top management [11]. A year later, NASA issued another important handbook entitled *Project Management Information and Control System (MICS)*, the content of which set the tone for the *Management Information Systems (MIS) that* later became so popular [12].

Efforts to be able to consider the schedule, financial, and technical project parameters as an integral unit also occupied NASA managers. Two authoritative documents have been produced on this subject, the Apollo document *PERT & Cost Correlation Technique (PACCT)* in which schedules and costs are integrated at the level of work packages [13] and the two-volume and computerized NASA document *Forecasts* and *Appraisals for Management Evaluation (FAME),* which deals in particular with the development, evaluation, and correlation of project trends (*schedule, cost, and technical indices*); see Sect. 9.4 [14].

NASA's pioneering developments in the field of project control methodology were made in close consultation with the industrial companies involved. Many of NASA's innovations were originally derived from industry studies. The constant dialogue between NASA and industry made it possible to test new procedures quickly and extremely broadly in industrial practice, to improve them wherever possible, and to disseminate useful procedures relatively quickly. Many methods were quickly transferred to other industries that did not cooperate with NASA, such as the automotive industry. Through a variety of publications, foreign countries also participated in NASA's PC developments.

9.1.3 Project Control—Expansion in Europe

In the early 1960s, about 5 years after the Sputnik launch, West-European countries also began to take a serious interest in space projects. In addition to national objectives, the main focus was on two European goals, the *European* rocket, for whose development, construction, and launch the *European Space Vehicle Launch Development Organization (ELDO)* was responsible, and European research satellites, for whose development, construction and operation the *European Space Research Organization (ESRO)* was responsible. Both organizations were founded in 1962 and, after several crises, merged in 1975 to form the *European Space Agency (ESA).*

It was clear that Western European space developments were planned independently of US programs, but that they benefited from US development experience. The first stage of the European rocket, the British *Blue Streak,* for example, was developed on the basis of the American *Atlas rocket,* and the development of satellites also looked to US projects. Particularly critical components had to be, and usually still have to be, procured in the USA and are often on the critical path of projects.

In the field of management, planning, and monitoring of projects, publications, management procedures, and cooperation with US partners were the basis for the development of management concepts. In particular, NASA's documents provided a constant stimulus for the development of its own methods and procedures. The author of this book has actively contributed to the transfer of knowledge and the creation of appropriate European standards.

In 1963, right at the beginning of the project work for the European Rocket, ELDO introduced the network planning technique developed in the USA as a project control instrument [15, 16]. The network planning technique was used at ELDO for project monitoring and also as a management information system. In 1970, ELDO produced a planning manual that laid out planning control procedures, and a standard industrial project control manual for the European launch vehicle program was issued that same year [17, 18].

The sister organization ESRO also worked diligently on the creation of uniform PC guidelines, and the first publication on this subject appeared in 1972 by Gehriger with the title: *Planning and Control for European Space Satellite Projects* [19]. This document represented for some time a most complete work on the subject of project control. It already contained the basic ideas of the *standard PC requirements* published by ESA in 1977 [20–23].

Between 1975 and 1977, ESA, which was founded in 1975, developed Project Control standard documentation based on the experience gained at ELDO and ESRO and in close cooperation with the West European space industry. It was understood that the methods and procedures to be used would have to be different in scope, depth, and detail for small, medium, and large projects. Thus, the PC documentation for large projects was created first, again distinguishing between definition studies and development and production activities [20, 21]. Then, after establishing the criteria for designating large, medium, and small projects, the PC guidelines for medium projects and small projects were developed [22, 23]. The standard PC guidelines issued by ESA in 1977 can be considered as a particularly noteworthy contribution to standardization in Europe and are applicable not only to space projects but also to other complex industrial projects.

Derived from the aforementioned standardization processes, ESA founded the "*European Cooperation for Space Standardization (ECSS)*" initiative in 1993 in cooperation with the national agencies and European industry. The ECSS documentation (*www.ecss.nl*) contains standards for the following three areas: "*Engineering,*" "*Quality,*" and "*Management.*" One of the aims of standardization is to ensure the cooperation and competitiveness of the European space industry. The application of standards is also becoming increasingly important in other areas, e.g., energy and transport projects in order to increase competitiveness. Therefore, a general transfer of know-how from space to complex industrial projects in other sectors involving ECSS is certainly appropriate.

9.1.4 German Company Handles Satellite Project According to NASA PC Standards

In 1965, Junkers Flugzeug- und Motorenwerke (*JFM*), Munich, today part of Airbus DS, won the tender for the development of the third scientific satellite of ESRO named HEOS*. HEOS was the first satellite contract won by a German company on international terrain, and its realization was an essential step in acquiring the necessary technical and managerial know-how for future satellite developments by MBB (*now Airbus DS*). As planned, HEOS-1 was launched into its designated orbit in December 1968 and functioned flawlessly for many years. In the realization of this project, the latest NASA methods and procedures were applied quite consistently. Scheduling and cost control were in one hand from the beginning. In order to make JFM a competent partner in the shortest possible time, the management decided to acquire the necessary management know-how through a cooperation with the US company Lockheed Missiles & Space Corporation (*LMSC*). LMSC had been instrumental in improving management methods in the USA as described in Sect. 9.1.2 and was included in the HEOS project team for a transition period. This ensured that the German colleagues were able to complete the complex HEOS project technically and financially in the shortest possible time through on-the-job training.

**Highly Eccentric Orbit Satellite*

With the realization of the HEOS project, it had been proven that the project control methods developed in the USA could also be successfully applied in Germany [24]. However, the HEOS project management had also quite consistently followed the advice of the LMSC consultants and implemented a self-sufficient, strong, and technically competent project management, in the context of which the Project Control team played a decisive role. The question of the correctness of the approach was answered by the successes of the HEOS project. The verdict was: technology excellent, costs not exceeded, and deadlines met. The members of the HEOS team, which included the author, were blessed with success and motivated for new, larger tasks. In 1968, the author published a report about the planning experience gained at the HEOS project [24].

Immediately following the HEOS project, the government of the Federal Republic of Germany and the USA agreed on the joint development of the HELIOS solar probe project, the primary objective of which was to bring a satellite for solar exploration to within one-third of the mean Sun–Earth distance. The task was scientifically very interesting, technologically extremely complex because of the great distance and the close approach to the Sun, and represented a challenge to the management capabilities of German industry. The German government included with this project a secondary goal " *... to (promote) the progress of German industry in managerial and technological sectors.*" [25]

Not least because of the HEOS successes, the government contract for the HELIOS project was awarded in 1969 to the MBB company, which took over the HEOS team as a result of the merger between the Messerschmitt, Bölkow, and Blohm groups. The HELIOS core team consisted of HEOS employees. Similar to the HEOS principle, a target and success-oriented project management was also implemented for HELIOS, in which the Project Control team, or as it was now called here "*Project Operations,*" played a significant role. Performance measurement, costs, and schedule were all in one hand. HELIOS also went down in Germany's space history as a successful project. In this context, other successfully implemented MBB projects can be mentioned, such as the research satellite COS-B, the German-French first communication satellite for commercial use Symphonie, and the large-scale project Spacelab, which was launched in November 1983 and developed by MBB/ERNO (*today Airbus DS*). For other future projects, MBB was able to prove that the company had the necessary management qualifications to successfully carry out international projects. Project Control was always at the forefront of the action in all these projects, but often also in the direct line of fire because the PC task involves considerable project responsibility.

9.1.5 Project Control—Developments in Germany

The proof of success in the handling of international space projects was reason enough to transfer the tried and tested PC methods to other branches of industry in Germany that were not yet familiar with PC methods because the problems of cost and schedule adherence are a problem everywhere. In the Handelsblatt of August 31, 1982, it could be read that "*the American-style controller has so far not really been able to establish itself in German companies.*" This statement certainly also applies to the Project Control function. It is true that network planning methods have been introduced in many offices and industrial companies, but only rarely as an integral part of modern project control procedures. In Germany, the rule still mostly applies that the planner is involved in the technical area and the cost control is carried out by the commercial area. However, effective PC activity requires the use of an integrated PC team, which sees the scheduling and control tasks and the cost planning and control tasks as a unified whole: see Sect. 9.1.1. The organizational separation of the PC task into schedule and cost tracking makes project execution quite difficult.

Fortunately, the concept of *project control* and, more importantly, the PC way of thinking has now gained a foothold in Germany among many companies. More and more German companies are moving towards using the PC methods described in this chapter to manage their projects. This is particularly the case in the realization of international projects, where cooperation with foreign partners is already a matter of course. Saynisch mentions, that a project management system was introduced at MAN (*New Technology*) in the early 1980s for a large international transport project in which the Project Control function played an important role [26]. According to Saynisch, the MAN PC team was responsible for following tasks:

- Planning and progress control
- Cost control
- Performance monitoring
- Configuration control
- Documentation control
- Interface management

The German Association for Project Management (*GPM*), founded in 1979, has played a significant role in the dissemination of knowledge of project management methods. GPM is a non-profit professional association for project management based in Nuremberg and is a member of IPMA (*International Project Management Association*). The association has a national capital representative office in Berlin and several specialists and regional groups and offers specialist seminars, workshops, certifications, and expert conferences. The GPM Forum is held annually with numerous lectures on theory and practice as well as workshops.

9.2 Significance of Project Planning for Management

9.2.1 Planning Requires Fantasy and Imagination

The word planning is widely used both as a standalone term and in a word combination. For example, there are government budgets, business plans, roadmaps, travel plans, city plans, etc., to list just a few examples. Planning processes of any kind are always about preparing future steps and courses of action. In any case, planning requires fantasy and imagination. This is just as true for the creation of a construction plan for a house as it is for the development planning of a technically complex system.

The preparation of a plan is based to a considerable extent on past experience. When drawing up a building plan, the architect refers back to previous projects and when developing industrial products, the technicians also refer to previously realized projects. However, if a completely new project is tackled, a new type of building or a new space device, increased demands are placed on the planner, because he can only draw on previous experience to a limited extent. Although some of the experience gained can still be extrapolated, it still often leads to major misconceptions. It is often not sufficiently taken into account that uncharted territory projects have to be planned particularly thoroughly due to a lack of know-how and that alternative concepts have to be developed at the same time.

Effective planning methods have been developed over the past decades in the context of aerospace research and development projects, taking into account the following aspects in particular:

(1) Precise definition of the project goal
   - Requirements
   - Specifications
   - Statement of Work *(SOW, see subsect. 13.2.2)*
(2) Creation of efficient basic planning documents
   - Project structure (*WBS, see subsect. 9.3.3*)
   - Development logic
   - Basic planning: performance, deadlines, and costs
   - Planning alternatives
   - Optimized use of resources
   - Risk identification
(3) Planning flexibility
   - Revision and/or adaptation of the workflow logic and planning
   - Inclusion of target changes and their impact on the project
   - Change Management.

9.2.2 Definition of the Project Objective

The basic prerequisite for planning is the definition of a clear and unambiguous objective in order to determine the tasks and results of a project as precisely as possible. A launch vehicle is developed with a very specific objective just like a new automobile. The goal of developing a new type of car is justified by the hoped-for sales, i.e., exploiting a gap in the market and generating a profit. In the development of an improved launch vehicle, e.g., Ariane 6, the primary objective is to create a new and marketable system for transportation of space payloads. A secondary objective may be justified by gaining technological and managerial expertise. It is therefore important that the objective of a project, including possible secondary objectives, is precisely specified; see Sect. 4.1.1.

With the definition of the project goal, the first important step has been taken. In a sense, it is the laying of the foundation stone of a project. This can be confirmed by the example of the successful satellite operator SES, based in Luxembourg. The key to success was based on realistic mission planning, for which sufficient time was taken. This was followed by the preparation of thoroughly prepared *Request for Proposal (RFP)* documents. The author was for more than a decade Senior Program Manager at SES and actively involved in the preparation of specifications, plans, functional specifications, and contract documents. When preparing the RFP, it is important that there are hopefully no or only very few open items such as TBDs (*to be defined*), because TBDs may cost money; see Sect. 14.2. However, customers are not always able to fully define all aspects of a project and there are always surprises, which then result in changes and usually additional costs.

In governmental and industrial development projects, especially in the IT area, where one often cannot draw on experience, it is usually very difficult to formulate clear objectives at the beginning. In the context of phase planning, a project goes through several iterations in the early phases (*Phases A and B*) and this can also lead to the objective being changed and/or re-specified; see Chap. 4. For this reason, performing technical-economic comparative analyses (*trade-offs*) in the early phases is an important responsibility.

Based on practical experience of the author, an example is described here. A country, several countries or an international organization intend to establish a global messaging system and announce following requirements:

1. A certain number of broadband communication channels will connect several countries and cities for the transmission of television data signals
2. The system must be expandable
3. The system must be able to operate largely without interference even in the event of equipment failures, interference with other systems, or operating errors
4. The existing ground facilities (*legacy*) shall be included
5. Trial operations and personnel training are scheduled to begin within the first 3 years and the system must be ready to start full operations 1 year later
6. An operator model including investment and operating costs shall be proposed.
7. The contract is to be awarded to an industrial company (*general contractor*) according to the rules of free competition

However, in order to be able to solve the task set, the global objective must first be divided into the following project phases; *see Chap. 4*:

- Phase A*:* to conduct a concept study to determine an optimal system taking into account all constraints listed below:
  - a terrestrial system (*cable network and/or directional radio*)
  - a satellite-based system or
  - a mixed system
  - When carrying out the concept study, all technical parameters, e.g., functionality, quality, service life, maintenance, etc., as well as all commercial aspects, development costs, operating costs, etc., must be taken into account. Their influence on the various systems shall be analyzed in detail; see Sect. 4.3.6 (Phase A)
- Phase B*:* Definition of the project with the aim of specifying all systems and subsystems in detail and developing precise plans; see Sect. 4.3.6 (Phase B).

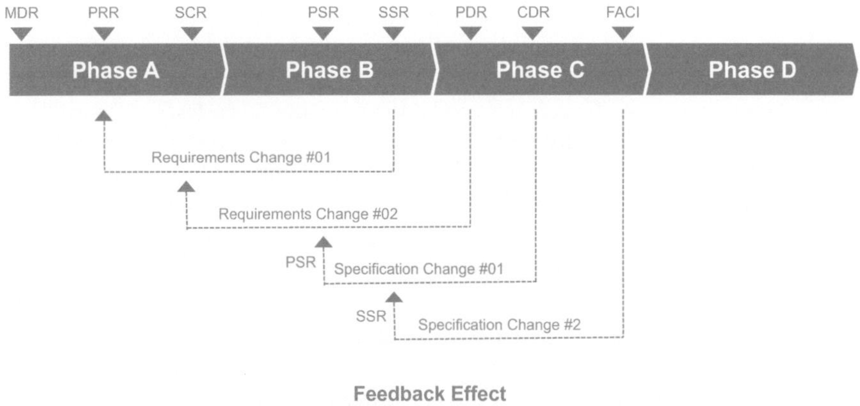

Fig. 9.3 Feedback effect of phases A and B

The performance of the first two project phases can lead to the modification of the original objective if, for example, the investment and/or operating costs are too high. It often happens that the internal or external client (*customer or stakeholder*) revises its requirements on the basis of the results of phases A and/or B. The customer would like to have a *gold-plated* solution but may be satisfied with the second-best solution when he sees the cost projections. Figure 9.3 shows the feedback effect.

With the successful completion of Phase B, the time has come to start preparing the tender documents for development (*Phase C*). Chapter 4 describes the results of Phase B and Chap. 14 describes the bidding processes. Once the Phase C/D bids have been solicited and evaluated, the C/D Phase work can begin.

- Phase C/D*:* Development of the specified system with the aim of being able to qualify the overall system. Production and installation of the system, start of operational use, and handover of the overall system to the customer; see Sect. 4.3.6 (Phase C/D).

The sub-goals mentioned for the individual project phases can be further detailed during the course of the project. Each objective, even that of the lowest level, must be defined as a measurable end product, e.g., as a milestone. The following criteria must therefore be taken into account when defining project goals, regardless of the level:

(1) Each defined target must be identifiable by an end product (*document, software, or hardware*). The end product (*deliverable item*) must be specified precisely and must be controllable in any case. Here are some examples:
   - *Report. The* control method is the content check; comparison of the desired and actually delivered result and the release of the report by a superior body

  - *Review.* Control method is the examination of the project result at a certain point in time; i.e., comparison of the specified and delivered performance and the preparation of a protocol (*minutes*) and determination of rework (*RID*)
  - *Device function. The* control method is the acceptance test; comparison of the specified and performed performance and preparation of a test protocol

(2) The time frame set for the defined goal must be clearly specified; deadline constraint.

(3) The dependencies (*interfaces*) and specifications (*inputs*) required to meet the objectives must be clearly defined.

The careful definition of project goals is an important preliminary work for the detailed project planning to be prepared subsequently. Project milestones can be derived directly from the previously defined project goals.

9.2.3 Why Planning Can Fail

However, it must be pointed out that planning very often fails because crucial basic requirements, e.g., sufficient professional experience of the planner or the information necessary for planning, are missing. In his book "*Management,*" Dale summarized five typical behaviors of planners that often lead to planning failures. The errors described by Dale are reproduced below [27]:

- The planner lacks the exact knowledge of the company
- The planner is not taking a realistic approach
- The planner tends to run after the facts rather than ahead of them
- The planner tends to make excuses rather than improve errors
- The planner plans in the short term

The flow of information plays a particularly important role in the execution of projects. A planner without information can contribute little to the project despite having great experience. In practice, it unfortunately happens time and again that the planner is not integrated into the project team, either organizationally and spatially. But precisely this is a basic requirement for promising planning in the interest of an optimal flow of information (*real time feedback*). To avoid failures, project planning must be firmly integrated into the project team, as shown in Figs. 5.3 and 5.4. This gives the planner rights, but also duties and responsibilities. In the past, there was often a trend for the planner to be available to the project as a consultant, without rights or duties. However, the successful planner should in no case earn his bread according to the staff position principle. In projects of a certain size, he or she belongs to the project with all his or her skin and hair, both organizationally and spatially. If he or she is also experienced, it is almost impossible for planning errors to occur for which he or she is responsible.

The former director of space programs at MBB/ERNO, Ants Kutzer, wrote very *aptly,* " *... the PM problems in space are typical of complex systems and open* *discussion of mistakes made can lead to the success of future projects.*" He then continues, "*... in fact, we cannot afford to repeat mistakes that have already been made, and this is not only true for spaceflight. However, in certain cycles, such as every ten years or so, mistakes are repeated*" [28]; see Sect. 3.2.

9.3 Planning Instruments and Their Use

9.3.1 A Whole Arsenal of Modern Planning Instruments

The project manager of the twenty-first century has a whole arsenal of modern planning tools at his disposal. However, he is repeatedly confronted with the following questions:

- What instruments are available (*overview*)
- Where do they exist and where can I read about them (*literature*)
- How can I make them work for me (*consulting*)
- Is use of IT recommended (*software*) and
- What planning effort is justified for a given project (implementation)?

Over the past 50 years, a variety of very effective planning tools have been developed, some of which have been disseminated worldwide, e.g., various planning techniques. Other tools, such as the Action Item Control System (*to-do lists*), became little known. In many cases, there was also a natural selection, so that instruments that were very promising at the beginning disappeared again very quickly, e.g., the PERT cost system propagated by NASA. In addition, today's project managers are hardly amazed by planning novelties, unless they are methods that have been tried and tested in practice and are really useful. Less but simple is better than more and complicated. KISS *(keep it simple stupid)* is a frequently mentioned formula.

Today, project managers are more dependent than ever on the use of planning methods that are effective and have been tested in practice. The days of experimentation are a thing of the past, because today hardly any project manager can afford the luxury of experimenting. Users must be convinced that the planning methods offered are useful to them. The following is a summary of the most important project planning methods that can be used in whole or in part:

(1) Requirements Specification
(2) Statement of Work *(SOW)*
(3) Project structure
   - Hardware/software tree structure (*product tree*)
   - Work Breakdown Structure (*WBS*)
   - WBS description *(WBS—Dictionary)*
   - Technical characteristics
   - Work Packages (*WP's*)

  - WP Descriptions *(WPD's)*

(4) Schedule and sequence planning
  - Milestone schedule (*milestone plan*)
  - Bar chart concept (*bar chart or Gantt chart*)
  - Network plans

(5) Personnel deployment and cost planning
  - Personnel deployment plan (*manpower plan*)
  - Cost plan
  - Cash flow plan (*resource plan*)
  - Payment plan

For most of the planning methods mentioned here, especially for scheduling and cost planning, the companies concerned have procured appropriate software packages; see Chap. 18. In addition, a large number of consulting firms offer appropriate support to companies concerned with the implementation of projects. If the use of software is not discussed in detail in the following remarks, this is not because the author does not attach sufficient importance to this topic, but solely due to the fact that the main concern of this book is to treat all planning and monitoring methods equally and to encourage the reader to critically review requirements from his own point of view.

Requirements →	**Specification Book**
Customer	**Project manager and Project team**
A document produced by the Customer containing all requirements on the deliverable items and services provided by the contractor ("Wish List")	A document produced by the Project Contractor, that contains Contractor developed implementation specifications. It describes the implementation of the Customer defined requirements.
Contains technical and content requirements, that the project shall accomplish. WHAT does the Customer want? WHY does the Customer need it?	Basis for the contractually binding Contractor services, WHAT will be delivered?
Important contents, that frequently arise with Specifications: • General project information, • Description of the Initial Status (IS-Condition), • Goals (SHOULD-Be_Condition), • Requirements and Framework conditions, • Estimate of effort →Resources, • Desceription of the Requirements, • Impact to other Projects, • Comparison with existing solutions, • Risks, • Release notice.	Important contents, that frequently arise with the Specification Book: • General project information, • formulated task descriptions, goals: Must and May criteria, • Results of the problem analysis, • Feasibility assessment, • Definition of individual functions, • Project organization, • Phase plan (Milestones), • Progress control, • Additional agreements, • Release notice.
General Requirements: current, complete, short, precise, clear (free from contradictions), feasable, verifiable.	

Fig. 9.4 Representation of the term's requirement specification and functional specification (Source: R. Fischer according to DIN 69901-5)

9.3.2 Requirements and Specifications—Statement of Work

The specifications and requirements are closely connected [29]. In the requirement document the client defines the expected supplies and services and in the specification the contractor describes the implementation of the supplies and services specified by the client. Alternative terms for requirement specifications are: Requirements specification, requirements catalog, product sketch or customer specification. In the English-speaking world, the term "Statement of Work *(SOW)*" is common and can be equated with the requirements specification. The specification, or SOW, describes in concrete form how the contractor will solve the requirements of the customer. Figure 9.4 shows the synoptic representation of the terms prepared by R. Fischer.

In the English language, there are a number of similar terms depending on the phase sequence, but these are not always applied consistently and uniformly; see Chap. 4. At the beginning of Phase A, the customer usually prepares a requirements document (*Mission Requirement*), to which all further documents refer in the course of Phases A and B. Subsequent documents are a fixed part of the contract when a project is awarded by the customer to a qualified contractor; see Sect. 13.2.1.

1. The Statement of Work (SOW) which describes the supplies (deliverable items) and services to be provided. NASA Guideline NPG 5600.2B, December 1997, is recommended for more in-depth reading.
2. The product specification as well as plans defining the future product (*functions, dimensions, etc.*) and the required tests. The contents of a system specification are described as an example in Sect. 7.2.1. In larger projects, individual specifications are created for the system, all subsystems and components. In addition, there are interface and test specifications to be delivered.
3. The quality requirements are usually described in a special QA specification.

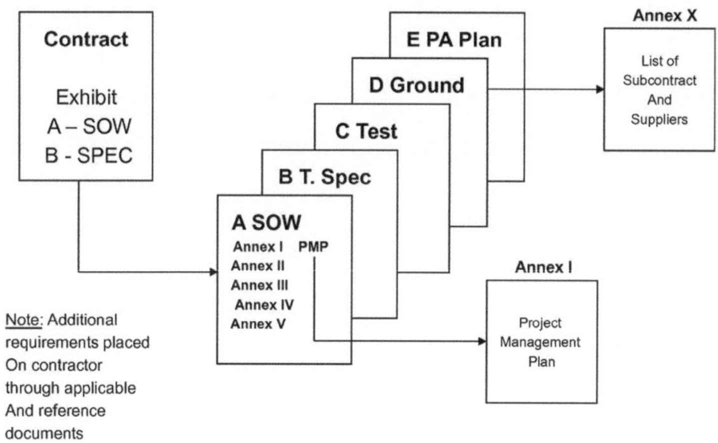

Fig. 9.5 Hierarchical structure of "*Contract, SOW and PMP*"

Figure 9.5 shows the hierarchical structure of the contract, the SOW and the PMP, as is typical for contract documentation. For smaller projects, the aforementioned documents may be combined in a single document with the appropriate structure.

9.3.3 Work Breakdown Structure

The topic of WBS is dealt with in great detail at this point because it is the mental entry into a project, since the WBS penetrates the project successively from A to Z, i.e., down to the lowest level, the work packages. The person who creates the WBS must automatically deal with the project contents and thus also coordinate with the responsible colleagues in detail.

9.3.3.1 Creating Transparency

In particular, large and complex projects, such as those that occur especially in the field of research and development, whose duration often takes several years and whose completion involves a large number of companies, institutions, and employees, require the project to be divided into individual construction and model phases, segments, assemblies, functions, work packages, etc., for reasons of efficient management. In other words, projects must be structured in a logical manner to ensure the transparency required for planning and progress control of the project. An effective instrument for this is the Work Breakdown Structure (*WBS*). It enables the clear assignment of companies, institutions and specialists to the respective project elements for which they are responsible.

For technical planning, the structuring of a project according to functions or individual assemblies is important because the processing of the individual project areas is usually carried out by very different specialists and, in the case of larger projects, often also by different companies. The WBS creates transparency! Here are two examples:

1. Example: a study project
   (a) Project objective: Study for a new mass transport system
   (b) Main Tasks:
      - Study of the passenger volume
      - Investigation of possible drive systems
      - Energy supply analysis
      - Tariff concept
      - Investment planning
      - Study Guide.
2. Example: Development of a rocket engine
   (a) Project objective: Development of a new propulsion system
   (b) Main Tasks:
      - System design
      - Component development

Combustion chamber
Turbine
Pump
Fuel supply
Servomotors
- Assembly and test
- Test facility construction
- Quality assurance
- Project Management.

The functional or system outline, resulting from the technical aspects and the logical procedure for processing a complex task, should always be the godfather when creating a work breakdown structure (*WBS*). In other words, the structure of the WBS should correspond as far as possible to the natural structure of a system. At this point, it should not go unmentioned that project structuring is also important for small projects and should not be forgotten. However, in most cases, small projects do not involve complicated structures, but rather a simple outline concept, e.g., in the form of a table.

9.3.3.2 System Structure and Work Breakdown Structure

So what is the difference between a pure system outline and the work breakdown structure? The answer is simple: in the objective! The system breakdown structure or the product tree (see Fig. 7.3) primarily serves the purpose of dividing the overall system, e.g., a production plant or a construction project, into its individual subsystems or assemblies so that the engineers or architects can structure their working documents (*specifications and drawings*) accordingly. The WBS serves the purpose of creating an integral instrument for project management. The project management, which is not only responsible for the purely technical but also for the

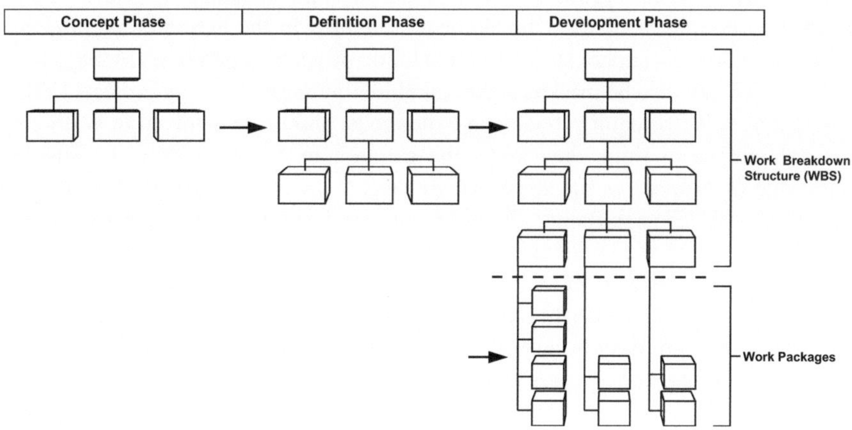

Fig. 9.6 WBS development steps for phase C/D

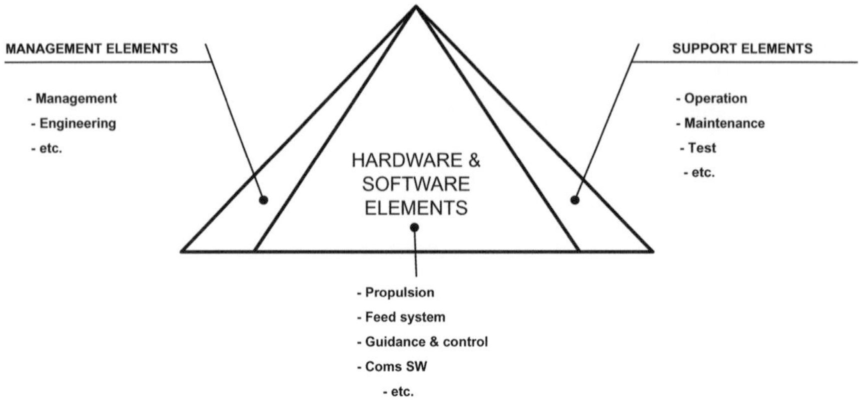

Fig. 9.7 Main areas of the WBS

economic and administrative project handling, associates a much broader objective with the WBS.

From this point of view, the WBS is the logical continuation of the system structure created in the early stages of a project (Fig. 7.3), starting with the upper project level. In this context, the creation of the WBS for the cost-intensive phase C/D is of particular importance. The preparation usually starts in the concept phase and is completed in the definition phase (*phase B*) so that a finished WBS including all work packages is available at the beginning of phase C/D. The system outline (Fig. 7.3) is an important prerequisite for creating the WBS for phase C/D. Figure 9.6 shows an example of the typical development steps of a WBS for Phase C/D [3].

In developing the WBS, it is important to remember that it is a management tool [30]. In the USA, an extensive guideline for the creation of work breakdown structures (*WBS*) was published in 1968 as part of the MIL standard series [31]. However, NASA also issued a detailed guideline for the preparation of work breakdown structures as part of its tasks [32]. European institutions and companies have also recognized the importance of work breakdown structures in the execution of projects. ESA has created a standard structure for satellite projects as part of its ECSS procedures, which can also be used as a general guideline for technology projects [33].

Creating a WBS requires hands-on experience, and those involved in WBS creation should never think they know the solution in detail up front, because as Burbridge (*a former Lockheed executive*) says: "*. ... there is no one best way to organize the WBS*" [34]. Rather, the WBS must be developed incrementally according to firmly established rules and in close contact with project engineers. The WBS is an important management tool. The question often arises as to whether the WBS should be presented in a hardware-oriented or function-oriented manner. As a rule, however, this is not easy to answer. In the vast majority of cases, a WBS for the development phase of a project will be a mixed structure. Very often, as shown in Fig. 9.7, the WBS is hardware-oriented at its core, but also contains function-oriented WBS elements. Software elements are treated like hardware.

The before mentioned principle was applied for the development of the WBS for the Europa III launch vehicle (*precursor of the Ariane launcher*). The major WBS elements are shown below [35].

1. Project management (*ELDO Secretariat*)
   - Project Administration
   - Project Control
   - Configuration control
   - Systems engineering
2. Systems engineering integration (*industry*)
   - System analysis and system design
   - Interface control
   - Test Management
   - Quality assurance
   - Special studies
3. Launch vehicle development (*industry*)
   - First stage
   - Second stage
   - Interstage
   - Electronic system
   - Equipment compartment (*control, etc.*)
   - Payload fairing
   - Ground equipment
   - Integrated test facility
   - Assembly and test
   - Test facilities
4. Launch facilities (*ELDO and industry*)
5. Launch Execution (*ELDO/CNES*).

9.3.3.3 The WBS Is a Central Management Tool

The WBS is the common backbone for all project areas and serves the following main purposes:

(1) Structure of the technical/administrative planning documents
   - Specifications*:* In the definition phase, the project specifications for the overall system and all subsystems must be created. In this context, a specification tree (*hierarchical structure of the specifications*) must be defined in accordance with the WBS; see Fig. 7.3 (Product tree).
   - Statement of Work *(SOW):* The project activities to be carried out are to be defined in the respective SOW of the companies involved. By assigning the company responsibilities to the WBS, it is possible to link the SOW to the WBS.
   - Schedules*:* The schedules to be prepared shall be structured to ensure consistency with the WBS.

– Cost plans*:* The total project costs are to be broken down according to the WBS, so that the planned costs are clearly and unambiguously defined for each top level WBS element down to the lowest level and/or work package. A further breakdown by cost categories (*personnel, material, IT, external service costs, etc.*) and time intervals shall be made per WBS element or per work package.

(2) Project progress control: progress control is based on the planning documents mentioned in (1) above. The clear assignment of the technical/administrative planning documents to the WBS ensures integrated progress control (*performance/dates/costs*). At regular intervals (*weekly, monthly, quarterly*), the project status, i.e., the technical, schedule and financial status and any deviation to plan, shall be determined by a comparison between plan and actuals; see Sect. 9.4.

(3) Reporting: The quality of reporting depends very much on the possibility of summarizing information. Detailed information from the lowest project levels must be summarized hierarchically upwards in accordance with the WBS, so that overall reports at the level of the overall project are possible without loss of information; see Sect. 11.2.

(4) Documentation structure: The importance of well-managed project documentation should not be underestimated. For this reason, a documentation system should be set up as early as possible. The clear assignment of all documents in reference to the WBS is of utmost importance. The documentation key (*numbering system*) should therefore be linked to the WBS code; see Sect. 12.1.

(5) Industrial work-sharing: Based on the WBS, the work packages of the project are to be shared among the participating companies and their departments.

(6) Building project databases: Based on the WBS, actual values (*technical, scheduling, and financial*) shall be recorded in databases to be specially created for future projects. It is therefore recommended that organizations and companies develop their own product-related standards based on the previously mentioned WBS standards.

9.3.3.4 WBS Standard and Diversification

As experience in WBS creation progressed, the call for standardized work breakdown structures became stronger in many industries. This was mainly due to the following reasons:

- Creation of a standard structure for project evaluation (*post-calculation and comparability of projects*) and
- Provision of a standardized WBS for future projects (*avoidance of expensive new developments, because the wheel does not have to be reinvented again and again*).

For this reason, the European Space Agency (ESA) commissioned an industry study in 1979 to develop a standardized WBS concept.

The author supported the project as part of his MBA studies. Fifty different WBS configurations from European industry were examined for their practicality in defining a standardized ESA WBS concept [36]. The very different WBS configurations could be summarized as follows:

– Configuration I:	Hardware and company oriented,
– Configuration II:	Hardware[a] and function oriented
– Configuration III:	Function and test model oriented.

[a]*Version II correlates best with the product tree shown in Fig. 7.3*

In order to arrive at a useful result, the three configurations were tested for their usefulness using the following management criteria and an evaluation scheme [36]:

- Management and organization
- Scheduling and monitoring
- Cost planning and monitoring
- Product assurance, documentation, and configuration control
- Contract and procurement management
- Systems engineering and test coordination.

The evaluation was made by an industrial expert team using a point system and the result was as follows:

– Configuration I:	82%
– Configuration II:	91%
– Configuration III:	76%

It became clearly visible that there were only minor differences, which confirms Burbridge's statement that there is no absolute best way to create WBS [34]. However, Version II (*a mixed hardware and function-oriented version*) has prevailed at ESA; see Fig. 9.8. The WBS concept (*Configuration II*) has been selected as the most appropriate WBS configuration for future ESA projects [37, 38]. This structure is characterized in particular by the fact that certain functions, such as project management, product assurance, and systems engineering (*see Fig. 5.3*), which may recur several times in the various WBS levels, are arranged in mirror-image (*modular concept*). In addition to the WBS structure, a detailed WBS number key was developed, from which the individual WBS identifiers for the standardized hardware elements, work packages and company abbreviations can be taken.

As part of the creation of the ECSS procedures, the WBS standardization also finally found its place [33]. Four iterative steps are proposed for WBS development:

1. Creation of a function tree
2. Specification tree
3. Product tree
4. Work breakdown structure (*WBS*)

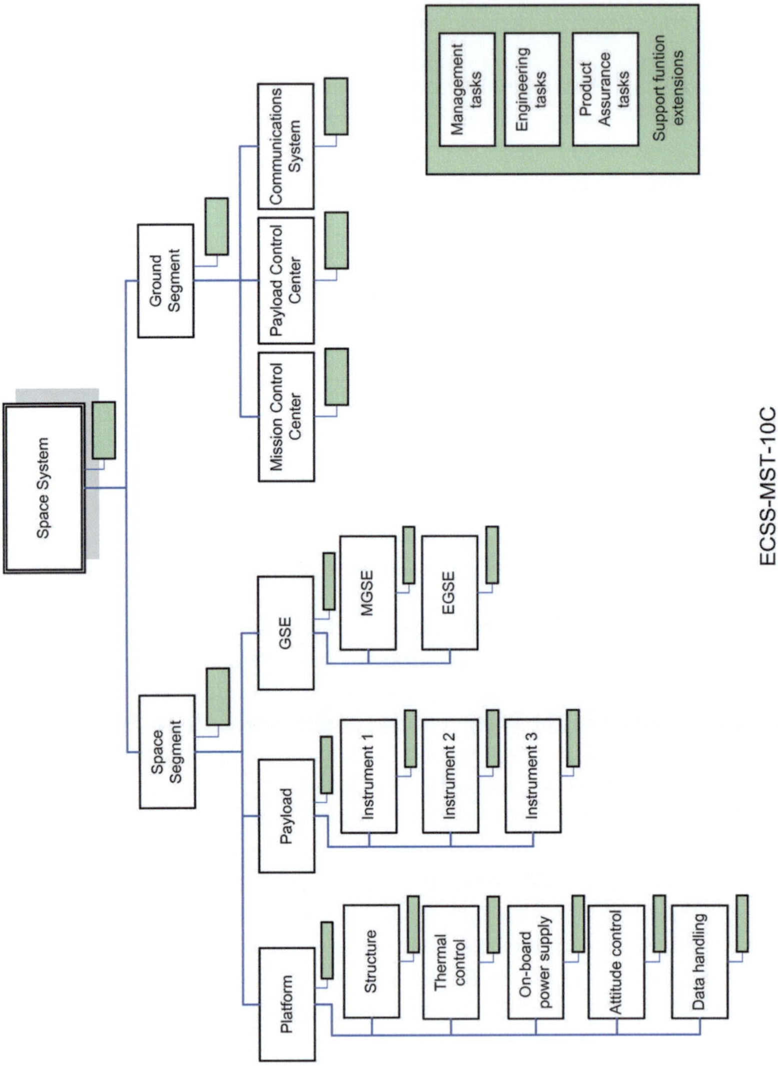

Fig. 9.8 Selected WBS—Configuration II (ESA)

Authorities and companies have come to realize that it is difficult or even impossible to plan and manage a project efficiently without a clear outline. Without a clear outline, misunderstandings, confusion, and sometimes downright chaos quickly arise, because organizational aspects are also linked to the project structure. The well-known "W-questions" arise: who, when, how, where and what. The work breakdown structure (*WBS*) is the PM backbone, as mentioned earlier because you can use it as a reference point. Many project managers have come to realize this. The question is, how do you create a WBS for a power plant or a ship project? The ESA experience described here provides a good reference.

Traditionally, many companies are used to structuring on the basis of assemblies. This corresponds to the product tree defined by the ESA and is also a good place to start. Often, one follows the bill of materials structure where functions such as project management, systems engineering, quality assurance, assembly, and testing are considered. From a commercial point of view, however, they must also be taken into account.

Especially for high-tech projects, it is essential to start developing a WBS at the beginning of the project, even if done by using a different name. Thus, the realization that systematic project structuring makes sense prevailed in many industries. Here are some examples:

(1) In 1982, with the active participation of the author, a major German company prepared a WBS concept for the construction of a power plant based on space experience. Major WBS elements are reproduced here [39]:
   - Project management (*function*)
   - Quality assurance (*function*)
   - Planning of the entire plant (*function*)
   - Arrangement planning (*function*)
   - Machine technology (*hardware*)
   - Control technology (*hardware*)
   - Power engineering (*hardware*)
   - Construction technology (*hardware*)
   - Site/Assembly/Commissioning (*Hardware/Function*).

(2) Airbus Industries also developed a standardized WBS in 1982 for the Airbus successor A320, which corresponded to the WBS concept shown in Fig. 9.8 [40].

(3) In 1982/1983, MBB conducted a WBS study on behalf of the German Government (*BMFT*) for the project "*Ship of the Future,*" in which, among other matters, the previously described WBS philosophy was also the inspiration [41].

(4) In principle, structuring concepts in transport technology and the automotive industry do not differ from the approach taken in aerospace. There, too, the products (*hardware*), such as engine, chassis, doors, flaps, exterior, etc., are initially at the center of the structure, flanked by elements such as documentation, quality, and management tasks; see Fig. 9.7.

(5) In 1977, MBB supported the procurement activities for the Caracas Metro as part of a systems and management consultancy, where the author was responsible for the introduction of the management concept, including WBS [42].

To avoid misunderstandings, it should be mentioned that, in addition to the WBS standard for technologically complex development projects shown in Fig. 9.8, other forms of structuring can also be useful depending on the industry. In the case of service projects, the focus may be only on functions. For a project study, the top WBS level could also contain only functions such as:

- Project management
- Engineering
- Process planning
- Plant design
- Economic view

It is important that the PM makes sure that in each case a WBS is created in which the tasks are clearly expressed because the WBS is a reference point for the scheduling and sequence planning and the cost and resource planning.

9.3.3.5 WBS Levels and Model Philosophy

For large projects, the WBS structure is usually made up of several levels, as shown below using the Apollo program as an example [43]:

– Program (Apollo)	Level 1
– Projects (launch vehicle, etc.)	Level 2
–Systems (levels 1, 2, n, etc.)	Level 3
– Subsystems (drive, etc.)	Level 4
– Sub-subsystems (engine, etc.)	Level 5
– Components (nozzle, etc.)	Level 6

The German Aerospace Center (*DLR*) plans to develop a supersonic plane supported by rocket engines; named SpaceLiner (SL). The SL flies at several times the speed of sound to traverse a suborbital trajectory from Europe to Australia in 180 min. Figure 9.9 shows the WBS for the C/D phase of the project. The structure consists primarily of hardware elements to which functions, such as PMO, SE, PA, PC, and AIT, are assigned at all levels, as well as ground facilities (*Ground*) and operations planning (*Operations*) [44]. The chosen principle corresponds to the philosophy of ESA; see Fig. 9.8.

In the development of technologically complex systems, such as transportation and aircraft projects, several test models and prototypes are usually created for testing purposes, which must also be recorded in the WBS. This is explained below using the SpaceLiner as an example. As part of the WBS element 5000, the "*model philosophy*" of the SpaceLiner project is defined; see Fig. 9.9.

Fig. 9.9 SpaceLiner WBS concept (Source: Trivaillo, March 2015)

Table 9.1 WBS table example for cost tracking

WBS item [Nr. & Title]	Responsibility [coordinates]	Budget [1000 €]	Actual [1000 €]	CTC [1000 €]	CAC [1000 €]	Balance [1000 €]
100 Project Management Office	Name, Tel. Nr. EMail, Address	100	70	40	110	−10
200 Mechanical Subsystem		800	500	275	775	+25
300 Electrical Subsystem		1200	950	350	1300	−100
310 Power Generator		800	600	200	800	0
320 Power Converter		400	350	150	500	−100
GRAND TOTAL OF THE PROJECT		2100	1520	665	2185	−185

- 5200 MU/BBM components—test models
- 5300 STM structural model
- 5400 EQM Engineering/Qualification Model
- 5500 PFM 1 First proto-flight model
- 5600 PFM 2 Second proto-flight model

In the first step, the models are used to qualify all components and the overall system (*5200–5400*) and thus also to minimize risks. On the basis of the test results then available, the construction of prototypes (*5500–5600*) can be initiated or completed in the second step in order to start flight testing after completion of the first prototype. The number of test models and prototypes serves to create a safe means of transportation, but it is also a considerable time and cost factor. Therefore, the decision of the model philosophy is an important system engineering task of the project management, which has to be done in close cooperation with the system engineering department: see *Chap. 7*. In order to save costs, the prototypes (*PFM*) are often used commercially after the completion of the flight test [45].

In addition to displaying the WBS as a tree structure, a table can also be selected, which is easier to handle in most cases. The WBS example for cost tracking is shown below in Table 9.1 can easily be created as an Excel table and can also be changed quickly. The important thing is that the WBS corresponds to reality and is officially adopted by the PM. The table can then be expanded as needed to include other project information. The table can become an important control element in the hands of the PM.

9.3.3.6 WBS Tables of Contents

During WBS development, it happens again and again that project members assign different contents to the individual WBS elements. If the WBS creation was even several months or years ago, the WBS contents are often interpreted completely differently. It can happen that the contractor has different ideas for a certain WBS element than the client. Many a conflict is already pre-programmed as a result. To

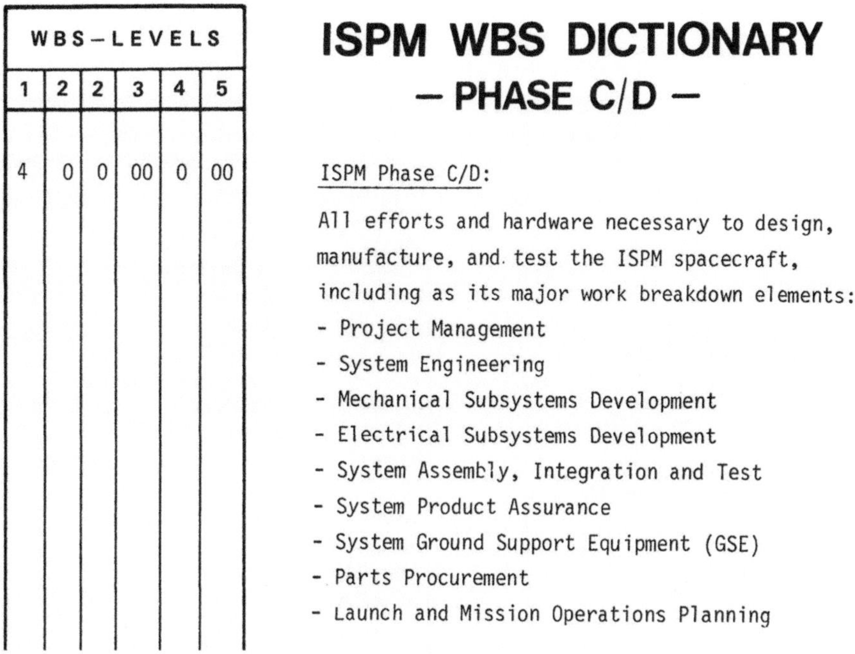

WBS-LEVELS					
1	2	2	3	4	5
4	0	0	00	0	00

ISPM WBS DICTIONARY
– PHASE C/D –

ISPM Phase C/D:

All efforts and hardware necessary to design, manufacture, and test the ISPM spacecraft, including as its major work breakdown elements:
- Project Management
- System Engineering
- Mechanical Subsystems Development
- Electrical Subsystems Development
- System Assembly, Integration and Test
- System Product Assurance
- System Ground Support Equipment (GSE)
- Parts Procurement
- Launch and Mission Operations Planning

Fig. 9.10 Example of a "WBS Dictionary" (*Source MBB*)

avoid such problems, the McDonnell Douglas company introduced a detailed WBS dictionary as part of the Space Shuttle program [46]. The procedure was later partly adopted by MBB; see Fig. 9.10.

In addition to the *WBS dictionary*, it has proved useful in practice to add the technical characteristics for the hardware and, if necessary, for the software elements of the WBS. An example of a "Technical Characteristics List" is shown in Table 9.2 [47]. This is not an exotic creation but a procedure for creating a basic reference for technical performance control (*TPC*); see Sect. 9.4.7. In other sectors, too, such as transport technology and automotive engineering, it is a question of achieving the project objectives and specifically of technical parameters such as dimensions, weight, battery volumes, etc.

Both documents, the WBS table of contents and the list of technical characteristics for the hardware elements of the WBS, are ideally suited as supplementary documents to the WBS to bring more transparency to project execution. The WBS table of contents shall be prepared in accordance with the bill of quantities (see Sect. 13.2.2) and the list of technical characteristics and in accordance with the specification; see Sect. 7.2.1. This may seem to be an additional burden for the reader; however, it should not be forgotten that the management of large complex projects requires the highest level of transparency. The WBS, including the additional information described earlier, is an important information and control tool. The planning effort that goes into the careful preparation of the WBS documents represents a

Table 9.2 Example of a technical characteristics list

Technical characteristics list				
Identification number	WBS identification	WBS level	Units of measure	Characteristics
100-00-00-00-00	Spacecraft	3	Pounds	Weight
			Days	Mission Duration
			N. Miles	Apogee
			N. Miles	Perigee
			Pounds	Gross Weight (Wet)
			Pounds	Weight (Dry)
			Feet	Diameter
			Feet	Length
100-01-00-00-00	Module	4	Ft^2	Wetted Area
			Ft^3	Crew Quarters Volume
			Number	Crew Size
			Ft/Sec	Reentry Velocity
			Number	Number of Flight Functions
100-01-01-00-00	Structure	5	Ft/Sec^2	Maximum "g" Loading
			Type	Type of Structure
100-01-03-00-00	Propulsion	5	Pounds	Thrust
			Lb/Sec/Lb	I_{sp} Actual
			Lb/Sec/Lb	I_{sp} Ideal
			Lb-Sec	Total I_{sp}
			Ft/Sec	Delta V Requirement
			Number	Number of Burns
			Pounds	Inert Weight
			Seconds	Maximum Burn Time
			Days	Operating Life
			Ratio	Throttling Ratio
			Ratio	Expansion Ratio
			Pounds	Weight of Propellant
			Specify	Solid or Liquid
100-01-05-00-00	Reaction Control	5	Pounds	Thrust per Engine or Thruster
			Number	Number of Thrusters
			Lb-Sec/Lb	I_{sp} Actual
			Lb-Sec/Lb	I_{sp} Ideal
			Lb/Sec	Total Thruster Impulse
			Pounds	Weight of Propellants
			Seconds	Minimum Impulse Bit
			Pounds	Inert Weight
			Seconds	Total Burning Time
			Deg/Sec	Minimum Attitude Change Rate
100-01-05-00-00	Electrical Power	5	Watts	Total Power Available
			Watts	Maximum Power Level
			Watts	Nominal Power Level
			Specify	Source Type (Chem, Solar, etc.)
			Watts	Average Power Level
			Pounds	Subsystem Weight
			Days	Power Profile, Storage
			Hours	Allowable Source Down Time

worthwhile investment for the definition of work packages and the preparation of the sequence and cost planning, as well as for the monitoring that builds on this. The WBS should not be seen as an administrative piece of busy work. WBS creation is a creative process in which the system engineer and planning engineer should work closely together.

9.3.3.7 Work Packages and WP Descriptions

Only after the WBS has been fully developed can the definition of work packages be started. Each work package (*WP*) represents a real task in the sense of work, in contrast to the end-product-oriented WBS elements (*hardware, software, or functions*) [3]. Typical work packages are:

- Design
- Construction
- Manufacturing
- Assembly
- Test
- Analysis etc.

Work packages are tasks that can be performed independently by a specialist department of a company. The development and production of an engine cannot usually be carried out by a single department but requires the involvement of several departments (*design, production, and test departments*). It therefore makes sense to divide up the end-product-oriented WBS element "*Engine development*" according to the work packages just mentioned. The work packages must be defined in the specific case in accordance with the project organization. Figure 9.11 shows the incorporation of the work packages at the lowest WBS level.

Once all work packages (*WP*s) have been defined, the WP tasks and results must be described in detail. All work packages together represent the entire scope of a project. For this reason, the WP description is particularly important. One cannot leave the definition of the WP contents to the WP managers alone, because then the project management would not be sure to get what is needed to fulfill the overall project goal. WP descriptions (*WPDs*), which are tantamount to mini-performance specifications, must be prepared in close coordination with the project management. In a sense, they are the link between the project management and the specialist departments; see Fig. 5.11. Detailed schedules and cost plans are drawn up on the basis of WP descriptions.

Figure 9.12 shows a form developed in practice for a Work Package Description (*WPD*) [3]. The WPD is a link between the project and the specialist department and in principle acts like a "company-internal contractual agreement." See the following practical example: both parties (*project and functional department*) must agree about the following:

1. A service to be provided, including deliveries (*outputs or end results*); e.g., deliverable products, documents, or services

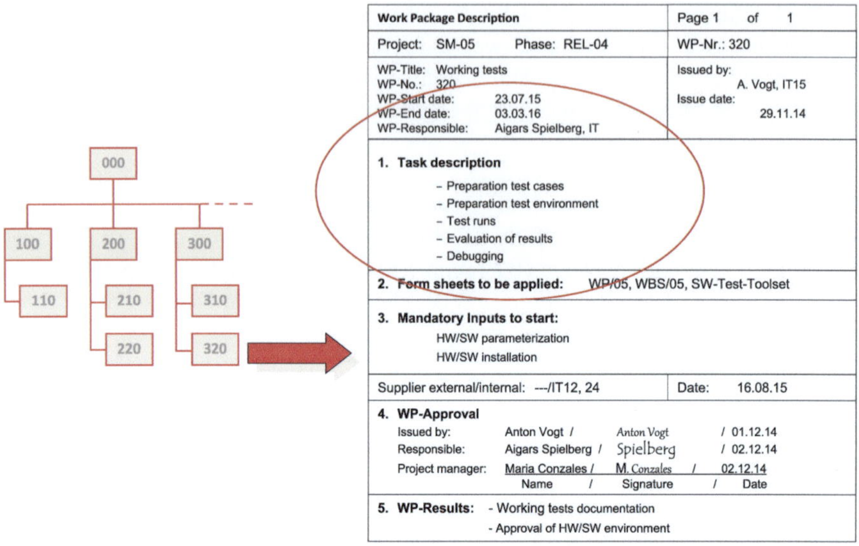

Fig. 9.11 Work package identification in the WBS (Source: FMT, Bremen)

2. Documents to be applied (*inputs*), e.g., specifications, plans, procedures, regulations, etc.
3. Provisions (*inputs*), e.g., components, software, machines, test equipment, etc.
4. WP—Approval/release by: PM and specialist department
5. Attachments (*as required*).

9.3.3.8 WBS Versus Project Organization

In practice, misunderstandings often arise between the WBS and the project organization. Both structures are indeed organizational instruments, but they serve different purposes. The WBS serves the process organization (*what has to be done*) and the organization chart serves the organizational structure (*who has to do it*).

The differentiation between the organization and WBS is very important. NASA explicitly points out this problem in its WBS manual and stipulates that "The WBS shall not be used as an organization chart." [32] The levels of hierarchy shown in the WBS are not identical to the levels of organizational structure, even when similarities exist.

It must be ensured that a particular organizational unit (*OU*) defined in the organization chart is responsible for one or more WBS elements or a work package. Duplication of responsibilities must be avoided. On the other hand, an OU can, of course, be responsible for several WBS elements and/or work packages at the same time; see Fig. 9.13 [48, 49]. The relationship between the project and the department is also illustrated by the matrix organization in Fig. 5.11.

WORK PACKAGE DESCRIPTION	ISSUE: DATE:	SHEET:
PROJECT:	WP:	
W.P. TITLE: CONTRACTOR: START EVENT: END EVENT:	 Planned date: Planned date:	
W.P. MANAGER:		
INPUTS: TASK DESCRIPTION: TASKS SPECIFICALLY EXCLUDED: OUTPUTS FROM WP: • Hardware • Software • Documents		

ESA PSS A20

Fig. 9.12 Standard Work Package Description (WPD)

9.3.3.9 WBS Number Key

The clear and unambiguous identification of the individual WBS elements and work packages requires the assignment of a WBS numbering system. Since many project documents have a direct reference to the WBS, WBS coding takes on special significance. In the quest for integrated project monitoring, the WBS is increasingly being used as a central management tool, as mentioned at the beginning of this subsection. The WBS number key as an integrating project coding can be used to unify other numbering systems, e.g., the documentation numbers, the cost key (*order number*), or the equipment designation (*nameplate*). In contrast to simple numbering without structure, the WBS number key should be used as far as possible as a reference for all other numbering systems in the interest of efficient project monitoring.

However, practical experience has shown that it is hardly possible to create a universal WBS number code that meets all project needs. Nor is this necessary, since it is quite sufficient if the core structure of the WBS code, as shown in Fig. 9.14, can be partially reconciled with the other codes, e.g., for cost control, etc.

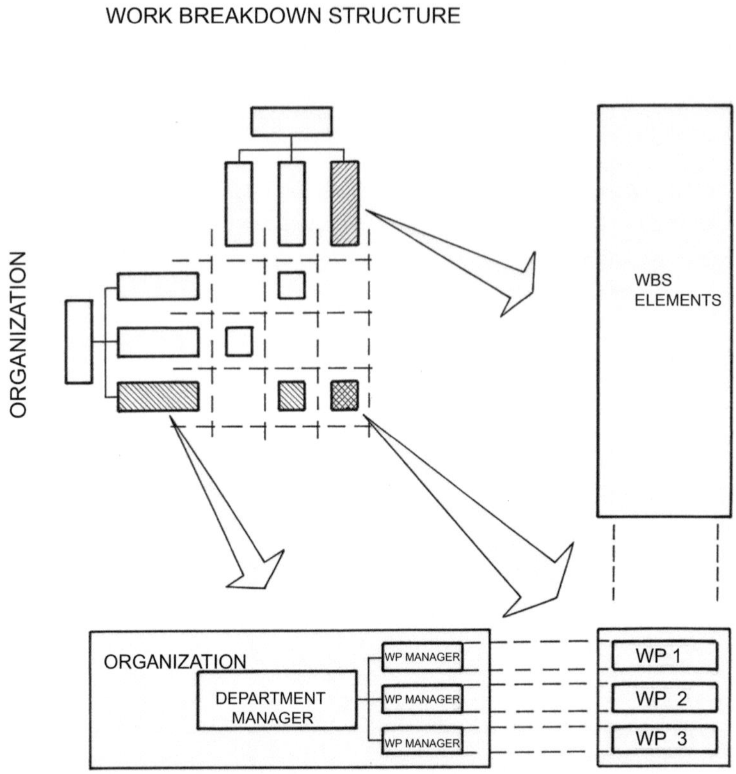

Fig. 9.13 WBS and organizational unit (Source: ECSS)

9.3.4 Scheduling and Sequence Planning

Planning is a main task of project management and is responsible for creating the following: Definition of project objectives, work breakdown structure (*WBS*), process and schedule plans, and resource and cost plans.

- Planning leads to the targeted use of resources
- Plans are the basis for determining deviations
- Plans are an information tool
- Plans represent a decision visible to all stakeholders

"*Time is the scariest resource and unless it is managed, nothing else can be managed*" [50]. It can be seen time and again that engineers and managers are usually not enthusiastic planners and are happy to hand over this work to "*someone*" who agrees to take on the laborious task. The creation of a process and schedule plan is a very creative system task, because behind the planning there is pure technology, and every plan must be approved by the PM before it is distributed to the team.

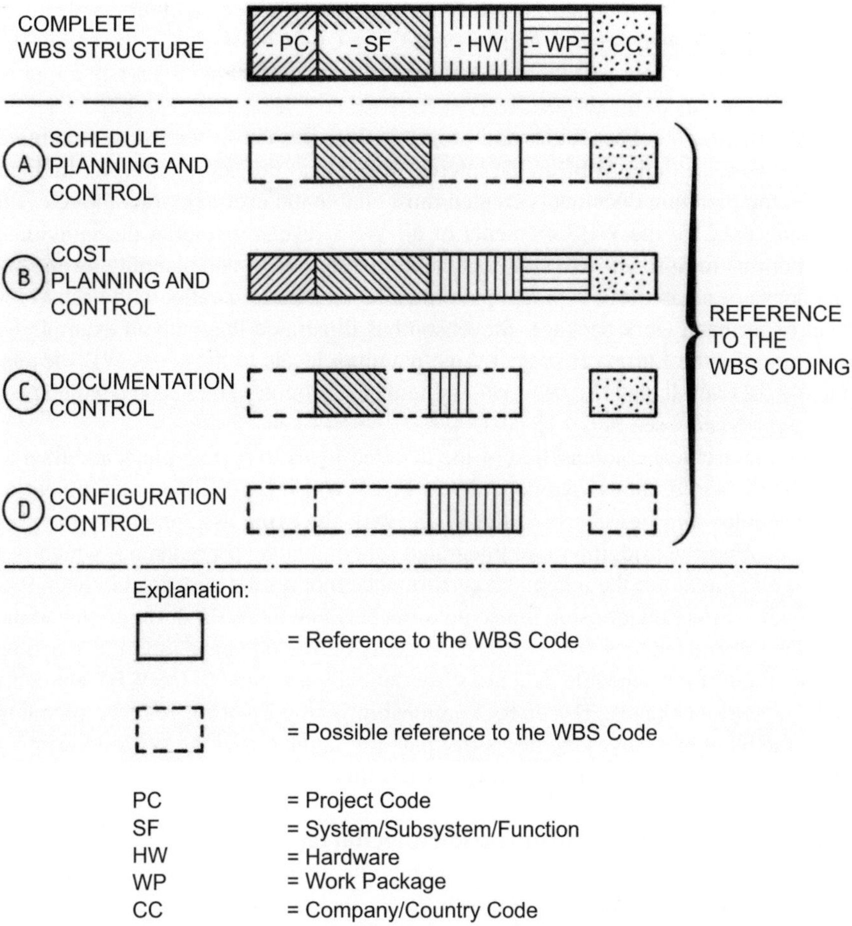

Fig. 9.14 WBS numbering system

9.3.4.1 WBS and Scheduling

Scheduling and process planning is an important management tool for project control. It involves the thorough planning of individual activities that are coordinated in terms of time and the definition of project milestones. It is the basis for detailed cost and resource planning and the basis for effective status determination. The schedules and progress plans must be drawn up in close accordance with the WBS and on the basis of the WP descriptions. "*A well-designed WBS makes it easy to assign any project activity to one and only one terminal element. ... It (the WBS) unifies engineering and management planning, scheduling, cost estimating, budgeting, contracting, configuration management, and performance reporting*" [51]. Just like the structure of the WBS, the creation of the schedules and flowcharts can only be done in individual steps. This means that the level of detail of the planning documents for the development and production phases must be slowly increased in the early phases of the project (*concept and definition phase; see Chap. 4*), so that at the end of the definition phase (*Phase B*) there is a complete and detailed plan corresponding to the specifications of the project.

The preparation of schedules and progress plans for a project must be structured in accordance with the WBS in the interest of efficient project monitoring. In other words, the planning documents created must enable the user to extract the relevant planning data for the WBS elements of all WBS levels, including the individual WP. For this reason, it is essential to synchronize the WBS and planning levels, but this does not mean that a separate plan must be created in parallel with each WBS level or for each work package. Stevenson has illustrated this with an example by assigning for the Europa II project three planning levels to the seven WBS levels, namely the overall plan (*ELDO level*), system plan (country level), and detailed plan (*company levels*); see Fig. 9.15 [17].

The hierarchical condensation of the detailed plans to system plans and then to the overall plan is shown symbolically in Fig. 9.16 [52]. The direct assignment of the schedules and flowcharts to the WBS, with the same assignment of the cost plans, enables the bridging of an integrated schedule and cost planning, which is a basic prerequisite for the integrated performance monitoring described in Sect. 9.4. However, a direct relationship is also possible between the WBS description and the technical characteristics for hardware elements on the one hand, and the schedule on the other, if the schedule data are systematically assigned to the WBS elements and the work packages. The direct relationship is important for realistic planning since the planner receives the necessary planning inputs from the WBS description and the work packages when creating the schedule.

9.3.4.2 Creation of Schedules and Flowcharts

Project planning is usually not based on only one planning method. In many cases, schedules and progress plans are drawn up on the basis of bar charts, which many planners and especially project managers find most convenient. Top management, i.e., the executive board or the client, often only needs limited milestone information to have a rough idea about the project. On the other hand, very detailed bar charts and/or network plans should be prepared for the project and specialist departments involved, from which the interrelationship of the project processes can be derived in detail [3]. Network plans or interconnected bar charts should be combined to form summary plans, but this involves effort and often fails. In general, it can be said that the choice of planning methods depends on the size and complexity of the project as well as on the experience of the planner and the project manager. Figure 9.17 shows an example of the development process for schedules in conjunction with the WBS.

9.3.4.3 Bar Chart Applications

The Gantt chart, respectively Bar Chart, is probably the most widespread planning instrument for representing project processes in their chronological sequence. Gehriger writes about this: "*Until the 1960s, projects were planned and monitored*

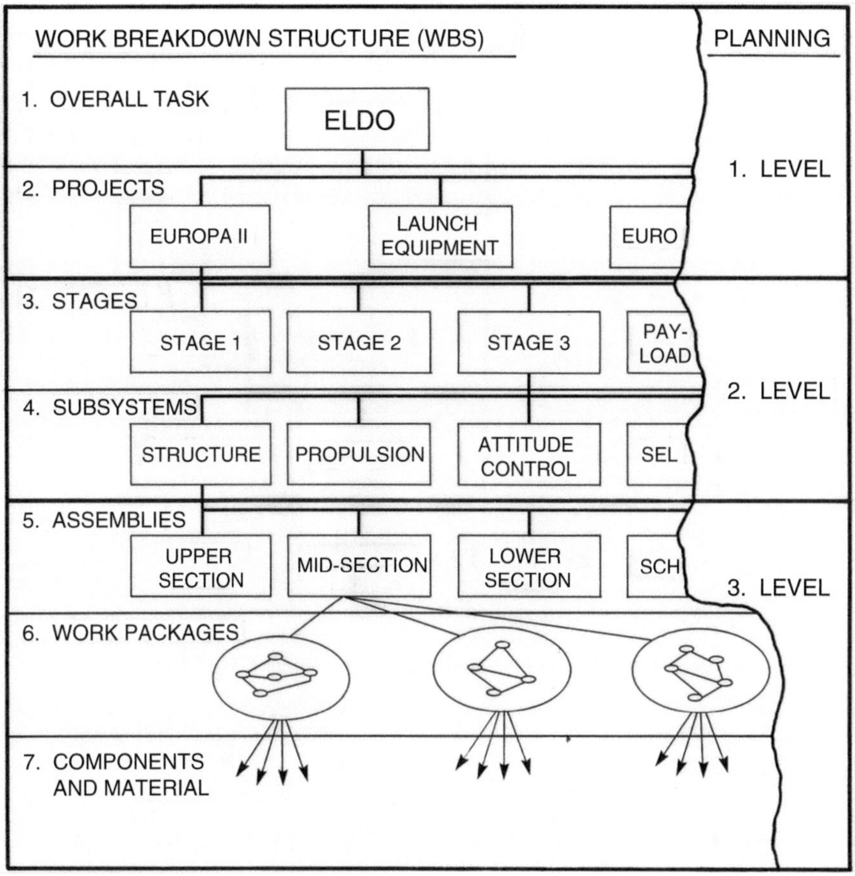

Fig. 9.15 Synchronization of WBS and planning levels (ELDO)

almost exclusively with schedule tables and bar charts, and even today these are still the predominant planning and monitoring methods in numerous companies and public institutions." [1] In many cases, the aforementioned "*today*" *of* 1980 is still a reality in 2023, because the Gantt chart can also be created by employees who are unfamiliar with planning without much prior knowledge and thus experience a particularly high level of acceptance; every employee can create a simple Gantt chart without training and instruction. However, this is dangerous because it often results in unrealistic schedules that no one then adheres to.

Particularly in the case of Gantt charts that are not supported by detailed plans, a particularly high requirement and, above all, an appropriately high level of accuracy must be achieved on the Gantt chart. Bar plans that are drawn up quickly and not in detail often overlook important details, resulting in illusionary schedules and progress plans. This in turn leads to unrealistic plans, which above all also have a negative effect on the schedule-related cost plans.

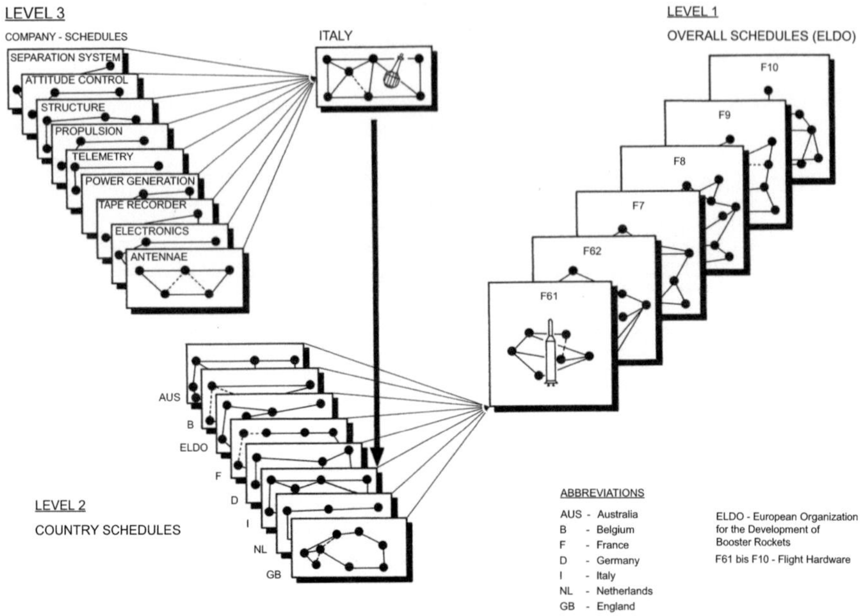

Fig. 9.16 Hierarchical summarization example of project plans (Source ELDO)

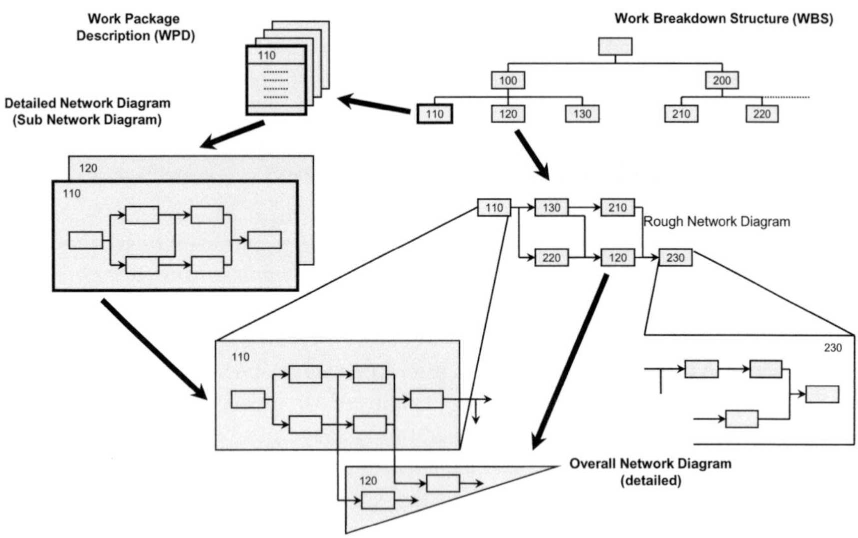

Fig. 9.17 Development process of schedule sequence plans (Source: FMT, Bremen)

One shortcoming of the simple Gantt chart is the lack of a precise flow logic and networking of activities. Gehriger: "*However, these methods of representation (schedule tables and bar charts) are seriously flawed. Dependencies and interdependencies of individual operations (processes, activities) are neither duly considered nor adequately reflected by them*" [1]. Many practitioners have therefore moved to a synthesis between network and Gantt charts, networking the Gantt charts that have been created. The networked Gantt chart is not a substitute for the network, but it can be a useful compromise solution; see Fig. 9.18 [53]. Henry Gantt introduced the method of networked Gantt charts in the early twentieth century.

Figure 9.19 shows a practical example of a *master bar chart* of a multi-stage project to be created based on detailed plans [54].

9.3.4.4 Planning Logic and Network Planning Technology

When creating a plan, it is important that the project activities are presented in the correct order. Flow charts can be very helpful in this respect. They are particularly useful for large and complex projects in order to better understand the process flow [55]. This is the top level of the project. Of course, it is useful to create flow charts at all levels of the project. In this context, it is important to point out the advantages of the network plan (NP) technique, because the NP is the best way to show the process logic and all the dependencies. Figure 9.20 shows a simplified NP. It is recommended to apply the advantages of the NP in addition to the flowchart already mentioned at the beginning of the project. The author has had several years of practical NP experience and knows the advantages of the network diagram technique, but also the disadvantages.

Here is a quote from Wille et al.: "The representation of the project flow in a network plan offers the following advantages according to our experience:

(1) Drawing up a network plan means that all those involved in the project must think through its course in detail and reach agreements and decisions at an early stage. This plays a particularly important role when different companies or departments of a company work together.
(2) As a graphical representation, the network plan provides an excellent overview of the project and makes it possible to clearly identify the planned project sequence for all participants.
(3) The advantages mentioned in (1) and (2) facilitate control over the completeness of planning." [56]

Since many publications have already appeared about network planning technology, this book will refrain from going into further details of network planning technology, such as the calculation of the critical path. In 1982 the new edition of the book *Netzplantechnik (network technic)* by Groh and Gutsch was published, to which special reference is made here [57]. The third edition of the book is didactically very logically structured and highly recommendable for newcomers as well as for practitioners. It is interestingly written and therefore well suited for self-study. In contrast to many other books on the subject of network planning technology,

ID	Task Name	Duration	Start	Finish	% Complete	Predecessors
1	**Solar Array Project (5 panels per wing)**	**414 days**	**Mon 03-11-03**	**Mon 30-05-05**	**67%**	
2						
3	**ENGINEERING**	**190 days**	**Mon 03-11-03**	**Tue 20-07-04**	**100%**	
4	**MS#01 Kick-Off Meeting Completion**	**2 days**	**Mon 03-11-03**	**Tue 04-11-03**	100%	
5	**DVT activities**	**75 days**	**Wed 05-11-03**	**Mon 16-02-04**	**100%**	**4**
6	Prelimin. DVT lay-out def. + Drawing Substrate	75 days	Wed 05-11-03	Mon 16-02-04	100%	4
7	DVT Interface Drawing Cells & Wiring	5 wks	Tue 13-01-04	Mon 16-02-04	100%	6FS-5 wks
8	Cup/Cone for DVT Sample	3 wks	Wed 24-12-03	Tue 13-01-04	100%	6FS-2 wks
9	**PDR Phase**	**124 days**	**Wed 05-11-03**	**Thu 22-04-04**	**100%**	
10	Prel.Design & Analyses + PDR Datapackage	17 wks	Wed 05-11-03	Mon 01-03-04	100%	4
11	Substrate production drawings	69 days	Tue 20-01-04	Thu 22-04-04	100%	6FS-4 wks
12	**MS#04 Preliminary Design Review (PDR)**	**2 days**	**Thu 18-03-04**	**Fri 19-03-04**	100%	52FS+3 days;10FS+2 wks
13	**CDR Phase**	**89 days**	**Mon 22-03-04**	**Tue 20-07-04**	**100%**	
14	Detailed Design & Analyses + CDR Datapackage	82 days	Mon 22-03-04	Fri 09-07-04	100%	12
15	Review Datapackage CDR	2 wks	Mon 28-06-04	Fri 09-07-04	100%	14FS-10 days
16	**MS#09 Critical Design Review (CDR)**	**2 days**	**Mon 19-07-04**	**Tue 20-07-04**	100%	15FS+5 days;55FS+5 days
17						

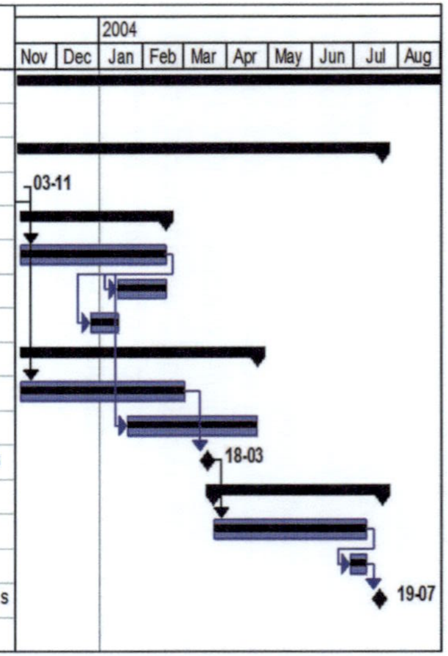

Fig. 9.18 Networked bar chart (Source ESA)

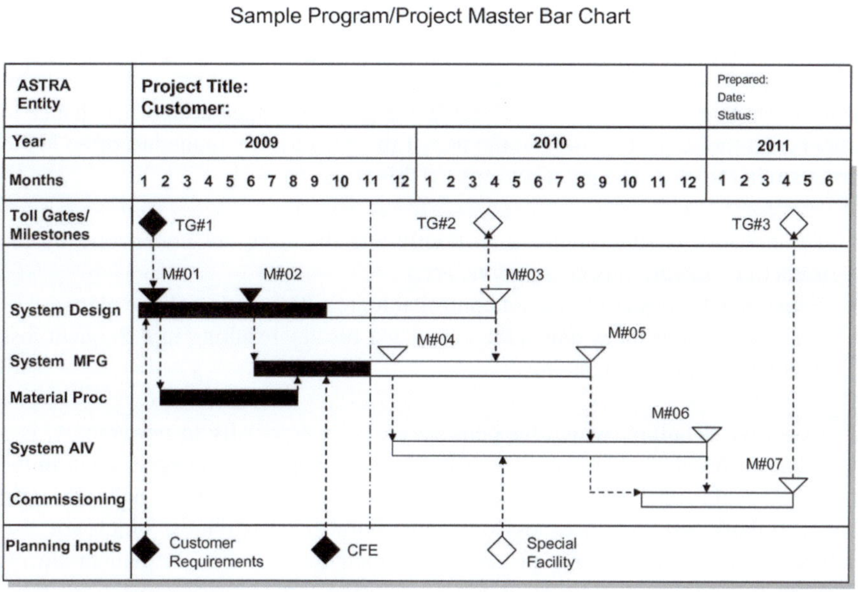

Fig. 9.19 Sample Master Bar Chart showing actual status

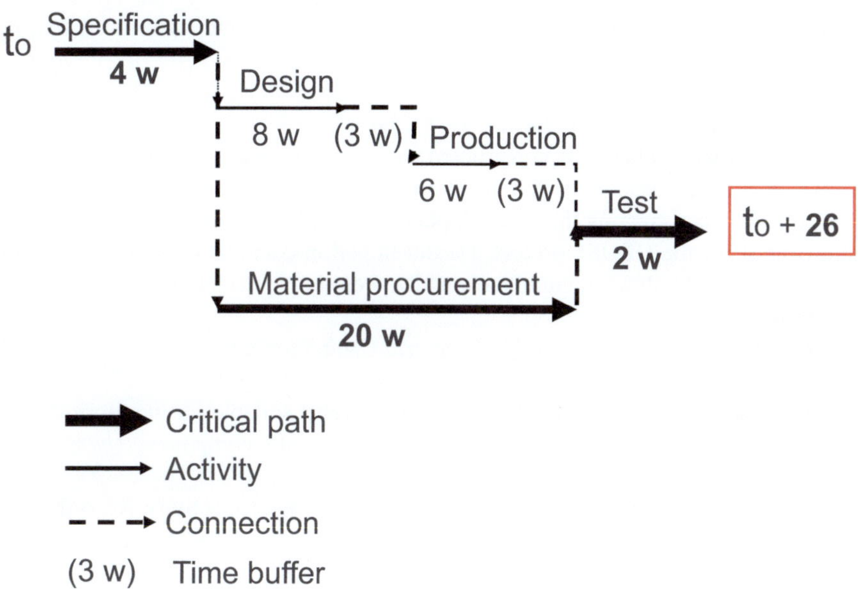

Fig. 9.20 Network Plan showing critical path

Groh and Gutsch deal primarily with user problems in practice and not so much with network planning theory. Particularly noteworthy is also the equal treatment of the previously mentioned network planning methods. Clear graphics and the inclusion of bar charts based on network diagrams give the reader an optimal chance to understand the subject. However, the use of the network technique has often led to criticism in the past. The main reasons for this are as follows:

- The network is structured too abstractly and therefore not understood by the users (*technicians, business people, etc.*).
- Network activities are often not controllable, resulting in unrealistic plans.
- Only the planner understands the calculated results, resulting in a frequent loss of confidence in the planning.

Excessive detailing, which happens very easily, especially to newcomers, is a well-known problem. Many a planner has become hopelessly bogged down in the excessive zeal to anchor as many detailed activities as possible in the network, right down to daily actions. An excessively large number of activities, which are also subject to a very strong change process due to their detail, lead to the planner always running behind things. During a NASA hearing, an industry project manager responsible for a major Apollo subsystem was asked the question: why did he introduce and then discontinue the PERT* planning system several times? The person asked summarized: "*I think if we had to do it all over again, we would certainly use PERT*, but certainly not to the detail that we tried to do, and I also think that in addition we would introduce bar schedules much earlier, because people understand them more easily*" [58].

**Program Evaluation and Review Technique*

Often, network plans are constructed in a very abstract and unrealistic way. However, it is very important that the planner only considers activities that are also controllable when creating the network plans. The author was in charge of planning and control for the HEOS* project. He developed in cooperation with the project management the following guidelines, which were implemented to ensure realistic planning:

**ESA Research Satellite (Highly Eccentric Orbit Satellite)*

- The network shall contain only activities or events suitable for monitoring, i.e., controllable, e.g., a pre-specified product, satellite component (*hardware*), report, or drawing.
- A responsible department is to be assigned to each defined activity. According to the organization chart, the activities have been assigned to the responsible workgroups or companies.
- The level of detail of the activities is to be determined in such a way that sufficient status monitoring is ensured. It should be avoided that subordinate activities, such as in-house processes, are included in the network.
- Planning interfaces of the subnetworks must be clearly defined. Existing planning interfaces of the subnetworks which define the deliveries of information or

components, etc. (*software and hardware*) from one company or department to another work group have been marked by milestone networking to obtain a clear delineation of all connections.
- The definition of activities is to be based on a uniform terminology that is equally valid for all companies and departments involved. This measure forced the use of a uniform and binding language after preceding term clarification, in order to avoid subsequent misunderstandings [59].

To ensure that the network contains all important activities, the planner should prepare an appropriate checklist. A corresponding example is given below [3]:

- Are all project reviews shown in the plan?
- Is it ensured that all project reviews are logically linked in the plan?
- Is there a design phase shown for each subsystem?
- Are test plans and procedures shown?
- Are activities for facilities included?
- Are subcontractors and suppliers identified?
- Is information about components with long delivery times included?
- Are the major tooling provisions included in the plan?
- Are all planning interfaces defined?
- Are design clearances shown on the plan?

As already mentioned, the use of planning software is an absolute prerequisite for large projects; see Chap. 18. But even the use of planning software must be approached with moderation and purpose. Gehriger aptly writes: "*Expensive operations research specialists were hired who obtained sophisticated computer programs, used them to block computer systems for hours, and produced tens of network printouts that no one understood.*" This is to be avoided at all costs, however, or else one day we'll have sayings like, "*To hell with the arrows and circles brigade!*" [1].

Gehriger's statements are from the past. But they are still relevant in 2023. A measured application of management software is the order of the day, not least in order to save costs through planning. The network planning technique is an excellent planning instrument for project management if it is used correctly and with moderation and purpose. Gehriger once again makes the following plea: "*The network planning technique brings the project participants closer together, it is their common language and enables a dialog on a common denominator. This is especially important in internationally interwoven projects. With the help of a computer, a realistic network plan allows the project manager to simulate various decisions, making their influences on the project visible before the final decision is made.*" [1]

9.3.4.5 Importance of Time Estimation

Time estimation is a very important task of project management. Very often, assumptions are made at the beginning of the project in cooperation with the specialist department, which later cannot be adhered to. This results from the carelessness with which the problem of time estimation is approached. In many cases, time estimation is shifted to the end of a long chain of project preparation activities and the engineers requested to estimate the time for a particular activity are quite often not prepared for this kind of activity and most of them do not have any related experience in this field. It is therefore no wonder that scheduling under such conditions is often far from reality.

On the other hand, realistic planning is an important foundation for cost estimation and the determination of binding contract dates. Deadlines that are too tight, but also too broad, increase costs; see Fig. 10.13. In the first case, planning changes and, as a consequence, cost changes, and in the second case, *time is money*. In addition, contractual penalties are the rule if deadlines are not met. There is also a psychological aspect. A lack of credibility leads to a lack of discipline, as people no longer believe deadline information due to bad experiences and increasingly ignore it. Therefore, as much attention as possible should be paid to a careful estimate of deadlines.

An important prerequisite for arriving at solid schedule estimates is the establishment of a detailed flow logic. As already mentioned, networks are an ideal basis for planning. Once the exact flow logic has been drawn up and individual activities identified, a good basis for a solid schedule estimate is already in place. With the help of an experienced planning engineer, it is possible to analyse in detail the time required to complete an activity. Past experience (*bench-marking*) should of course be taken into account. Whenever possible, the estimate should include *margins* for contingencies. Unfortunately, schedule analyses of completed projects are rarely prepared in this regard [60].

In the context of the network technique (see Fig. 9.20), the "three-time estimation" is a good method to determine the most probable time in critical cases using a known statistical formula. For very innovative projects, this method can be very useful. The estimator is requested to provide three estimates as defined by Miller [61]:

- *"Optimistic time (to):* Estimate of the shortest possible time required to perform a given activity. A time that is achieved only if one is exceptionally lucky and *everything goes right the first time.*
- *Probable time* (tw): An estimate of the time normally required to perform a similar activity. A time that will be required repeatedly if the same activity is repeated several times in an interdependent manner.
- *Pessimistic time (tp):* Estimate of the longest time required to complete an activity. A time that is needed only in the case of extraordinary problems. It should include the possibility of initial errors or even new starts but should not be affected by such things as catastrophic events, strikes, etc., unless these risks are directly related to the activity."

The mean time (tm) can be determined according to the beta distribution using the following formula:

$$\mathrm{tm} = \left(\mathrm{to} + 4\,\mathrm{tw} + \mathrm{tp}\right) / 6$$

However, this method was rarely used in the past. The reason for this is probably the relatively large estimation effort and the fact that the use of the network technique has declined in many industries. However, the three-time estimation can also be used sensibly without the application of the network planning technique and should therefore not be completely shelved.

In addition to a detailed time estimate, the planning analysis for a new project should also include an analysis of the critical project areas, which is best achieved with the help of the network planning technique. The critical paths can be clearly determined based on the network plan. However, the analysis of schedule-critical project areas also requires that precautionary measures and alternative plans be drawn up; it is not enough merely to point out schedule-critical areas. Planned countermeasures that can be initiated in the event of an emergency include:

- Increase in personnel
- Short-term scheduling of overtime and weekend work, if applicable
- Bringing forward particularly critical tasks
- Early ordering of components with long delivery times (*long lead items—LLIs*)
- Initiation of parallel work
- etc.

The analysis of critical areas and the timely planning of countermeasures are so important because schedule bottlenecks, if known in time, can often be overcome by relatively simple and relatively inexpensive countermeasures, whereas failure to overcome schedule bottlenecks as a result of downtime of large teams or machines then often leads to enormous cost overruns. Often, overcoming schedule bottlenecks in a timely manner also eliminates the obligation to pay contractual penalties due to late delivery. A realistic schedule estimation and analysis of schedule-critical areas is of utmost importance for the execution of a project. Schedules and time estimation of activities must be realistic.

9.3.4.6 Milestone Definition

The definition of project milestones, in addition to the planning techniques described above (*bar chart and network plan*), is an effective method of controlling results (*the term "key event" or "gate" is used as an alternative*). After all, the activities expressed in a network or bar chart only say that very specific work is to be carried out, but nothing about the results. In the network planning technique, a distinction is made in principle between operations (*activities*) and events (*results*), but often in a very abstract form. Only through the definition of controllable project milestones, which must be firmly anchored in both the bar charts and the network plans, do the network and bar charts acquire an increased connection to reality. Milestones or key events must be defined in such a way that effective performance monitoring is possible. Milestones can be defined according to the following criteria:

(1) Start and finish events (*releases and final products*):
  - Overall project (*project start and end*)
  - Project phases (*phase start and end*)
  - Subprojects/subsystems (*start and end*)
  - Work packages (*start and end*)
  - etc.
(2) Test and delivery events (*test completed and/or product delivered*):
  - Hardware
  - Documentation
  - Software.
(3) Planning interfaces (*major planning links to other plans*)
(4) Project reviews; see Sect. 7.3:
  - Requirement reviews
  - Design reviews
  - Acceptance inspections.

However, all project milestones may not have the same importance for project management. There are very important and less important milestones. Grouping the milestones according to their importance is therefore highly recommended for large projects. Dividing the project milestones into individual milestone levels, which are defined according to the WBS, has proven very effective. Staggered milestone lists must be created for all milestone levels. The milestone concept for a large-scale project is reproduced below from practical experience [62]:

- *Level 1 milestones*—Milestones to be controlled by the client
- *Level 2 milestones*—Milestones to be controlled by the main contractor
- *Level 3 milestones*—Milestones to be controlled by subcontractors and WP managers.

The list of the major milestones (*level 1*) of the HELIOS solar probe are given below [60]:

01 Contract conclusion and KO meeting
02 Implementation of the PMP (*see model PMPin Chap. 22*)
03 Ordering HIREL components
04 Completion of structural model (*STM*)
05 STM vibration test completed
06 Vibration test experiments completed
07 Telecommunication system first compatibility tests delivered
08 Antenna alignment tested on a rotating body
09 Availability of the thermal model for thermal tests
10 Completion of all EM function tests
11 Completion of all development tests
12 Start prototype integration

13 Prototype production completed
14 Completion of prototype testing
15 Start integration of the first flight unit
16 Acceptance of the first flight unit
17 Start of integration of the second flight unit
18 First spacecraft ready for shipment
19 Acceptance of the second flight unit
20 Second flight unit ready for shipment.

Figure 9.21 shows a milestone example for a fictitious project "*Robot cell with laser*" [63].

For company and project managers, the introduction of milestone control lists is an ideal control instrument. Without having to go into details in the network or the Gantt chart, they can be informed about the project status by means of the milestone tables.

Of great importance in the preparation of flow charts and schedules is the question of controllability. Plans must be designed in such a way that they are controllable. This means that the plan must not only contain horizontally arranged bars that can be changed at will but must be completed by controllable milestones. This is a basic requirement for later status control. The project status can be measured on the basis of a realistic schedule. Therefore, in principle, every activity shown in a plan must end with one or more concrete results. This must be the basic premise of every planner, because a certain amount of effort (e.*g., time and money*) is required to complete an activity (*task*). This principle applies at all levels. For smaller in-house projects, it may sometimes be sufficient for the team to agree on the outcome. For large and especially very complex projects, it is negligent not to clearly define the deliverables (*end products*) through milestones.

However, defining milestones is also an important task for the planner, but one that he or she cannot do alone in silence and without continuous discussion with colleagues on the team. From the project manager's point of view, an important question is: what will be the results (*deliverables*) within the assigned time and cost budgets? In other words, what are the specific *deliverable end items of* a project? The development of a new product consists largely of documents, e.g., reports, plans, specifications, procedures and drawings. The financial value for this can amount up to 30% of the total project budget. But what does the final end product look like in terms of content and quality? Does the report or specification meet the project management's expectations? Have the documents been reviewed and approved? In this context, the documentation management described in Chap. 12 is important. The same requirement for defining milestones also applies to hardware and software deliveries. The milestone in question must be precisely defined, i.e., "frozen," as a basis for subsequent status determination.

Based on the deliveries assigned to a milestone, the PM or the PC can check task completion. All hardware and software deliveries must be assigned to specific milestones. An example of an abbreviated hardware and software delivery list is shown

	Milestone	Measurement and Measureable Results
00	**Project Initiation**	o Project Description o Technical and Commercial Requirements o Project Manager named o Project Team named
0	**Project Kick-Off**	o Project Objectives and Strategy o Project Organization o Internal Project Order o Project Results Structure o Milestone Plan
1	**Start Design**	o Work Breakdown Structure (WBS) o Draft Schedule and Capacity Plan o Cost Plan/Calculation o Risk Analysis o Technical Concept + Draft Layout
2	**Start Development**	o Approved Layout o Manufacturing Guidelines o Manufacturing Plans o Control Concept o Pre-Calculation o Detailed Manufacturing Plans o Supplier and Purchased Parts List
2	**Start Manufacturing/Procurement**	o Long-lead Items Ordered o Assembly Drawings o Circuit Diagrams o Mechanical and Electrical Parts Lists o Detailed Assembly/Mounting Plans o Supplier Contract Framework
4	**Start Assembly/Installation**	o Parts Available for Assembly o Assembly Plan and Guidelines o Software Drafts
5	**Pilot Facility ready for Operations**	o Functional Facility o Software Documentation o Parts available for Test Operations o Functional Acceptance (Protocol)
6	**Completion of Pilot Operations**	o Test Operations Documentation (Protocol) o Optimized Facility o Pilot Parts o Production Acceptance (Protocol)
7	**Acceptance**	o Complete Documentation o Final Customer Acceptance (Protocol)
8	**Project Completion**	o Final Calculation o Release for Service o Project Review and Report o Project Organization Disbanded

Fig. 9.21 Milestone example for a fictitious "*robot cell with laser.*" (Source: Hab and Wagner [63])

in Fig. 9.22. Inspections must be performed to determine whether the products provided or delivered also meet the specifications (*DOC reference*); i.e., that they have been tested and accepted. Corresponding test and/or acceptance documents must be submitted; only then is the milestone considered to be completed.

The same procedure applies to document deliveries. Project documents must be completed and delivered to the PM or customer at the times and milestones specified in the schedule. Figure 9.23 shows an example of an abbreviated *Document Requirements List (DRL)*. The contents of the documents to be created are to be defined either in the associated paragraph of the SOW or in a special *Document Requirement Description* (*DRD*). Figure 9.24 shows a typical example of the documents to be supplied for performing a *design review*; see item 11 in Fig. 9.23.

The described procedure for "*hardening*" of milestones may seem complicated and laborious at first. However, it is an intelligent piece of hard work that is worthwhile, because sooner or later you will have to do the work anyway. However, if you don't prepare it systematically early enough, chaos very quickly ensues and, in the end, experience shows that this leads to considerable additional work. Conclusion:

- For large and/or complex system tasks, the previously described effort to systematically plan and harden milestones is clearly a MUST.
- For smaller and less complex projects, the methods described should also be valid, tailored on a reduced scale.
- For very small tasks, e.g., a study or a single work package, the methods described should serve as a guideline. However, it is also true that the end product is a component, report or software whose completion is to be checked by the PM.

Deliverable Assemblies	QT	MS	Date	Documentation Reference
01 Drive Unit Base	1	07	To+11	Drawing XXXXX
02 Drive Unit	1	08	To+12	Specification XX 01
03 Control electronics	1	15	To+14	Specification XX 02
04 Control software	1	15	To+14	Specification XX 10
05 Transformers	2	12	To+13	Specification XX 25
06 Cabling	5	14	To+13	Wiring diagrams
07 Machine buildings	1	17	To+10	Construction Plans
Etc.				

To – Effective Date of Contract
QT – Quantity
MS – Milestone

Fig. 9.22 Example of a *deliverable items list (DIL)*

Deliverable Documents	QT	MS	Date	SOW Reference
01 Project Management Plan	E + 5	01	To	SOW - Paragraph
02 PMO Organization Plan	E + 5	02	To+1	SOW - Paragraph
03 Project Schedule Plan	E + 10	01	To	SOW - Paragraph
04 Project Cost Plan	E + 2	01	To	SOW - Paragraph
05 System Specification	E + 5	07	To	SOW - Paragraph
06 System Test Specification	E + 2	15	To+5	SOW - Paragraph
07 System Test Plan	E + 2	18	To+8	SOW - Paragraph
08 Interface Control Document	E+2	07	To+5	SOW - Paragraph
09 Test Report (Test results)	E+2	22	To+10	SOW - Paragraph
10 Software Control Plan	E	05	To+5	SOW - Paragraph
11 Design Review Package	**E+10**	**20**	**To+12**	**SOW - Paragraph**
12 Quality Control Plan	E+2	07	To+5	SOW - Paragraph
13 Risk Management Plan	E+2	07	To+1	SOW - Paragraph
14 Status Report (Schedule/Cost)	E	--	M	SOW - Paragraph
15 Acceptance Protocol/Certificate	10	31	To+15	SOW - Paragraph
etc.				

To - Effective Date of Contract
Qty - Quantity
MS - Milestone
E - Electronic Version

Fig. 9.23 Example of a *Document Requirements List (DRL)*

9.4 Methods of Integrated Project Monitoring

9.4.1 Significance of Integrated Project Monitoring

"*The role of management is to respond intelligently to changes*" is a quote from former US Secretary of Defense Robert McNamara [64]. Transferred to the field of project monitoring, this means that the project manager or a person authorized by him in the project team, e.g., the project controller, must react to measurable project changes based on plans drawn up in advance in order to fulfill the project objective once it has been set, using the fewest possible resources. One can compare this process quite well with the mode of operation of a control loop. Following Fayol's management control loop (see Fig. 9.1), the eight main project control steps are shown in Fig. 9.25 [3]. The figure also shows how the successive steps follow a logical order; see below a description of the project control steps.

- *Step 1: Creation of the contract basis*. The contract basis, in which the required Performance (*P*) is anchored on the basis of the statement of work and system specification, the main or key schedule dates (*S*) and the project costs (*C*) is the foundation for the following management process. Project changes are measured against the contract baseline.
- *Step 2: Planning and work organization*. Detailed plans and a detailed work organization are to be built up based on the contractually defined project goals.

Sample List of Design Review Data Package	
1	Proof of compliance with the system specification, including environmental considerations
2	Final test and verification plans
3	Updated reliability analysis
4	Updated program risk analysis
5	Identification of all critical items, critical points, limited-life and age-sensitive parts and materials with back-up information as to their margins
6	Functional descriptions of equipment being delivered
7	Complete system and subsystem block and level diagrams and design description
8	Analysis of the impact of operating the maximum number of transponders per beam throughout the Satellite operating life time (satellite only)
9	Satellite alignment budgets
10	Final stress analysis and margins of safety for all structural elements (satellite only)
11	Qualification analyses for all components, from unit up to system levels
12	Final antenna patterns and performance data (satellite only)
13	Final predictions for worst-case propellant line propulsion subsystem.
14	TT&C Link Budgets: nominal transfer orbit, on-station cases and emergency situations such as Earth lock loss shall be included (satellite only)
15	Mass properties budget (satellite only)
16	TT&C list update (satellite only)
17	Power budgets (satellite, SGTs, …)
18	Propellant budget and delta-v budget (satellite only)
19	Antenna pointing budget (satellite only)
20	ICD Information (satellite/launcher, others TBD)
21	Overall System reliability analyses down to subsystem and unit level of the Space, Civil Ground, Commercial Ground and Management Segment
22	Civil Ground Segment network design and network operations
23	Civil Ground Segment interfaces to terrestrial networks
24	Communications services, service handling
25	Anchor station and ground terminal design
26	Link budgets

Fig. 9.24 Example of delivery contents for a "*Design Review*" Package

As a rule, the planning documents (*WBS*, *WP descriptions, schedules, workflows, and cost plans*) are already drawn up in the bidding phase and updated and finally adopted after contract (*letter of intent, preliminary contract or main contract*) has been awarded.

- *Step 3: Assignment and work approval.* The *Authority to Proceed (ATP)* is, in a sense, the official start of the monitoring process that must be performed on a regular basis. To prevent work from being started without authorization, an effective release procedure must be put in place. Motto: *Where there is no approval there can be no overruns! A problem arises, however, when it is neglected to issue a release in time!*

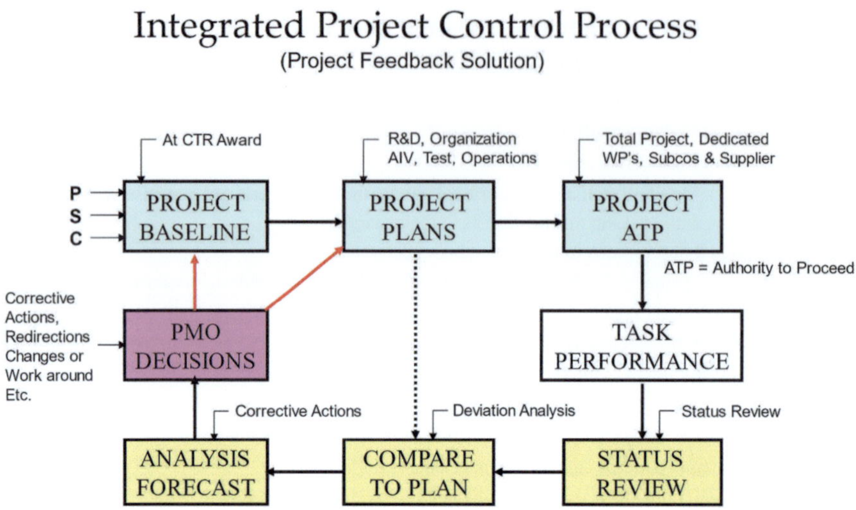

Fig. 9.25 Project management control loop

- *Step 4: Work Execution*. Project work should be carried out at all levels and by all project participants based on clear and unambiguous planning documents and in conjunction with an official work release.
- *Step 5: Performance measurement*. Performance measurement (*work progress, funds used, and changes*) is performed at regular intervals. For costs, the status is usually checked monthly and for deadlines weekly and/or monthly.
- *Step 6: Deviation analysis and corrective actions*. The measured performance (*Step 5*) can now be analyzed with reference to planning and work organization (*Step 2*). Planning deviations must be identified and their impact on the targeted project goal must be assessed. If necessary, appropriate corrective measures are to be defined and forwarded to management.
- *Step 7*: Management decision. The corrective measures resulting from the deviation analysis and forwarded to management as a proposal are to be implemented by management in planning changes or, if necessary, rejected. The management decision is to be regarded as an active control element in the management control loop. In practice, project management often initiates further analysis to be able to better assess the effects of planned decisions.
- *Step 8:* Changes. Two situations are possible when initiating changes:
  - Contract amendments and
  - Internal project changes.

It is very easy to see that the control loop shown can only work if all those involved adhere to the rules. First of all, it is important that the specifications are clearly and unambiguously defined in the first step (*1*). For example, if contract negotiations are slow, work may begin on the basis of a preliminary contract or *letter* of *intent (LOI)*. This is not unusual. Just make sure that the terms of reference

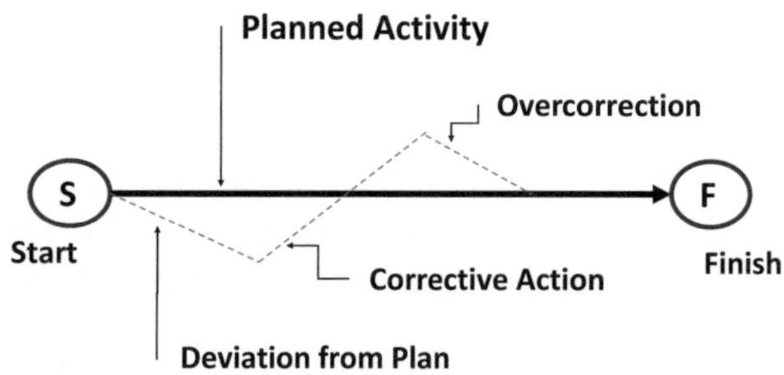

Fig. 9.26 Principle of project control

(*performance, time, and costs*) are then clearly defined, as in the final contract. Factors that are not clearly defined and can only be clarified at a later stage, the so-called TBD issues (*to be defined*), can become expensive.

In the second step (*2*), the preparation of detailed planning, bar and/or network plans and milestones is mandatory for an orderly flow of all work. The planning serves as a reference for status control. The third step (*3*) is about the project and work release, which is discussed in more detail in Sect. 9.4.3.

In the fifth step (*5*), the status of the work performed in step (*4*) is determined (*performance, time and costs*) with reference to the control date (*time now*). In the sixth step (*6*), a target/actual comparison is then performed. Minor deviations from the plan are usually negligible. However, if serious deviations occur, such as significantly delayed milestones, deterioration of performance data, e.g., an anticipated increase in weight and/or significant additional costs, then the reasons must be determined in the seventh step (*7*) and proposals for corrective measures must be developed.

Established deviations from the plan may have been caused by internal or external influences. Externally by delayed customer inputs (*deliveries or provision of specifications/machinery etc.*), which can lead to contract changes (step *8a*), or internally by the delayed completion of work packages (*step 8b*).

The control mechanism or steering process of a project can be compared very vividly with the steering of a passenger car; see Fig. 9.26 [3]. In many cases, effective project steering fails because the information required for the steering process is not available with the necessary transparency. Often, planning data is actually not structured clearly enough, sometimes only imprecisely defined or very incompletely completed. Another problem results from inadequate performance measurement and unclear variance analysis.

Another factor that is of utmost importance for efficient project monitoring (*control*) is the response time. The speed of the monitoring process, i.e., the time from becoming aware of a deviation (*steps 5 and 6*) to implementing the change (*step 8*), is a decisive factor for the efficiency of project monitoring. Here, too, the analogy with the steering of a passenger car is relevant. Excessive reaction times are problematic in both cases. A vehicle (*or project*) driven into a ditch can only be gotten out again with great effort. The importance of fast action is impressively demonstrated in Fig. 9.27 [14]. If an occurring problem is recognized in time and the

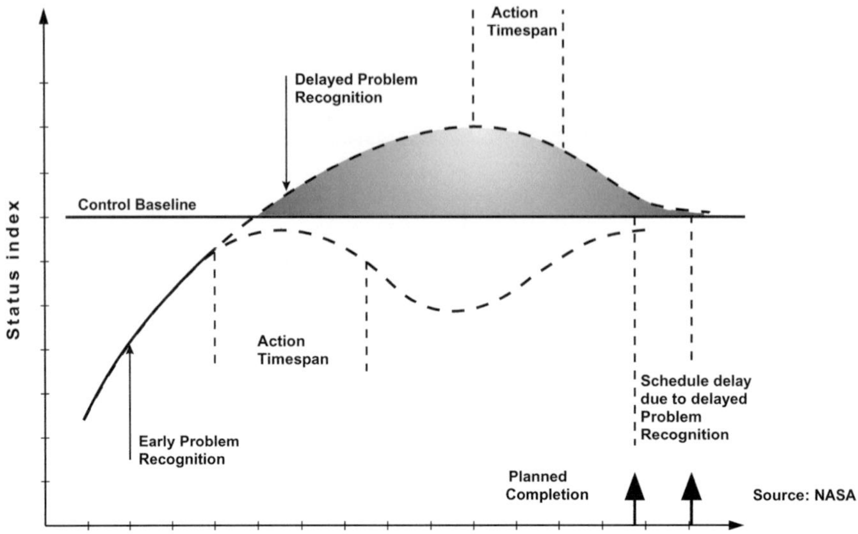

Fig. 9.27 Importance of response time for project monitoring (NASA)

necessary corrective action (*countermeasures*) is initiated in time, time and/or costs can be saved in contrast to corrective actions initiated late.

An objective assessment of the project status can only be made on the basis of an integrated view of the parameters time deadlines, costs, and function (*technical performance*). If only one of the three parameters is the focus of the investigation, i.e., in isolation from the other two, one-sided and non-optimal project decisions are too easily made.

The project manager is constantly faced with the task of considering or evaluating the different project aspects of task, costs, and deadlines not in isolation but as a whole. However, it is often difficult to compare the different factors with each other, and the project manager therefore often finds himself in the situation where he can only establish a relationship between the individual parameters with difficulty. The following considerations are therefore to be seen under the following premise: "*A relatively rough correlation of different project parameters using a systematic approach is better than a simple, unsystematic estimation of the ratios.*" [13] Efforts to capture the performance triangle of *performance, schedule, cost,* in figures has led to two basic procedures:

1. the *task-oriented* performance monitoring (*Task Performance Control*) and
2. of *technically* oriented performance monitoring (*Technical Performance Control*)

Task Performance Control is the monitoring of the performance status of the service specified in the Statement of Work (*SOW*), (*see Sect. 9.3.2*) and Technical Performance Control is the monitoring of the technical parameters (*performance specifications*) described in the specification; see Sect. 9.4.7.

9.4.2 Project and Work Release

Effective project monitoring can only be carried out on the basis of detailed and complete planning documents; see Sect. 9.3.4. Another important prerequisite for efficient project monitoring is the release procedure. In accordance with the planning specifications, work releases must be issued for the entire project, for individual WBS elements, work packages, lots and/or activities. In addition, a further subdivision of the release can be made according to time aspects, e.g., for particularly long-term work packages. Either a temporal delimitation is possible or a delimitation according to milestones [65]. Depending on the type of project (*complexity, etc.*), the type of contract (*cost reimbursement, fixed price, etc.*), and the controls provided (*WP control, milestone control, etc.*), the level of detail of the project/work authorization and the responsibility for the release must also be specified. In many companies there is a corresponding procedure and a standard format (*project/work order*).

For large projects, it may be useful to create a release plan that clearly identifies the individual releases and the responsibilities for the releases. The introduction of a release plan ensures that the teams/contractors involved in the project or scheduled for the project initiate the planned activities on time (*i.e., on schedule*) and according to their necessity. Figure 9.28 shows an example of a release schedule (*ATP Schedule*) that clearly indicates when the individual releases are to be made for the project teams, the company departments involved and the subcontractors.

To start project work, a release of funds is required for the entire project or for subsections of the project. After this step, the project management can issue detailed work releases "*project orders.*" In most companies, the release of work is done through the previously mentioned project order; for this purpose, there are usually appropriate templates for recording planned hours, travel expenses, etc. and requisitions for the procurement of materials and other purchased parts. The company's internal project numbering system should be brought largely into line with the WBS numbering; see also Fig. 9.14.

Subcontractor approvals are usually identified in the subcontractor contracts and specific details of the approval procedure may be described in the WBS. The release mode for subcontractors essentially depends on the contract type, project length and complexity of the project.

When analyzing the critical path of a project, the negative or almost negative time buffer is critically examined in order to be able to protect the final deadlines of the project. However, the analysis of the positive buffers is also of particular importance for the efficient operational control of a project in order to be able to optimally determine the work releases. If, for example, large positive planning buffers are identified for some project areas, the project manager will endeavor to set the releases not at the earliest possible time but at a later time; see Fig. 9.29 [3]. The example shows that, taking into account the positive buffer, the release can be postponed in order to optimize the cash flow plan. This means that the capital required to carry out the project does not have to be made available until later and can be invested profitably in the meantime. However, it should be mentioned in this context

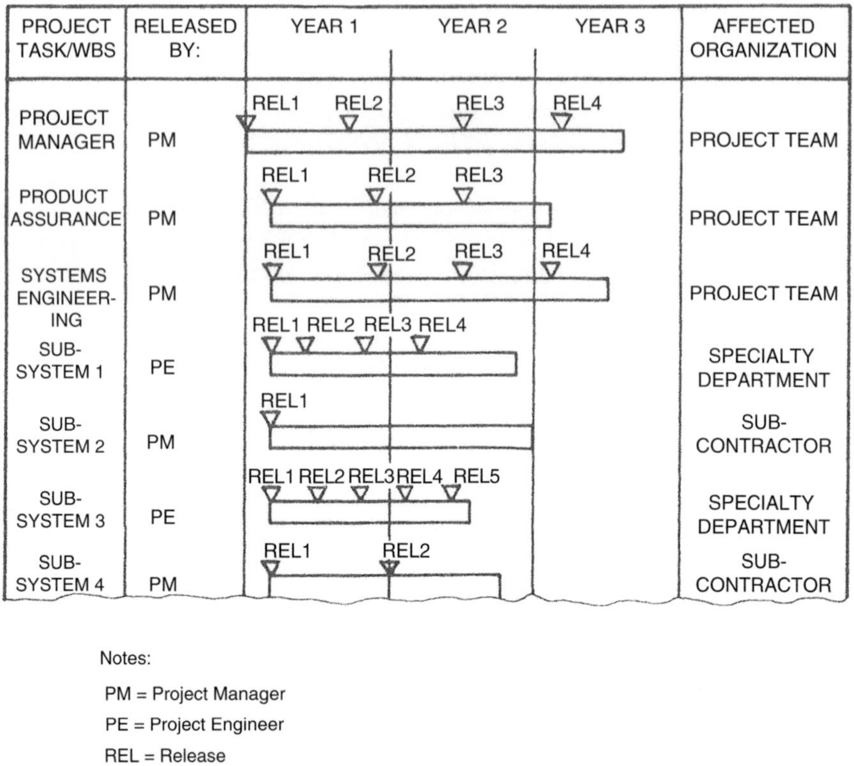

Fig. 9.28 Example of a release plan

that such considerations are not carried out for many projects because the planning statement on the positive analysis is not trusted, detailed and credible planning does not exist at all or other aspects, such as utilization issues, are more decisive.

9.4.3 Project Progress Control

The regular determination of the project status in terms of deadlines is an important measure of project management for successful project monitoring. On the basis of previously prepared and released plans, the progress of the project must therefore be measured at the lowest planning level of the project. The method of control depends on the type and nature of the plans available. Network plans provide the more accurate basis for progress control, while progress control based on bar charts allows room for inaccuracies.

When determining project status using bar charts or networks, determine at regular intervals (*i.e., weekly, monthly, etc.*) which activities have been started, completed, or are still in progress and whether they can be completed within the planned time. Following the status determination, a new planning analysis shall be

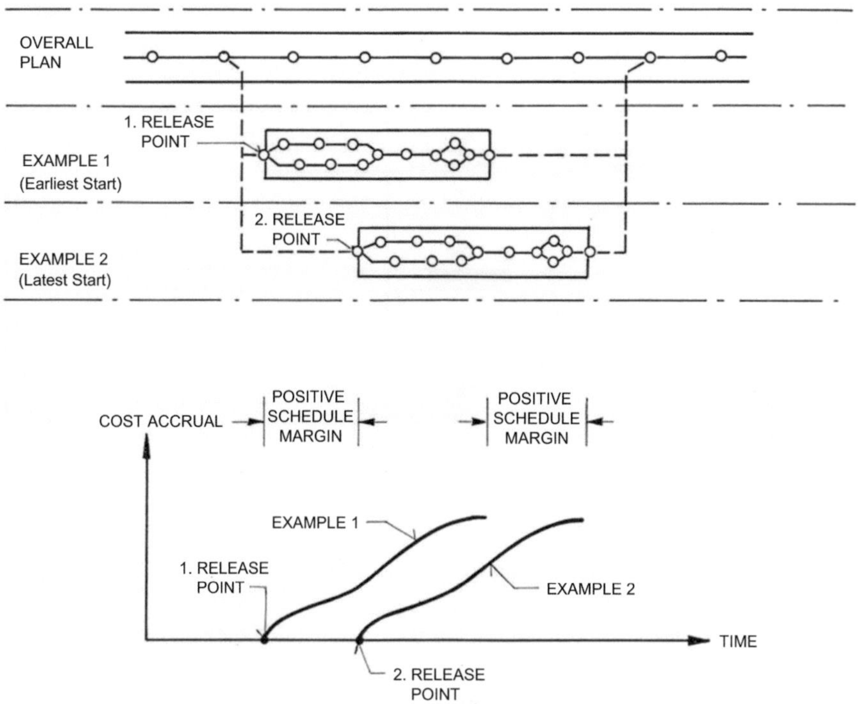

Fig. 9.29 Determination of work releases depending on the buffer analysis of the project plan

performed to determine the impact of the new situation, e.g., planning changes, reported delays, etc., on the overall project. If necessary, corrective measures are to be initiated in order to achieve the set project goal within the established time frame. Changes to *the* plan (*e.g., logic or estimated time*) shall also be recorded as they may have an impact on the end date(s).

On the basis of the regularly performed planning analyses, planning reviews are to be carried out in cooperation with the teams involved in the project, in which the actual status is compared with the planning (target) in order to initiate planning corrections if necessary. Planning corrections can only be considered in connection with other project parameters, such as the technical effects and the cost situation; see Sect. 9.4.5. An example of a status overview based on a Gantt chart with controllable milestones is given in Fig. 9.30. The strict tracking of milestones is of utmost importance for a future-oriented status monitoring.

Practical findings have shown that regardless of whether bar charts or networks are used, there is effective control of the project status only on the basis of controllable, i.e., "*hardened*" milestones, as described in Sect. 9.3.4. Based on hard milestones, the project manager can ask the following questions in the project status meeting:

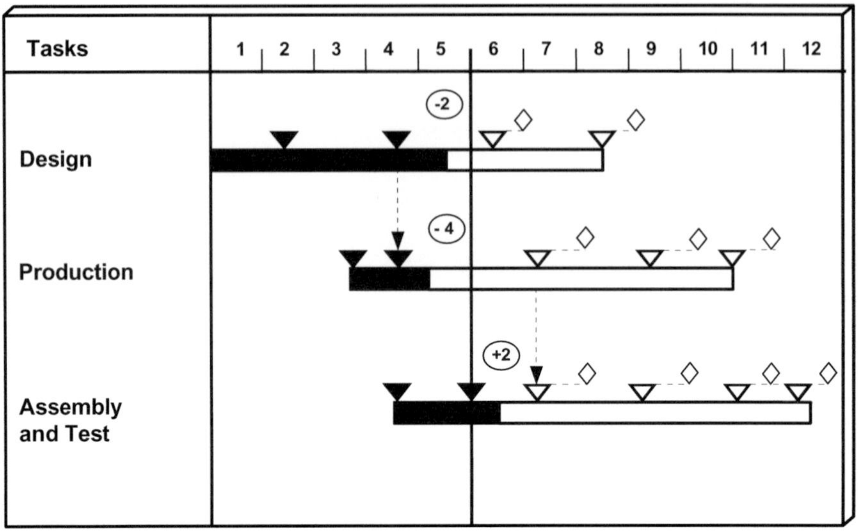

Fig. 9.30 Example of a monthly status report (Source NASA)

(1) Has the milestone(s) been completed by the scheduled date and the answer can only be "*yes*" or "*no.*"
(2) If the answer is *"yes,"* evidence must be provided.
(3) If the answer is "*no,*" i.e., if the milestone is not yet completed because the associated activities are still in progress, the question "*when is completion expected?*" must be added; i.e., what is the required time to completion *(TTC)* at time now must be provided.

Answers such as "*yes*" or "*in principle, yes*" are not acceptable because they are, in principle, just obfuscation. Project managers and their teams must learn to ask precise questions without creating a cold climate as a result. Precise questions about status are by no means pedantry, because the PM cannot achieve the specified goals if colleagues do not provide correct progress information to the PM. After all, the PM is obliged in the interest of the company to ask precise questions and colleagues and/or partners must respond accordingly. In the case of rigid milestone control, the introduction of symbols can be helpful. The symbols shown in Fig. 9.31 must be adapted accordingly when using planning software.

After the project status has been accurately recorded, a planning analysis is to be carried out on the basis of network or bar plans in order to determine the critical path or critical area. The changed dates are to be checked for compatibility on the basis of the planning documents in close cooperation with the technical departments, i.e., the influence of the new situation on the completion of the project and on personnel deployment is to be determined.

In the buffer analysis, the activities that lie on the critical path and their links must be examined with regard to the estimated times. By changing dependencies or

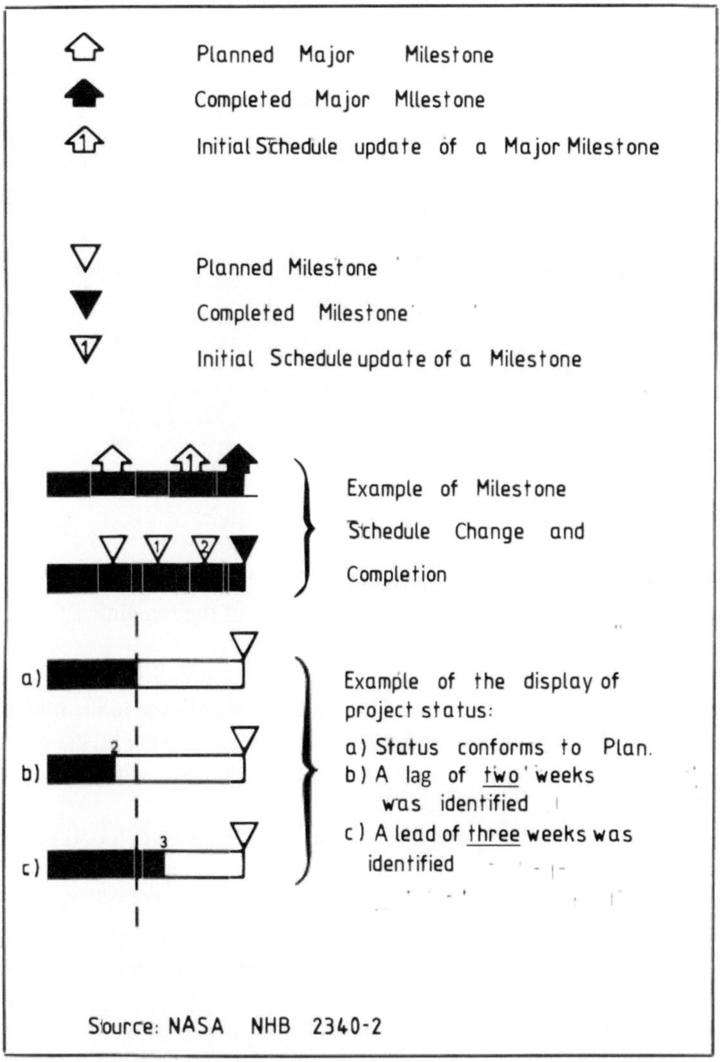

Fig. 9.31 Example milestone symbolism (Source NASA)

utilizing any built-in schedule reserves, an identified negative buffer can often be eliminated. Planning changes that have an impact on the project schedule must be reviewed and approved by the project management before they are incorporated into the overall project planning.

For the continuous tracking of project deadlines, deadline trend charts are an important tool, because they quickly enlighten even the most indifferent and disbelieving manager about a looming deadline problem. Often, a scheduling catastrophe only becomes apparent after a reporting period of several months. It is not uncommon for the first planning analyses to be optimistic, and even then, when the

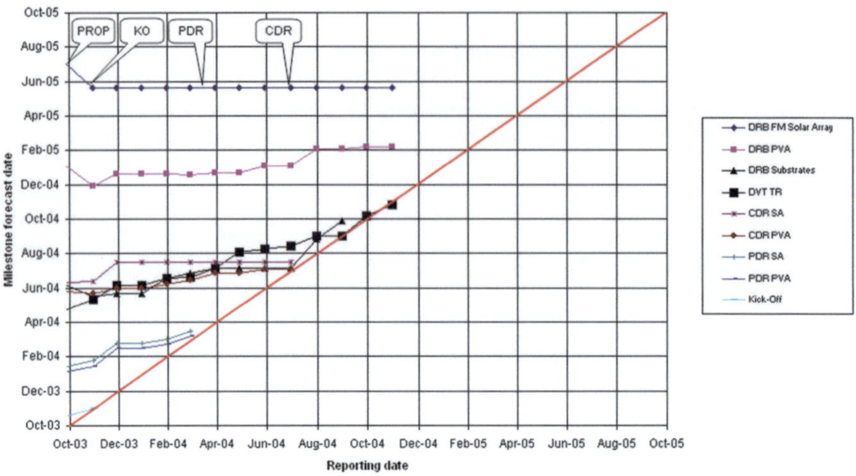

Fig. 9.32 Early problem detection using a deadline trend diagram

deadline of the first milestone can already no longer be met, the subsequent milestones are often presented too optimistically in view of the remaining time until the planned completion; see Fig. 9.32 [49].

The *trend chart* shows future problems with unerring clarity and provides easy recognition of all the approaching deadline misery. You can see in the middle of the chart that action must be taken. In most cases, however, there are no planning tricks, such as shortening deadlines or overlapping activities, which do not have consequences.

Trend charts represent a top management tool, because here, too, time *(deadline overrun)* is *(means)* money.

An interesting alternative is the traffic light system, where *"red"* can be used to indicate for example a very critical activity/milestone, "yellow" a critical and "green" a non-critical activity/milestone. A precise examination of all critical situations is required.

So, what are the causes of deviations from the plan? Burghardt names three causes: personnel, technical, and organizational. He then distinguishes between avoidable, hardly avoidable and unavoidable deviations from the plan; see Table 9.3 [66].

It is clear that one should first focus on the avoidable deviations from the plan, such as demotivation, incorrect planning and bottlenecks. It becomes more difficult with the deviations that can hardly or not be avoided.

9.4.4 Action Control

In addition to regular planning, the action item (*AI*) control system also known as a "*to-do list,*" is suitable for the detailed control of *ad hoc* defined actions that were

Table 9.3 Causes of deviations from plan (Source: Burghardt [66])

	Personnel causes	Technical causes	Organizational causes
Avoidable	• False motivation	• Planning failures	• Bottlenecks with RZ and test rigs
	• Inadequate training	• Prone to errors	• Unclear competency
	• Misunderstandings	• Incomplete test data	• Staffing bottlenecks
	• Overloading	• Inefficient tool use	• Production problems
Seldom avoidable	• Problem cases	• Performance Problems	• Supplier changes
	• Variations	• Overwhelmed test technology	• Priority changes
	• Limited competence	• New requirements	• Spatial division
		• Uncertain system baseline (e.g., operating system errors)	• Schedule pressure
		• Lack of support	
Unavoidable	• Illness	• Technical challenges	• Higher level organization issues
	• Pregnancy	• Defective parts	• Contract changes
	• Resignations	• Missing parts	• Supplier bankruptcy

not shown in the plan but must be fulfilled at short notice. The defined actions must be understandable and controllable and should not represent activities already foreseen in the plan or taken for granted as standard operating procedures. The AI control system has proven to be extremely useful in managing projects. Its implementation is simple and its success is very effective. Particularly urgent actions that need to be completed as quickly as possible in the interest of project progress are defined as additional actions that are necessary for the project. Here is an example from practice:

- During the weekly status meeting, it is determined that unacceptable tolerance problems with high-precision fitting parts constantly occur during prototype production under laboratory conditions.
- The experts uniformly conclude that this is due to temporary temperature differences caused by malfunctions in the air conditioning system.
- The resulting measure is: overhaul/repair of the air conditioning system during the following week, in which prototype production is interrupted anyway due to planned retooling work.

The situation described here certainly has many parallel cases in industry. But urgent actions such as the overhaul or repair of the air conditioning system described here are not always handled with the necessary vigor and deadline consistency, and unnecessary delays may occur in the project. The action described here may require the immediate initiation of ordering actions for spare parts and weekend work by those responsible for the laboratory. If necessary, spare parts must be procured by courier from the nearest warehouse. If one compares the costs incurred for the special action with the additional costs incurred for delays in the project, a delay in the repair action by only 1 or 2 days is often not justifiable from an economic point of view. The author is aware of cases in which the prevention of delays by only a few days justified an ad hoc business trip to the USA in order to find an urgently needed component, the procurement of which was not possible in Europe, from partner companies.

At this point, one should be reminded once again of the character of the project work: the highest priority is the fulfillment of the project goal with the least possible means. The AI control system is to be seen in this sense and gets its special importance by the frequency of special actions to be initiated. The practical examples described here, which represent extreme situations, are intended to underline the importance of the AI control system. In everyday work, it is usually much fewer spectacular cases that fill the AI register. But even the many small actions should be pursued with enthusiasm and not put on the back burner. It is always an astonishing experience to note from practice how urgent actions, if pursued with vigor, can be completed on the same day through increased effort, while they often drag on for days if the deadline is less consistently monitored. This statement is in no way intended to imply that urgent work can only be completed if the project manager exerts increased pressure. That would not be in keeping with the image of good project management. It is only meant to emphasize that the acceptance of a realistic deadline due to the realization of the importance of one's own activity as a link in the deadline chain (*critical path*) often leads to the fact that one gives oneself a jolt and is motivated to complete the started activity on the same day. The next day means a fresh start, often reorientation to get back to the same point of yesterday and the routine of the new day often tempts one to postpone again.

An AI control system (*to-do-list*) should therefore not be missing from any project. Especially in large projects it is a necessity. However, the definition of actions must follow certain rules. Actions are usually defined during regular status meetings, e.g., weekly progress meetings, reviews or other meetings. To be defined are:

- the action, including a detailed description,
- the deadline and
- the *actionee,* i.e., the action taker.

Table 9.4 shows an example of an AI control list. For the placing of actions, the following rules must be followed:

- The actions must be described clearly and unambiguously. They must be realistic and feasible.
- The ad hoc actions determined must not replace or repeat activities that are already included in the official schedule or are to be added.
- The actions should not be long-term activities, because then they also belong in the schedule.

Table 9.4 Example of an AI control list

Action control (example format)				
Status: 31.03.20XX				
Action	Description	Responsible	Date	Status
001	Motor adjusted	Maier	12.03.20XX	Open
002	Text	Mueller	01.02.20XX	01.03.20XX
003	Test	Huber	15.01.20XX	15.01.20XX
004	Text	Maier	20.01.20XX	15.02.20XX
005	Text	Maier	31.03.20XX	Open
006	Text	Maier	15.03.20XX	Open
007	Text	Maier	12.03.20XX	Open
008	Text	Maier	10.04.20XX	Open

- Routine activities, such as preparing the monthly status report, are not actions to be specified. They are defined in the management plan or should be defined there. Assigning actions requires acceptance by the actionee. Actions set anonymously (surprise for the actionee) are usually rejected directly or indirectly. The action date must be realistic.
- The fulfillment of the defined actions must be monitored regularly, for example weekly.

Systematic monitoring of the defined actions is crucial for the implementation of an AI control system. If action items are set but not tracked, the whole AI control system is in bad shape. The specified actions will not be executed in a very short time and the system will fall asleep. To ensure a successful implementation of the AI control system, the following actions should be taken by the project management:

(1) *Nomination of a project employee to monitor the action items* (*AI*): If the project manager does not perform this task himself, he can assign it to the project control manager or the project planner, for example.
(2) *Maintenance of an AI control list:* Actions shall be numbered in the order they are defined (*AI-0001, AI-0002, etc.*) and recorded in an AI control list *(register)*. The AI control list shall contain an accurate AI description, the name of the actionee (*action taker*), and the scheduled completion date. When defining the actionee, an individual should always be named, if possible, rather than a department.
(3) *Regular review of AI:* The AI status is to be reviewed at regular intervals. Actions that have been successfully completed are to be closed *(closed action items)* and for open action items that have not been completed on time, new deadlines or special measures (*for example, management involvement, etc.*) are to be initiated.

The following points should be given special attention during the AI review:

- The regular AI review is most appropriately conducted in conjunction with the regular project meetings
- The Project Controller or Project Planner responsible for AI control must attend the regular project meetings
- Closure of an Al must be verifiable, for example by citation of a report, notice, acceptance report

For smaller projects, it does not cause any difficulties if the person responsible for AI control performs the monitoring by keeping a logbook or a logbook-like AI control list. In many cases, this is the most economical method for AI control. However, for large projects or projects with many sites, the point is quickly reached where the use of appropriate software is appropriate. The control capability of the Al must always be kept in mind.

9.4.5 Cost Monitoring

In addition to progress control, cost control is another important pillar of project monitoring. It concerns the thorough recording of the actual costs and the commitment status at the time of control (*time now*). For this purpose, it is important to start cost control at the lowest level; i.e., at the WP level.

9.4.5.1 Cost Control per Work Package

Detailed cost control is usually based on the smallest unit of the WBS, i.e., at the work package (*WP*) level. For each WP a Personnel and Cost Plan (*PCP*) must be prepared; see Fig. 10.3. The PCP provides vertically the budgeted manpower (*hours and costs*) and other direct costs for material, travel, purchasing of items, services, etc. and horizontally the distribution of manpower and costs over time. The PCPs provide the basis for efficient detailed manpower and cost control. For many projects, personnel costs are the largest cost element. This is especially true for research and development (*R&D*) projects with up to about 80% of the project total. In addition, there is usually a direct relationship between the schedule (*Fig. 9.19*) and staffing (*Fig. 10.3*). For better understanding, hours can easily be converted into man-days or man-months. Applying the hourly rate, one can determine the personnel costs which are directly related to the number of hours (*hours* × *hourly rate* = *personnel costs*). When determining personnel costs, the first step is to determine the hours of a work package broken down by cost category. The responsible WP manager is responsible for the following three aspects of the WP, just like the PM of the overall project:

(1) Task or performance (*see WP description, Figs. 9.11 and 9.12*),
(2) Dates (*see schedule, Figs. 9.18 and 9.19*)
(3) PCP (*see Fig. 10.3*).

For cost control, it is important to record the actual costs incurred by the particular WP as quickly and regularly as possible in order to compare them with the costs planned for the controlled period. Deviations from the plan must be identified and evaluated; see Fig. 9.33. Simultaneous control of the current status of financial obligations (*commitments*) must also be taken into account; see Fig. 9.34.

The cost monitoring cycle should be kept as short as possible in order to be able to recognize planning deviations quickly and to initiate corrective measures as soon as possible. Since cost control is sensibly to be carried out in accordance with operational accounting and operational accounting of most companies are usually carried out on a monthly basis, a monthly control cycle is therefore also recommended for project cost control.

As already mentioned, detailed cost control is based on the individual work package cost plans; see Fig. 10.3. The project manager or his cost controller must compare the personnel, material and other direct costs incurred with the planned monthly costs as shown in Fig. 10.3. In principle, a similar procedure must be followed for monitoring of the commitment status. When identifying variances, e.g., cost overruns, first investigate whether there is an actual overrun or a possible posting error. Project debits are in principle only possible on the basis of the approved planning documents (*Fig. 10.3*) and a previously issued release; see Sect. 9.4.2. Intra-company project and work orders released by the project manager and the project controller as well as demand reports are the detailed release instruments for project funds. Nevertheless, due to incorrectly used project numbers, indifference or intention, incorrect bookings occur again and again, which may have to be rebooked by the project control officer. After the cost unit has been cleaned up, a true target/actual comparison can now be made. Cost overruns must be reported to the project management immediately. Since the personnel hours are directly related to the WP schedule, a simplified cost control is in many cases possible for a quick cost

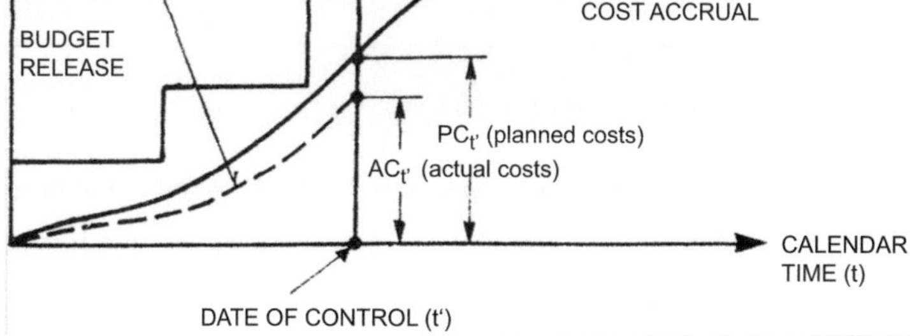

Fig. 9.33 Comparison of target/actual costs

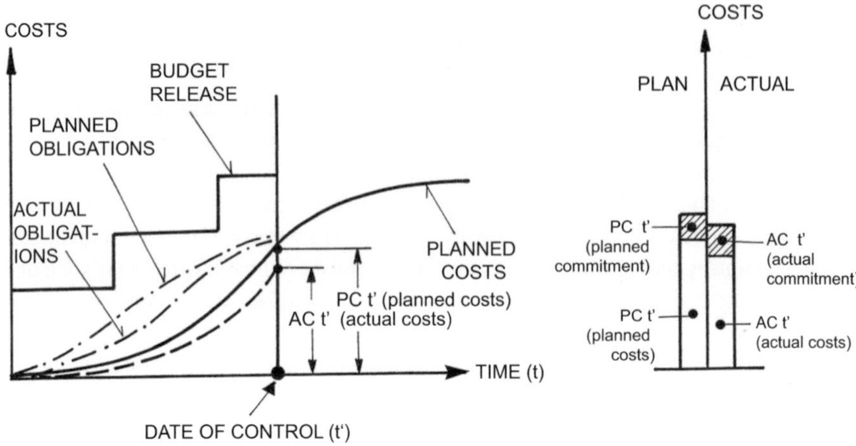

Fig. 9.34 Cumulative cost/commitment target/actual *comparison*

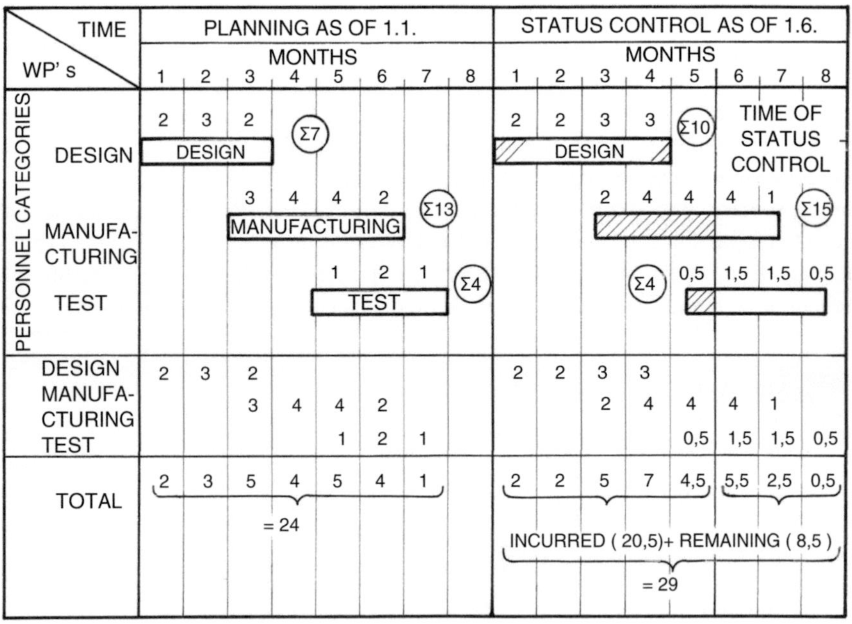

Fig. 9.35 Simplified resource and cost control scheme

overview by establishing a simple relationship between the activity and the planned personnel deployment; see Fig. 9.35.

9.4.5.2 Cost Control at Project Level

The cost control of the total project is based on the cost control of the individual WPs. Therefore, it is important that the responsible WP managers carry out a thorough control. A project summary can be prepared in accordance with the WBS at all levels up to the overall project level. The highly aggregated overall result can be presented in exactly the same way as for a WP, i.e., personnel hours and costs over time; see Fig. 9.36. The cost trend over time is also an important information source for:

- the preparation of payment schedules
- for the company's internal workload planning
- for Earned Value Analysis; see Sect. 9.4.6.

The determination of the total actual costs of a project for a certain month is achieved by summing up the actual costs in the period concerned (*accumulated actual costs of all work packages*). In order to be able to quickly analyze the causes in case of overruns, not only a simple summation of the actual costs as a total sum, but a summation broken down by cost categories (*personnel, material, etc.*) should be carried out. In this way, it is easier for the project manager to analyze the causes of cost overruns on the basis of a presentation of actual costs broken down by cost categories. A simple example (*see Fig. 9.36*) demonstrates how a 10% cost overrun can be quickly localized by looking at the cost categories. The 125% overrun shown for purchased parts can be identified as the cause of the cost overrun, while the other cost categories have only insignificant variances that are of no particular importance for cost control. For better understanding, the values in Fig. 9.36 are only symbolic. In the case of the 125% overrun for purchased parts, it may quickly become apparent, that it is only a possible overrun, since the costs causing the overrun (*125%*) are

	PLAN/IS COMPARISON		DEVIATION ANALYSIS	JUDGEMENT
	PLANNED	ACTUAL		
– PERSONNEL	70	72	–2	Small Overrun (~3%)
– MATERIAL	12	10	+2	17% Savings
– PURCHASED ITEMS	8	18	–10	125% Overrun
– EXTERNAL SERVICES	2	1	+1	50% Savings
– RENT	3	3	—	—
– TRAVEL	5	6	–1	20% Overrun
– Etc.	—	—	—	—
TOTAL	100	110	–10	Total 10% Overrun

Fig. 9.36 Example of a detailed cost analysis

not scheduled according to the plan until the following month, for which coverage is available.

However, an objective assessment of the actual cost development requires an examination of the cumulative target/actual analyses; see Figs. 9.35 and 9.36. The trend towards cost overruns can only be identified on the basis of a target/actual comparison carried out over a longer period of time. An overall assessment of the project situation, i.e., cost underruns or cost overruns, determined in the target/actual comparison, can only be compared with the schedule situation (*schedule underruns or overruns*) and the technical performance status; see Sect. 9.4.6 (Earned Value Management—EVM).

9.4.5.3 Integrated Schedule and Cost Control

An absolute prerequisite for the implementation of integrated control is the linking of cost planning with scheduling. This means that for each work package, a technical performance description, a coordinated schedule and a detailed cost plan are created. The recording of actual data for dates and costs must be controlled in such a way that a clear allocation is ensured. The work breakdown structure is at the center of this process. The use of the common WBS number key for both schedule and cost control is a prerequisite for the joint assignment of schedule and cost monitoring and the resulting performance evaluation. Figure 9.37 shows the process of integrated schedule and cost control [67, 68].

9.4.6 Earned Value Management

The "Earned Value Management—EVM" method is not new. It was developed as early as 1966 as part of the "Cost/Schedule Control Systems—C/SCS" concept [69]. The method is a management achievement, but its application causes great difficulties for many managers. Why?

First of all, it requires that the project manager has reliable schedule and cost data and that these are regularly updated. This means that the progress control described in Sect. 9.4.3 and its forecasts must be based on hard facts. The same applies to the cost control described in Sect. 9.4.5. Figure 9.38 shows a typical example of an EVM representation.

The budget curve (*planned costs*) is shown in clear relation to the schedule. At the control date (*time now*), the situation is as follows:

– The accumulated budget	325,000 Euro
– The actual costs incurred	250,000 Euro
– The labor value[a]	140,000 Euro
– Delay	2 months

[a]*BCWP, see Fig. 9.38*

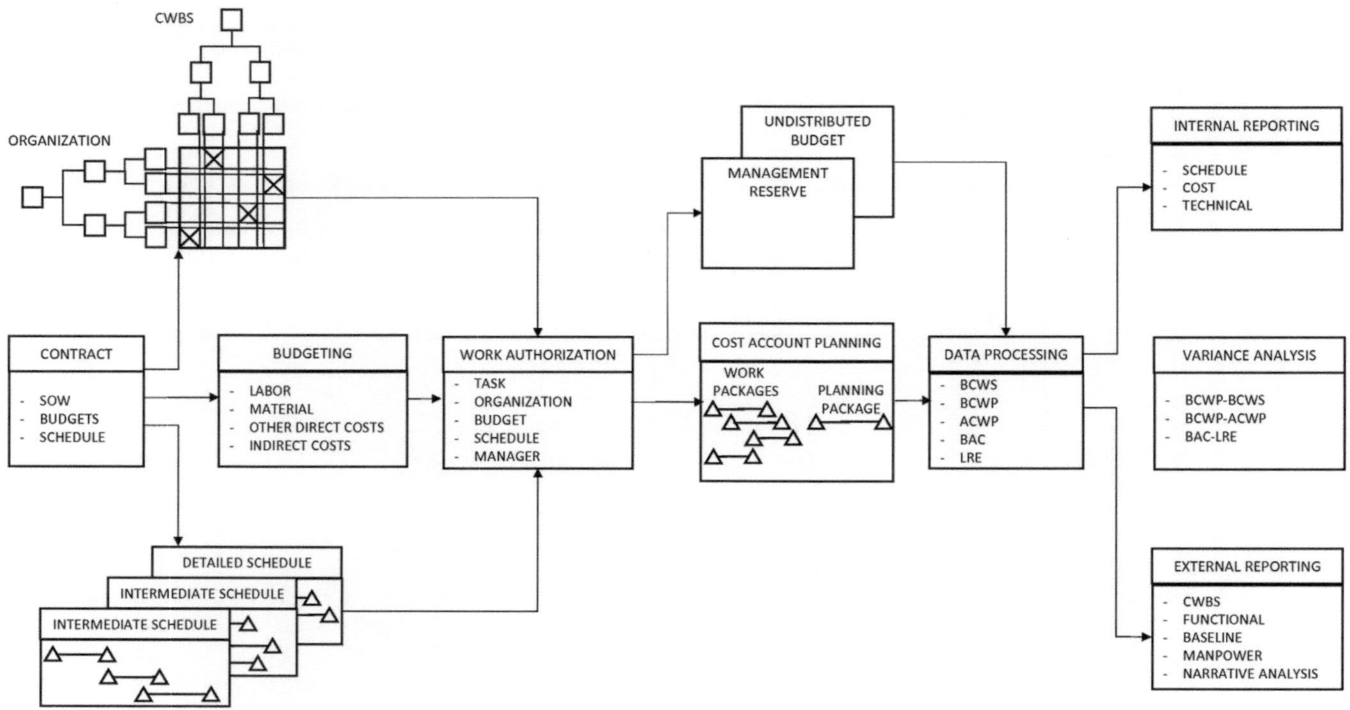

Fig. 9.37 Example of an integrated schedule and cost control (Source: TMSA seminar document)

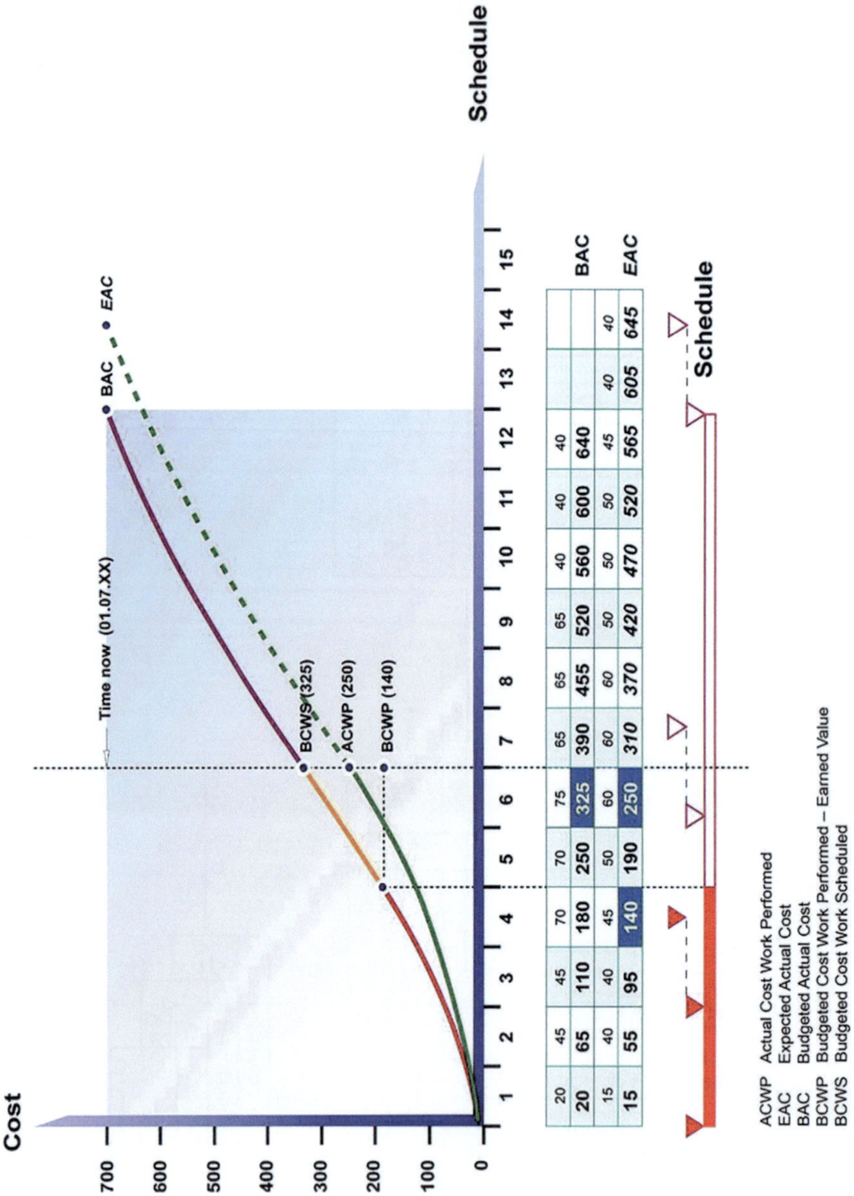

Fig. 9.38 EVM—example

That is, although less money was spent than was budgeted (*325 − 250 = +75*), too much was billed (*140 − 250 = −110*) compared to the established work value (*140*). This means that looking at costs without including the work value calculated from the 2-month delay is misleading. At this point, it is now necessary to investigate whether the schedule delay can be recovered as quickly as possible and what cost to completion (*CTC*) is still required. It is very important that the CTC estimate is as realistic as possible. In this context, it is also useful to consider possible savings. The bottom line, however, is that a realistic and, above all, credible CTC value must be determined. However, this process should not be a flash in the pan, but should be repeated depending on the situation, e.g., monthly or quarterly. The EVM methodology is a very effective management tool. In practice, many managers find it too complicated and burdensome because it requires rigorous schedule and cost monitoring. If project planning and monitoring is poor, EVM will not help.

Often, a simple examination of the cost situation of a project can be helpful. Project managers, but also WP managers, may ask themselves at regular intervals, e.g., at the end of the month, about the cost situation, including a residual cost analysis (*CTC*) and a final cost analysis (*Cost at Completion—CAC*). The following is an example (with fictitious values) of a simplified cost review at the time of control (*Time Now—TN*).

1. Amount of the agreed budget (*CAC #0*)	+100 T€
2. Approved changes (additional costs)	+20 T€
3. New total budget (*CAC #1*); 1+2	+120 T€
4. Actual costs + commitment at the time of control	−80 T€
5. Estimated remaining cost requirement (*CTC*)	−60 T€
6. Cost balance (*expected losses*); 3 − (*4* + *5*).	−20 T€

The original budget (*CAC*) was increased by 20 T€ to 120 T€. This is usually based on an approved change request from the customer. The new total budget (*CAC #2*) of 120 T€ is offset by incurred costs and liabilities of 80 T€ and a calculated remaining cost requirement (*CTC*) of 60 T€, resulting in a final cost balance of 140 T€. This means that the forecast result is negative (*−20 T€*). In the next step, it must now be investigated whether an expected negative result can be avoided through savings and/or other measures. It must also be investigated whether the negative result is due to additional services and whether the customer (*client*) will assume these in full or in part.

9.4.7 Technical Performance Monitoring

9.4.7.1 TPC Analysis

Technical performance control *(TPC)* is a useful supplement to schedule and cost monitoring. Integrated project monitoring can only be carried out successfully with the help of functioning technical performance monitoring. In other words, the PM must also deal with the so-called *metrics* in addition to schedule and cost monitoring.

Technical Performance Monitoring or Technical Performance Control *(TPC)* is a system for continuously demonstrating and predicting the current or expected performance status of selected technical goals/objectives of a system, subsystem, etc., and is intended to enable project management to respond in a timely manner as problems arise [70]. Typical TPC parameters include system weight, operating time, trajectory, diameter, power consumption, etc. However, monitoring all values listed in the system specification would be too time-consuming and therefore would not bring the desired success. Therefore, only a few particularly important parameters are usually selected. The selection of parameters for regular monitoring of technical performance data depends on various factors. The following criteria are of importance:

- Contractual receivables
- Mission critical data
- Critical development parameters
- Key data for a premium contract

For each of the selected parameters, a special control sheet with the following information should be created:

- Parameter designation
- Selection criterion
- WBS Reference
- Specification reference
- Organizational responsibility for the parameter
- Parameter Status Profile

Table 9.1 shows a practical example of a list of various technical parameters *of* the system, subsystem, etc., which can be clearly identified by a WBS reference. They can therefore be easily used for comparative analysis with other project parameters, such as costs and project schedules. Similar to the tracking of schedule, cost and performance parameters, the formation of indices is advantageous for the control of technical parameters.

Figure 9.39 shows a practical example of a TPC performance diagram for the power supply of a satellite project [71]. Based on the TPC information, the development engineer can determine whether the technical requirements (*specification values) of* the relevant assembly can be met or whether there is a conflict.

Developers repeatedly encounter problems in complying with specified weight limits. The situation is equally critical in vehicle and aircraft construction as well as in space travel. The example of the Mercury capsule clearly shows the increase in weight during development; see Fig. 9.40. Misjudgment can result from several problems. One very important aspect is that a heavier device or vehicle no longer meets customer requirements or, for example, that the energy consumption of a passenger car increases too much. In space travel, a conflict can easily arise as can be seen in Fig. 9.41. The example shows in a forecast that the capacity of the rocket to be developed is constantly decreasing and the weight of a satellite to be developed is constantly increasing. The forecast shows the resulting problem at an early stage.

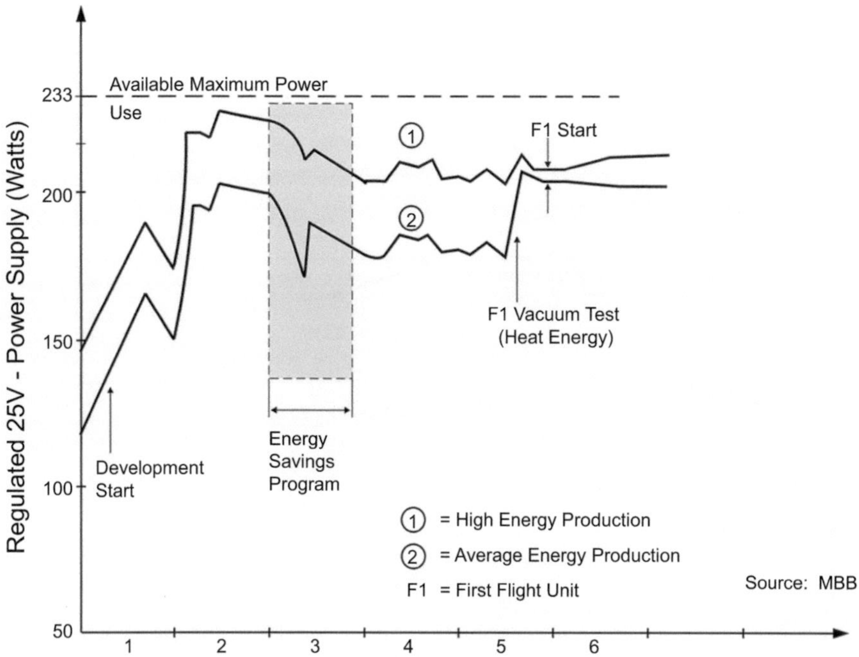

Fig. 9.39 Practical example of a TPC power diagram for the power supply of a satellite project (MBB)

Fig. 9.40 Weight increase of the Mercury capsule (Source NASA)

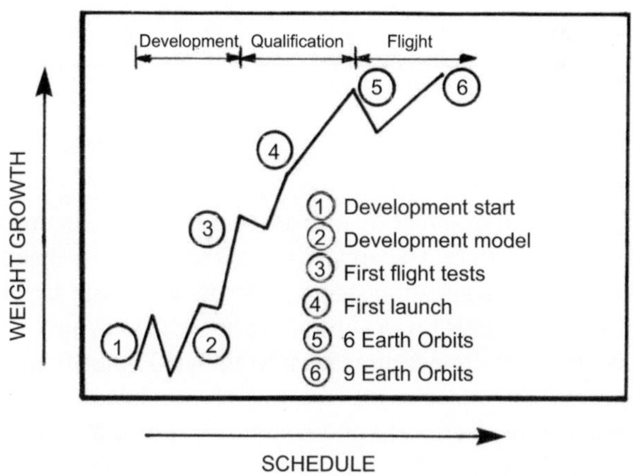

It is the task of project management to identify problems in good time and forecasts (*trends*) are an important tool for this. It is therefore important that projections of the masses are made at regular intervals. Incidentally, the same applies to other parameters such as the energy balance; see Fig. 9.39.

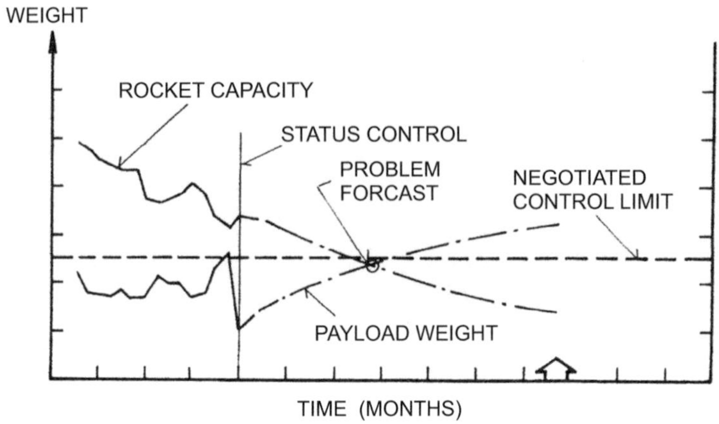

Fig. 9.41 Early detection of the conflict situation (Source NASA)

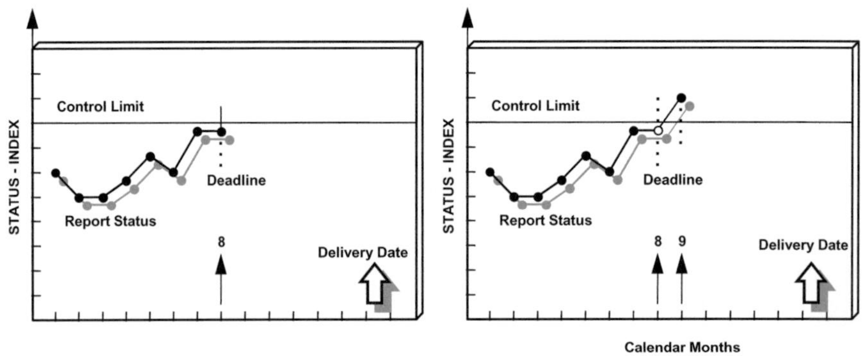

Fig. 9.42 Trend analyses—worst case considerations (Source NASA)

Figure 9.42 shows in addition that corresponding trend analyses must be carried out continuously. The importance of trend analyses becomes completely clear in the context of the situation shown, since they indicate an impending conflict at an early stage. This illustration shows that in order to avoid a conflict, in this case it was a weight conflict, measures must be initiated immediately. So how is a trend analysis created? It is usually difficult to create an exact mathematical method, because in practice there are seldom enough measurement points necessary to establish the shape of the curve (*regression analysis*). However, if we consider a situation such as that shown in Fig. 9.42, it is not difficult to determine the trend in terms of a worst-case analysis indicating the situation as shown in month 9.

9.4.7.2 Key Performance Indicators

The definition of particularly distinctive technical and programmatic key performance indicators (*KPIs*), in conjunction with the TPC analysis described in Sect. 9.4.7.1, is a valuable management tool. The KPI process provides accurate measurement of technical and economic performance delivered at all project levels. It is a simple and understandable method that can be applied in many areas of a company. Possible areas of application include:

- Management
- Research and development
- Project management
- Production and test
- Purchasing and materials management
- Marketing and sales
- Customer service
- Workforce development

"*Performance measurement is a principle of management*" [72]. For project management, the use of KPIs is recommended to measure the respective progress of the project (*degree of fulfillment*). The following KPI examples are provided for this purpose:

- System weight
- Dimensions
- Volume
- Fuel consumption
- Energy balance
- Transmitting power
- Delivery date
- Lifetime
- Project costs
- Profit expectation

Krause and Arora have produced a comprehensive publication in German and English on the subject of "Controlling—Key Figures." [73]

9.4.7.3 Technology Readiness Level

The manager of a development project is expected to be able to determine the status of the development process at any time. This is not only about the schedule status, but also about what has been technically achieved at the time of the question. In other words, "what is the technology readiness level *(TRL)* at the time of the survey? To answer the question precisely, you need a reference, i.e., a scale.

Organizations such as ESA, EU, DOD, DOE, NASA, and the oil and gas industry adopted a nine-level TRL procedure to measure the maturity of development efforts. A scale from TRL 1 to TRL 9 is reproduced below:

1. Establishment of the basic principles
2. Formulation of the technology concept
3. Analysis and testing of critical functions and/or characteristics

4. Validation of *breadboard* components under laboratory conditions
5. Validation of *breadboard* components under environmental conditions
6. Functional testing of subsystem test models or prototypes under operating conditions
7. Functional testing of system prototypes under operating conditions
8. Completion and qualification of the system and qualification tests
9. Completion, acceptance, and testing of the system in operational use

Figure 9.43 illustrates the stage of development maturity by means of a scale analogous to a clinical thermometer. If the TRL definitions are related to the phase planning described in Chap. 4, TRL 1 to TRL 4 belong to the range of early phases A and B, and TRL 6 to TRL 8 to development, i.e., phase C. TRL 9 is reached after all prototype tests are completed and the system is fully qualified. TRL 9 is the prerequisite for the start of production (*phase D*). In Fig. 9.43, however, it can also be seen that there is no sharp demarcation, and that overlaps are discernible.

Serry has published a detailed TRL definition in addition to this, which is too extensive to reproduce in this book [74].

9.5 Project Control Management

9.5.1 Implementation of PC Management

The project control (*PC*) task is an important key function of modern project management and, as shown in Fig. 5.3 and Table 5.1, can be classified organizationally

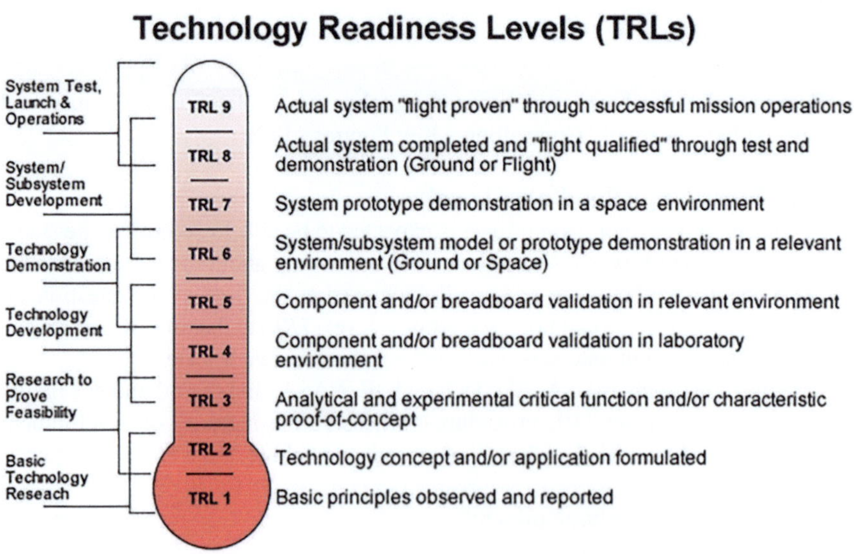

Fig. 9.43 TRL scale (Source NASA)

in the second project management level. In smaller projects, the PM must also take over the PC tasks in personnel union. In the PC task, the individual functions of planning and the specialist disciplines of scheduling and costs are combined to form an overall function. What is special about the PC idea is that one does not only try to consider the planning and monitoring of schedules and costs as an integral whole by merely comparing schedule and cost data, but rather, starting from a certain project size, one creates a corresponding organizational unit from the outset, in which performance monitoring (*performance control*) is in the foreground.

Traditionally, it might have been sufficient for the project manager to be assisted by a businessman and a planner. The project manager then had to create the necessary synthesis between the technical planning data and the business information himself. Often, however, the PM was even deprived of the competence for the financial project control, which inevitably had to lead to unbalanced decisions, because technical decisions usually cause a financial influence, which, however, often does not appear immediately, but only with a certain time lag. Adamowsky wrote as early as 1968: "*The fact is, however, that in practice economic decisions are made to a considerable extent by people with engineering training, while many trained business-people merely act as manual data processing technicians.*" [75] The interweaving of technical, scheduling and financial parameters of influence in the realization of large and complex projects is an absolute and probably indisputable necessity in order to be able to intervene in a controlling manner in the project events. This fact-based constraint leads to the implementation of project control management.

9.5.2 Tasks and Competences of PC Management—Trust But Verify!

The primary task and competence of PC management is to plan and monitor the scheduling and financial progress of a project; see Table 5.1. This requires qualified experts, e.g., such as project controllers, planners, and project managers. Figure 9.44 shows a summary of the planning and monitoring process based on the planning and monitoring packages to be created, as described in detail in this chapter. The monitoring process is based on the motto: *Trust but verify* [76]. In practice, in addition to the original PC tasks (*schedules and costs*), PC management is increasingly being assigned additional functions, such as documentation and configuration management, etc., which considerably extend the scope of PC management; see Table 5.1.

According to Figs. 5.3 and 5.4, the tasks of the project manager are extensive and the financial outlay is not insignificant. Experience shows that the proportion of the responsibilities described for managing large technology projects is between 7 and 15%; the lower value can be used for standard projects and the upper value for new developments. This may seem high, but experience shows that it is not too high for efficient project management. It should be borne in mind that a budget estimate of 10% on average is a good investment for project management if it can prevent a cost increase of several hundred percent, for example; see Sect. 3.2.

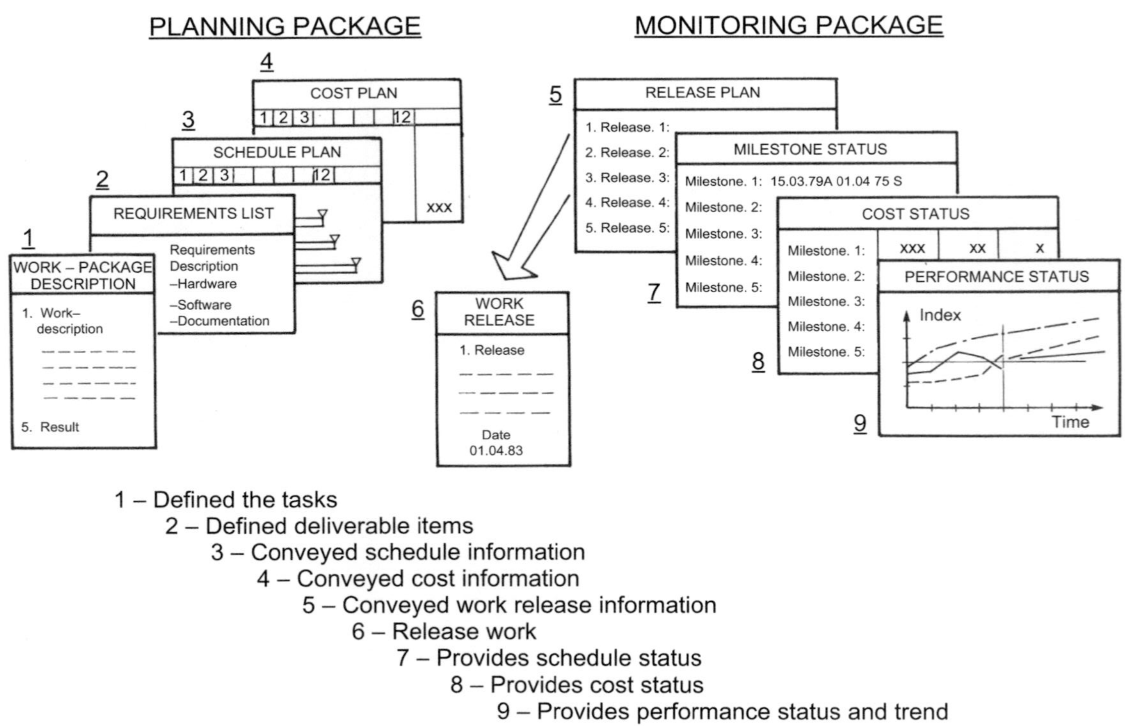

Fig. 9.44 Overview of typical planning and monitoring instruments

For project managers of small projects, e.g., a work package, the situation is usually somewhat different, and the previously mentioned rules cannot be transferred completely. To manage a small project the following tasks must be performed as a minimum with regard to Fig. 9.44:

- Creation/adoption of the work package (1)
- Schedule creation (3)
- Preparation of a cost plan (4)
- Appointment status control (7)
- Cost status control (8)

9.5.3 Integration of Schedule, Costs, and Technical Performance

9.5.3.1 The Software Supported Dashboard

This book contains many procedures and methodologies required for efficient PM. The author confirms that they are proven in practice and can be considered mature and as state-of-the-art. In other words, it is a solid basis for future project management assignments and only minor improvements are required.

An important improvement is the establishment of a close link between some key project characteristics. For example, between schedules (Fig. 9.19) and costs (Fig. 10.3) to provide an integrated view of schedule and cost performance as shown at the EVM presentation (Fig. 9.38); see also [77]. A further step to expand the control system is through the inclusion of technical parameters (TPC metrics) such as power, mass, and frequency (Figs. 9.39–9.42) in connection with schedule and cost data (Fig. 9.19 and Fig. 10.8). Figure 9.41 shows a practical example of an impending conflict between a predicted rocket capacity and the payload mass increase. It is an example of an early warning of a negative trend (crisis). In many cases immediate risk identification and mitigation is required; see Chap. 19.

For improvements of Project Management in the future, the author suggests the use of a software supported Dashboard, which would provide the PM and his team with a real-time information overview of all KPIs and their interrelationships.

9.5.3.2 Collaboration Tools

The following is a contribution to the above topic by Dr. Thor Möller, Bremen; see chap. 18. Microsoft has released MS Teams, a collaboration software that can increasingly be used for work in projects, especially in cross-organizational projects. Many users are currently only familiar with MS Teams as a tool for online conferences. What is special, however, is the extensive and low-threshold, cross-organizational collaboration coupled with various programs from the Microsoft 365 program range. This means that distributed teams from different organizations and locations can share, view, edit, and exchange essential and current information about the project on a common platform at any time. There are, among other things, project-related chats and calendars, shared access to files and various tools for planning and control.

The project-related files can be linked or synchronized with your own Windows Explorer, so that editing shared files is very convenient and the data can also be securely backed up. Microsoft tools such as Planner, ToDo, OneNote, Outlook, and Project can be integrated within a team. Tasks that are planned in the Planner and assigned to people, for example, synchronize with the organization programs mentioned. The interaction of these individual tools has improved enormously and made it easier for users. Specific advantages include significant increases in effectiveness and efficiency in teamwork and self-organization through quick orientation, always up-to-date data and the avoidance of redundancies, email traffic, and much more.

9.5.3.3 Artificial Intelligence

The term artificial intelligence (AI) continues to hover over everyone in society, the economy and does not stop at the project economy. Many people see AI as a kind of magic bullet that solves their problems. But caution is advised here. A successful application of AI requires certain prerequisites and does not replace human expertise and creativity. Well-structured processes are required for sufficient digitalization of the organization. Based on poor processes, neither digitalization nor AI will be able to bring any particular benefit.

Since AI reproduces based on reams of data and information, the quality of the results is usually good, nothing more, nothing less. The background is the connection of what already exists and not the creation of really new information. Very good or even excellent results and innovations have so far only been achieved through appropriate human expertise. However, to make the experts' work easier, they can have AI generate a preliminary solution, which they can use as preliminary work or inspiration and, if necessary, only need to add the final touches. The same applies to creativity. Real innovations still require people and their expertise. AI can provide excellent support, but the final touches for real innovations have not yet been possible without human intervention.

9.5.4 PC Checklist

The following PC checklist has been developed and applied in practice [78]:

(1) Has the Turnkey Contractor *(TKC)* developed a planning and monitoring philosophy for systematic monitoring and control?
(2) Have the proposed planning and monitoring methods been tested on other projects?
(3) Can the PM rely on the support of specialist departments, e.g., IT, when introducing monitoring methods?
(4) Are qualified planning and monitoring personnel available for the TKC?
(5) Has a WBS (*work breakdown structure*) been prepared by the TKC that includes all project areas and all TKC and subcontractor work packages?
(6) Can the WBS breakdown be used to determine the work shares of all project partners (*TKC and subcontractors*)?

(7) Can corresponding work packages be correlated to the WBS?
(8) Has the WBS been designed in such a way that the specifications (*specification tree*) and project documents (*document tree*) can be clearly assigned?
(9) Can the WBS be used as a basis for creating the cost structure?
(10) Have realistic and credible schedules and timelines been developed in consultation with departments and subcontractors?
(11) Was the structure of the schedules and flowcharts based on the WBS?
(12) Has the TKC created an overview "master bar chart" including all key milestones?
(13) Have all contract and/or payment milestones been clearly defined?
(14) Has the TKC conducted a critical path analysis in collaboration with subcontractors?
(15) Have all schedule interfaces been clearly defined (*input/output*)?
(16) Have all provisions (*tools, equipment, etc.*) been clearly defined?
(17) Do the schedules and flowcharts contain planning margins?
(18) Has the TKC determined how and at what time intervals the project status will be performed?
(19) Are TKC procedures in place to address planning overruns, e.g., work-around solutions and corrective actions?
(20) Have reporting formats (*templates*) been defined by the TKC?

Literature

1. Gehriger, Helmut: Overview of the network planning technique as a central control instrument, ZFU Management Seminar, Zurich, 1980, 5. 1, 27, 29
2. Madauss, Bernd-J., ISU: Lecture Notes M1-MGB, Eng-L03, 2016.
3. Madauss, Bernd-J.: Planung und Überwachung von Forschungs- und Entwicklungsprojekten, AlB-Fachliteratur, Bad Aibling, 1978/82.
4. Madauss, Bernd-J., ISU: Lecture Notes M1-MGB, Eng-L04, 2016.
5. Ostdik, Marion A. and Estes, Richard T.: Cost/Schedule Control System Griteria: An Analysis of Managerial Utility, Thesis, Air Force Institute of Technology, Report No. SLSR-15-75B, 1975.
6. Wright, Donald, D.: Cost/Schedule Control Criteria: A Case Study, (Thesis) Air Force Institute of Technology, Report No. SLSR-65-71B, 1971.
7. DIN 69900, 1987
8. DOD: Performance Measurement for Selected Acquisitions, DOD Instruction 7000.2, 1977.
9. Starch, Stephan: The Use of Cost and Schedule Data by Program Managers under G/SGSG, Study Project Report, Defense Systems Management School, Program Management Course Glass 75-1, 1975.
10. DOD/NASA: DOD and NASA Guide PERT/Cost, 1962.
11. NASA: Program Scheduling and Review Handbook, NHB 2330.1, 1965, p. 31.
12. NASA: OSSA/OART Project Management Information and Control System- MIGS, NHB 2340.2, 1966.
13. NASA: PERT & Cost Correlation Technique (PACCT), SC006-003-1A, 1966, p. 13.
14. NASA: Forecasts and Appraisals for Management Evaluation, SP6009, pp. 1-6.
15. Stevenson, Ian: The Use of Critical Path Analysis Methods in the European Space Vehicle Launcher Development Organization, BMWF Symposium on Project Management, 1968.

16. Madauss, Bernd-J.: Die Netzplantechnik im Einsatz zur Projektüberwachung und als Management-Informationssystem bei der europäischen Organisation zur Entwicklung von Trägerraketen, in: Astronautik, 2/1970, S. 88.
17. ELDO: Procedures Manual for ELDO Planning and Progressing System, CPPS, 1970 (Draft).
18. ELDO: Project Control Procedures Europe 1/11, CPP 100, 1970.
19. Gehriger, Helmut: Planning and Control for European Space Satellite Projects, ESRO SP70, 1972.
20. ESA: Project Control Requirements and Procedures - Systems Definition, ESA PSS-32, 1977.
21. ESA: Project Control Requirements and Procedures - System Development and Production ESA PSS-33, 1977.
22. ESA: Project Control Requirements and Procedures - Medium-Size Contracts, ESA PSS-38, 1977.
23. ESA: Project Control Requirements and Procedures - Small Contracts, ESA PSS-39 1977
24. Madauss, Bernd-J.: Experiences with the PERT Planning Procedure in the HEOS-A Satellite Project, in: Luftfahrttechnik + Raumfahrttechnik, 7/8, 1968.
25. MBB: HELIOS solar probe - final report, research and development contract QC-90-00 VP 6/70, 1976.
26. Saynisch, Manfred: Project Management System for a Large International Project (a real case), Proceedings of the 7th INTERNET World Congress, 1982
27. Dale, Ernest: Management, Econ-Verlag, Düsseldorf, 1972, p.335.
28. Kutzer, Ants, Projekt-Management in der Raumfahrt, ERNO Raumfahrttechnik GmbH, Bremen, 1987.
29. DIN 69901-5, 2009
30. DIN 69900, 1993
31. Military Standard: Work Breakdown Structures for Defense Material Items, MIL-STD-881C, October 3, 2011.
32. NASA: Handbook for Preparation and Implementation of Work Breakdown Structure, NASA Document NHB 7120.5, May 1994.
33. ESA: ECSS-M-ST-10C Rev. 1, March 2009, p. 16.
34. Burbridge, Keith: Final Report, Low-Cost Program Practices for Future NASA Space Programs, LMSC Report D469858, Volume II, LGGP17, 1975, p. 15.
35. Madauss, Bernd-J. and Orye, Raymond: Guideline for the Establishment of the Europa III Work Breakdown Structure and Work Package Plan, ELDO Document EIII/PGO1 - 60/BM, June 11, 1971.
36. Madauss, Bernd-J.: Management Advantages of the Standard Work Breakdown Structure proposed by the European Space Agency for Space Satellite Projects - Master Thesis, MBB Report UR-446/80 0, July 17, 1980.
37. ESA: General Instructions for the Establishment of Standardized Work Breakdown Structures for Technological Projects, ESA PSS-44 (Draft), December 1979.
38. ESA: Requirement Specification for Preparation and use of a Standardized Phase G/D Work Breakdown Structure for Satellite Projects, ESA PSS-53 (Draft), December 1979.
39. Project Structure Working Group. Standardization and Rationalization of Engineering in Power Plant Construction June1982 (not published).
40. Airbus Industrie. Industrial Feasibility Study A320, AI/P N 082 (Book 5), February 3, 1982.
41. Hopf, Hans-Dieter and Madauss, Bernd-J.: Schiff der Zukunft - Gliederung für das Transportsystem Schiff, MBB-Report, UR-61 1/83, 1983.
42. MBB, Rapid Transit System Caracas/Litoral, Proposed Technology System Management, June 1977.
43. Source 12, p. 6-1
44. Trivailo, Olga: Innovative Cost Engineering Approaches, Analysis and methods Applied to SpaceLiner - an advanced, Hypersonic, Suborbital Spaceplane Case-Study, PhD thesis, March 2015, p. 69.
45. Madauss, Bernd-J., Trivailo, Olga: SpaceLiner System - Engineering and Management Approach, DLR SART Workshop, July 2015,

46. McDonnel, Douglas: Space Shuttle System - Phase B Study Final Report, MDC Report E0308, Part III-1, June 30, 1971.
47. NASA: Standard Data Requirements Descriptions, NASA Document MF002M, 1969.
48. Q. 8
49. ESA, ECSS-M-ST-60C, July 2008
50. Aucoin, B. Michael: From Engineer to manager mastering, the transition, Artech House, Inc., Boston - London, 2002, p. 231
51. Seery, Bernhard D.: Project Planning, in Larson Wiley J. (managing editor), Applied Project Management for Space Systems, McGraw Hill Companies, 2008, p. 88.
52. Stevenson, lan: L'Analyse dc Reseau, in: ESRO/ELDO Bulletin, August 1968, p. 15.
53. ESA: ECSS-M-ST-60C, July 2008
54. Source 3
55. General Electric: BMBW Management Study for Europe III, September 1970
56. Wille, H./Gewald, K. and Weber, H. D.: Netzplantechnik, R. Oldenbourg Verlag München-Wien, 1972.
57. Grob, Helmut and Gutsch, Roland W: “Netzplantechnik” - Eine Anleitung zum Projektmanagement für Studium und Praxis, VDI-Verlag, Düsseldorf, 1982, p. 127.
58. NASA: Apollo Program Management - Staff Study for the Subcommittee on NASA Oversight, July 1969, p. 46.
59. Source 24, p. 174
60. Source 25, p. V-16
61. Miller, Robert W.: Zeit-Planung und Kosten-Kontrolle durch PERT, R.V Decker's Verlag, G. Schenck, Hamburg-Berlin, 1967, p. 51.
62. MBB: Spacelab Offer, Volume II, April 1974.
63. Hab, Gerhard, Wagner Reinhard: Projektmanagement in der Automobilindustrie, 4th ed. Springer Gabler Verlag, Wiesbaden 2013, p. 94
64. Servan-Schreiber, J. J.: Die amerikanische Herausforderung, Hoffmann und Gampe Verlag, Hamburg, 1968, p. 94.
65. MBB: Measuring system Temes offer, April 1975.
66. Burghardt, Manfred: Introduction to Project Management, Siemens, 4th ed., 2002, Fig. 5.5
67. TMSA, Advanced Program Control Technique, Seminar Paper, 1983.
68. Expenditure Profiles, SSGAG Working Group, letter from Mr. L. J. Taube, Rockwell International, April 7, 1981.
69. USAF System Command: C/SPCS Symposium Summation, 10-11 August 1966.
70. Miller, A. E.: Technical Performance Measurement Guidelines for a Compliant System, in: GE Report (no. and issue date).
71. MBB: Project GOS-B, Monthly Progress Report, No. 12, 1973
72. Weber, Al, Thomas Ron: Key Performance Indicators, IVARA Cooperation, Canada, 2005
73. Krause, Hans-Ulrich: Arora Dayanand, Controlling-Kennzahlen - Key Performance Indicators, Oldenbourg Verlag, Munich, 2nd ed. 2010.
74. Serry, Bernard et al, NASA: Project Planning, published by Whily J. Larson, Introduction to Applied Project Management for Space Systems, Mc Graw-hill, 2008, p. 83.
75. Adamowsky, Siegmar: Die zweite organisatorische Revolution, in: Rationalisierung, 19th ed. 1968-5, S. 102
76. Proverbial Russian saying; Büchmann, Georg: Geflügelte Worte, Verlag Ullstein, Frankfurt/M--Berlin-Wien, 33rd edition, 1981, p. 353.
77. Duffy, Jim: Improving Project Management Performance, Dissertation, George Mason University, March 2021
78. Madauss, Bernd-J., Project Management Checklist, Management and Organizational Concept for the Metro Medellin Project, September 1983 (unpublished), “*Project Planning and Control,*” p. A.5.

Finance Control

10

Contents

B. -J. Madauss, *Project Management*,
https://doi.org/10.1007/978-3-662-69057-4_10

Abstract

Prior to the start of a project, the availability of adequate resources and financial means must be ensured in addition to the goal setting and planning according to the well-known motto "No Money, no Honey." At the start of a project, companies must investigate whether they have qualified specialist personnel, the necessary equipment, and the appropriate internal funds or external financial support from banks and investors. To overcome bottlenecks, projects are therefore often carried out in cooperation with partners. In addition to the provision of a cost budget, efficient cost management, at the heart of which is cost determination, is an important prerequisite.

Finance control (FC) is a PM support function; see Fig. 5.4.

At the initiation of projects, the focus is on three aspects: the service to be provided (*performance*), the target dates (*schedule*), and the financial resources required for the project (*budget*). Cost estimation and, above all, estimation accuracy are becoming increasingly important in this context, as misjudgments repeatedly give rise to fierce criticism. Particularly in the realization of public projects, there have been frequent reports in recent years of cost increases in which, among other factors, the quality of the cost estimate has always been called into question. The most important reasons for cost increases mentioned by Hudock at the beginning of the 70s still apply in 2023 [1]:

(1) Excessive optimism in cost estimation: more realism is therefore needed. The confidence of the cost estimate must be taken into account when awarding the contract.
(2) Changes during the development and production phases: this results in the requirement for better definition, elimination of nice and desirable but not necessary requirements, strengthening of configuration control, and implementation of changes only after prior correct cost estimation.
(3) Inadequate identification of risk areas: At the start of the project, i.e., before the contract is awarded, all risk areas must be identified and the potential impact on project events thoroughly analysed.

In the interest of better economic results, costing must be carried out systematically and with appropriate methodology. It is necessary to be able to rely on the input parameters of a business plan, which includes the cost estimation of a project. Tried and tested methods must be used and the project staff concerned must be prepared for this difficult task through training. This also includes analyzing and evaluating completed projects as thoroughly as possible from the point of view of "lessons learned" in order to pass on experience gained.

Professional cost management is a basic requirement for companies and individual projects. According to Stibbe: "*Cost management is aimed at a goal-oriented design of costs. The main goal ... is to optimize the cost/benefit ratios in the company.*" [2] There is a direct relationship between the service to be provided and the costs required for it. This becomes particularly clear during project implementation because changes in the objective usually lead to changes in the costs. Projects often involve very complex technical and economic interrelationships that have to be solved in detail. The project manager (*PM*) must therefore be able to rely on professional cost management. In the case of complex large-scale projects, it is usually necessary for cost management to be organizationally integrated into the project management office (*PMO*) during the project term; see also Sect. 5.2.5.

10.1 Cost Estimation Methods

10.1.1 Forecasting—A Difficult Art

The "estimation of costs" * is to be regarded as a forecast of expected expenditures for very specific operations. But both the word "estimate" and the word "forecast" already contain a great deal of uncertainty. It is a truism that one can of course misestimate, that is, one has either overestimated or underestimated. It is quite similar to a forecast, which is a prediction based on estimates. Von Weizsäcker writes: "*Economic forecasting, like any forecast, is an eminently difficult art.*" [3]

**In the following descriptions, the terms cost and price are used equally. The difference between the terms cost and price is shown in Fig. 10.5.*

This is also why many people have a deep aversion to estimates and try to avoid the request for an estimate. This is especially the case whenever the estimate is one that has feet of clay because from experience most people know that once an estimate has been pronounced, it obligates them in some way, even if they have explicitly pointed out the low level of knowledge. It can be observed that someone who is asked to make a quick estimate about a process they do not know intimately is instinctively looking to gain time to gather more background information first. In order to make a qualified cost estimate for a new project, one needs:

- corresponding background information about the respective process
- sufficient time for estimation and
- Experience values from projects already completed (*benchmarking*).

This distinguishes *estimation* from *guessing*. Of course, the experience of the estimator and the estimation methodology used also weigh heavily. But even under the most favorable conditions, an estimate remains what it is, namely an estimate, even if one chooses to use the word "determination" for it. The estimated cost of a project predicts, with some degree of probability, how much the project, or a subproject or phase, will cost. Unforeseen problems in development and increases due to inflation can also be estimated, but their actual impact cannot be accurately predicted. Other factors, such as strikes and construction stoppages, usually cannot be predicted but it is good to have a financial reserve margin.

Every cost estimate for a project must be seen under the previously mentioned boundary conditions. However, one often has the impression that the estimated value once stated is seen as an absolute project commitment and is also handled in this way. This then often leads to unnecessary disputes between the project and the responsible project sponsor. However, to avoid any misunderstanding, it is explicitly stated here, and this is the main concern of this chapter, that a qualified cost estimate should be made with high credibility. All risk areas should therefore be clearly and unambiguously identified from the outset. On the other hand, the project team should be prevented from introducing fear surcharges into the estimate, thereby worsening the chances of competition. Involving the project manager in the sales and competitive process is the best way to avoid such actions.

In the case of fixed-price contracts, project reserves must of course be provided for unforeseen problems in consultation with management and the contract must be drafted wherever possible in such a way that factors beyond the project's control, such as inflation-related price increases, are not to be borne by the project. Efficient project monitoring also helps to identify deviations from the planned values at an early stage so that appropriate countermeasures can then be initiated as quickly as possible to avoid project losses; see also Sect. 9.4.

It should also not be concealed at this point that project costs are often underestimated both knowingly and unknowingly: knowingly under the premise "*win the contract first, additional costs will be claimed later*" and unknowingly due to lack of information, carelessness, etc. In both cases, this usually leads to very serious financing problems and sometimes even to project disasters, including project cancellation. This is often associated with a significant financial loss as well as damage to the image of the companies and/or authorities concerned.

10.1.2 Estimation Methods and Their Application

In principle, various methods can be used to estimate the development and manufacturing costs of a product. Probably the best-known and most widely used method is the detailed estimation method (*bottom-up*), in which the quantities required for specific tasks or activities, such as planned engineering and production hours, material quantities, and machine usage, are estimated and then multiplied by the hourly rates and overhead rates specified by the company. However, the detailed estimating method assumes that a detailed project outline and plan have already been prepared that clearly identifies in detail the project tasks to be completed.

But this is precisely the problem in the early project phases, when detailed planning documents are not yet available and cannot yet be available because the project is not yet conceptually mature and defined. Any attempt to make detailed cost estimates on the basis of incomplete planning documents will therefore lead to unrealistic results. In addition, the relatively large expense of a detailed cost estimate at this stage is not justified.

The decision-maker for a new project, who in many cases has to decide not only on the start of the project in question but often also on alternative projects, needs cost information that is as reliable as possible, especially in the early phases, on which to base his decisions. Pfohl and Wübbenhorst point out the dilemma between early decision-making and inadequate information: "*The general dilemma now is that the most important decisions have to be made at the beginning of the life cycle, but the level of information is lowest.*" [4]

Therefore, one should in early project phases alternatively resort to key figures that do not allow very detailed cost forecasts, but are sufficient for an initial preliminary cost statement. This does not refer to a so-called *thumbnail estimate* but to an approach based on technical parameters, such as weight specifications and performance data. Wildemann therefore speaks of an ... "*economically expedient forecast accuracy,*" [5] and Seldon mentions, in the context of considering the life-cycle costs of a project, mutatis mutandis: "*If one cost estimate is 50 percent higher than the other, this is of considerable importance for a decision, while a difference of 20 percent should be considered significant only if it is a sound estimate and a conventional project with a solid background; differences below 20 percent should first be carefully investigated before a decision is made.*" [6]

The various methods by which estimates are made have been summarized by Jones and Niebisch as follows [7]:

(1) Judgmental Methods
  - Expert Opinion
  - Qualified Guess (*Educated Guess*)
  - *Rough-Order-of-Magnitude* (*ROM*) estimation
  - Empirical
  - Analogies

(2) *Parametric Methods*
  - Cost Estimating Relationships (*CERs*)
  - Statistical
  - Probability
  - Mathematical Models

(3) Detailed methods (*Detailed or Bottom-Up Method*)
  - Work Package Costs
  - Engineering Estimates (*Industrial Engineering*)
  - Other detailed estimates (*Grass Roots*)

PMI proposes a three-point estimate which is little known in practice but can also be applied in a similar form for time estimation; see Sect. 9.3.4.5. The mean cost (*Cm*) can be determined according to the Beta distribution using the following formula [8]:

$$\mathrm{Cm} = \left(\mathrm{Cop} + 4\,\mathrm{Cpr} + \mathrm{Cps}\right) / 6$$

- Optimistic costs (Cop)

- Probable cost (Cpr)
- Pessimistic costs (Cps)

The decision as to which estimation method should be used in a specific case essentially depends on the current project phase and thus on the level of knowledge about the project. In the pre-project phase, i.e., before the concept phase (*see also Chap. 4*), initial cost estimates are often used on the basis of assessments, e.g., expert opinions and estimates of the rough order of magnitude (*ROM*) of costs. However, parametric estimation methods should be used as early as the start of the concept phase. The use of detailed methods can usually be initiated only in the second half of the definition phase, since sufficient planning documents, e.g., specifications, work plans, schedules, and the work breakdown structure, are available only then. Cost quotations for the subsequent phases are usually prepared based on detailed cost estimates.

METHOD/CRITERIA	APPLICATION PREREQUISITES	APPLICATIONS AREAS	APPLICATIONS BOUNDARIES (NOT RECOMMENDED)
I. JUDGEMENT			
• EXPERT OPINION • EDUCATED GUESS • ROM	• EXPERT/EXPERIENCE • ROUGH PRODUCT DEFINITION • ANALOG MATERIAL	• EARLY STAGE • NO-RISK SITUATION • INDEPENDENT CROSS-CHECKS • BUDGET ESTIMATES	• SUBJECTIVE • UNDEFINED ACCURACY • NOT USABLE FOR DETAILED PRICE NEGOTIATIONS
II. PARAMETRIC			
• CER • STATISTICAL • MODEL • COST FORMULA	• HISTORICAL DATA • REGRESSION ANALYSIS • CER MATERIAL	• CONCEPT COMPARISONS • BUDGET PLANNING • PROPOSAL EVALUATION • INDEPENDENT CROSS-CHECKS	• EXTRAPOLATION FROM DATABASES AND MODELS OFTEN DIFFICULT (IF DEFINITION MISSING) • ESTIMATE ACCURACY QUESTIONABLE
III. DETAILED			
• WP MANAGER ESTIMATE • WORK PREPARATION ESTIMATE • COST FORMULA	• SCHEDULE PLANNING • SOW & SPECIFICATION • DETAILED TECHNICAL MATERIAL • PRICE PROPOSAL	• HIGH-RISK SITUATIONS • PRICE NEGOTIATIONS	• EXPENSIVE AND TIME CONSUMING (DON'T USE IN EARLY STAGES) • LIMITED FLEXIBILITY • CAN LEAD TO COST INCREASE (IF TOO DETAILED)

Fig. 10.1 Estimation methods and their application according to Jones and Niebisch [7]. WP = Work Package; CER: Cost Estimating Relationship; ROM = Rough Order of Magnitude

Figure 10.1 shows the possible applications and limitations of the estimation methods. In practice, it is recommended to use both the parametric and detailed cost estimation methods in parallel for a certain time. The cost estimation models described in Sect. 10.3.2 also provide a good way to perform intra-project cost reviews. The parametric estimation methods, also referred to as "*top-down-estimation methodology*" and the detailed estimation methods, also referred to as "*bottom-up-estimation methodology,*" are two complementary methods that project managers should not neglect.

10.1.3 Detailed Work Package Cost Estimates

At the start of project phases C/D, detailed (*bottom-up*) cost plans must be available for each work package, because this is the prerequisite for efficient cost control; see also point III of Fig. 10.1. The costing of a work package requires that the quantity estimates of the cost categories of personnel, material, purchased parts, travel costs, etc. are first determined. Using factors such as hourly rates and overhead rates, the next step is to calculate the unit costs of a work package.

The preparation of a detailed project cost plan requires that the work breakdown structure (*WBS*) is defined down to the lowest level and that the relevant *work package descriptions (WPD)* are defined; see also Fig. 9.12. Furthermore, all technically necessary requirements, e.g., specifications, and completed schedule and sequence planning must be complete and available prior to the start of the estimation process. It should be noted at this point that work packages are, in principle, like small projects and must therefore be managed in accordance with the usual PM rules. On the basis of the aforementioned documents, an experienced project engineer (*WP manager*) should be able to estimate the required quantities (*hours, material, travel, etc.*) and the related costs. Figure 10.2 shows a schematic overview of the WP inputs and outputs.

In practice, the term "*Manpower and Cost Plan—MCP*" is used. Figure 10.3 shows the standard MCP template applied for ESA projects [9].

The MCP contains the personnel and cost categories vertically and time intervals, such as months, horizontally. The personnel and cost categories mentioned

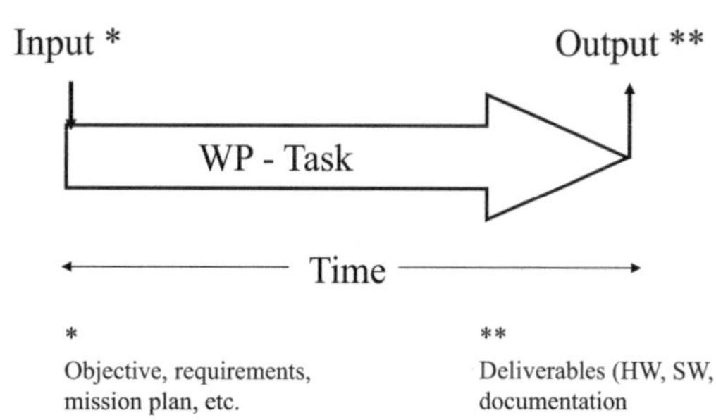

Fig. 10.2 WP input and output

PROJECT: LV SM-05	WORKPACKAGE MANPOWER AND COSTPLAN													WP-Titel: Working Tests	
PROPOSAL-Nr.: 4711	National Currency:	UAH				Issue date:	Oct 10, 2014							WP-Nr.: 320	
COSTCATEGORIES		Time periods after contract award (Labour Hours)												Conversion Rate	
		2015						2016							
		Jul	Aug	Sep	Oct	Nov	Dec	Jan	Feb	Mar	Apr	May	Jun	NC/€	0,05961
LABOUR EFFORT														HOURS	
Engineering		80	50	150	200	300	250	100	150	80				1.360	
Design					20	80	60							160	
Manufacturing				80	80	120	50							330	
Test Support		60	60	100	180	180	120	60						760	
Group 5														0	
Others														0	
TOTAL LABOUR		140	110	330	480	680	480	160	150	80	0	0	0	2.610	
LABOUR COST	Hourly Rate													NC	€
Engineering	1.250,00	100.000,00	62.500,00	187.500,00	250.000,00	375.000,00	312.500,00	125.000,00	187.500,00	100.000,00	0,00	0,00	0,00	1.700.000,00	101.337,00
Design	1.050,00	0,00	0,00	0,00	21.000,00	84.000,00	63.000,00	0,00	0,00	0,00	0,00	0,00	0,00	168.000,00	10.014,48
Manufacturing	1.470,00	0,00	0,00	117.600,00	117.600,00	176.400,00	73.500,00	0,00	0,00	0,00	0,00	0,00	0,00	485.100,00	28.916,81
Test Support	940,00	56.400,00	56.400,00	94.000,00	169.200,00	169.200,00	112.800,00	56.400,00	0,00	0,00	0,00	0,00	0,00	714.400,00	42.585,38
Group 5		0,00	0,00	0,00	0,00	0,00	0,00	0,00	0,00	0,00	0,00	0,00	0,00	0,00	0,00
Others		0,00	0,00	0,00	0,00	0,00	0,00	0,00	0,00	0,00	0,00	0,00	0,00	0,00	0,00
TOTAL LABOUR COST		156.400,00	118.900,00	399.100,00	557.800,00	804.600,00	561.800,00	181.400,00	187.500,00	100.000,00	0,00	0,00	0,00	3.067.500,00	182.853,68
NON LABOUR COST															
Material			300.000,00	200.000,00										500.000,00	29.805,00
External Services					100.000,00	100.000,00								200.000,00	11.922,00
Travel & Subsistance														0,00	0,00
Insurance						500.000,00								500.000,00	29.805,00
Others														0,00	0,00
														0,00	0,00
														0,00	0,00
														0,00	0,00
														0,00	0,00
TOTAL NON LABOUR COST		0,00	300.000,00	200.000,00	100.000,00	600.000,00	0,00	0,00	0,00	0,00	0,00	0,00	0,00	1.200.000,00	71.532,00
TOTAL COST (NC)		156.400,00	418.900,00	599.100,00	657.800,00	1.404.600,00	561.800,00	181.400,00	187.500,00	100.000,00	0,00	0,00	0,00	4.267.500,00	
TOTAL COST (€)		9.323,00	24.970,63	35.712,35	39.211,46	83.728,21	33.488,90	10.813,25	11.176,88	5.961,00	0,00	0,00	0,00		254.385,68

Note: Only Standard Hourly Rates to be applied including Overheads, but without Profit and Risk Margins

Fig. 10.3 Work package personnel and cost plan with fictitious figures (Source: ESA)

below are to be understood as examples. As a rule, companies define the categories on an industry-specific basis.

(1) Personnel manpower and costs
  - Management
  - Development
  - Work planning
  - Manufacturing
  - Test
  - Others
(2) Other costs
  - Material
  - Purchased parts
  - IT and software
  - Services
  - Facility rental
  - Travel
  - Others
(3) Total costs of (1) and (2)

The definition of the cost categories must in any case be carried out in accordance with the respective cost structure of the company so that the individual categories can be clearly assigned to the hourly and overhead rates used in the company. If, for example, mixed hourly rates are used for cost calculation for the individual departments in a company, it may be appropriate to list these individual departments in question under the main category of personnel costs. In other companies, it may be customary, for example, to differentiate the hourly rate according to junior and senior engineers, and this should be taken into account when determining the cost categories.

From the horizontal breakdown, the costs per time interval can be obtained. This is an important prerequisite for subsequent cost and performance control; see also Sect. 9.4.5. In addition, the time breakdown of costs is an ideal basis for developing the cash flow plan of a project. The time breakdown of costs should correspond to the cost control cycle. For example, if cost control is performed monthly, a monthly cost breakdown is also appropriate. A quarterly cost breakdown would not be sufficient in such a case. However, it may make sense not to immediately break down costs into monthly time intervals for work packages that, according to the schedule, will not be started for 12 or more months. Rather, one can initially make a quarterly or semi-annual breakdown and then change this to a monthly breakdown at a later date when the level of knowledge is greater. This also corresponds to the management principle of the "*rolling wave.*"

The representation of WP costs (WPC) over the time axis leads to the well-known term "*manpower mountain*" and this naturally includes both men and women. If one aggregates all WPCs in accordance with the WBS, one obtains the manpower mountain for the entire project at the top project level; see Fig. 10.4. The

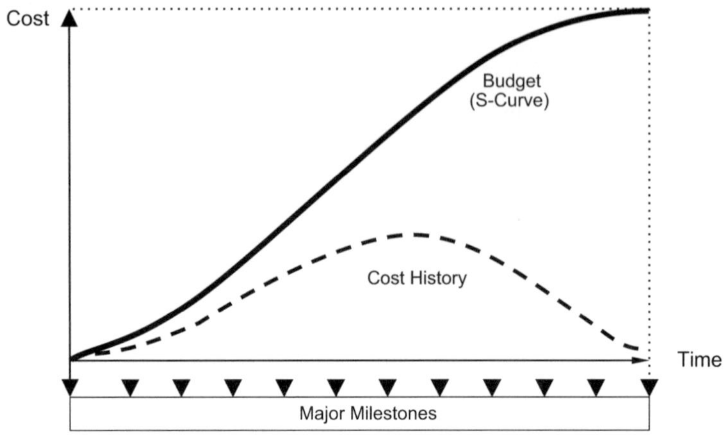

Fig. 10.4 Representation of the accumulated project cost budget

upper S-curve shows the accumulated total costs (*i.e., the funding profile*) of a project. The S-curve is also needed in conjunction with the main milestones for EVM analysis; see Fig. 9.38.

In summary, the representation in Fig. 10.4 is the prerequisite for the following aspects:

(1) Preparation of payment and financing plans; see also Sect. 10.1.5.
(2) Information on the company's workload planning; see also Fig. 17.3
(3) EVM basis; see also Sect. 9.4.6 and Fig. 9.38

The process of detailed cost estimation is usually divided between engineers and business people. The engineers estimate the required quantities of resources and the business people multiply them with the customary cost and overhead rates and calculate the total costs for a work package. Figure 10.5 shows an example of the calculation of the hourly rate including all allocations.

If possible, gross hourly rates may be used, i.e., hourly rates to which the company's internal surcharges have already been applied. This has the advantage that the project manager, and/or the business manager of the project are in the position to carry out full cost accounting at the work package level. Taking into account the profit markup, the (*sales*) price can already be determined at the work package level; see also Fig. 10.5. Competition often forces reduction measures and in this context a price comparison with external suppliers can then be carried out. In this regard, the term "*cost*" is used internally and the term "*price*" is used externally.

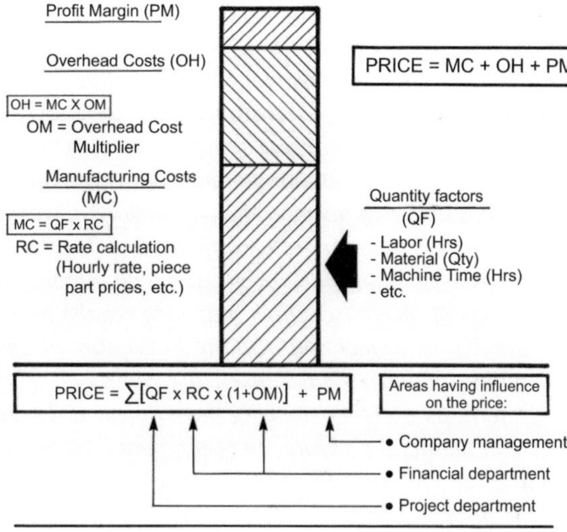

Fig. 10.5 Composition of the hourly rate

10.1.4 Target Costs and Basic Rules

Companies are usually in competition with other companies, increasingly so in a global environment; see also Chap. 20. This usually leads to target cost agreements: *target costing*. For example, if the product is new, target costs must be defined for development and production costs. The sales price achieved includes the unit production costs and the development costs allocated to them, as well as the profit markup. Parametric costing is a good way of determining target costs; see also Sect. 10.3. In this context, it is also pointed out that increasing development costs (NRC) may reduce production costs (RC) if new cost-saving production technologies can be developed during the development phase. This will be explicitly discussed in the context of DTC considerations in Sect. 10.5.3. The total target costs determined must be broken down into the subsystems or components of the project.

The basic rules of cost or price determination include the following 10 factors that must be considered in whole or in part:

(1) The year in which the estimate was made (*cost/price basis*): *The price or cost basis determined (baseline cost) is in principle only valid for the year in which the estimate was made because in the following year, there is usually a Cost/ price increase due to inflation.*
(2) Application of a price escalation clause (*escalation formula*): *Inflation-related cost/price increases can be determined by an appropriate index; see also Sect. 10.1.5.*
(3) Technology year (*use of old or new technology*): *Quite significantly, costs are also influenced by the state of the technology. Old technology is usually more expensive. The pocket calculator is an apt example. Within only one or two decades (70s to 90s), there was a drastic price reduction.*

(4) Currency: *The agreed currency is an important factor—it should be as stable a currency as possible. If several partners are involved in the project, an exchange rate arrangement must be made.*

(5) Exchange rates: *Often a strong lead currency is agreed upon for budget planning and subsequent cost control. Other currency packages are then converted using the official exchange rates. However, officially available exchange rates do not take into account the purchasing power of the country concerned. The problem is therefore often difficult to solve. One way out can be a compensation transaction (bartering), in which mutual claims are settled in money by means of offsetting. Often, this also results in a barter transaction (goods for goods). In connection with the International Space Station (ISS), the participating countries have taken the approach of manufacturing their modules internally and then docking them in space. This means that there is no significant financial exchange between them, except for the in-orbit assembly activity. The prerequisite, of course, is that all interfaces between the modules must be perfectly defined.*

(6) Bonus/malus regulations (*contractual penalties and bonus payments*): There *are corresponding regulations in many contractual agreements. Contractual penalties may be incurred in the event of technical deviations from the specification or schedule delays. The same applies to cost reimbursement contracts in the event of cost overruns. Conversely, such provisions may also apply to a premium arrangement. However, it depends on the contractual provision, because early delivery is not necessarily an advantage for the customer.*

(7) Planned reserves (*margins*) for unforeseen cases: No planning is precise enough to exclude errors. This is why a planning reserve is necessary. However, failures can also occur with technologically demanding tasks. For this, too, it is good to have a buffer, e.g., 10%. An ironclad schedule reserve for "*work around solutions*" is overtime and weekend work in exceptional cases.

(8) Task distribution key: *In joint projects, it is customary to agree on a task and contribution key at the start of the project. The competence and political interests of the companies and/or countries involved play a decisive role here. The distribution key may also be expressed in the structure of the budget.*

(9) Contract conditions (*Cost reimbursement, FP, FFP, other*): *Contract types are also an important factor in cost determination. For a fixed price (FP) or firm fixed price (FFP), the requirements are usually much tougher than for a cost reimbursement price because errors in estimating can result in losses. This is especially true if the contract and the specifications do not allow any room for deviation.*

(10) Cost management: *The project manager must ensure that the cost estimation is carried out by experienced persons with knowledge of the methods and that all specified rules of the company or customer are adhered to.*

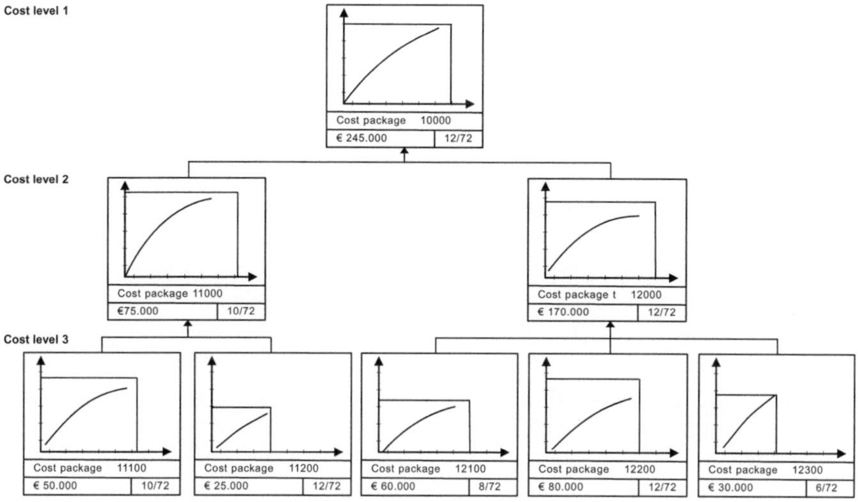

Fig. 10.6 Overall projection (Source: Schmitz)

10.1.5 Cost Breakdown and Pricing

The MCP shown in Fig. 10.3 represents the smallest element of the project cost planning. If a project consists of several work packages, the individual WP cost plans (*MCPs*) of a project are combined in stages based on the WBS. This results in a corresponding summary of the individual budgets at each WBS level, i.e., at the project, system and subsystem levels of the WBS; see also Fig. 10.6. For large projects, the use of appropriate software is helpful, since changes in particular would otherwise degenerate into a manually time-consuming calculation activity.

The detailed staffing and cost planning described in Sect. 10.1.3 is the basis for preparing cash outflow, payment, and financing plans. Based on the detailed WP costing, a cash outflow plan shall be prepared showing the monthly cost requirements. For externally funded projects, a financial plan shall be prepared in addition to the cost plan. The financial proposal for a project is based on cost estimates prepared in detail. However, in the financial offer, by adding the expected profit rate, the project price is the decisive variable; see also Fig. 10.5. The costs summarized in the total cost plan of a project (*Fig. 10.6*) represent the financial means necessary to carry out the planned project performance (*execution of project activities*). The funds required in terms of time (*material, purchased parts, etc.*), as well as the general services of the company can be taken from the MCPs of WPs; see also Fig. 10.3. The profit margin to be added is usually determined by management. In the case of public contracts, the profit is often fixed in advance. In the case of financial offers, the following aspects must be given special consideration:

- Pricing
- Profit surcharge

- Payment mode (*plan*)
- Funding
- Price escalation clause (*inflation index*)

Pricing is of great importance in the execution of a leveraged project. The contractor often takes a large risk on customer projects and will make the profit markup and any cost reserves dependent on pricing. The topic of pricing is discussed in detail in Sect. 13.1.2. The payment schedule is an important factor for project execution because the P&L (*profit and loss*) consideration for the project is strongly influenced by the payment schedule. For example, if payments are made annually or only after completion of very specific milestones, the interim obligations (*salaries; material procurement, subcontracting, etc.*) may need to be pre-funded. The costs of pre-financing and the possible procurement of external funds (*interest*), must also be taken into account in the financial offer.

The financing of large and complex projects is often another problem faced by project managers and the finance departments of project-processing companies. Project financing is closely related to the mode of payment already mentioned. Especially in turnkey construction projects *(turn-key contracts),* the selected manufacturer (*contractor*) usually includes only a down payment (*e.g., 5–10% of the contract volume*) when the contract is signed. The remainder is reimbursed in installments (*milestone payment*) or after project acceptance and handover to the customer. Often, the reimbursement is made less a final payment, which is paid only after an agreed period of time has elapsed (e.g., *acceptance plus 1 year*). Such long-term project financing, often extending over several years, requires quite substantial financing costs. In this context, however, there are often other issues to be considered, such as credit security for foreign transactions (*guarantees*) and currency fluctuations that may occur.

The situation is completely different in cases where the monthly costs can be billed immediately (*as incurred*). In the case of technologically very complex and lengthy projects in the range of 10 years, which involves a particularly high risk, government clients may grant industrial contractors the option of staggered contract conditions and subsidies.

Price increases are another plague of project management that must be taken into account in the financial offer. Any price offer can only be bindingly determined at the hourly and material rates valid at the time of the cost estimate. The price development of the following year or years cannot be predicted with absolute certainty for any country. However, if the customer insists on an absolute fixed price *(firm-fix-price),* a corresponding inflation surcharge must be added to the estimated price, which is based on the price of the year in which the cost/price estimate was made (*bid year*). In such cases, however, the risk of inflation-related cost increases lies fully with the contractor. If the inflation surcharge applied increases, the contractor will incur correspondingly high losses. However, many contractors are not readily prepared to take on such a risk. Particularly in the case of large and long-term international projects, where the inflation rate on the one hand and fluctuations in the exchange rate on the other are factors that are difficult to calculate, the contractor must hedge accordingly. In many such cases, therefore, a fixed price with an escalation or price escalation clause is agreed between the client and the contractor in order to be able to invoice inflation-related price increases separately. Hoss describes the standard formula for inflation adjustment for ESA projects as follows [10]:

$$Pn = Po\left(a + b\,X\,Gn / Go + c\,X\,Mn / Mo\right)$$

It means:

Pn = revised price for billing year or billing time "n."
Po = agreed price in the contract year
a = non-inflationary price share; quantified as 10% (0.1) in the example.
b = price share for salaries; in the example quantified as 60% (0.6)
c = Price share other direct costs, e.g., materials, supplies, etc.
Sn/So = Salary index
Mn/Mo = Material index

For ESA member states, generally, the salary and material index defined by the official Statistical Offices of the member states is accepted. Although escalation equations are used in slightly modified forms in the various sectors due to different tasks, the basic principle is clearly described by the aforementioned formula. In order to avoid risks due to exchange rate fluctuations, the safest solution is to make a contractual arrangement whereby the project is offered and also invoiced in the customer's national currency. Alternatively, offers in stable foreign currencies or insurance against exchange rate fluctuations are common.

10.1.6 Cost Estimation—Responsibility

In addition to the specified performance and schedules, costs are also an important factor to be considered; see also Fig. 9.2. All project decisions are in the area of conflict between these three factors. The complete fulfillment of the project objective is achieved when the promised product (*deliverable item*) can be handed over to the customer in good quality (*form fit and function*) and on time and within the planned cost framework. Realistic cost planning therefore plays a very special role.

In practice, the overall responsibility for project cost estimation is often assigned to a project manager who carries out the estimations of required resources in close cooperation with technicians, taking into account empirical values from completed projects (*benchmarking*). The project manager should therefore have basic technical knowledge in addition to commercial training. Sometimes the estimation of project costs is led by the PM or an experienced technician, who should have basic commercial knowledge, e.g., accounting knowledge. A double or postgraduate degree is advantageous in both cases. The third and most optimal variant is when the responsibility of cost estimation is assigned to an experienced industrial engineer(s).

The employees responsible for cost estimation should, if possible, also participate in the analysis of completed projects in order to be prepared for future tasks in the area of cost estimation. For large projects, there is often a move to integrate the individual project staff into a team complex, i.e., to a project management office (*PMO*); see Sect. 5.2.5. This usually leads very quickly to genuine teamwork, in which employees from different disciplines, in particular, e.g., engineers and economists, work closely together and exchange their experience; see also Chap. 15.

10.2 Cost Analyses and Databases

10.2.1 Evaluate Completed Projects

Cost analysis is an important prerequisite for cost estimation. Although cost analysis and cost estimation are closely related, they are fundamentally two different processes. Cost analysis means reviewing and evaluating cost data from projects that have already been completed, and building on this to create a cost database. On the basis of a benchmarking process, reference values can then be determined that serve as a reference for cost estimation. Cost estimation is a prediction of the costs likely to be incurred to perform a very specific task. The process of cost analysis involves reviewing and evaluating cost information from all projects performed in the past. Quotations, cost reports, and the results of post-calculations serve as the basis.

The author gave a presentation in September 2011 in which he presented the problem of cost increases using the example of 10 randomly considered projects [11]. There were increases up to 10 times the estimate. The following is a list of what caused the cost increases:

- Underestimated complexity
- Management and planning errors
- Political squabbles and construction stop
- Lengthy approval processes
- Increased cost of materials
- Poor cost determination and control.

But what is to be done when such drastic deficiencies are identified? The Berlin Airport and Hamburg Elbe Philharmonic Hall projects also show significant cost increases; see also Sect. 3.2. The following is a summary of recommendations [11]:

- Cost-conscious project management
- Precise project definition
- Systematic and thorough planning
- Reliable cost determination
- Professional cost management
- Use of proven cost estimation tools
- Good training and preparation of cost engineers
- Lessons-learned measures

- Implementation of cost databases (*empirical values of completed projects*); see also Sect. 10.2.3.

Cost analysis is a very important and informative task, which in practice, however, often turns out to be very problematic, as access to useful data material is often very difficult. Above all, it is important that the cost analysis is not carried out according to purely accounting aspects (*commercial post-calculation*), but in close connection with the development history of the project and lesson-learned reviews with the involved engineering teams. Changes in project objectives and planning must also be given special consideration here. However, if such changes have not been documented thoroughly enough by the project, cost analysis becomes extremely difficult and the results are often very unreliable.

10.2.2 Analyses Improve the Quality of Cost Estimation

Careful cost analyses of completed projects provide the cost estimator with the necessary information to reliably estimate the costs of future projects. The cost estimation of new and very complex systems also increases in quality if one can fall back on historical information. In most cases, such new systems consist largely of existing components, or analogies can be made to similar components. A detailed cost analysis is therefore of the utmost importance for the future. For this reason, it is extremely important for authorities and companies to create corresponding cost information in the form of a database to capture long-term information.

For future cost estimations, it is important to have empirical values from projects that have been carried out in the past. Experience values can then be used to develop *Cost Estimation Relationships* (*CERs*). This provides a solid basis for future cost estimates. CERs are particularly well suited to cost estimation during the early phases of a project because, in the early phases of a project, the estimator has little information available.

The development of CERs is industry-specific and the results are usually not publicly available. Thus, only the principle can be explained here, which of course also applies to other industries. For example, D. Kölle has published a comprehensive manual for space transportation systems, from which the CER example shown in Fig. 10.7 for the calculation of the development costs of a rocket stage was taken [12]. In 2014, Kölle demonstrated at a DGLR workshop in Augsburg the use of CERs which he had developed for the Ariane 6 launch vehicle program, consisting of a combination of liquid stages and solid motors.

Batchelder et al. point out that "*weight changes alone, however, do not always lead to useful cost results, and often additional variables must be included.*" [13] They illustrate this with a practical example, contrasting cost, weight, output power, and frequency for 10 different aircraft radios, but failing to arrive at a useful, generally applicable cost estimating equation as a function of any of the factors. It is only by classifying the devices into VHF, UHF, and HF frequency families that three different, but simple, linear cost estimating equations can be derived.

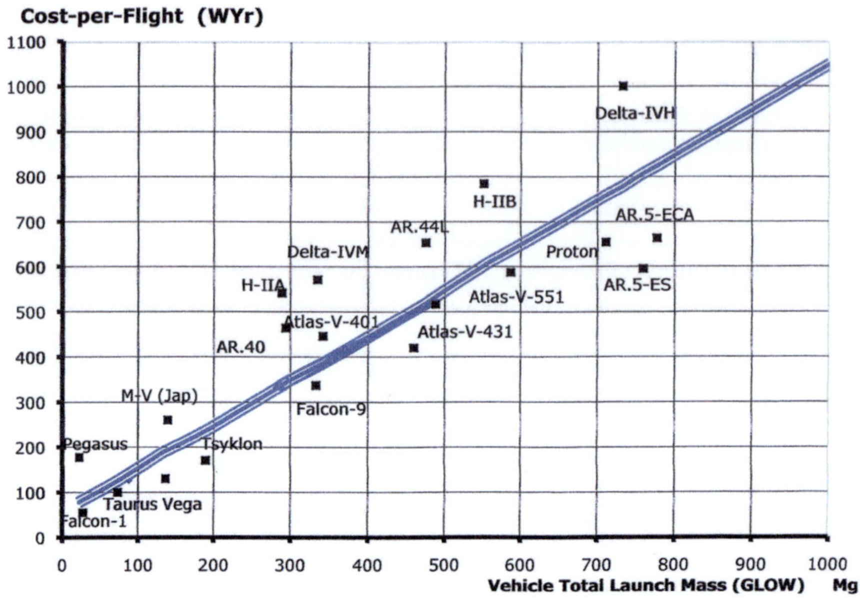

Fig. 10.7 Development costs CER for rocket stages (*Source: Dietrich E. Kölle*)

The *Unmanned Spacecraft Cost Model of* the USAF, on which several companies have collaborated, holds a larger collection of cost estimation equations for professional cost estimators of satellite projects; see also Fig. 10.8 [14]. When using these equations, however, the boundary conditions must be carefully observed, because the CERs are only applicable if the characteristics of the new device to be developed or produced match the characteristics of those devices that were the inspiration for the establishment of the CER equations. In addition, the respective year of validity must be taken into account as well.

The above data sheet allows the determination of the specific costs in thousands USD (*Y-axis*) depending on the reference value EPS x BOL Power (*X-axis*) as well as the boundary conditions specified below:

1. The equation is valid for the specified year of development
   a) EPS, power supply unit weight (*lbs*)
   b) BOL, electrical power of the power supply unit at the start of operation (*watts*)
2. 17 Comparison projects
3. Static standard deviations
4. Variable deviations
5. Notes
6. Y-axis: Development costs in FY79 USD
7. X-axis: system weight x BOL power
8. BOL—Beginning of Life in orbit

ELECTRICAL POWER SUPPLY
SYNCHRONOUS ALTITUDE OR ABOVE

NONRECURRING COST CER

1. EQUATION: $Y = 2419.43 + .02295\ X^{.97}$
   Where: Y = Nonrecurring FY79 $ in Thousands
   X = EPS Weight (lbs) x BOL Power (watts)

2. SAMPLE SIZE: 17

ATS-F	ATS S/G (D,E)	ATS S/S (B,C)	DSCS II
DSP I	FLTSATCOM	I-3	I-4
IDCSP (DSCS I)	IDCSP/A	MARISAT	NATO III
P78-2 (SCATHA)	SMS	SYNCOM	TACSAT
VASP			

3. MEASURE OF STATISTICAL FIT:

Coefficient of Determination (R-Square):	.66
Standard Error of the Estimate (SE):	2620.73
F Statistic:	29.46

4. RANGE OF INDEPENDENT VARIABLES:

$228.40 \le X \le 912824.00$ (lb-watts)
$9.10 \le$ weight ≤ 619.60 (lbs)
$25.10 \le$ power ≤ 1640.00 (watts)

5. REMARKS: Vela and Lunar Orbitor were deleted from the sample because each had a design life of 6 months or less.

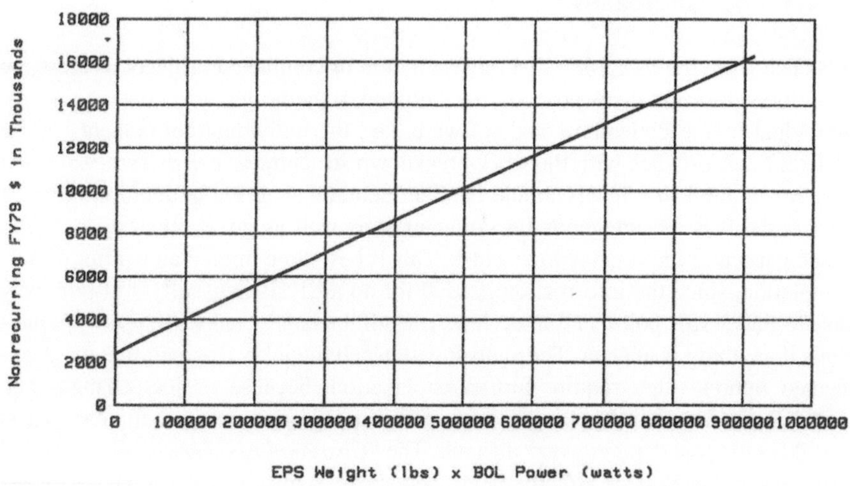

Fig. 10.8 CER for the evolution of the power systems of a satellite in an Earth-synchronous orbit of 36,000 km altitude (Source: SSCAG)

This makes it clear that costing on the basis of CERs is not black magic, but a sound method. However, the prerequisite is that the CERs consulted are based on a systematic analysis of completed projects, and this requires a corresponding investment in time and money. The aerospace CERs developed in the USA are generally only accessible to certain individuals and organizations for competitive reasons [14, 15]:

Therefore, if necessary, companies should carry out in-house regression analyses based on their products in order to develop corresponding CERs.

The establishment of cost estimation relationships (*CERs*) is, of course, not limited to the space sector, but lends itself to many areas. The specification of prices per kg is also a common praxis. One just has to be careful not to compare apples with oranges. For example, the specific car-kg price for a luxury-class vehicle is naturally much higher than that for a small vehicle. In addition, the boundary conditions (*see Fig. 10.8*) play an important role. When analyzing costs and developing CER equations (*regression analyses*), one must pay particular attention to the product description and possibly also the make. In addition, CERs for development (*non-recurring costs*) and production costs (*recurring costs*) must be estimated separately.

CERs are the prerequisite of cost estimation models. The development of cost estimation models began in the 1970s. The best-known and most widely used cost estimation model was developed by RCA in 1957 and later marketed under the name PRICE; see also Sect. 10.3. In the aerospace industry, special estimation methods have long been used to determine the development and production costs of a system or component as a function of very specific technical parameters.

10.2.3 Cost Databases

It is clear from the foregoing that the evaluation of completed projects increases in importance if the analysis process is not limited to the project level but also covers individual building elements and software, i.e., the individual components of the system. In accordance with the work breakdown structure (*see also Sect. 9.3.3*), all elements at the lower levels should be identified and analysed in detail in terms of their costs. It is advantageous for companies, as well as government agencies, to attach particular importance to this task. This is best done immediately after project completion, since the information needed for an analysis is usually still relatively easy to find at this point, and interviews can still be conducted with project participants if anything is unclear. The analyst will usually quickly discover that the documented figures often require further explanation, because project changes and occasionally incorrect bookings often make an objective post-calculation that is useful for the cost database very difficult. The "*Cost Analysis Handbook*" of the US Army prescribes that, as already mentioned above a, information on the characterization of the respective analysed construction unit must of course be recorded in addition to the cost data [16]. According to this, the information is to be divided into four groups:

(1) *Product characterization:* description of the component, including all technical and physical characteristics
(2) *Schedule:* Key milestones, schedule, and delivery dates

(3) *Expenses:* Summary of all expenses for development, testing, production, and operation of the component described in 1 (*quantities and/or costs*)
(4) *Additional information:* Description of quantity/cost documents evaluated; are they actual or bid costs; under what contract terms was the project completed (*fixed price, cost reimbursement, etc.*)?

The European Space Agency (*ESA*) has started to build a satellite component database that will collect technical and financial data for the system, subsystem, component, and individual part levels [17]. In this context, the minimal scheme for cost databases defined by Schelle should be mentioned [18]:

(1) Short title
(2) WP or component description
(3) Name and address of the provider, if the service was purchased
(4) Start date of the work package
(5) WP duration
(6) Place of execution and executing organization
(7) Quantity of purchased units
(8) Hours spent by category
(9) Cost of materials by category
(10) Equipment costs
(11) Other costs
(12) Special terms and conditions (*type of contract, reliability requirements, etc.*); price for purchased parts
(13) Other data, e.g., weight, volume, and general performance data

10.3 Parametric Cost Estimation Models

10.3.1 From CERs to Cost Estimation Models

Cost estimation is prediction based on experience. For a high-quality cost estimate, it is important that the estimator uses proven methods and that he has product knowledge. Past experience (*database*) is another important requirement. For parametric cost estimates using CERs (*Cost Estimating Relationships*) and/or cost estimating models, the availability of historical data plays a particularly important role. The use of CERs requires that a relationship exists between the component for which a cost estimate is developed and the reference components used for CER development. In other words, the related relationship must first be found; see also Fig. 10.8 (*item 2 seventeen comparison projects*). In this case, the weight and electrical power are the input parameters that also define the validity range.

Weight and electrical power are the best-known parameters for estimating costs on a statistical basis. However, it is obvious that other parameters, such as year of technology, project duration, and a variety of commercial conditions also play a role. For example, CERs that were created on a price basis of the year 2000 must be

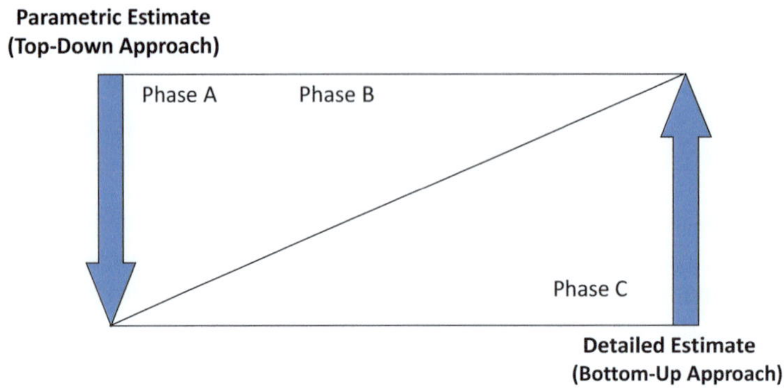

Fig. 10.9 Application of parametric and detailed estimation methods

adjusted to the respective year of application by means of an inflation index. Since technologies also change rapidly in some cases, the CER equation inevitably changes as well.

The parametric cost estimation methodology has gained considerable importance in recent decades, firstly because it is the only viable method available in the early phases of projects, and secondly because it can be used more and more as an independent estimation method to verify detailed cost estimates in the subsequent phases; see also Fig. 10.9. The detailed estimation method (*bottom-up*) is also called *Baseline Cost Estimate* (*BCE*) and the parametric estimation method (*top down*) is called *Independent Parametric Cost Estimate (IPCE)* [19]. The IPCE information is collected with a special format that shows the physical and technical data of the component; see also Schelle [18].

Parametric costing can be used in all project phases with different intensity. As already mentioned, a detailed manpower and cost plan (*see also Fig. 10.3*) must be prepared at the beginning of a particular project phase. However, it is usually not possible to prepare a detailed cost estimation for phases C and D at the beginning of phase A. This is due to lack of preliminary technical work and a mature work breakdown structure (*WBS*). On the other hand, however, preliminary LCC estimates are needed at an early phase for preparation of a ROI analysis. Therefore, parametric cost estimations are of particular importance at the beginning of the project. There is then a gradual transition to detailed *bottom-up* (*BU*) costing. But also, in the further course of the project, accompanying parametric cost determinations can be useful for the verification of internal BU cost determinations and offers from subcontractors and suppliers. Figure 10.9 shows how a successive transition from parametric to detailed cost estimation can be made. Conclusion: parametric costing is not a competing but a complementary estimating method and sometimes it is still actively used in phase C.

In 1982, the *Defense Contract Audit Agency (DCAA)* in the USA decided to conduct price negotiations on the basis of IPCEs [20]. Of course, this is only possible if all agencies and companies involved have proven estimating methods and experts.

It also indicated that contractors must consider the following six criteria in future bids:

(1) The cost estimation relationships (*CERs*) used must be logically structured!
(2) The CER information used must be verifiable,
(3) There must be a sufficient statistical correlation between the parameters used!
(4) The CERs used must be applicable and accurate!
(5) The parametric estimation systems used (e.*g., CERs or a SW-based cost estimation model*) must be easily verifiable!
(6) Before submitting an offer, there should be complete compliance with the above criteria!

The aforementioned six criteria provide a useful guideline for the development and use of CERs.

10.3.2 Universal Cost Estimation Models

In 1979, several engineers and business economists joined forces in the USA under the leadership of Frank Freimann, an industrial engineer and former RCA employee who developed what is probably the best-known cost estimation model (*RCA-PRICE*), and founded an international association for parametric analyses, the *International Society of Parametric Analysts* (*ISPA*), in 1979. The main topics of the association were the developments and/or improvements of parametric cost estimation models, risk analysis, life cycle cost (*LCC*), and *design-to-cost* (*DTC*) methods. The author chaired the *European Chapter for* several years. Since 1990, there has also been a German chapter. Members of the German ISPA are employees of authorities and companies in the sectors telecommunications, automotive industry, aerospace, information technology, and defense technology. ISPA merged with SCEA (*Society of Cost Estimation and Analysis*) in 2012 and is now known as ICEAA (*International Cost Estimating and Analysis Association*).

With the ongoing refinement of parametric cost estimation methods and the availability of powerful IT assets, the development of SW-based databases and cost estimation models was obvious. In the meantime, many technology companies worldwide apply such models. But also, government agencies use cost estimation models and/or databases. While many cost estimation models are only applicable to very specific product groups, known as *data-full* models, there are also the universally applicable models, known as *data-empty models*. "*Data-Full models*" are based on regression analyses of existing data (*weight, energy consumption, etc.*), while data-empty *models* have standard equations that need to be enriched and calibrated with real data.

Most cost estimation models can only be used for very specific product groups. Therefore, in the early 1970s, the RCA company successfully dedicated itself to the development of the first universal cost estimation model.

As mentioned before, the former RCA employee Frank Freimann was the initiator of the PRICE programs. "*The basic idea and philosophy of the PRICE models was that they should be universally applicable and easy to use, and at the same time guarantee fast access*" [21]. Here is a model selection:

- PRICE-H: A model for determining the hardware costs of individual components
- PRICE-L: A model for determining the life cycle costs of a project
- PRICE-M: A model for determining the cost of microelectronic components
- PRICE-S: A model for determining the cost of IT software
- PRICE-SL: A model for determining the lifecycle costs of IT software.

In the past decades, further parametric cost estimation models have been developed in the USA and in Europe; see also Sect. 10.3.4.

For several decades the programs developed and updated by RCA have been available on the market. RCA PRICE developed the aforementioned universally applicable PRICE cost estimation models based on knowledge of cost estimation acquired in practice [22].

In principle, the CERs that have been used in US industry for some time (*see Fig. 10.8*) were integrated to form a cost estimation model. A significant extension is that additional factors, such as the area of application (*platform*), the complexity for development and production, the year of the technology and deadlines, have been included in addition to the previously known CER factors. The initiator Frank Freimann, who had already been working on this task since 1957, initially introduced the following application areas (*platform categories*): Ground Equipment, Vehicles, Aircraft, Space Equipment, and Manned Spaceflight Equipment. Ground equipment was given a code of 1 and manned spaceflight equipment the highest code, namely 2.5. For the remaining categories, codes between 1 and 2.5 were defined. This provided the basis for evaluating different system requirements (*specification level*). The rating of 1 to 2.5, which is by no means linear but exponential in the calculations resulted from the different system requirements.

A uniform measure of the complexity of development and production was then defined in corresponding complexity scales; see also Figs. 10.10 and 10.11.

There are two parameters that determine the development costs: a), the required development effort (*scope of design effort*) and b) the experience of *the* personnel appointed (*experience of personnel*). The complexity factor derived from these parameters ranges from 0.2 to 3.1. It is logical that it will be very expensive to assign a highly complex task to an inexperienced team, which corresponds to factor 3.1. The opposite is true if a very experienced team is assigned a very simple task, which corresponds to a factor of 0.2. On the other hand, if you assign a team with average experience to a routine task, you are in the nominal range (*factor of 1 to 1.2*). Only in the case of very complex tasks does the factor increase beyond the nominal range. Even if the factors developed by Freimann should deviate in individual cases, the fact remains that the effort and the personnel qualification must be considered together.

EXPERIENCE OF PERSONNEL / SCOPE OF DESIGN EFFORT	Extensive experience, with similar type designs. Many are experts in the field, top talent leading effort.	Normal experience, engineers previously completed similar type designs	Mixed experience, some are familiar with this type of design, others are new to job	Unfamiliar with design, many new to job
Simple modification to an existing design	.2	.3	.4	.5
Extensive modifications to an existing design	.6	.7	.8	.9
New design, within the established product line, continuation of existing state of art	.9	1.0	1.1	1.2
New design, different from established product line. Utilizes existing materials and/or electronic components	1.0	1.2	1.4	1.6
New design, different from established product line. Requires in-house development of new electronic components, or of new materials and processes	1.3	1.6	1.9	2.2
Same as above, except state of art being advanced or multiple design path required to reach goals	1.9	2.3	2.7	3.1

Fig. 10.10 Definition of development complexity (Source PRICE)

In the case of manufacturing complexity, a distinction is made between mechanical and electronic complexity. The values vary between 4.75 (*ground antenna*) and 10.848 (*VLSI of a manned space flight transmitter*). The scale used is open upwards, since new technologies may well exceed the values given in Fig. 10.11. These are table values which have to be adapted to the respective product by calibrations.

Important input data are, for example, weight, volume, number of prototypes, quantities of the series as well as start and end dates for development and production. Other factors can optionally be added or determined by the program. Figure 10.12 shows an input scheme used in practice.

Cost estimation models are used by many technology companies and government agencies in the US and Europe, such as ESA, the UK MOD, and NASA. Government agencies, such as the DOD and supporting organizations, also apply cost estimation models with success. A particular advantage of IT-based cost estimation models is their speed. Even complicated technical changes can be reviewed for their financial impact in a very short time, often in minutes.

The Berlin-based company 4cost-aces has developed another cost estimation model that can also be used universally and is successfully applied in various industries in Europe. This model must also first be adapted to the environmental

Electronic Equipment Description	Types of Components	1.0 Ground	1.4 Mobile	1.8 Airborne	2.0 Space	2.5 Manned Space
ANALOG:	Discretes	6.264	6.857	7.854	9.597	10.063
	IC's	6.415	6.950	8.090	9.780	10.098
Receivers, OP Amps, Audio, Video,	LSI	6.742	7.274	8.267	9.894	10.184
RF, Servo Drive, etc.	Hybrids	6.919	7.451	8.446	10.076	10.365
	VLSI	7.096	7.629	8.626	10.259	10.547
DIGITAL:	Discretes	6.032	6.742	7.820	9.309	9.710
	IC's	6.182	6.888	7.940	9.420	9.857
Gates, Registers, Buffers,	LSI	6.474	7.174	8.232	9.564	9.963
Counters, etc.	Hybrids	6.648	7.348	8.411	9.743	10.144
	VLSI	6.823	7.523	8.590	9.924	10.325
DISPLAY WITH CRT:	Discretes	5.920	6.681	7.662	8.871	9.664
	IC's	6.094	6.824	7.771	9.008	9.801
TV, Terminals, Radar Consoles,	LSI	6.410	7.139	8.087	9.184	9.936
Test Units, etc.	Hybrids	6.582	7.312	8.262	9.361	10.113
	VLSI	6.755	7.486	8.438	9.537	10.290
DISPLAY NO CRT:	Discretes	5.801	6.535	7.638	8.841	9.527
	IC's	6.000	6.681	7.727	8.985	9.638
L.E.D.'s, Liquid Crystal, Indicators,	LSI	6.300	7.000	8.019	9.165	9.744
Controls, etc.	Hybrids	6.473	7.173	8.197	9.344	9.924
	VLSI	6.648	7.348	8.375	9.524	10.104
TRANSMITTER:	Discretes	6.470	7.218	8.090	9.692	10.252
	IC's	6.650	7.368	8.245	9.813	10.369
TV, Radar, Communications,	LSI	6.801	7.516	8.397	9.930	10.481
NAV AIDS, Laser, etc.	Hybrids	6.979	7.695	8.579	10.112	10.664
	VLSI	7.158	7.874	8.761	10.296	10.848
POWER SUPPLIES:	Discretes	5.391	5.978	6.941	7.527	8.494
	IC's	5.548	6.289	7.196	7.642	8.602
	LSI	5.678	6.415	7.368	7.971	8.732

Equipments	Typical Examples	WSCF	1.0 Ground	1.4 Mobile	1.8 Airborne	2.0 Space	2.5 Manned Space
Antennas	Small, Spiral, Horn, Flush, Parabolic	4	4.75	5.39	5.64	6.55-7.04	6.92-7.44
	Scanning Radar 10-40' Wide	8	5.3	5.4	5.5	—	—
	Phased Arrays (Less Radiators)	6-8	5.9	6.2	6.4	7.0	7.2
Engines & Motors	Automobile - 100 to 400 H.P.	25-35	—	4.30	—	—	—
	Turbo-Jet (Prime Propulsion)	25-35	—	—	6.6-7.9	—	—
	Rocket Motors	14-15	—	—	6.1-6.5	6.4-7.3	7.2-8.2
	Electric Motors	75-100	4.47	5.08	5.3	5.4-6.3	5.4-6.3
Drive Assemblies	Machined Parts, Gears, etc.	7-10	5.11-5.24	5.5	5.8	—	—
	Mechanisms w/Stampings (Hi Prod)	12	3.33-3.73	—	—	—	—
Microwave Transmission	Waveguide, Isolators, Couplers,	11-20	5.4-5.6	5.4-5.6	5.5-5.7	5.5-5.9	5.5-5.9
	Stripline Circuitry	9	5.7	5.8	5.9	6.0	6.1
Optics	Good (Commercial)	70-90	5.1	5.4	6.3	6.7	7.3
	Excellent (Military)	70-90	5.4	5.8	7.3	7.8	8.0
	Highest (Add 0.1 per 10% Yield)	70-90	5.9	6.8	8.0	8.3	8.5
Ordnance Fuze	Automated Production	14-20	—	4.3-4.65	4.3-4.65	—	—
	Small Production-Min. Tooling	14-20	—	5.11-5.33	5.11-5.33	—	—
Servo	Mech Drive & Coupling Networks	65-75	5.63	5.63-5.7	5.7-6.26	5.7-6.86	5.7-6.86
Tools	Machine Tools	25-30	4.45-4.52	—	—	—	—
Printed CKT Cards (Boards Only)	Paper Phenolic	83	4.1-4.3	4.1-4.3	4.1-4.3	4.1-4.3	4.1-4.3
	Glass Expoxy, Double Sided	110	5.3	5.3	5.3	5.3	5.3
	(Add 0.2 for 3 Layers & 0.05 for Addn'l)						
	Add 0.1 for Plated-Thru Holes						
Cabling	Multiconductor w/MS Connectors	40	4.9	5.0	5.0	5.1	5.2
	Same w/ Hermetically Sealed Connectors	40	5.1	5.2	5.2	5.3	5.3
Battery	Lead Acid	68-125	4.47	4.49	4.61	4.8-5.4	4.9-5.5
	Nickel Cadmium	75	5.39	5.83	6.73	7.63	8.38
Gyro	Inertial Platform Type	79	6.01	6.56	6.8	6.9-9.1	7.0-9.4
Laser Module			7.6	8.5	9.4		

Fig. 10.11 Definition of manufacturing complexity (Source PRICE)

Fig. 10.12 Typical costmodel input scheme

I. General A
   a. Production Quantity
   b. Protype Quantity
   c. Total Weight
   d. Volume
   e. Type (Mechanical or Electronic)
II. General B
   a. Quantity Next Higher Assembly
   b. Integration Level (Electronic)
   c. Integration Level (Mechanical)
   d. Application (Platform)
   e. Price Basis (Year)
   f. Year of Technology
III. Mechanical/Structure
   a. Mechanical Weight
   b. Technology index
   c. Percent New Development
   d. Percent Design Repeat
IV. Electronic
   a. Electronic Weight
   b. Packing Density
   c. Technology Index
   d. Percent New Development
   e. Prozent Design Repeat
V. Schedule
   a. Start Development (Phase C)
   b. First Prototype Complete
   c. End of Development
   d. Development Complexity (see. Fig. 10-10)
VI. Production
   a. Start Production (Phase D)
   b. First Unit Complete
   c. End of Production

conditions of the respective company. Chapter 24 demonstrates the effectiveness of cost estimation models using the example of 4cost-aces.

The author is a certified user of the cost models PRICE and 4cost-aces and has gained practical experience with both models.

10.3.3 Use of Cost Estimation Models at MBB

At the beginning of 1979 Messerschmitt-Bölkow-Blohm GmbH (*MBB*) in Munich applied the hardware model of PRICE for some space projects [23]. The results were not convincing at first, since, e.g., the assignment of the appropriate complexity factors caused considerable problems. On the other hand, however, complexity had a direct effect on costs, since it is an important influencing factor. Successful

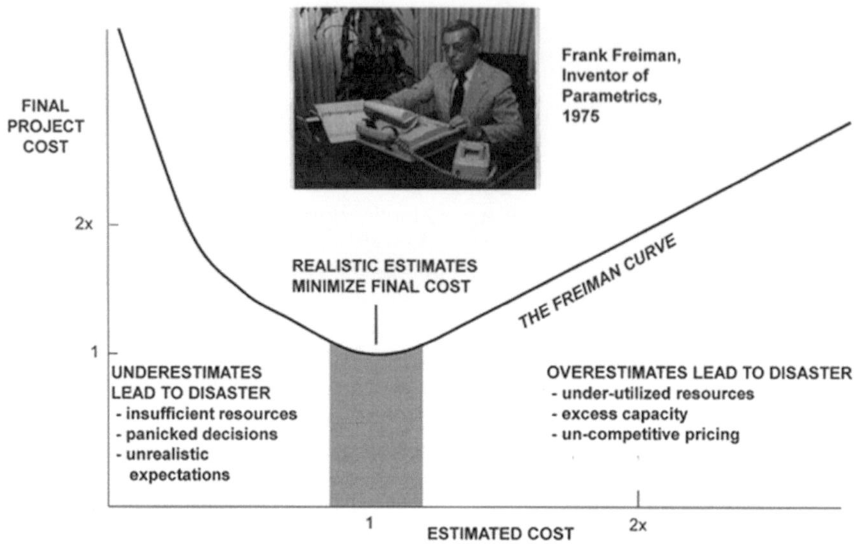

Fig. 10.13 Schedule-cost dependency (*Source: RCA-PRICE*)

calculation with the model requires calibration work over a period of several months. In other words, the model first had to be tailored to the MBB environment. During a preparation period of several months, completed projects were evaluated by inputting the incurred costs of completed projects to the model and performing backward runs: ECIRP runs, i.e., PRICE backward. The result was now not costs but complexity factors.

It is common knowledge that there is a relationship between the project deadline and costs, and that there is an optimal deadline when costs are lowest. Deadline pressure usually leads to additional costs. But even a very relaxed schedule entails additional costs; see also Fig. 10.13. This phenomenon is plausible and usually known to project staff. Frank Freiman, father of the PRICE model *(see Sect. 10.3.2)*, has demonstrated this under the term "*Freiman curve,*" sometimes called the Bathtub Curve; see also Fig. 10.13.

The problem, however, is to determine where the optimum is. Can this be determined intuitively or is there a clear connection—a plausible *relationship*? With the help of cost estimation models, it is possible to prove the relationship with concrete figures. Hans-Dieter Hopf, a former work colleague of the author, transferred the relationship between development complexity *DC*), manufacturing complexity (*MC*), and development schedule to first prototype (*DS*) into a ray diagram to better see the relationships; see also Fig. 10.14 [24].

Here are five examples of electronics development for an LSI analog receiver for human spaceflight—Platform 2.5:

- first example
  1. MC 10.2 (*rounded*); Fig. 10.11
  2. DK 1.6 (*new development, team has average experience*); Fig. 10.10

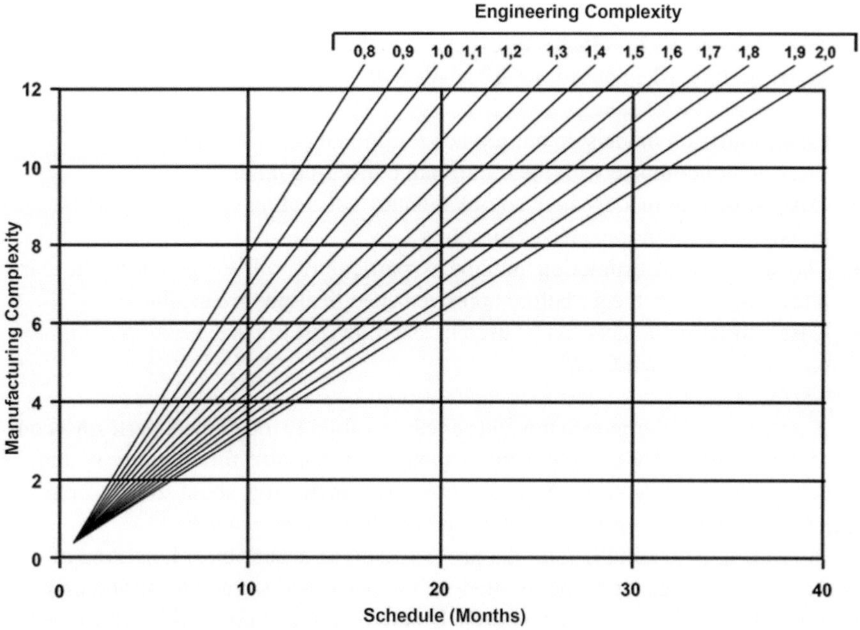

Fig. 10.14 Relationship between EC, MC, and schedule (first prototype) (Source: H. D. Hopf)

3. DS approx. 28 months for the first prototype

- second example
  1. MC 10.2 (*rounded*); Fig. 10.11
  2. DC1.9 (*new development, team with more mixed experience*); Fig. 10.10
  3. DS approx. 32 months for the first prototype
- third example
  1. MC 10.2 (*rounded*); Fig. 10.11
  2. DC 1.3 (*new development, team with a lot of experience*); Fig. 10.10
  3. DS approx. 23 months for the first prototype
- fourth example
  1. MC 7.3 (*rounded*); see Fig. 10.11
  2. DC 1.3 (*new development, team with a lot of experience*); Fig. 10.10
  3. DS approx. 18 months for the first prototype
- fifth example
  1. MC 7.3 (*rounded*); see Fig. 10.11
  2. DC 1.0 (*new development, team with a lot of experience*); Fig. 10.10
  3. DS approx. 12 months for the first prototype

The five examples clearly show the influence of the platform, the manufacturing complexity (*MC*) and the development complexity (*DC*) on the schedule. The schedule varies between 12 and 32 months. What is interesting now is that the model can also be used in reverse by specifying a target date and adjusting the development complexity (*DC*). Specifically, this could mean, as any development manager knows, increasing team experience to meet the given schedule. At this point, frequently asked questions are answered:

- Cost estimation models do not replace traditional cost estimation methods
- They do not compete with the traditional bottom-up *method*
- Cost estimating models are a complementary estimating system that can be used to validate conventional cost estimating
- The use of cost estimation models is particularly advantageous in the early phases of a project, e.g., before approved specifications are available
- Cost estimation models can be used to examine cost impact variables in minutes; e.g., performing *trade-offs*

A trained and experienced model user can, as already mentioned, perform a complete cost analysis for a component, a subsystem, but also for an entire system or project within a very short time. By correctly answering about 20–30 technical questions, the model expert can query the result in a few minutes.

In order to be able to handle the program and to master the whole field of cost analysis based on cost estimation models, the person concerned must first undergo training. Here, above all, the approach and communication with colleagues is trained. Often, the data provided is either not detailed enough or structured completely differently from what the model user needs. Problems of this kind often lead to the user throwing in the towel prematurely. In order to prevent this, the training course specifically addresses this point. At the end of the training, participants are expected to take a test to determine the costs of a system on the basis of a few details, e.g., photos and sketches.

10.3.4 Parametric Cost Estimation Models in Use

The following information was provided by Ronald Larson (*USA*) and Pierre Foussier (*France*). Both were professional colleagues of the author and model users with several years of experience. Larson worked for several years as a cost engineer at MBBs space division, Munich and at NASA HQ, Washington and Foussier was for many years an employee of the French space agency CNES and founded the consulting firm "*3f*" in Paris.

The application and acceptance of parametric cost estimation methods and corresponding software-based models, has steadily increased in importance over the last four decades, especially for scientific and commercial high-tech projects. The following list shows the currently universally applicable parametric cost estimation models:

- PRICE ® True Planning® Cost Estimating Framework (*PRICE Systems, LLC*), USA

  - An integrated suite of parametric models including hardware, software, IT, assembly, and program management
  - The origins of PRICE date back to 1969 when it was developed for internal use at RCA. In 1975, PRICE Systems began making models available to the public
- SEER by Galorath (*Galorath Incorporated*), USA
  - Galorath began as a consulting firm in 1979, and has developed parametric models for Software, Hardware, Electronics and Systems, Manufacturing, IT, and Systems Engineering.
- MAP-H by 3f, France
  - A universal cost estimating model for development, production, and maintenance for any kind of hardware
- 4cost-aces Software GmbH, Germany
  - Generalized all-purpose model for mechanical and electric components and, software development*; see also Chap. 24

**Author's addition: including LCC analyses involving a special learning curve application.*

The above models use CERs developed through intensive analysis of historical data. Using inputs that define the respective product, this allows meaningful cost predictions to be made based on model calculations.

There are also other CER-based cost estimation models that are used to determine costs and schedules, e.g., for the development, production, testing, integration and operation of hardware and software projects. All models are based on relevant historical data from similar projects. NASA has historically evaluated, normalized, and analysed historical space equipment cost data with appropriate effort, incorporating the technical benchmarks. In addition to the aforementioned universal cost estimation models, NASA uses specialized CER-based models, such as (*R. K. Larson [2017]*):

- PCEC (*Project Cost Estimating Capability*)
  - PCEC utilizes much of the data contained in NAFCOM *(NASA Air Force Cost Model*)
- NICM (*NASA Instrument Cost Model*)
- COCOMO (*Constructive Cost Model*) for software projects
- QuickCost
- TCASE (*Technology Cost and Schedule Estimation*)
- QTIPS (Quantitative Techniques Incorporating Phasing and Schedule)
- EstimLab by 3f, France (*a "deductive" which incorporates data selection and analysis while generating CERs*)

Currently, models are being developed to quantify the interrelated confidence of cost and schedule forecasts to better understand cost and schedule risks. Model examples are:

- Polaris (*developed by Booz Allen Hamilton*) and
- JACS (*Joint Analysis of Cost and Schedule*), an add-in to Microsoft® Project developed by Tecolote Research, Inc.

The NASA Cost Estimating Handbook (www.ceh.nasa.gov) is another source of useful information regarding cost models and cost estimation in general.

10.4 Estimation Accuracy

10.4.1 Credibility of Cost Estimates

In principle, every project member can estimate costs. The only question is, however, how accurate and useful is the estimate. Therefore, the topic of cost estimation cannot be left to laymen. But very often this is exactly what happens, because there is frequently a lack of professionals. The subject is at times left to employees who happen to be available at the moment. Companies with a cost calculation department are of course in a better position to produce reliable cost estimates, because they can draw on the experience of completed projects and professionals. However, it always becomes difficult when accurate estimates have to be made in the early stages, when no reliable planning documents are available.

In practice, it is usually not enough to present a cost estimate without supplementary additional information about the accuracy of the estimate. How reliable is the estimate. Is there any information on the accuracy and/or trustworthiness of the cost estimate made? When conducting cost studies, investigating the accuracy of the estimate is usually very important to substantiate the statement. However, knowledge about the quality of the cost estimate also plays a major role in contract negotiations and related pricing. For example, it is very risky to contract projects or project areas for which costs have been estimated with questionable confidence with a fixed price, unless appropriate reserves have been included in the fixed price quote.

It must be remembered, that a cost estimate reflects available knowledge about the project. If project definition is incomplete or misrepresented, then the cost estimate cannot be considered reliable and/or accurate.

Now, on the other hand, it is of course also certain that the estimation accuracy is phase-dependent, i.e., the estimates made at the beginning of the project are usually much less accurate than those estimates that are available after the definition phase has been completed. In the case of development projects, this is mainly due to the many unknowns that cannot yet be estimated well enough, especially at the beginning of a project. In this context, Norman Augustine speaks of two different categories of uncertainty that are a plague for almost every project: the "*known unknowns*" and the "*unknown unknowns*" [25]. He writes in this regard that the known unknowns, e.g., the exact soil characteristics of the lunar surface at the site where the first Apollo landing was to take place, could still be relatively well accounted for in the planning. The second category of uncertainty, the "*unknown unknowns,*" on the other hand, cannot be identified and planned for in advance, but its existence can be predicted according to the scheme of an insurance statistic. Augustine cites the example of Apollo XII, in which a lightning strike hit the rocket shortly after launch to the moon, and concludes: “Somehow, in every great project, lightning strikes somewhere; it cannot be predicted where it will strike, but it is certain that it will strike [25]. This refers to the project mishaps that occur again and again, often out of the blue. Augustine evaluated 38 major projects in the USA between 1962 and 1976 and concluded that cost overruns would not have occurred on the projects analysed if the original estimated costs had been multiplied by a correction factor that already accounted for the unknown unknowns. The correction factor Augustine developed, by which he believed each estimate should be multiplied, is as follows [25]:

$$\text{Correction factor} = 1 + \frac{0.8}{1 + 8t^3}$$

CF = correction factor, t = development date level (*0 = start, 1 = full completion*).

This means that the costs for the entire project estimated at the beginning of the project would have to be multiplied by 1.8 to be realistic, according to Augustine. The extrapolation made at the end of the development phase still needs to be increased almost 10%. If the formula is applied to public or private projects, the previously mentioned correction factor still seems rather too low.

Of course, this does not solve the problem of estimation accuracy. It is also not the author’s intention to offer this to the reader as a patent remedy. However, Augustine’s notes do an excellent job of describing the problems that project staff have to deal with in practice. At the beginning of a project, it is very difficult to make an accurate cost estimate because of the many known and unknown unknowns, to use Augustine’s words. With parametric estimation methods, one can get a better grip on the problem, but even then, there remains a significant uncertainty in the estimation accuracy due to a concept that is not yet fully developed.

Moreover, a reliable estimate at the beginning of a project is sometimes undesirable. Boeing Company President W.M. Allen wrote in 1964: “*I can think of a lot of programs in the Boeing Company where, if the estimate had been realistic, you wouldn’t have had the program. And that is the truth.*” But it is a fallacy to take such thinking as a motto. For new developments, it may sometimes be true that the “*unknown unknowns*” lead to an unexpected cost increase. But it is not acceptable for commercial projects.

When estimating costs, one should ideally also provide information about the quality or trustworthiness of the estimate. However, expressing confidence in the estimate requires a detailed analysis of how the estimated quantities and costs were arrived at. Determining the confidence in a cost estimate depends on several factors. The most important factors are [26]:

(1) *Estimating conditions:*
   - Estimation time (*time available*)
   - Estimating team (*experienced, partially experienced, inexperienced*)

Estimating Confidence	Estimating Conditions		
	Estimating Time	Product Knowledge	Estimating Method
High	Too long	Excellent (expert knowledge available)	Proven methods available
Medium	Just right	Partly available (expert knowledge partially available)	Some methodology available
Low	Too short	Not available (no expert knowledge available)	No methodology available

Fig. 10.15 Scheme for determining the estimation accuracy

- Prerequisites (*economic conditions: Hourly rates, expected price increases, profit rates, etc.*)
- Assumptions (*made by project, the company, others*).

(2) *Type of product:*
- State of the art (*established and known, new, only little known or completely new previously unknown technology*)
- Production experience (*available, partially available, not available*)

(3) *Product Description:*
- Specification status (*available, partially available, only rough data available*)
- Drawing status (*existing, partially existing, only existing in draft*)

(4) *Estimation methods and comparative data:*
- Methods (*detailed or parametric estimation methods*).
- Comparative data (*available, partially available, not available*).

With the help of these factors, a scheme for assessing trustworthiness (*confidence*) can be created. Figure 10.15 shows a practical example of a scheme for determining the trustworthiness of a cost estimate. In conjunction with the criteria listed, three levels of trustworthiness, (*L*) low, (*M*) medium, and (*H*) high, have been defined. The estimate of each defined component is then assigned a flag (*L, M, H*) for the trustworthiness of the estimate according to the scheme shown in Fig. 10.15.

This means that, in accordance with the example shown, work packages or components with a high degree of confidence in the estimate (*here class H*) are marked with an estimation accuracy of, for example, ±5% of the nominal value. If a nominal value of 100 Euros was estimated for a component, this value may fluctuate between 95 and 105 Euros, for example, due to unforeseen events. The same procedure is to be followed for the estimation categories "*M*" and "*L*." It should be mentioned at this point those statistical methods for determining the accuracy of estimates, provided they are well-founded, are a useful supplement to cost forecasting. Figure 10.16 emphasizes that estimation accuracy is valid only within a certain statistical range [27].

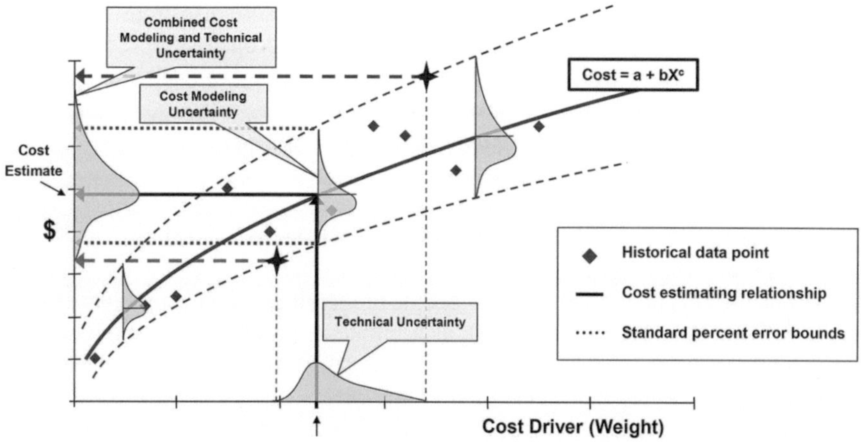

Fig. 10.16 Range of estimation accuracy (*Source: Larson*)

A case described by R. K. Larson: *Let us consider for a moment a hypothetical situation to provide a framework for assessing cost model accuracy. Establish two identical teams based on education, prior experience, etc. and provide them with identical resources (office, laboratory, infrastructure, etc.). Call the two teams together and provide them with a task such as "design and production of a prototype New Product with these performance characteristics" and ask them to report back in 12 months with a working model. Twelve months later the two teams report back to demonstrate their New Product devices. They may not look the same, and there might be some differences in appearance, configuration, etc. When we look at what each team expended in terms of money, labor hours, parts procurement costs, etc. would we expect the expended resources to be identical? Team A may have gone down a path which turned out to be non-productive and some of the work may have had to be repeated. Team B may have had a brilliant stroke of intuition which led to an optimum solution with minimum design/redesign. There are an unlimited number of scenarios which could have impacted the paths to project completion. Would we be surprised if there were a difference of 15 or 20% in bottom-line costs between Teams A and B? Probably not! So, if there can be a range of outcomes for actually performing project work, is it not reasonable to accept a range of outcomes for a cost estimate?*

Of course, the experience of the estimator and the estimation methodology used also weigh heavily. But even under the most favorable conditions, an estimate remains what it is, namely an estimate. The estimated cost of a project predicts, with some probability, how much the project will cost. Unforeseen problems in development and increases due to inflation can also be estimated, but their actual impact cannot be accurately predicted. In contrast, other external factors affecting the cost of a project, such as strikes and construction stoppages, can hardly be predicted.

Cost estimates in the early phases must be considered preliminary and in the context of the planning that exists at the time. Cost estimating is an iterative process that runs in parallel with the planning activity. It is therefore disastrous if, in order to avoid unwelcome justification in terms of possible project delays, the project team already includes excessive cost reserves and thereby exceeds the specified target costs. "*In order to avoid a business disaster and to guarantee business success, a professional cost estimating attempt must be implemented.*" [28]

10.4.2 Time Pressure Jeopardizes Cost Estimation Quality

Project managers and project business people are familiar with the problem of having to provide detailed and, above all, as accurate as possible cost estimates for a new project in the shortest possible time, because the future client usually does not allow enough time for the preparation of the offer. This then regularly leads to rushed actions in the preparation of the cost offer, which, as is well known, can only be started after the technical offer documents have been completed. However, the quality of the cost quotation depends to a large extent on the knowledge of the product, the experience of the estimator and the data material available to him from previous projects, as well as on the time available; see Fig. 10.15. Consultations with specialist departments and subcontractors are therefore often extremely brief. It can then easily happen that important details are forgotten and facts are underestimated.

If the internal or external client also requires that the costs be broken down in great detail, the project manager often can no longer see the forest for the trees. On the other hand, the quality of the cost estimate is not necessarily improved by greater detail, as is well known among experts. However, the desire for greater quality in cost estimating is very great. Especially with fixed-price contracts, which have recently been increasingly used for development contracts, the quality of the cost estimate plays a very important role! And time? Time for cost estimation will continue to be extremely tight in the future.

However, national and international competitors set the pace, as a German saying goes: "*first come, first served*" or "*he who is late is punished by life*" [*Gorbachev, 1989*]. Now, at the latest, it becomes clear how important it is to have a tool with which cost estimates can be carried out faster and more accurately, and these are the parametric cost estimates with corresponding models described in Sect. 10.3.

10.5 Cost as a Design Parameter

10.5.1 System Comparisons Including Costs

System comparisons in the early project phases cannot be overestimated in their importance, because the future system operator does not want to buy just any system, but the *best system.* Through system comparisons (*system-trade-off*), one tries to identify the most favorable solution through a combination of subsystems, assemblies, and components that lead to the optimal performance of the system. First of all, technical system comparisons have to be carried out in order to then combine them with economic aspects. The goal is a technically and economically optimal system configuration or architecture. Many companies are in global competition and therefore have to offer at market prices.

It is not about the best possible technical *nice-to-have solutions*, which tradition-conscious German companies often tend to do, but about a customer-oriented good-enough *solution* at compatible market prices. In summary, it is about the following three main parameters: (1) technical performance including QA (2) schedule, and (3) costs. In the early phases of the project, the three parameters can still be varied, as shown in Fig. 10.17. This is illustrated by the arrows. In so-called *trade-offs,* the most optimal solution is identified.

Technically, a lot is possible today, but on the other hand, one quickly comes up against financial limits. Cost-optimal, what does that mean, lowest cost or reasonable cost? In this context, two processes are particularly noteworthy:

- The consideration of the life *cycle* costs *(LCC)* and
- the *Design-To-Cost* (DTC) method.

Projects must pay off, i.e. they must be profitable. This includes a profitability analysis that takes into account a) investments for system development and production, b) system operating costs and c) expected *revenues*. The *return on investment* (*ROI*) ratio can then be used to measure the profitability of the project. However, a positive ROI can quickly slip if the costs for a) and b) increase and the revenues are slow to materialize or do not generate the expected amount at all.

With regard to cost for a) and b), the subject of this chapter, the operator should make every effort to obtain a credible cost estimate because the decision-making processes should be based on solid foundations; see also Sects. 10.1–10.4. Many *business plans,* although formally presented in a convincing manner, are often not based on robust estimates and assumptions. Projects go through a life cycle as described in Chap. 4. Sometimes the life cycle takes several years or even decades before the predicted end result (*profit & loss*) can be confirmed. In the early phases of a project, strategic decisions must be made with a focus on the end result. This is

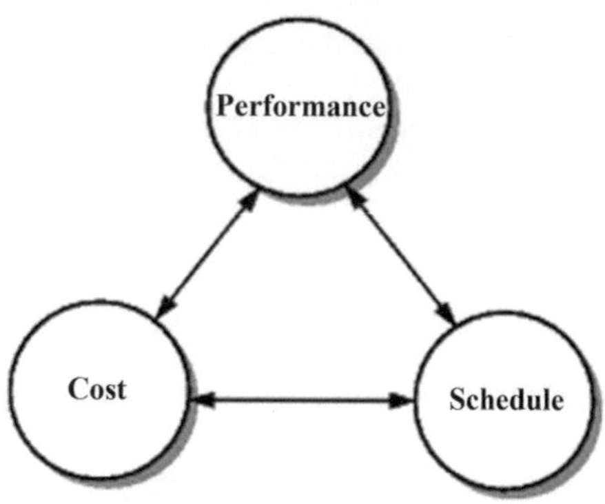

Fig. 10.17 Flexible relationship of the factor's performance-schedule-costs

the difficulty at the beginning of any project. Therefore, it is important that experienced engineering and business teams lay the foundation as accurately as possible during Phases A and B (*see also Chap. 4*).

10.5.2 Life Cycle Cost Consideration

As early as the beginning of the 1960s, LCC methods were developed in the USA because it was recognized that in the procurement of large systems, it was not only the level of research and development costs that had to be assessed but increasingly also the follow-up costs, i.e., the costs of production and operation. In other words, a system with relatively low development (*R&D*) costs does not necessarily result in the lowest total project cost. Bethke has published ratios on this subject that illustrate the importance of the LCC consideration particularly dramatically [29]. The procurement and operating costs (*phases D and E*) are often many times higher than the development costs; see also Fig. 4.3. This clearly shows the importance of LCC considerations. The early introduction of cost reduction measures during phases A and B influences the level of the subsequent phases.

All project phases must be taken into account; i.e., costs must be determined for all phases of the life cycle. This is usually done initially using top-down *methods* based on empirical values. The temporal distribution of life cycle costs (LCC) is shown in Fig. 10.18.

10.5.3 Design-to-Cost: Cost as a Design Parameter

The basic principle of *Design-to-Cost (DTC)* is not new and has been successfully applied by many producers for commercial products for quite some time. Due to the competition in the market, the mass goods industry has always had to be particularly careful to design only what could be sold, i.e., the designer always had to include cost as an important parameter when developing new products. In many major projects, especially in the aerospace sector, this principle was not always followed, sometimes leading to excessive cost increases.

It is of course understandable that increased requirements also entail increased costs. This connection cannot be eliminated by the DTC method either and where increased requirements are absolutely necessary, appropriate budgets must be made available for them. The DTC method is always useful when you have to realize a project within a tightly limited budget, but at the same time want to ensure that you get the best possible product for the funds provided. This goal is achieved by balancing the planned performance of the product being developed with the available cost and schedule in a continuous iteration process. Cost thus becomes an important design parameter.

The US Department of Defense (*DOD*) issued a directive in 1975 on the application of the DTC methodology, which set forth the regulations and guidelines for implementing the DTC processes to manage government projects. The major provisions of the DTC approach are listed below [30]:

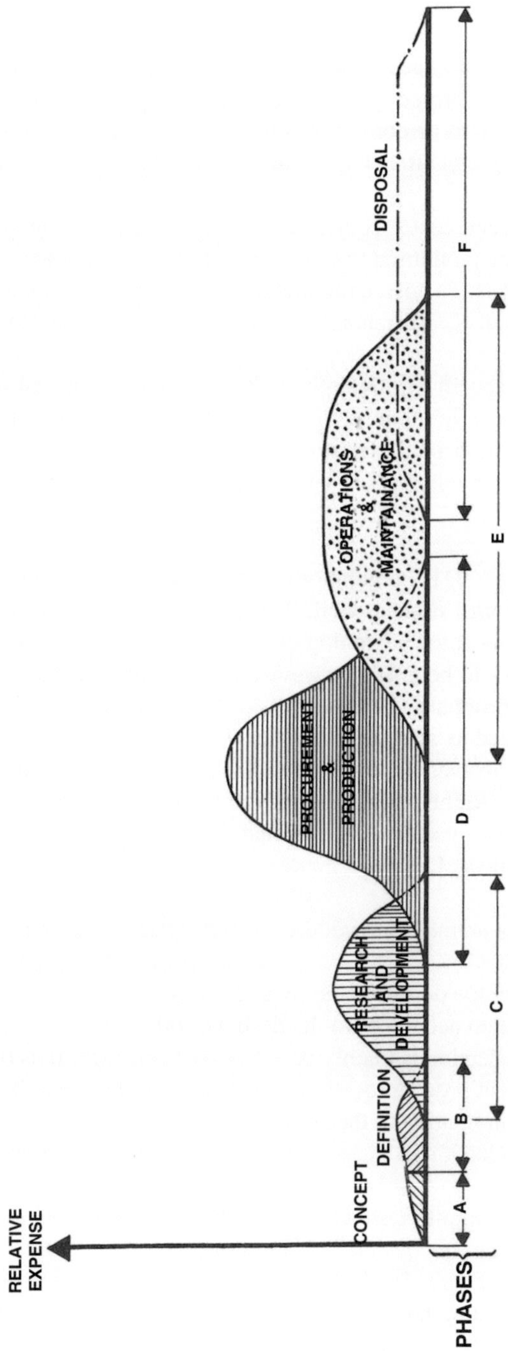

Fig. 10.18 Time distribution of life cycle costs (LCC)

- The DTC concept requires that cost be included as a design parameter during the development phase to improve cost discipline during the procurement and operation phases.
- For each project, cost targets of the entire life cycle are to be defined and subdivided into the main project phases, development, procurement, and operation. The cost targets are to be converted into fixed cost specifications as the system definition progresses, after which the system is developed and the costs controlled.
- During development, cost specifications (*target costs*) and compliance with target values must be monitored with the same rigor as compliance with technical performance targets. To ensure the lowest possible costs for the entire life cycle, system comparisons (*performance, costs, and deadlines*) are to be carried out continuously.
- The defined cost specifications apply to the development and subsequent phases. The production costs are to be controlled according to the cost specifications.
- After implementation of the system, monitoring of costs for personnel, spare parts, rework, etc. in comparison with the established cost targets for the operating phase.

Although the above DTC regulations were intended for US companies, it is a useful guideline for other users as well. DTC methods should be introduced as early as possible, preferably at the beginning of the project, i.e., in the concept phase; see also Fig. 4.13. It should be noted, however, that the introduction of DTC methods does not require additional DTC managers or DTC teams; because DTC management is to be regarded as an integral part of the already existing management and concerns everyone involved in development [31]. The use of DTC methods does not create new positions. Rather, it is a new way of thinking. A comparison between the procurement procedure used in the past and the DTC procurement philosophy is best described as follows [26]:

(1) Traditional procurement procedure—*Fixed system requirements and variable costs:*
   - Adaptation of the design to the requirements
   - Cost calculation according to the design effort
   - Contractor selection is mainly based on cost considerations (*lowest price*)

(2) DTC procurement procedure—*fixed cost and variable system design*
   - Best design in relation to the specified costs
   - Contractor selection is based primarily on performance (*best design*)

Breadly defines design-to-cost as "*Provide the most for the money (best buy)*" or "*Eliminate the gold plating design.*" [31]

The introduction of the DTC method often requires a joint approach by the client and contractor. As a rule, the client sets the DTC cost target in the first step on the basis of preliminary investigations. There are various methods for setting the DTC cost target; here are some examples from the R&D sector [32]:

- The cost of the system design shall not exceed ... Euro
- the cost of the design work shall not exceed ... percent of the total anticipated price; or
- the design effort shall not exceed TBD man-hours

In the second step, the DTC cost target specified by the client is to be broken down by the contractor into the main elements of the project and/or the subcontractors involved. In the next step, cost studies are to be carried out to check the feasibility of the specified DTC cost targets. As a result of these cost studies, in the fourth step, detailed cost specifications are to be established for all main assemblies of the system, which are then reflected in the cost plan for the project. Based on these targets, integrated cost/DTC tracking is then performed. The individual implementation steps for a typical DTC program are defined in Fig. 10.19 [32].

10.5.4 Optimization of Life Cycle Costs

The life cycle cost (*LCC*) of a system can be defined as the total cost of a project for all project phases mentioned in Fig. 4.2.

- A Concept
- B Definition
- C Development
- D Production and procurement
- E Operation
- F Disposal

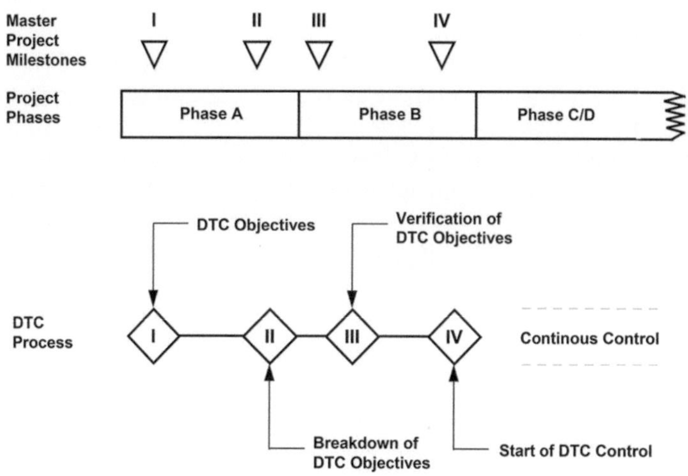

Fig. 10.19 DTC implementation plan (Source: K. Burbridge)

In recent decades, *Life Cycle Cost (LCC) Consideration* has become increasingly important for many industrial and public projects. Why? Increasingly, it is not only the development and procurement costs that cause headaches for the purchaser or operator (*client*) of a product, such as a new machine, plant, or large facility. Above all, it is also the consequential costs or operating costs for operation, maintenance, repair, etc. that cause headaches. With the increasing complexity of modern systems, operating costs have also skyrocketed. As mentioned before, a high-performance weapon system has already resulted in ratios of 1:5:15 for the shares of development, procurement, and operation [29].

The phenomenon of extremely high operating costs is not only a problem of so-called high-tech programs. Buyers of industrial products and municipalities have also been affected by this problem, especially in recent years. Stapel writes, for example, in connection with a consideration of the less expensive purchase of a compressor: "*After all, the energy and operating costs (including maintenance) of an industrial compressor can considerably exceed the original purchase price over the course of its life.*" [33]

Of course, any car buyer can have similar experiences when he examines the various car offerings not only in terms of their purchase price but also in terms of operating costs. Lenk and Lang show that follow-up expenditures on municipal administration buildings for 1979 averaged 47.5% of the 1978 capital expenditures [34]. Although this information is not fundamentally new evidence, the magnitude of consequential expenditures, on the other hand, is quite surprising. The defense budgets of many countries are so stretched because of the high operating costs of new weapons systems that increasingly fewer funds are available for new developments. Coble writes, "*We are constantly reminded that operating costs consume two out of every three dollars budgeted for defense.*" [35]

In the past decades, it has become clear that the consideration of life cycle costs is important, because the follow-up costs are often a major burden for manufacturers and operators. But disposal costs are also part of the total cost consideration. Slowly, the realization is gaining ground that the impact on subsequent costs is greatest in the early phases. According to Coble, after completion of the concept phase, 70% of the follow-up costs have already been determined, and after completion of the definition phase, 85%; see also Fig. 10.20. Blanchard [36] and other authors come to similar conclusions. This means that after completion of the definition phase, only about 15% of the follow-up costs can be influenced. However, the effects are often only visible after several years. The follow-up costs are already frozen or fixed after completion of phase B; see also Fig. 4.9. Kerzner confirms the percentage values mentioned [37].

This means, however, that correspondingly high importance must be attached to early phases A and B. They must be carried out thoroughly; see also Sect. 4.3.4 (*Roadmap*). For a better understanding, it is necessary to keep the project sequence clearly in mind. The development of a modern combat aircraft for example, takes many years, and the concept is defined at the very beginning of the development. Once development has been successfully completed, production (*procurement*) is initiated and the operational phase comes into play with the delivery of the aircraft.

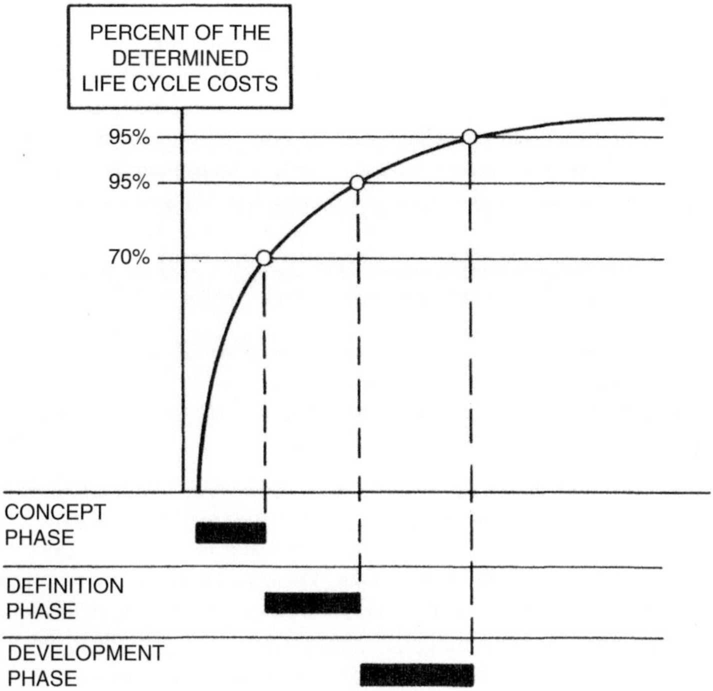

Fig. 10.20 LCC determination (Source: Coble)

At least 10 years lie between the concept definition and the operational phase. It is therefore not uncommon for the problems of the operational phase not to be considered thoroughly enough in the concept design. There are certainly many reasons for this. Cook writes, for example, that the following logistics activities should be carried out in the concept phase with a view to an optimally running operating phase, although a period of several years may lie in between [38]:

(1) Logistics engineering:
   - Development of preliminary logistics concepts
   - Analysis of preliminary logistics requirements
   - Identification of logistics activities for the entire project duration
(2) System maintenance:
   - Development of preliminary system maintenance concepts
   - Development of preliminary criteria for system maintenance
   - Participate in comparative weight, reliability, and system safety studies
(3) Maintenance (*development of a preliminary maintenance concept*).

The early start of this planning work (*Phases A and B*) is so important because the development engineers must incorporate the information into the individual specifications and design documents in good time. However, if this is neglected, or if things are left to chance, the development engineers will define their designs without considering the concerns for the operational phase and the operational concept will have to be based on the conditions defined by the developer. But this can be expensive.

It is often very difficult to properly assess the issues of the operational phase during the concept phase. It is simply too far away in time to plan the things that will have to be carried out in 10–20 years. This is also true for cost estimating. Broder mentions in this regard "*Estimating the cost of unbuilt (space) systems is very difficult, and cost estimates have consistently been too low.*" [39] To improve the situation, the following things need to be systematically resolved:

(1) System architecture	Systems analysis, trade-offs, etc.
(2) Program planning	Major milestones of all phases
(3) Life cycle costs	LCC model calculations

Often, however, managers, technicians, or politicians can distinguish themselves better through short-term solutions than through far-sighted project planning. Sometimes, however, one can get the impression that it is not very popular to know the total costs. But that is window dressing. There is, however, a general problem that needs to be solved in companies and authorities, namely the determination of the overriding responsibility for all phases. At the highest level, this is the company manager or a senior official. Since there are also many other tasks to be solved at this level, in the case of large and lengthy projects, a program manager should be appointed from the outset, who can then delegate the overall task to subordinate project managers for handling individual phases. It goes without saying that other solutions are possible for smaller projects, but the principle of overall responsibility must be maintained in order to keep continuity and responsibility on one hand.

Blanchard, in his book *Design and Manage to Life Cycle Cost*, points out the importance of understanding LCC [36]. Often, as he says, very significant areas of LCC are completely overlooked at the beginning of the project—only the tip of the iceberg is seen; see also Fig. 10.21. *In* this context, Laube states, "*Now in the past we all often made the mistake of looking at the costs of a project statically and partially, i.e., we always saw the costs only in the phase or even only in the financial year in which the project was at that moment.*" [40] Laube also points out the mistake often made (*saving at the wrong end)* by saying, "*The less money spent on development and procurement, the higher the material maintenance costs inevitably become! The higher the material maintenance costs become, the less money is left for development and procurement! We must have the courage and insight to break this vicious cycle*!" Following Earles, [41] the most important cost elements for determining life cycle costs are summarized below:

$$LCC = A + B + C + D + E + F$$

A. Concept and Feasibility
B. Definition and specification

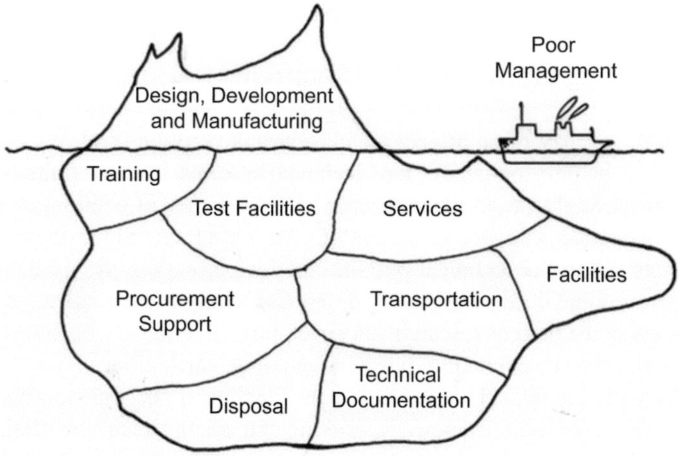

Fig. 10.21 Tip of the LCC iceberg

C. Research and development (*R&D*):
   - Development Teams
   - Tools and fixtures
   - Prototype production
   - Documentation
   - Test and evaluation
   - Project Management
   - Training
   - Facilities
D. Procurement
   - Production
   - Tools and fixtures
   - Modification costs
   - Facilities
   - Acceptance tests
   - Project Management
   - Documentation
   - Installation
   - Training
   - Spare parts
E. Operation and maintenance
   - Personal
   - Materiel consumption
   - Maintenance
   - Modification
   - Operation
F. Disposal

10.6 Efficient Methods for LCC Reduction

10.6.1 Avoiding of *"Nice-to-Have" Concepts*

The decisions to carry out a project are closely linked to the level of expected life cycle costs. In the development of new technical systems, the "*cost*" factor is occupying an increasingly broad place, because it is a matter of economic efficiency. Therefore, a precise knowledge of the LCC is important. Since there are often hardly any limits to the technical possibilities for perfection, the available means usually represent the limit of the possibilities. Especially in the realization of large-scale projects in the field of research and development, it quickly becomes apparent that everything that is technically feasible cannot be carried out in a cost-covering manner. Instead, it is necessary to weigh how a technical goal can still be achieved within the specified cost framework, leaving out all the nice and desirable but expensive extras (*nice to have*) that the engineer would like to offer if the cost limits were not observed.

In the developments of aerospace equipment, wishful thinking has been systematically reduced in recent decades in order to reduce expensive desirable design proposals to more cost-effective solutions. In the context of generally increasing budget constraints, this is a significant economic measure. Especially for very complex projects—a trend of our time—the search for cost-effective solutions is of utmost importance. In past decades, NASA initiated the race for *low-cost* solutions with the slogan "*faster, better, cheaper.*" However, this has also led to mishaps as quality defects have emerged. Conclusion: *low costs* yes but not at any price.

Cost reductions should not lead to quality losses, because a system failure is often more expensive than the savings. In other words, cost reduction must be planned in the development office. Ehrlenspiel says: "*Cost reduction is an increasingly important task in every company. Costs are most likely to be influenced in the areas where they are fixed.*" [42] The approach proposed by Ehrlenspiel et al. is right on target because the development engineer and the designer hold a key position in influencing costs. They are responsible for 60–70% of the influenceable portion of manufacturing costs. Without a doubt, the developer also influences the follow-up costs for maintenance and operation, which have become an oppressive burden for many project sponsors. In this context, reference is made to system availability (*MTBF*) as an influencing factor of LCC; see also Sect. 8.2.3.

Cost reduction measures should be initiated as early as possible. It has proven useful to carry out reduction measures at regular intervals, e.g., in connection with the *design reviews,* similar to the regular weight reduction measures, because mass as well as schedules and costs are subject to creeping inflation. Regularly initiated reduction measures—forcing the customer to slim down—reveal correlations that would otherwise not come to light clearly enough. The cost reduction measures to be initiated (*analogous to weight reduction*) usually set a positive cost-benefit process in motion.

10.6.2 Developments Under DTC and LCC Specifications

The DTC analysis is a procedure for identifying undesirable developments as early as possible and, as a consequence, introducing appropriate countermeasures. The *design-to-cost (DTC)* philosophy therefore assumes, that costs are to be regarded as a design factor. By limiting costs, the designer is encouraged to innovate and develop low-cost solutions through new ideas such as standardization, use of standard parts, streamlining, etc. Inspiration and inventiveness must be brought to bear; see also Sect. 7.5. The early inclusion of life cycle costing is therefore particularly important for preventing undesirable developments, because very often there are cost increases due to complexity and planning extensions; see also Fig. 10.22 [43].

Another factor that should not be neglected is the multi-nationalization of projects, which can also lead to cost increases. It is often additional requirements and tasks that lead to additional costs. There are also different manufacturing processes, technology differences and different training standards. Management problems, laws, exchange rates and different inflation rates can also lead to cost increases. Brents gives an example of this; see Fig. 10.23 [44]. However, there is no question that the additional costs of globalization mentioned by Brents can be reduced by training and systematic preparation in order to be competitive.

The processes of the project's early phases are quite crucial, because they are meant to thoroughly analyse and plan the future, i.e., what will happen in the subsequent phases. As shown in Fig. 4.2, the development process starts on the basis of a clear requirement, namely the *mission requirement definition (MRD)*. However, there are often several paths that can lead to success. However, not all paths are optimal from the LCC point of view.

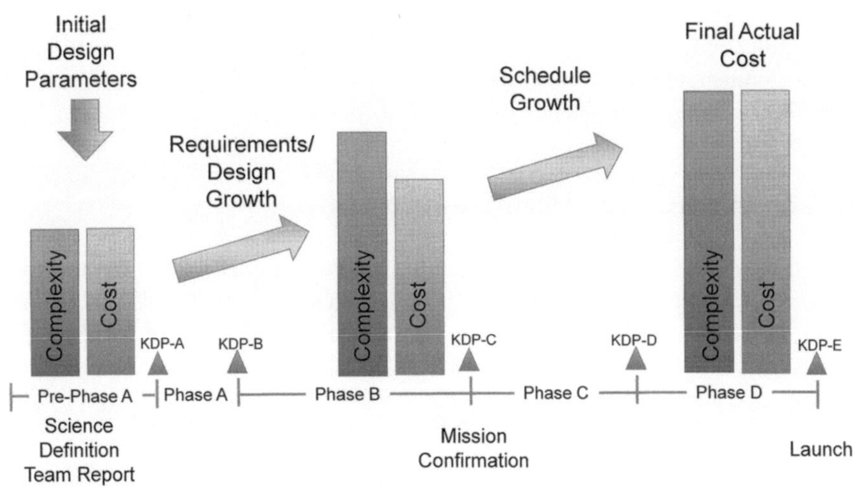

Fig. 10.22 Cost increases due to complexity [43]

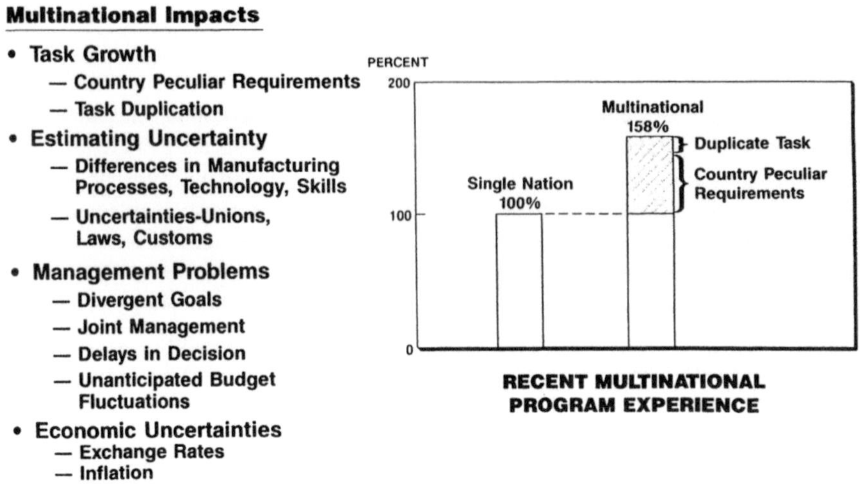

Fig. 10.23 Cost increases of multinational projects

The use of effective measures to reduce project costs is possible in principle in every phase (*A to F*). However, the cost comparison in Fig. 4.3, which is indicated in principle, shows that cost reduction measures for phases C to F are of particular importance for the reduction of LCC. Effective cost reduction measures should therefore be initiated as early as possible. This means that a measure that is to be effective in Phases C through F should be prepared or initiated in Phase A and/or B. In the interest of a cost-saving project solution, it may make sense to increase funding for the early phases in order to develop effective cost reduction measures for the more expensive follow-on phases. The use of effective measures to reduce LCC should be undertaken in the following steps:

- Identification of significant *cost drivers*
- Development of effective measures for cost reduction and
- Introduction of cost reduction measures.

10.6.3 Identification of Significant Cost Drivers

Cost drivers in this context are those technical and/or administrative project factors that have a significant influence on the LCC. The identification of these factors is an important step towards cost reduction. Ten typical cost drivers are listed below. However, this list can only be considered as an example, since the actual factors causing cost increases of a project have to be determined on a project-specific basis [45].

- Poor concept formulation
- Poor project definition

- Overspecification
- Tendency to technical perfectionism
- Late introduction of changes
- Excessive documentation requirements
- Insufficient use of standard constructions and equipment
- Delayed project decisions
- Specification of unrealistic planning deadlines
- Poor performance control

Burghardt also mentions the following technical factors that have an influence on LCC [46]:

- Number of pieces (*learning curve*)
- System complexity
- Availability
- Component quality
- MTBF—Mean Time Between Failures; see also Sect. 8.2.
- MTTR—Mean Time to Repair
- HW and SW test effort
- Number of SW errors
- Incoming inspection of the components
- Final testing of the product
- Diagnostic equipment
- System redundancy (*active, passive*)
- Mounting capability

He supplements this with the following social determinants:

- Qualification of the developers (*little to significant experience, see also Fig. 10.10*)
- Training of the operating personnel
- Qualification of the maintenance personnel
- Improvement proposal system (*see also Sect. 7.5.2, innovation*)

10.6.4 Development of Efficient Measures for LCC Reduction

Once the cost drivers have been identified, the second step is to look for effective measures to reduce LCC. Although no specific measures can be proposed for individual cases at this point, five practical examples of cost reduction are given below. The author refers to personal experiences from the R&D sector and related publications:

- Development of project goals
- Carrying out compatibility studies

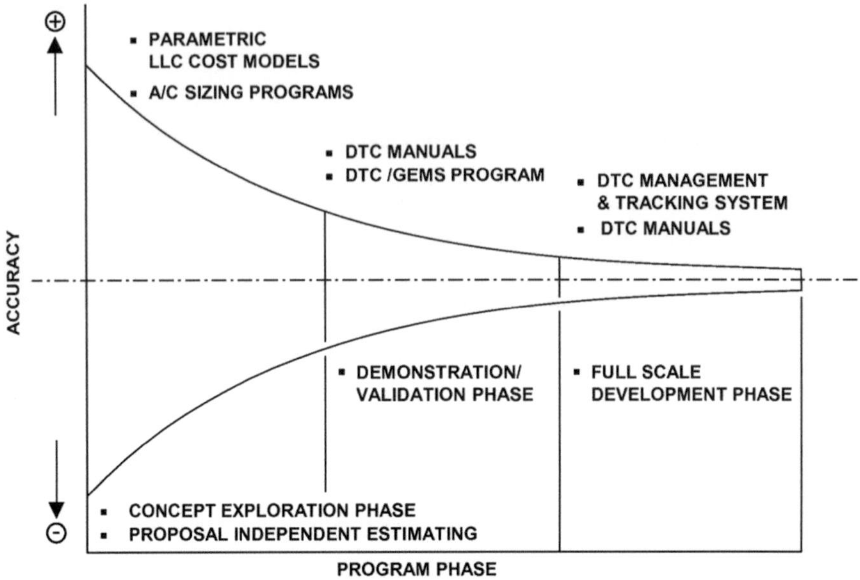

Fig. 10.24 LCC/DTC optimization process

- Identification of research and technology requirements
- Development of overall project plans (*deadlines, costs, etc.*)
- Carrying out comparative analyses (*trade-offs*)
- Identification of advantageous and disadvantageous development factors.

Phase A, Systematic concept formulation (*see also Fig. 4.13*): The concept definition usually represents a fundamental decision for the system to be developed and is thus an important starting point for all further development steps. At this point, the course is set for all subsequent steps. If carried out correctly, the concept phase concludes with the formulation of a clear concept and includes the individual activities shown in Fig. 4.12 [47]. The execution of *trade-offs* should be based on the highest possible estimation accuracy, see also Sect. 10.4.1. In the early phases, many technical parameters are not yet defined and can still be varied or optimized, see also Fig. 10.24 [48]. It is clear that the degrees of freedom of open parameters, e.g., material selection (*steel, aluminum, or fiber composites*) must be selected taking into account technical factors, schedule, and costs.

A poorly executed or not clearly completed concept phase often contributes to increases in LCC due to transformation of unsolved problems into the more cost-intensive subsequent phases. Delays or additional costs for the purpose of additional parallel work in phase A in favor of a better concept solution before the start of the more personnel- and cost-intensive subsequent phases can therefore lead to cost savings for the overall project. This becomes even more understandable if one considers the cost ratio of Phase A to Phase C, as shown below; see also Fig. 4.3:

$$\frac{\text{Cost of Phase A (approx. 3\% of Phase C)*}}{\text{Phase C costs (100\%)}}$$

**Experience value for technology projects [26].*

Phase B, systematic project definition: After concept selection, the definition phase (*see also Fig. 4.14*) is of particular importance, since the following conditions must be met without fail before the design and development phase (*Phase C*) begins [49]:

- Creation of final and realistic specifications
- Definition of all interfaces and responsibilities
- Identification of risk areas
- Presentation of system comparisons and alternatives
- Selection of the best overall solution
- Formulation of realistic manufacturing and operational concepts
- Preparation of realistic schedules and cost plans
- Setting the stage for future fixed and/or premium contracts.

A poorly executed or not clearly completed definition phase usually contributes to significant increases in the LCC due to transformation of unresolved problems into the considerably more cost-intensive subsequent phases. Delays or additional costs for the purpose of additional parallel work in phase B in favor of a more thorough project definition before the start of the considerably more personnel and cost-intensive subsequent phases can therefore lead to a reduction in the LCC. This argument can be justified by comparing the cost ratios for Phase B and Phase C, as shown below:

$$\frac{\text{Cost of Phase B (approx. 10\% of Phase C)*}}{\text{Phase C costs (100\%)}}$$

**Experience value for technology projects [26].*

In the context of Phases A and B, it is pointed out to avoid over-specification and technical perfectionism. When designing technical system architectures, there is a tendency to over-specify, mostly resulting from the operators' tendency to introduce nice-to-have *requirements*. However, over-specification is also often used simply because existing documentation is relied upon for simplicity.

In recent years, however, considerable efforts have been made for economic reasons to simplify the generally accepted design specifications for aerospace equipment wherever possible or to adapt them specifically to requirements in order to avoid any over-specification. The new slogan was *Design-To-Cost,* by which means that the technical concept of a system must be adapted to the specified cost targets in order to remain competitive [39, 40].

In the 1970s, the aerospace industry began taking targeted action to reduce costs by minimizing system requirements. A study conducted for NASA by Lockheed Missiles & Space Company (*LMSC*) disclosed opportunities to reduce costs by modifying (*reducing*) 10 of the client's authoritative program requirements [50].

The proposed changes resulted in an approximate 10% reduction in total cost for two projects on the *Mariner Venus Mercury* project.

Complex projects, e.g., technical plants, traffic systems, aircraft, satellites and IT networks, are expensive. System failures can result in extreme financial loss. The development philosophy of space flight has therefore traditionally provided for the greatest possible redundancies and extensive test programs to ensure mission success.

Over the past few decades in spaceflight, and specifically in commercial spaceflight, people have been looking at how to reduce LCC by reducing conventional system and test requirements. This is an example that applies to other high-tech projects as well. Eaton describes the successful completion of a European satellite project that was developed based on a reduced model and test philosophy and refers to this concept as the "*good-enough philosophy*" that is consistent with design-to-cost *ideas* [51]. European high-tech and aerospace companies have conducted cost reduction studies in recent years, the results of which have then been adopted for new projects [52–54].

10.6.5 Avoidance of Excessive Documentation Requirements

A certain degree of project documentation is absolutely essential for smooth project execution, especially when projects of a certain scale and complexity are involved and cannot be handled by one company/organization alone. However, as much as an appropriate level of project documentation is necessary on the one hand, since the economic situation of a project depends quite significantly on the quality of project planning and documentation, on the other hand the danger of excessive documentation requirements, which can then in turn lead to cost increases, is also very great. Public customers in particular often impose increased and sometimes very bureaucratic documentation requirements on industry, which then result in cost increases. The generally observable trend toward excessive documentation requirements is a byproduct of perfection.

In its search for an appropriate documentation cost, NASA published a report in the early 1970s, immediately after the end of the Apollo program, summarizing the findings of a working group that evaluated the costs of past projects [55]. This recognized that the actual costs of project documentation were not readily identifiable, because the clearly identified costs of 1–4% were only the costs of typing and copying. Other costs for documentation effort are included in other items, such as PM, QA, and systems engineering. A close examination of all proportions revealed a much higher amount of up to 15% of total project costs for the preparation of project documentation, with an average of 3% for duplication and distribution.

A questionnaire survey conducted in this context resulted in the following suggestions for cost reduction, which certainly have general validity:

- The documentation distribution list is often too extensive (*distribution according to the watering-can principle*) and should be kept as small as possible without withholding information for important recipients (e.*g., decision-makers*)—the *need-to-know* principle applies here. It was also suggested that standardized distribution lists be used, that microfilming be intensified, and that a number of documents produced in industry be sent to the client only upon special request. This is now implemented by paper-less, electronic distribution.

- Contracting agencies should make more use of existing in-house reporting procedures, as they have in the past, and not impose a new requirement on firms for each project.
- Clients should critically examine the requested documents in terms of their relevance, as many documents, e.g., reports, often only contain historical data that is no longer relevant for project execution.
- Increased costs are often incurred due to the client's delays in the approval process and a more efficient approval process is therefore required.
- Accelerating project decisions: "*The role of management is to respond intelligently to change*" is a quote from Robert McNamara, reproduced in the book "*The American Challenge*" by J. J. Servan-Schreiber [56]. This is especially true for projects where changes occur that must be responded to intelligently so as not to jeopardize the project goal. In most cases, changes in the project require a decision because they represent a deviation from the planned state. Therefore, any decision-making process requires a detailed analysis of the project operations, commensurate with their scope, as well as a prediction of their potential impact on the project. The time required for this is not infrequently the cause of delays in decision-making, especially when several bodies are involved and the flow of information between the various partners is poor.

Literature

1. Hudock, R. R: The National Scene - DOD Sets New Policy on Weapon Acquisition, in: Astronautics and Aeronautics, August 1970.
2. Stibbe Rosemarie, Kostenmanagement, 3rd edition, Oldenbourg Verlag Munich, 2009.
3. Von Weizsäcker, Carl Friedrich: Wege in die Gefahr; Carl Hanser Verlag, Munich 1976, p. 47
4. Pfohl, Hans-Christian and Wübbenhorst, Klaus L.: Zum Konzept der Lebenszykluskosten, Technische Hochschule Darmstadt - Arbeitspapiere Fachgebiet Betriebswirtschaftslehre, Darmstadt 1982, p. 32.
5. Wildemann, Horst: Kostenprognosen bei Großprojekten, C. E. Poeschel Verlag, Stuttgart, 1982, p. 160
6. Seldon, M. Robert: Life Cycle Costing: A Better Method of Government Procurement, Westview Press, Inc, Boulder, Colorado, 1979, p. 248.
7. Jones, Ray D. and Niebisch, Klaus: Cost Estimating Techniques, INTERNET Expert Seminar on Cost Control in Project Control, Zurich, 1975.
8. PMI, Guide to the Project Management Body of Knowledge (PMBOK Guide), ANSI/PMI 99-001-2013, 5th ed. p. 205.
9. ESA Cost Formats, ECSS-M-ST-60C, July 2008
10. Hoss, G.: The Role of Price - Revisions Forms in ESA Contracts, in: ISPA-News, No. 2, Spring 1980.
11. Madauss, Bernd, Project Costs: 4Cost User Conference, Capital Club, Berlin, September 2011.

12. Kölle, Detrich E.: Handbook of Cost Engineering and Design of Space transportation Systems, Rev. 4 including TRANSCOST 8.2, Report TCS-TR-200, TVS-TransCostSystem, Ottobrunn (dekoelle@t-online.de), p. 185.
13. Batchelder, C. A./Boren, H. E./Campbell, Jr. H. G./Dei Rossi, J. A. and Large, J. R: An Introduction to Equipment Cost Estimating, The Rand Corporation, Memorandum RM-6103-SA, December 1969, 5. 33, 34.
14. Unmanned Spacecraft-Cost Model - Fifth Edition, ST-TR-81-45, USAF, June 1981, pp. V-29, V-40.
15. Beltramo, Michael N.: The Use of System Level Cost Estimating Relationships to Establish Initial Design-to-Cost Targets for Rotorcraft, Fixed Wing and Hybrid Aircraft, SAWE Paper No. 1225, May 1978.
16. Army Material Command: Cost Analysis Handbook, January 1968.
17. ESA, Satellite Databank (SATELDATA), Cost Analysis Group ESA/ESTEC, March 1982.
18. Schelle Heinz, Projekte zum Erfolg führen, 7th ed., Deutscher Taschenbuch Verlag, Munich, 2014; p.167
19. Department of the Army Pamphlet, No. 11-2, May 1976.
20. Starrett, Charles O., Jr.: Parametric Cost Estimating - An Audit Perspective, in: ISPA Journal of Parametrics, Vol. 1, No. 4, Spring 1982, p. 3.
21. RCA-PRICE-Systems: PRICE an automated Parametric Cost Estimating/Predicting Family of Models, GC 1614, 1979.
22. RCA-PRICE-Systems: Users-Manual for PRICE Hardware Model, 1981
23. MBB: PRICE - Model for cost estimation in test at MBB - Greatest experience currently at UR, MBB Aktuell/Intern, 1/83, p. V.
24. MBB: Field report "*Greatest experience currently at UR*", MBB intern, 1/83
25. Augustine, Norman R.: Augustine's Laws and Major System Development Programs, in: Astronautics& Aeronautics, April 1980, p. 36.
26. Madauss, Bernd-J.: Planung und Überwachung von Forschungs- und Entwicklungsprojekten, AlB Fachliteratur, Bad Aibling, 1978/82, pp. III-54, 56, 10.
27. Larson Ronald K. NASA HQ, Washington, Cost Estimating Relationships (CERs) & Cost Models, ISU lecture, Strasbourg, 2012.
28. Madauss, Bernd, Cost Estimating Methodologies for Space Products and Services - A Critical Observation, PMT, Bad Aibling, September 2002.
29. Bethke, Rodin: Modern Material Conservation for Missile Systems, in: Wehrtechnik 3/81, p. 74.
30. Department of Defense: Design to Cost, DoD Directive No. 5000.28, May 23, 1975.
31. Bradley, Fred E.: Design to Cost, McDonnel-Douglas Astronautics Company N75 20414, 1975.
32. BFH Parametrics: DTC Proposal Inputs, letter from Keith Burbridge to MBB, 12/5/1981.
33. Stapel, Arnold: Kosten für Kompressor vor Kauf abschätzen, in: Handelsblatt, 28. 7. 1982
34. Lenk, Reinhard and Lang, Eva: Herstellkosten und Folgelasten öffentlicher Investitionen, Ifo-Institut für Wirtschaftsforschung e. V., Munich, 1981, Vol. 1, p. 216.
35. Coble, Fred W.: LCC Analysis, A Must for Designing Affordable Systems, RCA PRICE Systems, Hollywood, 1981.
36. Blanchard, Benjamin S.: Design and Manage to Life Cycle Cost, M/A Press, Portland, Oregon, 1978.
37. Kerzner, Harald: Projekt-Management, mipt Verlag, Bonn, 2003, Fig. 14.14
38. Cook, Ward H.: Logistics Planning for Phased Programs, Logistics Spectrum, Spring 1973, p. 4.
39. Broder, Melvin A., Bearden David A.: Lifecycle Cost Estimating, in Larson Wiley J. (managing editor), Applied Project Management for Space Systems, McGraw Hill Companies, 2008, p. 315.
40. Laube, Gerd: Kostenprobleme in der Nutzungsphase von Waffensystemen, Wehrtechnik-Seminar, Bonn, 7. 4. 1981.
41. Earles, Marry Eddins: Factors, Formulas, and Structures for Life Cycle Costing, Edding-Earles, 89 Lee Drive, Concord, Mass. 01742, 1978/81.

42. Ehrlenspiel, K./Kiewert, A./Lindemann, U.: Produktkosten senken - eine Aufgabe der Konstruktion, Konstruktion No. 30, 1978.
43. Bitten, Robert E., et al: Cost Estmating of Space Science Missions, Aerospace Report ATR-2013-00108, April 2013.
44. Brents, Tom; Cost Groth, GDC Fort Werth Division, 1988.
45. Madauss, Bernd: Plan to Avoid Catastrophe, Industriemagazin, March 1982
46. Burghardt, Manfred: Introduction to Project Management, Siemens 4th ed. 2002, p.153
47. NASA: Phased Project Planning (PPP) Guideline, NHB 7121.2, 1968.
48. Grumman: Design to Cost, Program History
49. USAF: System Program Office Manual, 375-5, 1964.
50. LMSC: Low-Cost Program Practices for Future NASA Space Programs, LMSC Report D469858, 1975.
51. Eaton, D.: ISEF B - A Minimum-Model Project, in: ESA Bulletin No. 12, 1979.
52. MBB: EXOSAT Study (Optimization of Planning) - Final Report, ESRO Tender AO/557, MBB, 1974
53. CNES: Cost Reduction - An Ariane Production Phase Objective, Report DLA/RC No. 536, 1978.
54. Siemann, H., Wagner, H.: Study on effort optimization in satellite and spacelab experiments, Dornier System GmbH Friedrichshafen, 1978.
55. NASA: Historical Costs Panel - Report to the NASA Space Cost Evaluation Program, 1973.
56. Servan Schreiber, J.-J.: Die amerikanische Herausforderung, Hoffmann und Campe Verlag, Hamburg, 1968, 5. 94

Information Control 11

Contents

Abstract

Information is an important aspect of project management. "Project managers use many communication channels, informal conversations (face-to-face), email, phone calls, meetings, and workshops." (Barker, 2000). Information exchange in the project is the "salt in the soup." Failure to communicate inevitably leads to project problems. A primary responsibility of the appointed proj-

B. -J. Madauss, *Project Management*,
https://doi.org/10.1007/978-3-662-69057-4_11

ect manager is to ensure that the required flow of information between the project team members is continuously maintained.

Information control (IC) is a PM support function, see Fig. 5.4.

The efficiency of project management depends quite significantly on the way information is processed and exchanged between the project team members. Project work, which is principally nearly always under time constraints, can only be optimally planned and controlled if it is ensured that the required project information is passed on to the appropriate recipients as quickly as possible and in a comprehensible and clear form. Information is not least an important prerequisite for decision-making, and project managers and their teams often have to make a disproportionate number of decisions as quickly as possible compared to other areas of a company. Martin states in this context, "*A good project manager makes significantly more decisions than a normal manager at the same organizational level.*" [1] This requires ensuring a purposeful flow of information between those involved. "*The objective of Information/Documentation Management during a space project life cycle is to ensure that every actor has ready access to all the information he needs in order to perform his task.*" [2]. For small projects, it is often sufficient to pass on information verbally for the most part. However, important information should also for small projects be recorded in writing; sometimes a notebook is sufficient for this purpose. Large projects, on the other hand, are far more dependent on information being passed on in a more formal, written form.

It's not just about information you would like to have but also information you need to know for completion of the task at hand. "*The flow of information inside a project is of paramount importance for the achievement of success ….*" [2] The issue of project information cannot be left to chance, because the unstructured use of new media, such as the Internet, smartphones, tablets, etc., which in principle can be considered modern achievements, can also create a chaotic variety of information that can be confusing and not helpful, for project management.

For highly interdisciplinary and technologically complex projects, the rapid processing of information is of particular importance and here the previously mentioned media are helpful, provided they are used in a structured manner. This becomes understandable when one considers that the processing of projects with very high priority usually involves the participation of several companies and/or institutions. Using the example of development programs such as Airbus 380 or Ariane 5, which often involved several hundred engineers, technicians, scientists and managers from government agencies and industrial companies at peak times, one can well understand the importance of information management in the project. This is also true for non-aerospace projects.

Emerging problems in a subordinate part of a large project that has a direct impact on the overall project often require immediate action by the project management to prevent jeopardizing the project goal, the final deadline, or cost overruns by taking timely initiated measures. This requires precise information about the technical context, schedule and cost at the current point in time. Information quality is directly related to timeliness. Servan Schreiber defines it this way, "*The revolutionary advantage in the new way of passing on information is that ideas (information) can be used in a reasonably timely manner.*" [3]

Large project undertakings, for example development projects in facility engineering, the automotive and aerospace industries depend on a well-functioning management information system (*MIS*). However, the basic MIS principles as described in this chapter also apply to smaller projects. However, all too often, in the assumption that the colleagues or interested parties involved will somehow learn what is necessary, a functioning MIS for the project is unfortunately dispensed with. This often leads to misunderstandings and problems due to a lack of information. "*… The questions of who, why, when, about what and how to inform require clear and generally binding answers in the form of organizational regulations.*" [4] Information that is too broadly distributed but not targeted is usually not expedient either, since the "*watering can principle*" often comes to nothing because no one feels addressed and therefore no one reacts. This form is only suitable for passing on general information.

Information management in the project is a serious task of the project management, and the information flows should therefore be defined in an appropriate information procedure before the start of the project. However, care should be taken to ensure that the means and procedures used are appropriate to the project type, size, and complexity. The implementation of a PM information system (*MIS*) does not always have to be expensive. It is a matter of proportionality, but discipline and clarity are important prerequisites for success. Fig. 11.1 shows the different ways in which information can be passed on and the intensity of reception by the recipients, see sect. 11.3.

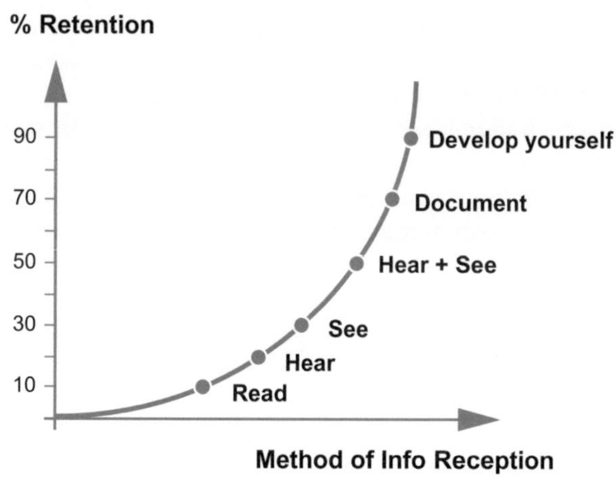

Fig. 11.1 Information transfer and reception by the recipient (Source: Schrader, Munich)

11.1 Project Management Information

11.1.1 Information and Reporting Systems

The management information systems (*MIS*) to be applied in the project are to be adapted to the respective needs. The term information is to be understood here in the sense of a definition by Wahl: "*Information is defined as the communication and utilization of the knowledge that a task owner must have in a specific situation in order to be able to fulfill a defined task.*" [5].

It is important that the formal information channels, including the organizational structure, are defined by the project manager right from the start of the project. This will ensure the necessary exchange of information in the project. In this context, it should be mentioned that one should strive to adapt the formal information channels to the informal information channels as far as possible. The informal information relationships, for example friendships and acquaintances, are often a realistic picture of the interpersonal relationships that should not be underestimated [6]. *In* this regard, an appropriate statement by Aucoin: "*In-person, face-to-face communication typically offers the richest content of the various types of communication.*" [7] The sender/receiver relationship can be classified according to the following criteria:

1. Information that does not require a response from the recipient(s).
2. Reporting information.
   - which are transmitted on the basis of recipient requests or,
   - where recipients are prompted to respond (e.g., *comments or actions*).
3. Formal instructions to which a response must be made in each case.

Of course, in a project where people meet every day, there are unofficial communications in addition to structured information flows. Bußkamp writes, ", *you just get caught up in communications.*" [8] Unofficial conversations between project members are, of course, an important source of information for project work.

However, broadly scattered information, where the author hopes that those affected will react, should be avoided as far as possible. This is not very helpful and is reminiscent of mailboxes that are constantly flooded by direct mail. Recipients do not feel they are being directly addressed and usually do not respond to information from a watering can. If one examines the reason for such distributed information, it usually quickly becomes apparent that this means is always resorted to when the responsibilities in the project are not clearly identified. Often, the author in question tries to protect himself against accusations of inactivity. In the interest of efficient communication, communications and reports should always be targeted.

On the other hand, it is also understandable that information about the project is provided in general terms. "*Information is not only a factual prerequisite for the functioning of an operation, but also satisfies an essential basic human need.*" [4] Project employees have a right to general information, even if it is not directly related to their field of work. Reasonable sharing of information about the overall context of the project, such as information about the project goal, greatly enhances collaboration. Jack Welch, former CEO of General Electric says, "*Whenever a good idea was born, it has to be announced within the whole company.*"

Statement by Ronald K. Larson (20 September 2023): *The challenges encountered during the COVID-19 pandemic in recent years have changed many of the traditional views regarding the necessity of having face-to-face communications within the project team. The evolution of desk-top video conferencing made it possible for team members to exchange information, provide direction, and receive instructions when in-person contact was severely limited. There has been pressure from employers to have staff members return to their classic offices (at least for a portion of each work week) and resistance from employees to retain the flexibility to work from their home offices. There are valid arguments on both sides and the verdict is still out on the final blend of in-person and remote work.*

11.1.2 Verbal and Written Information

Good communication is an important prerequisite for project success. The following types of information are available to the project management [6]:

1. Verbal information.
   - Individual meetings.
   - Reporting to the project manager.
   - Meetings—regular and/or ad hoc.
   - Internal or external presentations.
   - Project reviews.
2. Written information.
   - Memos/Emails.
   - Minutes and protocols.
   - Technical and status reports, etc., see also Chap. 12
   - Interim and final reports.
   - Documents and publications.

The decision about which information type to use depends on the particular needs or requirements of the institutions and staff involved, as well as on the size, type, and complexity of the project. Often, however, all information types are in play. The following examples show how they can be focused on in different projects: [6].

- For large and complex projects, the use of written information, including the written recording of the most essential results of discussions, plays a special role.
- Similarly, for projects with external client requirements or industrial partners, written information transfer will be preferred.
- In the case of internal or smaller projects, on the other hand, verbal information will be preferred to the more time-consuming and less flexible written communication of information. Nevertheless, essential project aspects, for example deci-

sions and minutes of meetings, should be recorded in writing even for small projects.

In any case, the exchange of information in the project takes place to a very large extent in a verbal form. It is above all the many informal discussions mentioned above that contribute to the enormously important team spirit. Written information creates commitment that no project, even a small project, can do without. Collaboration is best fostered through daily conversation. "*Gestures and actions convey additional information, in many cases give further emphasis to the spoken word, and underscore what is being said.*" [8] This plays a particularly important role in interdisciplinary groups.

In large space projects, it is common for project staff to be physically located in a separate office complex, the project office or Project Management Office (*PMO*). Interdisciplinary teams consisting of technicians, planners, financial and legal experts, etc., are housed in adjacent offices. This promotes communication and leads to better quality information. For example, when a project planner (*engineer*) and a project business person are housed in the same office complex, there is inevitably a better understanding between them, as over time there is a mutual transfer of knowledge between their different areas of expertise. This avoids misunderstandings. According to Bußkamp, "*An important reason for misunderstandings lies in people's different communication styles.*" [9] Business people and engineers have a different language culture due to their special backgrounds. However, an optimal solution of technical—economic problems forces the close cooperation of both disciplines.

11.1.3 Information Tools

When passing on information verbally and in writing in the project, it is important to use modern methods, such as email, tele- and videoconferencing, etc., in addition to the usual media, such as letters, telephone, and e-mail. As mentioned at the beginning, the quality of information also depends very much on its timeliness. In the project, information carriers must be used as quickly as possible. The most frequently used media and forums are:

- Telephone (*individual & fast*).
- Internet & e-mail (*fast*).
- Power Point (*graphics*).
- Excel (*spreadsheet*).
- Telenet Forum.
- Webinar, WebEx, and Zoom.
- Signed letters (*binding*).
- Report (*status*).
- Project software (*see Chap. 18*).
- Meeting and control rooms & Teleconferences.

- Video Conferencing & Skype.
- Fax (*only rarely*).
- Database connections.

The media of telephone, internet, letter etc. are, as already mentioned, part of the communication standard of every company. The creation of reports is another very important information tool in the project, which requires a systematic approach. Reports should not be created in an uncoordinated and haphazard manner, so that they suddenly appear as a surprise, so to speak, but should fit into a scheme set by the project management, see subsection 11.2.5.

Regular and ad hoc meetings, which are also important project information means, should follow specific guidelines and should never begin without an agenda or end without minutes. Agenda and *minutes of meeting* (*MOM*) can be prepared very quickly and unbureaucratically if necessary, e. g. also handwritten. Unfortunately, it happens again and again that neither meeting points and objectives, nor meeting results or measures or actions to be carried out by certain employees are formulated and specified clearly enough, which then leads to ambiguities, see also subsection 11.2.1.

Information systems, such as tele- and video-conferencing, lend themselves to the implementation of large and complex project undertakings involving several international partner companies. Equipping a special conference room to conduct video-conferences is a further enrichment of project information systems, as it allows the rapid exchange of important project information even over long distances. Video- and teleconferences save time and travel expenses. As a substitute, Skype can also be mentioned. The pioneer for video conferencing was NASA. As early as 1970, the author was able to visit a video conferencing facility at NASA Houston Texas.

11.2 Project Reporting

11.2.1 Reporting According to Ordered Guidelines

Project reporting is an important source of information. The project control to be carried out by the project control manager in the team is dependent on a well-functioning reporting system because no well-founded decision is possible without precise knowledge of the interrelationships. In addition, reports must reach the recipient regularly or according to a schedule and in an understandable form. Information that arrives by chance and is not fully understood by the recipient often cannot be assessed and may be worthless. Project reporting must therefore follow orderly guidelines, see subsection 11.1.2.

It must be determined who must report, to whom, about what, and when. The reporting flow must in any case be brought into line with the organizational structure and the defined responsibilities. The example in Fig. 5.8 can be used to illustrate the flow of information from the project to the management. It should be noted, however, that information should not only be directed upwards to superiors but also in the opposite direction as well as horizontally to neighboring areas. Lateral information exchange between the project and the specialist departments is a "*must*" in order to realize the matrix concept, see also Fig. 11.2. For system projects, lateral communication to project partners and the specialist departments is a necessity.

The report content is, of course, based on the areas of responsibility of the corresponding job holders. For example, the project controller will report on schedule and cost status, while the systems engineer will report on technical status. Report content should be drafted in such a way that statements are related to planning, any deviations are clearly highlighted, and any resulting or anticipated problems are presented.

The reporting frequency is not to be determined in general, but depends quite decisively on the following factors:

- Type of project (*development, production,* etc.)
- Contract terms (*externally funded projects*).
- Information type (*deadline status, cost status, actions,* etc.)
- Information priority (*bug report, staffing level,* etc.).

It is understandable that a complex development project usually requires a shorter reporting cycle than, for example, a less complex production project. In the first case, the project manager is dependent on the fastest possible information in order to be able to react very quickly to the changes that are common in the development sector. In projects with an external client, the project manager is usually required to keep the client management regularly informed about the project status. In research and development projects, there is often a contractual obligation to report at relatively short intervals (e.g., *monthly*) on the technical, schedule, and financial status of the project.

Information on the financial status of a project is only retrieved on a monthly cycle in almost all companies, as this should be consistent with accounting since monthly reporting is common practice in most companies. Information about the schedule status is also usually recorded monthly, but in schedule-critical cases also fortnightly or weekly, and passed on to the project management. In particularly critical situations, for example, when preparing a rocket launch or schedule-critical work on a construction site, there may also be brief daily reporting. In contrast, a quarterly reporting sequence is often sufficient in certain situations.

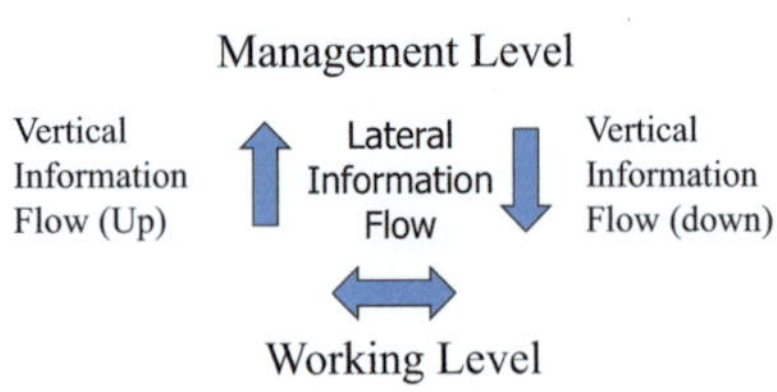

Fig. 11.2 Schematic representation of the information flows

In addition to regular reporting, employees must send an "*exception report*" to the project management within a very short time in exceptional situations. The exception report can be made informally (e.g., *by e-mail*) but should be marked as such and used only in exceptional situations. Regularly held status meetings, in conjunction with the previously submitted written reports (*handouts*), are an important information tool, see also Sect. 11.3.

11.2.2 Fast Information Transfer

Passing on information as quickly as possible is important for project management. Decisions can only be made on the basis of up-to-date information, which means that the project manager must be able to assess the respective situation and the possible consequences before making decisions. In principle, the ideal situation would be achieved if all information could be collected and passed on without delay. With the help of networked IT tools, it is possible to create and pass on information in *real* time. Unfortunately, the reality is often different because project information is usually not available without prior review by the responsible employees. Therefore, the introduction of fixed reporting cycles and meetings is necessary.

11.2.3 Integrated Reporting Systems

"*Regular reporting of status information to project management has become an integral part of modern project management.*" [10] For projects with a multi-level project organization or an external client, subcontractors, and suppliers, regular reporting is particularly important, and therefore the reporting flows, report contents, and reporting frequency should be defined at the start of the project as part of the contract negotiations. In addition, in multistage projects, it must be remembered that detailed information can be usefully condensed into top management information. Fig. 11.3 shows an example of planned information flow between project

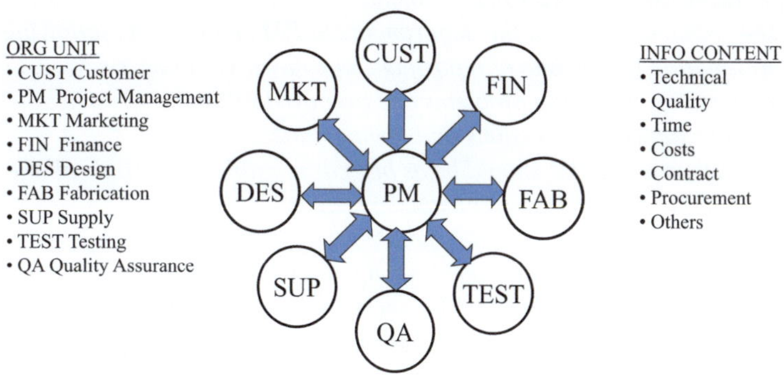

Fig. 11.3 Planned information flow

management (*PM*) and the functional departments. The content can be technical, scheduling, financial, and other information.

An eminently important task of project management is to ensure that communication works both within the team and between the actual project team members and the wider stakeholder group [11].

Project status reports must be prepared at all levels defined in the work breakdown structure. At the lowest level, the WP managers must forward their results to the subsystem managers, who in turn report to the project management level (*system level*). It is advantageous if the report content is identical at all levels. Although the topic of "*reporting*" is not new, in many companies there are no consistent standards in which a reporting structure is clearly defined.

11.2.4 Standard Status Reports: Examples

Project managers are responsible for regularly determining the project status and forwarding the status information to management and/or the customer in the form of a written *status report*. Five practical examples of status reports at different levels are given below. In principle, status reports contain technical, scheduling, financial, and contractual information. Subsection 11.2.5 defines the contents in detail:

A. Project management to company management (CM):
   1. *The status report from the PM to CM contains the following status information at the time of reporting (time now):* (*1*) *order backlog,* (2) *technical status, including problems and proposed solutions,* (*3*) *schedule status,* (*4*) *cost status,* (*5*) *decisions and open actions.*
   2. *The report should be kept as brief as possible and limited to approximately 5 to 10 pages. Additional details can be provided later by request.*
   3. *Reporting and meeting cycle: monthly to quarterly, depending on priority.*
B. Project management to customer: *identical to A., if no special requirements are specified by the customer. The report structure can be taken over from A, but the contents are to be checked and changed if necessary.*
C. Departments to the project management:
   1. *The status report from the departments to PM contains the following status information at the time of reporting (time now):* (*1*) *order backlog,* (2) *technical status, including problems and proposed solutions,* (*3*) *schedule status,* (*4*) *cost status,* (*5*) *decisions and open actions.*
   2. *The report to the PM should be as brief as possible but include all the necessary details for a subsequent status meeting.*
   3. *Reporting and meeting cycle: monthly.*
D. Subcontractor to project management *identical with C.*
E. Work package owner to the next higher organizational unit:
   1. *WP status reports are to be sent to the respective supervisors in the department with a copy to PM with the following contents:* (*1*) *work order,* (2) *technical status,* (*3*) *schedule status,* (*4*) *cost status, and* (*5*) *open actions.*

2. *The report should be as brief as possible but include all the necessary details for a subsequent status discussion.*
3. *Reporting and meeting cycle: should be synchronized and depending on the priority (weekly or monthly).*

11.2.5 Standard Contents: Report Contents

Good communication is an important prerequisite for project success. In order to standardize information management, NASA developed a special information system as early as 1966, which was published as a manual under the name *Project Management Information and Control System (MICS)* and is still regarded as a *path maker* today [11]. Since the procedure created is very systematic and simple in structure, the author has retained the outline concept in this edition. The MICS reporting methodology can be used at all project levels since the number of pages is limited [12]. General Electric (*GE*) developed a procedure for regular project review in 1969 based on MICS under the title "*Program Appraisal and Review*" *(PAR)* [13]. MBB adopted the system of regular project review in 1975 as part of a cooperation with GE [14]. The author was involved in the PAR implementation at MBB, see the reporting structure shown in Fig. 11.4.

The main advantage of an integrated reporting or information system is the targeted consolidation of all technical and commercial project parameters into a single

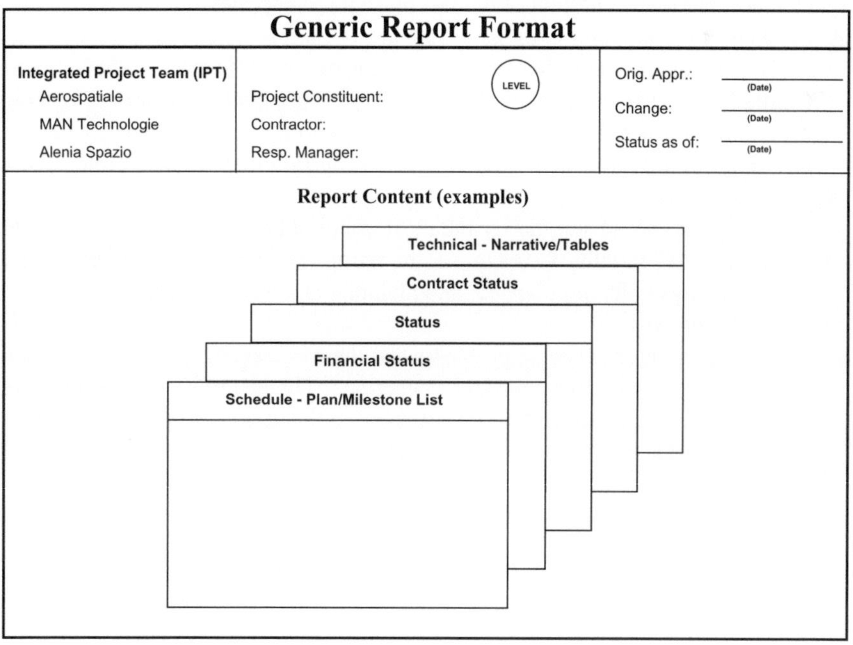

Fig. 11.4 Example of a report structure

reporting package. The project managers and their teams, but also the company management (*CM*) and, if applicable, also the client (*customer*) are provided with an instrument that provides information on all project areas. Using the PAR reporting package as an example, the report contents of a multi-page status report to the management and/or the customer, as is usual for large projects, are shown below:

1. *Title page and main project data.*
   - Project name.
   - Client (*internal or external*).
   - Order number.
   - Type of contract (*fixed price, cost reimbursement,* etc.)
   - Duration of contract (*start and end*).
   - Order volume (contract value).
   - Contract Changes (*applied for, approved*).
   - Deadline status (*on schedule, behind schedule*).
   - Cost extrapolation (*final costs*).
   - Report status (*date*).
   - Reporting date.
   - Reporting by ….
2. *Project organization (see Fig 5.3).*

   Reproduction of the project organization valid at the time, including the interfaces to the customer, subcontractors, and internal company departments. In addition, information on key personnel and their main tasks. If there are no changes, the representation of the last presentation applies.
3. *Configuration description of the technical system (see Sect. 12.2).*

   The actual product, for example hardware or software, is shown here: Configuration sketches, exploded views, or functional diagrams. The brief description must reflect the task, technical requirements, main data (*for example, dimension, performance, and weight*), design alternatives, and project risks.
4. *Important events.*

   This sheet should summarize the most important events of the last 2 months and the future 3 months, for example, the status of key milestones, work results, work meetings held and planned, and planning changes implemented.
5. *Critical problems.*

   Here, current or expected problems and their influences on costs, deadlines, and technical performance are to be described. The resulting actions that have either already been initiated or are still to be initiated, as well as expected decisions by management, must also be listed.
6. *Action Control Overview.*

   This is the place to list opinions, results, and completion notes for actions (*action items*) defined in the past, see subsection 9.4.4.
7. *Work breakdown structure* (*see subsection 9.3.3).*

   Only the upper levels of the work breakdown structure (*WBS*).
8. *Project schedule and flowchart* (*see subsection 9.3.4.3 and Fig. 9.30).*

The main schedule is to be inserted here, from which both the planned and the actually achieved deadlines as well as a deadline deviation analysis must be visible.

9. *Project costs (see* section 10*).*

   Summary of the hours and costs incurred up to the reporting date in comparison with the planned targets. In connection with the schedule analysis (8), the results of the final cost calculation (*cost-to-completion—CTC*) and any suggestions for possible savings must also be reported.

10. *Integrated schedule/cost analysis (see subsection 9.4.)*

    On the last page, an integrated performance analysis, that is a comparison between the schedule and cost situation, must be performed; example *Earned Value Analysis—(EVA), see Fig. 9.38.*

For medium and small projects, the reporting package must be reduced accordingly, see definition of project size at Table 1.1.

The creation of a monthly status report is of course associated with effort, although most documents only need to be recreated when there are changes. For large projects, the introduction of such a reporting system is, therefore, recommended in any case. For smaller projects, a condensed report may be sufficient. The previously mentioned version can then be used as a guideline for designing a simpler system. The author was an employee of SES Astra and responsible for the implementation of an integrated project control and reporting system. In an effort to further simplify the reporting system, a one-page weekly report, "*the one-page report*," was introduced at the department head level. All department heads who participated in the weekly status meeting were required to submit a one-page report to the upper management 1 day before the weekly status meeting, see also Fig. 11.5.

One-Page-Report

1. Goals and Strategy Points	**2. Important Results from Last Week**
Short description of the most important goals in a specified time period (e.g. week)	Description of the achieved objectives, deviations and special results; e.g. Decisions, completed reports, important calls or business travel.
3. Problems and Recommendations	**4. Planning for Coming Week**
Technical or programmatic problems encountered and recommendations for resolution or measures already implemented	Planned activities for the coming period (e.g. week). Tasks to be completed, travel, and their scheduling.

Fig. 11.5 Abbreviated weekly status report

11.3 Project Meetings

The ability to remember information depends strongly on the ability to take in information, to read, to hear, to see, to document or to acquire it oneself [15]. According to this, hearing and seeing in interaction reach a value of retention of more than 50% in comparison to "*self-worked out*" with approx. 100%, see also Fig. 11.1. Practice confirms that presentations including pictures are more informative than the spoken word. "*A picture is worth a thousand words*" (*proverb*).

11.3.1 Exchange of Information through Meetings

Regularly and ad hoc scheduled working meetings are an important source of information for the project management. The main advantages are the fast exchange of information, the possibility of immediate reaction, and the personal contact of all meeting participants. Identifiable problems can be analyzed and dealt with immediately. Since the time required for meetings and the associated costs are sometimes very large, each meeting must therefore be thoroughly prepared. This includes, above all, the presentation of an agenda, the determination of the circle of participants, the appointment of the meeting leader (*chairman*), and the writing of meeting minutes (*minutes of meeting*). Careful definition of the group of participants helps to avoid unnecessary attendance and unnecessary personnel and, if necessary, travel costs.

It goes without saying that meetings to solve technical problems cannot be conducted in a comparably short time. However, the principle that a well-prepared meeting (*agenda, minutes of results,* etc.) and a carefully selected group of participants to increase meeting efficiency also applies here.

11.3.2 Regular Status Meetings

Regular status meetings, which take place weekly, fortnightly, or monthly, represent a special form of meeting. As already mentioned, the main objective of status meetings is to record the respective project status and to pass this information on to the project management. The project management is to be informed about the technical performance achieved, the schedule status, and cost situation compared to the planed budget, see subsection 9.4.5. The importance of regular status meetings cannot be emphasized strongly enough. But at the same time, the quality of the status meetings to be held is also important. Unfortunately, many meetings are held without the necessary prerequisites. It happens again and again that no or only insufficiently detailed project documents are available, and the status statements are often completely inadequate. The question about the status of the project is often answered with the recurring standard answer "*everything is going according to plan.*" At a closer look, however, there are sometimes no valid plans available, and the project status is not known in detail either.

It is easy to see that the project manager cannot carry out any exact project control on the basis of such unclear information. In any case, the information about the project status must be clearly quantified and transparently presented on the basis of an approved plan, see Fig. 9.19. Status meetings should therefore be held on the basis of previously available status reports, see also Sect. 11.2. Status reports, also the weekly one-page report, are the basis for the status meeting. A procedure that has been tried and tested in practice is to be suggested, in which the status report is distributed for example 1 day before the status meeting. The facts shown in the status report, for example, the schedule status compared to the original schedule and any scheduling problems that may have arisen, provide a reasonable and, above all, strongly objectified basis for discussion. Here is a tabulation of the advantages and disadvantages of status meetings:

- Advantages:
  - Fast exchange of information.
  - Immediate reaction.
  - Personal contact.
  - Recognizable problems can be treated immediately.
- Disadvantages:
  - Time required (*many participants may be required*).
  - Costs (*meeting hours*).
  - Room requirement (*-planning*).
- Prerequisites:
  - Good preparation.
  - Agenda.
  - Meeting leader.
  - Tight conversation.
  - Result minutes & actions.

Project meetings have a very different character. While status meetings are basically about announcing the state of affairs or, in other words, establishing facts, development and design meetings, on the other hand, are innovative in nature. Status meetings are, therefore, much easier to time. One example is the standard agenda developed by the author:

- Review of the last minutes.
- Action list review.
- Scheduling and status review.
- Resource and cost review.
- Technical status and problems.
  - Identified problems.
  - Potential problems.
  - Proposed solutions.
  - Initiated measures.
- Others as required.

In-house status meetings should be held on critical projects "*as often as possible but briefly*"; for example, weekly for a maximum of one-to-one and one-half hours. The following points should be observed:

- Weekly status meetings should be held on the same day each week without gaps, so that it becomes a "*jour fixe.*" All those concerned should take part. In case of unavoidable absence, a well-prepared representative must be sent.
- An agenda shall be distributed by the meeting chair to all meeting participants at the previous day.
- Meeting participants should send a one-page status report (*one page report*) to the PM on the previous day, see Fig. 11.5.
- The meeting time should not exceed one or up to a maximum of 2 h. In the interest of keeping the meeting time tight, it should be avoided that problems that cannot be solved immediately are clarified during the status meeting. For issues that cannot be resolved immediately, a separate meeting with different participants must be scheduled by the meeting chair.
- *Minutes of Meeting* (*MOM*) must be prepared and distributed quickly, recording decisions, action status, etc.—no verbatim minutes are to be prepared.
- New actions (*action items*) are to be included in the action list.

11.3.3 Project Reviews: Reviews

Conducting project reviews is an effective management tool for project management, see Sect. 7.3. However, one must distinguish between two review categories:

- Technical reviews, for example a development review (*Critical Design Review—CDR for example*) and.
- Regular status reviews.

Technical reviews, such as a CDR, are used for a detailed review of the technical development status at all project levels (*system, subsystems, and components*) and take place at specified dates (*milestones*) of the life cycle, see also Fig. 4.2.

In addition, there are regular status reviews, which should take place, for example, at monthly or quarterly intervals. For critical projects, a monthly meeting certainly makes sense and for less critical projects it is often sufficient to meet for a status review only every 6 weeks or even only quarterly, see also Section 9.4.

It is important that there is efficient communication between the different management levels. The Apollo program is a good example of how communication took place between all levels of the program [16]. Regular status meetings were used to directly exchange technical and administrative information. Status reports *(see also subsection 11.2.5*) are the basis of a status review *(meetings*). Wernher von Braun aptly mentions, "*We think this success (Apollo) has something to do with our review procedure-specifically, the introduction of a management advisory board.*" [17]

The aim of the official status reviews is a critical discussion of the services provided, the problems and possible solutions identified, the schedule and cost situation, and a strengths and weaknesses analysis (*SWOT*). In parallel to the aforementioned official status reviews, supplementary internal company meetings should be held in critical situations. In addition to the monthly status meetings, NASA also conducted regular meetings at the individual centers [18]. A special feature was daily but very brief "*stand-up meetings*" where people could talk about the day's upcoming actions over a cup of coffee in a relaxed and collegial atmosphere at the beginning of the day. As an "*equal among equals,*" the boss could join in the unofficial discussions.

11.3.4 The Results Protocol: Minutes

Project meetings do not serve an end in themselves but have a very specific task to fulfill, which must be clearly and unambiguously defined in the context of the agenda. Consequently, project meetings should not be concluded without a record of the results. What is meant, however, is not a verbatim record, but a record describing the meeting results and measures to be initiated, actions to be taken, and decisions made.

It is advantageous if the meeting minutes are prepared during the meeting (*handwritten, if necessary)* and read out to the participants immediately so that corrections can be made on the spot. The minutes can then be approved and distributed immediately. Especially in meetings with other companies, clients, etc., it is convenient if the meeting result is available immediately and can then be signed by the respective partner representatives. The minutes are then a binding project document. Of course, there is no question that occasionally one would like to avoid precisely this in order to coordinate internally beforehand in the case of serious decisions. In such cases, the protocol will not contain any firm agreements, but it will contain a binding promise that the decision in question can be achieved within a certain period of time.

For many readers of this book, the obligation to take minutes is certainly self-evident, but in practice one can repeatedly observe that time-consuming and costly meetings are held without a concrete result having been defined, see also Sect. 3.1. Admittedly, especially in meetings in which very conflicting views clash, it is often not possible to crystallize a clear result at first glance. However, the meeting leader has the duty to steer the meeting in such a way that the opposing views can be clearly formulated and measures for solving the problem can be identified. Thus, it can be generalized that taking minutes of the results is important in any case, even when there is disagreement, because the minutes make any problems transparent and provides documentation. Project work is almost always highly correlated to schedules and costs and heavily dependent on meeting results. Poorly conducted and unrecorded project meetings often contribute to significant project problems.

11.4 Modern Information Systems

11.4.1 The Project Control and Information Center

In the handling of large and complex projects, such as the MRCA Tornado project or the SPACELAB space laboratory, the establishment of a central control and information center, where all project information comes together, has proven its worth. However, the establishment of such centers is also recommended for projects that do not have the dimensions of the above-mentioned projects. After all, it is possible for a company to use such a control and information center for several projects at the same time or for the entire company. The best solution is therefore to set up a central Project Management Office (*PMO*) where all the threads come together, see also subsection 5.2.5.

The basic idea of a project control and information center is not new and is common, for example, in military command centers. Probably almost every manager will have secretly thought about the introduction of an information system at some time or another when he was sitting in a sparsely equipped conference room and very specific information important for the meeting was either not available to him or the documents available to him were not visible or clearly prepared for everyone. This thought, which certainly arises frequently, leads directly to the project control and information center. The importance of such a center is directly related to the speed with which decisions have to be made by management. Bilstein writes in this regard, "*NASA's Project Control and Information Center, which became known as the Program and Control Center, served as the basis for all program decisions by project management.*" [19]

It had been recognized early on how important it was to combine the different information flowing together from all program areas, for example costs, deadlines, technical parameters, organizational aspects, etc., into an overall summary. This was also the main reason for setting up the control and information center described above. The Apollo Management Manual for the Marshall Space Flight Center (*MSFC*), released by Wernher von Braun, states, "*The Apollo Program Control Center provided management with an integrated and detailed view of cost, schedule, technical performance, and program status related to plan data.*" [20] The focal point of the information and control center is the control room with corresponding equipment, such as the projection and telephone installations, and IT software tools [19], see also Chap. 18.

11.4.2 Bridging Distances

The realization of large projects, such as development work on aircraft or the construction of power plants, almost always involves many companies. But it is certainly not limited to very large projects, because smaller projects, for example studies, are also increasingly being carried out jointly by cooperative ventures of companies. For this reason, business trips are now more than ever the order of the day, and this often means a great expenditure of time and money. Video and teleconferencing can be used for status meetings and other routine meetings because they can reduce costs.

11.4.3 Communication Via Video/Teleconference

The video/teleconference procedure, which turned out to be appropriate and in which several people participate at the same time, foresees that the participants of the conversation send their presentation documents in advance and present them simultaneously using appropriate display technology in all participating conference rooms. Of course, holding a video/teleconference requires increased discipline. In this regard, Wetmore states that participants who ask a question or make a comment should identify themselves in advance by giving their name and location [21]. He adds that the prepared graphics force a certain discipline. The procedure described earlier is now used worldwide.

However, whether meeting in the office or via video or teleconference, effective communication should consider the following five points, according to Aucoin: [22].

- *Appreciate the audience*—The recipients should be considered as customers.
- *Establish mind share*—The communicator (*sender*) should formulate his statements in a comprehensible way, so that the receiver understands and remembers them.
- *Deliver simply and* clearly—information should be formulated simply and clearly.
- *Prompt the change*—An important reason of communication is to influence the recipients to change things.
- *Be human*—Although business communication is mostly about passing on very dry information, human aspects should not be disregarded if possible.

11.4.4 MIS Checklist

In preparation for a Metro project to be performed by a general contractor (*GC*), the following checklist was prepared by the author as part of an RFP:

1. Has the turnkey contractor *(TKC*) developed an information and reporting plan?
2. Are the reporting paths between the TKC, subcontractors, and suppliers defined?
3. Do standard distribution lists exist?
4. Has the TKC ensured the provision of technical communication equipment such as telephone and video- and teleconferencing equipment?

5. Are there standard guidelines for meetings and project reviews?
6. Have reporting guidelines (*status reports, action lists,* etc.) been established by the TKC?
7. Has the TKC established a schedule for meetings and reporting to the client and/or company management?
8. Have rules been created for special and urgent reports (*emergency reports*)?
9. Has the TKC established a firm-wide and/or project-specific action monitoring system?

Literature

1. Martin, Charles C.: Project Management - how to make it work, AMACOM, 1976, p. 49
2. ESA, ECSS-M-50A, 1996, P 7
3. Servan-Schreiber, J.-J.: Die amerikanische Herausforderung, Hoffmann und Campe Verlag, Hamburg, 1968, p. 107
4. Work Reports of the Committee for Social Company Design at the Confederation of German Employers' Associations - Information for Company Management No. 6, October 1979, p. 2
5. Wahl, Manfred P.: Grundlagen eines Management-Informationssystems, Luchterhand Verlag, 1969, p. 123
6. Madauss, Bernd-J.: Planung und Überwachung von Forschungs- und Entwicklungsprojekten, AIB Fachliteratur, Bad Aibling, 1978/81, p. VI-3/VI-4
7. Aucoin, B. Michael: from engineer to manager mastering the transition, Artech House, Inc., Boston - London, 2002, p. 200
8. Bußkamp, Werner: Kommunikations-Management, Wirtschaftsverlag Langen Müller/Herbig. Munich 1998, p. 16
9. Source 8, p. 51
10. Source 6, p. VI-11
11. Hobbs, Peter: Professionelles Projektmanagement, (German edition), mvg-verlag im verlag moderne industrie, Landsberg am Lech, 2000, p. 75
12. NASA: OSSA/OART Project Management information and control system (MICS), NASA Handbook 2340.2, 1966
13. General Electric, Space division, Program Appraisal and Review System (PAR), August 20, 1969
14. MBB: Program Appraisal and Review, R W 1-75/6, 1975
15. Schrader, Einhard, Straub, Walter G.: Darstellungstechniken und Techniken zur Auswahl und Verdichtung von Informationen, Munich
16. NASA: NASA Apollo Program Management, Vol. I, December 1967, pp. 4–29
17. Committee on Science and Astronautics: Apollo program management staff study for the subcommittee on NASA oversight of the committee on science and astronautics, VS House of Representatives, Ninety-First Congress, First Session, Serial C, July 1969, p. 10
18. Q. 16, P. 4–31
19. Bilstein, Roger E.: The Saturn management concept, NASA Report CR-129029, June 1974, p. 37
20. NASA: NASA-Apollo Program Management, Vol. III, December 1967, pp. 4–15
21. Wetmore, Warren C.: Boeing network cuts Apollo work costs, aviation week and space technology, February 10, 1969, p. 43
22. Source 7, p. 206ff

12 Documentation and Configuration Control

Contents

Abstract

Documentation and configuration monitoring is an integral part of project management. The development results and the respective configuration status must be documented by the project management from the very beginning during all project phases.

Documentation & configuration control (DC) is a PM support functions; see Fig. 5.4.

B. -J. Madauss, *Project Management*,
https://doi.org/10.1007/978-3-662-69057-4_12

This chapter was prepared with close collaboration of Heinz-Herrmann Müller, Airbus Safran Launchers, Bremen.

This chapter addresses two important project management tasks that are closely intertwined:

(a) documentation management and
(b) configuration management.

Responsibility for both tasks is, therefore, usually in one hand during project implementation. However, long-term archiving should be assigned to a central archiving office of the respective company, as there is usually a long-term retention obligation.

Professional documentation management is the basis for a well-performing configuration management. Experienced project engineers know the correlation between good documentation and configuration management and the costs of a project. Even when laying the foundation stone of a new project, one starts from very specific prerequisites, such as the documented project requirement (*objective*) including all technical, economic, and legal boundary conditions.

An initial rough cost estimate can then be made on this basis; see Sect. 10.1. The project requirement (*mission requirement*) specified by the internal or external customer should already be documented in writing. Changes to the configuration defined in the project requirement and any resulting technical, scheduling, and financial effects should also be documented. This is the beginning of documentation and configuration management.

In small projects and in the realization of the preliminary phases of large projects, the task of documentation and configuration management often has to be performed by the project manager in personnel union. This is, of course, a major burden. That is why it is even more important that companies, whether large or small, should implement a documentation management system (*DMS*) because project managers and their teams often spend many hours on useless redundant work searching for documents if there is no central archiving available in the company.

12.1 Documentation Management

12.1.1 Importance of Documentation Management

The definition of a document according to ISO 8613-1 is as follows: "*A document is a defined and structured set of information that can be managed as a unit and exchanged between users and systems.*" The execution of projects is always associated with the creation of more or less extensive project documentation, consisting of specifications, plans, procedures, drawings, etc. The early planning phases (see Chap. 4) of a project are essentially focused on creating plans, specifications, procedures, etc. The results of the planning phases consist largely of corresponding documents since the entire development effort is reflected in paper and/or electronic data, for example in CAD files. Here is a quote from Götzer: "*Document management includes … all processes, procedures and responsibilities related to the administration of documents.*" [1]The mere fact that employees working on large project undertakings often need several years to develop a future project underscores the importance of document management. It can happen under certain circumstances that a development team works intensively for several years on the preparation of a new project, but in the end the results are not available as a complete and verifiable documentation package due to the lack of well-structured documentation, but only fragmentary, incoherent individual documents without a recognizable structure. Gröpper mentions concrete examples of mismanagement in the documentation area: [2]. "*Companies have to cope with ever-increasing amounts of data and information and manage the challenges associated with it. According to estimates,*

- *85% of the information is unstructured,*
- *30% of working time is spent searching for information,*
- *and 40% of work time is attributable to handling paper documents alone.*"

The cases of inadequate documentation are unfortunately not rare. Especially in the case of small projects carried out under great time and cost pressure, it is often neglected to create consistent and comprehensible documentation. This can have catastrophic consequences when the project is accepted by the customer, as it often leads to necessary rework. In most cases, this also results in problems during subsequent repair or maintenance work, for which good documentation is also necessary.

An international company, which was entrusted with the implementation of a multibillion-dollar power plant financed by the public sector, confirmed this in a letter with the following words: " … *however, we can only point out again at this point that no overview list of the project documentation exists for the project.*" [3] Carsten Eckart rightly mentions that document management is treated "*stepmotherly*" in most projects, and the reason most often given is: "*We don't have time for it.*" And he adds to this, " … *that a large part of the information that is not set down in writing is lost to memory again within a very short time.*" [4]

Documentation management ensures that the documents in circulation and their issue status as well as the distribution key are recorded and controlled. This means that, via a documentation overview or distribution list, it should be possible at any time to display the respective status of the changes and the distribution key in the project for all to see via all documents belonging to the project and officially issued.

It is important that all project documents, their issue status, and distribution to project participants, that is customers, subcontractors, and project staff, are systematically recorded. This ensures that all project partners are provided in good time with the documents they need and that they are kept up to date, and that the planned project work can be carried out on the basis of up-to-date documents. Götzer addresses the technical requirements of document management in the public authority environment, where the focus is on ordinances and rules, for example, as well as the technical environment, which involves complex documentation structures such as technical drawings and CAD documents [5]. In the course of tightening liability issues and high-quality requirements, the proof of and procedures followed in the case of documentation changes are becoming increasingly important [6].

12.1.2 Main Tasks of Documentation Management

"*Data and documents are important resources for a company. In many cases, they are the only evidence of the fulfillment of obligations arising from the actions, processes and products of a company or institution.*" [7] Early in the project, it is necessary to be clear about which documents are required for project execution and the importance of these documents to the project. Furthermore, it must be determined who is responsible for the creation of the respective documents and how the release procedure should work. Thorough planning of these processes prevents important actions from being left to chance. The responsibilities and required processes are usually written down in a "*document management plan.*" Here is a rough outline:

- Identification of all required documents.
- Creation of a documentation requirement list (*DRL*).
- Document requirement description (*DRD*).
- Determination of the document distribution.
- Document status monitoring.
  - Documents planned.
  - Documents in preparation.
  - Documents distributed.
  - Documents in the review or change process.
  - Reissue of the amended document.

Performing these activities ensures that the right documents are routed to the right recipients at the right time. The significance of this statement can be easily understood with an example. For the construction of a large industrial plant, several thousand individual documents have to be created, released, and distributed. However, many of these documents have to be changed several times during the course of the project, and this means that the documents in circulation have to be replaced by new ones. It is easy to imagine what happens when different outputs are used in different places: nothing fits together anymore.

For large projects with the involvement of multiple subcontractors and suppliers, documentation control becomes very complex and costly. It is, therefore, necessary to apply appropriate SW tools to support this process. There are a variety of SW tools and databases that can support documentation management efficiently. An overview of the tools offered is provided by the Softguide platform http://www.softguide.de.

There are good reasons that companies introduce a centralized document management system (*DMS*). Götzer points out that the introduction of a DMS is a complex organizational and technical task. The requirements and goals of one's own organization determine the outcome of the project. Therefore, the focus is not on buying and implementing software but on building a specific document management solution for the company [8]. Companies where many projects are running in parallel are encouraged to create a central DMS organizational unit for the following reasons:

- Implementation of DMS standards.
  - General DMS guidelines.
  - Standard formats and templates.
  - DMS software.
  - Numbering systems.
  - Testing and release.
- Management of a central archive.
  - Warranty requirements.
  - Legal requirements for the duty to store.
  - Creation of a know-how database.

During project execution, it may be that project-specific archives are also created to ensure fast access. At the same time, however, copies of all documents should always be stored in the central archive of the DMS.

12.1.3 Identification of the Document Types

A first important measure for the identification and classification of the project documents is the definition of the different document types. This provides a basic overview of the diversity of project documents. A typical example of documentation types required for the creation of a list of required project documents (*Document Requirements List—DRL*) is provided below:

- Requirements.
- Regulations.
- Specifications.
- Plans.
- Procedures.
- Reports.
- Deviation or error reports.
- Acceptance documents.
- Manuals.
- Contract documents.
- Construction documents (*drawings, parts lists,* etc.).

These document types can be further subdivided according to their specific purpose, as demonstrated by examples:

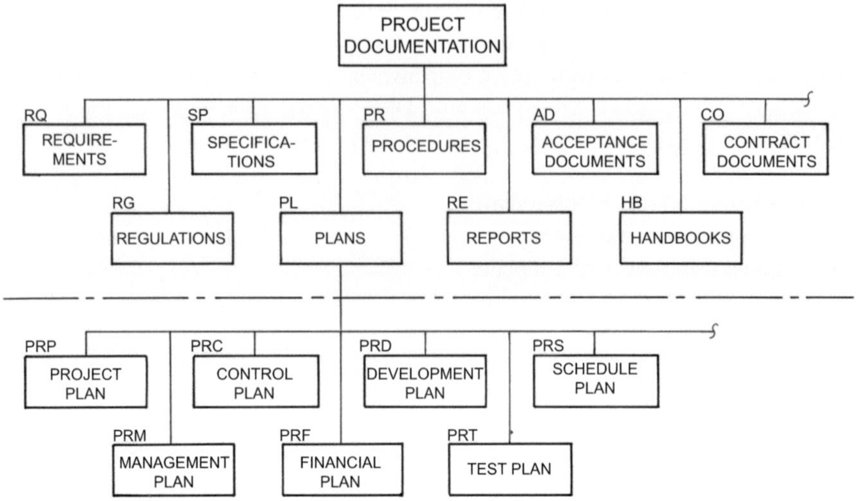

Fig. 12.1 Example of a documentation tree

- Specifications (*see Sect. 7.2*):
  - System specification.
  - Design specification.
  - Manufacturing specification.
  - Test and acceptance specification.
  - Interface Specification.
  - etc.
- Plans (*see Sect. 6.1*):
  - Project plan.
  - Management Plan.
  - Control plan (*configuration, time, cost,* etc.)
  - Financial plan.
  - Development Plan.
  - Test plan.
  - Facility plan.
  - etc.

For better understanding the representation of the different document types in the form of a tree, the "Documentation *Tree,*" has proven useful; see Fig. 12.1.

12.1.4 Determination of Documentation Requirements

The identification of documentation types is only the first step in determining the required project documentation. It represents a rough classification and serves as a guide. The definition of documentation requirements builds on the identification of types and must be carried out in close cooperation with the development management. For each project element defined in the work breakdown structure (*WBS*), the required project documentation must be determined. In other words, in cooperation with the respective person responsible for a WBS element or a work package, it must be determined which specifications, plans, drawings, etc. are required for his work.

In the early phases of a project, in which, with a few exceptions, no hardware is yet created, the final results of the respective phase are clearly summarized in the documentation requirements list (DRL). Incidentally, this also applies to studies. In other words, the final product of an early project phase or a study, which ultimately consists only of documents, can be excellently presented using the DRL. For third-party orders, the DRL becomes part of the *Statement of Work (SOW)* and thus also part of the contract; see Sect. 13.2.

As part of the NASA Apollo program, General Electric (*GE*) developed in the 1960s a procedure for standardizing project documentation. Initially, the focus was on the documents to be created in a project, the *Document Requirements, which* are usually *summarized* in a Document/Data Requirement List *(DRL)*. In order to provide guidance to authors on how to create documents, NASA published 1970 a manual with several hundred proven *standard Document/Data Requirement Descriptions (DRDs)*. This resulted on average in a cost reduction of $500 per DRD [9].

The DRL should include the following information:

- Documentation name (*title*).
- Documentation number.
- Document creator (*company/department/function*).
- Required issue date; milestone (*first issue*), if applicable.
- Requested updates (*status*).
- Review & approval (*by whom*).
- planned distribution.

The DRL, which should not be confused with the documentation status list to be created during the course of the project, is an important planning tool. Table 12.1 below shows a practical example (*excerpt*) from a DRL for the management area. In reality, the DRL often contains technical and programmatic delivery contents on several pages. The completion or delivery date is defined by the respective milestone.

The DRD should be seen as a supplement to the documentation requirement list, and the creation of a DRD is extremely important for project management. This ensures that the documentation content is what the project management expects. The importance of this tool should not be underestimated because a clear description of the desired content ensures that there are no surprises. Without such a specification, the following surprises could occur:

- the document is not detailed enough—*too little content*,

Table 12.1 Example of a documentation requirements list (DRL)

Document Title	*Quantity*	*Due Date*	*SOW Chapter*
• Project Management plan.	5	PDR	6.9
• Documentation and configuration management plan.	5	PDR	6.12.2
• Monthly status report.	1	Monthly	6.2.6.1
• Quarterly report: Technical, management, cost, and contract.	5	Quarterly	6.2.6.2
• PDR data package.	10	PDR—2 weeks	6.10.2
• CDR data package.	10	CDR—2 weeks	6.10.3
• Software Control plan.	10	PDR	7.3

- the document did not address the current situation or was too general and the facts do not apply—*bypass the problem,*
- the document is unstructured and/or too voluminous—*too many details.*

Therefore, a sure way to avoid the above-mentioned problems is to provide a detailed specification of the content or a DRD pattern. In this context, one should keep in mind that the creation of detailed plans, specifications, etc. for large and complex projects often takes several months and a large amount of manpower. The development phase, as mentioned above, is often focused exclusively on the creation of documents regardless of whether they are paper documents or electronic media or both. Precise planning of documents, that is the creation of requirements descriptions, is therefore worthwhile. Using a project control plan as an example, project management should create a DRD with at least the following specifications:

1. *Purpose of the document:* to provide procedures, techniques, guidelines, and examples for planning, monitoring, and assessing the project.
2. *Responsible organization:* Project Control Management.
3. *Reference to Other Plans:* Project Plan, Management Plan, Work Breakdown Structure, Schedule, and Finance Plan
4. *References:* company's Planning Manual, company's Organizational Instructions, etc.
5. *Creation instructions:*
   (a) Introduction

   Brief summary of how the plan works and how project requirements are met for project planning and monitoring.

   (b) Planning requirements.

   Interpretation of the set requirements

   (c) Proposed planning strategy.

   The planning strategy must be described in detail and comprehensively, for example, the individual functions for early detection of problems trend analyses and how corrective measures are initiated

(d) Planning tasks.

Description of the individual tasks and their combination into an integrated statement

–Schedule monitoring.

–Cost monitoring.

–Integrated power measurement.

–Control of subcontractors.

(e) Reporting.

– Report contents.
– Weekly and monthly status report.
– Quarterly reports.
– Other reports.

In the meantime, the DRD method has become a standard for handling technology projects, especially in the aerospace industry. The modular structure of both the documentation requirements list (*DRL*) and the requirements description within a Document Requirement Description (*DRD*) is a great relief for the project. With the introduction of this system, however, another positive effect occurs. Already existing DRDs can be reused for a new project without much effort. Often, only minor changes are required. For this reason, NASA as well as ESA developed appropriate standards at an early project stage. The space industry and ESA followed this development. Corresponding standards are available in the ECSS system of ESA [10]. The following Fig. 12.2 shows a typical example of a DRD for a project control plan.

12.1.5 Documentation Numbering System

Efficient documentation control requires a unique documentation numbering system, which is easy to imagine with a documentation volume of several thousand documents for a large project. The mistakes often made in practice go in two opposite directions. Some apply an extremely simple system consisting only of a counting number, which is at most suitable for archiving, and others invent a universal super-code, which is too complicated for practical use. However, both mistakes should be avoided at all costs. The documentation numbering system should basically fulfill three conditions:

- Creation of a system of order.
- Unique identification of a document and,
- Identification of the documentation status.

The creation of a sorting system is particularly important for establishing search criteria for project documents. For example, if you want to get an overview of the existing specifications or to know how many plans the client has created, this overview can be easily determined using sorting criteria with the support of a sorting program. The sorting criteria most frequently used in practice can be summarized as follows:

DOCUMENTATION REQUIREMENTS DESCRIPTION

1) Title PROJECT CONTROL PLAN	2) Number 3) Date September 11, 1970
4) Use To provide program management and contractors with a system of procedures, techniques and responsible organizations by which work is planned, controlled, measured and assessed.	5) Organization Program Office
6) Interrelationship Program Plan Master Program Schedule Accounting System	7) References NASA NHB 2340.2

8) Preparation Instructions

1.0 INTRODUCTION

Provide a brief summary of how the Project Control Plan as presented herein will satisfy the project control requirements imposed on the contract.

2.0 REQUIREMENTS FOR PROJECT CONTROL

The Contractor should describe his interpretation of the requirements which make certain control aspects necessary:

1 To provide effective project control over the areas of Work Performance, Schedule Progress, and the Consumption of Resources.

2. To provide control of interfaces between the project and other elements of the overall program.

3. To cause the project to function as an integral part of the overall program.

4. To provide an action mechanism for problems/critical project areas.

Fig. 12.2 DRD-example for a project control plan

- Document type; see Fig. 12.1.
- Documentation creator, for example company, department, or author.
- WBS assignment; see Fig. 9.14.

The nine-digit document key for the classification system shown in an example in Fig. 12.3 can be simplified even further if shorter individual codes are used.

The abbreviations for the document type must usually be redefined for each project. If applicable standards already exist, for example when using DIN or in-house

Document Identification	Document Number	Changes	
		Edition	Revision
DDD-FFF-WBS	0001	I	A

(1) Identification Block

DDD- Document type

FFF- Company ID (Originator)

WBS- Work Breakdown Structure

(2) Unique Numbering Key

0001, 0002, 0003, etc. to 9999

(3) Document Status

Edition (*issue*): I, II, III, etc.

Changes (*revision*): a, b, c, etc.

Fig. 12.3 Document numbering key

specifications, these should of course be used. For large projects, it is advisable to assign a code or abbreviation to the companies involved, which can then also be used to identify the documentation creator. WBS assignment does not require that the full length of the WBS be included in the documentation number; rather, WBS assignment should refer only to the absolutely necessary levels of the WBS. Usually, three to four WBS levels are sufficient for the assignment. In this context, one speaks of a "*speaking numbering system*", that is one can already recognize the type of document from the documentation number. Since document management is usually carried out with databases and corresponding software support, the documentation number is normally also assigned in the system and can be supplemented with corresponding attachments (*attributes*).

The documentation number is supplemented by the documentation status. As a rule, this is a two-digit code. The first digit indicates the issue, for example issue I, II, III, etc., and the second digit indicates the change status within an issue, for example a, b, c, etc. In case of a new issue, for example issue II, issue I becomes invalid. Changes to the new issue, for example Issue II, are then again designated by the change index, starting with "*A*."

It should then be pointed out that the documentation numbering system of a project should not completely disregard the numbering systems available in many companies, for example the standardized drawing numbering system or other standardizations. In the case of joint projects, it is often unavoidable, due to the various internal company numbering systems, that product documents are assigned a uniform project-specific documentation number on the one hand and the internal company documentation number as a reference on the other.

12.1.6 Documentation Release and Distribution

Each official document kept in the documentation list must be approved by the project management or by a person authorized by the PM before distribution. The importance of each document must be defined in the documentation tree according to its hierarchy. The release mode is adapted to the documentation hierarchy in most cases, that is a regulation is made according to which document group is released by the project management and which documents are released by downstream organizational units. The following principle applies: If the introduction of a change to a subordinate document does not result in changes to a superordinate document, the responsible organizational unit is authorized to release the document. If, however, the process results in changes to a higher level document, the higher level organizational unit must be involved.

Many documents still have to be checked by several bodies before they are released. For example, production specifications usually have to be reviewed by production management and the accompanying quality control department before they are released. The review and release mode must be defined in detail and unambiguously for the individual documents.

In addition to the official documents, which are created in accordance with the documentation requirements list (*DRL*), there are usually quite a number of unofficial project documents, which are only of an informational nature. These include, for example, special reports and analyses. The creation of these documents is a useful supplement to the official documents. They have the purpose of documenting very specific processes or situations and must therefore also be recorded by the documentation numbering system, document type, author, *WBS* assignment and sequential number. However, unofficial documents do not require release and status control. Their information content will, if applicable, be reflected in one of the official documents.

The distribution of project documents should not be left to chance and should also not be carried out according to the watering can principle. Random distribution results in employees being unaware of important working documents and broad distribution, on the one hand, often leads to weariness over the large flood of paper and, on the other hand, costs an unnecessary amount of time and money. Since each official document, as expressed in the requirement description, serves a very specific purpose and has pre-planned information, the distribution list can usually also be clearly defined. The targeted distribution of documentation according to the motto "*need to know*" also has the advantage that the recipients feel that they are being addressed directly.

In one case, however, broad distribution is appropriate, namely in the distribution of the documentation status list. If possible, this should be sent to all project staff or at least made easily accessible, so that it is known at all times which version of every document is the one that is now valid.

The responsible documentation manager of the project, in the case of small projects the PM, is responsible for defining the distribution key, which must be done in coordination with the project management. In addition to individual distribution, the definition of standard distribution lists has in practice proven successful. Today, the distribution of documents is also usually supported by an appropriate mailing system or information system, and distribution status can easily be traced.

12.1.7 Monitoring of the Documentation Status

Monitoring the documentation status is a function that should not be underestimated because everyone can imagine what happens when the lathe operator in one workshop carries out his work according to an old set of drawings and his colleague in another department or company machines the counterpart with a modified set of drawings. The situation is similar for specifications, plans, test specifications, etc. The document management system must, therefore, ensure that only the valid documents are in circulation. The following status situations are possible for all project documents defined in Table 12.1:

(a) Completion planned; see DRL.
(b) Created and released; see project schedule.
(c) Distribution to employees.
(d) Under review or change.

The change of a version-managed and already released document can only be executed via a change request or in the case of small projects informally via e-mail. The requested change must be coordinated by the documentation department or the documentation manager in cooperation with the affected project areas with regard to the resulting consequences before it can be officially executed. If the change request is approved and implemented, this leads to a new issue/release of the document. The distribution list must be modified with the new issue status of the document. Each change made is identified by the respective change index. The valid version is then stored in the document database and can be called up in each case by all those involved.

At regular intervals, the documentation status list, which shows the currently valid documentation status, must be published and accessible to all project staff.

12.2 Configuration Management

Configuration Management (*CM*) is a management function that provides technical and administrative management during the entire product life cycle with up-to-date configuration-related data, which includes at all stages of the life cycle identification and traceability of the product and maintains the achieved state of its physical and functional requirements and access to accurate information. Configuration management describes the "state" of all hardware and software objects (*end items*) defined in the WBS of a particular phase of the life cycle. For example:

- As-designed (*constructed*).
- As-built (*manufactured*) and.
- As-maintained (*revised*).

This results in configurations that are consistent and reproducible and can be distributed to other systems [11].

12.2.1 Significance of Configuration Management

Configuration management is an important management discipline to ensure that the specified configuration base of a product to be developed and/or produced and all changes to the configuration base approved during the course of the project and the resulting effects are identified and monitored. The first steps toward systematic configuration management were taken in the 1950s by the U.S. Department of Defense. The task of configuration management can be divided into the following main functions: [12].

- Configuration identification (*baseline*),
- Configuration monitoring (*status control*),
- Configuration changes (*change control*),
- Configuration status determination (*status account*).

Configuration management, which is closely linked to documentation management, involves the continuous recording of the basic configuration of the respective product, which in turn is directly linked to the project costs. Many a customer/buyer of the product initially wants a simple product with which he associates very specific cost expectations. The customer often does not stick to his original intentions, and the initially simple order then quickly becomes a less simple and consequently several times more expensive order. The capitol city airport in Berlin is a negative example of this; see Subsection 3.2.1 [13].

Configuration management (CM) is a field of management that focuses on establishing and maintaining consistency of a system's or product's performance and its functional and physical attributes with its requirements, design, and operational information throughout its life. For information assurance, CM can be defined as the management of security features and assurances through control of changes made to hardware, software, firmware, documentation, test, test fixtures, and test documentation throughout the life cycle of a system [14].

To ensure that the configuration of the developed or built system or product corresponds to the actual desired system, the project management must implement a strict configuration monitoring system. Configuration management is closely linked to documentation management (*Sect. 12.1*) on the one hand, and to systems engineering (*Chap. 7*) and project monitoring (*Chap. 9*) on the other. There are different ways of integrating configuration management into the project organization. In the interest of lean cost-saving management, it makes sense to combine documentation and configuration management (*DOC/COFI management*) in a single unit and there are various options for organizational embedding in the project organization:

(a) As an independent organization that reports directly to the PM because of its importance, a variant that is often found in complex and large projects.
(b) as part of the quality teams or,
(c) as part of the Project Control Team, see Figs. 5.3 and 5.4.

12.2.2 Historical CM Development

Configuration management methods and procedures were developed in the United States in the early 1960s. Schriever, a general in the U.S. Air Force, published one of the most comprehensive configuration management documents, *Configuration Management during Definition and Acquisition Phases,* in 1964 as part of the AFSCM Systems *Management* Manuals, which also became known as the 375 series [15]. In August 1969, NASA published the Saturn Configuration Management Manual and in 1970 the Apollo Configuration Management Manual [16, 17]. Both documents are based on the experience of the Apollo program and are also considered important standard configuration control documents today. Based on these documents, almost all major aerospace companies in the USA have produced their own in-house manuals on this subject. Slotkin wrote in his book 'Doing the Impossible': "*Implementing configuration management in the Apollo Program remained a key objective.*" [18]

In the European aerospace industry, configuration management became an important tool in project execution. ELDO (*now ESA*) published an instruction on configuration management in 1972 as part of the EUROPA III launcher program (*Ariane predecessor*), and ESRO (*now ESA*) issued a requirements document on configuration control in 1973 [19, 20]. In 1985, Saynisch published what is probably the most comprehensive work on this topic in the German-speaking Countries [21]. Today's requirements for project management for European spaceflight projects are defined by ESA's ECSS system. The configuration management is regulated in the standard "*ECSS-M-ST-40*" and for the other non-space industry in DIN ISO 10007. For the certification of aerospace companies, the system requirements are also defined regarding the configuration management processes in the DIN EN 9100. An overview of the relevant CM standards is listed below:

1. CM Standards.
   - NASA, Configuration Management Plan (*CMP*) Template, T2401, January 2011.
   - ANSI/EIA-649-1998 National Consensus Standard for Configuration Management.
   - EIA-649-A 2004 National Consensus Standard for Configuration Management.
   - ISO 10007:2003 Quality management systems Guidelines for configuration management.
   - Federal Standard 1037C.

- GEIA Standard 836–2002 Configuration Management Data Exchange and Interoperability.
- IEEE Std. 828–1998 IEEE Standard for Software Configuration Management Plans.
- MIL-STD-48 1B, July 1988, Configuration Control, AMSC N4478.
- STANAG 4159 NATO Material Configuration Management Policy and Procedures for Multinational Joint Projects.
- STANAG 4427 Introduction of Allied Configuration Management Publications (ACMPs).
- CMMI for Development, Version 1.2 CONFIGURATION MANAGEMENT.

2. CM Guidelines.
   - IEEE Std. 1042–1987 IEEE Guide to Software Configuration Management.
   - MIL-HDBK-61A CONFIGURATION MANAGEMENT GUIDANCE 7 February 2001.
   - ISO 10007 Quality management—Guidelines for configuration management.
   - GEIA-HB-649—Implementation Guide for Configuration Management.
   - ANSI/EIA-649-1998 National Consensus Standard for Configuration Management.
   - EIA-836 Consensus Standard for Configuration Management Data Exchange and Interoperability.
   - ANSI/EIA-632-1998 Processes for Engineering a System.

12.2.3 Configuration Identification

The basis of any effective configuration monitoring is the identification of the configuration baseline, a process that takes place step-by-step during the individual project phases. In accordance with the phase structure described in Chap. 4 (*see Fig. 4.2*), the proposed definition of the configuration basis in the following three steps makes sense: [15, 17].

1. Identification of project/system requirements.
2. Identification of the design requirements.
3. Identification of production requirements.

As shown in Fig. 12.4, the project/system requirements should be defined at the beginning of the definition phase and anchored in the first edition of the system specification. The definition of the project/system requirements is an important reference point for all further work. Configuration identification starts at this point.

As long as the system specification is not approved and frozen, all essential parameters hang in the air and the start of development work is hardly justifiable. In practice, however, it often looks as if the release of the system specification depends on only a few data. It is, so to speak, 99% ready, but cannot yet be finally released. In many cases, this problem can be solved in the interest of the progress of the work by temporarily excluding this data with the note "*to be defined*" (*TBD).* Often, the

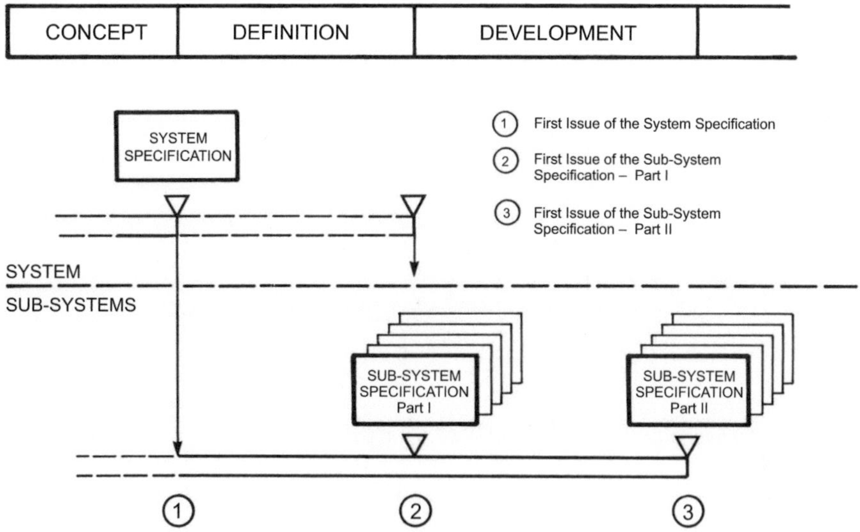

Fig. 12.4 Steps for defining the configuration baseline

system specification that has not yet been released is also issued as a *preliminary document.* However, both approaches should be only be used for a short time and extremely rarely. Sometimes it can be observed in practice that the release of the system specification is delayed for fear of an early definition. In view of the project delays and cost increases that often result from this attitude, a realistic and reasonable attitude to release is advisable. Later changes to the system specification that may have to be made by timely "*freezing*" are usually also associated with additional costs, but they are usually smaller compared to the cost increases caused by delays.

It is important to identify, in conjunction with the work breakdown structure (*WBS*), those elements (*configuration items*) of the project (*hardware and/or software*) that need to be subject to configuration control. These could be all elements of a system or structures classified as particularly critical, such as motors, transformers, voltage regulators, control electronics, hydraulic elements, or even control software. Fig. 12.5 provides an example.

At the end of the definition phase, all design requirements should be precisely identified and captured by relevant specifications; see Subsection 7.2. Based on the finalized specifications, detailed design of subsystems and their components may be authorized. The definition of production requirements is possible only after the most essential development work has been completed, and the results of prototype tests have been evaluated. The subsystem specifications mentioned above shall be expanded accordingly by the production requirements. The relevant U.S. Air Force and NASA regulations, therefore, provide from the outset for the division of subsystem specifications into two parts: Part I: Design Requirements and Part II: Production Requirements [16, 17]. In other sources of relevant literature, this process is also

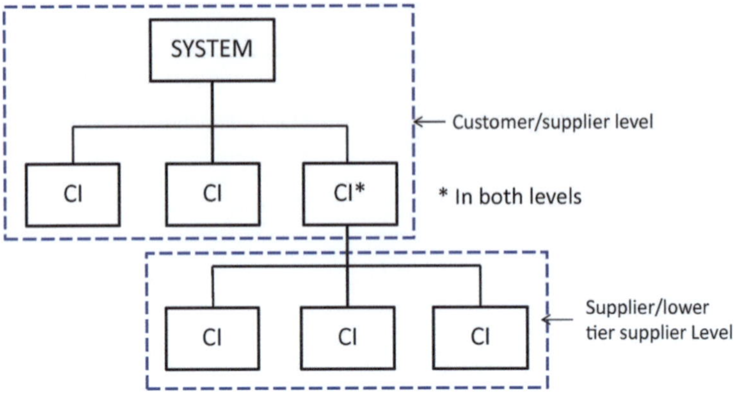

Fig. 12.5 Configuration Item (Source ECSS-M-ST-40C)

referred to as "*requirement management.*" For large-scale projects, it becomes a complex process that is handled in several cycles. The more parties that are involved, the more complex this process becomes because of interfaces between the customer, cooperation partners, subcontractors, and suppliers. However, since a clear and universally accepted specification of the product is a mandatory prerequisite for the successful completion of a project, this process should be handled very carefully, transparently, and consistently. There are also very helpful tools for this. The tool "*DOORS,*" which is frequently used in industry, should be mentioned here as an example.

An important measure for proper identification of the configuration base is the implementation of a specification tree in which all specifications, including their hierarchical assignment, are recognizable. Figure 7.12 shows a typical specification tree. Technical drawings and bills of material are also part of the configuration basis and are therefore also subject to configuration identification.

12.2.4 Configuration Monitoring

In a DOD, directive configuration management is defined as follows: "Configuration management is a discipline that uses technical and administrative monitoring methods,

1. Identifies the functional and physical characteristics of systems, subsystems, and components,
2. Changes in the characteristics tracked and,
3. Records the change and implementation status." [22].

Laine wrote in this context in 1966, mutatis mutandis: "With thousands of technical changes, which are normal in the development of complex systems, how can the purchaser and the developer of a product be sure that the final product actually corresponds to what had been agreed upon? This new discipline (configuration management), a method for up-to-date and accurate documentation, is designed to ensure that." [23]

Therefore, there must be a clear goal of configuration management such as described by ESA: [24].

- Know at any moment the technical description of a system and its components, using approved documentation,
- Control the evolution in the system's technical description,
- Provide traceability of the evolution of the system's technical description,
- Facilitate the consistency of the system's components (*control of external interfaces*) and the products that make up these components (*control of internal interfaces*),
- Verify that documentation is and remains the exact image of the products it describes,
- Identify the current baseline and the as-built configuration, in order to deal with discrepancies detected during production, delivery, or operation of the product,
- Enable any user to know the operational possibilities and limitations of each product item and, in case of non-conformance, to know which items are affected.

Continuous monitoring of the configuration base once it has been established is an extremely important management measure that cannot be overstated in terms of its significance because money is at stake. Many projects flounder financially because of uncontrolled changes to the configuration baseline. Companies often have the problem that their project engineers, working with the client's engineers, adopt or propose and initiate design changes based on new ideas without these changes being evaluated for their schedule and/or financial impact on the project, formally recorded, and approved by project management. Changes to the configuration basis, and every private home builder knows this particularly well, are in most cases associated with additional costs. Every new idea, no matter how good, must be coordinated with the configuration baseline and approved as a change by the project management before it is implemented.

The following Fig. 12.6 shows the information flow to configuration management. The main information comes from management (*Chap. 6*), systems engineering (*Chap. 7*), and quality assurance (*Chap. 8*).

Conversely, there is of course the flow of information from configuration management back to management, systems engineering and quality assurance, see Fig. 12.7.

12.2.5 Configuration Changes

The monitoring of the configuration baseline starts with the release of the system specification at the beginning of the definition phase. In order to make a technical change, the following process must be followed: [25].

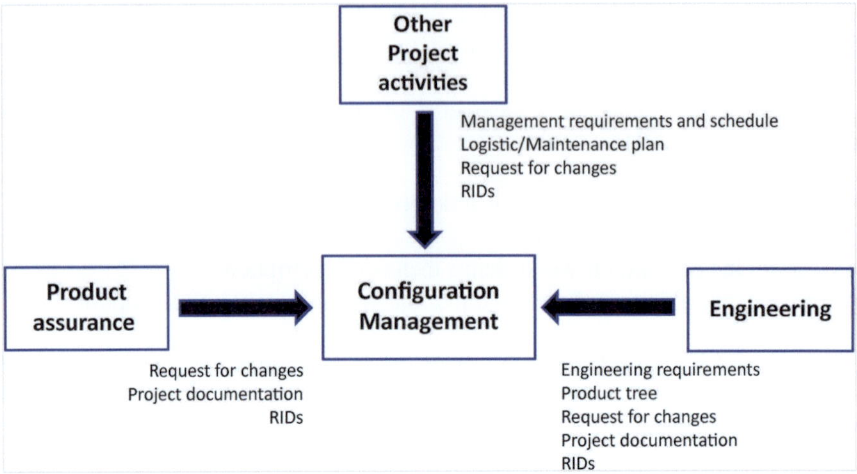

Fig. 12.6 Information flow for configuration management ECSS-M-ST-40C

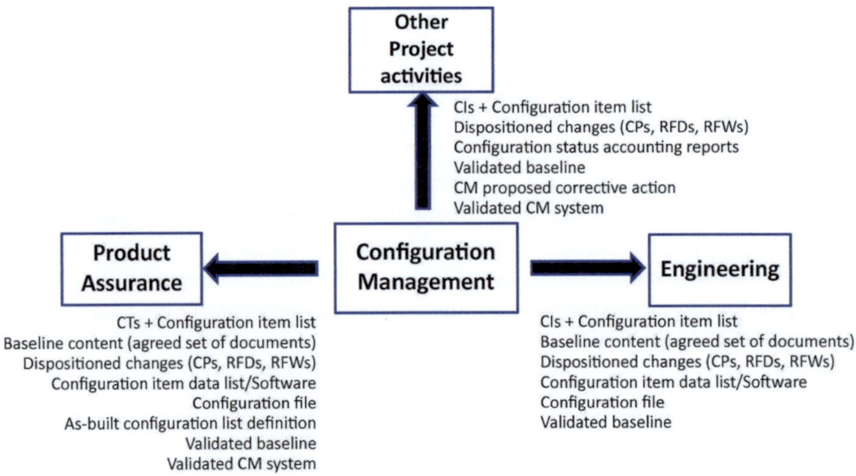

Fig. 12.7 Information flow for configuration management ECSS-M-ST-40C

- Justification of the need for the technical amendment.
- Determination of the change class by the person requesting the technical change.
- Preparation of an application for a change.
- Review of the resulting consequences of the change (*technical, schedule and costs*) Submission of the change request to the PM (*contractor and/or client*).

- Acceptance/rejection or reclassification of the change, and
- Implementation of the change.

Providing a justification for a desired change is intended to prevent changes from being made too lightly. Many desired changes often fall into the *"nice to have but not essential to the task"* group. The establishment of change classes, as defined in the U.S. Military Standard, MIL-STD-480, is an important prerequisite for classifying changes according to their importance to the system [26]. Document No. 480 provides for two classes, Class I and Class II changes. Changes are always to be assigned to Class I if they affect the configuration basis anchored in the system specification in terms of form, fit, or function. All changes that cannot be assigned to Class I, for example pure document changes or hardware changes that have no influence on the system, are to be assigned to Class II.

Change Management: according to ESA rules all required changes must be classified as class I* or class II*. All class I changes, for example, are changes of the system specification, must be approved by the system-level parties [24].

**Note: at some organizations called class A and B.*

- Class I changes must be clearly described and approved by the joint customer/contractor Change Control Board (*CCB*).
- Class II changes are considered company internal and require approval of the company internal CCB.
- Class II changes may be re-classified as a class I change in case the overall system is affected by the change.

Changes to the system specification and the associated system drawings are to be regarded as first-level technical changes since they can have an influence on all subsystem specifications grouped together in the specification tree. Changes to the system specification must therefore always be coordinated with the client. The prime contractor is usually responsible for monitoring changes to specifications and drawings of the subsystem and component levels. It should be noted, however, that technical changes to the subsystems and components can only be approved by the contractor if they are Class II changes. However, if it turns out that a Class II subsystem change has an impact on the system specification, the Class II change request must be converted into a Class I change request to be reviewed by the customer/prime contractor.

Technical changes have an impact on the functionality and quality of the product: the schedule, the costs, and the contract. Therefore, in addition to the change description and justification, each change request must, above all, contain an analysis of the technical consequences resulting from the change, including schedule, cost, and contract impacts. The assessment of technical changes by the respective project management of the customer or contractor requires the involvement of a *change control board (CCB)* in which the relevant experts for the assessment of technology, quality, deadlines, costs, and contract are represented. The change control board (*CCB*) must, therefore, be staffed with the following project personnel:

I. Change Committee Management.
   - Project manager and/or,
   - Systems engineering (*deputy*).
II. Standing Committee Members.
   - Systems engineering.
   - Product assurance.
   - Planning and monitoring.
   - Contract Management.
   - Configuration management (*must write minutes*).
III. Additional committee members (*as may be required*).
   - Manufacturing Management.
   - Integration and test management.
   - Operation.
   - Other.

The Change Committee should be convened at regular intervals and on an ad-hoc basis as needed.

The Configuration Control Board (*CCB*) shall approve or disapprove any Engineering Change Proposal (*ECP*). Permanent members of the CCB will be the key people of the PM team. Additional specialists from functional departments and/or partner companies may be invited to participate in the CCB. All CCB members, whether permanently assigned or specially invited, must be fully authorized to take decisions on behalf of their respective organization, and to accept actions placed upon the organization they represent [24].

In addition to the possibility of making technical changes via the formal change request route, arrangements are common in practice that allow temporary deviation from the specification in the interests of speedy project progress without making a general change to the system. Two procedures are involved here:

- Approval of construction deviations in special cases *(request for deviations—RFD).*
- Approval of construction variances that have occurred and been accepted *(request for waiver—RFW).*

In the first case, this is a request to be allowed to deviate from the specification in order, for example, to be able to keep to the schedule by using a different, non-specified but equivalent material or production process. In the second case, the issuing of waivers, the aim is to check whether production deviations that have occurred can be tolerated and whether the component can be accepted by the customer despite the deviation from the specification. Both procedures serve the purpose of checking minimal deviations and their effects on the overall system in an unbureaucratic manner in the interest of rapid project progress and, if necessary, approving them without changing the basic requirements. This allows that minor deviations in certain cases will be tolerated but will not be recognized as a fundamental technical change for subsequent production batches. Temporary deviations are to be carefully and conscientiously examined for their effects by the *material review board (MRB)* to be set up for this purpose.

12.2.6 Configuration Status Determination

Every development and production process is sooner or later affected by changes. Precise recording of the change status is particularly important in the case of large and complex projects, in which it is not uncommon for many companies to be involved at the same time. Documentation of the configuration status must be done systematically, so that the identified configuration baseline and the approved changes to it can be clearly identified at any time. Furthermore, the project management must be informed about the configuration status at regular intervals. This is best achieved with the known methods for monitoring the documentation status. The production and integration of a product usually starts with a so-called "*frozen*" configuration status of the *"baseline."* This configuration status is also referred to as "*as designed*" or "*as ordered.*" However, the status of both versions can also be different. All changes that become necessary after this and are incorporated in the product after appropriate testing and approval are documented and recorded in the "*as built*" configuration at the end for delivery. The *change* history must be recorded in the change record. If changes/rework are still carried out after completion and acceptance of the product, which often occurs in larger and complex projects, these changes must of course also be documented. The final configuration status is designated as *"as delivered."*

It is the responsibility of the configuration management to examine changes of the baseline configuration of a product, the "*Configuration Baseline*" in co-operation with the respective project specialists regarding their necessity and the time schedule and financial effects most carefully, in order to counteract the creeping inflation of often very expensive, however, not always necessary, technical extras in time and efficiently. With contracted projects, necessary changes are to be coordinated with the customer and documented accordingly and determined if necessary, all effects; see Subsection 12.2.5. Finally, a summary of the tasks and outputs of the configuration management are shown below: [24].

1. Configuration Baseline Control.
2. Configuration Item Data List (*CIDL*).
3. As Designed Configuration Data List (*ADCL*).
4. As-Built Configuration Data List (*ABCL*).
5. Non-Conformance Report (*NCR*).
6. Request for Deviation (*RFD*).

7. Request for Waver (RFW).
8. Engineering Change Proposal (*ECP*).
9. Contract Change Notice (*CCN*).

There is a close interface between configuration management, documentation management, and requirements management. All three functions are very extensive and complex tasks in larger projects. For this reason, it is now standard in the industry to support these processes with integrated SW tools to accompany the process. As already mentioned, there is a multitude of corresponding SW tools/databases that can support configuration management very efficiently. An overview of configuration management tools is provided by the website: (http://www.softguide.de/).

12.2.7 Documentation and Configuration Control Checklist

The following is a checklist for the completeness of documentation and configuration control developed for a TKC (*general contractor*) of a METRO project: [27].

1. Has the *turn key contractor* (*TKC*) included a documentation and configuration management position in the organizational chart?
2. Do the personnel deployed for documentation and configuration monitoring have the required qualifications?
3. Have all project documents on the *document requirement list* (*DRL*) been recorded?
4. Were the completion date and author (*or company*) of each document specified in the DRL?
5. Has the content of the project documents been clearly described in the *document requirement description* (*DRD*)?
6. Has a uniform numbering system been specified by the TKC in conjunction with the WBS?
7. Does the documentation number include a change index?
8. Has the TKC created a procedure for status control and approval of documents?
9. Has the completion date and author (*or company*) of each document been specified in the DRL?
10. Has the TKC documented the hardware and/or software configuration, e. g. in the associated specification or drawing?
11. Has a change control procedure been implemented by the TKC?
12. Has the TKC formally established a *change control board (CCB*) that meets regularly or when there is an urgent need to approve or deny requested changes?
13. Have the permanent CCB members been nominated?
14. Does the TKC have a formal contract change procedure?
15. Has the change procedure from the TKC described how to handle construction deviations (*request for deviation—RFD and request for waiver—RFW*)?

Literature

1. Götzer, Klaus, et al: Dokumenten-Management, dpunkt.verlag, 5th edition, Heidelberg, 2014, p. 2
2. Gröpper, Meinolf et al: Einführung eines Dokumentenmanagementsystems, VDMA Verlag, Frankfurt a.M., 2012, p. 43
3. Letter from a major international company, copy available to Aitor
4. Eckhardt Carsten, WEKA Media
5. Source 1, p. 13
6. Source 1, p. 19
7. Source 1, p. 35
8. Source 1, p. 257
9. NASA: Data Requirement Descriptions, Document DM002–001-1, March 1970
10. ESA: ECSS - Documentation of ESA; ECSS Standards CD and ECSS database download
11. CM ISO 10007
12. Madauss, Bernd: Configuration Management, ISU/ACT - Training at the Royal Army of Oman, HQ-SAF, 2010
13. Berlin Capital Airport (BER); see subsection 3.2.1
14. Source 12
15. USAF: Configuration management during definition and acquisition phases, AFSCM 375–1, June 1964
16. NASA: Saturn Configuration Management Manual, MM8040.10, August 1969
17. NASA: Apollo Configuration Management Manual, NHB8040.2, January 1970
18. Slotkin, Arthur L.: Doing the impossible, Springer N.Y., Heidelberg, Dordrecht London, 2012, p. 50
19. ELDO: configuration management functions and preliminary requirements, EUROPA III, Document 20.02.00/1, March 1972
20. ESRO: Configuration Management Requirements, QRA-06, July 1973
21. Saynisch, Manfred: Konfigurationsmanagement - Entwurfssteuerung, Dokumentation, Änderungswesen, TÜV Rheinland, Cologne 1984
22. DOD: Directive 5010.19, p. 2
23. Laine, Morton J.: Configuration Management, Space/Aeronautics, Vol. 46, No. 6, November 1966, p. 74
24. ESA: ECSS-M-40A, P. 15
25. Madauss, Bernd-J.: Planung und Überwachung von Forschungs- und Entwicklungsprojekten, AIB-Fachliteratur, Bad-Aibling, 1978/81, p. V-2
26. Configuration Control- Engineering changes, deviations and waivers, MIL-STD-480, October 1968
27. Madauss, Bernd-J.: Project Management Checklist, Management and Organizational Concept for the Metro Medellin Project, September 1983 (unpublished), "*Documentation and Configuration Control*," p. A.8

Contract Management 13

Contents

Abstract

For customer projects, the cooperation between the parties is usually regulated by a contract. Project contracts are intended to set out the agreements reached between the client (customer) and contractor on the scope of supplies and services, deadlines, and payment arrangements.

Contract management (CM) is a PM support function; see Fig. 5.4.

B. -J. Madauss, *Project Management*,
https://doi.org/10.1007/978-3-662-69057-4_13

Public and private organizations that cannot accomplish project tasks on their own due to lack of resources, required know-how, experience or for other reasons look for appropriate project partnership. The awarding of project tasks by the client (*customer*) to a competent company usually requires a competitive process to ensure that a contractor (*supplier*) is found that best fulfills the specified tasks; see Chap. 14. After successful contract negotiations have been concluded, a binding contract defining all the rights and obligations of the partners can be implemented.

The tasks of a project manager are, as described in Chap. 5 (*see Figs. 5.3 and 5.4*), very diverse. They are not limited to technical tasks but also to administrative and contractual tasks, for example, *contract administration.* This does not mean, however, that the PM should solve the contractual task without legal assistance. Well-structured project contracts contain in addition to the formal legal aspects (*major sections*), in annexes to the contract information about the supplies and services to be provided; see Fig. 9.5. Typical annexes are, for example, a Statement of Work *(SOW*), specifications, time schedules, and cost plans. The preparation of a project contract requires very close cooperation between the project manager and his team together with the legal department.

It must be avoided, for example, that contradictory delivery dates are mentioned at different places in the contract. The principle applies, to remain with the previously mentioned example, that deadlines are only contained in the schedule, and delivery dates only in a special *Deliverable Item List (DIL*) and that reference must be made to these in other places in the contract. This also applies to the *Document Requirements List (DRL*); see subsection 12.1.4. A second reason for strict compliance with this rule is that, in the case of amendments, an amendment need only be made at one place in the contract.

13.1 Contracts as a Link between the Partners

13.1.1 Project Contracts

The purpose of concluding the contract is to specify the deliveries and services of the contractor (*supplier*) on the one hand and the remuneration and possible provisions by the client on the other hand. The contract represents the link between the project partners. "Contracts regulate the relations between two or more parties" [1]; see Fig. 13.1. It sets out the rights and obligations of both parties, as well as the legal framework, for example, applicable law, arbitration procedures, breaches, withdrawal clauses, and termination of the contract.

Projects pose a very special problem for cost and price determination since precise forecasts of deadlines and the reliable determination of expected costs are always accompanied by a greater or lesser degree of uncertainty. This results from the difficulty of making credible forecasts for tasks that lie in the future; see Chap. 9.

The plans (*processes, deadlines,* etc.) available when the contract is concluded ideally represent a well-founded conception of the possible operations, that is, a conception based on experience with similar projects in the past. Based on plans, required quantities for personnel, materials, machine hours, etc. are determined. In

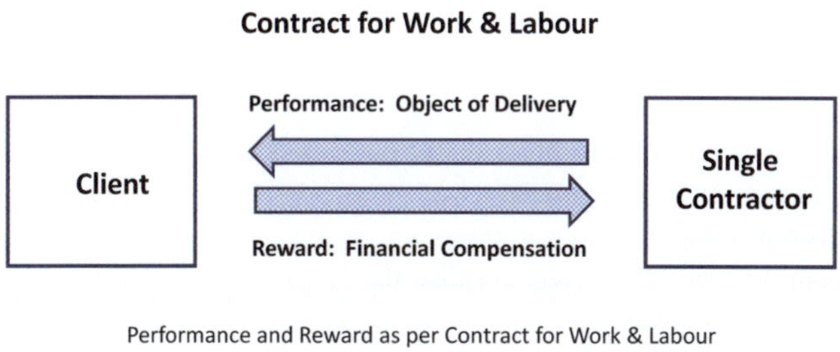

Fig. 13.1 Performance and consideration in the contract for work and services

contrast to the mass production of goods, every project carries an inherent planning risk due to its novelty because no two projects are alike.

Research and development projects involve a particularly high planning risk due to their innovative nature. For this reason, projects should be planned in accordance with the guidelines described in Chap. 4. In the course of the preparatory phases (*concept and definition phase*), the planning prerequisites for a well-founded cost estimation process are created; see Chap. 10. It should be noted, however, that the cost estimate made for a project must be regarded as a forecast subject to errors, as mentioned above, which is important for pricing. Either the customer or the supplier must pay for errors in estimation that result in additional costs (*prices*), depending on the contractual arrangement.

13.1.2 Contract Types and Pricing

The contract and pricing structures are interlinked. In principle, there are three basic types of prices for projects: (1) *fixed price,* (2) *cost-reimbursement price, and* (3) *other pricing arrangements* (e.g., *with maximum limits and premium arrangements).* The preference for fixed prices is a key pricing principle of the Regulation on Prices in Public Contracts [2]. That is, in public contracts, a fixed-price agreement should be sought wherever possible. This applies equally in the private sector. In the case of long-term contracts, price reservations in the form of price escalation clauses should be negotiated; see subsection 10.1.5. A cost-reimbursement price arrangement, which must be based on the contractor's reasonable costs, may only be agreed in exceptional cases [3]. Reasons for exceptions are lack of certainty of the price and limited competition. In the case of public contracts, cost-reimbursement prices are subject to price verification.

The determination of the price type to be applied in the project depends on various prerequisites. Some typical price types and their application criteria are mentioned below: [4].

1. Fixed prices.
   (a) *Firm Fixed Price (FFP).*
       Prerequisites:

- Well-defined specifications and bills of quantities.
- Realistic cost estimates.
- Adequate competition.
- Well-reasoned cost or price data that allows for price comparison.

(b) *Fixed* Price with *Escalation.*

Prerequisites:

- Unstable market or wage conditions.
- Long-term contracts.
- Otherwise as (a).

(c) *Fixed price with incentive* clause.

Prerequisites:

- Cases where there is estimation uncertainty and opportunity for cost reduction and/or performance improvement.
- Otherwise as (a).

(d) *Fixed* Price with *Redetermination.*

Prerequisites:

- If a realistic price can only be set for the early stages but not for later periods of time.
- For the early phase (a) applies.

2. Cost-reimbursement prices.

Cost-reimbursement price basic forms are cost price, target price, and cost-reimbursement price. Typical forms are:

(a) Cost-Reimbursement Price plus Profit.

Prerequisites:

- The agreement of a cost-reimbursement price is allowed by the contracting authority only if the price determination is not otherwise possible.
- A performance guarantee by the contractor is not given.
- Costs exceeding the agreed upper limit are not reimbursable.
- Costs shall be disclosed and reimbursed as incurred.

(b) Cost *Plus Fixed Fee (CPFF*).

Prerequisites:

- In the case of orders whose bill of quantities is not yet fully known at the start of the project. Possible later additions do not automatically lead to an increase in profit.
- Otherwise as (a).

(c) Cost *Plus Incentive Fee* (*CPIF*).

Prerequisites:

- Projects where a premium scheme can lead to efficiency improvements in project execution. Wherever possible, premium incentives should be used to improve technical performance, schedule adherence, and cost savings. Conversely, deteriorations in performance naturally lead to contract penalties.
- Otherwise as (a).

3. Other cost/price arrangements.

(a) Grants (*cost sharing*).

Prerequisites:

– An eligible project that is fully (100%) or partially (< 100%) funded by the contracting agency.
– The costs are to be disclosed.

(b) Time and Material Contracts.
Prerequisites:
– Time Limited provision of personnel and/or materials at agreed costs/ prices (*hourly/material rates*).
– The costs are to be disclosed.

The decision as to which price type should be applied depends on the type of project and the current project phase. Government contracts, for example, research and development projects with a high degree of innovation and long durations, are often handled under cost reimbursement terms with premium-malus rules because of the high risk involved, while production and procurement projects are predominantly awarded under fixed-price terms. Study contracts in the early stages of a project, on the other hand, are preferably awarded under fixed-price conditions due to the relatively low risk involved. In the past, many projects awarded under cost-reimbursement conditions have resulted in particularly excessive cost increases because the contractor had few incentives to minimize costs. An alternative to cost-reimbursement is the aforementioned cost-reimbursement with a bonus arrangement, which promises the contractor appropriate incentives (*bonuses*) if savings are made. In return, however, there may then also be a malus regulation.

There is an increasing tendency to also handle development tasks more and more under fixed-price conditions, for example as a fixed price with a price escalation clause; see subsection 10.1.5. The German Ministry of Defense (*BMVg*) announced this in a decree of 19 February 1982, also for development contracts in the defense sector [5]. The European Space Agency (*ESA*) has also started to handle development projects under fixed-price conditions. However, this is often only possible for some parts of the project. It should be mentioned in this context that the award of a development contract at fixed-price terms requires a precise task description (*functional specifications, specifications,* etc.). Therefore, comprehensive execution of Phases A and B is very important; see Chap. 4. The contracting of private satellite operators to augment the satellite fleet, such as SES, is usually done on FFP terms.

It is recommended to make more use of differentiated pricing than in the past. That is, fixed prices can be agreed for project areas for which reliable cost estimates are available, and cost-reimbursement contracts with premium regulations for projects with increased risk. In contrast, the subsequent phases, for example production and operation should be handled under fixed-price conditions or under time and material contracts. Fig. 13.2 explains the principle of different pricing for the various project phases.

Peeters and Veld point out in the context of ESA projects that it is almost impossible to find a company (*contractor*) willing to take on the financial risk of an FFP for high-tech projects with long durations and one therefore usually resorts to CPFF (*cost plus fixed fee*) or CPIF (*cost plus incentive fee*) contracts in such situations [6]. They explain this with the following equation:

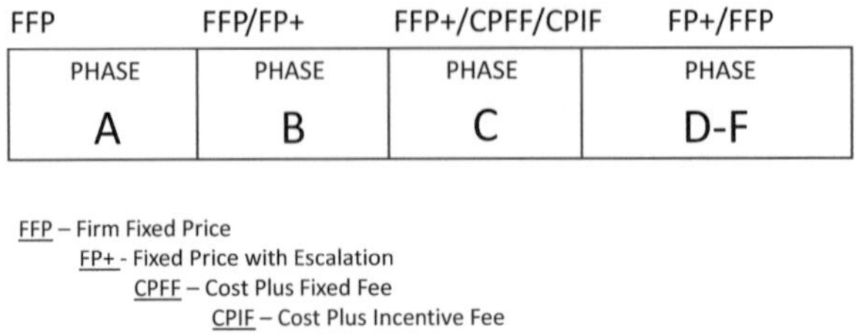

Fig. 13.2 Phase-dependent application of different price types

$$P = BF + S(TC - AC).$$

P = Profit *(absolute profit)*
BF = Basic Fee (*basic profit*)
S = Sharing Ratio (risk sharing)
TC = Target Cost
AC = Actual Cost

Thereafter, "*S*" is the determining factor to determine the type of contract (CPFF, FFP, or CPIF):

S = 0 (the company does not assume any risk)	CPFF
S = 1 (the company assumes the full risk)	FFP
0 < S < 1 (the company participates in the risk)	CPIF

13.1.3 Fixed Price Versus Cost Price with Premium Regulation

In addition to the fixed-price arrangement, the *cost-plus-incentive-fee (CPIF)* contract is a viable contracting alternative for research and development projects with significant risk areas. Kahn writes that ESA, for example, seeks fixed-price contracts wherever possible, and in cases where a fixed-price arrangement cannot be reached immediately, efforts are made to establish a target cost price *(ceiling price)* that can be converted to a fixed price as early as possible—an arrangement that is also consistent with German pricing law. If it is not possible to agree on a ceiling price due to an increased risk, a cost-reimbursement arrangement can be considered [7]. Kahn further states that a fixed price or a temporary maximum price regulation is always appropriate if the project risk can be determined with some certainty and does not exceed approximately 10–15% of the order volume. In addition, a fixed-price arrangement also leads to simpler project management in which client intervention with the contractor is largely excluded [7].

For projects with a large or not clearly determinable technological risk, a fixed-price arrangement does not make sense. The contractor would be forced to provide for very large risk reserves or not accept the order. Therefore, in such cases, a cost-reimbursement or target cost price is preferable. However, in the case of cost-reimbursement contracts, the contractor is obliged to disclose the planned costs to the customer (*client*) to ensure full cost transparency. This leads to higher management costs for the client and especially for the contractor. As already mentioned, in practice cost-reimbursement contracts often lead to increased costs, as there are no incentives to minimize costs, because the costs incurred—insofar as they are in line with expenditure—are charged to the customer, including all agreed surcharges and profits. Far more dramatic, however, is the emergence of a cost-reimbursement mentality among employees. Companies with cost-reimbursement experience are then often no longer a match for the competition.

In this context, however, the impression should not be created that the cost- reimbursement contract encourages dishonest actions; that would be wrong. What is correct, however, is that the cost-reimbursement does not motivate an economically efficient way of thinking to the same extent as the fixed price and for this reason, as already mentioned, should only be used in exceptional cases, for example for projects for which an exact cost estimate is not possible due to a high degree of novelty and large risk areas.

Cost overruns in cost-reimbursement contracts were in the past the reason to apply the CPIF contracts already mentioned, to ensure that the contractor shares in the loss in case of possible additional costs through reduced profits, while he benefits in case of a cost reduction by an increase of the nominal profit (*incentive*). CPIF contracts were a viable solution especially for large projects with long durations for aerospace, where there was no way around cost-reimbursement solutions. As early as 1967, NASA produced a manual on the premium scheme, which was a good reference for their European counterparts [8].

From the point of view of cost-effectiveness, in the early phases of the project, that is in the concept and definition phases (see Chap. 4), intensive cost studies must be carried out in order to be able to provide qualified cost estimates and as definitive a statement as possible on the accuracy of the estimates; see Sect. 10.4. Based on relatively accurate cost estimates, the *cost-plus incentive fee (CPIF)* is a viable alternative to the fixed price. For example, if a project is estimated at 10 million Euros and a nominal profit margin of 5% is agreed upon, the contractor would be able to settle for 10.5 million Euros after the project is completed and the costs are proven to be 10 million Euros. However, if less or more costs are incurred, the contractor will participate in the saved or additional costs in accordance with the participation formula specified in each case. The following model case illustrates this:

(1) *Basic data of the premium contract.*

– Target costs	10 M€
– nominal profit (5%)	0.5 M€
– cost sharing of the contractor	10%

(2) *Example 1* Target cost overrun (±20%)

– Incurrence of costs	12 M€
– nominal profit	0.5 M€
– cost sharing by the contractor (2 M€ × 10%)	0.2 M€

– Incurrence of costs	12 M€
– real profit of the contractor (0.5–0.2)	0.3 M€

(3) *Example II* Target cost underrun (−20%)

– Incurrence of costs	8 M€
– nominal profit	0.5 M€
– cost sharing by the contractor (2 M€ × 10%)	0.2 M€
– real profit of the contractor (0.5 + 0.2)	0.7 M€

This model case shows that the contractor (*cost sharing 10%*) participates with 10% in the additional costs in case of cost overruns and thus reduces his nominal profit accordingly or participates with 10% in the cost savings in case of cost underruns and can increase his nominal profit accordingly.

It is easy to see that the defined participation formula (*customer/contractor participation*) is of decisive importance for the effectiveness of the contractor's behavior. If the contractor participation is set too low, the motivation to save funds is often not effective enough despite profit cuts, for example, if the contractor is willing to accept profit losses in whole or in part in favor of personnel employment due to capacity utilization problems. This means that the *customer/contractor* participation value must be carefully selected. In Fig. 13.3, four participation models are explained in detail, based on a paper by Kahn [7]. Case A) provides for a 15% (*85/15*) and Case B) a 30% (*70/30*) contractor participation in cost overruns and underruns. In case C), a two-stage participation formula was defined. Under this formula, the contractor has a 15% share of the savings (*85/15*) and a 30% share of the cost overruns (*70/30*). Case D) also shows a tiered arrangement, 85/15 for cost savings and 60/40 for cost overruns, but with the additional introduction of a neutral zone. Kahn writes: "*It is sometimes appropriate, in the case of cost overruns due to technical uncertainties, not to have the contractor share immediately in the additional costs, but to introduce a neutral zone.*" [7] Kahn further elaborates on the procedures used in ESA that when nominal costs are exceeded into the neutral zone, only the costs incurred are reimbursed but without profit markup. The neutral zone should be between 5 and 10% of the nominal costs. If, however, costs are exceeded beyond the neutral zone, the costs must be reimbursed in accordance with the example in Fig. 13.3 (*Case D*), 40% must be paid by the contractor.

The determination of the cost-sharing formula depends very much on the quality of the cost estimate, the type of project and the qualification of the contractor, and should therefore be made individually for each project, taking all factors into account. In this context, reference should also be made to the common practice of setting upper and lower profit limits. In extreme cases, such a regulation could mean a doubling of the nominal profit (*maximum*) or a one hundred percent loss of profit (*minimum*).

The determination of the upper and lower profit limits clearly defines the range of flexible profit management (*premiums or penalties*). A regulation exceeding the

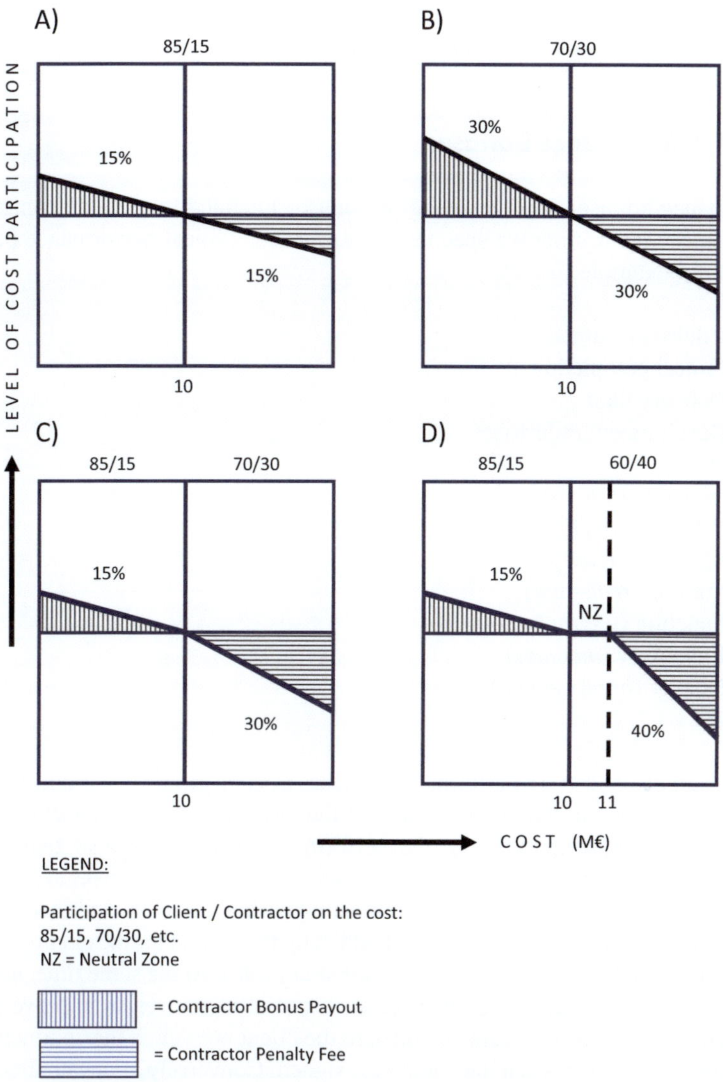

Fig. 13.3 Participation model for premium contracts (Source: Kahn, ESA)

mentioned range is not in the sense of a premium contract since larger cost deviations indicate that the target costs were not estimated realistically enough. In this context, it should also be borne in mind that profit maximization achieved through drastic cost underruns raises the suspicion of questionable favoritism on the part of the contractor and excessive cost overruns can lead to losses for the client that he is no longer able to bear on his own, which in the worst case leads to the project objective being jeopardized. A premium regulation only makes sense in connection with a well-founded estimate of the *target costs* and a reasonable premium range (i.e., *definition of reasonable maximum and minimum profits*).

13.1.4 Performance Bonuses

Performance bonuses are the setting of bonuses intended to give the contractor a special incentive to improve specified performance. Typical performance parameters are, for example:

1. Schedule fulfillment:
   - Project completion.
   - Delivery data.
   - Development milestones.
   - etc.
2. Technical parameters:
   - Weight (*reduction*).
   - Speed (*increase*).
   - Volume (*reduction*).
   - Function (*improvement*).
   - Reliability (*increase*).
   - Quality (*improvement*).
   - etc.

The purpose of setting performance bonuses is to maximize the required performance. Analogous to the determination of the target costs, when concluding contracts for development projects, the performance parameters also represent an achievable compromise due to many yet unknown problems and risks and in accordance with the estimated costs. The contracting parties agree on a performance target (*specification*), for the fulfillment of which by the contractor the client is obliged to pay the agreed price (*target cost plus nominal profit*). At the same time, however, by suspending premiums, the client creates special incentives to improve performance in those areas that seem desirable to the client but which the contractor cannot guarantee to meet when the contract is signed. Conversely, many contracts also contain corresponding contractual penalties if performance deteriorates. In this context, however, it should be mentioned that the reduction in performance naturally comes up against lower limits, that is, a range is specified. If, for example, a target speed of 200 km/h is specified for a vehicle to be developed and an increase in this speed is considered desirable without an upper limit, the lower limit may be set at 195 km/h, for example, and further deterioration may lead to refusal of acceptance by the customer.

Often, however, a performance improvement that is not required or desired, for example, the reduction in weight of a component beyond the designated minimum value is not necessarily eligible for a premium. If there is no discernible benefit to the client then the assessment of the performance premium should be directly related to the benefit achieved for the client. If the early completion of a project or the increase in the performance of the product results in a measurable benefit for the client, this benefit should serve as the benchmark for the assessment of the premium. Conversely, the assessment of contractual penalties should also be adapted to the actual damage incurred by the client and not represent a fictitious figure. Of course, performance bonuses must also be within a reasonable range of magnitude in relation to the contract value. For example, they should not exceed the magnitude of the agreed profit, because the purpose of the performance bonus is to create incentives, but not to cover risks or generate additional profits. For a European satellite project, the following premium scheme was established: [9].

- Target cost (TC).
- Nominal profit (NP).
- Maximum incentive (MI).
- Maximum penalty (MP).

As a result, the nominal profit (NP) can be doubled in the best case (NP + MI = 0.14 TC) and put at risk for contractual penalties in the worst case (NP—MP = 0.0 TC).

According to this scheme, the contractor could double the nominal profit *(7%)* under ideal conditions and had to give up any profit in the worst case. The possible premium or penalty was divided according to the main parameters of cost, design, engineering (*0.35–0.35 - 0.3*) and corresponded to the particular circumstances of this project. However, the mentioned ratio is not a general rule but is valid only for this case. For fixed-price contracts, for example, the cost premium would be omitted, and any premium system would relate only to design and critical technical parameters. The interested reader of this section is specifically referred to the following publications: *NASA Incentive Contracting Guide* [8] and *DoD Incentive Contracting Guide* [10].

13.2 Project Contracts and their Structure

13.2.1 Modular Contract Structure

The project contract is the link between the client and the contractor, and it is the main reference document between the partners. It defines the services and obligations of both parties as well as the legal regulations. The drafting of project contracts requires close cooperation of specialists, for example, lawyers, technicians, managers, and business economists, as it is an interdisciplinary task. This cooperation requires that the specialists introduce their respective points of view at the appropriate place in the contract. The contract should therefore be structured accordingly. Particularly in the case of larger projects, specifications and requirement specifications should therefore be included as appendices; see Fig. 9.5.

The modular contract structure provides in the main part all legal provisions, and the detailed technical documents shall be included as an annex to the contract. This has proven successful in practice. In contrast, a contract structure that jumbles technical, legal, business, and managerial facts is difficult to handle and often very confusing. "*The modular structure of contract documents has the advantage that the respective specialists can negotiate independently without the entire team having to negotiate at the same time.*" [11] In other words, the individual contract elements can be worked on separately by the respective specialists and it is not until the overall compilation and negotiation of the contract documents that all specialists come together to jointly approve the contract document. According to this principle, a project contract could be structured, for example, as follows—minimum contents: [11, 12]:

Main part of the contract

1. Definitions.
2. Deliveries and services (*reference to the specifications*).
3. Price and terms of payment.
4. General arrangements (*rules to be applied, language,* etc.)
5. Special guarantees.
6. Legal arrangements (*applicable law, taxes, and duties, right of withdrawal,* etc.)
7. Project implementation (*general conditions, task assignment, subcontractors,* etc.)
8. Project acceptance.
9. Other conditions (*patents, licenses,* etc.)

Annexes to the contract

1. Service specifications; see subsection 9.3.2.
   - General requirements.
   - Supplies and services.
   - Project management.
   - Dates.
   - Project documentation.
   - etc.
2. System specification; see subsection 7.2.1.
3. Other (*special equipment according to the respective requirements*).

Below are examples of contract structures from different industries. [13]:

1. Architect contract.
   (a) Definitions.
   (b) Subject of the contract.
   (c) Services.
   (d) Fees and incidental expenses.
   (e) Terms of payment.

   (f) Object monitoring.
   (g) Changes.
   (h) Duty to provide information.
   (i) Confidentiality agreement.
   (j) Liability of the architect.
   (k) Copyright.
   (l) Contract termination.
   (m) Applicable law.
2. Procurement contract for major projects.
   (a) Definitions.
   (b) Subject of the contract.
   (c) Project dates.
   (d) Supplies and services.
   (e) Price and payment conditions.
   (f) Inspections and acceptances.
   (g) Patents.
   (h) Release of confidential documents.
   (i) Warranty services.
   (j) Force majeure.
   (k) Changes.
   (l) Contract termination.
   (m) Management and key personnel.
   (n) Reporting.
   (o) Confidentiality agreement.
   (p) Language and communication.
   (q) Applicable law.
   (r) Arbitration.
   (s) Attachments.
      i. Statement of work (*SOW*).
      ii. System specification.
      iii. QA specification.
      iv. Test and acceptance plan.
      v. Schedule.
      vi. Cost plan (*optional*).
3. Procurement contract for a small/medium project.
   (a) Subject of the contract (*goals, tasks, results*).
   (b) Terms of contract.
   (c) Applicable law.
   (d) Commercial conditions.
   (e) Communication.
   (f) Confidentiality agreement.
   (g) Changes.
   (h) Contract termination.

13.2.2 The Statement of Work

The *statement of work (SOW)* is also an important part of the project contract, in addition to the product or system specification; see subsection 9.3.2. The specification expresses *what* the final product of the project should look like and how it should perform, while the statement of work contains information about *what* the contractor must do, *when, where*, and *how*. In other words, the SOW describes all the contractor's tasks and duties that he must perform under the project contract. However, this description should not be understood to mean that all individual activities that the contractor must perform must be described in the SOW. Rather, only the main tasks and duties and the end products to be delivered are to be described (i.e., *the DIL and DRL for example*), while the details are to be taken from the individual work package descriptions and planning documents. The following guidelines should be followed when preparing the SOW [14]:

- *Avoidance of misunderstandings.* A precise description of the tasks and obligations as well as the deliverable items of the contractor and a clear demarcation from the customer is required. Customer Furnished Equipment (CFE) must also be precisely defined.
- *Use of clear wording.* For example: "*The contractor shall perform the XYZ test*" and not "*An XYZ test will be performed.*"
- *Clear definition of deliveries.* It is not sufficient to describe only the tasks and duties but also deliveries of the hardware/software and documents (*see Sect. 12.1*) as well as the exact date of the delivery and the place of delivery must be described in detail; an effective instrument is the *Deliverable Items List (DIL)*. Furthermore, the delivery conditions must also be specified, that is free on board (*FOB*), cost insurance, freight *(CIF)*, etc.
- *Inclusion of existing procedures.* Reference must be made to existing and applicable procedures.
- *Avoiding over-specification and multiple entries.* A very detailed task description (*similar to a detailed WPD, see subsection 9.3.3*) must be avoided at all costs. Likewise, multiple mentions, for example, a delivery date at different places, must be avoided. For reasons of efficient change control, important data (e.g., dates) should, if possible, only be mentioned once at a central place. Reference can then be made to this point in the individual chapters.
- *Inclusion of reference documents.* Reference to documents shall be made only once at a central place of the relevant specifications.

The SOW should be structured by modules similar to the contract structure. The advantage of the modular structure is in the independent processing of the SOW by the different specialists. The modular structure of a SOW can be as follows [15–17]:

Main part of the bill of quantities

1. Overview.
2. Project goals.
3. General requirements.

4. Tasks.
   - Project management.
   - Development.
   - Production.
   - Test.
   - etc.
5. Documentation.
6. Supplies (*documentation, hardware, software*).

Attachments (examples)

1. Applicable Documents (*DRL*).
2. Documentation to be delivered (CDRL); document number and date of delivery.
3. Documentation Requirement Descriptions *DRD*); see Sect. 12.1.
4. Hardware deliveries: list of items to be delivered and delivery dates (*DIL*).
5. Software deliveries: list of programs to be delivered and delivery dates (*DIL*).
6. Provisions by the client; list of provisions, place, and date.
7. Work breakdown structure; see subsection 9.3.3.
8. Milestone dates; list of major milestones including dates.

An important aspect in hardware deliveries is the delivery terms as defined in the international rules for the interpretation of customary forms of contracts, that is the Incoterms. In this regard, the previously mentioned terms FOB* and CIF**, the application of which is also reflected financially, because in the first case the freight costs up to the delivery point defined in the contract, including insurance, are to be borne by the customer, and in the second case by the supplier.

*Free On Board

**Cost, Insurance, Freight

13.2.3 Special Contractual Arrangements

- Letter of Intent. Project deals are often initiated very quickly and sometimes there is not enough time to draft a thorough contract. This often results in a memorandum of understanding (*MOU*) or a letter of intent (*LOI*) to express interest in a business agreement. In both cases, the MOU or the LOI, care should be taken to ensure that they are legally binding. Often, a financial limit (*Limit of Liability (LOL)*) is also stipulated in connection with the LOI.
- Barter. An interesting contractual arrangement is barter in which the partners exchange goods or services with each other. A special variant of barter is that a part of the project costs consists of the delivery of products. However, the problem is that it is necessary to agree on the goods offered in exchange because the recipient of the goods must be able to use or resell the offered products in a useful way. Sometimes the customer makes barter a condition and a criterion for awarding the contract. The author became acquainted with the barter model during a satellite project that was paid for in part with goods from the country in question.

- Intergovernmental agreement (IGA). A special form of bartering is cashless settlement, known as "*bartering.*" The partner countries of the International Space Station (*ISS*), for example, have agreed on a bartering concept to avoid the exchange rate problem within the framework of the IGA. Farand writes that the ISS contract structure consists of three levels: [18].
  1. The Intergovernmental Agreement (*IGA*).
  2. Memoranda of Understanding (MOU) between NASA and the four cooperating space agencies.
  3. Implementation agreements between NASA and cooperating space agencies.
- Buy-back. Another special form of contractual agreement is the *buy-back agreement*, in which a certain portion of the products generated by the newly built facility is offset as compensation. Another example, in which the author was involved, is the financial compensation of a satellite manufacturer with revenues generated by the communication satellite launched into orbit.

13.2.4 Consortium Agreements

As described in subsection 5.3.3, companies often form a consortium because they cannot solve the task alone. Cooperation within a consortium requires that a consortium agreement is concluded by the companies. The content of the contract is described in subsection 5.3.3; see also Steinrücke [19]. Within the sixth Framework Program, the EU has prepared a very detailed model consortium agreement, including definitions and checklists, for cooperation which can serve as a general guideline for consortium agreements [20].

13.2.5 Producer Liability

"*The global tightening of producer liability has reached a dimension of risk that demands fundamental action from a growing number of ventures.*" [21] The producer, as well as the designer or developer in question, is liable for defective goods and damages caused to persons using those goods. The producer's liability does not concern the defect of the product itself because that falls under warranty claims, but around those damages caused by a material defect [22]. In detail these are

- Design flaw.
- Manufacturing defect.
- Instruction errors.
- Development errors.

All those involved in the creation of the product, from the project manager to the developer, can be held liable for the aforementioned damage if they can be proven to have made the aforementioned errors in the event of damage. It is the task of the project manager, as far as his competence is concerned, to monitor this aspect as well. "*Liability exists for anybody who has put a product on the market which is defective or harmful and has by doing so caused an injury. A contractual relationship does not have to exist.*" [23]

13.2.6 Contractual Risks

Every contract also harbors risks, even if they are not clearly visible. Hamann cautions: "*Risks must not be overlooked; they must be countered.*" He lists the following contract risks: [24].

- Contractual risks from the project environment.
  - Country risk.
  - Customer risk.
  - Consulting risk.
  - Cooperation with third parties.
- Contractual risks from project services.
  - Licensing and know-how assignment.
  - Engineering.
  - Ordering.
  - Material deliveries and transports.
  - Construction services and assembly.
  - Commissioning.
  - Training and customer personnel.
  - Staffing/cooperation for maintenance and repairs.
  - Financing risks.
  - Settlement risks/exchange rate risks.
  - Special risks.

13.3 Contract Management

13.3.1 Contract Negotiations

An important task of contract management is the negotiation of contracts. In the process, serious mistakes are made time and again because basic rules of negotiation are violated. "*After all, every contract negotiation is always about ignoring or disregarding certain basic rules that have a decisive influence on the course and outcome.*" [25] If the customer (*client*) must choose between several suppliers, he will endeavor to draw up a ranking list, starting with the best offer, in order to be able to make a preliminary selection. In subsection 14.4.2, a procedure for evaluating bids is explained by means of a practical example. The aforementioned ranking *list* is also known as a *short list,* whereby the term *short list* already indicates that not all bids are shortlisted.

In principle, the negotiation process can be divided into the following three stages:

1. Compilation of facts that are available for consideration.
2. Conducting preliminary negotiations.
3. Conduct final negotiations and determine agreements.

From the customer's point of view, the compilation of facts is essentially based on the available offers and the associated evaluation results for those suppliers who have achieved the first goal, namely to get on the *short list*. "*Without conscientious preparation, the negotiator, whether acting as an employee of the company or as a freelance agent, cannot perform the task assigned to him*." [26] After comparing the bids to the project requirements (*see Chap. 14*), the *short list* is drawn up. Bids that do not meet the *RFP requirements* are eliminated from the negotiation process. The negotiation rounds can now begin. In practice, it has been found that it makes sense to divide the negotiating teams of the client and the contractor into four to five major groups according to the bid structure; see Chap. 14. Here is an example:

- Compliance list.
- Technology, if necessary, further subdivided into subsystems.
- Management.
- Costs.
- Contract and SOW.

The thorough preparation of the tender documents (*see subsection 14.2.2*) is a first step to be performed by the customer. Step two is the careful selection (see Sect. 14.5) and the third step is to make any necessary corrections.

Suppliers should also be well prepared for this process. Firstly, by analyzing the customer requirements contained in the tender package as precisely as possible and responding to them in their offer; see subsection 14.3.2. The latter does not mean, of course, that all requests must be met unconditionally, especially since a customer request may also be unfulfillable or may be fulfillable only in conjunction with very high costs or by means of an alternative solution. Once the offer has been delivered, the question-and-answer game often begins very quickly, that is, the aforementioned preliminary negotiation. The customer needs clarifying additional information and sometimes also insists on further concessions. This post-offer phase is often just as important as the offer phase itself and suppliers are well advised to have qualified personnel on *standby* during this phase. Of course, it is always necessary to assess which concessions can be made and, above all, at what price. But if one wants to stay in the game, this process must be followed.

As described in Sect. 14.2, a draft contract (*sample contract*) should be prepared as early as possible, on the basis of which the contract negotiations are then conducted. In many cases, a sample contract is already supplied by the customer as part of the invitation to tender for projects, and the bidders must then comment on it. Meins says in this context: "*Before the contract discussion with the other side takes place, a draft contract should be available*" and he adds: "*It is obvious what advantages arise if the draft is prepared by one's own side* [27].

Projects are almost always under some element of time pressure. The customer will therefore be tempted to shortlist the first two or three suppliers and invite them to preliminary negotiations. Through the previously described compilation of facts relating to the technical, administrative, and legal components of the contract (*see Sect. 14.2*), it should be noted:

(a) where there is already agreement,
(b) where it is felt that compromises can be achieved and,
(c) where one is of the opposite opinion.

This is to ensure that the above-mentioned points b) and c) are dealt with consistently and that the areas that have already been adopted are no longer brought up for discussion. In this context, Meins recommends adherence to the following principles: [28].

1. No reopening of positions once they have been conceded.
2. no contradictory behavior,
3. no violation of the dictates of fairness.

In practice, and in connection with the usual time constraints, it often looks as if compromises are reached and then, at the end, a few differences of opinion remain, which can then lead either to the customer conducting further rounds of negotiations with the competition or to the seemingly insurmountable differences being dealt with in connection with the price negotiation. The negotiation of prices, payment terms, contractual penalties, and delivery guarantees should only take place after the results of the negotiations on technical, administrative, and legal details have been achieved. In this regard, once again a quote from Meins: " ... *it is best to talk about money last*." [29] This is also the experience of the author.

With the results of the preliminary negotiations, which are not infrequently conducted in parallel with two or more companies at the same time, the final negotiation is then to be called by the customer. At this point, the competing companies are often asked for a final price reduction *(best and final—BAFO),* which can be decisive for the final selection. In the case of joint projects (*consortia projects*), the outward price reduction often leads to an in-consortium cost reduction in which all participants must make a percentage reduction *(royal cut).*

Due to time constraints, the final negotiation is often conducted with only one supplier, and if the negotiations fail, the second-best supplier can then be invited for negotiation. The purpose of the final negotiation is to incorporate all the agreements reached into the already pre-negotiated contract documents (*contract and annexes*) and to make the contract ready for signing, page by page. In projects with a high degree of complexity, it therefore makes sense for the relevant specialists (*engineers, managers, and lawyers*) of the customer and the contractor to negotiate and seal the contract components for which they are responsible temporarily in separate rooms. "*Only when there is agreement on all points, i.e., when the reading has also taken place, is the time right for final agreement on the few but crucial points of dispute. This will include, if necessary, the price.*" [30] The sealing is done by initialing the individual pages. This makes the contract document ready for signature by the official representatives of the contracting parties. In many cases, the signing is a ceremonial act which is accordingly honored and sometimes also takes place in public. Here again a quote from Meins: "*Whether with a great signing ceremony or summarily with the signature by both negotiators—at least a drink, if not a banquet, should follow. The contracting parties thereby assure themselves of business friendship and bow to the psychological phenomenon of closing loyalty.*" [31]

13.3.2 Contract Implementation

The signing of the project contract by the client and the contractor on both sides is the prerequisite for the contractor to start the project work. In practice, however, this process does not always proceed smoothly. In many, if not most cases, the contract is not signed immediately for a variety of reasons. Often it is the contract negotiations that drag on for a long time before a final agreement can be reached, because the offer, usually the basis for awarding the contract, is in many cases not clear enough for the client on all points. Nevertheless, due to the deadline situation, the project work often has to be started immediately after the bid has been evaluated; see Sect. 14.5. In these cases, the client very often makes do with a Letter of *Intent (LOI),* as already mentioned, which can bridge the period between the start of the project and the conclusion of the contract. However, the LOl must be based on detailed documents describing the range of tasks to be covered (*service description, deliverables, schedule,* etc.) and the associated preliminary order volume *(Limit of Liability—LOL).* In practice, it is usually done in such a way that the client separates out the work packages to be performed based on the offer and makes them part of the LOI. A similar procedure is followed when determining the LOL for the activities to be brought forward.

Especially during the contract implementation phase and/or during the implementation of the LOI, close cooperation between the project manager and the company contract management is very important. For large projects, it may even be necessary for the person responsible for contract management to be seconded to the project permanently or for the implementation phase. As a direct member of the project management team, he has much closer contact with the project team and gets to know the project issues better.

The official start of the project work takes place after the signing of the contract or the LOI combined with a Kick-Off Meeting (*KOM*). The KOM is an important milestone and should be well prepared. At the KOM, the tasks, schedules, and teams are presented and introduced.

13.3.3 Contract Amendments

Changes initiated by the customer or contractor can occur in all project areas, which may affect technical, schedule, and/or financial aspects, for example; see Sect. 12.2.

If, for example, the customer wishes to extend the range of services, this will lead to a corresponding change which will usually also have a schedule and/or financial consequence. A different situation arises if the contractor requests changes in the schedule, for example due to delayed provision of CFE by the customer.

Requests for changes to the contract must be indicated by a formal change notification (*change request*). The European Space Agency (ESA) has developed a special change procedure and form for notifying change requests. The *Contract Change Notice (CCN)* triggers the change process and approval of the *contract change* is usually subject to the findings of a *Change Control Board (CCB)*. The applicant must provide following information: [32].

- Proposed classification of change; see subsection 12.2.5.
- Intended implementation date for the contract amendment.
- Description of change.
- Justification for change.
- Impact of the change on the schedule, costs, or on other contractual aspects.
- Technical impact and consequences of the change on the performance of the system or subsystems.
- Identification of the work packages affected by the change.
- Estimated cost of change.
- Signatures of the project and the contract managers of the contracting parties.

Approved amendments to the contract become part of the contract in the form of an addition, deletion, or revision.

"*A contract change notice (CCN) is the means by which changes to the business agreement are made,* i.e.

- *Any change with a financial or schedule impact on the project is submitted by the supplier to the customer.*
- *The change is notified to the customer with accompanying documentation defining classification of the change, technical, schedule, and cost impacts in accordance with the specified contractual regulations.*
- *The customer's decision is given after a proper impact assessment (technical, cost, and schedule) is performed.*

The agreement reached between the customer and the supplier is documented in a CCN. The implementation of the CCN results in a new baseline agreed for the project, against which cost control, analysis and reporting is performed." [33]

In certain industries, the term "*claim management*" is also used instead of change management. Both the customer and the contractor have the possibility of amending the contract; see Doetsch: [34].

1. "*CM (claim management) has its origins in Anglo-Saxon. In the U.S. and the U.K., CM has been practiced successfully for many years.*"
2. "*Analysis of expired projects showed that typically at least 15% of the project total is 'claim suspect.'*"

CONTRACT CHANGE NOTICE

Contractor: Prime Contractor	Contract No.
Title of area affected (work package, etc.):	WP Ref: Initiator of change:
Description of change:	
Reason for change:	
Price breakdown (currency):	
Effect on other contract provisions:	Start of work: End of work:
CONTRACTOR	
Project Manager Contracts Officer	Date Date
CUSTOMER	
Technical Officer Initiator Contracts Officer	Date Date Date

Fig. 13.4 Contract Change Notice (Source ESA)

3. "*As a rule of thumb, it can be noted: The more concrete the contract, the lower the claims and*
4. *The fuzzier the contract, the more necessary the claim management.*"

Conclusion: A precise contract preparation which begins with the invitation to tender (*see Chap. 14*) reduces the number of contract amendments and claims. In addition to a clear contract structure, this presupposes that all legal formulations and the contract annexes, such as the specifications (*supplies and services*), plans, and specifications, have been described without gaps and negotiated between the partners. If possible, the contract including annexes should not contain any open or insufficiently defined points, that is the so-called TBDs (*to be defined*). A Contract Change Notice is shown at Fig. 13.4.

Literature

1. Weber, Kurt E.: Vertragsgestaltung und Vertragsmanagement bei Projekten - Juristische Grundlagen und Probleme, Attorney at Law, Kanzlei Weber, Munich, IIR Frankfurt, August 1994
2. Ebisch, Hellmuth and Gottschalk, Joachim: Preise und Preisprüfungen bei öffentlichen Aufträgen, Verlag Franz Vahlen, Munich, 4th edition, 1977, p. 57
3. Source 2, p. 117
4. Madauss, Bernd-J.: Planung und Überwachung von Forschungs- und Entwicklungsprojekten, AlBFachliteratur, Bad Aibling, p. VII-19
5. Armament Decree of 19 February 1982 (Rü VIII 4-Az: 78–60–40-01), Wehrdienst 841/82, 29 March 82
6. Peeters W. et al: The Use of Alternate Contract Types in Europe as Protection against Overruns, National Contracts Management Journal, Vol 23, summer 1989, Issue 1, p. 23
7. Kahn, G.: Partners in Risk - Cost Incentive in Development Contracts, in: ESA Bulletin No. 26, 5. 48
8. NASA: Incentive Contracting Guide, NASA Handbook NHB 5104.3, 1. 8. 67
9. Source 4, pp. VII-23
10. Department of Defense: Incentive Contracting Guide, DoD Handbook FM 38–34, 19 Jan 65
11. Source 4, p. VII-15 and VII-16
12. ESA: "General Clauses and Conditions for ESA Contracts", ESA-Document ESA/C/290, Rev. 2
13. Madauss Bernd-J, Contract management/ execution of major international projects, Bonn, August 1994 (not published)
14. Source 4, p. VII-17
15. US Air Force: Work Statement Preparation, AFSCM 70–5, 1. 2. 68
16. NASA: Statement of Work Handbook, NASA Handbook NHB 5600.1, 6. 10. 66
17. NASA: statement of work space shuttle system program definition, RFP No. 10-8423, February 1970
18. Farand, Andre: Legal aspects of the international Space Station (ISS), ISU Strasbourg, 2003
19. Steinrücke, Bernhard: Das Konsortialgeschäft der deutschen Banken, Drucker & Humbolt, Berlin, 2nd edition, 1973, p. 11
20. Consortium Agreement for Integrated FP 6 Projects
21. Brendl, Erich: Der Produzenten-Haftung gewachsen sein: Ein Maßnahmen-Programm, Management-Zeitschrift io 51(1982)
22. Federation of German Industries, Producer Liability, BDI Printed Paper No. 182, p. 9
23. Thoma, W.: Product Liability: Present Status, Trends and Perspective Measures, 2nd ESA Product Assurance Symposium, 1981, p. 31
24. Hamann, Michael: Vertragsmanagement und Nachforderungsmanagement beim Bau und Anlagenbau, IIR, Frankfurt 1994
25. Meins, John: The negotiation of contracts, 2nd edition, Schäffer-Poeschel Verlag, 1993, p. 11
26. Source 25, p. 11
27. Source 25, p. 15
28. Source 25, p. 49f
29. Source 25, p. 58
30. Source 25, p. 64
31. Source 25, p. 75
32. ESA: Project Control Requirements and Procedures for Major Procurement Actions, Phase C/D System Development and Production, ESA Document PSS-33, Issue 1, 3. 1. 1977, pp. 8–2
33. ESA: ECSS-M-ST-60, 2008, P 28
34. Doetsch, Wilhelm: Claim-Management, in Lange, Dietmar (Ed.) Schäffer Poeschel Stuttgart, 1995, p. 104ff

Acquisition and Proposal Management 14

Contents

B. -J. Madauss, *Project Management*,
https://doi.org/10.1007/978-3-662-69057-4_14

Abstract

Today more than ever, companies find themselves in global competition. In order to survive, competitiveness must be maintained and, if possible, increased. This includes the development and marketing of new market-driven products. Project offerings are an effective instrument for achieving this.

Acquisition and proposal management (APM) is a PM support function; see Fig. 5.4.

The economic system is based on the principles of the free market economy, in which supply and demand are the decisive regulative. The handling of projects, like any other industrial or governmental task, is also subject to the laws of the market and must conform with them. The same conditions also apply to research and development projects. These conditions result in the compulsion to compete and to be selected. One of the supporting pillars of competition is the preparation of bids.

Whenever the rules of the free market economy are violated by special measures, sooner or later unwelcome market distortions inevitably occur. Official subsidies undertaken for politically motivated reasons, the justification of which is not to be questioned here, are a good example of this phenomenon. In the past, for example, public projects were sometimes handled without regard to free market principles, which in some cases led to project disasters. The avoidance of any competition or competition-like conditions usually leads to considerable delays in deadlines, cost increases, and functional and quality losses.

The strength of an economic system depends to a considerable extent on the performance and competitiveness in national and international markets. One must therefore also face up to national and international competition in the realization of research and development projects. The results of this competition not only lead to products that are more in line with market requirements but also provide society with a positive selection in the area of human resources, which is of great importance for the future. Talent can only develop optimally in a competitive climate and derives many necessary impulses for creativity from the competitive situation. Just as the athlete's award is derived from the results achieved in competition, the demonstrable personal achievements of project employees are clearly visible in the competitive climate.

"*Thus, contrary to the opinion of its critics, competition is, under the conditions of the modern world, the best means yet known for achieving precisely the solidarity of all. Competition is more solidarity than sharing.*" [1]

14.1 From the Vision to the Competitive Product

14.1.1 Looking Far into the Future

"*Standing still is taking a step backwards*" is a well-known saying. In order not to fall behind, companies must dare to look into the future. It is not enough to stand still with the established product range—you have to think about future possibilities, because "*the competition never sleeps*" is a well-known saying. It sometimes happens very quickly that you lose market potential because you bet on the wrong horse. Even large traditional companies can lose their traditional market because the market or the environment has changed. Well-known examples are the German brand companies Nord Mende and Telefunken or RCA in the USA. Collins addresses this phenomenon in his book "*How the Mighty Fall,*" and mentions the decline of *Bank of America* in the 80s and attributes this to bureaucratic bank managers who had not recognized the trend of the times with new businesses and markets; see Fig. 14.1 [2].

In this context, Collins describes a four-stage process of the rise and fall of companies; see Fig. 14.2 [3].

- Level 1: Overestimation of self, based on successes.
- Stage 2: Undisciplined striving for more because things have been going well so far.
- Stage 3: Ignoring hidden business risks because the existing business is still going well.

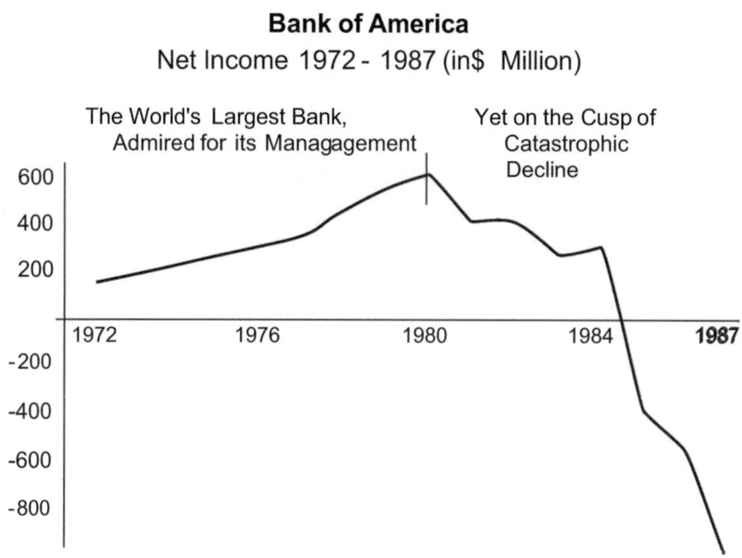

Fig. 14.1 Decline of Bank of America in 1980 (Source: Collins)

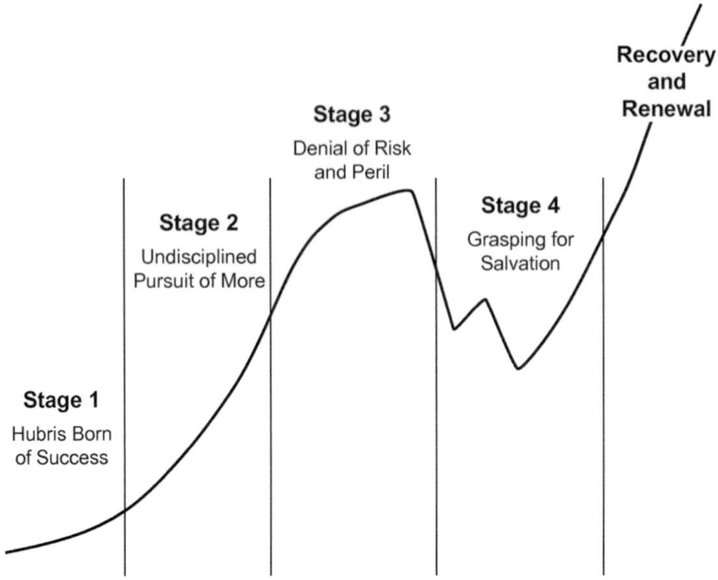

Fig. 14.2 Rescue by an appropriate thrust (source: Collins)

- Stage 4: Search for rescue possibilities, what can be done to still save the business. The longer you stay in Stage 4, that is delay the problem, the harder the downward fall becomes.

Of course, there is also the possibility of counteraction, as Collins mentions on page 113 of his book [2]. There is hope if counteraction is taken accordingly (*recovery and renewal*). Examples are cooperation with potent partners, capital injections, and the timely development of new competitive products. In October 2012, the author accepted an invitation from the Estonian Leadership Conference in Tallinn to give a presentation on "*Why Rockets Fall,*" alluding to corporate bankruptcies and downfalls [4]. Under the motto "*what goes up must come down*" (*title of a song by Tom Petty*), the trajectory of a satellite rocket was compared to the life course of a company. The rocket stage comes back to earth due to gravity unless it receives an additional thrust to stay in orbit. The same is possible if a company that is in the previously mentioned Stage 4 of Fig. 14.2 receives an additional thrust, that is through a cooperation or a new product. "*Companies or projects, especially the so-called "high flyer undertakings" may start like a rocket but soon get out of tune with the environment (technology change, global competition, political impacts, etc.) and therefore fall like a rocket falls by its end of life.*" [4]

The well-being of companies depends heavily on their innovation potential. The rise of MBB, a German aerospace company, that also became known as the "*Munich think tank,*" is a very good example of this. In his book, Bölkow, CEO of MBB, mentions several innovation projects, such as the very successful Bo 105 helicopter with a hingeless rotor, the VJ 101 vertical takeoff aircraft, of which a prototype model is shown in the Deutsches Museum in Munich, the successful Airbus aircraft developments, the German-American solar probe project Helios (*two flight models orbited around the sun*) and the development work of magnetic levitation technology [5]. Bölkow also mentioned in this context that it was always his principle to achieve constructive cooperation with the customer.

Iacocca writes in his book "*An American Career*" that all economic operations can be reduced to three words: People, Products, Profit and that people come first because if you don't have a good team, you can't do much with the other two terms [6]. This is also what Bölkow was aiming for, building and maintaining highly skilled and motivated development teams to turn ideas into new products.

Regarding idea generation, Kotler writes: "*The development process begins with the search for ideas. These should not be left to chance. The company should define focal points for its products and markets. It should set target priorities for new products, be it high cash flow, large market share, etc. It should determine how much energy may be spent on developing original products, modifying existing products, and copying competing products.*" [7]

The ASTRA satellite project of the private operator SES in Luxembourg is an excellent example of how an idea became a globally successful commercial enterprise. The vision to use space satellites commercially can be found in rudiments in a 1923 publication by space pioneer Hermann Oberth under the title "*Die Rakete zu den Planentenräumen*" [8]. In the 70s, the Grand Duchy of Luxembourg investigated the possibilities of using the still young satellite technology for the European-wide transmission of radio and television programs.

On 1 March 1985, the satellite operating company *Société Européenne des Satellites* (*SES*) was founded. The contract to build the first TV satellite, designated ASTRA 1A, was awarded to the U.S. company RCA and launched into 36,000 km geostationary orbit by an Ariane 4 launch vehicle in December 1988. The author was appointed project control manager of ASTRA 1A. The project was fraught with great risks, as technical failures could occur and it was not known whether the leasing of the satellites would be successful. But the project became a "*success story*" in a short time. Success quickly led to demand for more satellite capacity, and by 2016, SES had a global fleet of more than 50 geostationary satellites in orbit.

A key success factor of SES is rooted in its professional procurement management and satellite operations. The procurement process is a decisive factor for the business model because "*space business is a risky business*". However, this is not only true for space projects but also for projects in other industries and especially for so-called "*high-flyer-undertakings.*" The experience gained from space projects can therefore in principle be transferred to other business areas.

14.1.2 Procurement Processes in the Project Environment

The procurement process for a new product has its origins in business planning—the *business plan*. The business plan for a new product (*hardware, software, or service*) should include the following aspects, for example:

1. National/global market potential.
2. Business segments.
3. Existing market segments (*marketing grid*).
4. Customer information.
5. Product requirements.
6. Legislative aspects.
7. Competitor analysis.
8. Price analysis—benchmarking.
9. Time window.
10. Existing resources.
11. Business risks.
12. Marketing strategies.
13. Return on investment (*ROI*).

The creation of a business plan with the previously mentioned or similar contents applies equally to a hotel expansion, software development, or to the procurement of new satellites. In the latter case, both the satellite operator (*customer*) and the supplier must create a business plan. The hotelier as well as the satellite operator cannot perform certain tasks by themselves due to a lack of competence and/or resources. Therefore, they resort to one or more qualified suppliers. This is the beginning of the procurement process, which can be divided into four sections; see Fig. 14.3.

The processes of the four steps are described in the following. During his professional career, the author has had several opportunities to become acquainted with the procurement processes of complex systems as an employee and in a managerial capacity at manufacturers and customers.

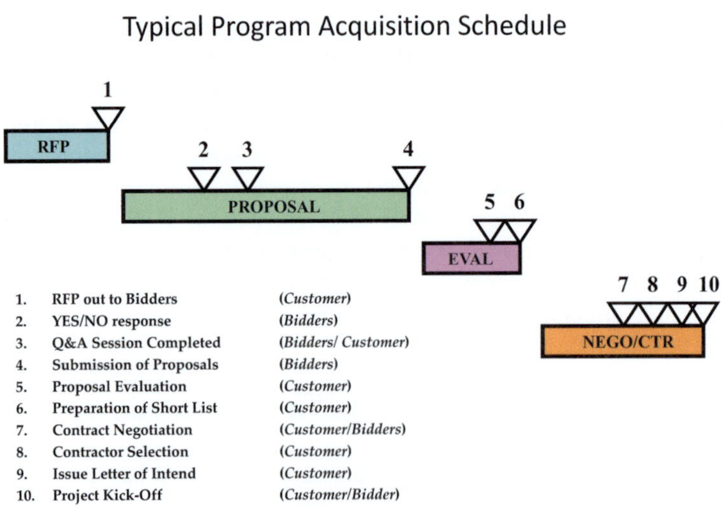

Fig. 14.3 Example of the procurement process

- The first step is the creation of the *request for proposal (RFP)* by the customer. This is a task that requires a lot of experience. In practice, it has been shown that the precise elaboration of requirements is extremely important. The customer should try to define realistic requirements and not wishes that cannot be fulfilled or are very expensive. It is necessary to decide between *"nice to have"* and *"good enough."* For this reason, the customer must first conduct a thorough *feasibility analysis*; see subsection 4.3.1. Although this takes time, it is a valuable investment for the success of the project. When preparing tender documents, customers usually draw on empirical values. In addition, customers often obtain prior to the RFP-specific information from potential suppliers via a *request for information* (*RFI*). This can be helpful for both sides. The RFP is then drawn up on the basis of the RFI information obtained. A detailed description of the *RFP package* is given in Sect. 14.2. Words such as "*should*" must be replaced with the word "*shall*" in the RFP.
- The second step deals with the creation of offers and the process sequence by the provider from the RFP evaluation to the submission of the offer to the customer and the subsequent question and answer phase. The creation of offers can be compared to a project where the result is a product that convinces the customer "*we are the best.*" It must be done fairly, as in a sports competition, although some readers may grin at this point, it is true. In the offer, all aspects of the customer's requirements must be fully answered. Many customers therefore demand proof that all requirements have been answered—that is suppliers are asked to include a *compliance matrix in* the quotation. In addition, the quality of the offers (*presentation and layout*) is an important factor because "*ugliness sells badly*" says Raymond Loewy [9]. Further details are described in Sect. 14.3.
- The third step describes the tasks and responsibilities of the customer in evaluating the offers received, which must be done in fairness. The providers have a right to fairness. Public and private customers must pay particular attention to this aspect in order to avoid possible legal consequences because the providers have often incurred considerable costs in preparing the offer. A fair evaluation is also important for the customer because he does not want to buy a pig in a poke. This underlines the importance of a systematic selection because cheating can also have negative consequences for the customer. The result of the evaluation is usually summarized in a list, and the best offers appear on a *short list*. Companies whose bids are on this list have a good chance of being invited to negotiate a contract; see Sect. 14.7.
- The fourth step deals with *contract negotiation* (*NEGO*). This is the final round. As a rule, the customer invites the companies with the best offers to contract negotiations, referring to the aforementioned *short list*. In practice, it has proven effective for technicians, business people, and lawyers to conduct separate negotiations for larger projects. At the end of the negotiations, all negotiators meet with the project manager for an overall analysis and decision on how to proceed. After the technical and programmatic bid evaluation is completed, the financial bids (*price and payment schedule*) are then evaluated. Questions about a price

reduction, the best and final offer *(BAFO)*, are common, and the answer should be well considered by the offeror; see Sect. 14.7.

14.2 Call for Competition

14.2.1 Tender Documents

The award of a project contract is usually preceded by the preparation of a proposal, on the basis of which the client can then place the expected order. The preparation of a proposal is based on a request for proposal (RFP) from the customer or client, in which all requirements and general conditions are described. The client must therefore carry out extensive preparatory work in advance in order to be able to provide the contractor (*bidder*) with the RFP documentation. If it is a development project, for example, the project usually goes through several phases: concept, definition, development, production, operation, and decommissioning; see Chap. 4. In the early phases, the customer must acquire the necessary knowledge about the future market demands and the development trend of new products. For that, he may start together with some well-known expert consultants a research project in order to gather the required information prior to the preparation of a Phase A RFP. Sometimes customers initiate a pre-Phase A prior to the start of Phase A; see Sect. 4.2. During the concept and definition phases, required system and economic trade-off studies must be performed in order to reach a technically and economically optimal project solution; see Sect. 10.5.

In the aerospace industry, for example, it is common for public-sector clients to involve potential suppliers or experienced consulting companies in the processing of concept and definition studies. This has the advantage that a wide range of expertise flows into the future project, and industry has the opportunity to prepare for the new project challenges and can form appropriate industrial teams. This principle is illustrated in Fig. 14.4. The eight Phase A bidders (*A to H*) are to be down selected to four Phase B bidders and one company is finally awarded the development contract: that is Phase C. Since the selected company is perhaps unable to handle the order for a major project using its own resources and capacity or is required to involve other companies for political reasons, the competitors from Phases A and B may be involved in the project as subcontractors.

For each of the Phases A, B, and C shown in Fig. 14.4, the customer must prepare a Request for Proposal (*RFP*), which typically includes the following elements: [10–12].

- Tender letter (*cover letter*).
- Attachments to the cover letter, consisting of:
  - Product specification.
  - Service specifications.
  - Evaluation criteria.
  - Terms of contract (*sample contract, if applicable*).

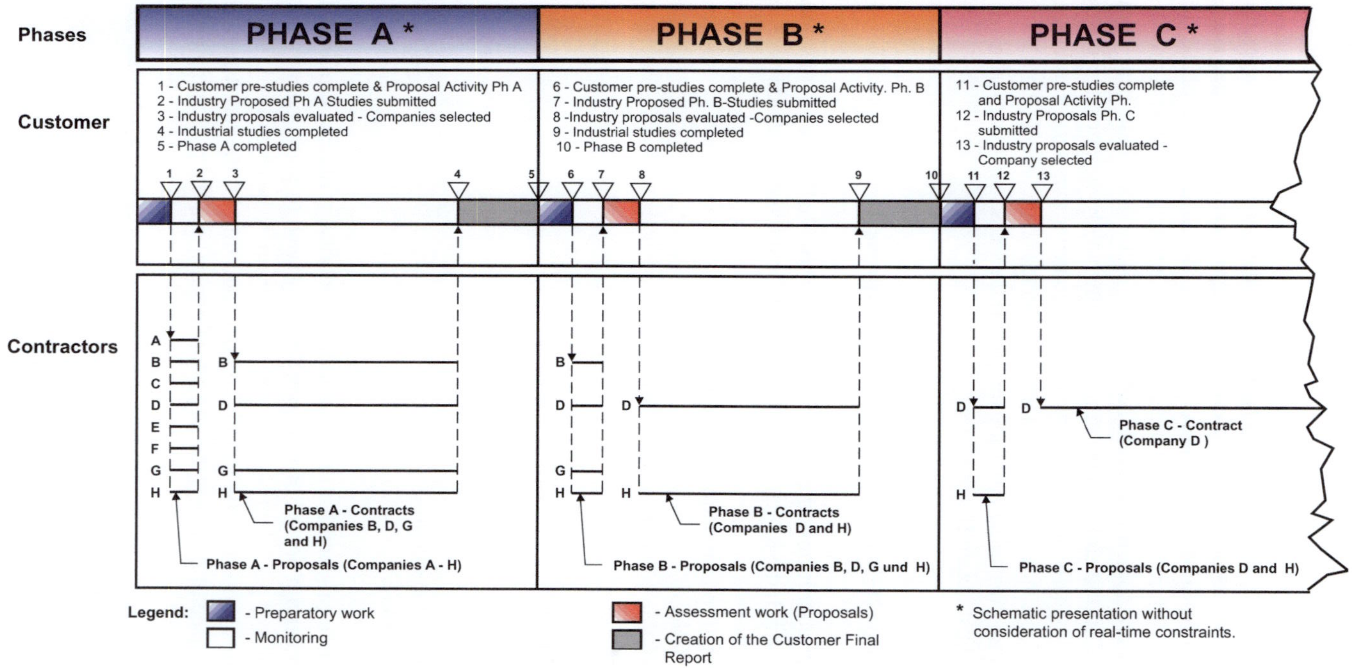

Fig. 14.4 Phase-related bidding process

- general tender conditions and,
- in the case of international government projects, the condition for the return of funds.

With reference to Fig. 14.4, it is pointed out that the tender packages for Phase A, B, C, or D bids naturally differ from each other in terms of content. This applies in particular to the SOW, which must be prepared for the individual phases in accordance with the different tasks. In order to accelerate the procurement process, individual phases are sometimes combined; see Fig. 4.11. ESA industrial studies that are carried out in the early stages are usually financially supported. However, private customers usually do not support the RFI contributions shown in Fig. 14.3. In tenders, clients often refer to groundbreaking standard documents from the country or the contracting organization, for example on the ECSS standards of ESA, which the provider must obtain if necessary.

14.2.2 Tender Requirements and Conditions

In the tender conditions, the contracting authority shall specify the conditions according to which the tender shall be issued. In detail, these include

- Declaration of participation of the provider.
- Terms of delivery.
- Instructions for the preparation of the offer.
- Language.
- Pricing.
- Taxes and duties.
- Currency (*Euro, USD,* etc.)
- Binding period.
- Offer costs.
- Questions for the client.
- Other conditions.

It is common practice for the contracting authority to require suppliers to indicate within a certain period of time whether they wish to participate in the tender. The declaration of participation is a basic requirement to be able to bid. The submission conditions proscribe the submission date and time of the latest possible submission, the place of delivery, the number of copies, and the packaging and sealing modalities as well as the transmission of e-versions in a binding manner.

The contract officer must also provide offerors with detailed information regarding system requirements, such as the following:

- Service specifications in which all supplies and services are specified; see subsection 9.3.2.

- System specification defining all required performance parameters and interfaces; see subsection 7.2.1
- Test specification of the overall system describing the required development and acceptance tests.
- Quality assurance specification describing the required system reliability, quality standards, and safety requirements; see subsection 8.1.1.
- Criteria that the customer will use to evaluate the offers.
- Financial specifications: price type, payment schedule, etc.
- Terms of contract: model contract.

The instruction on the preparation of the offer usually extends to the structure of the offer. For technical products, it is common to divide the quotation into the following volumes or chapters:

Volume I:	Executive Summary
Volume II:	Technical offer
Volume III:	Management offer
Volume IV:	Cost, Price, and contract offer.

Often, the client also explicitly specifies the maximum number of pages (*page limitation*), the font to be used, and the line spacing. For international tenders, the language to be used naturally also plays a decisive role. International customers sometimes stipulate that the bid or parts of the bid, such as the summary and all contract and price documents, must be written in the local language.

Sometimes, the client also specifies the type of price, for example a fixed price or a fixed price with a price escalation clause. The taxes and customs duties to be taken into account as well as the currency to be used are usually also part of the terms and conditions of the offer. In addition, each price quotation must generally be accompanied by a binding period, which is also usually stipulated by the contracting authority. In order to avoid misunderstandings, the customer usually points out that the costs incurred in connection with the preparation of the offer will not be reimbursed by the customer.

It is normal practice for the customer to grant the bidders the right to make inquiries. Often, the tender documents are not clear on some points and require further clarification. Normally, therefore, the client sets a specific time period by which additional questions must be submitted in writing. For reasons of competition, all questions posed by bidders to the customer, as well as the answers provided by the contracting authority, should be communicated to all bidders. This is to ensure that all bidders have the same level of knowledge. Many clients also offer a joint bidders' conference, where all bidders have the opportunity to ask additional questions, which are then answered directly by the client. However, caution is advised here because the enemy (*competitor*) is listening in!

14.3 Offers: An Important Marketing Tool

14.3.1 Offer Strategies in the Project

For companies, offers are the prerequisite for future cooperation. Through them, new business relationships are initiated in the event of selection, and at the same time, they are the figurehead of a company. If Majaro's outline of the marketing mix is followed, the following market elements must be taken into account when preparing offers: [13].

- External, uncontrollable elements:
  - Environmental conditions.
  - Conditions of competition.
  - Institutional conditions and.
  - the legal situation (*legislation*).
- Internal, controllable elements:
  - the product,
  - the price,
  - the advertising,
  - the personal contact and,
  - the future sales conditions.

The external uncontrollable market elements, which, with a few exceptions, can only be influenced to a minor extent, set a fixed framework for the supplier. The environmental conditions are understood to be the political, cultural, and economic conditions of the countries in which one applies for a project task. The competitive conditions are also mostly unchangeable due to existing competition and must be accepted as a given. Similarly, the institutional and legal requirements of a country, for example, the acceptance of technical products by an official institution, legislation, and tax and customs regulations, are also unchangeable conditions for the competitor.

Internal controllable market elements, on the other hand, are the product to be offered and the price. Advertising, personal contacts, and sales planning are other important market elements that a company can control. A company's project management has a significant role to play in developing offering strategies for customer projects. In contrast to mass products, the product to be sold is not yet finalized but only exists as an idea or product description. The project group must therefore try to offer the future customer a product that does not yet exist, including all external uncontrollable market elements in the early stage, which involves risks. Nevertheless, the goal of the project group must be to create an offer with a competitive advantage.

In most cases, the product price is of equal importance to the product itself and is the focus of competition. If the product offering is the same, the price is usually the decisive selection criterion. The future project team must therefore be involved in the cost and pricing process from the outset since the decisive parameters for cost determination, for example system function and quality, are to be influenced by the project management. During the bidding process, it is often essential to weigh the technical options against the associated costs; see Sect. 10.5 (*costs as design parameters*). The task of the project engineer is therefore not only to design a technically competitive product but also to design a product with the most favorable price/performance ratio in the market. In this context, Hinterhuber writes: "*The engineer who is not able and willing to realize a trade-off between additional performance and additional cost with respect to a given price target is in reality, in our time, practicing only half of his profession; the good engineer must, in other words, help decide which combination of performance and cost will yield the most market-competitive product.*" [14]

It is extremely important to point out the correlation between project costs and the determination of the sales price. On the basis of detailed project plans, the project team determines the quantities of personnel, materials, machines, etc. required to carry out the project and multiplies them by the fixed rates defined by the company (*hourly rates for personnel, cost allowances for materials and machines,* etc.); see subsection 10.1.3. The sum of all individual costs must then be added together as shown in Fig. 14.5 and supplemented by surcharges for administrative overheads and the target profit margin in order to determine the selling price. Fig. 14.5 illustrates the strong influence of project management and the technical departments on pricing. In order to achieve a competitive price, all parties involved in the project share responsibility by not increasing but rather realistically estimating the quantity estimates. The technical-economic comparative analyses (*trade-offs*) described in subsection 10.5.1 are helpful in finding a solution through optimization processes with which one can secure a competitive edge in terms of technology and price. In addition, the management has the possibility to achieve the expected market price through profit variations.

The previous explanations make it clear how important it is for the future project management of larger projects to assume a responsible role for the future project as

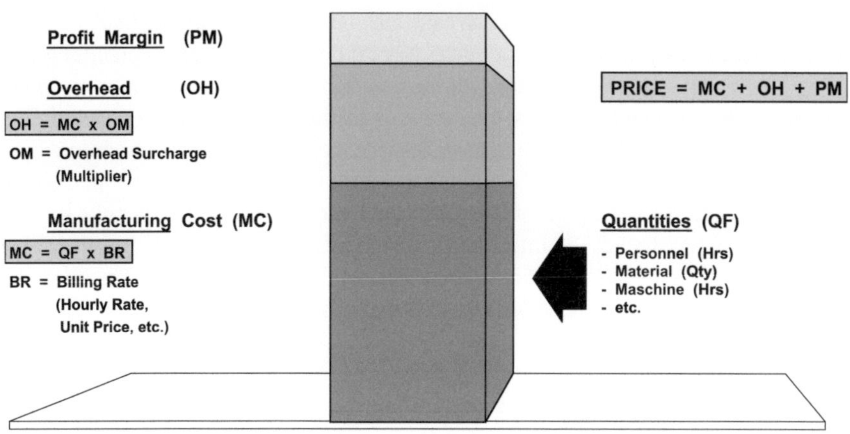

Fig. 14.5 Breakdown of the sales price

early as the bidding stage. In most cases, the intended project team already has the relevant know-how or has already been involved in preliminary phases. In any case, the preparation of a quotation requires a thorough analysis of the customer's requirements, the technical possibilities, and the capacities of the company itself before the work of preparing the quotation can begin. However, the assessment of the competition and the competitive opportunities that can be derived from this must also be taken into account in good time in order to avoid submitting an offer that is lost from the outset because offers cost money.

Taking all the known facts into account, an offer strategy must be defined which, as far as possible, takes into account the strengths of the supplier's own company and the known weaknesses of the competition. In many cases, for example, it has proven successful if the bidder offers the customer options or alternatives in addition to the required basic offer. On the one hand, this demonstrates the flexibility of the provider and, on the other, offers the customer greater decision-making potential. In the case of large projects, it is naturally very difficult for the customer to decide between two equally good proposals. He will therefore look for ways to split the decision into several individual decisions. He may want to approach the problem slowly. In such cases, a modular concept from the bidder, in which project execution is offered in phases (*see Chap. 4*), would certainly have competitive advantages. The customer sees a way out of the "*all or nothing*" decision, and the supplier has the opportunity through this strategy to create an optimal starting position for subsequent phases. Here is a quote from Hinterhuber: "*Strategic decisions are characterized by the fact that, in contrast to tactical decisions, they are not directed toward achieving a specific goal, but toward finding an optimal position from which specific goals that can be specified once certain elements of uncertainty have been eliminated.*" [15]

14.3.2 Proposal Preparation

The preparation of a proposal is not an easy task but usually a very difficult one that involves a lot of effort, because it requires a great deal of innovative potential if one wants to win. This is especially true for the creation of offers for complex or newly developed products and services that are still in the planning stage. It requires a thorough analysis of the customer's wishes, the general boundary conditions, the technically and economically justifiable possibilities as well as the personnel, mechanical, and financial prerequisites (*resources*) of the bidding company. Even if it can be assumed that all the information just mentioned is available in some form or other, the company should first clarify in a preliminary analysis the following points—Silver refers to this process as self-analysis: [16].

- Are there chances of winning,
- Does the product fit into the company strategy and,
- What are the costs for bidding?

Based on this preliminary analysis, a decision must be made as to whether or not the company should participate in the competition. If the decision is positive, the bid strategy must now be developed. Ammon-Wexler and Carmel refer to it as the plan for winning a particular contract and define the following as the main elements of the bid strategy: [17].

- Detailed analysis of customer requirements,
- Analysis of the possible targets of the competition,
- Determination of one's own possibilities and,
- The determination of an offer plan with the highest possible chances of winning.

Based on his many years of relevant industrial experience, the author suggests that one should ask oneself the following question in order to arrive at a strategy and decision from it. Namely, how can one differentiate oneself from the competition in one's own favor? [18].

- Through precise responses to customer requirements and wishes—*be fully compliant,*
- By the quality of the products and/or services to be offered,
- Through convincing presentations of one's own abilities,
- Demonstrating professional excellence of the team,
- Through proven success in the past—*heritage,*
- By showing their own strengths. Here are some examples:
  - Extensive experience at all project levels (*system, subsystems, components,* etc.)
  - Available backup teams at all levels.
  - Demonstrated mastery of all required developmental areas.
  - Maximum flexibility of deployed teams (*including subcontractors*).
  - Full acceptance of project staff by the customer and project partners.
  - International project connections.

The creation of a bid strategy requires the following actions that are critical to the success of a profitable bid:

- Appointment of a bid manager.
- Formation of a proposal team.
- Preparation of a proposal plan and.
- Approval of proposal funds.

Fig. 14.6 shows the detailed sequence of the bid preparation phase. The appointment of an experienced bid manager, who is responsible for all technical and administrative matters, is very important. This has been proven in practice, especially for large and complex projects with a high degree of novelty. This ensures that the proposal preparation process is carried out with great efficiency. The bid manager is the central contact for all bidding activities.

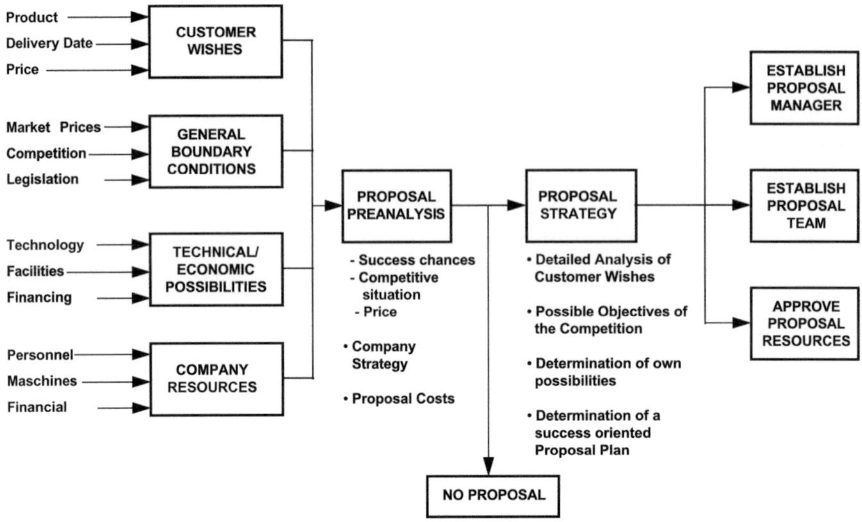

Fig. 14.6 Offer preparation phase

Some companies may have a central proposal department that automatically coordinates all proposal activities. There is no question that this is a convenient situation. However, the central proposal department must cooperate closely with the future project department when preparing a proposal since system-related bids contain many technical documents. Often the proposal responsibility is in such cases transferred to a technician, who is responsible for the technical proposal part with the support of the central proposal department. The central proposal department is also responsible for the preparation of uniform guidelines, which the proposal manager has to follow. This includes the preparation of the financial proposal.

The standard activities based on routine, such as layout design, editing, and printing, are another area of responsibility of the central proposal department. Fig, 14.7 shows this situation using the example of a typical proposal organization for large projects. The bid manager is a temporary organizational unit that is assigned only to a specific bid. When appointing the bid manager, the following principle must be observed: "*The best person to manage the future project is often not always the best person to manage the bid work.*" [19]

The preparation of a proposal for a large and complex project requires that employees from a wide range of technical and administrative departments participate with their knowledge and experience because the future customer expects a clear technical elaboration, the presentation of a flawless management concept as well as a mature planning and, based on this, a convincing cost and price offer. In the case of larger bid campaigns, where, for example, more than 10 employees work closely together over a longer period of time, it is advisable to temporarily consolidate the team physically. If, for example, there is an opportunity to temporarily free up a larger office or conference room for a certain period of time, this should definitely be taken advantage of.

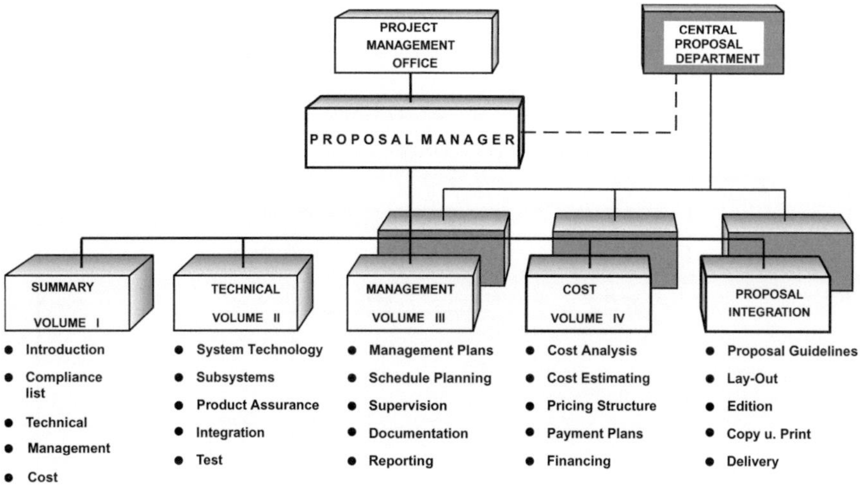

Fig. 14.7 Interaction between the offer department and project team

The preparation of a detailed schedule, in which all actions as well as the respective responsibilities are precisely defined, is very important for the preparation of the offer. After all, it is extremely difficult to prepare a brilliantly written offer comprising several 100 pages in a relatively short time—often only a few weeks are available even for complex projects. For example, it is extremely important to have a detailed proposal outline at a relatively early stage and to have the first drafts of the individual chapters ready by halfway through the total time because the main work is still to come. The chapters have to be coordinated, contradictions have to be eliminated, and the final copies that are then to be prepared still have to be edited. About 20% of the proposal creation time must be reserved for editing, compilation, duplication/printing, and delivery. A quotation schedule is shown in Fig. 14.8. The time segments are based on an evaluation of several technology offers prepared by the author. According to this, planning should allow approximately 20% each for bid planning and bid integration. Individual tasks are shown in Fig. 14.8.

The cost of preparing proposals is not a negligible factor, and only in the rarest of cases will the customer be prepared to bear some or all of the proposal costs. It is not possible at this point to give guideline values for proposal costs since the costs to be incurred are industry- and product-specific and most industrial companies are probably not prepared to provide information. In addition, a company will certainly be prepared to allocate more resources to achieve a new market position than for a routine task. In any case, for many companies, the cost of preparing a proposal is a significant item. Reducing the cost of bidding without sacrificing quality is therefore certainly a welcome measure. In Sect. 14.4, this is dealt with in particular.

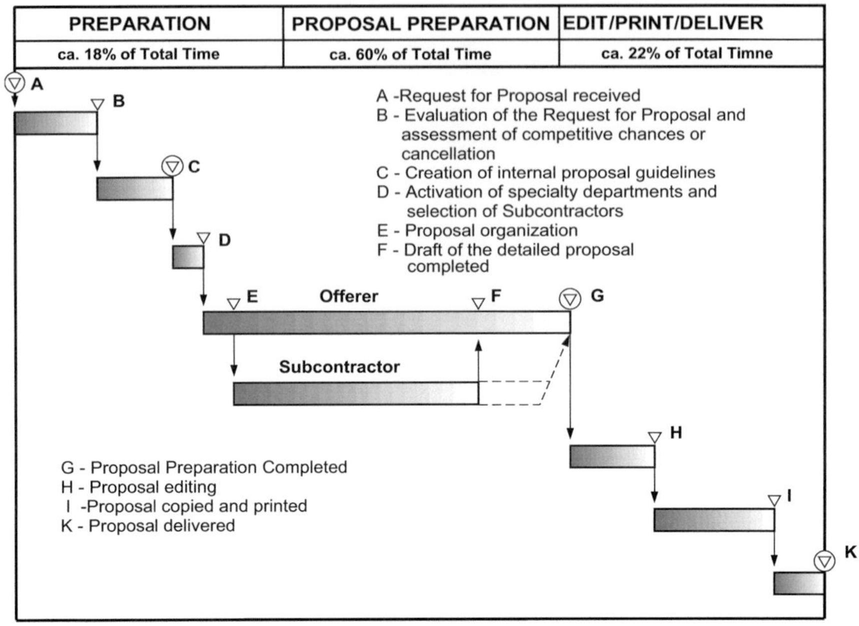

Fig. 14.8 Typical proposal preparation schedule and durations

14.3.3 Features and Options

The bidder should endeavor to point out the advantages of its product and the benefits of cooperation with the customer through its offer. As already mentioned, at the self-analysis many companies will not infrequently come to the conclusion that there are hardly any competitive advantages over the competition if the situation is honestly assessed. There are no technological advantages, the market price can hardly be undercut, and there are no other competitive advantages to be discovered.

Under these circumstances, how can the supplier still gain a significant competitive advantage over the competition? In many cases, highlighting features and defining options or alternatives to the basic offering is a promising measure. Of course, these must be realistic, credible proposals that improve the offering. A verbose but unfeasible additional promise would tend to damage the basic offering.

Underlining the features included in the offer, for example the application of a simpler management process—lean management approach—as a characteristic feature of the project management draws the attention of the client's evaluation team to the advantages of the offer. Other examples might include the modular and easily modifiable nature of the product and maintenance advantages. Often, these points are considered in the design but not explicitly discussed at any point. New customers are usually receptive to a vendor offer for training options. Possible options should be recorded in a catalog of features to be provided at the beginning of the bidding phase and included in the bid at the appropriate place.

Any bid strategy should also include the early definition of project options and alternatives. In most cases, it can be assumed that the client has a clear idea of the product to be purchased but will respond positively to options that allow for further wishes and to alternatives that provide for a choice. The offer is, as it were, the continuation of the principal's thoughts. If the offeror deals now with the thoughts of the client, by providing first a basic offer as exactly as possible fitting to the customer demand, including additional options and alternatives, which equals a qualitative improvement, then the client will often honor this positively. Of course, options and alternatives also have their price. They can improve the product and usually cost more. However, it is also possible to make suggestions aimed at reducing costs; see Sect. 10.6. However, this topic is too extensive to be dealt with exhaustively here, but it should not be forgotten in any bid.

14.3.4 The Roles of the Customer and the Contractor

Larger industrial or public projects can rarely be carried out by the future user (*customer or client*) alone. Often the capacity (*personnel, machines, and equipment*) or the special experience is missing. In many cases, therefore, one or more companies are contracted to carry out parts of the project; sometimes, however, the implementation of the entire project is entrusted to a general contractor; see Sect. 5.3. Often, however, the selected company cannot carry out the task alone and must involve partners or subcontractors, which in turn may depend on suppliers. Fig. 14.9 illustrates this relationship and the resulting client–contractor relationship up to the third stage.

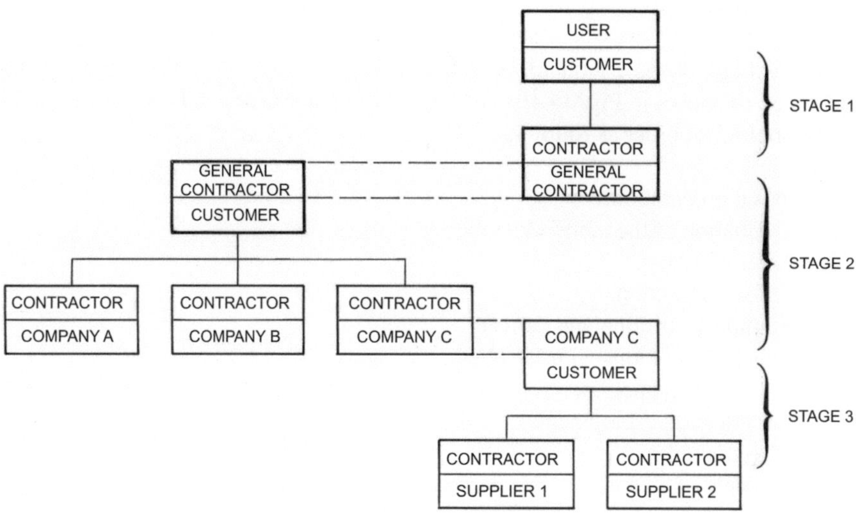

Fig. 14.9 Three-stage customer-contractor relationship

Awarding task packages to a contractor requires thorough preparation and planning on the part of the client before he approaches external companies with a request for a bid (*RFP*). Above all, he should know as precisely as possible what he wants and—to speak in the language of the private home builder—have a rough construction plan.

The role of the customer can be summarized in its chronological order as follows:

1. Preparation of the tender documents:
   - Definition of project requirements (*product, goals, constraints,* etc.)
   - Preparation of the specifications and SOW.
   - Determination of the cost ceiling.
   - Setting of deadlines.
   - Description of the offer conditions and guidelines.
   - Preparation of the bid evaluation criteria.
2. Selection of the contractor:
   - Bid evaluation,
   - General assessment of applicants (*company capacity, past performance,* etc.)
   - Collateral (*for example, bank guarantees*).
   - Other aspects (*for example, political aspects*).
3. Contract negotiations with the selected contractor; see Subsection 13.3.1.
4. Supervision of the contractor during project implementation:
   - Status meetings.
   - Project reviews.
   - Change monitoring.
5. Product acceptance:
   - Quality checks during the takeover.
   - Acceptance tests.
   - Incoming inspection.

The actions of the contractor are closely intertwined with the activities of the customer, as shown in Fig. 14.10. The role of the contractor can be summarized in its chronological order as follows:

1. Proposal preparation:
   - Evaluation of the tender documents.
   - Offer planning.
   - Offer completion.
   - Summary, printing and delivery.
2. Contract negotiations with the client; see Subsection 13.3.1.
3. Project Implementation:
   - Design and development.
   - Manufacturing.
   - Assembly and test.
   - Quality control and acceptance.

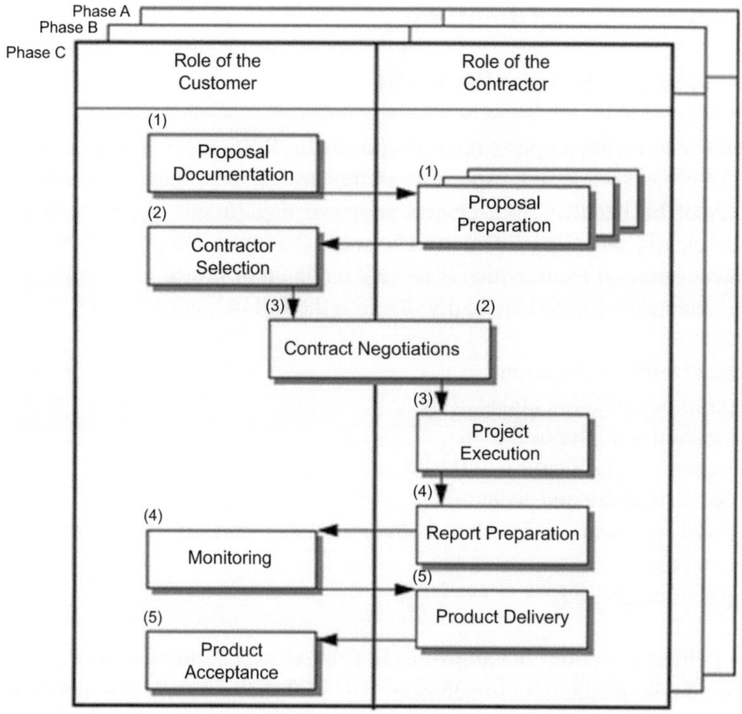

Fig. 14.10 Linking the client-contractor actions

4. Reporting to the client:
   - Technical reports.
   - Status reports.
   - Cost and financial reports.
5. Product delivery:
   - Packing and loading.
   - Transportation.
   - Delivery (*handover to customer*).

The relations between the customer and the contractor are usually defined by contract. The contract concluded between the contracting parties, consisting of the main legal part and the additional contract elements, such as specifications, SOW, performance descriptions, etc., regulates the relations between the two partners; see Sect. 13.2. The thoroughness and depth of the implementation of the aforementioned client and contractor actions contribute quite significantly to the success of the project.

14.4 Proposal Coordination

14.4.1 Proposal Process Planning

Without question, the preparation of proposals for R&D projects can be considered as part of project work and, especially in the case of large and very complex projects, cannot be handled by a central proposal department of a company alone. Fig. 14.7 clearly shows the interaction between the central proposal department and the project proposal team, which is largely made up of project personnel. The proposal preparation process can be divided into the following main tasks:

- Detailed tender evaluation.
- Proposal organization and staffing.
- Preparation of a proposal plan.
- Questions for clarification to the client.
- Proposal structure and scenarios.
- Proposal preparation by the main contractor and, if necessary, including subcontractors.
- Proposal Integration.

The individual actions are shown in Fig. 14.11 as a continuation of Fig. 14.6 in the form of a logic diagram. In addition, Fig. 14.8 shows a typical schedule for preparing a quotation. A study conducted by the author shows that the preparation time of a quotation takes about 20%, the actual quotation preparation time about 60%, and the quotation integration (*finalization*) about 20% of the total time [20]. In other words, the time allocation for "*preparation: elaboration: compilation*" behaves like 1: 3: 1 to each other. Experience has shown that it is extremely important to set up an effective proposal organization in the shortest possible time and to assign tasks to all proposal staff with the help of the proposal plan. Another key point that should

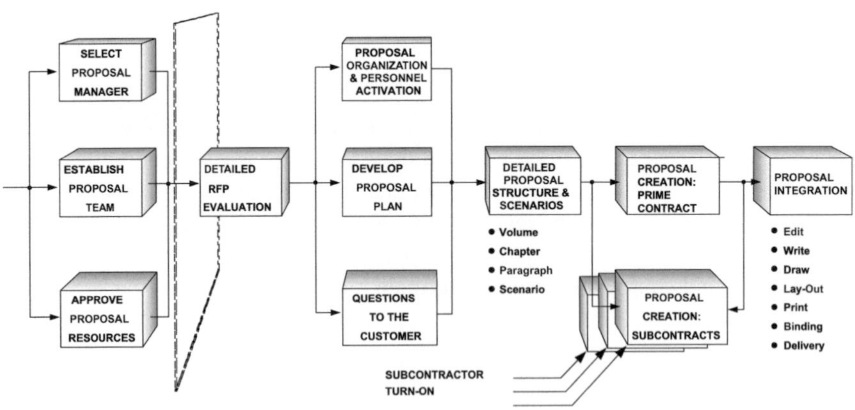

Fig. 14.11 Typical proposal peparation sequence

be taken care of during the preparation period is the creation of a detailed quotation structure.

14.4.2 Tender Evaluation

Successful bid work requires the bid team to analyze and evaluate the client's tender documents in detail. The client associates a very specific idea with the tender, which should be recognized and taken into account by the supplier. As a rule, the tender document is also compiled by a team, and therefore the requirements and wishes are not always described and coordinated clearly, distinctly, and transparently enough. Related requirements are often found in different places within the tender or are expressed only indirectly. For this reason, it is highly recommended that the bidder draws up a checklist of requirements and wishes, which he can then address in a targeted manner.

Drawing up a checklist is also recommended for another reason. It is easy to lose track of the requirements and wishes contained in the tender document. It happens again and again that at the very end of the tender preparation, someone points out an important demand of the client, which has not yet been discovered and processed by anyone. So the checklist has a very special purpose here, too.

14.4.3 Supply Planning and Organization

One of the first tasks of the bid manager is to define a detailed list of tasks and assign individual responsibilities. The offer work can be divided into the following five main tasks and responsibilities, as shown in Fig. 14.7:

1. Executive Summary (*Volume I*).
2. Technical offer (*Volume II*).
3. Management offer (*Volume III*).
4. Cost/price/contract offer (*Volume IV*).
5. Proposal Coordination and Integration (*Volumes I through IV*).

In practice, it has proven effective for the bid manager to delegate the responsibility for preparing bids for large projects to specialists in the main tasks mentioned above, for example, to four *volume leaders* or *book captains* and the bid coordinator. For smaller project bids, the bid manager assumes this responsibility. The team organization for bids described by Stewart confirms the proposed division of labor [21].

For bids of smaller projects, efforts will be made to place the individual bid sections in a single volume (*binder*) with chapter divisions (I) through (IV). One reason for placing the four sections mentioned above in separate volumes is the specialization of the proposal preparation and evaluation teams. Dividing the offer into separate volumes or sections has another advantage. The offer leader can delegate the responsibility for the offer preparation according to the individual volumes, and the persons responsible for the individual volumes (*volume leaders)* can pass on detailed tasks of their volume to the appropriate experts according to the individual chapter structure.

Proposal work must be carried out by the company's most capable experts in the interest of competitive opportunities. This statement also applies to the involvement of subcontractors. However, the best proposal staff of the company often do not have their office in the same building and sometimes not even in the same city. When subcontractors are involved, the case is usually even more complicated; sometimes the favored company is even located in another country. However, with today's media diversity, distances can be easily bridged. In any case, however, it is important for the proposal manager to prepare a detailed proposal plan and guideline with the following content:

1. Proposal breakdown: volume, chapter, paragraph, etc.
2. Information about the proposal organization: proposal manager, volume/chapter responsibility (*volume leader*).
3. The volume leaders are responsible for further distribution into individual chapters and subchapters.
4. Schedule and key dates: detailed outline, first draft, second draft, review, final manuscript, etc.
5. Writing and graphic instructions: font types, line spacing, etc.
6. Terms and abbreviations to be used.
7. Scenarios: descriptions.
8. Page number limitation.
9. Coordination of subcontractor inputs.

With reference to Fig. 14.8, the above instructions (1)–(9) should be completed and distributed to the relevant employees during the first-time segment (*20%*). It is important that this deadline is met to prevent chaos. For status determination, it is essential to prepare a detailed schedule that also includes the reviews; see Fig. 14.12. For critical phases of proposal preparation, short daily status determination is necessary. Important milestones are the reviews, for example the *red team review* (*RTR*), by which time all bid work should be completed. The RTR is attended by the respective authors and experts who were not involved in the bid in order to prepare an independent assessment. The RTR can also be described as an independent critical review before the customer reads it. In addition to the schedule, it is advisable to show the degree of completion of each chapter as shown in Fig. 14.13.

In the first section of the proposal plan, the preparation of scenarios should also be completed because scenarios are very important, especially for very large proposal tasks with many employees from different locations. This largely ensures that the proposal sections created by the individual authors fit together in terms of content and that the individual topics are balanced in terms of content. Scenarios are the scripts of the offer, similar to a script for the preparation of a cinema film; see Fig. 14.14.

PROPOSAL CHAPTER	CHAPTER RESPONSIBLE	PROPOSAL REVIEW AND INTEGRATION PLAN (Example for a large and complex proposal) FINAL PHASE BEFORE PROPOSAL SUBMITTAL (Fictional dates)											
		M 19	T 20	W 21	Th 22	F 23	WE	M 26	T 27	W 28	Th 29	F 30	Sa 1
1. TECHNICAL	Gerhard	IR				FR			DR				
2 MANAGEMENT	Bernd		IR			FR				DR			
3. COST	Volker				IR			KR			DR		
4. CONTRACT	Hans				IR			FR		DR			
5. OPTIONS	Peter					IR			FR		DR		
6. EXECUTIVE SUMMARY	Victor			IR					FR	DR			
INTEGRATION	Horst											PACK	SEND

Legend:
DR **Printing**
FR **Final Review** by external review team that was not involved in proposal preparation (also called "Red Team Review" and which can be seen as "Customer Simulation"
IR **Internal Review** after work completion (Text, Figures and Tables)
CR **Cost/Price-Review** (by Company Management)
MGMT Management
WE Week End

Fig. 14.12 Example of a detailed proposal milestone schedule

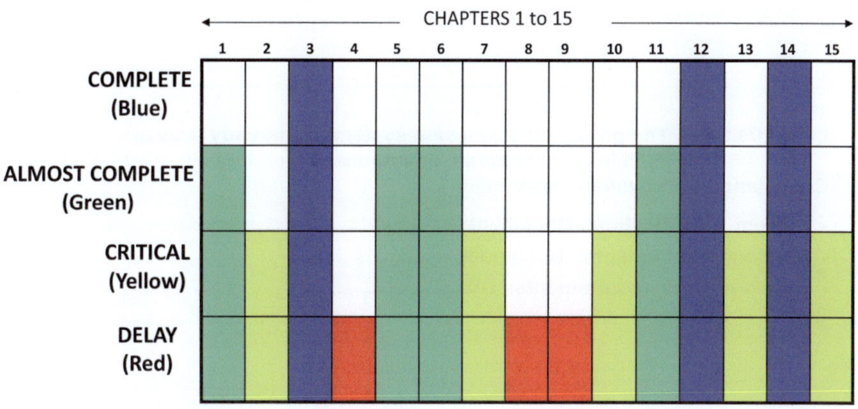

Fig. 14.13 Degree of completion of the individual chapters

By the time the proposal chapters are finished, there is usually little time left for redesign and customization. The company *Procurement Associates, Inc.* has developed a special *scenario technique* under the name *STOP-Technique*, in which, similar to the production of a film, a script is first created [22]. Then the entire bid is pre-planned page by page according to the detailed outline in model form. Fig. 14.14 shows an example of a scenario. Scenario creation may seem bureaucratic or cumbersome, but it leads to better coordination and a better offer. Of course, the authors must be involved in the creation of scenarios.

If there is a large spatial separation of the offer staff, the creation of scenarios is also a cost-saving measure if carried out correctly. This can be easily seen in the example in Fig. 14.15 in comparison with the conventional method. The conventional method (*case A*) requires the teams to be in constant contact with each other

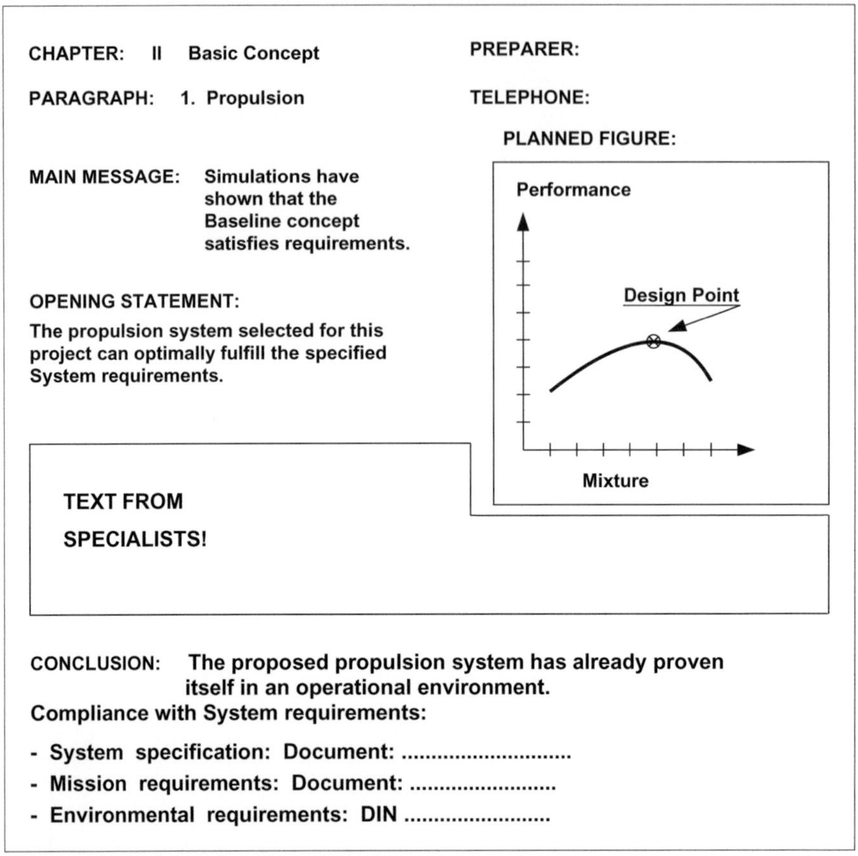

Fig. 14.14 Example of a typical proposal preparation scenario

and is associated with high travel and telecommunications costs as well as many redrafts. At the very end, when proof-reading, the so-called last-minute changes regularly occur. Often the proposal team then must be considerably reinforced in terms of personnel and sometimes work has to be done on weekends.

The modular scenario method (*case B*), on the other hand, introduces a more planned approach. A relatively small group of the bid team creates the scenarios in the preparation phase (*see* Fig. 14.8), which are then presented on boards or walls (*the storyboard*). The next step is then scenario correction. All employees involved in the offer now have the opportunity to critically analyze the scenarios in comparison with each other. The main objective here is to ensure that the interfaces and information are aligned. Once the scenario correction has been completed, the teams can now return to their home locations and complete their chapters.

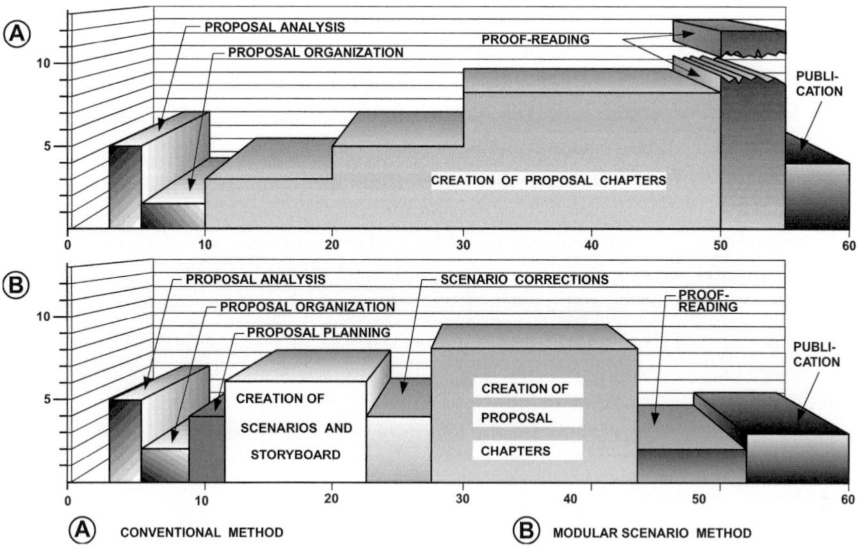

Fig. 14.15 Cost savings through different proposal preparation scenarios A or B

14.4.4 Proposal Preparation

Proposals are a sales instrument and, in this respect, also a self-presentation of the bidder. The question often asked by technicians in particular as to whether proposals have to be so extensive, detailed, and ultimately expensive cannot be answered in a general way with a yes or no. This is a question that can only be answered with a single premise. Particularly when one is trying to *win* international contracts, such questions are subject to only one premise, and that is: *everything necessary must be done to win the contract—the competition never sleeps.* The result is that everything appropriate to the circumstances must be invested in terms of bidding effort.

Regardless of the importance of the offer for obtaining a contract in the event of an award, it is also an ideal document for the future implementation of the project. Often, the bid's execution is relied upon, as it is a compilation of the company's know-how at the present time. Even lost bids still have a significance for the company, as they can also be seen as a competitive yardstick. If the client has carried out a thorough evaluation based on points and published the results (*see Sect. 14.5*), the weak points of the company's own bid will be clearly revealed and this will provide important input for future bidding work. The reference to an objective bid evaluation is not intended to give the impression that bid evaluations are always carried out according to objective criteria. In business life, of course, subjective evaluation criteria often play a major role, which are then described, for example, by the term "*politics.*" Nevertheless, much can be learned from lost bids: *lessons learned.*

The preparation of proposals must be carried out in the spirit of positive and convincing drafting. Criticism of the customer, even if justified, is not a means of winning a contract. Therefore, the following principle should be followed: "*Make your proposal a Winning Proposal by responding positively to the demands of the customer or client.*" [23] This principle should be followed even if it is difficult to answer or fulfill the customers' demands since criticism can also be expressed in other forms, for example, through additional proposals (*alternatives and options*). An experienced client is aware of the problems associated with project execution and is usually sympathetic to a partner with good ideas.

Each offer volume and their chapters should be logically structured, understandable, and easy to read without adding additional documents. According to the motto "*A good picture expresses* as *much as a thousand words,*", illustrations and tables are a very useful and, in many cases, a necessary supplement to the text, especially when a complicated situation is explained. However, too much accumulation of figures and tables should be avoided, as well as the use of very complicated graphics that are difficult to understand.

Typical outline suggestions for the offer volumes are reproduced below:

- *Volume 1: Executive Summary*
  1. Preface and introduction.
  2. Compliance list—*compliance* (*requirement/fulfillment*).
  3. Product presentation and system description (*overview only*).
  4. Development planning and deadlines (*overview only*).
  5. Management plans and organization of the industrial team (*overview only*).
- *Volume II: Technology*
  1. System requirements and design strategy.
  2. System design.
  3. Sub-system design.
  4. Manufacturing concept.
  5. Assembly, integration, and test.
  6. Facilities and tools.
- *Volume III: Management*
  1. Management strategies.
  2. Introduction of the industry team and division of tasks.
  3. Organization and key personnel of the main contractor.
  4. Organization and key personnel of subcontractors.
  5. Management procedures.
- *Volume IV: Costs*
  1. Commercial conditions.
  2. Price definitions; see Chap. 13
  3. Price and cost overview.
  4. Estimation methods used.
  5. Payment plans and financing.
  6. Contract and SOW.

In addition, it is always necessary to prepare extensive lists, such as cost tables, detailed schedules, and a detailed SOW, which often also contains individual work package descriptions. To prevent unwelcome interruptions in the bid texts, it is advisable to place these bid sections in special appendices to the individual volumes.

The *compliance list* mentioned in Volume I is an important tool for demonstrating the completeness of the tender submitted. The supplier has the possibility to demonstrate there the most important offer requirements in comparison with his answers once again in a clear form by listing the corresponding reference sources. Here is a brief example of a *compliance* list to determine whether the offer meets the customer's requirements:

Customer demand RFP - paragraph	Contractor offer Offer - paragraph
1.1 delivery schedule	1.1 integrated schedule
1.2 definition of milestones	1.2 Main milestones
1.3 etc.	1.3 etc.

It is recommended that the proposal breakdown be aligned with the RFP breakdown as much as possible.

In addition to the bid package, in this example consisting of Volumes I–IV, a signed cover letter is usually prepared by the bidder, which includes the official handover of the bid. It is recommended that this letter should address the main points of the offer, for example the merits of the proposed technical solution or the cost-saving opportunities offered.

14.4.5 Importance of Proposal Integration

The quality of an offer depends not only on the content but also on the presentation quality. The font and drawing quality, the offer layout, editing, and printing are also decisive quality features. Here it is advisable to call in specialists if necessary.

The bid coordinator or proposal manager also has the task of monitoring bid integration, packaging, and delivery work, usually under extreme deadline pressure, because "*the last one to bite is the dog.*" As mentioned at the beginning of this Chapter, many customers are very strict about deadlines for competitive reasons. A minute's delay means non-acceptance. If a proposal consisting of four volumes and some appendices, for example in quintuplicate and at the same time in electronic form (e.g., *by e-mail or CD*), has to be sent to a distant country, it is essential to allow sufficient time for transport and customs formalities. Often, however, authors have already used up all their deadline reserves. For this reason, it is absolutely recommended to have the offer integration carried out by specialists who are not simultaneously entrusted with offer work. To reduce the risk of late delivery due to unforeseen mishaps, it is advisable to plan time reserves for particularly important offers and, if necessary, to send a special courier who can deliver a reserved copy in good time by a second route, for example by air.

14.4.6 Preparation of Proposal Activities

The time to prepare a quote is in most cases very tight. Customers often have a problem completing the request for proposal (*RFP*) in time, so suppliers receive it late. At the same time, however, the offer is expected to be prepared in the shortest possible time. In a competitive environment where all vendors are in the same situation, there is often no alternative but to accept a tight deadline. Therefore, it is advantageous to prepare accordingly in advance of an announced tender, that is before the customer has sent out the tender. Especially if one already knows the customer, there are usually good opportunities to prepare accordingly. Silver, therefore, recommends that one should complete the bid planning in a draft form in advance of tender receipt [16]. He also suggests that under certain circumstances, one should already prepare a draft of the proposal. For example, if one is well acquainted with the customer's working methods and knows what he particularly values, one can prepare many sections that the customer routinely expects. About 80% of the requirements, in this author's experience, are easy to predict and can be prepared in a rough draft; called a "*straw man.*" The author has successfully put Silver's suggestion to start preparatory work even before receiving the tender documents into practice in several cases; see Fig. 14.16 [16].

Initiating the pre-bid activities summarized below for a future bid, it can be very important to spend the time immediately after receipt of the RFP to address critical tasks and focus on special areas with the aim to produce a high-quality bid:

1. Determine offer manager and team.
2. Define profit strategies.

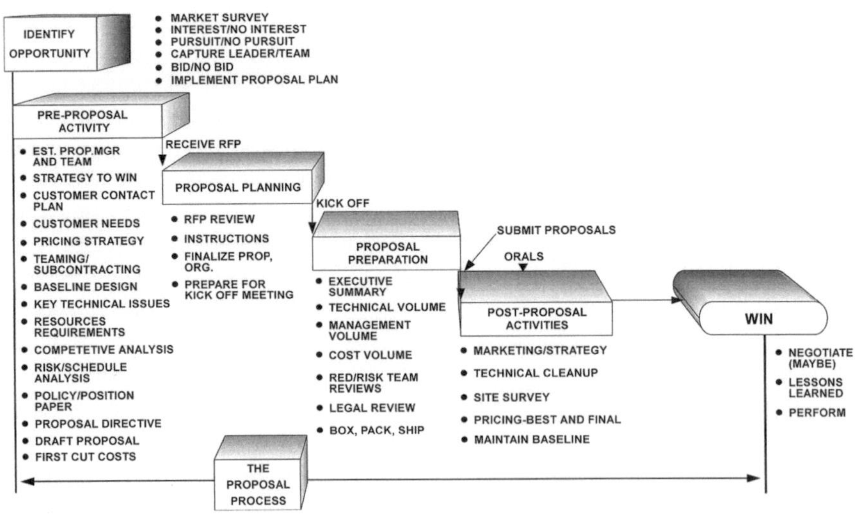

Fig. 14.16 Offer flow including preliminary activities (source Hyman Silver)

3. Develop a customer contact plan (*who knows whom at the customer*).
4. Cost and pricing strategies.
5. Subcontractors and partners.
6. Basic technical design.
7. Investigate key technologies.
8. Determine resource requirements.
9. Perform competitive analysis.
10. Perform risk analysis.
11. Create draft proposal plan.
12. Create a draft offer (*straw man*).

14.5 Proposal Evaluation

14.5.1 Evaluation Criteria and Standards

In order to be able to carry out a fair evaluation of bids, the contracting authority (*customer*) must apply evaluation criteria and quantifiable benchmarks. He also protects himself in this way, because with regard to future cooperation, he cannot have any interest in selecting a non-qualified bidder. In addition, he thereby avoids possible protests by bidders who did not win. A quantifiable selection procedure is also highly recommended in the interest of unintentional distortions of competition. Kahn writes in this regard, "*A proper and correct evaluation of offers is fundamental to good and professional procurement.*" [24]

For industry, a fair selection system is an important benchmark that does not raise false hopes. Despite any criticism of the existing procedures that may arise, the methodology of quantifiable bid evaluation should therefore be endorsed in principle.

In its proposal evaluation guidelines, the European Space Agency (*ESA*) has defined an evaluation scale ranging from 0 to 100 points, which is reproduced below: [25].

Points	Evaluation
100	Perfect elaboration
90	Excellent elaboration
75	Very good elaboration
60	Good elaboration
50	Sufficient elaboration
40	Just acceptable value
0	Worthless elaboration

The further procedure is that the contracting authority draws up a catalog of criteria to be evaluated, in which, for example, each main point of the bid is evaluated and the result is assessed using a point system, for example 0 to 100. The criteria catalog is usually sent to the bidders together with the bidding documents. The following is a typical list of criteria: [26].

- *Executive Summary (Volume I)*:
  - The executive summary is usually not evaluated because it contains a proposal summary.
- *Technical (Volume II)*:
  - Consideration of the given objective.
  - Observance of external boundary conditions.
  - Conformity of the design proposal with the system requirements.
  - Thoroughness, comprehensiveness, and credibility of the design analyses performed.
  - Identification, clarity, and completeness of interfaces.
  - Planned design margins (e.g., weight, power, etc.) and flexibility of the system design.
  - Quality of the design options.
  - Thoroughness, comprehensiveness, and credibility of the proposed scheduling.
- *Management (Volume III)*:
  - Clarity, consistency, and credibility of the proposed industry organization.
  - Clarity and credibility of the project organization and structure.
  - Distribution of responsibilities between the companies and departments involved.
  - Proposed key personnel.
  - Evidence of experience of the proposed team with the implementation of similar projects.
  - Applicability of the proposed management procedures.
- *Commercial offer (Volume IV)*:
- Interpretation of the given project requirements.
  - Comprehensiveness, thoroughness, and credibility of the cost/price proposal.
  - Quality of the options offered (costs).
  - Explanations of the terms of the model contract.

14.5.2 Weighting of the Results

The second important step in bid evaluation is the weighting of the individual evaluation results since not every point is of equal importance. In most cases, the technical part is rated higher than the management part, for example. For the weighting of R&D projects, here is an example: 50, *30*, 20 for the technical, management, and commercial bid areas, respectively. The weighting must then be subdivided into individual areas [27]. A practical example of the management proposal (*Volume III*) is shown below.

CRITERIA	W[a]
(1) clarity, consistency, and credibility of the industry organization.	5
(2) clarity and credibility of the project structure	3
(3) distribution of responsibilities between the companies involved.	3
(4) proposed key personnel	**8**
(5) evidence of experience of proposed project teams.	**8**
(6) applicability of the proposed management procedures.	3
TOTAL MANAGEMENT (volume III)	30

[a]Weighting

The weighting given indicates that items 4 and 5 of the table are particularly important. It corresponds to the general experience that the quality and the experience of the key personnel for the realization of a project are more significant than for example the quality of the management procedures (*point 6*). However, the above-mentioned evaluation of the main areas 50:30:20 is composed differently for fixed price offers. Silver gives the following example: 60:40:0 for technology, management, and price [16]. In this case, the price is not an evaluation criterion, but an offer value that is compared with other offers. Table 14.1 shows an example of an evaluation scheme.

14.5.3 Evaluation Team

The evaluation task for a proposal is an important responsibility. Therefore, the evaluation team must meet two basic conditions:

- It must have professional competence
- be independent in its decision.

Table 14.1 Example of an Evaluation and Ranking Table

Evaluation Criteria	BIDDER A	BIDDER B	BIDDER C	BIDDER D	WEIGHTING
Organization. Aspects	**70** 210	**75** 225	**65** 195	**60** 180	**3** 300
Subcontractors & Suppliers	**55** 165	**75** 225	**80** 240	**65** 195	**3** 300
Key Personnel	**70** 420	**75** 450	**75** 450	**75** 450	**6** 600
Related Experience	**65** 390	**70** 420	**75** 450	**80** 480	**6** 600
Management Tools	**65** 130	**75** 150	**75** 150	**70** 140	**2** 200
Total • **Un-weighted and** • **Weighted**	**67.8** 1,315	**75.2** 1,470	**77.4** 1,485	**72.0** 1,445	**20** 2,000
Ranking	**65.8%** **4**	**73,5%** **2**	**74,3%** **1**	**72,3%** **3**	**100%**

For this reason, it is necessary that the evaluation team must be carefully selected. It must consist of experts from a wide range of specialist areas and be free from external influence, for example from the direct supervisor, for the duration of the evaluation. Among other things, the evaluation guideline must be strictly observed for any bid evaluation action: [28].

- Establishment of an evaluation committee, which should normally have no more than six members.
- The evaluation committee shall be chaired by a customer executive, such as a principal department head.
- No member of the evaluation committee shall be under the influence of, or accountable to, his or her immediate supervisor in his or her evaluation work.
- Decisions of the evaluation committee are made according to the principles of majority voting.
- The advisory committee shall perform the following tasks:
  - Release of the tender procedures.
  - Determination of the evaluation criteria.
  - Determination of the weighting for the individual evaluation criteria.
  - Determination of the individual evaluation groups.

The assessment teams must have the required qualifications and be familiar with the project. During the evaluation work, they should not be under the influence of or report to their immediate supervisor in connection with the evaluation work. One member from each individual assessment group shall be appointed as the spokesperson of the group. The collective result of the assessment group shall be arrived at by a majority vote. Fig. 14.17 shows the principle of the ESA's bid evaluation system.

In his role as SES business manager, the author successfully applied in 1992 a tailored version of the ESA proposal evaluation guideline [10] for proposal evaluation of SES satellites ASTRA 1C and 1D.

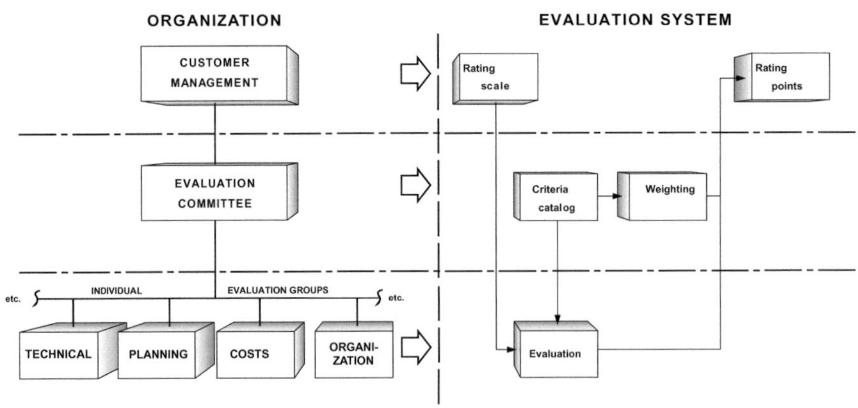

Fig. 14.17 ESA bid evaluation system

Kahn mentioned that the main reasons why a formal evaluation process is necessary are as follows: [24].

- to ensure that the best product is procured,
- to make a fair and honest selection,
- to create an evidential basis for the involved decision-makers,
- to have written documentation available to demonstrate at any time that a fair selection process has taken place.

14.6 Chances of Winning

14.6.1 To Win or Not to Win, that Is the Question Here

Burmeister, formerly director of RCA PRICE Systems, writes: "*Eighty to 90% of the bids we prepared are lost. Name me any activity of yours where you would settle for a ninety percent failure rate; I know of none. In most cases, we bid to lose.*" [29] True, if one knew one's odds better, one would naturally bid only when there was a relatively high probability, that is, a probability greater than 10–20%, of succeeding. On the other hand, it is clear that there cannot be a 100% chance of winning because then there would no longer be any competition. So, it is a question of a qualitative improvement of the self-analysis, that is, a more precise statement of the chances of winning [16].

Very often, the decision to participate or not is made without a thorough analysis of the chances of winning. There are two main reasons for this:

- Wishful thinking is all too often taken as fact.
- there is a lack of reliable analytical methods.

It is understandable that the technical departments of a company are easily persuaded to take part in a competition involving a product for which their own company in principle has the ideal prerequisites. The feeling arises that one cannot lose here since one has already developed and built something similar. Often this feeling is additionally strengthened by statements of the marketing experts, who allegedly have first-class contacts to the client and express *we have the best chances*. How can the management still decide negatively there? But nevertheless, it happens again and again that a supposedly safe offer misses. Mark Burmeister has summarized the typical comments after the loss of a bid that was considered safe as follows [29]:

- They (*the competition*) have influenced the client!
- This time they will lose (*competitor*) head and neck!
- Technically we were the best!
- We probably could have reduced the price!?
- They (*the competition*) have already been selected!

These comments contain part of the required analysis process that should have taken place before the expensive bidding process. Thus, one can counter with the following questions: Could we also have influenced the client? Would we also have risked our necks at that price? If we were technically the best, were we perhaps too good and therefore too expensive? If we could reduce the price, why didn't we? Why didn't we know beforehand that the competitor had already been selected?

The chances of winning the project award must not be considered one-sidedly, for example, only with regard to the technical bid (*Volume II*). The bidder competes with all factors against his competitors. This means that it must have a foreseeable competitive edge in both technology and management, and the price must be right.

14.6.2 Difficulties with the GO-NOGO Decision

Bids carry a twofold risk: first, the risk of not winning and thus spending bid funds unnecessarily, and second, if winning, taking a project at too low a price. Both risks are directly related to each other. Burmeister very aptly says, "*You can't lose money on a contract you don't have.*" [30]

For this reason, the analysis of the chances of winning, which ultimately leads to the GO-NOGO decision, must begin as early as possible. Many bidding experts, therefore, repeatedly recommend maintaining intensive contact with the client as early as possible in order to know better what the client actually wants and to respond more precisely to the client's wishes. Silver points out that information about the client's plans and also the budgets available to him are usually not a secret but are often made known through publications [16]. An early feasibility analysis, combined with initial cost estimates based on cost estimation models (*see Sect. 10.3*), therefore often provides information as to whether or not a bid is worthwhile. In the case of public-sector research and development projects, there is usually the opportunity to participate in the early Phases A and/or B of the project; see Fig. 14.4. This automatically gives the bidder detailed knowledge about the project and puts him in a better position to estimate his chances of winning. In this regard, Laine says, "*If you haven't already been involved in the concept phase, don't bother bidding on subsequent phases.*" [31] Silver further recommends influencing the design of the bidding documents and suggests the following checklist for a GO-NOGO decision [16]:

- Does the company have appropriate know-how in all areas?
- Has sufficient preliminary work been done with the client that influenced the bidding documents?
- Have preliminary technical investigations and bid preparation work been carried out in accordance with the plan?
- Does the client consider the bidder to be one of the potential winners?

- Have sufficient funds been set aside to provide a good service?
- Does the order not involve too much risk, and/or does it lead to lucrative follow-up orders?
- Have personnel, facilities, and management requirements been provided for successful competition?
- Are credible strategy benefits evident?

Silver believes that if these questions can't be answered almost 100% in the affirmative, you shouldn't submit a bid.

It is also recommended for competitive reasons, as Silver states, not to start the bidding process when the RFP is received, but considerably earlier; see subsection 14.4.6 [16]. He recommends preparing a dummy RFP for certain areas prior to the expected official RFP and preparing a *straw man* on this basis. This gives one a head start in terms of time and allows one to focus on specific issues and the quality of the bid during the main bid preparation phase. However, the described procedure presupposes that one already knows the customer and the future project in its outlines due to previous projects or preliminary studies. After officially receiving the RFP, one can then adjust the preliminary offer already prepared accordingly. In most cases, according to Silver, the replacement tender already corresponds to a high percentage to the official tender. In Fig. 14.18, with reference to U.S. companies, an experience curve for the effort of bid work, including preliminary and follow-up work, is reproduced [16].

To avoid misunderstandings, it is pointed out at this stage as a precautionary measure that the processes described above must be adapted to the respective circumstances. Employees who frequently create quotations usually fall back on existing standard texts and quotations created in the past. This enables them to react quickly. This is also true for small projects. Hyman Silver mentioned early

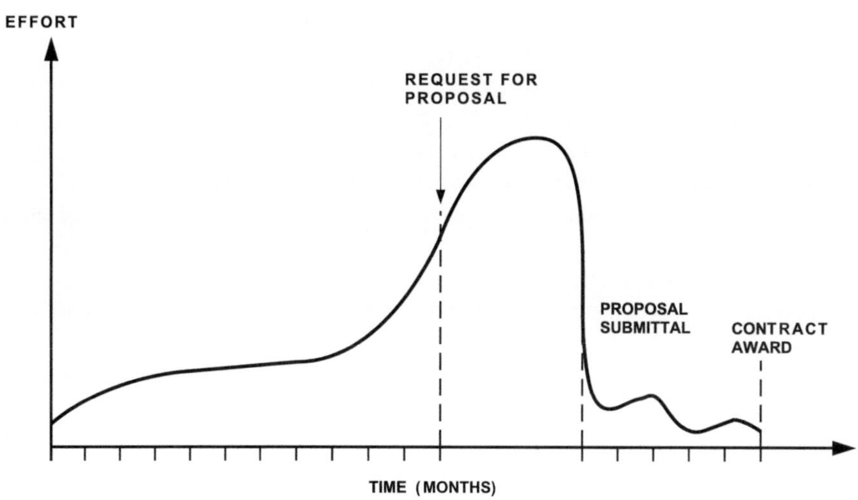

Fig. 14.18 Bidding effort before and after receiving the request for proposal (source: Silver)

preparation for an expected tender and the creation of a straw man is advisable and not exotic; see Figs. 14.16 and 14.18 [16].

The following list contains typical questions for proposal evaluation of communication satellites: [32].

1. What would be the best strategy for market penetration—what kind of business do you think is profitable?
2. Is the business plan acceptable?
3. How should we start defining/verifying the system objectives?
4. What would be the best replacement strategy—your experience?
5. Is the pricing strategy competitive?
6. If appropriate, how should government support be used to best advantage?
7. Are the project time schedules credible?
8. Are all of the bidders (*bidders list*) qualified and are their factory allocations and resources available?
9. Do all bidders show a clear determination to win?
10. Considering international project management and marketing, what are the main risks?
11. How should we start to secure frequency coordination?
12. Considering satellite size and insurance cost, what is the best launcher strategy?
13. What is the best policy for using, or not using commercial-grade Satellite ground terminals?
14. Are all bidders fully compliant with major requirements of the RFP and have all completed the compliance matrices?
15. Are those bids that contain US components potentially ITAR compliant?
16. Do all bids include quotation for launch and in-orbit insurance and how has it been calculated?
17. Are the quoted prices firm fixed (*FFP*)?
18. Have bidders proposed payments by milestones or by progress?
19. Have all bidders included bank guarantees as part of their proposals?
20. Have bidders identified, leased, or purchased land and/or buildings for the required ground stations?
21. Do the proposals include the design and delivery of equipment, including software, for the satellite, payload, and network control center?
22. Do the proposals include the design and delivery of all required Satellite ground terminals?
23. Have bidders included sufficient training concepts and plans in their proposal?
24. Has system acceptance been clearly defined–in-orbit including ground and Satellite ground terminals?
25. Have bidders clearly defined the role and responsibility of the turn-key contractor *(TKC)* and the management interfaces to the client?

14.7 Contract Negotiations

If the supplier is shortlisted and invited to negotiate the contract (*NEGO*), the goal of winning the contract is closer; see subsection 13.3.1. It now depends very much on whether and how one survives the contract negotiations. Negotiation experience is important and this applies equally to both sides. Typical contract negotiation aspects are described below:

- Negotiation goal: Achieve full compliance of technical and programmatic offerings with customer requirements (*example*):
  - System function.
  - Quality and reliability.
  - Availability.
  - Implementation schedule.
  - Life cycle costs; see Sect. 10.6.
  - Payment schedule and milestones.
  - Acceptance of the contract conditions.
- Negotiation strategy:
  - Parallel negotiations with two (*maximum three*) suppliers.
  - Negotiation of open technical points.
  - Negotiation of open contractual points.
  - Negotiation of open financial points.
  - Identification of non-compliances and show stoppers.
  - Price reductions—best and final offer (*BAFO*).
- Negotiation plan:
  - Invitation of the selected provider (*short list*).
  - Negotiation Management.

    Planning of separate meetings (*technical, management, deadlines, costs, and contract*).

    Final selection of the best bidder.

    Signature of all contract pages including annexes.
  - Summary of NEGO results.
  - Contract signing and ceremony.

Contract negotiations do not always proceed in a straight line but often in stages. If the first round of negotiations (*NEGO*) between the customer and the shortlisted bidders has not been successful, there is often a time break; see Fig. 21.7. Negotiating teams sometimes ask for a consultation break (*caucus*) so that they can prepare for further action. This is not always in line with the NEGO plan, but if both sides have an interest in working together, they will put up with it. Victor Billig has described a detailed NEGO process using the example of a turnkey contract; see Subsection 21.3.4. Billig and the author have repeatedly experienced that a resulting delay can usually be accepted in the interest of a better result.

Literature

1. Pierer, Heinrich von, Lohmann Karl, Lübbe-Wolff, Gertrude: Zwischen Profit und Moral, Carl Hanser Verlag, Munich Vienna, 2003, p. 42
2. Collins, Jim: How the mighty fall, Harper Collins Publishers Inc., New York, 2009, p. 9
3. Source 2, p. 20
4. Madauss, Bernd-J.: Slide presentation "*why rockets fall*", Estonia leadership conference, Tallinn, October 2012
5. Bölkow, Ludwig: Committed to the future, Herbig Verlag, Munich 1994
6. Iacocca, Lee: An American Career, Econ Verlag Düsseldorf and Vienna, 1985, p. 214
7. Kotler, Bliemel: Marketing-Management, C.E. Poeschel Verlag Stuttgart, 7th ed., Stuttgart, 1992, p. 494
8. Oberth, Hermann: Die Rakete zu den Planetenräumen, Verlag Oldenbourg, Munich, 1923
9. Statement of the French-American industrial designer Raymond Loewy
10. ESA (ESRO): Manual of Competitive Tendering Procedures in the European Space Research Organisation, March 15, 1974, p. 4
11. Madauss, Bernd-J.: Result Report of the First Practical Seminar: Planning and Monitoring of Research and Development Projects; MBB Education Program, MBB Report, p. 4
12. Kahn, Stephan: Advanced technology projects and international procurement: the case of the European Space Agency, 2 public procurement law review, Sweet & Maxwell, Andover, 1993, p. 19
13. Majaro, Simon: International Marketing - a strategic approach to World's markets, George Allen and Unwin, London, 1978, p. 30
14. Hinterhuber, Hans H.: Strategische Unternehmensführung, Walter de Gruyter, Berlin/New York, 1980, p. 185, p. 24
15. Source 14, p. 24
16. Silver, Hyman: Proposal preparation and source selection, State of the Art Seminars, Rome, Munich, Stockholm, 1981
17. Ammon-Wexler, J. and Carmel, Catherine: How to create a Winning Proposal, pp. 2–5
18. Madauss, Bernd-J., Preparation of offers, MBB report UR-589/82, 1982
19. Madauss, Bernd-J.: Planung und Überwachung von Forschungs- und Entwicklungsprojekten, AIBFachliteratur, 1978/81, pp. VII-6
20. Source 19
21. Stewart, Rodney D. and Stewart Ann L.: Proposal Preparation, John Wiley & Sons, Inc, 2nd. Ed., 1992, p. 81ff
22. STOP-Technique, Procurement Associates, Inc, Covina, Cal. USA
23. Madauss, Bernd: Preparation of Offers Special Offer Techniques, MBB Report UR-589/82, Part 2
24. Source 12, p. 24
25. ESRO (ESA) Manual of Competitive Tendering Procedures …, p. 19, 3/15/1974
26. Source 19, p. VII-9
27. Source 25, p. 18
28. Source 10, p. 19
29. Burmeister, Mark: Bidding to lose, RCA Corporation, 1982, p. 1
30. Source 29, p. 5
31. Laine, Morton J.: Configuration Management, in: Space & Aeronautics, Vol. 46, No. 6, Novembre 1966, p. 76
32. Madauss Bernd-J.: contract negotiation, training for satellite amd Management, Royal Army of Oman HQ SAF, December 2010

Project Personnel 15

Contents

Abstract

Project management is first and foremost a leadership task because without leadership a project wallows like a ship without a captain. Therefore, successful project management requires professional leadership. The project manager must master the methods of project control in theory and in practice. In addition, he or she should have the qualifications to lead employees. Appropriate preparation, training, and motivation are required for project managers.

B. -J. Madauss, *Project Management*,
https://doi.org/10.1007/978-3-662-69057-4_15

It is not uncommon for authorities and companies to think that project management is essentially a method but not a leadership task. Furthermore, since there is often a lack of capable and, above all, experienced project managers and project teams, attempts are sometimes made to compensate for the problem with new methods and software solutions. Often, however, the use of new terms or buzzwords without knowing their meaning creates confusion without solving the actual management problems. First and foremost, qualified project managers with people management skills are needed. Barker and Cole say: "*If you can't work well with people, you'll never be a brilliant project manager. As a project manager you won't be brilliant without the respect and support of your team and other key individuals around you. There are many effective techniques that are simple to use … grouped into three topics:*

- *Building a project team—recruiting and inducting new team members.*
- *Motivating the team—techniques for improving your team's productivity.*
- *Providing leadership—setting direction and building loyal following* [1]."

The methods and procedures for planning and monitoring projects have been continuously improved over the past four decades with regard to the manifold practical requirements. Management efficiency can be significantly increased by PM methods, such as procedures for earned value analysis, cost estimation models, and risk management; see Chaps. 9, 10, and 19. However, the available methods are often underutilized because their use requires implementation and training of the PM team members concerned. However, since this is too costly for many managers, they refrain from doing so and resort to methods they are familiar with, even if they are less effective.

Another problem is that the developers of new PM methods, and the associated software do not sufficiently consider the practical project requirements and often leave the project managers, who are frequently working under enormous time pressure, alone in handling the methods. It is a recurring experience that it is two different things to develop methods and then apply them. Project managers and their closest colleagues, who are busy with the problems of daily project work, are constantly faced with tasks that involve serious technical and/or administrative decisions, for which they in many cases do not have adequate training. This underlines the demand for simple and practical methods and corresponding implementation support.

However, methods and procedures alone are not enough. Just as with all other management tasks, the personal qualifications of the leader, that is, his leadership skills, his ability to integrate, in short, his personality, play a decisive role in project management. In view of the enormous difficulties involved in handling a complex project, it is of course necessary for project managers and their teams to be trained in the relevant project management procedures. However, it should be understood that simply attending training courses is not enough to become a project manager. Like any type of management, project management is first and foremost a leadership task. However, beyond that, the following additional qualities are highly beneficial to future project managers and their teams: Flexibility, logical thinking, creativity, imagination, commitment, willingness to work in teams, and the ability to make mental analogies. Barker and Cole list the following qualities that a PM should possess: [2].

- Excellent knowledge of management techniques.
- Ability to deal with people.
- Insight to continuously educate yourself.

15.1 Motivation: a Key to Success

"*Motivation always means making a difference, first to oneself, to one's environment, and then to one's fellow human beings*" says Lejeune [3].

15.1.1 Success, Recognition, and Development Opportunities

A project represents a small self-contained unit in which qualified employees of a company can best realize themselves and maintain their uniqueness. The limited autonomy and the pronounced goal orientation are a source of motivation for many project employees. This is true for all project levels that is, from the work package to the overall project. When the delegation principle is applied correctly, employee success becomes measurable. The manager of a work package (*WP manager*), as described in Sect. 9.5, has principally an identical responsibility for a WP as the PM for the entire project. The successful completion of a work package is an important component of the overall project. Measurable success (*see Sect. 9.4*) and recognition by the supervisor of the work performed contribute to motivation.

Herzberg describes the six most important motivational factors that lead to job satisfaction as follows: [4].

- Success.
- Recognition.
- Independent work.
- Responsibility.
- Progress.
- Development Opportunities.

In practice, however, frustrations and resignations of highly qualified and relatively well-paid employees unfortunately occur again and again since they hardly find a chance in their company to even come close to enjoying the motivational factors mentioned by Herzberg. Höhn draws attention to the well-known problem of internal resignation and refers in this context to the following misconduct of many superiors:

- *"Directives that go right down to the details and leave no room to maneuver whatsoever*
- *Lack of information with corresponding strain on the relationship of trust.*
- *Specifying targets without first discussing them with the employee.*
- *Control with demotivating effect,* i.e., *lack of constructive criticism, instead of hurtful rebuke.*
- *Lack of recognition for services rendered,*
- *Disinterest of the supervisor in the ideas of the employee.*
- *Supervisor's annoyance with employee's ideas that are not his or her own.*
- *Adoption of the employee's ideas by the supervisor, who passes them off to others as his own."* [5]

Aucoin mentions the reasons for employee dissatisfaction mentioned by Adele Lynn [6]. Lynn quantified the dissatisfaction after a survey of 1,000 people from different organizations as follows:

- 61% mention that their superiors do not attach particular importance to employees and their work
- 57% say leaders do not care about employees' personal issues
- 88% of employees miss recognition of their performance by supervisors
- 62% shared that the workload is unfair.

However, as Barthel writes, motivation in the company (*or project*) is a key for success "*With the motivation of its employees achieved, a company is enabled to harness many previously untapped human resources.*" [7] And Aucoin adds, "*As a direct manager, you will depend heavily on the productivity of the people who work for you.*" [6]

Probably almost every employee in a company has at some time thought about his or her advancement and associated it with considerations of promotion. Promotion, that is, advancement within the hierarchy. However, the hierarchy only allows for a laborious climb toward the top, and there, the vacant places become rarer and rarer as one moves up the ladder. With an average supervisor/employee ratio of, for example, 1–5 (*span of control*), it is easy to calculate how much or how little opportunity there is for advancement along the hierarchy levels; see Fig. 5.7. Particularly in companies with low expansion and fluctuation, it is very difficult and, above all, tedious for many a capable employee to climb the ladder. But in many companies, it is only the climb to the top that brings the longed-for success, recognition, independent work, responsibility, further advancement and development opportunities and, finally, a higher income; see Herzberg [4].

Strong hierarchical thinking is in many companies usually associated with a very strong centralism. "*At our company, everything goes over the table of Mr. or Mrs...*" was one remark, expressing the fact that the employees have no responsibility or decision-making power whatsoever, but that everything is decided by the boss personally. This expresses the conditions in many authorities and companies very aptly.

On the other hand, the creation of hierarchies is definitely a very useful and, of course, necessary measure of business management. The abolition of hierarchies would simply make it impossible to manage a company. Instead, it is more important to better and more adequately distribute tasks, responsibilities, and powers between top and bottom or to delegate them downwards. Today's increasingly complex tasks require decentralized solutions and the associated decentralized responsibilities and decision-making powers to a much greater extent than in the past. Bernhard cites an important leadership element described by Pascale and Athos (*The Art of Japanese Management*) as follows: "*Organizational structures tend to be broad, decentralized, with few centralized functions and with few hierarchical levels to allow information to flow as directly as possible in both directions. Matrix relationships for planning, supervision, and support are used to integrate the independently operating units…*" [8] In this context, reference is made to subsection 5.2.4.

15.1.2 Project Management: Motivation and Career Path

There are usually several levels of hierarchy in a company. In a small company, there may be only one level below the company management and in a large company there may be several. For example, Fig. 5.8 shows a company organization with seven levels of hierarchy (*decentralization levels*). Terry writes: "*The matrix organization, also known as the lattice work organization, has proven to be a viable answer for solving ever more complex and larger undertakings (projects), for the realization of which more flexible and, above all, systems-oriented organizational structures are imperative.*" [9] The matrix in Fig. 5.8 shows the project core team (6) together with the work package managers (7). They represent a temporary integral and fully functional working group in which capable employees of a large company can express themselves. The temporary limited autonomy and goal orientation are a source of motivation for many employees. Heintel and Krainz put it this way: "*Project groups therefore often become oases of motivation in otherwise barren hierarchies.*" [10] Project work aims at the broad and decentralized organizational structure in the company addressed by Pascale and Athos [11]. For project workers, the vertical top-bottom relationship is often less meaningful than the horizontal or lateral and interdepartmental left-right relationship.

In many respects, work in a project is similar to work in a small business, where one still knows one's direct co-workers personally (*village community situation*). Especially, when the project team (*engineers, business people, lawyers,* etc.) is housed in a common office complex, teamwork is most likely to occur. The common objectives and the daily living together, door to door as it were, lead almost automatically to pulling together in the same direction. In the project, that is, at the base of the action, detailed knowledge and expertise are an absolute prerequisite. Many business decisions are and can only be made here. There is no question that project work in a team is best suited to motivate employees. This is where success can be clearly measured, where recognition is most likely to be achieved, where independent work is a prerequisite, and where responsibility is commonplace. Successfully completed projects also ensure the advancement and development opportunities of project employees. Martin writes in this context, "*Project experience is an excellent prerequisite for personal career*" and he continues, " *… the project manager learns to think like a company manager … is there a better situation of 'on-the-job' training?*" [12]

Of course, the opportunities for personnel development within the framework of project management depend very much on the status that project management has within the company. However, it can be assumed that project management will gain enormously in importance in the future and thus the chances of many capable junior employees to plan their personal success within the framework of project management will grow. However, it is recommended that in the future, in addition to the realization of new projects, companies should also view project management more than before as a training school for junior staff, on-the-job training.

15.2 Project Managers and their Teams: Where do they Come from and Who Trains them?

15.2.1 Shortage of Project Management Specialists

The preparation and training of project personnel is still a dilemma for public authorities and companies. There is a lack of qualified project managers, controllers, system engineers, etc., in short: project personnel are in short supply. Hansen writes in this regard: "*In the area of large public tenders, providers often have to offer project segments where they do not have the sufficient number of specialists from the outset.*" [13] Thorough training of project managers and project staff should therefore be a priority in the future because many management errors result directly from inadequate staffing of project positions and inadequate management training of technicians assigned to project management tasks. In 1981, the German Society of Project Management (GPM) conducted an evaluation of 250 job advertisements for project management personnel as well as numerous interviews with project managers in industry.

According to this evaluation, project management personnel were sought for the following PM-tasks: [14].

1. *Planning, implementation, monitoring, and control of project activities.*
2. *Fulfilling the coordination function in technical, scheduling, and economic terms.*
3. *Planning and scheduling of resources.*
4. *System analysis.*
5. *Ongoing contact and communication with internal and external authorities.*
6. *Offer preparation and offer evaluation.*
7. *Contract negotiations.*
8. *Leading project staff.*

9. *Employee training.*
10. *Financial and cost control.*
11. *Documentation control and archiving.*
12. *Configuration management.*
13. *Risk management.*

Although this is not a new evaluation of GPM, the demand for positions 1–9 has not changed significantly according to the author's observations. An additional demand for specialists to cover functions 10–13 is discernible. The fulfillment of these project management tasks requires extensive specific expertise, corresponding practical experience, and certain personal behaviors.

With reference to the organizational structure for a technology project shown in Fig. 5.4, examples of various tasks to be performed by the project team are shown below:

1. *Planning and Monitoring (Project Control):*
   - Project administration: information management, contract administration, mail/filing, etc.
   - Schedule monitoring: WBS, WP descriptions, schedules, status checks, etc.
   - Cost monitoring: manpower and cost schedules, payment schedules, work release, staffing and cost control, etc.
   - Configuration and documentation control: documentation outline, specification outline, release, change service, status control, etc.
   - Reporting.
2. *Product Assurance:*
   - Reliability: analyses, monitoring, etc.
   - Quality Assurance (QA): QA monitoring (design or development), component identification and selection, inspection, cleanliness control, material control, etc.
   - Security control.
   - Parts and materials supervision: specification, control, etc.
   - Reporting.
3. *Systems Engineering:*
   - System analyses.
   - System design.
   - System specification.
   - System engineering: design supervision, reviews of technical documentation, etc.
   - System integration.
   - Reporting.
4. *Subcontractor Control:*
   - Evaluation/selection of subcontractors.
   - Introduction of project guidelines at the subcontractors.
   - Preparation of bills of quantities.
   - Subcontractor coordination: deadlines, costs, interfaces, meetings, etc.

- Status controls.
- Action controls.

5. *Production Control:*
   - Manufacturing engineering: methods, processes, tools and fixtures, make-or-buy, etc.
   - Manufacturing support: material supply, warehousing, transportation, etc.
   - Production monitoring: quality, tolerances, safety, changes, etc.
   - Reporting.
6. *Integration and Test (Integration and Test):*
   - Integration planning: procedures, system planning, etc.
   - Test planning: procedures (functional, environmental and acceptance tests), facility planning, etc.
   - Test execution,
   - Integration and test documentation: specification, evaluation, protocols, etc.
   - Reporting.
7. *Operational:*
   - Operating requirements.
   - Maintenance requirements.
   - Operation and maintenance planning.
   - Personnel training.
   - Logistics.
   - Spare parts.
   - Reporting.

It is easy for the reader to see that this is an example of a larger and complex project. For a smaller project, certain functions may be omitted, for example, item (4) if there are no subcontractors. The same is true for items (5) and (7) if manufacturing, integration, and operations are not included. While for large projects there are corresponding positions, for example, "*Head of Planning and Supervision*", for small projects some of the mentioned functions are performed by the PM in the personnel union. The following example of a task reduction for a small project illustrates this:

1. *Planning and Monitoring (Project Control):*
   - Contract Administration.
   - Schedules and status control.
   - Manpower and cost plans and cost control.
   - Configuration and documentation control.
2. *Product Assurance:*
   - Quality assurance (QA).
   - Security check,
3. *Systems Engineering (SE):*
   - System design.
   - System specification.
   - System reviews.

The task and overriding responsibility of the project manager is that of a leader. Capps and LeRoy mention in this context that project managers should adopt a leadership philosophy in which the following aspects play a role:

- Inspiration: for example, ethical and moral standards.
- Intellectual stimulation: actively involving employees in innovation and strategy discussions.
- take into account individual concerns of employees: Needs, personal and professional goals, talents, etc.

Conflict in a team is best resolved by fostering personal cohesion among team members because when you achieve that they care about each other, the willingness to deliver excellent performance also increases [15].

15.2.2 Importance of Project Management Training

Key positions in the project must be filled on time and correctly to prevent serious project mishaps from occurring. But "*where to take from and not to steal*" says a German proverb! Project personnel must be thoroughly prepared for their future tasks through training or on-the-job training. However, many project staff are thrown into the water without preparation and without training. After a few years, they then naturally have very specific experience but often unfortunately without the necessary theoretical background. The everyday life and the routine hardly grant them the necessary time to familiarize themselves comprehensively with the very complex topic of project management. They are forced to muddle through. Yet it is precisely the interdisciplinary and broad-based expertise of the overall management topic that is a prerequisite for a methodical and efficient approach to the project. In this context, Oelsnitz writes: "*Quite a few people think that management is a profession without training.*" … "*In fact, in few professions is training as lacking as in management. And this despite the fact that wrong business decisions today have much more serious consequences than in the past.*" [16] It should also be mentioned that there is a particularly serious shortage of personnel to fill the key positions of project management, planning and monitoring, product assurance, and systems engineering.

In today's world, the manager of a technical project urgently needs an extended interdisciplinary education in addition to a basic technical degree in order to be able to cope with the tasks assigned to him. Systems engineering, planning, and monitoring, organization, contract management, and personnel management often play a greater role for him in everyday life than detailed technical expertise. However, the career path to becoming a project manager or project employee certainly does not lead to the goal through studies alone since practical experience in the company and personality structure play just as important a role as method training. A viable path for future project managers of technical products, perhaps the most promising, is practical experience building on an engineering degree, combined with postgraduate management studies. In the USA, many project engineers take the following path:

- Engineering degree (*BS or MS*).
- Five to 10 years of hands-on experience as a development engineer, production engineer, etc.
- Completion of a second-degree program, for example, Business Administration (*BA or MBA*).

In many U.S. companies, related university courses or company training courses recognized by universities are offered as part of the company training system. The courses held as part of MBA programs are not limited to business subjects alone but also include topics such as quality control and systems engineering. However, PM training has now also achieved a high status in Europe.

15.2.3 Sustainable Training to Become a Project Management Specialist

There are only a few educational institutions where engineers, business economists, and business people can learn the methods and procedures of project management as well as people management in projects. In this context, the question also arises to what extent project management can be learned in courses alone. This question is extremely important for the conception of a thorough training as a project management specialist. In fact, while project management can be learned to a large extent, for example, the correct application of procedures much has to be experienced in practice. In addition, a high level of commitment and talent is also necessary.

The learnable part is essentially a thorough study of methods as well as training in personnel management and teamwork. The PM professional must have a very broad knowledge. This includes all the topics in this book. In particular, planning and monitoring, systems engineering, and contract management are at the forefront of the methodology. Future PM employees should receive comprehensive training. Chapter 25 outlines the complete training program to cover all knowledge areas. Optionally, training can also be provided in individual steps. A comprehensive training course for project management developed by ISU for ESA is provided in Chap. 26. Training on personnel management and teamwork is not project-specific and can therefore also be completed outside PM training.

Commitment and talent are two decisive factors for the career planning of future project employees. Commitment is a basic requirement for working on a project and needs no further explanation. But what about talent? Project work requires specific character traits and skills, namely:

- A strong sense of flexibility, that is, quick adjustment to new situations,
- A view for the whole,
- A limited willingness to take risks,

- Contact and enthusiasm skills.
- Inventiveness and creativity.
- Cooperation and team spirit.
- Willingness to make decisions and the.
- Ability to convince (*inspire*) and lead people.

Charismatic qualities are a great advantage for project managers. In a nutshell, future project employees, especially project managers, should have at least some of the essence of an entrepreneurial personality in addition to a thorough knowledge of methods because project managers and their employees are basically always on the front lines and, in order to be successful, must prove themselves like small business owners. The play of forces (*market, technology, feasibility, deadlines, costs,* etc.) is much more noticeable in the project than in other parts of a company.

On-the-job training is an important aspect in addition to the various training opportunities for project employees, to get to know the practice. It is an additional training module for future project managers. For example, one should avoid entrusting an experienced engineer without any management experience with the management of a large and complex project without any transition. Rather, one should proceed in stages here as well. Martin describes the possible path of an engineer as follows: [17].

- Development Engineer.
- Group leader development.
- Systems engineering.
- Department Manager Systems Engineering.
- Manager of configuration control in the project.
- Head of systems engineering of a medium-sized project.
- Project manager of a medium sized project.
- Project manager of a major project.

In summary:

- Project staff and especially project managers require thorough project management training in any case.
- However, training alone is not enough to be successful on the project team; future project staff must also have the appropriate level of commitment and talent.
- On-the-job training is another important measure for a project management career.

What conclusions can now be drawn from this for project management training, because training is, as already mentioned, very important? A possible model for the training of project staff could be that the various disciplines of the universities and technical colleges offer a special teaching program on project management as part of their training curricula in order to provide students with the basic project

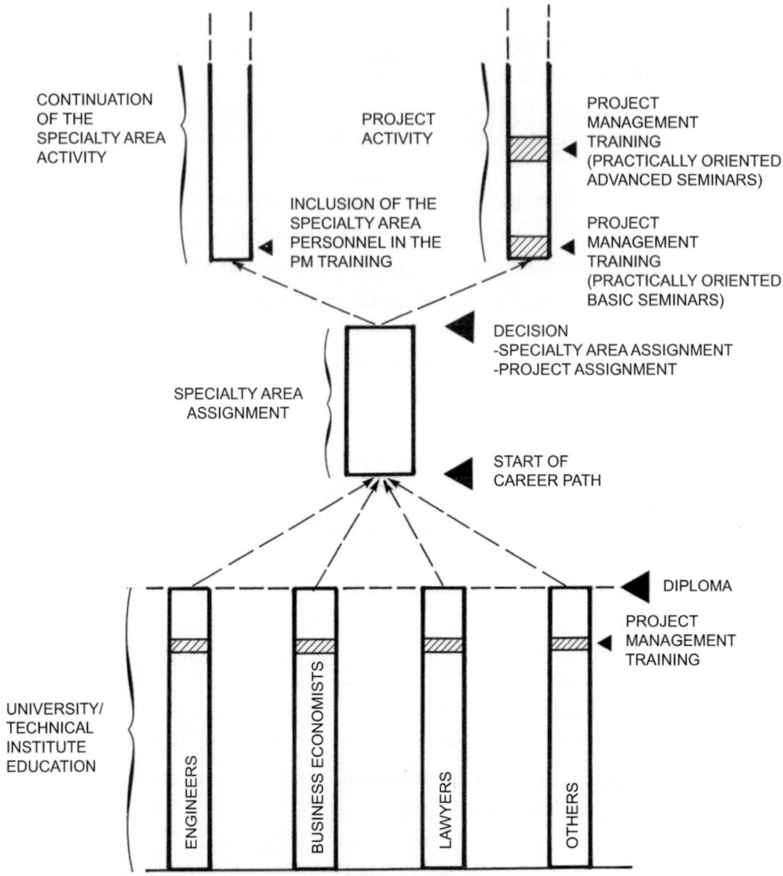

Fig. 15.1 Project Management Training Concept

management knowledge necessary for practice; see Fig. 15.1. A PM training program should include at least following topics:

- Definition of project goals.
- The life cycle of a project.
- Decision theories.
- Project organization and leadership behavior.
- Planning and monitoring.
- Cost estimating and bids.
- Systems engineering and configuration control.
- Product assurance and quality control.
- Contract management and change control.
- Documentation control and reporting.

Sustainable training also includes case studies. Therefore, 4–5 days are required to teach each of the above topics. It is also advantageous if future engineers, business administrators, business people, lawyers, etc. are offered a uniform basic program which is supplemented by in-depth training courses; see Chaps. 25 and 26. In addition, an interdisciplinary cooperation in the field of universities and technical colleges in this respect is of great benefit for the project management idea (*preparation for practice*). The case studies with examples from practice to be included in such a program must be solved by students from the individual faculties in teamwork. By involving lecturers from industry, the practical relevance can be ensured.

Only rarely does the professional career of a university or technical college graduate begin in a project. This is essentially a result of the complex project management requirements, which presuppose corresponding experience. In order to be able to perform a project activity, the project employee should have a correspondingly large wealth of experience, which he or she has acquired over several years of specialist department activity, for example, 5–10 years; see Fig. 15.1. Insufficient practical experience in the specialist department repeatedly turns out to be a handicap in project management practice. Project management is a cross-cutting function with many internal and often also external interfaces, for which the project employee should be able to draw on thorough practical experience from the line. This experience gives him the judgment necessary for the project work as well as the necessary confidence to make decisions. This applies equally to engineers, administrators, business people, and lawyers. In this context, the job rotation principle can also play a significant role. In summary, it does not make sense to hand over a project task to an employee without sufficient specialist knowledge. In fact, companies recruit their junior project management staff for the most part from the line and not directly from college or university.

If possible, entry into a project activity should be combined with accompanying project management training measures (*basic seminars*); see Fig. 15.1. These courses should be practically oriented project management training courses that build on the university and technical college programs also proposed. The practically oriented basic seminars should be supplemented by advanced seminars in which further PM special knowledge, such as *earned value* or risk management, is taught. It is also recommended that specialist department personnel also receive training to improve communication. However, project staff are often not thoroughly prepared or trained before taking up their responsible project activities. But even those companies that consider a training measure to be necessary can usually only decide on a project management seminar lasting a few days. Here is an example from MBB (*Messerschmitt-Bölkow-Blohm*) in Munich, which offered its employees a two-week project management course with in-house and external instructors as part of its central training program, supplemented by additional special courses.

The German Society for Project Management (*GPM*), founded in 1979, has been conducting project management seminars since 1980 and defined the requirements for project management education and training as follows: [18].

- Efficient use of training time.
- High practical relevance.

- Not only getting to know the *"tools of the trade" of* a project manager, but also practicing and routinely mastering them.
- Use of active teaching methods to ensure active participation of participants and transfer of know-how.
- Promote efficient group work.
- Coordination and motivation of working groups.

For the training program, GPM involved a pool of speakers consisting of selected experts from practice and academia. Great importance is attached to the spread of industries and fields of activity, a balanced treatment of material in the seminars and the possibility of really using the "*expert*" in the individual topics. GPM writes: "*Participants should have a technical or business education and have already worked in practice for about 3 to 5 years. The events are particularly suitable for participants who are to be prepared by their companies to take on project management tasks.*" [19] The learning objectives of the five-day seminar modules are as follows:

Module 1: Project management basics, project preparation phase:

- Understanding and familiarity with project and systems thinking.
- Knowledge of project preparation tasks and training in a multi-stage case study.
- Understanding of and ability to establish project structure, organization, and flow charts.
- Familiarity with assessment and risk analysis techniques.
- Knowledge of organizational embedding of project management in the organization of a company.
- Know-how transfer for daily project management work.

Module 2: Project planning techniques and tools:

- Deepening the knowledge in project structuring.
- Familiarity through practical training with methods of sequence and time planning as well as for cost and financial resource planning.
- Knowledge of procedures for specifying technical project performance objectives.
- Insight into project information and reporting systems.
- Know-how transfer for daily project management work.

Module 3: Project control:

- Knowledge of the requirements but also the limits of project control.
- Comprehensive knowledge of the interrelationships between project planning and project control, in particular the possible control methods and procedures, both for schedule and cost control, including examples.
- Insight into the tasks and procedures for quality control and quality assurance.

- Basic knowledge of the use of project-related information and documentation systems.
- Specially tailored case studies are conducted for all individual topics in order to familiarize the participants with the issues. Participants can help define the boundary conditions, which ensures that the case studies can be tailored precisely to the needs of the participants.
- The overall case study to be conducted at the end of the course additionally gives participants the opportunity to test the applicability of what has been covered on the actual example and clearly shows advantages and disadvantages of certain procedures depending on the task, size, and duration of a project.
- Know-how transfer for daily project management work.

Module 4: Collaboration in and on the project:

- Recognizing and training cooperative behavior in project teams.
- Development of decision understandable presentations of work results, reports, and protocols.
- Recognize conflicts in project teams and consider possible solutions.
- Insights on team and conference facilitation.
- Recognize and use limits and possibilities of project management.
- Know-how transfer for daily project management work.

Although the GPM Academy Program for Project Management did not cover the topics of systems engineering, product assurance, and tender preparation, or at least not in sufficient detail, the training program developed by GPM in 1980 and its further development represent a milestone and an important step in the right

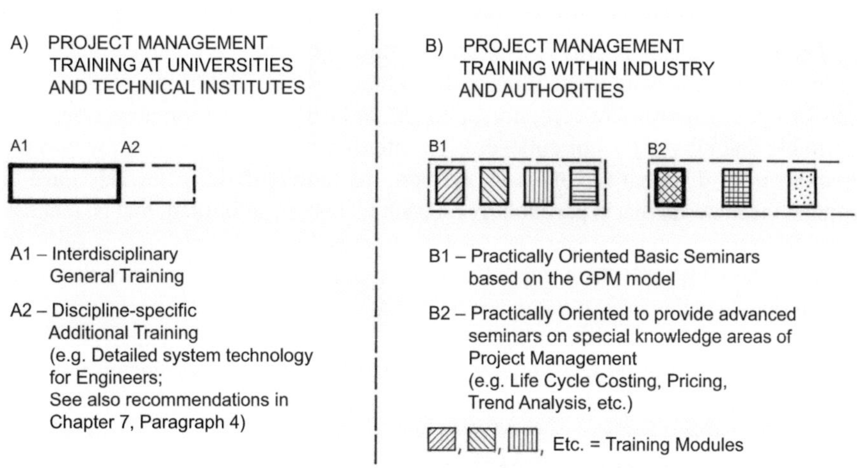

Fig. 15.2 Proposed Project Management Training at Universities, Colleges and Companies

direction because with the best will in the world, such a complex and extensive topic as "*project management*" cannot be covered thoroughly enough in 1 or 2 days. One- or two-day seminars on the subject of project management can only serve the purpose of providing information and orientation. Under this premise, short seminars are, of course, also very useful.

However, 30 years after the GPM push, it can be seen that the academic program of several weeks, although well thought out and useful, is often implemented only hesitantly by the industry and the authorities. This is certainly because organizations are not prepared to invest the effort (*time and money*) required for it. Therefore, GPM, just like most organizers, also successfully offers two-to-three-day courses on various PM topics, which is better than nothing. In this context, reference is also made to the modular training concept developed by the author; see Chap. 25.

15.2.4 Suggestions for a Project Management Training Concept

Project management is an interdisciplinary activity and therefore also requires the use of interdisciplinary teams. In the future, this fact should be taken into account more than in the past at universities and technical colleges and in industry and public authorities. However, since the decision as to who can or wants to be part of a project team only becomes apparent in the context of an industrial or governmental activity (*see* Fig. 15.1), it seems sensible, as already mentioned, to impart project management knowledge on two levels (*see* Fig. 15.2), namely:

- through basic project management training at universities and technical colleges and,
- in-depth and hands-on project management training for industrial and government employees who want to pursue a project management career.

The basic education provided at universities and technical colleges makes the most sense when it is interdisciplinary, that is, across faculties. For this purpose, one can imagine a matrix concept similar to the industrial matrix organization. It is desirable that universities or colleges offer interdisciplinary PM lectures within the framework of different faculties. In addition, the individual faculties may have to expand the material that is particularly relevant to them in additional, that is, faculty-specific, training courses; see Fig. 15.2. For example, it certainly makes sense to offer engineering students in-depth lectures on systems engineering since this topic is of particular relevance to them; see Sect. 7.4.

Practical training courses within industry and public authorities, which are aimed specifically at future project employees, should be sensibly coordinated with the training at universities and technical colleges or build on it. For this reason, universities and technical colleges as well as industry and public authorities must cooperate closely in the field of project management education and training. The previously mentioned GPM-training model, which provides for experienced speakers from theory and practice, can be considered a good example in this context. It is certainly useful to regularly bring in practitioners from industry as lecturers to teach project management courses at universities and technical colleges and to have university and technical college lecturers regularly attend industry seminars as guest speakers. As shown in Fig. 15.2, the practically oriented seminars should be divided into individual seminar modules in accordance with the modular training model described in Chap. 25 in order to avoid excessive employee downtime. This also applies to the practically oriented advanced seminars, which are important for imparting special project management knowledge. Here, the latest findings in the field of project management, for example, contributions on the subject of life cycle costs, and trend analyses, are taught in particular.

Due to the constantly growing importance of project management for the economy, the universities and technical colleges, the companies, and the authorities should become aware of the importance of project management training for young people and act accordingly. The potential for training certainly exists but the PM-organizations must give it the appropriate priority. Faculty and companies must develop joint training and seminar concepts. GPM has a leading role to play here. In this context, it is worth mentioning GPM already plays a very active role by offering practical PM seminars and that the annual conferences (*PM Forum*) have become the meeting place for PM professionals.

The opportunities that present themselves are certainly far from being fully exhausted. In the USA, but also in Europe, there is a lively exchange of experience between universities and industry in the field of PM training. For some time now, there has been a move to hold company-internal educational events, such as project management training courses, in close cooperation with universities. This gives course participants the opportunity, after passing an examination, to use the course they have taken as a building block in a university MBA-course for example. Engineers acquire an additional academic degree with the inclusion of in-house courses. The courses are often offered to the participants by their companies free of charge, but they have to make some of their free time available for this purpose.

The author has developed a model for training project managers and their staff over several weeks, which can be used to conduct smaller training sessions as needed; see Chap. 25.

15.3 Leadership in Project Management

Projects must be led to success because without professional leadership they can easily get out of hand. The PM must plan the necessary management processes, issue required instructions, make decisions, and lead the project in question with skill. This includes regular meetings, performance reviews, and personnel management. "*Leading is an art, something you learn over time, not merely by reading books*" says de Pree and continues, "*The first task of a leader is to define reality. The last is to thank employees. In between, the leader must become a servant and a debtor.*" [20]

15.3.1 Leadership Claim of the Project Manager

I can calculate the motion of celestial bodies, but not the behavior of human beings [Isaac Newton, physicist, mathematician, and astronomer].

In accordance with the organizational principles of project management (see Sect. 5.1), the project manager bears full responsibility for handling the project assigned to him. According to this understanding, there is no co-regent, for example, a person of equal rank responsible for the project costs; he is the (*flight*) captain. As shown in Fig. 5.3, he is the sole regent. In larger projects, however, he has a competent team at his disposal to solve the individual factual problems efficiently and in a correct manner; see Fig. 5.4. It also makes sense to appoint a deputy. However, the project manager is dependent on the fact that he is also equipped with appropriate powers and competences. From this, he can then derive the corresponding claim to leadership. It goes without saying that he must observe the general guidelines and rules of the company in which he works. The company's management can, of course, withdraw the delegated mandate from him at any time. Practice has shown that project leaders who were given full project responsibility, including the necessary powers and authorities, were able to demonstrate far better results than project leaders who were only partially mandated and/or authorized. The management of a project must converge at a central point, namely the project manager.

The determination to lead, that is, the project manager's claim to lead the overall task, usually has a very positive effect on project activities. This assumes that the project manager has a leadership style that promotes team spirit (*see subsection 15.3.2*) because leadership determination and teamwork are not mutually exclusive. "*Even a good team needs the captain's signals!*" With this statement, a journalist tried to paraphrase the importance of leading a team [21]. Team leadership plays a particularly important role in project management. Martin writes about this: "*Leadership is more difficult in a project than in a normal organization, because a project is inherently a freer form of work in which, moreover, changes occur more frequently … In a project, many employees are (initially) strangers (frequent personnel changes), many things are different and change more quickly, and the future is not always clear. Leadership here is a bigger and much more complex challenge.*" [22]

15.3.2 Leadership Style in the Project

Project management is a very interesting and motivating activity but as a rule also an extremely difficult and very hard one, if one thinks, for example, of the problem of meeting deadlines and costs. Martin says very aptly about this: " … *project management is particularly hard and can mean an enormous physical and mental strain for the project manager.*" [23] For the employees in the project, this statement is also true. In this context, motivation plays a particularly important role. "*Motivation is the magnet that attracts positive experiences and positive-minded people.*" says Lejeune [24].

Many project decisions have to be made very quickly and often represent compromise solutions. However, group decision-making, which is strongly favored in the literature, is only partially feasible in a project. It seems to be a misunderstanding. For factual reasons, it cannot be aimed at always bringing about group decisions in which the superior (*project manager*) only appears as integrator or coordinator. Small or large project decisions occur almost daily in the project. However, one must be aware that these decisions are assigned to different management levels and vary greatly in their importance and relevance.

As a matter of principle, factual decisions should only be made by employees who have the appropriate expertise and who also bear responsibility for them. This sometimes involves an authoritarian but fact-based decision for which the person concerned also personally vouches. Of course, he should consult with his colleagues before making decisions on his own responsibility. Management planning should also define the decision-making authority of senior staff. Making decisions requires courage and in retrospect, what one has decided has not always been right. On the other hand, a decision is better than a hesitant attitude. Napoleon Bonaparte is credited with the following statement in this context: "*Nothing is more difficult, and therefore more precious, than being able to decide*."

The leadership style in the project depends on several specific characteristics, which Grochla summarized as follows [25]:

1. *Task Characteristics*:
   - The task is of limited duration, with relatively precise objectives and usually a high degree of urgency.
   - The project task is relatively extensive, often highly complex and relatively unstructured.
   - The project task is usually associated with relative novelty and tends to have a high degree of risk (i.e., *also with generally high overall significance*) for the company.
2. *Personality traits*:
   - The project situation is characterized by the collaboration of numerous specialized experts.
   - Depending on the type of project organization (*total project management, matrix project management or influence project management*), the project manager has different decision-making powers; see Fig. 5.9.
   - Participation in the project requires good communication and cooperation skills as well as a high degree of conflict resolution ability and role flexibility on the part of the project team members.
3. *Organizational Features*:
   - A project requires the use of different resources from various parts of the company.
   - A project organization introduces an element that tends to be unstable into the organizational system that is designed for permanence.

- One project raises the issue of disciplinary subordination.
- In the course of the project, integration and motivation problems arise, as well as the problem of the "*back-home situation*".
- The project process takes place in phases (e.g., *project idea, project planning, project realization, project supervision and control*); see Chap. 4.

Grochla then summarizes: "*An exuberant propagation of the participative and employee-oriented leadership style as the sole-sanctifying variant of leadership behavior in project management certainly does not do justice to the problems shown. A task-oriented and also authoritarian behavior adapted to the specific situation is certainly just as conducive to problem solving in project management.*" [26] Differences in project management leadership style can be found especially in connection with the individual project phases. For this again Grochla: " *A guidance style differentiated after phases would be ideal.*" [26] Fig. 15.3 shows how the leadership style can change during the individual phases. In the early phases of the project, concept and definition (*see Sect. 4.3*), mainly study-oriented specialist teams are employed, while in the subsequent phases, development, production, and operation, mainly hardware-oriented development teams are employed. The change of tasks and personnel along the individual phases has an effect on the management style. Grochla draws from this the conclusion, easily verifiable in practice, that for the early phases of a project a participative, open, and employee-oriented leadership behavior is more promising, while in the aforementioned subsequent phases, due to a pronounced goal orientation (*technology, cost and deadlines*), a more task-oriented and authoritarian-seeming leadership style comes into question [26].

In their book *"The Managerial Grid"*, Blake and Mouton have broken down the leadership style into two variables, a) the person-oriented (*vertical*) and b) the task-oriented (*horizontal*) leadership style, and have presented these, provided with a scale, in a matrix diagram (*behavioral grid*); see Fig. 15.3 [27]. Blake and Mouton describe the five main leadership styles as follows:

- *Low task and personal interest (1.1).*
  - Leadership type: The disinterested type.
  - Style: inadequate leadership style.
- *Low task interest and high personal interest (1.9).*
  - Leadership type: The missionary type.
  - Style: person-oriented leadership style.
- *High task interest and low person interest (9.1).*
  - Leadership type: The autocratic type.
  - Style: task-oriented management style.
- *Sufficient task and personal interest (5.5).*
  - Leadership type: The type willing to compromise.
  - Style: balanced leadership style.
- *High task and personal interest (9.9).*
  - Leadership Type: The managing director type.
  - Style: best possible leadership style (*but difficult to reach*).

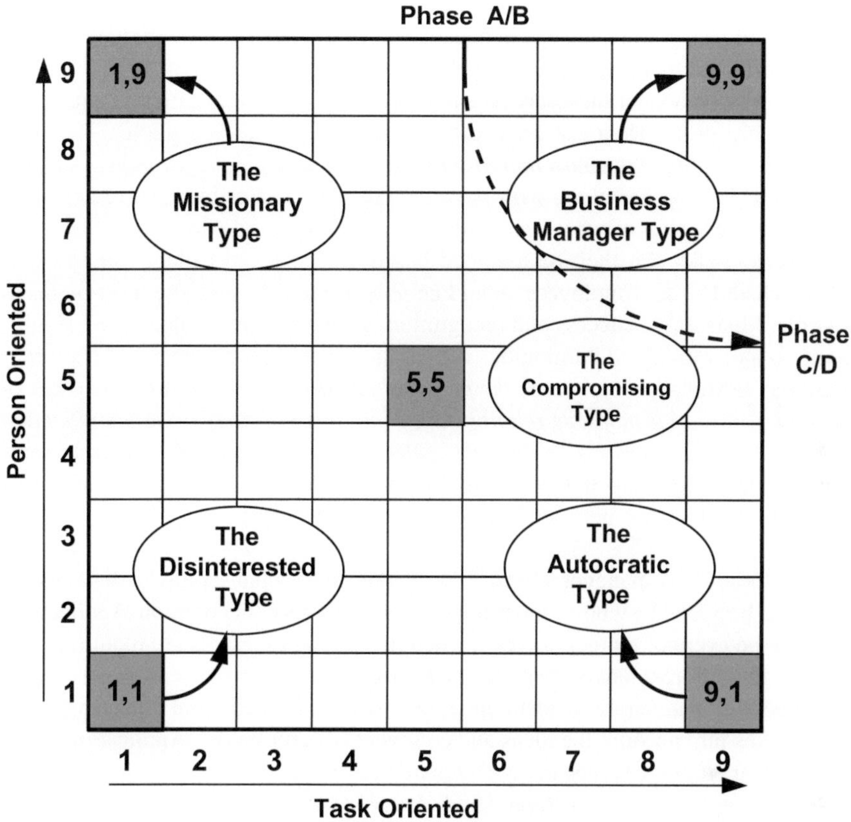

Fig. 15.3 Classification of Project Management Leadership Styles at different phasess based on the Behavioral Grid as defined by Blake and Mouton

Project management is on the one hand a strongly task-oriented function, but on the other hand very much a people-oriented function. The behavioral grid shown in Fig. 15.3 is suitable for illustrating the differences in project management leadership styles for the individual project phases. It is certainly obvious that the leadership style in the concept phase (*Phase A*), for example, is different from that in the production phase (*Phase D*). Grochla distinguishes in this context, as already mentioned, between the anticipatory and authoritarian leadership style [26]. This can be easily explained by the example of the task description for the individual project phases; see sect. 4.3. The tasks, employees, environmental conditions, etc. are very different for each phase. If one observes the behavior in typical Phase A teams and then, in comparison, in Phase D teams, this difference is clearly noticeable due to a research atmosphere on the one hand and a workshop or construction site atmosphere on the other. In Phase D, the production processes are in the foreground, while in Phase A the analysis processes are more pronounced. However, as shown in Fig. 15.3, the leadership style of project managers should lie between the ranges 5.5–9.9 in all phases because in many cases the project manager must behave like a managing director (*general manager of the project*). Differentiating between Phases A, B, C, and D, a gradual change in behavior from more personnel-oriented to more product-oriented management can be observed.

The project manager must rely on employees with often very different characters to fulfill his tasks. He must lead and motivate them. Aucoin quotes Theodore Roosevelt as saying, "*the most important single ingredient in the formula of success is knowing how to get along with people.*" [28] The leadership task requires the leader to have a plan of what needs to be done and to coordinate it with his or her employees. Tasks can then be assigned based on the project plan, consisting of WBS, schedule, etc. Employees should be able to identify with the tasks as far as possible. Measurable success and recognition, see Herzberg, [4] contribute significantly to motivation and harmonious cooperation. Kirchner comments: "*The art of conscious leadership is to merge the personal needs of the leader and the goals of the leader into a harmonious relationship. From this harmony arise identification and willingness to perform on the part of the led.*" [29] The project manager must decide which *leadership style* to use in the different phases and situations of a project. Rush distinguishes between four leadership styles: [30].

1. Dictatorial—the manager makes all decisions about what should be done, when, where, how, and by whom, and ruthlessly enforces his or her opinion. Application:
   - In the event of danger or crisis, when life-saving measures are required.
   - In the enforcement of disciplinary measures.
2. Autocratic—autocratic leadership style is similar to dictatorial leadership style but takes into account the ideas and comments of employees. Application:
   - When employees repeatedly disregard authority.
   - For new employees without detailed task knowledge.
   - In case of disregard of the organizational rules.
   - If the respective manager is the only one responsible with expertise to make decisions.
3. Consultative—this management style is based on the project plans and active participation of employees in decision-making. If necessary, the manager makes important decisions without consulting the employees. Application:
   - Implementation of planning tasks.
   - When solving creative tasks.
   - Training personnel to take the lead.
   - When there are many organizational tasks to complete.
4. Participative—in this style of leadership, which many managers do not like, authority is partially transferred to the employees. However, the manager remains the leader of the team. Application:
   - As employees become more competent at performing routine tasks independently.
   - During organizational planning processes.
   - During organizational evaluation processes.
   - In motivational actions of highly qualified employees through new tasks.
   - Whenever there are new creative and innovative tasks to solve.

In summary, there is no absolutely and universally correct leadership style, because it depends on the situation.

15.4 Staff Increase and Reduction in the Project

15.4.1 Overcoming Personnel Bottlenecks

Qualified project management personnel are usually in very short supply and the few available employees are often busy with another important project when they are needed most urgently. Many companies can only solve this problem with great difficulty. The nature of the project means that staffing levels are not constant over a long period of time but fluctuate greatly over relatively short periods. It can then happen those projects taken on by the company's management with a view to ensuring as continuous an order backlog as possible result in temporary personnel bottlenecks. This is a particularly serious problem in the area of key project management personnel (*project managers, systems engineers,* etc.) because there is often no adequate replacement for these employees due to their special qualifications. One then often hears from the affected employees "*I can't clone myself.*" The suggestion often made by outsiders that "*priorities have to be set here*" is of little help if the affected projects have to be handled with the same priority due to a tight schedule. Suggestions such as "*hire or train more personnel*" or "*take on fewer projects*" do not solve the problem either, since in the first case the specter of overcapacity looms (*projects have an end by definition*) and in the second case securing the order backlog is an important concern for every entrepreneur; see Sect. 17.2. How can companies overcome the problem of personnel bottlenecks? Hansen points out four alternatives for meeting staffing needs: [31].

- by existing personnel (*overtime and weekend work*),
- through temporary staff,
- by freelancers (*consultants*),
- by assigning tasks to external parties.

As a rule, companies always try to bridge personnel bottlenecks in the project with their own employees. This leads to the familiar overtime and weekend work arrangements. In this way, small and time-limited additional project tasks, that is, workload peaks, can be reduced in the short term. The situation is quite different, however, if more project management specialists are temporarily needed at the same time on parallel projects than are available in the company, and double staffing is not possible even despite overtime. In such cases, the project goal can often only be achieved by using additional temporary staff or freelancers, but this is not entirely unproblematic in terms of labor law [32].

Another option for bridging personnel bottlenecks is to outsource tasks. Hansen states: "*Of the three external alternatives for covering personnel requirements, only one remains that can be used in general,* i.e., *outsourcing of tasks* [33]." The external assignment of tasks by means of a contract for work and services seems to be suitable in many respects for overcoming the problem of personnel bottlenecks that occur permanently in project activities, provided that important know-how is not passed on. In this context, one can also well imagine that similar companies, may, if possible, support each other. Hansen states, "*It may be surprising to some that competing companies in many cases work very closely together … Alternately, a company may act as a client or as a contractor.*" [34]

15.4.2 Where to Put the Project Staff after the End of the Project?

The question "*Where to put the project personnel after the end of the project?*" is important, but in many cases, it is not considered critical. Since every project, even if it is a mammoth project, will end at some point, the end and the associated reduction in personnel must of course also be planned for in good time. Particularly in the case of major projects, the end of a project means for the project manager and his team "*saying goodbye*" to a phase of their sometimes years of working together and, in many cases, also to the familiar environment and the familiar team. From a business perspective, however, this is an absolute necessity.

However, it is also necessary to link the end of one project with the beginning of the other(s) as smoothly as possible. In many cases, management succeeds in creating a seamless transition. An ideal situation arises when the project team is retained in its entirety or at least in its core. Often, during the transition phase, the project managers concerned make their own efforts to bridge emerging funding gaps by temporarily taking on sub-tasks from another project, for example, studies, etc. This can be done both within the company and on a project-by-project basis. These tasks can be both internal and external to the company. The previously described outsourcing of tasks by a company with staff shortages fits into this model of thinking and is also realized by industry.

It should, however, not be overlooked that a smooth transitional solution does not always occur, and a successful project team has to be dissolved. The project manager and team members are then available as individuals for new tasks, which can lead to hardship. In these cases, the skill of management in making new appointments is particularly important. While the employees assigned to the project in question (*matrix organization*) can return to their regular departments, the process of dividing up the remaining employees now begins. Martin states in this context: "*The reassignment of employees at the end of a project is a challenging and important task. In the process, it should not be forgotten to thank the project employees for their commitment.*" [35] In this regard, Martin again states, "*At the end of a project, regardless of whether it was successfully completed or not, the staff's efforts must be appropriately rewarded, in writing as well as in words.*" [35] At NASA and to some extent also at ESA, it is for example customary for project staff to receive certificates (*awards*) for special achievements after completion of the project, which motivates and is a measure worthy of imitation.

Literature

1. Barker, Stephen, Cole, rob: brilliant project management, Pearson Prentice Hall, Harlow, 2009, p. 79f
2. Source 1, p. 149
3. Lejeune, Erich J.: Lebenswissenschaft Motivation, mvg Verlag, Heidelberg, 2005; p. 25
4. Herzberg, Frederick: Work and the nature of man, Staples Press, 1968, p. 95
5. Handelsblatt of 17. 1. 1983
6. Aucoin, B. Michael: from engineer to manager mastering, The transition, Artech House, Inc., Boston - London, 2002, p. 256f
7. Handelsblatt of 8/9. 10. 1982
8. Bernhard, Alfred: Japanese Management eine Kurzdiagnose, in: Management-Zeitschrift io (51)1982, No. 3, p. 106
9. Terry, George R.: Principles of management, Richard D. Irwin, Inc, Homewood, Illinois, 1977, pp. 268–274
10. Heintel, Peter u. Krainz Ewald E.: Projektmanagement - Eine Antwort auf die Hierarchiekrise?, Gabler Verlag, Wiesbaden, 2nd edition, 1990, p. 91
11. Pascale and Athos: 7S model, the art of Japanese management, Simon & Schuster, N.Y., 1981, 24, 83
12. Martin, Charles C.: Project Management - how to make it work, AMAGOM, 1967, pp. 57, 68
13. Hansen, Nils E.: Overcoming personnel bottlenecks in German companies among highly qualified carriers of temporary tasks, Dissertation, 1980, p. 35 (PSU, Los Angeles)
14. Brochure of the Gesellschaft für Projektmanagement (GPM) - Academy for Project Management, Munich, June 1982, 5. 4: GPM, Reitmorstr. 50, Munich
15. Capps, Charles and LeRoy, Robert J.: Project leadership and teamwork, in Larson Wiley J. (managing editor), Applied Project Management for space systems, McGraw Hill Companies, 2008, p. 531f
16. Oelsmitz, Dietrich von der: Management, Verlag C.H. Beck, Munich 2009, p. 99
17. Source 12, p. 59
18. Source 14 p. 8
19. Source 14, p. 10 to 16
20. de Pree, Max: Die Kunst des Führens, Camous Verlag, Frankfurt/New York, 2nd ed. 1992; p. 29ff
21. Author unknown (1982)
22. Source 12, p. 44
23. Source 12, p. 48
24. Source 3, p. 149
25. Grochla, Erwin: "Projektmanagement und Führungsstil", ZfU Management-Seminar, Zurich, 1980, p. 12
26. Source 25, p. 17 and 18
27. Blake, Robert R. and Mouton, Jane Srygley: "the managerial grid," gulf publishing company, Houston, Texas 1964
28. Source 6, p. 39
29. Kirchner, Baldur: Dialectics and ethics, Gabler Verlag, Wiesbaden 1991, p. 57
30. Rush Myron, Management: a biblical approach, Victor Books, Wheaton Illinois, 1985, p. 219ff
31. Source 13, p. 6
32. Source 13, p. 14
33. Source 13, p. 44
34. Source 13, p. 47ff
35. Source 12, p. 265ff

International Project Management 16

Contents

Abstract

For the successful execution of international projects, it is necessary to use project management that has been tried and tested in practice. If possible, project managers of international projects should have gained experience in the execution of national projects. In addition, they must be prepared for cultural differences.

B. -J. Madauss, *Project Management*,
https://doi.org/10.1007/978-3-662-69057-4_16

It is desirable that the project managers and their team members already have sufficient practical management experience in handling national projects before being deployed in the international environment and that they are additionally prepared for the new tasks by training measures. This is an important prerequisite for handling international projects successfully; see Sects. 1.8 and 20.7.

However, it alone is not enough to be well prepared for international cooperation. Indeed, it must not be disregarded that one's own management experience acquired in the national environment must be coordinated and harmonized with the experience of the partners. In aerospace, which almost always involves international business partners, appropriate standards have been created to standardize PM procedures. Particularly noteworthy are the ESA and NASA standards, already mentioned, which have been developed together with industry and are publicly available. ESA's ECSS procedures are also an excellent reference for non-space projects, for example, transportation engineering, facility engineering, and the automotive industry, for developing their own PM procedures; see subsection 9.1.3. The construction of the International Space Station (*ISS*) involved the space agencies of Canada, Europe (*ESA*), Japan, Russia, and the USA. All agreed on uniform standards as mentioned earlier.

Management is also to be understood as a leadership task in which interpersonal relationships and behavioral norms (*human behavior*) play a major role; see Sect. 15.3. This also applies to the role of project management. For the management of international projects, there is another important aspect, namely the consideration of cultural and linguistic differences; see Sect. 20.7.

16.1 Trend towards Internationalization

It can be assumed that the number of international project plans will increase continuously in the future. Authorities and companies which in the past were mainly entrusted with the handling of national projects are increasingly confronted with having to deal with the handling of international projects as well. The trend toward internationalization in the development, production, and operation of complex technical systems (*see Chap. 7*) is likely to intensify considerably in the coming years. A decisive reason leading to internationalization is likely to be the common interest as well as the shortage of funds of individual states. The opening of national borders within the EC, which was completed in connection with the Schengen Agreement from 1993 onward, is accelerating this process considerably.

However, project management in an international context often also leads to a number of complications that demand an increased level of management skills from the project leader. After all, project management is a difficult and multidimensional task from the outset, which can be compared with the task of a managing director. International project management means an extension of the multidimensional project management tasks and thus increases the degree of difficulty. In particular, cultural and political factors can strongly affect and jeopardize the handling of international projects, according to Herten, and he states, "*It (the political factor) must always be regarded as a strong, all too often unpredictable influencing variable in international management.*" [1] It is, therefore, advisable that employees designated to manage an international project be thoroughly prepared in advance by appropriate training measures; see Chap. 15.

The former CEO of General Electric, J. Welch, writes on this subject: "*It is a fact that there are sociological and educational differences in the various cultures. Appropriate education and training should therefore be a top priority for companies operating globally (internationally).*" [2]

16.1.1 International Governmental Joint Projects

One of the main reasons for the trend towards internationalization is probably the desire to increase market potential. Majaro states: ... "*that the internationalization process can pave new ways for a firm to achieve planned growth and greater prosperity, but this must be based on clear thinking and quantifiable marketing analysis.*" [3]

However, the development of international markets requires special attention from management, since in addition to the usual market influencing factors, there may be additional complicating factors that require considerable strategic thought. Majaro correctly states, "*Internationalization, like corporate mergers, acquisitions, and diversification, is a matter of corporate strategy.*" [4]

In many cases, national or international company expansions lead to the necessary larger units mentioned by Raidt: "*Technical progress forces larger units*" [5]; ... Bölkow put it this way: "*If enough intelligent people are allowed to work together, intelligence reaches a critical mass which automatically leads to a chain reaction of creative achievements.*" [6] The desired enlargement of market potential often initiates the merger of companies, even across national borders. "*The merger is anticipated natural growth,*" says Raidt, and continues, "*It is an artificial intervention—unfortunately often without a stroke of genius—a 'synthetic expansion.*" [7]

However, international cooperation also takes place within the framework of joint projects, which are handled by international joint ventures, consortia, or management companies. Such joint projects can easily lead to the merger of companies. Raidt again: "*If, for example, several companies in so-called joint ventures build a dam, develop a carrier rocket or a space vehicle, this is a so-called 'project merger' which, being temporary and partial in nature, can possibly lead to a total merger.* [8]

Common interest and the need to share costs are leading governments to more and more multinational joint tasks, the management of which is then entrusted to an international project management organization. In space, international organizations, such as ESA, provide for the implementation of an effective project management. In some cases, a joint multinational project management organization is established by governments and/or companies for the specific purpose of managing a single major international project.

16.1.2 Handling International Industrial Projects

Similar to multinational governmental cooperation, it is in most cases also of enormous importance for industry to enter into cooperations across national borders. On the one hand, in order to respond to the project requirements of international organizations with appropriate constructive proposals in international tenders, and on the other hand, in order to also be able to successfully handle self-financed bi- or multinational projects. Many customer markets are internationally oriented from the outset, which results in international industrial cooperations or the internationalization of companies.

The execution of multinational projects through industrial cooperation requires in any case the implementation of a joint, central project management, if one aims at achieving an efficient and profitable project execution. As shown in Fig. 16.1, the management of the established consortium could be entrusted to a control body, the *Board of Management* (*BOM*), which is composed of one representative of the management of each of the participating companies or of the responsible business unit; see Sect. 5.3. A spokesperson for the consortium is appointed in turn for a specific period of time, and all the threads of the consortium management are brought together in this spokesperson for the specified period. The consortium committee decides which of the companies will be selected as lead company for a future project. The lead company selected by the consortium committee then assumes the tasks of the main contractor within the consortium, while the other consortium members are actively involved as subcontractors.

International cooperation within the framework of a consortium with a selected lead company is a viable and proven technique to manage an international project. The involvement of a consortium is always useful when one company cannot assume the full risk and the establishment of a higher level management company is not desired. Tydemann and Kelm confirm the principle of international

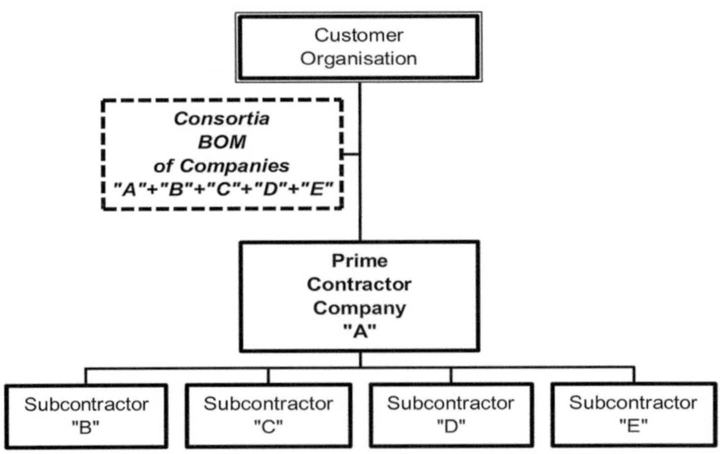

Fig. 16.1 Consortium organization

cooperation with a practical example of the production of television satellites by a European consortium of companies [9].

16.1.3 International Project Cooperation

As international joint projects in research, development and procurement tend to grow continuously and become technically and administratively more complex, project management must be adapted to these challenges. Therefore, whenever possible, the handling of international projects should be assigned to employees who have sufficient experience in handling national projects, are familiar with all common PM methods, and have a feeling for varied cultural aspects. Kutschker and Schmid describe different cultural fields in the international enterprise, such as the corporate and national culture as well as an industry culture; see Sect. 20.7 [10]. The latter is absolutely true for spaceflight, since, as already mentioned, people have been working together regularly for several decades.

International cooperation, including bilateral projects, is becoming increasingly important, forcing improved forms of management. One solution for the industrial handling of international projects is to create international project teams, for example, within the framework of management companies. It is, therefore, advisable to prepare the employees concerned for their task through intercultural training measures.

16.1.4 International Project Teams

Cooperation in the execution of international projects requires the use of international project teams. Frequently, this is also already a political requirement of the countries and/or companies involved, if, for example, proportionality requirements (*return of funds*) play a major role.

Particularly in the context of international organizations, such as ESA and NATO, but also in the case of international industrial enterprises, project management teams with international members are established. For bilateral projects, this applies in the same form but limited to only two nations or companies.

With the expected increase in international projects, the use of international project teams is also growing in importance. In recent decades, many international projects have been carried out in the context of cooperative ventures that have demonstrated the successful use of international project teams. Despite some failures, there is an overall positive trend in the use of international project teams, which can also be seen as a contribution to international understanding.

A special form of international project teaming is the integrated project team (*IPT*) or the integrated project group (*IPG*), whose members are recruited from the companies/organizations involved in a particular international joint project and then report to the management company; see Sect. 5.3.

National or international joint projects for which, due to certain reasons (e.g., competitive interests), no single company can be used at the top as the main contractor (i.e., system or lead company), often resort to a less efficient management concept, such as a joint venture (*ARGE*). This form of cooperation is not recommended for very complex system tasks, such as aerospace projects or other high-tech industrial projects, due to the lack of a lead company where system responsibility should lie.

The creation of an integrated project team (*IPT*) within a consortium can provide a remedy under certain conditions. Namely, if the companies involved in a joint project agree to assign system project management and responsibility to a higher level integrated project team created by themselves, to which they then jointly report. However, the IPT should then operate within the framework of an overarching, jointly established management company or consortium in order to give the IPT as neutral and overarching system character as possible. Practice has shown that industrial collaboration, as shown in Fig. 16.1, can be successful, provided that the participating companies follow the agreed-upon rules of the game. The IPT staffing in the context of a management company can be relatively small and limited to the company management, as the project team consists of the temporary seconded staff members of the parent companies. Although the IPT solution is not an ideal situation compared to the general contractor concept, it is, however, acceptable.

In order to be able to cooperate meaningfully in international projects, a clear agreement on the division of tasks to be carried out is required in advance. Not infrequently, failure to do so leads to considerable conflict between the partners concerned. The agreed division of tasks (*work sharing*) requires a willingness to compromise and trust on the part of the partners. It should also be noted that the consortium partners may well act as competitors in other projects. In the case of high-tech projects, task sharing often means giving up experience and strategic advantages. This is a frequently encountered problem of task sharing in international projects, which can naturally lead to considerable discord in cooperation.

This topic takes on further significance in that the division of tasks creates administrative and technical interfaces that must be carefully monitored. Time and again, interface problems arise in the handling of international projects, which can then lead to disastrous system errors. One example is the European Rocket program of the 1960s, which was then replaced by the successful Ariane program; see Sect. 3.8. In other words, when dividing up tasks, it is essential to ensure that responsibilities are clearly and unambiguously assigned so that complicated technical and/or administrative interfaces are avoided. This is a basic system engineering principle!

Finally, the division of tasks also has considerable influence on the financial return of international funds to the international institutions and companies of the individual member countries. This topic has often been given too little attention in research and teaching, mainly because it is a cross-cutting issue of the fields of management (*company strategy*), systems engineering (*interface solution*), and financing (*return of funds*). The division of tasks to be carried out is an extremely important management task in international projects, which should be defined as early as possible, that is in the early project phases. As a rule, this is accomplished during the competition phase, when the international customers, for example, ESA, make corresponding specifications (*quotas*) for the distribution of tasks.

16.1.5 International Cooperation Models

Collaboration always occurs when companies do not have the necessary know-how, personnel, facilities, etc. Other reasons can be strategic or political. Table 16.1 summarizes typical reasons for the formation of alliances [11].

Different forms of cooperation are possible. The VDI (*Society of German Engineers*) "Development-Construction-Distribution" distinguishes between two forms of project-related cooperation:

(a) the contract model with the characteristic feature of the exchange of services, and,
(b) the corporate model with the distinguishing feature of the exchange of services and a cooperation for a common purpose that goes beyond that [11]..

Under the German legal system, the contract model is also referred to as an exchange contract, bilateral contract, mutual contract, or collaborative synergy contract.

Table 16.1 Reasons for the formation of alliances (VDI 1991)

Grounds for building a bidding coalition
1. A single bidder alone cannot deliver all requested services.
2. Lacking in-house know-how.
3. Lacking in-house legal protection.
4. Lacking in-house prequalification.
5. Better chances with partners for successful orders and work flow.
6. Relationship with local partner in customer country (local content, local manufacturing).
7. Partners provided support (financing, acquisition, etc.)
8. Financing/export credit insurance from third country.
9. Partners can deliver at lower cost.
10. Possibility of specialization.
11. Reduction of project cost.
12. Risk and liability sharing.
13. A bidder cannot or will not cover contract volume alone.
14. Competition shutout.
15. Bidding under one's own flag is not possible.
16. Special customer wish for cooperation and/or specific bidding partners.

The best-known cooperation models are:

(a) General contracting.
(b) Management company.
(c) The open consortium,
(d) The silent consortium,
(e) Joint ventures.
(f) The working group (*ARGE*).

VDI: "*General contracting is the most common and important form of cooperation under the contract model.*" [11]

The execution of projects is usually characterized by a multitude of problems, as it is often challenging to achieve a set goal within the fixed deadline and cost framework. This is particularly true for innovative research and development projects. The associated management crises must be overcome by appropriate measures, such as clearly structured project organization and detailed, thorough planning measures.

Internationalization leads to further complications, which Majaro describes in the context of key cultural differences as follows: [12].

- Language differences.
- Cultural barriers.
- Consumer behavior.
- Various target groups.
- Socio-economic constraints.
- Marketing boundary conditions.
- Information sources.
- Financial problems.

These are problems that anyone involved in international cooperation must inevitably confront.

In many cases, industrial cooperation for the execution of international projects requires joint assumption of the project risk. The companies involved, therefore, often enter into joint liability within the framework of an international consortium. In such cases, the main contractor is the selected lead company of the consortium, whose project management is supervised by the consortium *board of management* (*BOM*); see Fig. 16.1.

International projects carried out by an international group of companies or consortium require that the lead company or prime contractor, respectively, general contractor assumes system responsibility and supervision of the subcontractors. The consortium organization proposed in Fig. 16.1 has in the past been used with success. However, this solution does not always occur. Frequently, the companies involved do not want to agree on a strong prime contractor and fall back on the easier variant of the Working Group (ARGE) solution; see subsection 5.3.2.

However, because of its lower assertiveness, the ARGE is not the best variant of international cooperation for projects where strong system leadership is important.

In recent years, the experience has been to establish a management company instead of a joint consortium management, which then assumes the overall management and acts as a contractual partner towards a customer. The establishment of a management company, which handles the tasks of project management and system management, entails a different legal form. When it comes to liability issues, the owners of the management company, who are also subcontractors of the management company, usually have to step in. Many people are suspicious of the concept of a management company. However, establishing a management company can be helpful in the smooth execution of an international project. It is the second-best solution after a general contractor concept; for further details, please refer to subsection 5.3.3.

16.2 International Financing Aspects

16.2.1 Geographical Distribution of Funds

The member countries of governmental international organizations are generally expected to recover the budgets used for the benefit of domestic industry within a reasonable period of time. The responsible international authorities or companies are, therefore, obliged to make the award of contracts dependent not only on the quality of the products offered and the most favorable prices but also on a recovery rate specified for the project. However, the recovery rates are usually provided with a corresponding bandwidth. This means that small deviations are temporarily tolerated. This procedure makes project awarding more difficult and is often at the expense of optimal project management.

The bidding industrial consortia must, therefore, prove in each competition that they comply with the prescribed payback ratios. If this is not possible for specific reasons, such as the lack of competitive bids, the bidding consortium must demonstrate that appropriate compensation can be achieved at a later stage, for example, in the subsequent phases of the project or in compensation with a parallel or follow-on project, because pedantic demands for exact compliance with the payback ratios can lead to poor solution approaches.

Vandenkerkhove, a program manager at ESA, points out that a financial distribution of funds in line with proportionality is undoubtedly a problem, but one that can be solved satisfactorily under certain conditions if the countries allow appropriate bandwidths [13]. Through clever allocation of contracts, the return quotient for the participating countries can then settle around the specified nominal value. This is possible primarily by requiring the respective prime contractors to structure their industrial team in such a way that the required return on investment can be achieved as optimally as possible.

16.2.2 Currency Problems

When handling international or bilateral projects outside the Euro zone, it is necessary to determine a lead currency in order to manage the various task packages efficiently. However, the shift in exchange rates often poses a financial problem because in most cases the project cannot assume the exchange rate risk for transactions in a foreign currency. Here, the support of the company management is required. Project contracts are, therefore, preferably concluded in the local currency.

16.2.3 Prequalification

The completion of projects often depends on whether the company in question fulfills the prequalification required by the customer, as many international customers require prequalification of suppliers as part of the tender process. "*Without providing such prequalification, participation in the tendering process and consideration of a bid by the customer is generally not possible.*" [14] The following information is usually required as part of the prequalification process:

- Name and size of the provider.
- Company statutes.
- Service program (*overview*).
- Published balance sheets/financial statements.
- Personnel data/curricula vitae of the members of the Executive Board.
- Personnel data/curricula vitae of project employees.
- Rough structure of the project (*scope of work*).
- Possible project partners.
- Project organization.
- Reference Facilities
- Evidence/assurance of national/local performance shares.
- local representative (*resident*).

The bidding company must ensure that the required prequalification is carried out in good time. The *bid bonds are* also part of the prequalification.

16.2.4 Project Financing

Financing often poses a further problem for the execution of international projects. "*In large-scale facility construction, the situation arises almost as a rule that the preparation of a financing concept, the raising and securing of capital is understood as an additional bid service to be provided by the suppliers and necessary for obtaining the order.*" [15] In other words, the supplier often has to provide financing at the same time as his product. For this purpose, in addition to equity capital, appropriate sources of financing, for example, banks, must be found that are also prepared to share the respective project risk. Herten explains: "*The special feature here is that all risks of a technical, economic and political nature are analyzed in much greater detail than in a conventional credit decision. The basis for such an analysis is the so-called 'feasibility study.* "[16] Heintzeler mentions the following financial sources: [17]

- International bank loans.
- Loans and guarantees under international export financing programs.
- Loans or guarantees from international institutions such as the World Bank, European Investment Bank, and other regional or national development banks, etc.
- National or international capital markets.
- Loans and public loans in the currency of the country in which the project is implemented.

According to Heintzeler, the activity of banks in project financing takes place in four phases:

1. Consultation with sponsors, analysis of the project, and development of an overall financing concept.
2. Preparation of legal documents and information to inform the future debt creditors.
3. Provision of external funds.
4. Ongoing control of project development and monitoring of loan agreements.

16.2.5 Offset and Compensation Agreements

The award of contracts for international projects is sometimes linked to a compensation requirement (*bartering*). For example, the client expects the bidder, a general contractor, or a consortium, to agree to bartering in that part of the payments are made by taking delivery of goods that must then be resold. However, this usually poses a significant problem for the project management, which in most cases cannot solve on its own, that is, without the support of the company management.

For industrial companies, compensation agreements entail the risk of not being able to sell the goods received in their own country at the expected price, especially if the goods are of inferior quality and/or difficult to sell. It should not be overlooked that a considerable amount of coordination is often required to handle compensatory goods. The decision as to whether corresponding conditions are to be accepted by a company therefore depends to a considerable extent on the level of the compensation requirement, that is, the percentage of total sales, and on the nature, quality, and sales situation of the goods to be expected.

In international business, it is often expected that local companies will be involved as subcontractors to the prime contractor with a task package (*local content*). However, the involvement of local industry in the customer country is often very problematic since the bidder must find competent, competitive companies, or partners in the customer country, which can be very difficult, especially in so-called developing countries. If it then becomes known that the bidding company or consortium is desperate to win the contract, contract negotiations are subjected to significant stresses. Here, too, it depends on the required scale of participation. A low percentage, for example, below or up to a maximum of 10% of the contract sum, is usually still easy to meet. The situation is simplified if only the less complex tasks, such as local construction projects, are carried out by companies based in the country, and all technologically more complex tasks remain with the general or prime contractor.

16.3 International Management Using the Example of Space

It is assumed that the management aspects described below can in principle be transferred to other non-space industries.

16.3.1 Project Requirements of the Privately Owned TV-Satellite Astra 1A

In 1985 SES, Luxembourg started the privately owned commercial TV satellite project Astra 1A [18]. The author was appointed project control manager and in this role, he was involved in the procurement action of the satellite, ground station, and launch service.

After a selection process, procurement contracts were signed with RCA, USA (*satellite and ground station* equipment), and with Arianespace, France (*Ariane IV launch service*). The RCA 4,000 series satellite was designated for direct-to-home reception of 16 television analog channels and was launched into geostationary orbit (*GEO*). This was absolutely uncharted territory at the time. ASTRA 1A was launched in December 1988 and was the start of a European success story [19]. In 2016, the Luxembourg-based company has a fleet of over 50 satellites in Earth-synchronous orbit of 36,000 km, covering almost all parts of the globe.

Approximately 30 years later, the satellite company YahSat, based in Abu Dhabi (*UAE*), decided to launch its own satellite program. With the support of the German companies DETECON and IABG, the satellite system was defined, and tender documents were then drawn up on that basis. After an international competition, a co-contract was awarded to Astrium (now *Airbus Defense and Space*) and Thales for the construction of two communication satellites. The first satellite, YahSat 1A, was launched in 2011 with an Ariane V rocket into its intended GEO orbit of 36,000 km.

A good example of international cooperation is also the International Space Station (*ISS*), which involved five space agencies; see Table 16.2. MOUs governed the cooperation [20].

Table 16.2 Space agencies involved in the ISS

MoU Partner	Contributions to Space Station	Start planned in:
CSA	– Space Station remote manipulator system. – Mobile remote Service Base system. – Special purpose dexterous manipulator.	– December 1998. – July 1999. – January 2000.
NASDA	– Pressurized module. – Remote manipulator system. – Exposed facility. – Experiment logistics module.	– March 2000. – March 2000. – March 2001. – March 2001.
ESA	– Automated transfer vehicle. – Columbus orbital facility.	– May 2000 (first Launch). – September 2001.
RSA	– Service module. – Universal docking module. – Docking compartment. – Science power platform segment 1. – Science power platform segment 2. – Russian research module 1. – Docking and storage module. – Russian research module 2. – Life support module. – Russian research module 3	– April 1998. – June 1998. – July 1998. – November 1998. – February 1999. – November 1999. – February 2000. – June 2000. – February 2002. – March 2002.
NASA	– Functional energy block. – node 1 – Z I truss segment. – US lab module. – S O truss segment. – S 1 truss segment. – Node 2, cupola. – P 1 truss segment. – P 3 truss segment. – S 3 truss segment. – Centrifuge module. – Habitation module.	– November 1997. – December 1997. – June 1998. – November 1998. – May 1999 – August 1999. – October 1999. – November 1999. – January 2000. – August 2000. – August 2001. – February 2002.

16.3.2 Project Organization

Fig. 16.2 shows the organizational structure for procurement and operation of a system project. The appointed project manager is responsible for the technical and administrative project execution and must be authorized accordingly. The PM of the prime contractor has direct contracts with the contractors (*suppliers*) of satellites, launcher services, insurance and the necessary ground facilities and must lead the project toward technical and commercial success. Conclusion: In the international environment, staffing of highly qualified experts, engineers, and managers from different countries and clear management structures, responsibility, and authority have proven their value in practice.

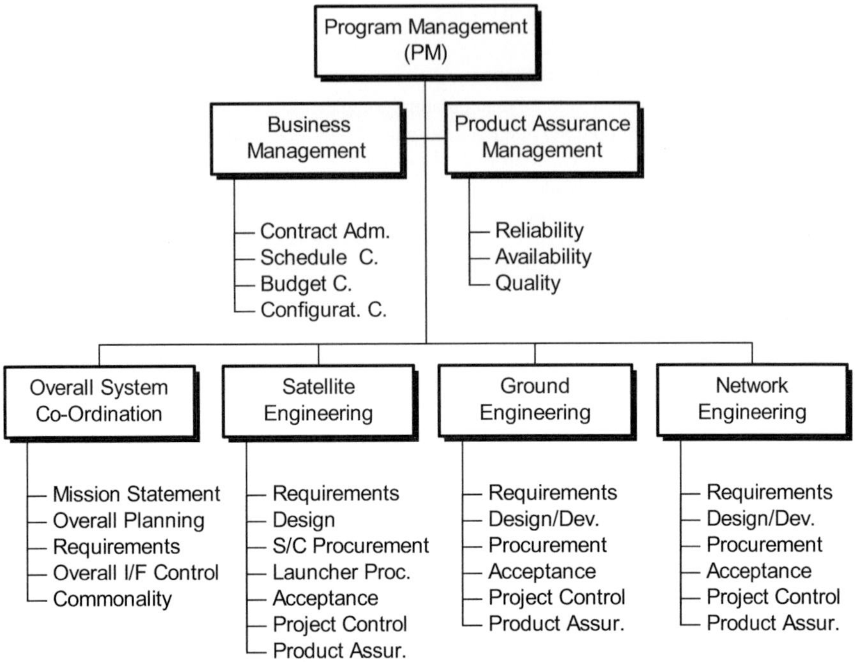

Fig. 16.2 Example of a System Project Management Organization

The project leads were responsible for the following tasks:

- Contract execution (*contract negotiations, specifications, contract amendments*).
- System design and specification.
- Quality monitoring.
- Scheduling and tracking.
- Budgeting and cost control.
- Documentation and configuration control.
- Reporting.

The project management was vested with the following powers to implement the project:

- Implementation of budget planning.
- Release of funds and audit.
- Adoption of the specifications.
- Monitoring of project costs.
- Central point of contact vis-à-vis the contractors.

In addition, Merz has reproduced the PM task priorities for a technical cooperation project of the German Society for International Cooperation (*GTZ*); see Table 16.3 [21].

Table 16.3 Main tasks of the project spokesperson

• Planning of projects and/or task orders.
– Support of the customer in preparation and proposal phases.
– Preparation of the project concept and definition including acquisition of qualified personnel from one's own department or specialty department or external experts.
– Establishing the scope of services of the GTZ for completion of the project/task order.
– Planning the table of quantities and manufacturing costs required for delivery of the planned GTZ performance scope.
– Elaboration and layout of the proposal in accordance with the appropriate guidelines.
• Preparation of the project realization/task order execution.
– Mobilization of the required expertise, including external sources, that is consulting.
– Cooperation in the search and selection of any foreign personnel.
– Providing directions to the GTZ partners.
– Support of project and team leads for detailed planning.
– Support for the customer.
• Cooperation in the achievement of project objectives and task performance.
– Support of GTZ partners on request.
– Control of project progress regarding results and schedules.
– Budget control concerning quantities and manufacturing costs complying with the approved guidelines.
– Informing supervisors regarding plan deviations (performance, quantities, costs, schedule).
– Establishment of accountability reports for the customer in accordance with valid guidelines.
– General support of the customer and preparation of change proposals.

16.3.3 Project Planning and Monitoring

The implementation of an international commercial project begins at an early stage with a market analysis, on the basis of which the project objectives are defined; see Chap. 4. The intended project management should be involved in this process as early as possible. This involves the following tasks in detail:

- Project structuring and budgeting.
- Determination of the main milestones and key dates.
- Preparation of specifications.

Project structuring is usually carried out in iterative steps in close cooperation with the project engineers involved in the technical departments. Based on the experience gained in the past, initial cost estimates are made in order to be able to make budget proposals.

In this context, the most important milestones, for example, the delivery dates of documents and equipment and the project reviews, are then also defined and the associated key dates are established.

In order to optimally organize a possible subsequent procurement and the subsequent order processing, a *statement of work* (*SOW*) must be prepared as early as possible in addition to the preparation of specifications; see Sect. 9.3. The preparation process must be carried out thoroughly and with the inclusion of internal project reviews. In the reviews, the project proposals (*technical, schedule, cost*) are presented, discussed and, if necessary, improved in terms of content. Comparative analyses ("*trade-offs*") and sensitivity analyses, for example, consideration of technical requirements in terms of required costs must be performed, and the financial impact of modified technical solutions discussed.

Fütstenberg points out the problems of international PM, which cannot be overlooked and are reproduced here [22].

- Spatial distance (*possibly different time zones*).
- Long-term risks and liability.
- Potential conflicts through international partnerships.
- Highly complex environmental structures and.
- Social–cultural communication barriers; see subsection 16.5.2.

General project planning and control processes are to be performed as described in detail in Chap. 9.

16.4 Procurement Process

Commercial national and/or international customers, for example, globally operating satellite operators, generally procure the products, satellites, launch vehicles, and ground stations required for their operational activities through international competition. The selection is made according to predefined criteria in the areas of technology, quality, deadlines, and price; see Chap. 14.

Experience has shown that fair international competition is advantageous for the customer. The customer plays a leading role here. It is up to the customer to shape the competitive process efficiently. One slogan could be: "*everybody wins.*" The Request for Proposal (*RFP*) must be well prepared. This includes the cover letter and, for example, the following key documents:

1. Proposal preparation guideline.
2. Statement of Work (*SOW*) including a delivery list (*hardware, software, documents, services,* etc.)
3. System or product specification(s).
4. Test specification(s).
5. Quality requirements.
6. Sample contract or contractual conditions.
7. Evaluation criteria and.
8. Compliance lists.

In order to carry out a correct and fair evaluation of the bids, certain rules must be observed. This is of particular importance for international projects. It is important, for example, that binding criteria are defined even before the bids are received, on the basis of which an objective evaluation is possible. In accordance with the principles of utility value analyses, a quantitative evaluation must then be carried out based on the previously defined evaluation criteria and their weighting. A corresponding guideline provides the evaluation team with the necessary specifications for a systematic evaluation of the bids:

1. The quotation guideline describes the most essential requirements of the customer regarding the preparation of the quotation, such as the deadline for submission, number of copies required, limit on the number of pages, language, quotation structure, price fixing, and currency.
2. The statement of work (*SOW*) describes the required obligations (*deliveries and services*) and contains corresponding information about the delivery quantity, time of delivery, and delivery conditions (*FOB, CIF,* etc.).
3. The design specifications are strictly separate from the functional specifications, as they describe the technical data of the system. Quality and test requirements are also included.
4. Each tender should be accompanied by a sample contract, which the bidder must include in his bid because the contract negotiations will later be conducted on the basis of this contract.
5. Evaluation criteria should show which aspects to which the customer attaches particular importance. His offer should be tailored to this.
6. Through compliance checks the customer wants to quickly determine whether the offer is complete and that all requirements can be met. This aspect is very important because the customer may conclude that the offer is excluded from further evaluation if its requirements are not met (*show stopper*).

The technical offer must be seen in conjunction with the dates and prices offered for the project. Extreme technical requirements have their price and over-specification can therefore lead to surprises. Care should be taken not to overstate technical requirements without justification. If possible, "*nice to have*" requirements should be replaced by "*good enough*" requirements.

The award of contracts to a JV and partners or subcontractors involved does not depend solely on the quality of the bid but very often also on strategic and/or political requirements, such as the geographical allocation of funds as described in subsection 16.2.1. Often, it is not only the given quotient that plays a role in task sharing but also whether the tasks are noble or less noble. In a high-tech project, for competitive reasons, each participant wants to be involved in the qualitatively challenging areas, for example, the latest technologies and software (*noble work*), and not in tasks that are not technologically challenging (*non-noble work*). This means that the PM has to perform a balancing act between best technology, geographically optimal task distribution, optimal deadlines, and an attractive financial offer. Details on bid evaluation and contract negotiations are described in Chap. 14.

16.5 International Cooperation

16.5.1 Experts with International Management Knowledge

International cooperation is not new in principle and is commonplace on both the international and European stage, if one thinks of the many international organizations such as the UN, EU, OECD, and ESA. But internationally oriented companies, such as IBM, Unilever, Mobil Oil, and SES, also have a multinational character.

Large institutions and companies usually have internationally trained specialist staff. Small- to medium-sized companies, on the other hand, usually do not have the appropriate internationally trained specialist personnel to handle international projects. They often have to rely on corresponding specialists or consultants.

The cooperation of international teams requires appropriate rules. The three most important ones are:

1. Use of specialists according to their qualifications and practical experience gained in the past, regardless of nationality. On the basis of a job description, the positions are to be advertised, and the best candidate should then be awarded the contract.
2. Staffing of management positions regardless of nationality. The only decisive factors should be qualifications and proof of experience.
3. As far as possible, use of an official language for internal and external correspondence in the project to improve communication with partners and foreign colleagues.

Global tenders for the procurement of new systems give full effect to the laws of the free market. Whenever possible, the procurement process should be preceded by an international invitation to tender, as described previously. Decisions should depend solely on factual factors such as functionality, quality, and reliability of the products as well as availability and price. This basic rule should be strictly observed in the procurement of all goods, hardware, software, studies, etc.

16.5.2 Cultural Aspects of International Management

When handling international projects, the aspect of cultural differences must not be disregarded, which can easily lead to unnecessary conflicts if not taken into account; see Sect. 20.7. Attention should be paid to

- What is the cultural background, taboos, rules of conduct, training, etc.?
- What is the legal situation, country-specific features, etc.?
- Are there cooperation partners with international experience in the partner country?

- Are common management structures in place?
- What are the likely issues in terms of exchange rates, inflation, investors, return of funds vs. GNP ratio, etc.?
- Are communication problems (*language*) to be expected?
- Does the partner country have technical, economic standards?

Handling international projects requires that the project partners agree on common goals and, when staffing the joint project office (*PMO*), ensuring that the selected employees are prepared for the cultural differences. Therefore, appropriate training is important; Welch refers to this as a top priority [2]. It is essential to take cultural peculiarities into account. This concerns both eating and drinking habits and consideration of religious regulations. In addition, legal and military regulations of the countries concerned may have to be observed.

The management of an international project should ensure that the communication between the project members is lively. Misunderstandings are often the cause of major project disasters. Therefore, an open climate and regular meetings as well as an open exchange of information are very important. Attention should be paid, for example, to culturally based expressions and hidden agendas. The international project management should ensure a positive management climate (*team spirit*).

Mole writes, "*When people of different nations and cultures come together in teams, meetings, or negotiations, they bring different expectations and perceptions of how to work together.*" [23] When people of other cultures do things differently, it is not inability, inattention, or incompetence, although it may seem that way. The judgment of what is right is often rooted in the tradition, values, and norms of the culture in question; see Sect. 20.7. Mole asks the following questions that have a cultural background:

- Do people in the company address each other by their first or last name? While in Germany you are usually addressed by your last name in the "Sie" form, in other countries, such as the USA, Luxembourg, and UAE, it is customary for people to address each other by their first names.
- Are humorous things said at meetings and presentations?
- Does the company communicate more verbally or in writing?
- Can the business relationship succeed without a good personal relationship?
- Is the social relationship and hospitality important?
- How important is punctuality—do all employees start work at the same time?
- Where do the most important conversations take place—in the office or elsewhere?
- Are meetings conducted using a detailed agenda, or are there only spontaneous discussions?
- Do all meeting participants participate equally, or does the boss dominate?
- Do all participants agree on decisions, or does the boss decide alone?

In this context, he continues with the following questions: (1) Will staff members from other cultures give the same answer? (2) What do they express about their inner motivations, and (3) How can you build and manage a team where the members have different answers to the previously mentioned questions?

This is indeed a difficult task because there are different rules, behaviors, and values between German colleagues for example and colleagues from neighboring countries. It is more difficult when working with colleagues from cultural circles with very different ethical and religious backgrounds.

Good preparation, for example, intercultural training, is an important prerequisite for cooperation in international teams. In addition, the relevant employees, and especially the managers, should be prepared to successfully complete the tasks of international project management with tact and sensitivity. Consideration and empathy are very helpful in this regard. The author was able to gather relevant experience in the realization of complex system tasks both in Europe, USA, and in other cultural circles, and experienced in practice that the interpersonal relationship level is of the utmost importance. It is, therefore, necessary that the employees of international projects approach each other with respect and understanding toward colleagues from other cultural circles.

16.6 Summary

It is no longer possible to imagine the handling of international projects without the project management methods that have been tried and tested in practice because international projects in industry or the public sector are constantly on the increase. It can be assumed that this development will intensify considerably in the course of time since there is an increasing international interest in joint projects and, above all, the ever-scarcer funds and resources are forcing internationalization. In connection with the Schengen Agreement, the opening of many national borders within the EC which took place from 1993 onward is a step forward.

However, international project management also leads to the difficulties already mentioned, which demand an increased level of management skills from the project manager. After all, project management is a difficult and multidimensional task from the outset, which can best be compared with the function of a managing director. International project management increases the demands of what is already a very complex task due to different cultural and country-specific characteristics. In addition, political factors can affect PM of international projects and jeopardize success. Herten mentioned, "*It (the political factor) must always be considered in international management as a strong, all too often unpredictable influencing variable.*" [1] It is, therefore, advisable that employees who are intended for a project management function of an international project have, as an important prerequisite, already successfully handled a national project.

Literature

1. Herten, H.-J.: Internationales Projektmanagement, Dissertation, RWTH Aachen, 1987, p. 45 ff
2. Welch, J., former CEO of General Electric, 2002
3. Majaro, S.: International marketing - a strategic approach to world markets, London, 1978, p. 13
4. Source 3, p. 14
5. Raidt, F.: The 'Inequality' fusion. Bad Harzburg, 1972, p. 174
6. MBB the nation's multipurpose forge. In: GEO, April 1982, p. 86 and 99
7. Source 5, p. 169
8. Source 5, p. 53
9. Tydemann, J. and Kelm, E. J.: New Media in Europe, London, 1986, p. 98
10. Kutschker and Schmid, Internationales Management, Oldenburg Wissenschaftsverlag Munich Vienna, 3rd ed. 2004, p. 670
11. VDI Gesellschaft Entwicklung, Konstruktion, Vertrieb (Ed.): Projektkooperation beim internationalen Vertrieb von Maschinen und Anlagen, Düsseldorf, 1991, p. 30 ff
12. Source 3, p. 70 ff
13. Vandenkerkhove, J. A.: "The experience of ESRO in managing international space projects, 3rd orbital international laboratory symposium," Konstanz, 5-6 October 1970 (Proceedings)
14. Source 11, p. 15
15. Source 9, p. 15
16. Source 1, p. 31
17. Heintzeler, Frank.: Internationale Projektfinanzierungen - aus der Sicht der Banken. In: Dülfer, E. (ed.): Projektmanagement International, Stuttgart 1982, p. 166 ff
18. Madauss, Bernd-J.: Internationales Projektmanagement am Beispiel von Europas ersten Privat-TV-Satelliten mittlerer Leistung ASTRA, GPM-Beiträge zur Jahrestagung 1986, S 229
19. Chris Forrester: The untold story of Astra, Europe's Leading Satellite Cpmpany, Springer-Verlag Berlin Heidelberg, 2011
20. Bosch Michael: Management internationaler Raumfahrtprojekte, Gabler Verlag- Deutscher Universitäts-Verlag, Wiesbaden, 1997
21. Merz Hans-Peter: Projektmanagement in der technischen Zusammenarbeit am Beispiel der GTZ. In: Dülfer, E. (ed.): Projektmanagement international, Stuttgart 1982, p. 83
22. Fürstenberg Friedrich: Problems of international project management from a social-organizational perspective. In Dülfer, E. (Ed.): Projektmanagement International, Stuttgart 1982, p. 179
23. Mole, J.: Mind your manners, 3rd Edition, Nicholas Brealey Publishing, London, 2003

Project Management Implementation 17

Contents

Abstract

The implementation of project management in authorities and companies usually represents a change in corporate culture. It requires a longer process of organizational development, which is primarily about training and convincing those involved.

B. -J. Madauss, *Project Management*,
https://doi.org/10.1007/978-3-662-69057-4_17

It is not always easy to apply the methods of project management consistently in practice, even if they are beneficial and proven; many obstacles usually stand in the way of the project. This is especially true for the permanent implementation of the project management idea in an organization because it usually leads to methodical and organizational consequences and thus to unrest within the authorities and companies; see Sect. 3.6. Therefore, Litke rightly says, "To *introduce project management in a company means: to initiate change processes!*" [1]; see Sect. 1.1. The courageous introduction of effective project management methods professionalizes project processes, which leads to increased efficiency in companies with project tasks.

The consistent use of project management leads to greater efficiency for companies and authorities involved in handling projects because technical/business expertise is placed closer to the scene of the action. In other words, project managers and their teams who understand their area of responsibility are prepared to assume full responsibility if they are given the necessary authority by the company's management and see themselves as entrepreneurs on a temporary basis. It can be deduced from this that rationalization in the project area results primarily from the potential of motivated project employees; see *Sect. 15.1.*

Although today it is understood quite well in principle that it is necessary to be able to handle projects rationally within the company, in concrete terms, it is a matter of preparing ways to introduce the project idea and the associated measures into the company effectively and with consistency. Project managers and their teams therefore need, even if they are only active as project managers or project staff for a short time, firstly sufficient preparation and training, and secondly the full support of the top management as well as organizational embedding within the company, e.g., in the company's project office (*PMO*); see subsection 5.2.5. Platz writes: "*In the discussion about the use of project management (PM), the discrepancy between the state of method development and the practical use of the methods is always noticeable*" [2]. He justifies this by disregarding the fact that PM is to be designed in such a way that it can be implemented in existing organizations simply and in a manner appropriate to the situation.

The behavioral norms of company employees and especially of project personnel also play an important role in project management implementation in a company because project management is not only a question of methodology, but also a question of leadership behavior, in which ethical aspects in particular play an important role. Here is a quote from Kirchner: "*Ethical leadership is creative leadership. It offers the leader and those led a new value view. Thus, the interaction becomes valuable.*" [3].

This chapter deals in detail with the obstacles to consistent PM introduction, the significance of project management for companies and authorities, and the paths to permanent PM introduction. Special emphasis in this chapter is placed on the aspects of ethics in project management with a view to the areas of friction that can arise from PM introduction in the company, as well as the accompanying personnel development. As a precaution, it is pointed out that the terms PM and project leadership (*PL*) are used with the same meaning.

17.1 Project Management: Yes! Change—No?

17.1.1 Situation Analysis

In recent decades, many companies have encouraged the introduction of project management. On closer inspection, however, one can very quickly see that the introduction of project management is often done only half-heartedly. The task, responsibility, and authority of the project managers, as well as the demarcation from the specialist departments, are not always clearly regulated. But when problems arise during execution, project management is often used as a scapegoat without any investigation into the true causes. This leads to the fact that sooner or later, at the latest, when the first problems become apparent, the efforts to introduce a functioning project management system are all too quickly abandoned.

"Project management," as is already clear from the combination of words, is of course always a good solution only where it is really a matter of project execution; see Chap. 2 (*project definition*). Project management requires the consistent application of very specific management methods (*planning, cost control, risk management,* etc.). Project leaders and their teams, and this is probably the most important point of all, must be given a home port through consistent organizational integration into the company, e.g., the PMO. It is also a fatal fallacy to assume that the problem of project management in the company can be solved by software solutions alone. Software, even if it is a product of first choice (*see Chap. 18*), does not solve any problems. Project management is first and foremost a management task, and only if it is understood as such can improvements in project execution be achieved, and intelligent SW systems, in the hands of competent project management, lead to increased efficiency.

17.1.2 Consequence: Project Management

The management of projects is part of the project, and conversely, the management of projects also presupposes projects. This statement is trivial, but nevertheless significant, since attempts are often made to apply project management methods even in those areas where there are no projects in sight far and wide. Companies should therefore first determine whether there are any real project tasks at all. In this context, reference is made to Chap. 2, in which a precise definition of the term project is reproduced.

Companies that are constantly faced with the problem of having to solve project tasks have, in principle, no choice but to press ahead quickly with the introduction of project management. Once this decision has been made, it is then necessary to consider how best to proceed. In many cases, the decision to introduce project management is based on the assumption that the implementation, i.e., training, software purchase, and implementation, can be completed very quickly in just a few weeks. However, this usually turns out to be a big and fatal misunderstanding. PM implementation always starts with the necessary organizational measures or organizational development, for which sufficient time should be taken, as it requires a certain amount of convincing. Namely, only when the organizational integration has been satisfactorily solved, e.g., by establishing a PMO, should methodologies and training be considered.

Today, more than ever, it is important that project managers and their staff, just like their departmental colleagues, are recognized for their importance and given authority appropriate to the tasks and responsibilities assigned.

17.1.3 Changes through Organizational Development

The introduction of PM leads to organizational changes and presupposes organizational development. Heintel and Krainz say: "*The introduction of project management is always the beginning of organizational development; you have to know this so that you are not taken by surprise* [4].

Changes in organizations, which are planned with regard to the imminent introduction of project management, must be carried out carefully and with the involvement of all relevant employees. This is a "must" if one does not want to risk causing unnecessary problems and turbulence in the company, as project management is often seen as a disruptive factor by the traditionally anchored functionaries of a company, and organizational instability can easily result; see subsection 1.1.3. Changes to the organizational structure and process organization in the company that are planned with regard to the introduction of PM should therefore be made with tact.

It often makes sense to take the path of a step-by-step plan, in which various PM organizational forms are then analyzed (*see Fig. 5.9*) before deciding on a final concept. A company that carries out a large number of projects in parallel will sooner or later certainly come across a variant of the matrix organization and then introduce it; see Fig. 5.11. However, it must be borne in mind that most companies are faced with the problem that the projects are not comparable to each other in terms of their size, complexity, and timing. This means that the management methods and the organizational embedding differ from one another. It has turned out to be a general guideline that, in such cases, a deliberate distinction should be made between small, medium-sized, and large projects. The question of project size naturally depends to a large extent on the size of the company in question. A large project of a small company can therefore be counted among the small projects of a large company. Table 1.1 shows examples of three companies of different sizes and the associated project size. The limits shown also clearly indicate the range for small, medium, and large-scale projects.

In many cases, project managers handle several small projects at the same time and make use of support from the business departments to a central PMO. In these cases of multi-PM, it is particularly important that the PM be supported by the central PMO. It is also particularly important that he can fall back on standardized methods and SW, because only in this way can he achieve the required efficiency. It is understandable that it is not always easy to implement this, but on the other hand, the limited resources available for the PM effort dictate the most efficient approach possible. One fact must be completely clear, no project can be managed at zero cost, even in cases of limited funds, regardless of whether PM costs are charged directly per project or by apportionment. Even in the case of small projects, appropriate funds, hours, travel expenses, etc. must be provided for project management. As a rule, the PM shares are calculated from the total order according to a key (*percentage based on experience*) and made available to the PM. It is important that appropriate funds are allocated for the PM task; otherwise, the door is opened for inexpensive PM, which in the end usually leads to the PM task not being performed at all or only being performed extremely poorly, which could then result in the well-known management crises. The problem of management crises must also be taken seriously in small projects. A poorly managed small project by itself certainly does not lead to a problem, but if many small projects get into trouble, it can become a problem. It makes sense for the managers of small projects to remain in the respective department organizationally since they usually also perform departmental work at the same time. In addition, proximity to colleagues is important because it promotes communication.

Let us now turn to the major projects of a company. Major projects often have a serious impact in terms of their financial volume. A major project, especially if it is carried out as part of an industrial team, has a financial magnitude that sometimes exceeds the financial dimensions of the lead company. It is then quite obvious that the quality of the project management is of great importance for the success of the project and, moreover, for the benefit of the company entrusted with the project management. In such a case, the project management must have qualified personnel with many years of experience and appropriate management powers. In this case, the matrix organization is suitable, in which the PMs and the PM team members working closely with them are grouped together in a special project organization, a PMO, because this also serves to improve communication; see Fig. 5.11. In this context, the temporary assignment of personnel from the specialist department to the PMO is often implemented.

The organizational unit for project management created in this way is given greater weight in comparison with the specialist departments, and a balance of competence is thus created between the specialist and project departments; see Fig. 5.10. The project leaders are thus in a better position to act appropriately, as they can act as an extended arm of the management (*see Fig. 1.1*), which corresponds to the tasks of a managing director on a reduced scale.

If one has now decided between the two forms of organizational integration of projects, namely between the small projects, which should always be located where the responsible employees (*project managers*) are located, and the large projects, for which the matrix is most suitable, then the medium-sized projects now remain. But this is the most difficult case. Here, one has to select one or another solution on a case-by-case basis. It is also conceivable, however, that mini-matrix solutions are found. This means, for example, that project management is established within a department that centrally handles all projects according to the matrix principle; see Fig. 17.1.

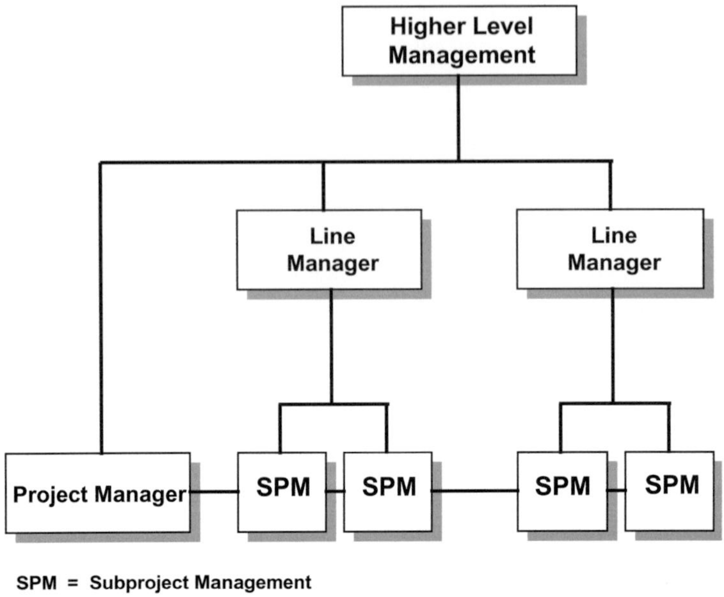

Fig. 17.1 Principle of a Mini-matrix for Small Projects in a Functional Department

In project work, in addition to the tasks specific to each project, one also encounters activities that are best assigned to a central project control center of the central PMO in order to carefully coordinate all of a company's planned and/or ongoing projects. The reasons for this are:

1. The definition and provision of resources must be centrally controllable.
2. The resulting priority regulation requires a centralized approach.
3. The management methods and procedures used by a company should be designed as uniformly as possible for all projects in order to achieve the following:
   (a) The reporting system for management can be standardized.
   (b) The methods and procedures are standardized to the extent that a necessary exchange of personnel is possible at any time.
   (c) A uniform PM training program can be easily introduced across the company and is acceptable to all employees concerned.

The central PMO management position should preferably be located in the program area or report directly to upper management; see Fig. 5.11. It is necessary to provide this position, which is partly of a service nature but must also perform active control tasks with appropriate powers. This position, for which there are very different designations (*central project monitoring, business management,* etc.), represents the link between project management and the specialist departments. This underscores the importance of creating a centrally located project department, namely the PMO, from which all projects of the company are centrally supervised;

A) Tasks	B) Responsibility	C) Authority
(1) Establish Management Standards	(1) Initiate and maintain current PM standards	(1) Coordination with upper management and project managers and Implementation of PM Standards
(2) Training of Project personnel	(2) Ensure adequate training measures	(2) Direction of PM training measures
(3) Control of Utilization planning	(3) Identification of bottlenecks and priority requirements	(3) Conviening and managing Priority discussions
(4) Planning support	(4) Provision of Planned capacity	(4) Control task assignments of Planning personnel
(5) Central Project reporting	(5) Regular provision of project reports to management	(5) Collect information and coordinate review meetings

Fig. 17.2 Tasks, Responsibility, and Authoroty of the Central Project Management Unit (Source: Heintel)

see subsection 5.2.5. The PMO has overall responsibility for all project management and is the central control center for all PM measures in the company. Figure 17.2 describes the tasks, responsibilities, and powers of the central project management office [4].

17.1.4 Project Management Costs

It should be pointed out once again that the PM services to be provided have their price. The cost aspect of project management is often controversial. Companies want to keep PM and quality assurance (*QA*) expenses as low as possible in the face of customer price pressures. Keeping all costs as low as possible is a worthy goal, but there is a lower limit. If you look at the workload of a project manager, as described in Chap. 5, it is clear that this is an important control task with corresponding responsibilities. If the work is reduced too much, it is comparable to a large ship with a rudder that is much too small. Many projects exceed the once-agreed budget, and there are two main reasons for this:

1. inadequate costing, and
2. poor project management.

Cost determination (*estimation*) requires thorough planning, sufficient time, and the experience of the estimators; see Sect. 10.1. It is obvious that it is easier to determine the costs for a standard project than for a new development. The latter case requires particularly thorough planning and recourse to empirical data and databases. The expense of thorough planning is worthwhile and represents a useful investment.

Professional project management is also a sensible investment to achieve a good project result. In this context, if one looks at the enormous cost overruns of several 100% for poorly managed projects, see subsections 3.2.1 and 3.2.2, it is easy to understand that project management for a complex project (*see definition in Fig. 5.4*) costs around 10 to 15% of the project total based on experience. This is a good investment if it means that high increases in cost can be avoided. This consideration does not take into account the fact that unforeseen events and external influences can, of course, cause cost overruns. However, a good PM is a guarantee that any problems that arise can be identified immediately and timely countermeasures can be implemented to avoid extreme cost increases.

17.2 A Place for Project Managers and Their Teams

17.2.1 Project Management: A Decentralized Business Unit

Projects comprise more and more of the business of today's companies. Especially for companies with high-tech character, project management represents a viable alternative to business management. The management of some companies has not yet recognized this but is more and more willing to include the project management idea in their considerations. In doing so, it is very often completely overlooked what motivational gain is possible through the consistent introduction of PM; see Sect. 15.1.

Saynisch writes very aptly: "*Projects will represent the form of work and action of the future, probably also on the global level of world politics—and the "project management principle" is the appropriate organizational form for this*" [5]. If one believes this statement, consistent action is necessary, but many companies have a problem with this, mainly because they often start it completely wrong.

It is necessary that large projects be viewed as independent *business units (BU)* and managed like *profit centers.* This idea runs like a thread through all the chapters of this book and also represents the author's many years of experience. In concrete terms, this means that the project managers act like managing directors, or managers of a business unit, and are monitored like a *profit center* of the company or their representatives. Through this measure, projects are better managed and the project managers are empowered to act independently as employees with management tasks, which ultimately leads to more motivation; see Sect. 15.1.

Projects must be managed in such a way that the set goal is achieved and the deadline and cost framework are adhered to. According to Zijl, "*The greatest challenge of project management lies in managing conflicts between cost and schedule goals on the one hand, and factual and quality goals on the other.*" [6].

One of the most serious problems in project management is probably the resolution of priority conflicts, whereby many a PM resigns because he cannot solve the problem alone. In fact, this is usually a difficult problem to solve, where the project managers are often left alone by the management or the PMs fail to involve the management in time.

If the PM is given the mandate to manage a BU, it must be ensured that they are well prepared for their task through training, that the company provides the appropriate tools (procedures and software), and that the resources required for project execution (skilled personnel, machines, materials, etc.) are available. However, this is exactly where the permanent conflict occurs! Resources are often in short supply, i.e., they are available in principle, but often at too late a stage, and then the wrangling begins because the project managers are in a tight spot.

So where are the causes? Who is not acting correctly here? The PMs, the CEO? When analyzing the matter in detail, it is easy to see that the conflict of priorities often results from taking on projects with short-term deadlines, for which resources cannot be made available at the appropriate pace. One has to understand that the CEO strives to secure as high an order backlog as possible and is therefore often forced to accept short-term deadlines, especially when the order books are not too full and long-term staff utilization is not guaranteed.

At the same time, it often causes great difficulties to be able to provide corresponding resources, employees, machines, materials, etc. in time. Here are some typical problems:

- Employees are tied to other projects that have not yet been completed.
- New employees can only be hired if long-term staff utilization is guaranteed.
- Employees, especially younger colleagues, must be prepared for new activities through appropriate training courses.
- The existing machines are fully utilized, and a new investment is not planned.
- Materials/components with long delivery times (*long lead items*) cause planning problems, especially if there are also delays.

From the situation described here, it is clear that the bottlenecks are preprogrammed, and a way out can only be found in close cooperation between PMs and the CEO. Figure 17.3 shows the case described above using an example involving the workload planning of departments "*I*" and "*II*." Due to the inclusion of projects "*A*" and "*B*," which are required to meet deadlines, department "*I*" *is* heavily overloaded, while department "*II*," provided there are no problems due to illness, etc., can complete the additional tasks in addition to the existing tasks by working overtime. Department "*I*" will inevitably experience the following problems:

- Priority wrangling between projects "*A*" and "*B*."
- Also between the new projects and those yet to be completed.

What should happen now? It must be prevented in the interest of the projects that the task overhang, "*the bow wave*," does not simply lead to postponing the deadlines. In practice, the situation is often not dealt with openly. Only if the project managers have a high degree of self-confidence, as required here, can the matter be tackled courageously, i.e., illuminated, together with the affected departments and

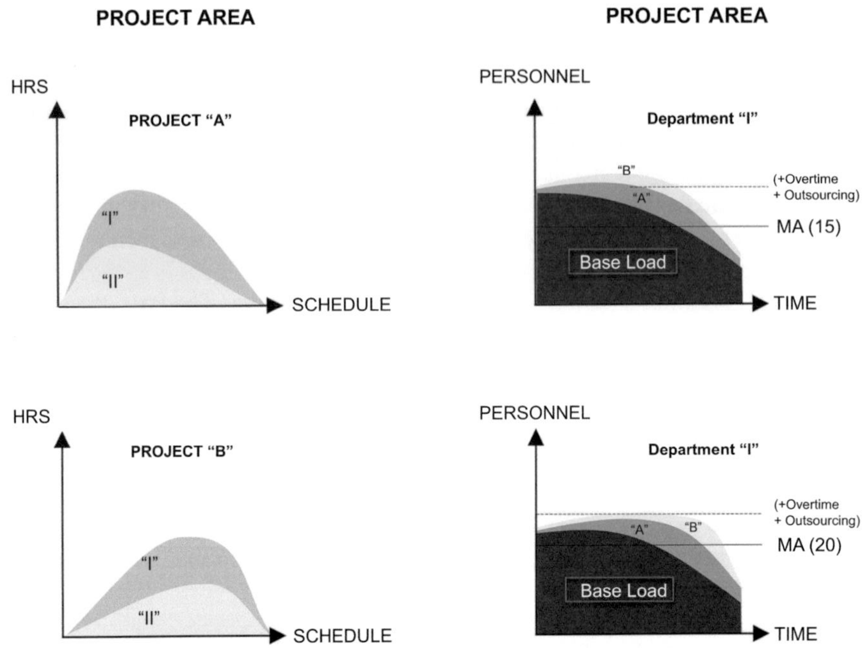

Fig. 17.3 Priority Problem of projects in a company

the GM. What options now remain in the example shown in Fig. 17.3? Here is a selection of possible measures:

- Additional overtime, including weekend work; problem: limited possibility, requires approvals from the employee council and, if necessary, supervisory authorities.
- New hire, possibly with fixed-term contracts; problem: deadline delay due to training.
- Acceptance of temporary workers; problem: Temporary Employment Legal Regulations; see Sect. 15.4.
- Checking whether a postponement is possible; problem: possible annoyance of the customer.
- Increase of motivation; problem: limited possibility of increase.
- Success fee; problem: distribution scheme.

Transparency and information are important supports for the manager, because in the situation shown in Fig. 17.3, it must be considered what is to be done in the interest of the company and not solely in the interest of an individual project or department. Therefore, all parties involved must come to the table to reconcile the aspects of the centrally operating CEO with the decentralized acting project managers and to find an acceptable solution in coordination with the departments. This also corresponds to Terry's statements with regard to the advantages of centralization (*CEO*) and decentralization (*projects*) in the company; see Fig. 5.8. Both poles of strength must be taken into account and are important because the PMs are or should be junior partners of the CEO.

How can the process of clarifying priorities work best now? PMs and especially CEOs usually have the problem, after all, that their own deadlines also have priorities, and if the priority clarification system is not solved professionally, success is questionable. The solution to this problem can be as follows:

(a) First, it should be noted that there are priorities at different levels. Not every priority is equally important and must be presented to the management.
(b) A mechanism needs to be established by which priority conflicts and their impacts become identifiable and assessable.
(c) Aside from ad hoc issues that arise, a forum needs to be created to prioritize corrective actions.

Based on a priority setting procedure in which the author participated, a priority setting concept is presented here [7].

1. Introduce a regular priority round to deal with top priority conflicts that cannot be resolved at a lower level, e.g., weekly, fortnightly, or ad hoc if necessary. The priority round can, of course, also be done as part of a regularly scheduled planning meeting.
2. Determination of the permanent members of the priority round: top-level representatives, second management level of the company, central planning, affected projects, and departments.
3. Weekly preparation of a short report (*maximum two pages*) by the department and project managers to the second management level (*see Fig. 5.11*) with the following content:
   - Project status/important events/actions.
   - Major problems.
     - Problem areas.
     - Introduced measures.
   - Outlook for the following week.
     - Planned activities.
     - Completion of milestones.
   - Travel and meetings.
   - Necessary support.

The introduction of such a system, where the PMs have the opportunity to solve problems they can no longer solve on their own because they first have to try to solve them together with the CEO in a consensual way, creates a new quality of management in the company and in the projects. It should be noted here that in larger companies with several hierarchical levels, the project managers can resolve priority problems not necessarily with the CEO, but with another authorized body, such as the technical directorate.

Finally, the importance of competent and authorized project managers (*junior partners of the company management*) should be pointed out again at this time, because the granting of corresponding authority is the prerequisite for successful project management in the company and corresponds to the trend of the times. In this regard, the following quotation from Heintel and Krainz: "*Project management brings hardly any advantages if its relationship to the hierarchy is not clarified, is redefined again and again, so that it constantly wears itself out in it or too much energy is tied up in it. A clear organization keeps the project groups on their toes.*" [8].

17.2.2 Project Manager at the Scene of the Event

"*Large companies can no longer be coordinated by the top (alone).*" [9] An apt statement, but certainly not valid only for large companies. Project management gets you right to the heart of the action, see subsection 1.4.1, which corresponds to the levels of decentralization described by Terry; see Fig. 5.8. This makes project managers much more powerful than they commonly assume. This is because whatever they perform, influence, or decide has not only technical but also, and above all, economic consequences for the project and thus also for the company. At this point, a quotation from Heintel and Krainz says: "*Project management is the insight, coagulated into organization, that certain upcoming problem solutions require prerequisites that are only given in small groups*" [10]. What are the advantages of such small groups with the character of a village community? Heintel and Krainz think: "*Small groups are the place of manageable direct communication and control*" [11]. They state the size of the small group as 7 to 15 members [12], which roughly corresponds to the team strength of larger projects. Now here are some important factors that make up the advantage of a small group:

- detailed knowledge
- great flexibility
- short communication channels
- rapid decisions
- good knowledge (assessment) of the employees among themselves
- good prerequisite for team building.

The project groups managed as BUs, whose project managers are the extended arm of the company (*see Fig. 1.1*), must on the one hand be allowed to act autonomously within the scope of their competence, but on the other hand must be firmly integrated organizationally in the company and regularly report the project results to the CEO or its representatives. Schrempp states, "*Decentralized responsibility leads to other, higher demands of controlling, which thus becomes one of the most important controlling instruments of top management.*" [13].

Organizational integration presents a special problem and is the subject of the following section. Reporting to top management should follow the procedures described in Sect. 11.2 and include the information described in Sect. 9.4. It is important that project information does not reach top management solely through reporting, but also through project reviews, because in addition to the communication of measurable facts (*performance status, deadlines, and costs*), there is often a large amount of information about the course of the project that is not measurable but that needs to be described. In addition, it is important for the project manager to receive *feedback* from top management.

17.2.3 Home Port for Project Managers

The quest for belonging and a home is a legitimate and completely normal concern for all people. The only question that arises is where this home port for project management should be. This is precisely the difficulty that companies and authorities struggle with and often fail to find a satisfactory solution. According to Schwarz, "*Transverse to the conventional pillars of the division of labor, for example, the project management structure was created: The stated mission is to complete a (seemingly!) definable task within a certain time frame.*" [14].

In solving this question, it is best to start with the project management mandate. If one sees the project managers as the extended arm of management (*see Fig. 1.1*), it is clear that they are working on behalf of management. In reality, however, a project-oriented company usually handles not only large projects but also medium-sized and small projects. This complicates the situation. To avoid a detailed discussion of the various organizational possibilities (*see Fig. 5.9*), the example of the matrix organization (*see Fig. 5.11*) is used to comment on this important point. The project sizes defined in Table 1.1 for small and large companies are used as a basis for all subsequent considerations. It certainly makes sense for large projects to be grouped together in a separate PMO area, see Fig. 5.11 (*left-hand side*), because of their importance for the company in question, since large projects often take up a considerable proportion of total sales. For the sake of completeness, it should be noted that the project office can either be subordinate to the CEO or to a corresponding divisional management, e.g., the central development area.

In this context, the author can refer to previous experience and describe the situation using the example of a large company with approximately 300 projects. According to the definition in Table 1.1, 10% were large projects, which contributed approximately 80% to the annual turnover. An optimal organizational solution for large and small projects has to be found, and the matrix organization described in detail in subsection 5.2.4 and the introduction of a PMO-like solution presented themselves. The home port for large projects was the central project area (PMO) of the company; see Fig. 5.11. The required PM culture could then develop in the PMO. But where should the home port be for the remaining projects, which also have to be handled professionally? They can be handled in the department, but the central PMO provides technical support on the basis of PM guidelines. The only exceptions are preliminary studies for customer projects or large-scale projects. They must be handled in the PMO.

Therefore, the managers of small projects usually remain at home in their respective departments, especially since project managers often oversee several small projects at the same time. In this context, however, it makes sense to transfer the PM functions to an experienced person who is particularly suited to this task in order to relieve the already overburdened specialists, who often have little or no PM experience. A mini-matrix organization can then be created, which has the following advantages; see Fig. 17.1:

- The appointed PM is a central point of contact within the specialist department.
- Only one person (*or only a few people*) in the department deal with PM tasks.
- Reporting to company management (*from the PM to the department head*) comes from a single source.
- There is a central and competent point of contact for the department head.

However, it is a prerequisite that a qualified person be responsible for the PM tasks, as is the case for large projects, because the success of the project depends decisively on the person assigned. It should not go unmentioned at this point that decentralized operating PMs should be professionally supervised by the central PMO.

Scholz comments: "*Project management stands and falls with the people involved* [15]. If the project teams for large projects have now found their home port in the project or program area and the managers of small projects in the respective specialist areas, they are nevertheless professional colleagues with principally the same tasks and problems. For this reason, it is important that the companies concerned ensure that a constant exchange of experience and collegial contact is maintained between the individual project islands (*small, medium, and large projects*) under the leadership of the central PMO. This primarily involves the following measures:

- Application of uniform standards (*procedures, guidelines, formats,* etc.), which primarily contributes to the continuous improvement of project work but also promotes communication.
- Workload planning and prioritization in order to be able to react in good time and correctly to emerging problems (e.g., *permanent or temporary shortage of resources*) for the benefit of the company; see Fig. 17.3.
- Regular exchange of experience within the framework of project reviews. It is recommended that despite the pressure of deadlines, PMs meet regularly, e.g., once or twice a year, for a meeting to exchange *lessons learned.* This is also important because some PMs of small projects are often put into the situation of having to manage a medium-sized or large project sooner than expected. Conversely, it is also useful for managers of large projects to learn from small projects, e.g., the application of simple solutions, sometimes precisely in order to regain a more realistic view of the small things in project life.
- Joint project manager and employee training is a basic prerequisite for efficient project management in operations; see Fig. 17.2.

In order to be able to implement the measures mentioned here effectively, the organizational embedding of a central project office (*PMO*) in the company organization is necessary. With reference to Fig. 5.11, the PMO should be at the same hierarchical level as the specialist departments in order to have equal competence in project decisions.

The question of the home port for project managers and their teams for small projects is now somewhat different: Disciplinarily, they undoubtedly belong to different departments, but in their PM function, they are closely linked to each other and to the central PMO. The PMO has policy authority for all projects in the company.

17.3 Project Managers: "*What*" and "*How*" They Should Be

17.3.1 Tasks, Responsibilities, and Authority

Zijl writes: "*Project management is a leadership discipline which is applied in all sectors of the economy and for all types of projects*" [16]. Project managers must first and foremost be able and willing to lead, which is one of the most essential characteristics of the PM job. Here is a quote from Scholz: "*Primarily, project management must see itself as people management....*" [17] The statements made here are indeed true for all projects and project types, no matter in which area they are handled or how large and complex they may be. Most project failures are caused by a lack of leadership and usually start with the planning, which is not carried out courageously enough; see subsections 3.2.1 and 3.2.2.

The task of the project manager is to lead the project, which is clearly expressed by the term "*project manager.*" Project managers must ensure through leadership that the project goal, once set, is completed on time and within budget. The project task can then be broken down into specialist and management tasks, because the project manager should be a specialist who has a good knowledge of project management methods, and he must also know how to successfully lead and motivate employees.

Project managers assume a high degree of responsibility in carrying out their tasks, because if the project goal is not achieved or only partially achieved, or if deadlines and/or costs are exceeded, the company for which they work can be damaged. They have the responsibility to avert or minimize possible damage to the company. In addition, they are also responsible for the employees entrusted to them or temporarily assigned by other departments; they have direct or indirect personnel responsibility. Thom defines responsibility as follows: "*Responsibility is expressed by the duty and the willingness to vouch for an action or omission with one's person or to give personal account for the fulfillment of the task*" [18]. Only those who also have a certain amount of authority (*competence*) are in a position to perform tasks and exercise responsibility. "The *task, responsibility and competence of each employee must correspond to each other. They form an inseparable unit.*" [19] So what do the powers of the project manager look like, or rather, what powers are the project manager actually entitled to? To answer this question, it is necessary to recall the objectives of PM implementation. Project managers are supposed to relieve the management or division managers, who are subordinate to the management, by taking over the project tasks, which by their nature represent principled cross-sectional topics. They are the extended arm of the CEO, the division management, or also the head of a department; see Fig. 1.1. The PM tasks, which represent an additional burden to the already existing accumulation of tasks for the responsible heads of companies, divisions, departments, etc., must be delegated to experienced project managers via the "*extended arm*" and in this context also be provided with appropriate authority. These include, among others:

- Budget responsibility within the scope of funds released by the company: i.e., project funds release and tracking.
- Deadline monitoring: i.e., holding status meetings with the departments involved.
- Conducting project reviews, i.e., determining whether the promised performance has actually been achieved, includes development reviews (*see Fig. 4.2*) but also acceptance tests. Participation in the procurement of special components and materials for the project by the purchasing department, i.e., representation of the project interests and veto right on the project-side if the project goal is endangered.
- Participation in the recruitment of personnel for the project, i.e., project-side participation and veto rights.

17.3.2 Project Manager Training

How does one become a project manager? There is no single universal answer to this question; as is well known, many roads lead to Rome. Martin, for example, describes an eight-step path for a development engineer to become the project manager of a large project: [20].

- Development Engineer
- Group leader development
- Systems engineering
- Department manager systems engineering
- Manager configuration control
- Systems engineering manager
- Manager of a medium-sized project
- Manager of a major project.

Of course, this is only one of the many ways to become a PM. Often, it is even much faster, but this also brings dangers with it. A PM who goes through the stages mentioned by Martin has good prerequisites for *on-the-job training*.

Now, of course, it is clear that future PMs should not only acquire their knowledge in practice according to the principle of "*trial and error,*" but must also be thoroughly trained to acquire PM methods and leadership knowledge. Companies that take this point seriously must be careful to provide thorough training for their PMs. This is a worthwhile investment for the future. However, it is important to understand how best to conduct such training. Section 15.2 discusses the training concepts of the German Society for Project Management (*GPM*) and the possibility of teaching basic PM knowledge at technical colleges and universities (*suggestions for a PM training concep*t). Chapter 25 describes a PM training program developed by the author, which consists of several individual courses that can be conducted independently of each other.

In practice, however, it often looks like employees are appointed project managers without being thoroughly prepared. People do not recognize the need and do not take the time. In connection with the eight-stage career path from development engineer to manager of a major project described earlier by Martin, a model career for a PM accompanying training (*see Chap. 25*) would be absolutely necessary.

What is the best way to solve the problem of training junior PMs (*or staff*) now? The best way is to create a career concept for PMs and their employees. It starts with the recruitment of potential candidates, who must already be suitable for project tasks in terms of their personalities. Character traits include above all:

- Pioneering spirit
- Prudent risk appetite
- Ability to work in a team
- Dynamics
- Serenity
- Flexibility and
- Personal stability.

The next step is to formulate the goals of PM candidates, and it should be noted that not all PM employees necessarily have the goal of becoming the leader of a large project, because leaders of large projects often, as one can easily imagine, have greater concerns and problems. Often, it is the small and medium-sized projects that are of particular interest, if one thinks of study projects or technology investigations, for example. Companies that urgently need junior PM staff should train project managers and their teams according to the principles described in Chap. 25. The first step is to impart basic PM knowledge. The course modules S.01 to S.22 provide the theoretical background, and the course modules P.01 to P.18 provide the closely related practical cases. It must be decided on a case-by-case basis whether all modules are required or only an absolutely necessary part. As shown in Fig. 15.2, the practice-oriented basic seminars (*B1*) must be supplemented by corresponding advanced seminars (*B2*) if necessary. The course durations (*number of teaching units*) also depend, among other things, on whether the participants already have prior PM training and/or practical experience, e.g., university courses or PM tasks in practice, or have not yet come into contact with the PM topic at all.

17.3.3 Dealing with Project Employees

If up to now the project manager has predominantly been in the foreground of the considerations, it must not be forgotten that it is not only the PM who is important but also his staff. They are just as important for the success of a project as the PM. In the case of small projects, it is not a question of employees who report directly (i.e., *disciplinarily*) to the PM, but also of seconded staff from other areas, departments, groups, etc., and who work on the respective project by project order. The situation is different for large projects. Here, the PM usually has a relatively large group of PM team members who are fully or partially subordinate to him in disciplinary terms; see Fig. 5.4. But here, too, other teams from the specialist department collaborate by project order. Regardless of the forms of cooperation described above, it must be taken into account that the goals and motivations of the project team members are also of decisive importance for the success of the project. If the PM does not succeed in motivating the project team members and aligning the project goals with the goals of the project team members, the necessary momentum and drive that project teams need in order to overcome the usually unforeseeable problems more easily will not be created. Motivated employees, whose personal goals are achieved in conjunction with the project goals, are more willing to accept inconveniences than employees who are not motivated and whose own goals deviate completely from the project goals.

Section 15.1 summarizes the motivational factors as defined by Herzberg. When hiring project staff, it is therefore essential to make sure that these factors are taken into account. What is the easiest way to do this? For almost all people, success, recognition, and independent work are probably the three most important motivating factors. Therefore, when dealing with project staff, it is essential to ensure that their efforts are made in the context of measurable success. This, in turn, means that the employee must also be given a real opportunity to prove himself or herself through clear objectives and clearly defined authority. Even if the tasks and goals are highly ambitious, an important step has been taken because success then becomes measurable.

Recognition (*praise*) is a strong motivating factor. We already know this from raising children. Of course, project employees should not be dismissed with infantile praise. Recognition must be well-founded and based on actual performance; otherwise, it becomes ridiculous. *Recognition* is an important motivational factor that people in the German-speaking world use only with great reluctance and with reservations. Basically, it is very easy to express appreciation and thanks for good performance and commitment without falling into exaggerated adulation. The important thing is that we don't just give recognition in private, but publicly. In many U.S. companies, it is customary to highlight special achievements by naming the person concerned "*Employee of the Month,*" for example. It goes without saying that this procedure must not be used for fraudulent purposes. NASA also honors its project employees with appropriate awards after successful completion of their work. Iacocca writes about this: "*If you want to praise someone, do it in writing.*" [21].

Now, in everyday practice, of course, an award certificate does not have to be presented on every occasion. In many cases, a "*thank you*" or a reference to who was responsible for the success is quite sufficient. Another important motivating factor is independent work. To a greater or lesser extent, everyone needs this. Last but not least, even people with secure jobs strive to become self-employed because they want to be independent and free, even though this is often a fallacy.

When recruiting project staff, ensure that the following aspects are considered:

- Training of employees and.
- Team composition.

It is assumed that the project staff have sufficient training with regard to their area of expertise and have the practical experience necessary for the respective task. With regard to PM training, the staff must in principle be given the same training as the project managers; see Chaps. 25 and 26.

Putting together project teams is a difficult task. This is especially true for employees who have to work closely together on a daily basis, because what matters most is not that each person performs satisfactorily on his or her own, but rather the team result. However, it would not be wise to combine several experts whose knowledge is important for the project into one team without checking whether they accept each other. That can go absolutely wrong. As Iacocca puts it, "*People who are on a team have to like each other. They have to really appreciate each other.*" [22] In other words, people who literally cannot stand each other should not have to work together on a team, if possible.

17.3.4 Ethics in Project Management

Rush says very aptly, "*What an organization will be tomorrow depends on how well it manages people's ideas today*" [23]. Those who perform management tasks should therefore always keep in mind that they must not ignore the ethical aspect when dealing with people. In this regard, Kirchner says: "*Ignoring opinions is immoral and un-Christian*" [24], and he defines ethics as follows: "*Ethics is the doctrine of moral willing and acting.*" [25] Moral willingness and acting provide the best guarantee for assembling successful and responsible project teams and then maintaining them as stable organizational units until project completion. Hamann states with regard to the ethical responsibility of project managers, "*The special power and ability of project management must be coupled with ethical responsibility derived from social values and moral obligations.*" [26].

The ethical responsibility of project managers and their teams must be regarded as an important socio-political factor since the individuals concerned often have a great deal of power. Their ethical responsibility can be decisive for the well-being of their colleagues' careers. Hamann therefore warns that "*project management must not be limited to the purely technocratic use of one's own tools and to the thoughtless implementation of project goals without evaluation. Without project evaluation and without an assessment of the consequences of one's own actions, the project manager degrades himself to a tool and does not do justice to his task.*" [26].

When the title of this subsection is "*Ethics in Project Management*," it is not implied that ethics in project management are different from those in corporate management or other human fields. It is about moral norms and behaviors. In his book "*Dialectics and Ethics,*" Kirchner has commented on essential aspects of human behavior with regard to their ethical standards of value. Project management is just as affected by this as any other area of our social coexistence. In the following, some guiding core sentences of Kirchner are reproduced, which are of great importance for the work in project management and should be considered especially by the executives [27]:

- "*He who wants to lead people must have learned to lead himself.*" [27, p. 52]
- "*Succumbing to the lust for power are especially those people whose self-esteem is deficient.*" [27, p. 55]
- "*True leadership is necessarily tied to the moral personality formation of those who lead.*" [27, p. 56]
- "*Whoever wants to become a leader may accept that in the end he cannot avoid control by the led.*" [27, p. 70]
- "*Most leaders in the present define their self-worth by their professional competence.*" [27, p. 63]
- "*But anyone who chooses technical competence alone as the basis of his entrepreneurial strategy will not remain stable in the long term.*" [27, p. 70]
- "*Attention is the most significant phenomenon in interpersonal experience. It is so significant because a person's mental flourishing is shaped by the manner and degree of attention.*" [27, p. 76]
- "*Moral competence as a quality of leadership prevents the selfish abuse of the led.*" [27, p. 93]
- "*In listening, the gate opens to the inner encounter with the partner.*" [27, p. 102]
- "*On the basis of tolerance, communication partners meet each other with dignity.*" [27, p. 107]
- "*Only those who are genuinely interested in their partners and their concerns can gain and transfer lasting trust.*" [27, p. 110]
- "*It is possible to exchange positive feelings only with the person who has them in the center of his being.*" [27, p. 133]
- "*Credible is only he who does what he proclaims.*" [27, p. 134]
- "*Where conversation dies, gradually every partnership dies.*" [27, p. 156]
- "*The outstretched, open hand builds a bridge to the partner.*" [27, p. 189]
- "*If you want to persuade your partners, you should appeal positively to feelings in them.*" [27, p. 192]

17.4 Project Management Implementation Plan

17.4.1 Project Management Strategy

The introduction of project management should be based on a clear and unambiguous strategy. If one merely follows a fashionable trend and hangs the mantle of progress around oneself in the process, sooner or later there will be setbacks and inevitable disappointments at all levels.

The decision to introduce project management entails considerable organizational consequences and should therefore be well thought out and planned and, above all, follow an overarching strategy. The strategic goals for introducing project management can be defined as follows:

1. Achieve greater efficiency in the execution of projects, i.e., improve results through more precise adherence to goals and greater adherence to schedules and costs.
2. Increasing employee motivation, i.e., improving results through project-oriented assignment of tasks, responsibilities, and authority to results-oriented employees (*entrepreneurs*).
3. Creating greater customer confidence and image enhancement through better compliance and a faster response to customer requests.

By setting clear and unambiguous strategic goals, the route for further action is unmistakably defined. All further steps can be measured against this, and project management does not become a flash in the pan. However, it is important to choose the right steps for PM implementation. But what are the right steps? Do we need new people, new methods, or new software? It is certainly a complex process, but it starts primarily with the employees, because project management is not just a method but above all a management philosophy that is sure to change the company culture by putting people (*employees*) and their ideas at the center. Rush: "*All management activities can be reduced to two basic functions: the management of things and ideas. Most managers focus on things rather than on ideas. However, all things begin as ideas. Therefore, the leader interested in progress should focus on managing ideas*" [28]. The introduction of project management today is still an experimental field, and therefore ideas should be highly welcome. However, it is recommended that validated findings be taken into account. In general, however, it can be stated that the PM procedures available today are much more mature than the procedures for project organization.

The focus of PM implementation is undoubtedly the human being (*employee*) and especially the committed employee—the entrepreneur. Stein says the following about the importance of the entrepreneur in middle management: "*What is needed today is the "Entrepreneur of middle management" who can and wants to make a difference. This species was "discovered"* in *the United States as early as the late*

1970s, and the realization was gained that companies with such middle managers can be much more efficient, provided that structures are in place in which they can also fully develop their entrepreneurial potential." [29] Project managers do not automatically belong to the middle management of a company, but they go through an instructive process that prepares them very well for advancement in the company; see Sect. 1.7.

In summary, the tasks of a project manager can be compared to a scaled-down version of the tasks of a company CEO, and they exert an attraction on employees with entrepreneurial traits. Employees with these inclinations and characteristics are attracted to PM tasks, and this is a double opportunity, first for the employees and second for the companies in question. PM can serve as a "*breeding ground*" for future managers, but at the same time as a filter for selecting suitable junior managers. In addition, as already mentioned several times, project management leads to more motivation if handled correctly. These are arguments that must be taken very seriously and are of great strategic importance because companies with project tasks are urgently in need of committed employees with entrepreneurial character. This type of employee represents an important part of the human capital of such companies. Companies in this situation must make a clear and unambiguous decision in favor of PM implementation if they want to survive in the national and international competitive process. Above all, this decision-making process must be timely. Iacocca points out the importance of timely decisions with the following words: " .... *because even the right decision is wrong if it is made too late.*" [30].

17.4.2 Introductory Measures: Start of a New Corporate Culture

Platz rightly mentions: "*Introducing PM is a project*" [31]. Once the decision has been made to introduce PM, the action must also be completed quickly. However, the approach must be based on a plan. To start with, the PM software is, as already mentioned several times, completely wrong.

First of all, it must be ensured that the decision to introduce PM has been initiated by the relevant body. As a rule, this is the CEO or a department head commissioned with this task, because the project managers are supposed to act as an extension of the CEO, as shown in Fig. 1.1. All too often, however, the decision to introduce PM is initiated by someone in the company who does not have sufficient authority. In other cases, the PM idea is introduced by employees at the lower management level, which usually leads to problems with implementation.

The project to introduce PM must be supported by management from the very beginning and must be based on a solid foundation. The PM introduction needs a strong lobby, and this should come from management, because PM is the beginning of a new corporate culture. After the PM introduction has been consciously and officially initiated, the project "*PM introduction*" can be started in earnest. First, an implementation plan must be created, which should consist of the following elements:

1. Inventory.
   - Identification of current and future project plans in the company
   - Project classification and types
     - Size (*small, medium, large*)
     - Customers
     - Industry (electronics, mechanical engineering, etc.)
     - Phase (*development, manufacturing, procurement,* etc.)
   - Area allocation
     - Marketing
     - Development
     - Production
2. Organizational Development
   - Investigation of different organizational concepts; see Fig. 5.9
   - Organizational contribution to the company
     - Major project
     - Medium-sized projects
     - Small projects
     - Project management center
   - PM key personnel
     - Available
     - PM potential
     - Junior staff
3. Development and implementation of PM procedures
   - PM systems (*schedule, cost, performance, risk,* etc.)
     - Selection (procedure)
     - Testing (company-*related*)
   - Policies/procedures and manuals
     - Creation
     - Review
     - Release
   - Implementation
     - Select pilot projects
     - Application of PM systems in pilot projects
     - Evaluation of PM systems, including the usefulness of policies/procedures and manuals
     - Modification/improvement of PM systems
     - General introduction of PM systems
4. PM software (*see Chap. 18*)
   - System requirements
   - Schedule monitoring
   - Cost monitoring
   - Performance monitoring
   - Documentation and configuration control
   - Action item control
   - System selection

- Market
- Performance comparison
- System compatibility
- Expandability
5. Implementation steps
   - Test phase (pilot phase)
   - Evaluation
   - Installation and test
6. Recruitment of PM personnel
   - Personnel requirements
   - Project manager
   - Other PM employees
   - Tenders
   - In-house
   - External
   - Candidate selection
   - Criteria
   - Interviews
   - Selection
7. PM training measures
   - Training concept (*see Chap. 25*)
   - Training materials
   - Hand-outs
   - Teach-ware seminars and training
8. Accompanying PM lobby
   - Announce the PM objective
   - In-house campaign
   - Information sessions

Figure 17.4 shows the time sequence of PM implementation. The selected implementation time, including thorough training for at least 12 to 24 months, can be considered typical and necessary for a technology company with 1000 to 2000 employees. Shorter time estimates are usually not realistic since it must be considered that the new concepts have to be planned, developed, tested, and implemented, but also accepted. Therefore, the implementation plan must allow sufficient time for reviews and training. Jay aptly states, "*If you want to create something permanent, the time factor is of the utmost importance.*" [32] In parallel, however, pilot cases are possible and even necessary in order to check whether the planned changes actually work.

Platz points out that project management must be organized in a stable way for the long term [33]. In fact, it often happens that the guidelines once introduced are not followed, the PM systems are not maintained, and the employees are not continuously trained. Therefore, two measures are of particular importance in the context of PM implementation:

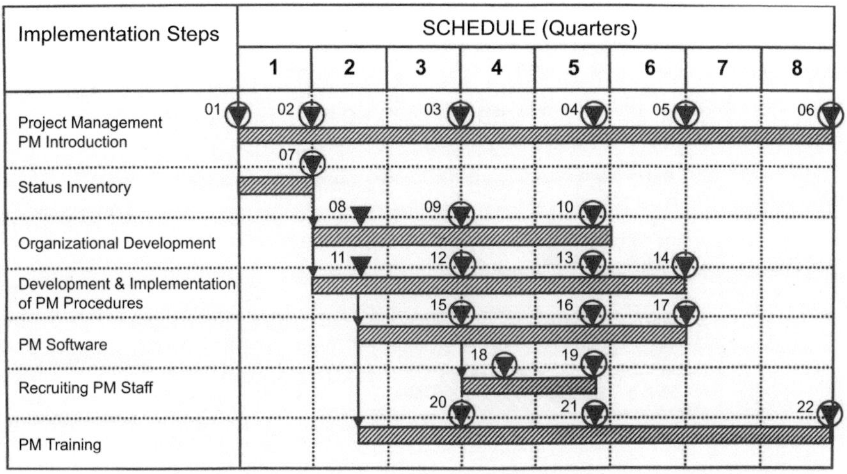

Milestone Definition	Documentation	Time (Months)	Milestone Definition	Documentation	Time (Months)
01. Kick-Off Meeting	Hand-Out	(0)	13. PM Concept Implemented	3. Review	(14)
02. Requirements Definition	1. Review	(3)	14. PM Handbook Complete	4. Review	(18)
03. Concept Formulation	2. Review	(9)	15. PM Software Specified	2. Review	(9)
04. Implementation Completion I	3. Review	(14)	16. PM Software Selected (Pilot case available)	3. Review	(14)
05. Implementation Completion II	4. Review	(18)			
06. Project Completion	5. Review	(24)	17. PM Software Implemented	4. Review	(18)
07. Status Inventory Complete	1. Review	(3)	18. PM Staffing Requirements Completed	Report	(11)
08. Organizational Concepts Available	Report	(5)			
09. Organizational Concepts Developed	2. Review	(9)	19. PM Staffing Selection Completed	3. Review	(14)
10. Organizational Concepts Implemented	3. Review	(14)	20. PM Training Concept Available	2. Review	(9)
11. PM Concept Available	Report	(5)	21. PM Training Material Available	3. Review	(14)
12. PM Concept Developed	2. Review	(9)	22. PM Training Completed	5. Review	(24)

Fig. 17.4 Proposed PM Implementation Plan

1. The firm organizational anchoring of a superordinate project management (*PMO*) as well as a sub-organization PM systems and software embedded in it; see Fig. 17.5.
2. The issuance of well-developed PM documentation, consisting of the following elements:
   - PM manual
   - PM policies and procedures

The anchoring of the higher-level project management (*PMO*) shown in Fig. 17.5 and the PM system control center are to be seen in connection with the matrix organizations shown in Fig. 5.11. Through the creation of these two positions, which in smaller companies can also be performed in personnel unions, the PM stability mentioned by Platz begins. The guideline according to which PM tasks are to be performed is anchored in the PM documentation. The handbook is the service instruction of every project manager. To ensure that the PM handbook (*HB*) is actually used, it must be simple in its structure and, above all, realistic. Wishes have no place here. One should be careful to make the explanations as simple as possible and include all necessary forms and formats (*templates*) in the HB. Experience

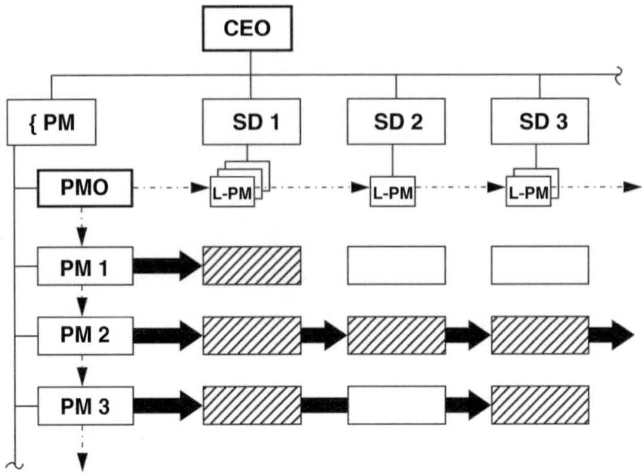

CEO	Corporate Executive Officer
SD	Specialist Department
PM	Project Manager
PMO	Project Management
L-PM	Local PM

Fig. 17.5 Anchoring of the central PMO and the deligated department PMs

teaches: the text must have instructional character; otherwise, it will not be followed. A complete outline of the PM handbook is given below:

1. Foreword
2. Organizational principles (*see Chap. 5*)
   2.1 Duties, responsibilities, and powers of the PM
   2.2 Organizational integration of project management into the company
   - Large projects
   - Medium projects
   - Small projects

   2.3 Project organization
   - Project functions
   - Interfaces to the departments
   - Other organizational interfaces

   2.4 Job descriptions
   - Project Manager
   - Project staff (*only for large projects*)

   2.5 Cooperation project
3. Project personnel (see Chap. 15)
   3.1 Qualification requirements

3.2 Training measures
3.3 Deployment and support
4. Planning and monitoring
4.1 Phase planning (see Chap. 4)
4.2 Project structuring (see Section 9.3)
4.3 Schedule planning (see Section 9.3)
4.4 Cost planning (see Section 10.1)
5. Documentation and configuration control (see Chap. 12)
5.1 Documentation requirements
– Customer
– In-house
5.2 Documentation control (*status*)
5.3 Configuration monitoring (*changes*)
6. Reporting to the CEO and/or the customer (*see Section 11.2*)
6.1 Report structure
6.2 Information content and formats
6.3 Reporting frequency
6.4 Review conversations
7. Project meetings (*see Section 11.3*)
7.1 Frequency and time required
7.2 Meeting content and minutes
7.3 Participants
8. Supply and competition (see Chap. 14)
8.1 PM Involvement
8.2 Request for quotation
8.3 Quotation
9. Contractual aspects (see Chap. 13)
9.1 Contract types
9.2 Contract structure
9.3 Annexes to the contract
9.4 Contract negotiations
10. Risk management (see Chap. 19)
10.1 Risk identification
10.2 Risk classification
10.3 Risk minimization
10.4 Risk reporting.

Although the outline of the PM manual reproduced here indicates a larger content, care should be taken to keep it as short and concise as possible. Detailed explanations of planning with a procedural character should be omitted because they belong in the guidelines yet to be prepared. The aforementioned 10 chapters of the PM handbook are accompanied by detailed guidelines; the author has participated several times in the preparation of a PM handbook. PM implementation is indeed, as Platz correctly says, a project [31]. But if one shirks from this effort, everything remains the same, or the urgently necessary measures are introduced only half-heartedly and incompletely. The result is often a failure. But the appropriate, i.e., customized PM introduction, must be seen as an investment in the future. Here is a quote from Eiff: "*Organizational design measures have investment character; they bring about production increases in the work, information and decision processes.*" [34].

The PM introduction must be planned and enforced in a general staff manner; see Chap. 26. However, mistakes often happen during the introduction, of which Platz records the essential ones. "*Typical PM introduction mistakes are for example:*

- *PM is operated*" on *the side*"
- *PM is prescribed, not elaborated*
- *No clear goals are pursued*
- *The affected parties are not involved*
- *The model is developed in isolation*
- *Resistance is not avoided*
- *Existing resistance is overlooked*
- *Management is not behind the new concept*
- *Alibi actions such as tool purchases or standard training.*" [35]

In order to avoid mistakes, one should, as already mentioned, handle the PM introduction as an independent project in the company. The most important measures for this are:

1. Development of an implementation plan (*see* Fig. 17.4).
2. Budget determination and release.
3. Appointment of a PM for PM introduction.
4. Conduct reviews on a regular basis.

17.4.3 Accompanying Personnel Development

PM personnel development should be understood here not only as training in the sense of courses, but quite generally as preparation for the new tasks. As already mentioned, PM is primarily a management task, often under difficult conditions. Training or preparation for PM does not only concern the project staff but all employees of a company who come into contact with projects. They all need to know what is involved and what is important. After all, it is not just the PMs who are expected to contribute to improving business results. All affected managers and employees of the company are required, the employees from the specialist departments as well as the PM specialists, because all employees are important when looking outwards to the customers. Project management must be supported by the entire workforce. Consequently, all affected employees, project and department personnel, as well as the CEO, must be systematically prepared for this new task. Project management must be learned and accepted by everyone involved.

Necessary PM training concepts have already been dealt with in various places; see Chap. 25 and in particular Sect. 25.2. There is hardly anything to add to this.

However, in connection with Fig. 17.4, reference is made to the individual measures that a company should take to prepare all the employees concerned sufficiently for the required PM tasks, not excluding the upper management. PM-oriented personnel development provides for the following individual steps:

1. Information events
2. PM seminars
3. PM workshops
4. On-the-job training
5. Coaching
6. Experience sharing.

Figure 17.6 shows the scheme for PM personnel development. During the PM implementation period, see Fig. 17.4, the six personnel development measures are to be carried out as part of the PM implementation and under the direction of the PM responsible for the PM implementation. After successful PM implementation,

Personal Development Measure	Event Management	Affected Participant			Comments
		Project Staff	Specialty Dept Staff	Company Management	
1. Information Session – PM Objectives – Measures – Problems – Advantages	o Company Management o PM Trainer – internal – external o Guest Speaker	*	*		During the two year implementation period, regularly scheduled information sessions were held; e.g. Quarterly
2. PM Seminars – PM Systems – Organization & Personnel – Cooperation – Supervisor behavior	o PM Trainer – internal – external o Experienced PMs o Central PM	*	*	(*)	Produce a PM Seminar program (see Attachment 6) in order to systematically train the affected Project and Specialty area personnel; Maximum course duration approx.. 2 weeks per class
4. Training on the Job – Training Program – Cooperation in the Project	o Experienced PMs o Central PM	*			Through job rotation in the next step, future PM candidates are trained on all activities through project personnel in Project; approx.. 1 year
5. Coaching – Cooperation in the Project – Individual counseling	o Experienced PMs o Central PM	*			The selected PM squad should be systematically prepared for their future assignments
6. Experience Exchange	o Central PM	*	*	*	Regular (i.e. monthly) sessions should be held to facilitate experience exchange.

Continuous participation, (~) Participation on Overview Themes

Fig. 17.6 Proposa for Development of PM Personnel (Source: Eiff, 1991)

everyday life should then return. However, it is important that PM personnel development continue on a reduced scale under the direction of the PMO manager; see Fig. 17.6.

17.4.4 Motivational Measures

In Sect. 15.1, the six motivational factors defined by Herzberg are stated, which should definitely be taken into account:

- Success
- Recognition
- Independence
- Responsibility
- Progress
- Development Opportunities.

Iacocca mentions the following point, which he believes also leads to more motivation: "*The only way to motivate people is through communication.*" [36] Indeed, good communication also causes an increase in motivation. Employees want to be informed and educated to remain capable of acting. Information thus becomes a privilege, and those who are informed are better prepared to make decisions and act on their own. Rush writes, "*Giving decision-making power is one of the most effective means of communicating trust.*" [37].

The most important motivating factors for successful project work are summarized below:

- Assignment of unambiguous, clear, and solvable targets (*PM as an extended arm of the CEO; see Fig. 1.1*).
- Ensuring vertical and lateral communication within the matrix organization; see Fig. 5.11.
- Granting authority to the PM.
- Recognition for services rendered and successes achieved.
- Planning professional development opportunities.

It is the task of higher-level project management (*see* Fig. 17.5) to create a climate conducive to motivation by complying with the aforementioned measures. "*The leader or manager is responsible for creating a positive work environment.*" [37].

17.4.5 Project Management: Handbook—Checklists

In the past, the author has been commissioned several times by industrial companies to introduce a PM handbook. In this context, he has developed a standardized checklist covering 11 aspects, which is reproduced below.

1. Project organization

	yes	no
– Is a project organization chart available?	☐	☐
– Does the appointed PM have the necessary qualifications?	☐	☐
– Has the PM been officially designated by his line manager?	☐	☐
– Have all positions (PM and team members) been nominated?	☐	☐
– Are the PM's tasks clearly and unambiguously defined?	☐	☐
– Are the responsibilities and powers of the PM's defined?	☐	☐
– Are the interfaces between the PM and departments known?	☐	☐
– Have the PM's reporting requirements been discussed and mutually agreed upon by all parties?	☐	☐
– Are the information flows to and from the PM defined and known?	☐	☐
– Is the signature authority for the PM available?	☐	☐
– Has the project organization been approved by the responsible line manager?	☐	☐

2. Project structuring

	yes	no
– Is there a coordinated and logical structuring concept?	☐	☐
– Are all projects included in the project structure?	☐	☐
– Is the service to be provided clearly defined by the project structure?	☐	☐
– Are all suppliers and vendors identified in the project structure (*make or buy*)?	☐	☐
– Are all responsibilities clearly and unambiguously defined in the project structure?	☐	☐
– Has the project structure been agreed with all stakeholders?	☐	☐

3. Project planning

	yes	no
– Are the main milestones of the project and the key dates known?	☐	☐
– Are the main milestones also anchored in the contract?	☐	☐
– Is there alignment between key milestones and payment milestones?	☐	☐
– Is there a complete schedule (e.g., Microsoft® project)?	☐	☐
– Was the company's standard format used?	☐	☐
– Are the previously mentioned key milestones (including reviews) included in the Gantt chart?	☐	☐
– Does the schedule show the exact planning sequence and the individual dependencies?	☐	☐
– Were the time estimates determined realistically and with the inclusion of empirical values?	☐	☐
– Has a planning analysis been performed and the critical path/area of the project been identified?	☐	☐
– Does the plan fully include all activities that are part of the project (*including subcontractors and suppliers*)?	☐	☐
– Have all project plans been coordinated between the PM and affected agencies and released as a binding project document?	☐	☐

4. Calculation (quantities/costs)

	yes	no
– Have all quantity approaches been taken into account for the realization of the project?	☐	☐
– Has a cash outflow plan been prepared?	☐	☐
– Has a payment plan (desired payment schedule) been established?	☐	☐
– Is the payment schedule consistent with customer deliverables?	☐	☐
– Can a capacity consideration be derived from the costing documents?	☐	☐
– Are there already identifiable resource bottlenecks compared to scheduling?	☐	☐
– Have all costing documents been coordinated between the PM and affected agencies and approved as a binding project document?	☐	☐

5. Deadline monitoring

	yes	no
– Has a status reporting procedure been established by the PM?	☐	☐
– Is status information regularly provided to the PM by the affected departments and subcontractors?	☐	☐
– Is a regular target/actual comparison carried out?	☐	☐
– Do regular status meetings take place in which all affected parties participate?	☐	☐
– Are trend analyses prepared by the PM for key milestones?	☐	☐
– Is an action list (*to-do list*) maintained by the PM?	☐	☐
– Are the scheduled reviews being conducted?	☐	☐

6. Cost monitoring

	yes	no
– Is the running costing used by the PM as a basis for tracking costs?	☐	☐
– Have all deviations from the plan (*target/actual comparison*) and their causes been identified?	☐	☐
– Have all changes/requests been reflected in the cost summary?	☐	☐
– Is the forecast for the costs still expected to be realistic?	☐	☐
– Does the receipt of payment correspond to the agreed payment targets?	☐	☐

7. Resource planning

	yes	no
– Has resource planning (*target, actual, forecast*) been carried out?	☐	☐
– Have resource bottlenecks been identified?	☐	☐
– Are there priority conflicts?	☐	☐
– Have potential/possible future priority conflicts also been considered?	☐	☐
– Have the causes of any priority conflicts been identified?	☐	☐
– Is there a proposed solution to eliminate priority conflicts?	☐	☐
– Have all alternatives (e.g., *outsourcing*) been considered to resolve priority conflicts?	☐	☐

8. Meetings

	yes	no
– Is there an invitation (*possibly standard*)?	☐	☐
– Have appropriate preparations been made (*room allocation, agenda, participants,* etc.)?	☐	☐
– Was the agenda distributed beforehand?	☐	☐
– Is the meeting process structured (*agenda*) and the time frame fixed (e.g., *one hour*)?	☐	☐
– Does the agenda (*possibly standard*) contain all important agenda items?	☐	☐
– Is the meeting leader known (e.g., *the PM*)?	☐	☐
– Are the participants known, and have they been invited?	☐	☐
– Are meeting minutes kept?	☐	☐
– Do the minutes contain all decisions and resolutions made?	☐	☐
– Is an action list (*to-do list*) maintained?	☐	☐

9. Information and reporting systems

	yes	no
– Are all required information flows to and from the project known?	☐	☐
– Have the information contents and formats been defined?	☐	☐
– Is the sequence of reporting fixed?	☐	☐
– Are the report contents used as a basis for project meetings?	☐	☐
– Does the PM inform the affected department's/subcontractors about important events in the project process?	☐	☐

10. Documentation Management

	yes	no
– Is there a customer requirement?	☐	☐
– Are the internal documentation requirements (list of requirements) defined?	☐	☐
– Is the documentation numbering system used?	☐	☐
– Has a documentation approval process been implemented?	☐	☐
– Are documentation changes systematically recorded?	☐	☐
– Is the documentation status regularly monitored (*in preparation, existing, changed*)?	☐	☐

11. Contract management (project issues)

	yes	no
– Has the contract/order been signed by both parties with legal force?	☐	☐
– Is the specification an integral part of the contract?	☐	☐
– Does the contract contain fixed information about delivery dates for the customer?	☐	☐
– Does the contract include fixed dates for customer provisioning/information?	☐	☐
– Does the contract contain a contractual penalty provision?	☐	☐
– Does the contract contain a provision on warranty periods?	☐	☐

	yes	no
– Does the contract contain a clear regulation about the payment modalities?	☐	☐
– Are the acceptance conditions clearly and unambiguously regulated in the contract?	☐	☐
– Has the contract been reviewed with the appropriate specialists?	☐	☐
– Has a contract review been performed (see QA specifications/plans)?	☐	☐

Literature

1. Litke, Hans-D.: Projektmanagement: Methoden, Techniken, Verhaltensweisen, Carl Hanser Verlag, Munich/Vienna, 1991, p. 171
2. Platz, Jochen: Projektmanagement erfolgreich einführen, Projektmanagement 2/92, Verlag TUV Rheinland, S. 5
3. Kirchner, Baldur: Dialectics and ethics, Gabler Verlag, Wiesbaden, 1991, p. 212
4. Heintel, Peter u. Krainz, Ewald E.: Projektmanagement - Eine Antwort auf die Hierarchie? Gabler Verlag, Wiesbaden, p. 1992, 42
5. Saynisch, Manfred: The Future of Project Management (Editorial), Projekt Management 2/92, GPM, Munich, S. 4
6. Zijl, Nico van: Projektmanagement heute, Die Orientierung, No. 92, Swiss People's Bank, p. 6
7. Madauss, Bernd-J.: Fortnightly meeting to identify problems and priorities, MBBKX04–096/88 (in-house)
8. Cf. source 4, p. 29
9. Cf. source 4, p. 21
10. Cf. source 4, p. 101
11. Cf. source 4, p. 72
12. Cf. source 4, p. 98
13. Schrempp, Jürgen E.: Top Manager's Address Controlling in High-Tech Companies, Controller Magazin, Management Service Verlag, Gauting/Munich, 4/93, p. 178
14. Schwarz, Reinhard: "Geheimauftrag: Seam Management, SZ No. 66, 20/21 March 1993, p. 75
15. Scholz, Christian: Projektmanagement ist Mitarbeiterführung, IO-Management Zeitschrift 60 (1991), No. 1, p. 72.
16. Source 6, p. 6
17. Source 15, p. 73
18. Thorn, Norbert: What does "bearing responsibility" mean in an institution? Association Management, 2/90, p. 8
19. "Organisationsbrevier", 6th edition Bern, 1974, p. 9.
20. Martin, Charles C.: Project Management - How to make it work, AMACOM, 1967, p. 59
21. Iacocca, Lee: Iacocca - An American Career, Econ Verlag, Düsseldorf/Vienna, 1988, p. 84.
22. Source 21-, p. 85
23. Rush, Myron: Management - A Biblical Approach, Victor Books, a division of SP Publications, Inc. Wheaton Illinois 60187, 1985, p. 20.
24. Source 3, p. 161
25. Source 3, p. 141
26. Hamann, Michael: Ethics des Projektmanagements, Zeitschrift Projekt-Management, Heft 3/92, Verlag TUV Rheinland, Köln, S. 2
27. Source 3, page number as given in the text.
28. Source 23, p. 30
29. Stein, Dieter: Mittel-Management - die gefesselte Leistungsreserve, Blick durch die Wirtschaft, July 1988, No. 127, p. 7.

30. Source 21, p. 78
31. Source 2, p. 7
32. Jay, Antony: Management and Machiavelli, Econ Taschenbuch Verlag, Düsseldorf, 1985, p. 156.
33. Source 2, p. 9
34. Eiff, von Wilfried: Organisation - Erfolgsfaktor der Unternehmensführung, Verlag Moderne Industrie, Landsberg/Lech, 1991, p. 11
35. Source 2, p. 13
36. Source 21, p. 81
37. Source 23, p. 46

IT Support in Projects: Project Management Software

18

Contents

Abstract

In principle, project management also works without IT support. But it would not be efficient. In the age of globalization, it is important that projects are managed from an economic point of view so that they are competitive, and with the support of suitable software, this is best achieved. However, it is important to use the appropriate PM software on the one hand and the right level of PM software on the other. With the explanations in this chapter, it is tried to give the reader appropriate assistance with the procurement of PM software.

B. -J. Madauss, *Project Management*,
https://doi.org/10.1007/978-3-662-69057-4_18

The author was able to enlist Dr. Thor Möller, Bremen, who has profound knowledge in the field of IT solutions for project management, to enrich this section with his expertise. Möller is a multiple author of PM standard works, such as the Project Management Lexicon and the practical handbook "*Managing Projects Successfully.*" He studied economics at the University of Bremen and holds a doctorate in business administration from the Institute for Project Management and Innovation (IPMI). As a consultant and trainer, he has appeared for many companies and universities worldwide. From 2004 to 2012, he was a member of the board and, in 2021 and 2022, President of the German Association for Project Management (GPM). Möller has been awarded several prizes. In 1999, he received the first Study Award for SME Research for his doctoral thesis. In 2011, IPMA honored him with the IPMA Otto Zieglmeier Project Excellence Award. Different faculties awarded him as teacher of the year.

18.1 Project Management Software

Project management is first and foremost an innovative leadership task, and project success depends very much on the leadership quality of the project manager and his team. Sophisticated procedures and IT tools cannot replace poor leadership quality. Experience teaches that a qualified PM team with only a little support from PM standards and software is still more successful compared to an inexperienced team with optimal IT support. This is also logical, because you should not drive a vehicle without a driver's license. To be successful, PM staff must have both good leadership skills and good knowledge of the PM procedures and IT tools mentioned before.

In order to survive in national and international competition, the companies concerned must ensure that the intended project teams have the necessary PM qualifications through training and practical experience and that suitable PM standards and IT solutions are available. A common mistake is to first procure PM-ITIT for PM out of ignorance, but not to be able to cope with it. However, the continuously increasing demand for rising management efficiency, such as precise schedule and cost forecasts and their monitoring, the establishment of know-how databases, and better communication between team members, requires the use of standardized PM methods and IT solutions.

Experienced project teams and suitable IT solutions are required to perform complex system tasks. First, however, the PM processes must be defined before targeting a suitable IT solution for PM. We are talking about PM processes as described in different chapters in this book.

The tasks to be mastered by the project manager are usually diverse and complex because they usually involve closely interlinked project parameters, e.g., the "*performance/scope*" to be provided, the specified "delivery *date,*" and "*cost management.*" They must be recorded, forecasted, and controlled at the various project levels over the entire course of the project. A synthesis of the three factors mentioned and structured project information are the basis for good project decisions. It is therefore obvious to use appropriate software to manage large volumes of data in order to be able to perform integrated project analyses thoroughly and, above all, quickly.

However, due to a lack of experience, companies and public authorities are often undecided about how to proceed when procuring PM-IT. There are often many important questions that need to be considered in the IT procurement process. For example, consider the question of the possibility of integrating scheduling and cost planning within a software package. Furthermore, there is the question of the interconnectivity of the IT packages used in multi-PM situations and of functioning workload planning. The interface to other IT programs of the company, e.g., to the commercial area, also plays an important role in this context. The many open questions make the decision to procure PM-IT difficult. For this reason, it sometimes comes to isolated, cost-effective stand-alone solutions, whose cost-effectiveness in the overall framework of the company, however, is not always optimal.

18.1.1 IT Support for Projects Over Time

This section begins with a brief outline of the historical development. In doing so, it becomes clear that project management IT (PM-ITIT) has sometimes also had a significant influence on the development of project management at various points. Current developments and perspectives are then highlighted. These are subdivided according to different dimensions of the application.

The birth of so-called modern project management in the 1950s was the development of the various methods of network planning technique; see also subsection 9.3.4. The calculation mechanisms of the network planning technique are basically simple, but because of the rapidly emerging complexity and dynamics of projects, they are very complex, require maintenance, and are susceptible to carelessness errors. Thus, the application of the network planning technique rarely took place outside of large-scale industries, such as the aerospace industry. It was not until the use of PCs that the first more convenient solutions were developed from the 1980s onwards. With sound planning and correct data input, computational operations could be performed at lightning speed and without careless errors. This was not only a significant pioneer of PM-IT but also led to a renaissance of network planning technology through IT. The applications of network scheduling can also be seen as the birth of PM-IT. At first, large-scale computer systems were used, which was very costly. Since then, continuous further developments have taken place, which have constantly expanded the performance spectrum of PM-IT solutions up to the present day.

The first PM-IT solutions could usually only plan individual projects and were also limited to scheduling. Basically, they were pure network planning programs. However, the range of services offered by PM-IT solutions was then expanded step by step. At first, capacity planning was added. A little later, cost planning was introduced. Then the planning calculations and presentations were increasingly extended by the input of actual data and the visualization of control. From then on, target/actual comparisons and further project-control representations supplemented the range of services. Gradually, further project management functions were added, such as risk assessments.

The next major development in project management in general and predominantly in parallel in PM-IT was multi-project management. The various PM-IT packages are now no longer focused on individual projects only but could aggregate project data into program and project portfolio data. On the other hand, important PM functions, such as capacity or risk management, could be handled across projects.

Another significant trend is the increasing integration of PM-IT with other IT systems in an organization over time. This overarching trend in IT development increases the precision and speed of data processing and provision and reduces duplication and redundant data storage, among other things. The internet then gave rise to new communication possibilities. Web 2.0 in particular brought with it a variety of dangers for projects on the one hand, but also enabled opportunities on the other; see also subsection 18.1.5.

18.1.2 Project Management with General Software

Overall, project management can be supported by general office solutions. The central tools for this are database and spreadsheet software. They are particularly suitable for checklists, scheduling, cost planning, etc. Other aspects of project management can be covered by word processing and presentation software. Presentation software is probably the least suitable, as data exchange in particular can be poorly automated. Paradoxically, however, it is precisely these applications that are widely used in practice. This can presumably be related to the inability of the users or to the widespread practice today that packaging is more important than the content.

In addition to the IT solutions already mentioned, accounting programs, certain deadline-tracking programs, etc. can also be used. Compared to professional PM-IT, all of these general solutions seem rather incoherent and not very professional, but depending on the initial situation of a company, they can make sense if projects are not part of the core business and are therefore not carried out in large numbers. In addition, professional PM-IT still does not cover the entire IT support of projects. Many project management activities must therefore still be supported by general software, especially word processing and presentation. These include, among others, goal definitions, and contract and *post-claim* management (*claim management, stakeholder management,* etc.).

When database or spreadsheet software is used as an alternative, several advantages and disadvantages arise. The advantages include the following in particular:

- No extra software needs to be procured and installed. Employees usually have the corresponding Office licenses at their workstations.

- The employees do not need to be trained in the software in general but already know the software well. They only need to be trained specifically for the application tool.
- One can start with very simple solutions and a low power spectrum. These solutions can then be elaborated on step by step.
- Thus, there should be no unused, superfluous functions. Thus, the tool can be kept consistently lean.
- There is a high degree of customizability.
- There is high traceability of the software operations if no complex macros are used.

However, solutions using general database or spreadsheet software are also subject to the following disadvantages in particular:

- A relatively high level of effort is required for "*greenfield*" tool development and maintenance.
- The user guidance is partly uncomfortable.
- Graphical limitations, especially in bar charts and network diagrams, reduce the overview and professional effect.
- Limited possibilities, especially in resource management, can only be implemented with disproportionate effort.

General software, especially a spreadsheet, can already support projects to a large extent. The author of this section has been using an Excel template for years, which contains a Gantt chart with an integrated work breakdown structure in the central view; see Fig. 18.1 [1].

The bars are shown at their planned periods ("*baseline*") and realized periods ("*actual").* Furthermore, the planned and realized costs are included in this view. Automated evaluations on work package, subproject, and project level, as well as per used time unit, lead to an overall evaluation of the current project status regarding performance, costs, and time (*magic triangle*). The overall evaluation is based in particular on the mechanisms of *earned value analysis* (EVA); see also subsection 9.4.6.

The tabular evaluations can be automatically supplemented by graphical representations. Other PM tools, such as cost trend analysis (CTA) and milestone trend analysis (MTA), can easily be added. Examples of these can easily be found on the Internet by entering the keywords "*milestone trend analysis,*" "*earned value analysis,*" "*stakeholder*" or "*risks*"—linked with the file abbreviation "*xls,*" or similar—into a search engine. In further spreadsheets, evaluations and updates of target achievement, stakeholders, risks and opportunities, contracts and claims, changes, etc. can be easily integrated. The central view can be duplicated, e.g., weekly and the corresponding spreadsheet can be named with the specification of the calendar week. This creates sufficient documentation of the process. A particular advantage of this approach is the extensive support for project management with only one file.

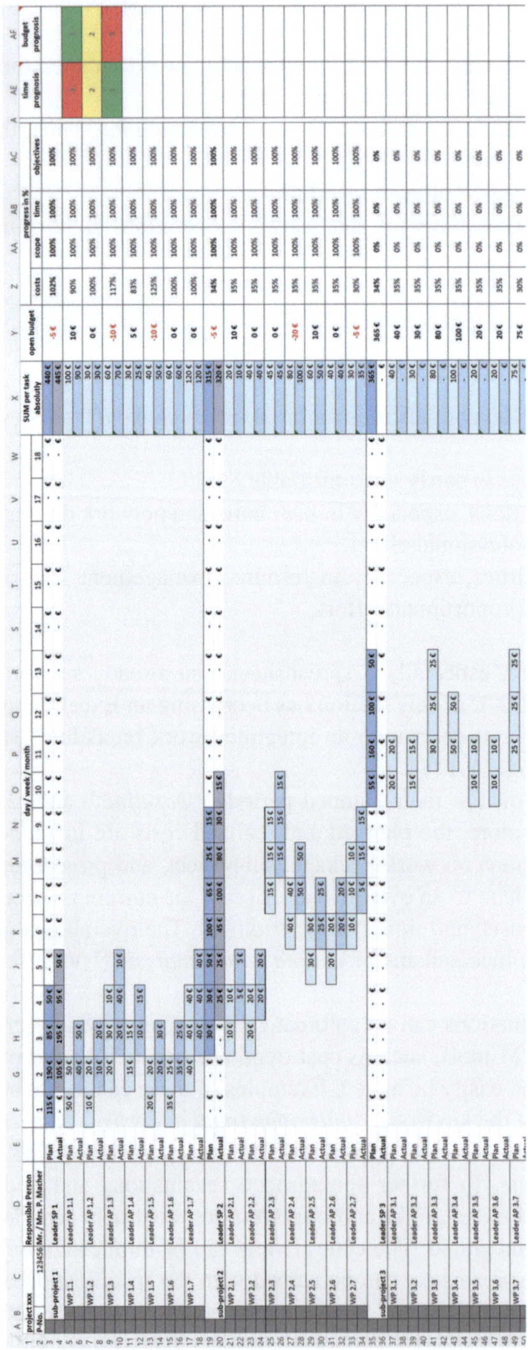

Fig. 18.1 Excel template for project control [1], download file at: http://www.pm-experten.de/index-Dateien/Downloads ProjektmanagementExperten.htm

Experienced users of spreadsheet software can develop such a file with relatively little effort. The arithmetic operations and presentation possibilities do not have to be reinvented for this. The Internet, especially in the English-speaking area, offers a huge amount of templates if you enter keywords in a search engine, such as "*earned value xls,*" "*bar chart xls,*" or "*milestone trend xls.*" However, as is generally the case on the Internet, utmost caution is advised and the accuracy and seriousness of these templates should be thoroughly checked. It is recommended to use these templates only as templates in order to develop a completely new, own template from them. In this way, you can also get a tool tailored to the company and project individually.

So, all in all, respectable solutions can be achieved with general software. However, as soon as a company aims at a larger number of projects and as a core of the management with many customer projects, professional PM-IT solutions are indispensable.

18.1.3 Freeware Solutions to PM

The Internet offers a wide variety of freeware solutions. Some can be downloaded free of charge from the Internet; most are now purely web-based applications. These freeware solutions already offer a somewhat more compact solution for IT support for project management. Although the possibilities are also limited, they are usually sufficient for the requirements of basic project management. Some project management activities can be supported much more comfortably here than with general software and office solutions. In addition, the overall graphical presentation in the screen view and in print looks much more professional.

Especially with cloud-based freeware solutions, however, one should refrain from entering sensitive data. So planning a company anniversary can certainly be done with such software, given the usual precautions on the Internet. When developing a new product, however, freeware is not advisable. Otherwise, the risk of sensitive data getting to unauthorized persons increases considerably.

18.1.4 Professional PM-IT Solutions

In addition to freeware solutions, there are, of course, a great many solutions that you have to buy or rent in the form of licenses. The spectrum of these solutions is enormous. Some solutions are more self-management tools that offer pure task tracking. Other solutions focus on financial aspects. The mass of PM-IT, however, largely offers complete solutions that can at least show the tasks, deadlines, capacities, and costs as well as risks and other aspects of individual projects. Since they can usually do more than freeware solutions and can often be part of an overall package, PM-IT packages are also quite widespread.

Due to a certain size and complexity of companies and projects, professional project management software is indispensable nowadays. But also, small companies should consider professional project management software if their core business consists of projects and basically several projects are carried out in parallel.

Many hundreds of different IT solutions for project management are offered in the PM-IT market. There are therefore sufficient tools for the most diverse requirements. The problem is therefore not finding tools but selecting the most suitable tool for the company's individual project management requirements.

Many PM-IT packages are offered by companies specifically established for this purpose. This means that it is often the only product or one of a few products offered by the company. The particular advantage of this is the extremely high degree of specialization. These companies offer their products and usually also support the implementation and integration of the tool up to the accompaniment of the initial phase or even constant support in the company. Furthermore, many PM-IT tools consist of a module solution within a large IT package. This applies, for example, to individual modules in integrated software solutions, such as SAP-PS at SAP, or to individual IT solutions in office packages, such as MS-Project at MS Office.

Many aspects, such as capacities, risks, etc., are nowadays managed across projects in companies and optimized company-wide. PM-IT solutions have therefore also been extended to the management of multi-project environments for many years. Key figure systems are an important basis for the aggregation of data. On the one hand, the same problems exist here as with KPI systems for companies as a whole. They are disliked by many employees, but the organization cannot do without them. And on the other hand, there are additional problems specific to project management when using KPI systems. IT support for multi-project management through PM-IT has made tremendous progress in recent years and provides services that can hardly be replaced.

IT systems for project management should also definitely be integrated solutions these days. This means that data entry and use are largely automatic via interfaces to other programs and databases. Redundant data entry and storage should be avoided as much as possible. This is the only way to realize the greatest possible benefit from IT systems. Unfortunately, many project management systems are still run as highly isolated systems with many redundancies to existing systems. User acceptance is usually correspondingly low due to these duplications alone.

Resource planning, in particular, should take place across projects. The current solutions for this are already quite well mature, but there are always conflicting situations resulting from this. Only with very clear rules and the consistent implementation of these rules can such conflict situations be avoided as far as possible. Furthermore, much more project information, especially risks, should be used across projects.

As a rule, companies and public authorities prefer ready-made IT solutions that may need to be *customized.* In-house developments are usually not economical. Ready-made packages for project management software (PM-IT) already contain sufficient reporting systems that have to be adapted to a greater or lesser extent by the organization. Due to these convenient default settings, many organizations forget or neglect a thorough analysis of what information is actually needed and how to obtain it for the report writers when implementing PM-IT. The proposed reports of PM-IT are largely adopted and usually little adapted without sufficiently critically questioning the usefulness of the information obtained. As a result, reports are often produced that require time-consuming data provision and do not invariably contain relevant information. As a result, the reports are hardly ever read. This is time-consuming as well. This should be avoided at all costs.

Software should always adapt to the requirements of an organization and thus serve the demands optimally. Particularly in the case of integrated software, however, it can also make sense to adapt the organizational structures to the software. The less "*customized*" software is, the easier it is to carry out repairs, updates, and migrations.

With the prerequisites mentioned in this section, an IT solution for projects also always requires the ability to aggregate projects into a multi-project management system. In the sense of a reporting pyramid, it must be possible to aggregate work package reports into subproject reports, project reports, program reports, and multi-project reports. The basis for this is a high level of technical and content standardization. For example, for data to be aggregated and meaningfully compared, it is essential that milestones are standardized and precisely defined across projects. All calculations at the various hierarchy levels should be automated. However, since each key figure usually also requires a qualitative commentary to increase its informative value, the possibility of a justification or comment must be provided at each level.

18.1.5 Project Management and the Web 2.0

The Internet has revolutionized the economy in two major waves, and, of course, the project business is not affected. In particular, the second Internet wave, Web 2.0, has had a significant impact on project work. On the one hand, it is a huge threat, as project opponents can find each other much more easily, organize better together, and communicate much more effectively. On the other hand, Web 2.0 also offers enormous opportunities for project work, since communication, especially toward today's diverse interest groups (*stakeholders*), but also overall internal and external communication, is given a large additional spectrum of possibilities. In addition, the communication channels of Web 2.0 can make the previously strongly characterized one-way communication take a big step toward dialog-oriented communication.

Certain developments in stakeholder behavior are forcing the use of these new technical possibilities. Citizens are once again taking a more active part in society's development. Whether "*angry citizens*" or "*courageous citizens,*" whatever you call them, our society has become recognizably more active again. It informs itself more and seeks and finds information and like-minded people, especially through the Internet. For example, opponents of the project can find each other on the Internet, organize themselves, and join forces to attack the project. Rights and other issues are increasingly demanded.

Current developments in stakeholder management reinforce these trends. The German Federal Ministry of Transport and Digital Infrastructure, for example, presented a "*Handbook for Good Citizen Participation in the Planning of Major Projects in the Transport Sector*" in November 2012 [2]. Modern communication and participation methods, such as *World Café* or *Deep Democracy,* also strengthen the involvement of stakeholders.

Companies and major projects, as well as all projects with a strong and/or explosive external impact, need Web 2.0 strategies. These should regulate, among other things:

- How employees may behave with regard to the project in Web 2.0 and what they may and may not communicate.
- How the project monitors Web 2.0 in terms of information and developments.
- How the project should respond to negative communication and developments regarding the project.
- How the project can actively use Web 2.0 for stakeholder work.
- How the project can also use Web 2.0 internally to support project work (think tank, internal communication (formal and informal), reporting, motivation, commendation, and more).

The tools of Web 2.0 are countless, and new possibilities are constantly developing. The literature now offers a lot of publications on how companies deal with Web 2.0. What one has to consider in general in project management and in particular in concrete projects with regard to Web 2.0 and what threats, opportunities, and possibilities there are is explored in greater depth in an extensive article by Möller [3].

18.2 Project for the Selection and Implementation of Project Management Software

18.2.1 Overview

The support of project work by IT is taken for granted today. However, the spectrum of possibilities is enormous, from self-developed templates or from the Internet to simple freeware solutions to complete professional solutions for project landscapes with corresponding expenditures of time and money for introduction and maintenance. The selection of the right solution always depends on the prerequisites and requirements of the company or user and their project. Most of the time, the tool selection is quite intuitive. In addition, the existing solutions are usually used as a starting point instead of openly searching for other—more suitable—solutions. Whichever solution is then used in practice, there are many exemplary solutions, but also countless examples of unstructured and thus mostly unsuccessful implementations. Furthermore, the implementation of the selected software is a complex task that should be implemented in the form of an organizational change project.

For a more structured selection of the right solutions, a wide range of background information is first provided. This enables the reader to make an informed decision regarding his approach and, as a result, the targeted selection of the appropriate tool. With a simple, self-developed Excel template, the requirements can be elicited. A well-founded solution conception could then use the experiences of the customized xls tool. Furthermore, many hints for implementation are given.

18.2.2 The Selection and Implementation of PM-IT Is a Change Project

Structured selection and application are certainly critical success factors for project teams in organizations. Inappropriate solutions can lead to significant risks. Important success-and-failure factors are outlined below. One thing that should not be done under any circumstances is that one should not introduce PM-IT without first introducing project management. The introduction of project management software in no way replaces the introduction of project management, but it should always be done downstream. The introduction of software cannot replace organizational structures or expertise but only supports them. A common mistake is the fallacy that the acquisition of software leads to the automatic completion of tasks and solutions to problems. The software can only be as good as its user. If you do not understand project management, you will not be successful with the corresponding software: *A fool with a tool is still a fool.*

PM structures and competencies must be in place! Before introducing PM-IT, the necessary structures and competencies for solid project management should be established. Only when these have been anchored to some extent in the organization can a PM-IT provide the necessary support. The higher the maturity level of project management in the organization, the more specifically an appropriate PM-IT can be selected and implemented.

For the introduction of PM-IT, an introduction strategy should be applied that is well thought out. The introduction strategy should possibly be done in stages. Small content-related steps can be used to test where and how PM-IT should support project work. Then pilot projects can deepen this experience and serve as a testing ground for different approaches. Only when sufficient experience is available as to which tasks are to be performed by the PM-IT can a targeted selection and implementation be made; see also Fig. 17.4.

18.2.3 Success Factors and Maxims in PM-IT

Key performance indicator systems are particularly suitable as a basis for successful planning, controlling, and reporting of organizational units or projects. They provide relevant information for those involved in the project and reduce the effort required to maintain data and create reports. Furthermore, they also reduce the effort for the report recipients at higher hierarchy levels and increase the usability and real use of the information in terms of quantity and quality.

However, for KPI systems to determine success factors, work effectively and efficiently, and be accepted by all users, many prerequisites are mandatory. The non-fulfillment of one of the prerequisites mentioned below can already be critical to success. In addition, the required users of KPI systems (donors and users of the information) are very sensitive. The consistent data maintenance and use required for success will only occur if as many users as possible are convinced of the usefulness of the system. An important prerequisite for this is that precisely these stakeholders recognize a significantly higher benefit than the expense of it. Thus, the success factors should be precisely identified at the beginning of the project. Examples of these success factors are described below. Furthermore, it is very beneficial for the success of the system introduction to define maxims [4].

The easiest way is to determine the needs of IT solutions that exist on the market and what services they offer. One can then check which of these services could also make sense for one's own organization. After selecting the tool, one could then adapt these performance features to the organization. However, this usually does not meet the diverse requirements of an organization, because one is limited from the outset to the options offered on the market. The better way is certainly to first determine what information the organization really needs for successful project management, and then select the right tool for the organization based on this.

As with any other software tool and with any other KPI system, the principle applies: *Reduce to the Max!* The effort required to determine and enter data must be reduced to a minimum in order to obtain the maximum amount of information that is really needed while consistently excluding all superfluous information. Information that is obviously interesting but not really relevant in practice must be excluded in order to avoid generating an unnecessary data graveyard that, in addition to the superfluous effort, also prevents a free and clear view of the really relevant information [5].

As in all other change projects, the active involvement of those affected (*stakeholder management*) is an important success factor. Successful implementation cannot be achieved with a detached team at the "*green table.*" Those affected must be turned into participants; see also subsection 17.1. Although this requires considerable additional effort, especially at the beginning of the project, it prevents uncontrollable dynamics and reactance on the part of the workforce, which can even lead to resistance. You have to get these stakeholders on board in a very targeted way. Only in this way will they have a real interest in ensuring that this boat does not sink. Most projects always have a change component. However, this varies enormously in its scope and explosiveness. The introduction of PM-IT has a large change in scope and explosiveness. This is especially true if a similar project has already been started once or several times but has not yet been successfully completed. When introducing PM-IT, it is therefore necessary to pay very close attention to and implement the mechanisms of change management right from the start.

The definition of maxims is very conducive to the success of tool selection and implementation. These maxims should specify the essential requirements for project reporting and consistently apply throughout the entire duration of the project as a hard guideline and meaningful benchmark for all participants. These maxims contain some of the success factors already discussed in advance. Stakeholders should definitely be involved in selecting and defining the maxims. Sufficient communication of these maxims to all those affected is also of decisive importance for the success of the project. This must by no means be a one-way communication. It must also allow for feedback and dialog in order to make the change process effective and to obtain important information.

Examples of this maxim are:

- The data generation and entry effort should be, to the greatest extent possible, a "*waste product*" of information that is already available or generated, in order to minimize the input to the system for report writers.
- Report writers should be able to use the report for self-direction as much as possible.
- There must be the possibility to explain the quantitative data with qualitative information, but without forcing it, but only when needed. However, the text length should be strongly limited to provide short and precise information.
- All data must be aggregable; the only exception is the qualitative supplements and notes, which remain at the respective reporting level.
- All operations and calculation paths must be easy to understand.
- The output should be as short as possible and contain only the really relevant information in a standardized structure.
- The system must be suitable for different project categories (A, B, and C projects).

18.2.4 Importance of Project Work in the Organization

The organizational requirements determine the requirements of the PM-IT. Only on this basis can the right tool be selected for the actual requirements. For this purpose, the current and future significance of project work in the organization must be assessed. The most important content-related question for the first rough assessment is, therefore, what significance and intensity the project work has for the organization. This can be represented on a spectrum leading from projects as a single situation to complex project situations as the sole business purpose; see Fig. 18.2.

The left outer point of the spectrum contains the following initial situation for the organization: The business purpose of the organization is, e.g., the production of or trade in goods, the operation of systems, or other ongoing services. Projects are not part of the core business purpose but are only used occasionally, for example, to

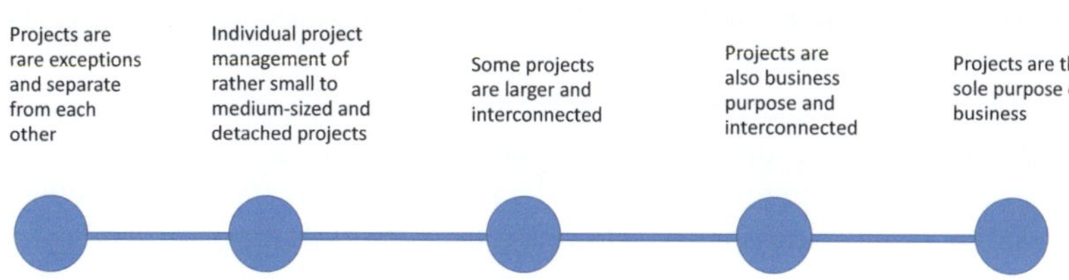

Fig. 18.2 Spectrum of the importance of project work in the organization

further develop the core business. Projects are thus rather rare special tasks for these organizations. For organizations in this initial situation, simple solutions for IT support of project management can and should be sufficient.

The right outer point of the spectrum is a situation where the sole business purpose of the company is the implementation of projects of all sizes as multi-project management (MPM). In particular, the business purpose of these organizations is the execution of projects. These projects are mostly carried out on behalf of customers, but may also be internal in nature. Therefore, the overall business activity is strongly characterized by projects, and thus project success contributes significantly to the company's success. The organization runs a larger number of projects of different sizes and also needs a solid overview of its entire project landscape. Thus, multi-project management is essential for successful business management. As mentioned earlier, when organizations reach a certain size and complexity, professional project management software is nowadays indispensable. Organizations in this initial situation must strive for comprehensive PM-IT solutions.

These two initial situations describe only two extreme starting points on the spectrum. As a rule, the initial situations of companies will lie somewhere between these two points on the spectrum. The further to the right they lie, the more significant the project success is for the company's success, and thus correspondingly more comprehensive solutions must then be sought.

In addition to the content-related, organizational, and technical requirements, other issues should be considered, such as:

- Are there any prevailing tools in the industry?
- Do central customers have specific tools?
- Can the exchange of project data with cooperation partners and customers only be done with certain tools?
- Is multilingualism required for the tool due to international work?
- What tools already exist in the company and group networks?

18.3 Project Management Software Selection

The market for PM-IT is diverse and confusing. Even if the company's own requirements for the PM-IT are already very well known, the selection of suitable software for this purpose appears to be possible only with great effort. If one has defined the requirements of the enterprise for the PM-IT (see Sect. 18.2), then one can mirror the PM-IT available on the market against it. First of all, there is the typical problem of self-promotion among providers of software in general: It is suggested to the potential buyer that the software will solve almost everything and every problem. This rather dubious, but unfortunately common behavior in this market and thus hardly avoidable for the providers is based on the assumption that non-existing elements can be added (*customizing*). However, the additional costs for the buyer for the additional programming are difficult to manage and are incurred again with each upgrade/release change. It is therefore advisable to consider only PM-IT packages in the pre-selection that actually fulfill the requirements for PM-IT to the greatest possible extent. It is not sufficient to rely on the manufacturer's specifications; a targeted review is necessary.

Some PM-IT solutions have been developed in specific industries and are still specialized in them in some cases. These then also generally use the procedures and technical terms typical of the respective industry. This can make the introduction of PM-IT much easier, both from a technical point of view and in terms of changes for employees. However, the strong focus of these solutions on typical industry projects also has a small disadvantage. General projects of the company are then also subject to this industry mode.

The first analysis of the market can be done via the Internet. Solid providers of specific software should also be easy to find via the Internet. Otherwise, you have to ask yourself whether this product is the right one at all. Many vendors are also quick to make on-site appointments at customer companies. This is convenient, but it can also quickly become very time consuming. One can also find a pre-screening of vendors at specific project management conventions. For example, the annual PM Forum of GPM (*Deutsche Gesellschaft für Projektmanagement e.V.*) includes a large exhibition area for vendors related to project management. Many well-known manufacturers of PM-IT present their products at such events. Furthermore, it can be a good advantage if the provider of the specific software has been on the market for a longer time. The reasons for this are assumed to be known.

For some years now, a comprehensive and neutral study by Meyer, Mey Marc/ Ahlemann, and Frederik [6], which is regularly updated, has offered considerable assistance. This study offers a variety of assistance and tips for the selection and implementation of PM-IT. In the first part, it guides the user through the selection and implementation processes, presents the basic architectures of PM-IT, and describes, with maturity levels, the different prerequisites as well as the common functions of PM-IT. In the second part, it provides a market overview with structured product evaluations as well as a summarized product comparison. Even if the price of this English-language publication appears quite high at first glance, the procurement and use of this study nevertheless enable a multiple of savings. Two renowned and product-independent experts present "*Requirements, Selection Process and Products*" in this study. Thus, an organization saves considerable effort in the selection and introduction of PM-IT, e.g., for the analysis and comparison of the most marketable PM-IT tools. Especially if this is to be done by external supporters, the procurement of the study is quickly amortized [6].

18.4 Implementation of Project Management Software

The implementation of PM-IT can basically take place through three different strategies:

- Pilot projects.
- Step-by-step technique or.
- Big Bang.

In most cases, pilot projects are preferred. While all other projects continue to be implemented according to the old system, the new system is tested on selected projects. In this way, individual system errors can be identified quickly and cause only limited damage. As soon as the "*teething troubles*" have been eliminated through several pilot runs and an initial degree of maturity has been reached, the new system can be applied to all projects to be restarted. This procedure is very safe but requires a lot of time.

With the step-by-step *technique,* one introduces the individual changes slowly, one after the other, but applies them immediately to all projects. A particular advantage is that the projects can slowly adjust to the new system. The individual modules can also be tested and optimized step by step. However, due to the many IT interfaces required nowadays, this is often not feasible, or only with far too much effort. This approach is also very safe, but it also requires a lot of time.

In the Big Bang approach, one immediately implements everything for all projects. The particular advantage is huge time savings, but this is only achieved at the cost of a rather high level of risk. If something goes wrong with the new system, all projects are usually affected immediately, and this can lead to considerable problems and additional work.

Which of these strategies is most suitable must be decided within the implementation project on a situational basis, based on the company's situation. Of course, many other implementation strategies can be developed based on these three basic strategies. In any case, it is important to ensure that sufficient test phases are planned. Through the test phases, many processes can be simulated, and thus some errors can be eliminated preventively so that they cannot even occur in the actual use case later. Furthermore, a parallel run is often created for a limited period of time. This ensures that if errors in the new system are too large, a *fall-back solution to* the old system can be quickly resorted to, thus limiting the negative impact on the project business.

The high proportion of change in these projects must be taken into account, especially during the implementation phase. Employees' fears and anxieties must be allayed through sufficient and open communication. A detailed communication concept must take into account all stakeholders and increasingly ensure dialog-oriented communication. It goes without saying that extensive training is also required for the employees concerned, i.e., the subsequent users. All users should be able to familiarize themselves sufficiently with the new system before they are expected to use it in the real world. IT experts often underestimate this requirement because, in line with their professional orientation, they rush into new systems with a great deal of curiosity. However, the majority of people have no particular affinity for IT. For them, changes to IT systems are complex stress situations that require a certain amount of time and a familiarization phase.

Only if the users of the future PM-IT feel sufficiently informed, considered, and prepared will they also sufficiently support the new system. Even in the event of minor or even major initial errors, they should then tackle the initial problems quite openly and in a solution-oriented manner and ultimately use the new system successfully. In practice, however, things often work out differently. Employees feel ambushed by the new systems because there is insufficient communication and preparation. As a result, they capitulate in the face of even the smallest problems and dismiss the new system as unusable.

A small but very effective and efficient measure is the four questions to the user. This measure can be used for every software implementation/upgrade and also for other change projects. The following four questions are asked of all users—or of selected users in the case of very large groups:

- Which elements of the previous (software) solution work well?
- Which elements of the previous (software) solution do not work well?
- Which elements of the previous (software) solution are superfluous?
- Which elements of the previous (software) solution are missing?

By answering these questions, the affected parties on the one hand already provide extensive information for the IT specification. On the other hand, they are actively integrated in the selection of the appropriate solution. Any respondent is thus deprived of the legitimacy to grumble about aspects they have not mentioned or even to boycott the work in any way. However, it goes without saying that this requires that the respondents' feedback be seriously evaluated and taken into account in the selection process. In addition, the aspects that have not been adopted in particular must be communicated to the respondents by means of sufficient feedback. This is particularly difficult if the respondents have an oversized "*wish list*" for the possibilities of the new system. In this case, realistic expectations must be tempered at an early stage.

During implementation, a direct hotline can provide immediate assistance and thus quickly solve any problems that arise, thus also improving overall acceptance. The early indication of this service can also relieve employees of many fears at an early stage.

18.5 Summary and Outlook

The use of computers in the management of projects is almost indispensable nowadays, but it must be well thought out. This applies to all formats: individual templates, office applications, freeware solutions, individual and group solutions, and multi-project solutions. No format replaces professional project management; but it can support it. Depending on the prerequisites and requirements of the company and the projects, the tool selection should be well-founded. The introduction of larger tools is also always a change project.

The presented template is rather aimed at single projects that do not have a large scale or complexity. The use of this template offers a lot of advantages, but its performance is limited quite quickly. To a certain degree, a professional PM-IT is indispensable.

One mega-trend for virtual work is collaboration platforms. Software like Microsoft Teams enables team members from different companies to share a platform to manage their projects. In addition to the sharing of files, calendars, chats, etc., they provide conference tools.

Another megatrend is the application of artificial intelligence (AI). Project managers will get more and more support from these services. Today, they can just use it for doing analyses, producing texts, etc.

Finally, the users of computers require more and more of a look and feel. PM-IT could be organized like a metaverse in the future, where avatars represent real team members who work together in a virtual place and have access to all information.

Note: Section 9.5.3 provides information about the application of SW Collaboration Tools and Artificial Intelligence for integrated project control.

Literature

1. Möller, Thor: Excel project folder for small projects, in: Managing projects successfully, chapter 05470, TÜV Media GmbH, Cologne
2. Federal Ministry of Transport and Digital Infrastructure: Handbook for Good Citizen Participation in the Planning of Major Projects in the Transport Sector, (download at: https://www.bmvi.de/SharedDocs/DE/Anlage/VerkehrUndMobilitaet/handbuch-buergerbeteiligung.pdf?__blob=publicationFile)
3. Möller, Thor; Nutzungspotenziale und Gefahren des Web 2.0 in Projekten, in: Projekte erfolgreich managen, Chapter 05450, TÜV Media GmbH, Cologne
4. Möller, Thor: Anforderungen an Kennzahlensysteme für Projekte und Projektlandschaften, in: Projekte erfolgreich managen, Chapter 10210, TÜV Media GmbH, Cologne
5. Möller, Thor; Wirtschaftlichkeit von Projekten konsequent nachhalten, in: Projekte erfolgreich managen, chapter 10130, TÜV Media GmbH, Cologne
6. Meyer, Mey, Marc/Ahlemann, Frederik; Project Management Software Systems, Requirements, Selection Process and Products, GPM Verlag, 8. edition 2014

Project Risk Management

19

Contents

Abstract

There are no projects without risk because a typical project characteristic is innovation, and that entails risks. Project managers must therefore always see themselves as risk managers. This means that project managers and their teams are responsible for identifying risks and their impact on the project in

B. -J. Madauss, *Project Management*,
https://doi.org/10.1007/978-3-662-69057-4_19

good time. The next step is to initiate efficient measures to minimize or eliminate them.

Risk management (RM) is a PM support function; see Fig. 5.4.

The execution of project tasks is always associated with entrepreneurial risk. There are no projects without any risk. This means that project managers and their teams must be prepared to deal with project risks, even if they are not really aware of them at first. Sometimes the risks may be small, but they exist and often lead to significant setbacks in the project's progress. The established project goals are not achieved; deadlines are not met, and cost limits are exceeded.

However, this situation is unfortunately not uncommon in projects. Thus, it can be assumed that it is often a matter of hidden and difficult-to-detect risks that lead to the previously mentioned setbacks. But how can project risks be recognized and mitigated in time? Are there any useful methods for identifying and combating risks that can be effectively applied in practice without major inconveniences, for example, through efficient risk management?

The occurrence of project risks is not only a phenomenon of modern times. Architects and construction experts have also been confronted with project risks in the past. Gutberlet describes very clearly how a construction or project disaster occurred, using the example of the choir vault of Beauvais Cathedral. The problem had its origin in the very high construction requirements: "*One wanted to build higher than ever before, but at the same time give the impression of weightlessness.*" [1] Twelve years after the consecration of the choir on November 29, 1284, the largest choir vault up to that time collapsed [2]. How could this have happened? "*The best expert on the history of Saint-Pierre de Beauvais Cathedral, Stephen Murray, blamed it on disagreements between the bishops (builders), who delayed further construction by bringing in new people in charge, which led to redesigns and thus put more and more strain on the statics of the building.*" [3] A lengthy reconstruction followed with increased demands again (*a new superlative*), resulting in another collapse.

In modern construction projects, too, the goals and expectations are often set very high (*nice to have*), which often leads to increased risks. But political squabbles and management errors can also lead to increased risks, as described in the examples of Berlin's new airport (*BER*) and Hamburg's Elbe Philharmonic Hall; see Sect. 3.2. Increased technical and/or commercial risks can also occur in high-tech industries, such as aerospace, power plants, or transport technology. The early identification, avoidance, or mitigation of risks through appropriate measures therefore makes efficient risk management an obligation. High standards of traffic safety and the avoidance of catastrophes are ensured by stringent functional and quality checks (*audits and reviews*) and extensive testing during development and production processes.

This chapter describes the need to combat risks at an early stage and the merits of a professional approach to risk minimization. The answer lies in the introduction of risk management, with the help of which all identifiable and hidden risks are identified at an early stage, combated, and, if possible, eliminated or otherwise hedged. By introducing effective risk management at all project levels, not only risks that have already been identified but also hidden risks can be effectively combated.

19.1 Risk Management

19.1.1 Definitions of Risk Management

Project risk is equated with a lack of certainty that the specified technical and/or economic target of the project will be achieved. This means that the achievement of the project objective is most likely to be jeopardized due to the existing risks. For a better understanding, a definition of the terms "*risk,*" "*risk management*," and "*risk potential*" is given below.

- Risk is the imponderability of the technical and/or economic success of the project. Because of existing risks, the project's success can be achieved completely, partially, or not at all. Certainly, there are situations that allow partial success as a result, while total failure is a technical and/or economic disaster. What is desirable, however, is full project success that is not threatened by risks and in which all specified objectives are fully achieved.
- Risk management refers to all the tasks and measures required to combat risk. This means that the range of risk management tasks includes the identification, assessment, and development of measures for reduction and continuous tracking of all actions as well as the results in connection with risk combating.
- The term risk potential is used to define the significance of the risk for the project in question, i.e., it is a large, medium, or small risk. For this purpose, the following description is provided:
  - Major Project Risk: Any identified and/or potential risks that, if not mitigated, are reasonably likely to result in failure to achieve the intended overall technical and/or economic objective of the project or system being built (*technical performance, reliability, safety, schedule, cost,* etc.).
  - Medium project risk: All identified and/or potential risks that, if not remedied, are reasonably likely to result in only partial achievement of the intended technical and/or economic objective of the project or system to be built (*technical performance, reliability, safety, schedule, cost,* etc.).
  - Low project risk: All critical and/or potential risks that, if not remedied, have a low probability of not achieving the intended technical and/or economic objective of the project or system to be built (*technical performance, reliability, safety, deadlines, costs,* etc.).

Lewis distinguishes between risk and threat; risks are unforeseen events that can happen, such as technical failures, accidents, inflation, etc., and threats, in his opinion, are competitors, business adversaries, etc. Although, as he elaborates, they are two different things, they can be combined into one overall analysis. If you can't avert the risk and threats remain, there is a risk that the project strategy will fail. "*Anything that can go wrong in a project will affect project targets.*" [4] Anything that can go wrong in the project process and jeopardize the project objectives is defined by Lewis as a risk. In this context, he suggests a SWOT* analysis to test the respective risk; see Table 19.1 [5].

* *SWOT—Strength, Weaknesses, Opportunities, Threats.*

19.1.2 Risk and Innovation

Innovation is the prerequisite for economic progress, e.g., the introduction of new materials, technologies, and manufacturing processes. For example, in the construction of the Airbus A380 ribs from aluminum-lithium composites, the wing from carbon fiber composite material and parts of the outer hull from GLARE*, resulted in significant weight reductions. The weight savings achieved can increase economic efficiency. Weight savings also play a major role in the automotive industry. Innovations in the communications sector (*internet and cell phones*) lead to more efficient business processes, and more electronics and software in cars lead to greater safety and comfort.

**GLARE = A glass fiber reinforced-hybrid material consisting of aluminum and a glass fiber laminate.*

However, innovation is always associated with risk because new products first have to be tested. Everyone is familiar with this from the automotive industry because when a new product is launched on the market, so-called teething troubles often occur despite extensive development tests during series production. But without a willingness to take risks, there can be no innovation. Gigerenzer comments: "*Risk aversion is closely linked to the fear of making mistakes. No risks, no mistakes, no innovation*!" [6] In order to be competitive with new products, the

Table 19.1 SWOT analysis

Strengths	Weaknesses
• Competence • Experience • Availability of qualified personnel • Availability of facilities • Good customer contacts • Secured financing	• New task • No or minimal experience • Lack of qualified personnel • Lack of facilities • Supplier problems • Lack of country knowledge
Opportunities	**Threats**
• Market leadership • Monopoly position • Long-term orders • Expansion possibilities • Full employment • High profits	• Competitive pressure • Technology advances • Absence of partner company • Financing problems • Currency fluctuations • Financial losses

developer must accept an acceptable level of risk-taking. Therefore, in addition to the development work, he must continuously assess the expected risk potential that can lead to the set goal not being achieved and evaluate the consequences. "*Calculating a risk is one thing, communicating it to people is another; risk communication is an important skill for lay people and professionals alike.*" [7].

Since a quantitative risk assessment is often not yet possible in the early phases due to a lack of data, all risk potentials should be recorded in the first step and described with the inclusion of empirical values. A quantitative risk assessment can then be carried out at a later stage; see subsection 19.4.2. However, it is crucial that the project manager does not lose sight of the topic of risk assessment and puts it on the agenda as an item for discussion at regular project meetings right from the start. Rules of thumb often suffice, says Gigerenzer, "*Experts often look for less information than novices and make do with heuristics.*" [7] "*In an uncertain world, complex methods of decision making that rely on more information and calculations are often worse and can cause harm because they raise unjustified hopes of certainty.*" In total:

1. "*RISK does not mean UNKNOWNNESS. The best decision under risk is not the best decision under uncertainty.*
2. *RULES OF THUMB ARE NOT STUPID. In an uncertain world, simple rules of thumb can lead to better results than sophisticated calculations.*
3. *LESS IS MORE. Complex problems don't always need complex solutions. Look for simple solutions first.*" [6].

Gigerenzer also points out that, as experienced project managers already know, the so-called gut feeling is definitely a phenomenon to be taken seriously. "*A gut feeling is neither a whim nor a sixth sense ... it is a form of unconscious intelligence. The assumption that intelligence is necessarily conscious and deliberate is a giant fallacy.*" [8] Psychologist Gigerenzer argues that intuition should be taken seriously and gives the following reasons: [9].

1. "*Intuition is neither a whim nor the source of all bad decisions. Intuition is unconscious intelligence that uses most regions of our brain.*
2. *Intuition is not inferior to logical thinking. Most of the time, both are necessary. Intuition is indispensable in a complex, uncertain world, while logic may suffice in a world where risks are known with certainty.*
3. *Intuition is not based on deficient mental software, but on intelligent rules of thumb and a lot of experience hidden in the subconscious.*

..., good decisions are made "*straight from the gut*," said Jack Welch, former CEO of General Electric, in 2009.

Intuition always plays a major role in project management when not all influencing variables can be quantified exactly. Particularly in the early project phases, project managers and their teams are permanently moving on unstable terrain, on black ice, so to speak. The situation at the start of the project is often chaotic, since many prerequisites, described as necessary in subsection 19.4.1, have not yet been established and, the start of the project cannot be postponed. Such a situation involves an increased risk, but its quantitative exact estimation at the time is difficult and unreliable, and therefore a simple qualitative assessment is preferable. "*Both simple and complex rules have their justification, depending on the state of the world:*" [10].

a) Simple rules of thumb	b) Complex risk models
Unstable, globally networked environment	*Stable and predictable Environment*
Many risk factors	*Few risk factors*
Small amount of data	*Large amount of data*

19.1.3 Risk Management: A Sub-Task of Project Management

For a long time, the term "*risk*" was not explicitly used in project management. Risk management was the domain of banks and insurance companies. But of course, depending on the complexity, risks always arose during project implementation, and they were often described with the word "*problems.*" In the meantime, "*risk management*" has become an integral part of project management because risk management is a proactive measure to prevent problems from occurring that were not identified in time. Early detection is important here in order to be able to initiate suitable measures to avoid or reduce risks, if necessary. In the aerospace industry, it was recognized early on that increased risks can be avoided by planned countermeasures, such as an extensive test program. However, this also comes at a price, and sometimes necessary budgets are cut due to profitability considerations, but this can lead to project mishaps and additional costs at a later date, sometimes even significantly exceeding the hoped-for savings.

The associated decision-making process is not always easy. It is understandable that one does not want to worsen the economic result by increasing risk budgets, but one wants to avoid the occurrence of a technical and/or schedule-related project risk. For innovative projects, the risk of a catastrophic failure can be particularly high. The aerospace industry has seen enough of this; see Sect. 3.8. Before a new system is ready for production, extensive qualification tests are carried out at all project levels during the development phase in order to eliminate as many risks as possible. This principle is also applied to high-tech projects, including facility engineering and the automotive industry.

Risk management is a task that falls within the area of responsibility of the appointed project manager. The project management must ensure that all identified and serious acute and potential risks that seriously jeopardize the success of the project are aggressively combated or eliminated. For projects of a certain size and/or complexity, it makes sense to assign risk management to a specialist reporting to the project manager, who is then exclusively responsible for risk management for a certain period of time. It is advisable for the respective risk manager to temporarily work in the core team of the project management; see subsection 5.1.3. The main tasks of risk management can be summarized as follows:

- Information on the status of all critical and/or potential project risks.
- Determination of the risk causes and consequences for the work package in question and for the project as a whole.
- Classification of all identified risks according to their importance for the anticipated project success.
- Preparation of a plan of action to combat risks.
- Introduction of regular reporting on risk status.

The deliberate integration of risk management into project management's area of responsibility is a compelling necessity for all innovative project tasks to avert damage to the project. Companies that have created a central office for risk management should ensure that there is a direct link to the practical realities of the projects. "*In few other project management areas do theory and practice diverge as widely as in risk management.*" [11].

19.1.4 Main Objectives of Risk Management

The main goal of risk management is to identify risks as early as possible and then develop adequate countermeasures. Here is a summary:

1. Identification of all current and potential risks.
2. Evaluation of the risks according to the criteria high, medium, and low; see Table 19.2.
3. Introduction of a risk elimination plan.
4. Regular reporting (*risk report*) on existing and potential project risks.

The purpose of the regularly prepared risk report is to inform the management of the company and/or the customer at certain intervals by providing an overview of the risk status. The report should contain the following information for each identified risk:

1. Risk title and brief description.
2. Creation of a risk register.
3. Description of the risk cause.
4. Impact on the function, safety, reliability, schedule, and cost of the project.
5. Risk potential (*high, medium, low*).
6. Planned risk avoidance or mitigation measures: changes, additional resources, insurance, etc.
7. Responsibility of risk management (*responsible person–risk owner*).

Table 19.2 Measures of the four main steps for risk prevention

Step	Individual Actions
1	**Risk identification** • Compile a list of all identified risks • Potential risk possibilities (experience) • Assemble prior experience from prior projects (lessons learned) • Review assembly and test anomalies from prior projects • Review identified operational anomalies • Generate a risk register
2	**Risk analysis and classification** • Evaluate identified risks • Classify risk impact (severe, moderate, minimal) • Continue the periodic evaluation of all risks • Expand the use of the FMEA/FMECA (failure modes, effects, and criticality analysis) methodology • Introduce measures for risk mediation
3	**Risk minimization or mediation** • Identification of the risk owner • Definition of goals for risk minimization or mediation • Performance of a risk analysis • Establish risk reduction measures • Establish a plan for the introduction of risk correction measures • Include team members in action item corrective measures
4	**Risk monitoring** • Monitor all risk mediation actions • Continuous evaluation of all open risks and introduction of corrective measures. • Evaluate all risk management information • Publish a risk status report

Risk analyses should be carried out in all project phases, e.g., in the early phases but also in subsequent phases up to disposal. The nuclear reactor disaster in Fukushima is a good example of how an early risk analysis would probably have helped to avoid the reactor disaster. In this case, it would have made sense if the operator had worked out appropriate *worst-case risk* scenarios when procuring the reactor plant and had provided appropriate earthquake-proof protection measures for the emergency power generators. This is a measure that would probably have resulted in additional costs, but the costs incurred as a result of the accident are many times higher. Similar incidents are also known from spaceflight, e.g., the Challenger disaster of 1986, where warning voices about increased risk were not heeded. In this context, Greenfield mentions that "*risk and risk management are a recurring theme and cites the failures*."

- *Insufficient risk identification and analysis.*
- *Insufficient risk mitigation and tracking.*
- *Weak systems engineering management.*
- *Inadequate application of risk assessment methods.*" [12].

Potential risk areas exist in many industries, e.g., transport technology and aerospace. Aircraft accidents and rocket crashes not only lead to spectacular press releases and thus a loss of confidence, but also to considerable liability and insurance costs, and in the worst case, future flights may be banned. Just as spectacular are the recalls of vehicles when defects are found and the associated substantial costs to the manufacturer. The most recent case is the VW emissions problem knowingly caused in 2015, which, in retrospect, must be regarded as a major financial risk.

In spaceflight, the issue of "*space debris*" and the associated risk of collision is particularly acute. Authorities and satellite operators are therefore calling on manufacturers to avoid or reduce the risk of collision by taking appropriate proactive measures. Another requirement of the operators is to ensure that no parts remain uncontrolled in orbit and that the satellites are returned to earth in a targeted manner after completion of the intended mission (*de-orbiting*) or, alternatively, placed in an orbit where they do not pose a problem. Of course, this risk-avoidance measure also entails additional costs, but liability for accidents would possibly be considerably more expensive.

19.2 Risk Management Process

19.2.1 Main Steps of Risk Management

The risk management process is generally carried out in several steps. Table 19.2 shows four main steps and a detailed description of the individual measures [13].

In addition, Fig. 19.1 shows the risk management process in a flow chart. In this context, reference should also be made to the well-known FMEA/FMECA procedure, a method that can be used to identify faults at an early stage and to make a quantitative assessment (*impact on the project*); see Sect. 8.1.

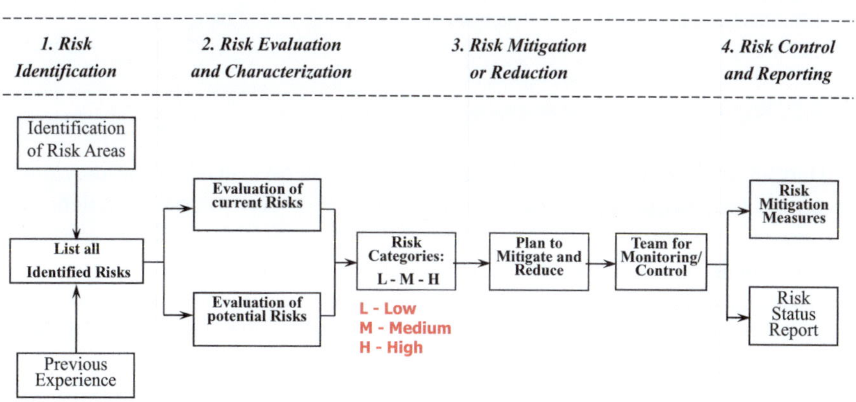

Fig. 19.1 Risk management process

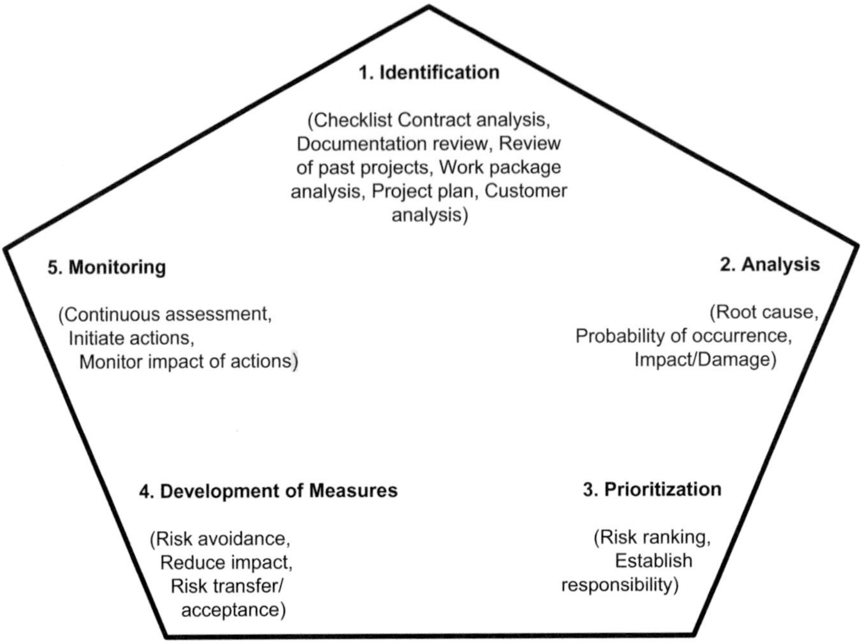

Fig. 19.2 Five steps of risk management (source: Harrent and Hemmrich, 2004 [14])

Table 19.3 Risk avoidance measures according to Michaels

1. Risk identification	2. Risk assessment	3. Risk monitoring
• Study objective • Employee indoctrination • Inclusion of other disciplines • User requirements • Implementation of the basic model • Identification of risks and dangers	• Establishment of the risk hierarchy • Quantification of risk measurement points • Selection of risk descriptions • Generation of a risk model • Calculation of risk impacts • Establish risk reserves	• Implement a risk organization • Budget to cover project risks • Application of best practices • Perform regular risk audits • Motivation measures • Respect risk avoidance

Harrant and Hemmrich have defined five main steps instead of the previously described four main steps, which, in detail, however, correspond in content; see Fig. 19.2 [14].

Looking at risks and opportunities together is another aspect of their book. "*Just as risks can have a negative impact on the project, there are also opportunities (positive impacts) inherent in every project, caused by different events.*" [15] In other words, "*Risk is inherent in opportunity …* " [16].

Michaels, in turn, breaks down the process of risk management into three individual steps by combining risk control and monitoring; see Table 19.3 [17]. He supplements his statement with a checklist containing the following four main questions:

- Is there specific technical risk assessment and reporting?
- Are formal reports produced regularly at all levels of management to provide information on the technical status, corrective actions, and potential impact on the project?
- Have technical risk indicators been developed for development, testing, manufacturing, cost, and management?
- Were all the technical requirements of the project (*top-level*) applied to the lowest development and testing level?

In summary, it is not so important whether the process is divided into three, four, or five stages, as long as the correct control measures are initiated. The author prefers the four-stage model described in Table 19.2.

19.2.2 Implement Risk Management at an Early Stage

With the urgency of the market for newer, better, and more economical products subject to the constraints of international competition, the technical and economic risk also increases. On the one hand, you have the technology, which is usually still in the development stage and not yet fully completed, but on the other hand, the demand is pressing (*time to market*). The recall actions of passenger cars are a good example of this. Also, in the field of space travel, one can hear the same story. The problems of the Iridium program (*satellite telephony by cell phone*) are another good example. Because the terrestrial mobile phone market developed very quickly, Iridium did not get off the ground as planned. *Simultaneous engineering* and *rapid prototyping are* still being used to improve the situation, but the pressure of deadlines also has its price.

On the one hand, you want to sell the customer the latest technology as quickly and profitably as possible, but on the other hand, you have the problem that the system is not yet technically mature and is afflicted with teething problems. The pressure of deadlines brings with it the danger of increased risk. However, since existing market conditions cannot usually be changed, it is therefore imperative that project management address the issue of "*risk management*" (RM) as early as possible. Klein describes the implementation stages of formal risk management as follows: [18].

1. Determination of RM content (*which WBS elements are to be monitored?*).
2. Which disciplines are to be involved (*PM, PC, PA, SE,* etc.)?
3. Identification of what (and what not) to monitor.
4. Determination of the required RM meetings.
5. Participants in the RM meetings.
6. Decision criteria for risk closure.

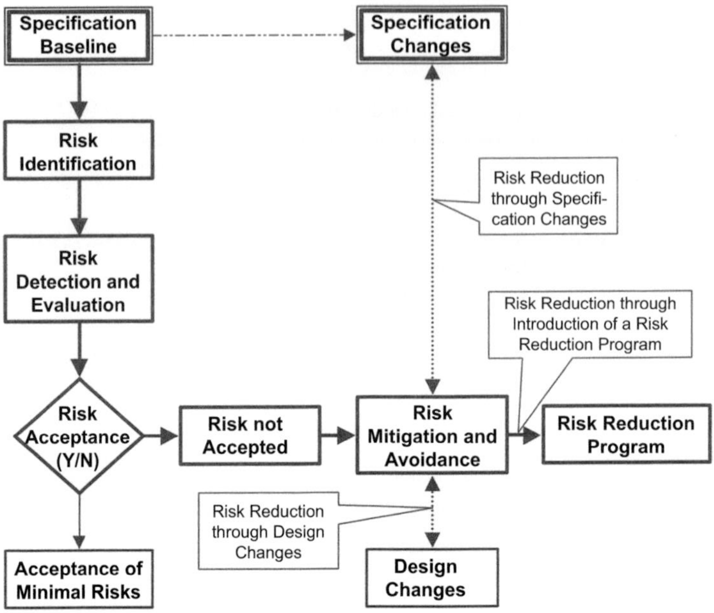

Fig. 19.3 Relationship between specification and project risk

He also mentions that the topic of risk has always been part of project management. However, it is only through the formal introduction of PM that regular information is provided to all project participants in the context of meetings and reports [18].

The logical connection of risk management with the specification process is shown in Fig. 19.3. It is well known to engineers and managers that over-specification (*nice to have*) can result in high risks. In aerospace, new technologies and materials are often included in the development process for competitive reasons. This includes the demand for weight savings. However, over-specification can lead to increased risks. In the case of the lightweight "*Centaur*" rocket upper stage, developed in the USA in the 1960s and powered by liquid oxygen and hydrogen, total failure initially occurred because the tank, which was stabilized only by internal pressure and had extremely thin walls, collapsed. Only the revised specification and the redesign of the upper stage based on it led to success, which continued in 2016. Figure 19.3 explains the relationship between specification and risk.

19.3 Risk Identification

19.3.1 Recognizing Risks in Time

Where and in which areas do project risks exist, and how can they be recognized? Because you can't fight what you don't recognize. Risks are often hidden, and you really have to go in search of the risks that are often difficult to identify in order to avoid negative surprises later on. The search for risks is often particularly problematic in highly complex projects because the interplay of interconnected assemblies prevents transparency. The former president of Lockheed Martin, Norman Augustin, has described this phenomenon as follows, "*There are the known unknowns on the one hand and also the unknown unknowns.*" [19] Against the known unknowns, the potential risks that one does not yet know, one can usually still take appropriate countermeasures relatively easily, e.g., plan for reserves. However, it is more difficult with the unknowns, the problems or risks that occur out of the blue and that no one has thought of, e.g., international or national catastrophes, political changes, or natural events such as earthquakes. Good advice is then expensive, and management is faced with an unexpected situation that is not always easy to cope with. It is difficult to give advice on this. However, if the project has been managed professionally and on a sound, contractual basis, the project situation can usually be successfully handled by working out positive solutions.

19.3.2 Systematically Identify Risks

Technical and commercial risks must be systematically identified through appropriate risk analyses. This means the engineers and managers, as well as the economists involved in the project, are required to pay attention to possible technical, scheduling, and/or commercial project risks and to identify them clearly. The following key questions arise:

- Do the planned project processes involve high, medium, low, or no risks?
- What is the past experience with the planned project processes: great, some, none?
- Have alternatives and/or planning reserves been provided for: yes, partially, no?

At the beginning of the project, the first step (*see* Table 19.2) is to identify all real identifiable risks and potential risk areas by referring to experience reports and recording their possible effects on the course of the project. During project meetings, all real risks and their effects on the course of the project must be recorded with the involvement of the departments, subcontractors, and suppliers. The information must be realistic and discussed within the teams. Identified risks can influence the performance target to be achieved, the schedule, and the estimated budget. Past experience (*lessons learned*) must be taken into account; see Fig. 19.1.

There are many areas of risk that can cause a project to falter. To avoid the risks, it is necessary for project management to regularly review the risk potential by making risk an important topic on the agenda during project meetings and consistently reviewing the identified risks and the status of their handling. It is an advantage if companies have developed and implemented risk standards. Creating a "*risk management*" organizational unit can also be useful. However, three aspects should be taken into account:

1. The standards of risk management should be simple and practicable. The possible application of special software should not be complicated (*KISS—keep it simple, stupid*).
2. The project manager must ensure that the project staff understands and accepts the need for risk management and the use of appropriate procedures.
3. The risk manager must work closely with project colleagues and the involved employees of the specialist departments at the working level.

A variety of risks can arise during project execution, which can be in the technical and administrative areas as well as in the financial and personnel areas. All risks can lead to the goal not being achieved within the specified targets. The most important risk areas are described below: [13].

- Technical risks: compliance with the given (*specified*) performance, such as function, test results, quality, and reliability, is at risk.
- Schedule risks: important milestones, such as design reviews (*PDR and CDR*), the completion of development, the start of production, acceptance tests, etc., are delayed, and delivery to the customer is at risk.
- Financial risks: The planned quantity estimates (e.g., *estimates for personnel and materials*) and calculated planned costs are not met. Contractual penalties for quality losses and late delivery are threatened, and there are financing bottlenecks.
- Personnel risk: the following four aspects are of particular importance for the risk assessment, namely the availability, qualification, commitment, and motivation, as well as the readiness of the team members.
- Legal risks: In the national but especially international environment, the applicable legal situation, official and internal regulations, standards, and procedures can represent a risk.
- Contractual risks: Contracts (e.g., *contracts for work and services*) are the link between partners (e.g., *customer and supplier*). The careful drafting of contracts is important for risk containment. Risk factors also include product acceptance, delivery dates, payment terms, and company insolvencies.

Batson has compiled a very detailed list of all technical and administrative risks that can be used as a checklist: [20].

1. Development risks
   (a) Development complexity
   (b) New technology
   (c) Relevant experience
   (d) Experience with similar projects
   (e) Resources required
   (f) Operating requirements
   (g) Required theoretical analyses
   (h) Differences in existing technology

2. Technical risks
   (a) Physical factors: Dynamic, Load, Thermal, Weight, Vibration
   (b) Material factors: Chemical, Pneumatic, Hydraulic, Atmospheric, Ionization
   (c) Radiation, etc.: EMI, Nuclear, Optical, Sound, Magnetic
3. Material availability
   (a) Sufficient stock?
   (b) Sufficient quality?
   (c) Alternative suppliers available?
   (d) Subcontractors and suppliers available?
4. Test and model risks
   (a) Facilities available?
   (b) Models available?
   (c) Interpolation/extrapolation required?
5. Integration and interface risks
6. Project Personnel
   (a) Personnel availability
   (b) Confidentiality agreement
   (c) Travel requirements
7. Software development
8. Security risks
9. Reliability risks
10. Critical error modalities
    (a) Reliability
    (b) Maintainability
    (c) Redundancies
    (d) Error detection
    (e) Error correction
11. Energy/environmental risks
12. Schedule risks
    (a) Technical problems
    (b) Material deliveries
    (c) Personnel availability
    (d) Plant availability (*production and testing*)
    (e) Test error
    (f) Communication problems
    (g) Requirement changes
    (h) Unpredictable events: weather, strikes, etc.
13. Cost risks
    (a) Technical problems
    (b) Delivery problems
    (c) Economic factors: Inflation, exchange rates, etc.
    (d) Political and social problems
    (e) Laws and regulations
    (f) Unpredictable events: weather, strikes, etc.
14. Support risks

Table 19.4 Example of an ESA risk register

<table>
<tr><th colspan="15">RISK REGISTER (Example)</th></tr>
<tr><td colspan="3">Project:
WBS Ref.:</td><td colspan="4">Organization:</td><td colspan="5">Source:
Controlled by:
Supported by:</td><td colspan="3">Date:
Issue:</td></tr>
<tr><th colspan="15">RISK SCENARIO and MAGNITUDE</th></tr>
<tr><td colspan="2">No.</td><td colspan="13">Risk scenario title:</td></tr>
<tr><td colspan="15">Cause and consequence:</td></tr>
<tr><td colspan="5">Severity (S)</td><td colspan="5">Likelihood (L)</td><td rowspan="2">Risk index</td><td>RED</td><td>YELLOW</td><td>GREEN</td><td>Risk domain</td></tr>
<tr><td>Negligible
1</td><td>Significant
2</td><td>Major
3</td><td>Critical
4</td><td>Catastrophic
5</td><td>Minimum
A</td><td>Low
B</td><td>Medium
C</td><td>High
D</td><td>Maximum
E</td><td>(*)</td><td>(*)</td><td>(*)</td><td>(**)</td></tr>
<tr><th colspan="15">RISK DECISION and ACTION</th></tr>
<tr><td colspan="7">Accept risk ❑</td><td colspan="8">Reduce risk ❑</td></tr>
<tr><td colspan="3">Risk reduction measures:</td><td colspan="3">Verification means:</td><td colspan="9">Expected risk reduction (severity, likelihood, risk index):</td></tr>
<tr><td colspan="6">Action:</td><td colspan="9">Status:</td></tr>
<tr><td colspan="14">Agreed by project management:

Name: Signature:
Date:</td><td>Risk rank:</td></tr>
<tr><td colspan="15">Notes
(*) Mark box as appropriate for the value of "R" (risk index), according to the criteria defined in the risk management policy.
(**) Indicate risk domain (e.g. technical, cost or schedule).</td></tr>
</table>

Notes

(*) Mark the box as appropriate for the value of "*R*" (risk index), according to the criteria defined in the risk management policy

(**) Indicate the risk domain (e.g., technical, cost, or schedule)

(a) Manpower Risks (*Client/Vendor*): Number, hourly rates, skills, training
(b) Plants: maintenance, simulations
(c) Transport problems
(d) System security
(e) Quality control: production capacity, spare parts, material availability
(f) Ambient Influence
(g) Data Management
(h) Alternative resources
(i) Configuration Management.

The identified potential risks and their effects on the project's progress are to be recorded in a risk register. An example of an ESA risk register is given in Table 19.4 [21]. The risk register should be kept as an integral part of the project documentation for continuous risk assessment.

Barker points out the importance of risk management: "*Risk and issue management helps you to pinpoint your plan's weaknesses and gives a useful insight into your project's overall health.*" … "*The risk management process is an iterative one. After the initial identification and assessment of risks and issues have been made, mechanisms should be put in place to keep them under regular review.*" [22] He describes the contents of a typical risk *and issue log* as follows:

1. Risk identification
2. Identification by [*name*]

3. Identification on [*date*]
4. Risk description
5. Probability that the risk will occur
   a. Very unlikely
   b. Unlikely
   c. 50/50
   d. Very likely
   e. The risk is most likely to occur
6. Risk impact
   a. Negligible
   b. Marginally
   c. Moderate
   d. Strong
   e. Catastrophic
7. Total evaluation (*probability x impact*)
8. Required measures for avoidance or mitigation
9. Risk handler [*name*]
10. Communication to senior management (*Y/N*)
11. Reference to previous similar risks
12. Risk status (open or closed)

Barker says in this context, "*A brilliant project manager avoids being a dictum of chance and circumstances.*" [22].

19.3.3 Consideration of Supplier Risk

An important aspect of the risk assessment in the project is the supplier risk because many fail due to the late or defective delivery of components or materials. Particular attention must be paid to parts with especially long delivery times, such as long lead items *(LLIs).* Orders must be placed as quickly as possible, and the delivery process must be continuously monitored. Stocking special components can be another way to meet deadlines. In the run-up to the order, the delivery reliability and product quality of the suppliers in question should be evaluated in order to avoid possible sources of risk. The purchasing decision should take the following criteria into account regarding possible risks: [23].

1. Company assessment: company image, know-how, management concept *(company/project),* organizational structures, qualification of the project manager and team members, subcontractors, and suppliers.
2. Bid evaluation*:* compliance, technical bid, management bid, commercial bid.
   (a) Technical offer: innovation, product, quality and reliability, warranty services, risk.
   (b) Management offer: organization, key personnel, scheduling, planning reserves, delivery risks *(bottlenecks),* contractual penalties.

(c) Commercial bid: price-performance ratio, identifiable benefits, design to cost (DTC), life cycle cost *(LCC)*, price bid *(FP, FFP,* etc.*)*.

In this context, reference is also made to the characteristic risk factors associated with the use of standard components off-the-shelf (*COTS*): [24].

1. Short-term asynchronous changes
2. Use of obsolete components
3. Use of protected data
4. Increased life cycle costs
5. Multiple configurations
6. Different quality practices
7. Fixed (*as is*) configurations
8. Commercial standard components
9. Temporary manufacturer support
10. Unprotected product information

19.4 Risk Analysis and Assessment

19.4.1 Prerequisite for Risk Control

The second step is risk analysis and assessment; see Table 19.2 and Fig. 19.1. This analysis and assessment are intended to help ensure that all possible effects of project risks or risks that can or could develop during project implementation do not go undetected. The process of risk analysis and the accompanying assessment of the identified risks are prerequisites for the development of measures to combat them.

It must be determined in detail what effects the identified risk, e.g., through the use of new materials, aggregates, or processes, will have on function, reliability, deadlines, and costs. Unrecognized or incorrectly assessed risks may lead to schedule delays and/or cost increases and, in extreme cases, to outright project disasters. Risks must also be seen in the context of legal requirements, e.g., producer liability; see subsection 13.2.3.

19.4.2 Risk Assessment

The elimination of discovered risks often leads to drastic changes in the project process. Often, this decision is very costly and time-consuming. If the project manager does not have appropriate project reserves, the decision-making process is particularly difficult, especially if the available risk analysis is not comprehensive enough to justify a budget increase. In fact, the offsetting calculation is the price and the project delay for the breakdown that has occurred, while hoping that it will not. But this is the crux of the matter because in many cases the disaster scenario is not clearly enough defined and supported with facts and figures to derive the necessary risk or defect elimination decisions.

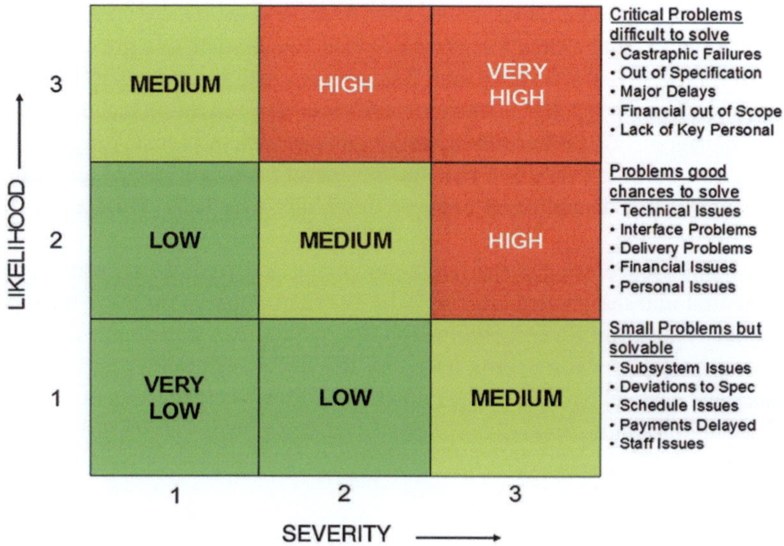

Fig. 19.4 Three-stage model (Y-axis: probability and X-axis: severity)

However, the question is often, "*Is it worth the technical and financial effort to correct the error?*" or should one take a chance? Often, the supposed problem does not actually occur, one has gotten away with it or has been lucky. In this context, reference is made to two catastrophes of recent years, the Challenger disaster of NASA in 1986 and the accident of the German Intercity Railway ICE (*Deutsche Bahn AG*) in 1998. In both cases, the damage to people and property was very extensive and could have been remedied with relatively few resources in relation to the amount of damage. In both cases, there were already indications of possible risks (sources of error), which then led to the disaster.

A method of risk assessment that has also proven itself in practice is considered:

(a) the risk class *(severity)*, see Table 19.2; and
(b) the probability *(likelihood)* of the risk occurring.

The three-stage model shown in Fig. 19.4, also referred to as the *"threat chart,"* distinguishes between

- Low (L)
- Medium (M)
- High (H)

Some companies also use a five-step model but, in most cases, three steps are sufficient.

Risk area 3–3 is a very high risk (*level 3*) that is very likely to occur unless mitigation measures are taken. In contrast, risk area 1–1 is low and unlikely to occur.

Only minor mitigation measures are required to address the residual risks. However, it should be noted that even a low class 1–1 risk can escalate to a greater risk. The aim is to achieve a 0–0 risk. Risk area 2–2 involves medium risks that require special attention from the project management. Crucial to good risk management is that project management continuously assesses risks as part of regular progress meetings (see *Sect. 11.3*). Class 3–3 risks do not permit any negligence and must be pursued with great discipline in order to eliminate or at least mitigate the existing risk.

As a quintessence, we can say that a systematic and credible assessment of the risk potential and the presentation of failure scenarios based on it, as well as the estimation of the effort for risk elimination and the costs for statistically determined project mishaps, are a compelling necessity for all risk-relevant project decisions. Madauss and Mennel emphasize the importance of project management in addressing risk, stating "*that professional project management is critical to minimizing risk.*" [25].

19.4.3 Internal and External Influences

Internal and external influences can have a significant impact on the project process and thus also on the project risk. Five areas are listed below that can be affected by internal and external influences. While internal influences can in principle be controlled, this is not true, or only to a limited extent, for external influences [26].

- Political Influence:
  - Internal examples: Reorganization, staff reductions, new partnerships, etc.
  - External examples: Legislative changes, environmental protection, currency changes, etc.
- Competition:
  - Internal examples: Streamlining, benchmarking, metrics, etc.
  - External examples: Market demands, competition, market prices, etc.
- Business aspects:
  - Internal examples: Rationalization, hourly rates, surcharges, etc.
  - External examples: Exchange rates, taxes, etc.
- Technology aspects:
  - Internal examples: New developments, production processes, tests, etc.
  - External examples: Legal requirements, regulations (e.g., *exhaust gas*), interfaces, etc.
- Operational aspects:
  - Internal examples: Standard procedures, operating instructions, etc.
  - External examples: Standards, safety regulations, etc.

Duffy mentions in this context that all internal factors (*previously mentioned examples*) are usually controllable, while external factors are mostly not; see Figs. 19.5 and 19.6.

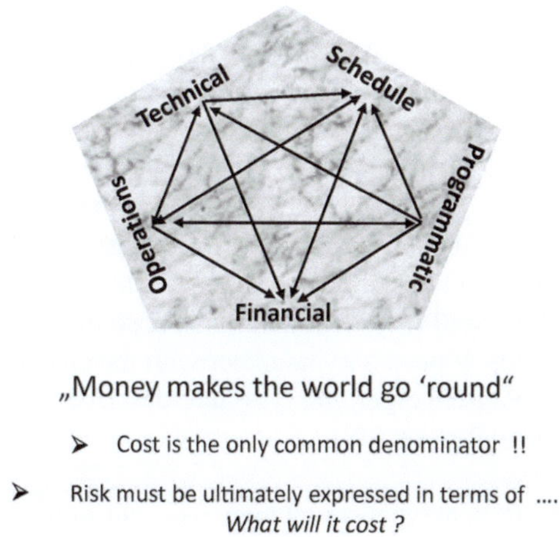

Fig. 19.5 Internal influencing factors (Source: Duffy, TeraSim Associates, 2002 [26])

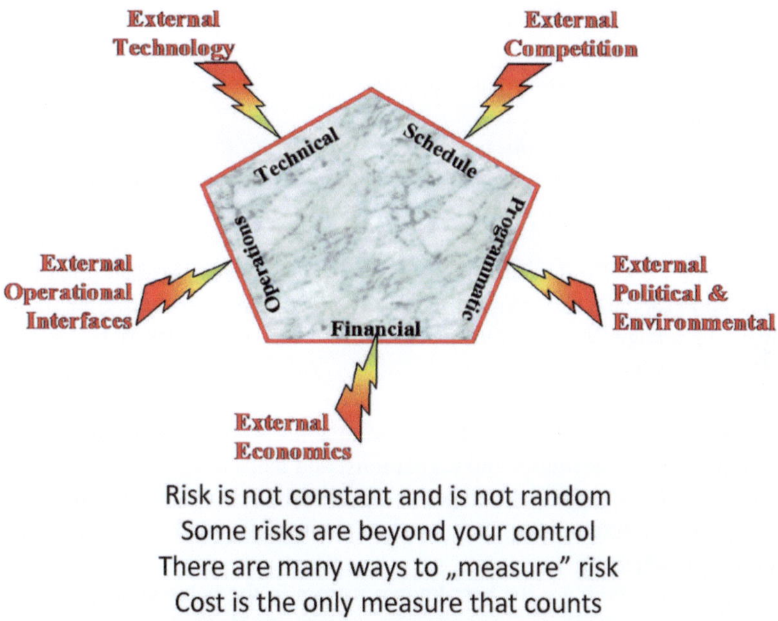

Fig. 19.6 External influencing factors (Source: Duffy, TeraSim Associates, 2002 [26])

19.5 Risk Reduction and Mitigation

19.5.1 Eliminate Risks in the Approach

It is best if all risk areas (*small, medium, and large*) can be identified and eliminated at the development stage. In concrete terms, however, this means that developers must strive to eliminate all potential risks during specification and design phases; see Fig. 19.3.

Of course, this is much easier said than done because customers are usually interested in innovative product development solutions; who wants a museum product? At the same time, however, competition is forcing companies to reduce schedules and costs while maintaining or even improving quality requirements. There is no question that this is a balancing act between innovation and risk; see subsection 19.1.2. But this is precisely why risk mitigation that begins in the specification process and/or in the development stage is most effective. To avoid project disasters, increased risk should not be accepted in favor of innovation. Therefore, a solid and fact-based risk identification is absolutely necessary to modify the project requirements, if necessary, together with the customer.

Pausenberger and Nassauer describe four management measures to avoid commercial risks: [27].

- Risk avoidance: *no high-risk contracts should be concluded,* e.g., *contracts with partners from politically unstable countries.*
- Risk reduction: *by reducing the probability of occurrence and/or limiting the potential for loss.*
- Risk transfer: *risk sharing with partners,* e.g., *the customer, subcontractors, or through insurance.*
- Risk limitation: *limit business risk through compensatory measures in the case of foreign businesses, for example, by taking out loans in the partner country at fixed interest rates at an early stage.*

Weber and Liekweg describe the different influencing factors and point out that benefits and risks are very closely intertwined because an increased or improved business advantage (*nice to have*) often also leads to greater risks [28]. They described six influencing factors:

- General external factors that affect all companies, such as GDP, climate change, natural disasters, new laws and regulations, and leaps in technology.
- Market factors change the competitive position of companies when marketing their products. Examples include product changes, supplier concentration, competitive changes, etc.
- Productivity factors are the best-known factors in the value creation process of companies, e.g., production interruptions due to accidents or strikes, unexpected research and development successes, transport damage, or the start of new product lines.
- Financial factors are often the most common factors that lead to risks but sometimes also to benefits, e.g., changes in exchange rates, profits, and bank interest rates.
- Organizational factors (*changes*) can lead to serious problems (*including fraud and deception*) at various levels of management, and thus add to risks.
- Legal factors play an important role in business life. Risks that arise are often a consequence of non-compliance with contractually stipulated conditions, which may lead to the payment of contractual penalties and liability compensation.

Table 19.5 Summary table for financial risk tracking (example)

Risk Nr.	Risk Title	Risk Reserve (€) Estimated Cost	Actual Cost	Open
R.01		35,000	–	
R.02		–	20,000	
R.03		–	–	???
R.04		10,000		
R.05			5000	
Total risk costs		45,000	25,000	
Reserve—Budget increase		70000.00 €		???

Rohrschneider [29], like Harrant and Hemmrich [14], points out that in addition to the recognition of risks, there is also the recognition of opportunities. He then states: "*Whoever wants to manage risks and opportunities in the project must first develop as precise an idea as possible of what the risks and opportunities are in the first place.*" In everyday language, the terms do not always have the same meaning. Therefore, a clarification of terms is necessary; see subsection 19.1.1. The identification of risks is particularly difficult because they are usually not obvious and are not recognized at all or too late. Rohrschneider suggests asking open-ended questions in defined search fields, such as: [30].

- Are there dangerous planning errors?
- Were binding commitments made without having accurate information?
- Were unrealistic specifications accepted?
- Were risks taken that were beyond one's control?
- Was overtime scheduled at the start of the project?

19.5.2 Financial Risk Assessment

Appropriate financial provisions must be made for risks that cannot be reduced or eliminated completely during the course of the project. Therefore, adequate technical, scheduling, and financial project reserves (*margins*) must always be planned for. In certain cases, insurance policies must be taken out to cover risks. Risk-conscious project managers will therefore ensure as far as possible that appropriate project reserves or insurance policies are planned. As a rule, any residual risks must be accepted. However, these risks should be monitored with particular attention by management. The financial provision can be well illustrated in a summary table using an example; see Table 19.5.

First, all identified risks, i.e., R.01 to R.05, are to be recorded. Then the estimated financial damage of the identified risks R.01 and R.04 is to be determined at 45,000.00 €. The already incurred additional costs of the risk areas R.02 and R.05 amount to 25,000.00 €. The project budget is to be increased in accordance with the example by 70,000.00 €, whereby it concerns estimated costs in the first case and

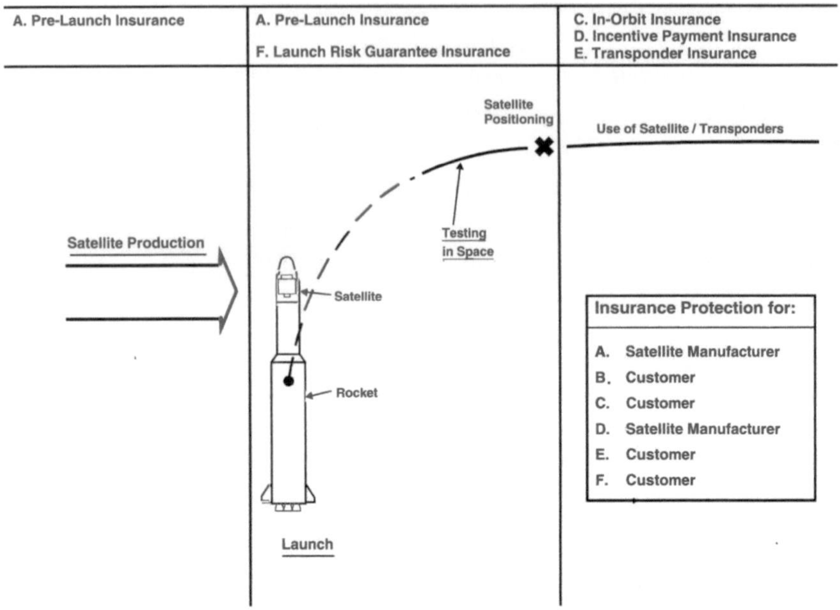

Fig. 19.7 Satellite insurance models

the final value can still change. The cost calculation for R.03 is still open and therefore an indication for the project management to estimate the cost impact.

Appropriate risk insurance is taken out for the still high-risk satellite launches and subsequent satellite operations since most commercially used launch vehicles only achieve a reliability of around 95 percent. In manned spaceflight, reliability reaches nearly 100 percent, but this results in significantly higher costs. In commercial spaceflight, the remaining residual risk is usually insured. Figure 19.7 shows an example of the various insurance packages. A statement by the Munich Reinsurance Company says: "*The use of space flight for commercial purposes has generated an increasing demand for tailor-made insurance solutions*" [31].

19.6 Risk Monitoring, Documentation, and Reporting

19.6.1 Continuous Risk Monitoring

The project manager is responsible for continuous risk monitoring. This means that project management must keep the topic of "*risk monitoring*" as a fixed agenda item. The existing risk potential of the project and the measures initiated in the meantime to eliminate or mitigate the risk must be recorded in the risk report, which must be updated regularly; see Sect. 11.2. In this context, checklists are a useful aid to project work, e.g., for recording all risks and their causes. Meier has reproduced a risk checklist in his book "Risk Management in Technology Companies" [32], which is a useful aid in accordance with the risk register (*see* Table 19.4):

- Risk designation
- Origin of the risk
- Risk target (*trend*)
- Risk impact
- Risk classification
- Risk networking with other risks

Greenfield, NASA, emphasizes that risk management is to be understood as a continuous process and that the project manager is responsible for the integral and coherent implementation of risk management [12]. In this context, reference is also made to ESA's comprehensive set of rules on risk management [21] and to Romeike's "Lexikon Risiko-Management" [33]. The NASA—Risk Guideline for Managers and Practitioners is also recommended [34].

19.6.2 Risk Documentation

It is the task of project management to document the risk management process. This includes the risk register in Table 19.4 and the overview table for financial risk tracking in Table 19.5. The current risk status is an integral part of the monthly project report; see Sect. 11.2. For joint projects with several partners or subcontractors and suppliers, the documentation should be adapted accordingly. The monthly risk status report should normally include the following information:

- The risk registers are valid at the current time.
- List of all open risks and the necessary planned and completed measures for risk elimination or mitigation, sorted by
  - Large
  - Medium
  - Small
- Reference to all relevant protocols on the subject of risk
- Risk avoidance decisions are made.

19.6.3 Risk Reporting

Based on the updated risk documentation, the respective higher organizational unit, e.g., the company management and/or the customer, must be informed regularly, e.g., monthly, about the current status of the project risk. This should preferably be done at the regular project status meetings; see Sect. 11.2.

Literature

1. Gutberlet, Bernd Igmar: Grandios Gescheitert; Bastei Lübbe, 2012, p. 37
2. Source 1, p. 55
3. Source 1, p. 58
4. Lewis James P.: The Project Managers Desk Reference, McGraw Hill, Boston, 1999, p. 298
5. Source 4, p. 300ff
6. Gigerenzer, Gerd: Risiko, C. Bertelsmann Verlag, 2013, p. 58ff.
7. Source 6, p. 41
8. Source 6, p. 46f
9. Source 6, p. 147
10. Source 6, p. 293
11. Steeger, Oliver: Risikomanagement - das "*Stiefkind*" in der Projektarbeit, Projektmanagement aktuell, 14. Jahrgang 2003 2 Quartal, S. 6.
12. Greenfield, Michael A.: Risk Management Tools, NASA Langley Research Center, May 2, 2000, p. 3.
13. Madauss, Bernd-J.: Risk Evaluation and Mitigation, ISU Lecture at the Royal Army of Oman, 12/2010
14. Harrent, Horst and Hemmrich, Angela: Risikomanagement in Projekten, Carl Hanser Verlag, Munich, 2004; p. 15
15. Source 14, p. 37
16. Aucoin, B. Michael: From Engineer to manager mastering the transition, Artech House, Inc, Bos ton - London, 2002, p. 147
17. Michaels, Jack V.: Technical Risk Management, Prentice Hall, 1995.
18. Klein, Gerald A.: Risk Management, in Larson Wiley J. (managing editor), Applied Project Management for Space Systems, McGraw Hill Companies, 2008, p. 471ff.
19. Augustin, Norman: Augustine's Laws and Major System Development Programs, As tronautics and Aeronautics, April 1980, p. 36.
20. Batson, R.G.: Program Risk Analysis Handbook, NASA TM-100311, 1987, Appendix B.
21. ESA-ESTEC: ECSS-M-ST-80C, Risk Management, 31 July 2008
22. Barker, Stephen, Cole, Rob: Brilliant Project Management, Pearson Printice Hall, Harlow, 2009, p. 30ff.
23. Madauss, Bernd-J.: Risk Assessment of the Customer - Evaluation of Suppliers and Offers, PMT, Bad Aibling, 1998
24. FAA COTS Risk Mitigation Guide, Rev 3.1, 2002, p- 3
25. Madauss, Bernd-J., Mennel, Volker: Project Controlling and Risk Management, Workshop, September 23, 1994.
26. Duffy, James B.: Identified Risk Factors, TeraSim Associates, 2002.
27. Pausenberger and Nassauer: Governing the Corporate Risk Management Function, in Frenkel Michael, Hommel Ulrich, Rudolf Markus, Risk Management, Springer Verlag Berlin Heildelberg New York, 2000, p. 273.
28. Weber, Jürgen and Liekweg, Arnim: Statutory Regulation of the Risk-Management Function in Germany, in Q.17, p. 285.
29. Rohrschneider, Uwe: Risikomanagement in Projekten, Haufe Verlag, Munich, 2006, p. 21
30. Source 29, p. 32
31. Munich Reinsurance Company: Space Flight and Insurance, 2006, p. 6
32. Meier, Peter: Risikomanagement in Technologieunternehmen, Wiley-VCH Verlag Weinheim, 2005, p. 30.
33. Romeike, Frank, Lexikon Risiko-Management, Wiley-VCH Verlag Köln, 2004.
34. NASA/SP-2011-3421, Probabilistic Risk Assessment Procedures Guide for NASA Managers and Practitioners, 2nd Ed, December 2012.

Project Management in a Global Environment

20

Contents

B. -J. Madauss, *Project Management*,
https://doi.org/10.1007/978-3-662-69057-4_20

Abstract

More and more projects in the 21st century are being carried out in a global environment. This has to do with the globalization of markets and the worldwide cooperation of authorities and companies. One corollary is the rise of global procurement and project cooperation. International project partners must therefore move closer together and share their different experiences in the interest of efficient project execution. This requires understanding each other and open communication.

Globalization is understood as a change in the world economy that leads to more transnational transactions. Globalization is also taking place in the cultural and social spheres. For example, many products are nowadays outsourced to countries such as China or India, which can produce them under more favorable financial conditions while maintaining quality standards. These include products such as textiles and simple toys, but increasingly also sophisticated goods. Globalization is a process through which markets and production in different countries become increasingly interdependent thanks to the dynamics of trade in goods and services and the movement of capital and technology. According to Müller, "*Beyond political opinion, globalization can be described as the spatio-temporal extension of social practices across state borders, the emergence of transnational institutions, and diffusion of cultural patterns*" [1].

Globalization can be felt in many areas through worldwide political, economic, and cultural networking. Globalization takes place at all levels. Between states, organizations, companies, societies, and individuals. Globalization is made possible above all by technical progress in worldwide communication and the well-functioning transport possibilities on the highways, by rail, by ship, and by air. This also affects the intensity of personal contact between members of different cultures. "*Globalization also causes cultural interdependencies and mutual influences to occur,*" Preyer writes [2].

Globalization is a major challenge for public authorities and industry. Competition, language, and culture are particularly worth mentioning in this context. Project management (*PM*) for the development and procurement of hi-tech products is also strongly affected by this. In the early 1960s, U.S. government agencies and aerospace companies developed and successfully used the management methods that are now commonplace; see Sect. 9.1. Building on this, a few industries already operating in the global environment, such as the European aerospace industry, adopted internationally recognized management standards in the late 1960s. They also successfully implemented new and efficient methods of project and procurement management.

For a long time, the aerospace industries in the USA and Europe were pioneers in the application of internationally recognized project management and procurement methods and successfully transferred them to other industries. In the context of advancing globalization, internationally proven project management methods and procedures can no longer be dispensed with. Special features of international project management are described in Sect. 16.1.

An intensive cooperation between American and European engineers and managers of aerospace was developed. To be successful, engineers and managers needed simple but efficient methods and procedures to master complex projects with technical and economic success. From a global point of view, the development and procurement management of the international space station ISS with the participation of Europe, Canada, Japan, Russia, and the USA, in which the management disciplines described above were applied, was a prime example of global cooperation; see Sect. 16.5:

Complex aerospace projects can be described as the hotbed of modern project management. In the meantime, proven PM methods are also successfully applied to non-space projects, such as in facility engineering and the automotive industry. In most cases, PM know-how cannot be transferred one-to-one but must be intelligently adapted. In principle, this should also be taken into account for the transfer to small projects, but in a reduced form, because one should not shoot at sparrows with cannons; see Sect. 9.5.

20.1 The Role of Project Management in Meeting the Future Challenges of the Global Marketplace

20.1.1 The Common Thread of Project Management

The project management methods used worldwide were developed and tested in the 1960s and 1970s, as mentioned elsewhere in this book. The 375 manual series of the USAFCM, which was available in the mid-1960s as a complete work and contained all aspects of modern project management, represented an important milestone in PM development [3]. Components of the 375 series included the management disciplines of systems engineering, procurement, production, configuration control, project monitoring (*schedules and costs*), testing, and quality assurance. The basic structure developed at that time is still a common thread running through aerospace project management procedures for complex systems. Initially, only major military and civilian aerospace and naval projects in the USA and later in Europe were affected. In the meantime, non-aerospace companies also benefit from it, and it is no longer the sole domain of large-scale projects but also of medium and small projects.

Project management knowledge is also disseminated in the global environment by international and national professional associations, such as the IPMA (*International Project Management Association*) based in Zurich. In the German-speaking world, there are national societies in Germany and Austria, such as the Gesellschaft für Projektmanagement (*GPM*). PM knowledge is also disseminated through training courses and publications. The PMI (*Project Management Institute*), founded in Philadelphia, USA, in 1968, has also contributed to the worldwide dissemination of PM knowledge with the PMBOK Guide (*Project Management Body of Knowledge*) and the training courses based on it; see Sect. 15.2.

Global project cooperation results in a worldwide distribution of proven PM methods. In this process, adaptation to the situation (*tailoring*) is always very important. Worldwide cooperation using proven PM methods also leads to the standardization of PM processes and, thus, an improvement in management efficiency. The ECSS standards of ESA are an important building block for this; see Chaps. 7, 8, and 9.

20.1.2 Globalization Requirements for Management

Institutions and companies operating in the global environment require, for example, efficient project management methodologies to design, build, and construct aircraft, facilities, transportation systems, communication networks, or pipelines. PM and procurement methods developed by aerospace organizations, e.g., systems engineering, configuration control, risk management, etc., have been fully or partially adopted by other industries.

The application of up-to-date PM and procurement strategies is a matter of course for European companies with project responsibility in order to be able to participate successfully in the global marketplace. Public authorities and companies that do not have a functioning project management system often resort to consulting firms when necessary. In the long term, however, they should create a PM position staffed with their own employees so that their own project managers can successfully handle the global project tasks assigned to them. It is essential to take appropriate organizational action. Project managers must be given the necessary powers of management and procurement. They must be familiarized with standard procedures, methods, and tools through appropriate training. This measure is particularly important for global projects.

It is extremely important for the European industry to be able to hold its own against its international competitors and, whenever possible, to take on leadership roles in major projects. This trend has become particularly clear in recent years with the examples of commercial aerospace and telecommunications.

20.2 Global Competition and its Effects

20.2.1 Examples of Globalization

Since the beginning of the 21st century, globalization has progressed further. One example is the worldwide networking of communications facilities between continents through cable and satellite links. These include, above all, fiber-optic links that enable the transfer of large volumes of data over long distances at the speed of light.

Global projects also include pipelines for gas and oil, international passenger and freight transport by road, rail, ship, and air, and worldwide production facilities and goods deliveries. An excellent example of globalization is the expansion of the Panama Canal, which opened to international shipping on June 26, 2016. The project was realized with the involvement of companies from many countries in a construction period of 9 years and allows cargo ships of the "*Post-Panama Class*" to navigate the canal with three times the container capacity; see Fig. 20.1.

Aerospace projects are almost always global projects. Many international companies are often involved in the development and construction of aircraft, e.g., Airbus or Boeing and their subcontractors and suppliers. In the planning phase, manufacturers must first obtain the appropriate approvals from the relevant aviation authorities, such as the Federal Aviation Office in Braunschweig, Germany, for example, or the U.S. Federal Aviation Administration (*FAA*), which is headquartered in Washington, DC. However, during the planning phase, the companies must deal with future customers, i.e., the airlines, in order to include their requirements. Another problem is financing the high development costs. Says Tathjen: "*The high development costs and the associated risks are forcing all manufacturers of aircraft, engines, and equipment, as well as airlines, to enter into cooperative ventures:*

Fig. 20.1 Expanded Panama Canal (Source: Creative Commons)

domestic, European and global. The buzzword "globalization" applies to aviation as it does to few other sectors of the economy" [4]. Figure 20.2 shows the Airbus A380 wide-body aircraft developed and produced by Airbus in cooperation with partner companies from four European countries. Over a period of almost 10 years (*October 2007 to March 2017*), aircraft were delivered to international customers.

The situation is similar in the space sector. When planning, there is no getting around official approvals, e.g., by the ITU (*International Telecommunication Union*), a special organization of the United Nations for information and telecommunications technologies, located in Geneva. When a satellite is launched, several countries, institutions, and companies are always involved, because, in addition to the launch site, many ground stations around the world are involved in the global network control system to track the flight path. In addition to the satellite constellations flying at low altitudes, *the* so-called LEOs (*Low Earth Orbit*), which have a flight path altitude of a few 100 km above the earth's surface, there are also other flight paths, e.g., the satellite systems with a geostationary orbit at an altitude of 36,000 km, the so-called GEOs (*Geostationary Earth Orbit*). Examples of global satellite systems are the satellites for message transmission between the continents of the company INTELSAT [5] and the Astra satellite fleet of the company SES Luxembourg [6]. Organizations and companies from different countries are involved in the realization of satellite systems. The initiative for the realization of global commercially operated satellites increasingly comes from private companies. The example of globalization described here on the basis of globally operating satellite

Overflight of the A30-800 in Airbus Factory Livery
Source: Creative Commons

- Nations participating in Development and Manufacturing: France, Germany, UK, and Spain
- First Flight on 27 April 2006
- Entered Service on 26 October 2007
- World-wide Delivery of 200 Aircraft to 13 Customers

Quelle: Statistica.com Stand Maerz 2017

Fig. 20.2 Globalization using the example of the A380 wide-body aircraft (Source: Creative Commons)

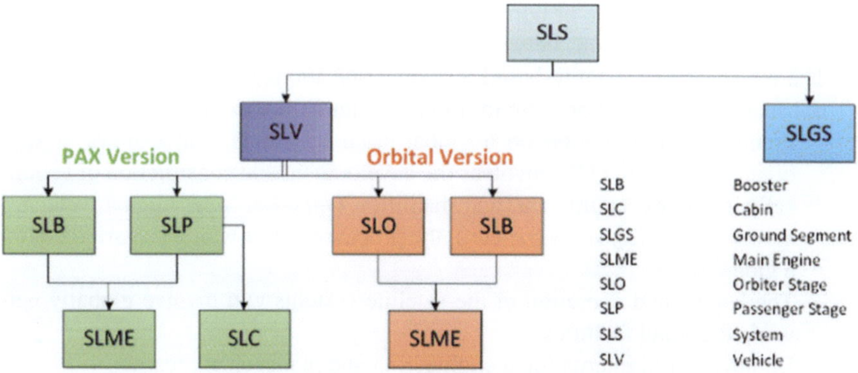

Fig. 20.3 SpaceLiner—system configuration (source: DLR SART)

systems can, in principle, be transferred to other branches of industry, such as facility engineering, because technical and economic networking is a characteristic of increasing globalization.

The German Aerospace Center (*DLR*) is investigating a concept for a hypersonic flight vehicle called *SpaceLiner,* based on rocket and aircraft technology, in order to be able to fly from Europe to Australia in 90 minutes, for example; see subsection 7.5.5 [7]. Figure 20.3 shows the system configuration of the two-stage SpaceLiner project, which consists of several elements. It is a future project with a time horizon for development, including flight testing, of about 20 years and is inspired by the technologies of rocket and aircraft developments by aerospace companies. SpaceLiner is undoubtedly a global project that will involve international and national authorities, industrial companies, and investors; see subsection 7.5.5.

Small countries, such as Luxembourg and the United Arab Emirates (*UAE*), are also reaching for the stars. The Luxembourg government announced a collaboration with the U.S. to extract special minerals from asteroids. As a first important step, Luxembourg wants to publish a set of regulations regarding exploitation rights. U.S. President Barack Obama signed the *Commercial Space Launch Competitiveness Act,* which gives all private U.S. companies mining rights on uninhabited asteroids" [8]. Globalization continues as Abu Dhabi's newly created Space Agency plans collaborations with the Japan Aerospace Exploration Agency (*JAXA*) and NASA to participate in a future Mars mission.

20.2.2 Globalization in the Telecommunications Sector

The situation can be well described using the example of satellite-based global telecommunications, which is characterized by the features described below:

1. Satellite-based telecommunications projects are usually global in nature from the outset because isolated solutions are usually not profitable.

2. The increasing opening of borders as well as the growing expansion of international economic relations between nations promote globalization.
3. The realization of satellite-based telecommunications projects usually requires transnational cooperation from the very beginning. For example:
   - Frequency coordination on a global scale by the ITU in Geneva, a sub-organization of the UN, involves the development and construction of technically complex equipment and facilities (*satellites and launch vehicles, network facilities, software,* etc.), the procurement of which is carried out on a global scale.
   - The launch and operation of the satellite systems will involve globally networked ground facilities.
   - Deployment of international engineering and management teams
   - International financing and insurance of satellite systems.

Anyone who is abroad on business or pleasure and then contacts family members or colleagues by telephone gets a good impression of the global networking of telephone systems. Fiber-optic connections are laid all over the world, and satellite-based mobile communications links make it possible to network all five continents of the world globally. Future globalization projects require project management to promote communication and close cooperation between specialist disciplines, such as marketing and financial management, in order to survive against technology-driven young competitors.

20.2.3 Selection of System Suppliers

The selection of system suppliers, e.g., industrial facilities, aircraft, etc., by the operating companies (*customers*) takes place within the framework of international competitive conditions. The customer usually selects the system supplier on the basis of appropriate criteria, such as technical performance characteristics, delivery dates, price, and contract conditions. The proposed management concept and the qualifications of the subcontractors and suppliers are usually also included in the evaluation; see Chap. 14. However, national interests also continue to play a decisive role. In some cases, global cooperation can also involve compensation transactions, e.g., when involving countries that have a financing problem on the one hand but possess high-performance products on the other.

20.2.4 Global Competition and Procurement

Globalization and the accompanying deregulation measures are leading to global competition in all areas of the value chain; see Chap. 21. As a result, all the companies involved are forced to adapt to global competitive conditions. Product developers, producers, and operators, as part of the value chain, must strive to maintain or even strengthen their global competitiveness. The products offered must therefore fully meet customer requirements, both technically and economically. Provided there are no political or other constitutional constraints, international customers make product selection dependent on the following 10 criteria, for example:

1. Does the product fit into the customer's corporate strategy?
2. Has the product already been tested and successfully used in practice?
3. Does the product meet globally recognized quality standards?
4. Does the manufacturer offer an acceptable product warranty?
5. Is a maintenance concept offered by the manufacturer?
6. Are the delivery conditions, delivery dates, and *Incoterms* acceptable?
7. Does the financial offer, price, and payment schedule have competitive advantages?
8. Does the manufacturer offer implementation support?
9. Is operational support provided by the manufacturer?
10. Does the manufacturer offer professional project management?

In global competition, where one has to prove oneself on the international stage, the previously mentioned criteria play a decisive role; see Chap. 14. Experience and trust are particularly important in the procurement process. Experience (*heritage*) can usually be demonstrated, but trust must first be gained, and it may take some time before a basis of business trust is established. This also applies to global business relationships, where the cultural aspect should also be taken into account; see Sect. 20.7.

The European aerospace industry has passed the challenge of global competition in the fields of civil and military aviation, helicopters, launchers, and satellites with flying colors. Founded in 1965, "*Airbus*" delivered a total of 9754 aircraft to nearly 400 customers in 42 years, from May 1974 to May 2016, and is also a global success in the helicopter sector. The Ariane 5 launcher, with a series of 115 successful launches (as of July 2023), which can carry satellites with a launch mass of 6000 kg into orbit, is a European success story.

Space companies have gained their competitiveness in part through national and international development projects. Contracts awarded by ESA, DLR (*Germany*), or CNES *(France*), for example, led to the acquisition of know-how in the areas of technology, procurement, and management and thus to an increase in competitive opportunities in the global environment. The development, construction, and utilization of space products for satellite-based telecommunications were carried out according to the rules of the global market. Development support from national and international governmental organizations has enabled the European space industry to succeed globally. This was important in order to be able to compete with globally competing companies technologically and at system prices in line with the market.

After significant advances in communications and satellite technology, private operating companies were established, e.g., INTELSAT in 1965 [5], to improve transatlantic telephone links, and SES Astra in 1985 [6] as the first commercial European television satellite operating company; see subsection 1.9.4. Both

Fig. 20.4 SES antenna park in Luxembourg (Source: SES)

companies have a large number of satellites located globally on almost all continents. Figure 20.4 shows the SES antenna park for global satellite monitoring.

In this context, however, the space developments in China, India, Japan, and Russia must not go unmentioned. There is worldwide cooperation with them. European operators, such as SES, have used the powerful Russian "*Proton*" launch vehicle several times in the past. China, too, has a powerful and reliable satellite launch vehicle, the heavy launcher Long March 3B. Globally operating company and project cooperation exists in many areas, e.g., also in the automotive and electrical industries.

20.2.5 Global Satellite-Based Communication Systems

The liberalization of telecommunications in conjunction with new technical developments, e.g., the use of very high frequencies and data compression methods, set the stage for digital global television. The attempt to create a global communications network with a constellation of 66 satellites in LEO was a failure for Motorola (*USA*). Terrestrial networks were powerful competitors. After bankruptcy proceedings, iridium is now used primarily for military purposes.

The planned projects of the international operating companies for telephone, TV, and data transmission services, which are mostly of a "*direct-to-home*" nature, are almost always of a global nature. Global telecommunications, however, presupposes that both standards and technologies are coordinated and harmonized jointly, i.e., worldwide, for example, by the ITU, e.g., by a common definition of interfaces.

The global satellite constellation project O3b (*the other three billion*) is intended to provide better service in the future to sections of the global population that do not have adequate communications links. In June 2013, "*O3b Networks*" sent the first four satellites into orbit, followed by eight more in 2014. The O3b Networks satellites deliver mobile communications and internet at fiber-optic speeds in developing and emerging countries in Asia, Africa, South America, the Middle and Far East, and the Pacific Rim. The reinsurance company "*Munich Re*" developed a customized system of coverage together with the operator and the broker to cover the failure of one or more satellites in the constellation, thus providing the necessary planning and decision security for operators and investors. Since 2016, O3b has been wholly owned by Luxembourg-based SES. Commenting on this, Karim Sabbagh, SES President and CEO, said, "*The move to take control of O3b is a game-changing acquisition and a major step in the execution of SES's differentiated strategy and complements SES's growth strategy.*" [9] The global environment is constantly undergoing changes that must be addressed in order to stay ahead of the competition. In an interview with Mark Holmes of Via Satellite Digital, Sabbagh makes a landmark statement, "*I am a believer of constant evolution*" [10]. Fig. 20.5 shows the SES and O3b satellites in GEO and MEO orbits.

There are a large number of national and international satellite operators worldwide that provide point-to-point communications links or television transmissions in different frequency bands. The following are four satellite operators operating worldwide with very high global coverage:

- INTELSAT, Washington
- SES Astra, Luxembourg
- EUTELSAT, Paris
- INMARSAT, London

Global satellite networks connect the continents and enable fast communication. HTS (*High Throughput Satellite*) is the name of a new global communication system consisting of satellites in different altitude orbits; LEO, MEO, and GEO; see Fig. 20.6.

20.3 Global Cooperation and Procurement Concepts with European Participation and Leadership

20.3.1 Global System Planning

Global telecommunications requirements presuppose a worldwide network. For reasons of economic efficiency, telecommunications networks should be planned and implemented from a global perspective. Terrestrial cable networks and communications satellites are available for this purpose. In the early phases of projects, technical-economic comparative analyses (*trade-offs*) must be used to evaluate how the system should be designed.

- Cable network
- Satellite network
- Interconnected Network.

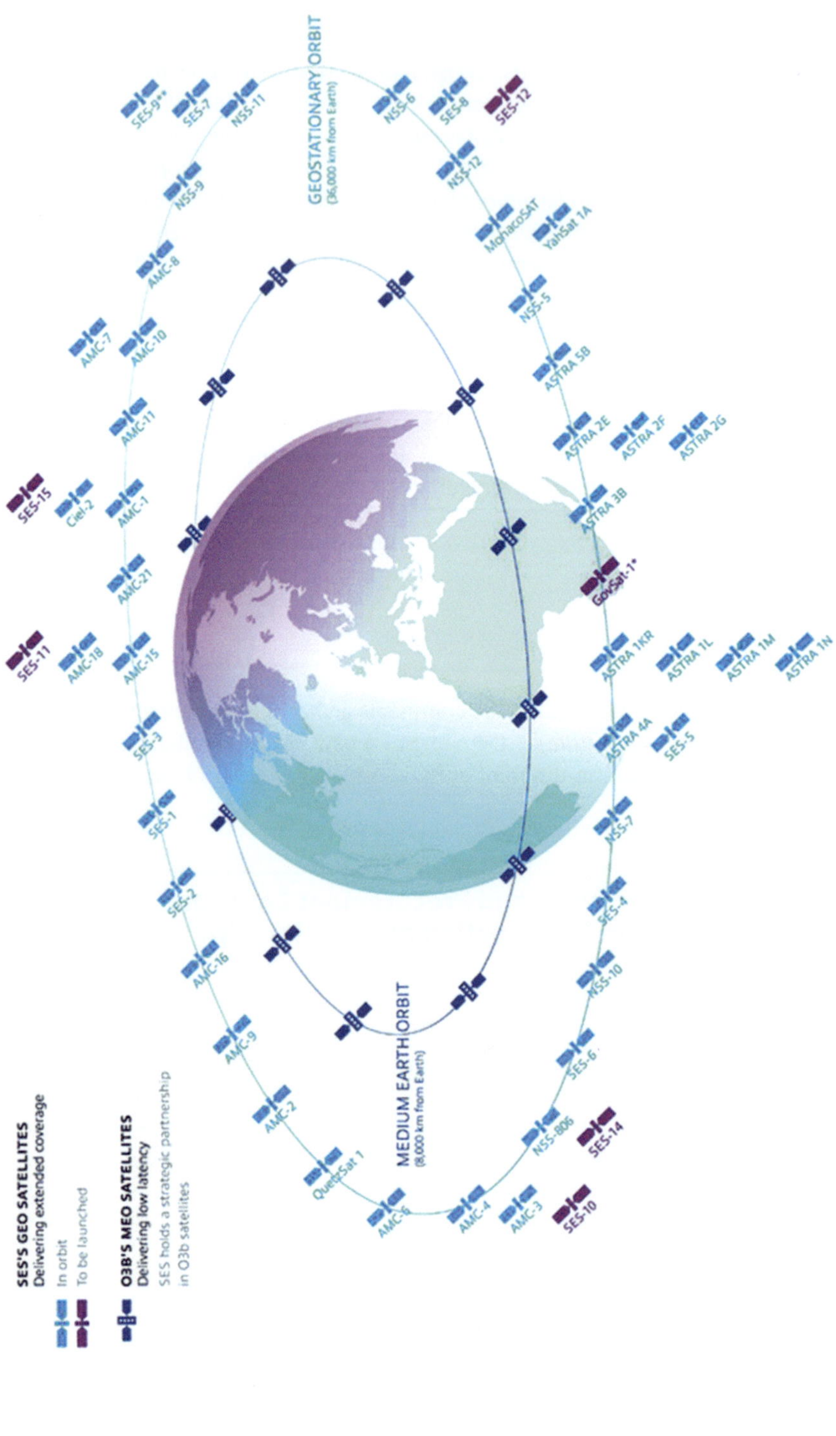

Fig. 20.5 SES and O3b satellites in GEO* and MEO** orbits (Source: SES). **GEO*, Geostationary Satellite Orbit at 36,000 km; ***MEO*, Medium Satellite Orbit at 8000 km

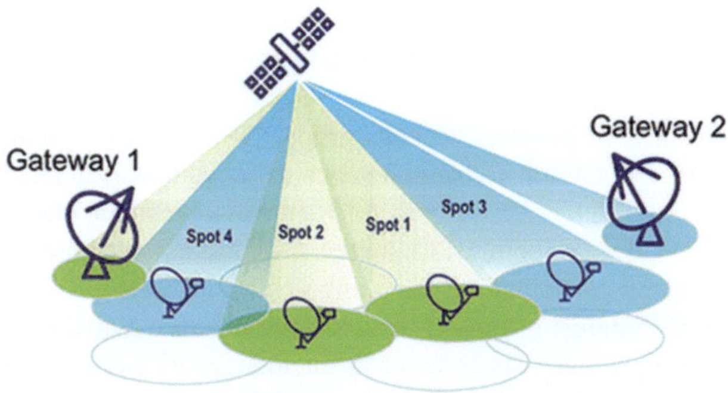

Fig. 20.6 HTS satellite constellation (Source: Creative Commons)

Compared to the familiar cable connections, including the very high-capacity optical line systems, there are one or two advantages and disadvantages, but the most essential advantage of satellite connections is their flexibility. Once plans have been made, they can also be changed again quickly.

Planning for new commercial satellite systems requires a globally operating project organization and industry participation. In many cases, the countries involved, which have sovereign responsibility for determining the operating frequency and position of the satellites, make participation by domestic industry a condition. This raises the question of technical and commercial competence, without which participation cannot take place or can only take place in the area of less interesting work.

20.3.2 Competence to Assume Management Tasks

However, if one wants to take on corresponding management tasks, technological competence, management ability, and economic efficiency play a decisive role in the awarding of contracts for demanding investments.

Most commercial space projects still originate in the U.S.; SpaceX is an example. Other countries, including those in Europe, are mostly involved in global projects but have rarely taken a major leadership role in the past. This is true for the procurement process (satellite, launch vehicle, ground facility, network) as well as for operations and services. Only in recent years has the situation changed somewhat in favor of European industry in the aircraft and aerospace sectors. However, competition has now intensified further, and American competitors are trying to win back lost customers by offering better delivery conditions.

20.3.3 Tight Deadlines and Adherence to Schedules Are a Decisive Competitive Advantage

The competitiveness of projects depends primarily on product quality, reliability, the price offered, and demonstrable adherence to schedules. Provided there are no overriding political influences, these factors determine whether international competition is successful. In principle, this applies to all industries, including aerospace.

Greater European participation, assuming greater leadership roles, is definitely to be sought in the interest of greater involvement in future global communications projects. The European aerospace industry, including the supplier industry, has in principle the necessary know-how in technology and management. In comparison with competitors, European suppliers often have the following disadvantages from the customer's point of view:

- Risk aversion
- High cost (*Euro currency*)
- Longer schedules in comparison
- Lack of adherence to deadlines (*past experience*)
- More cumbersome management (*lack of flexibility*)
- Weak project management
- Lack of international experience
- Lack of cultural knowledge (*customer country*).

In recent years, the situation has improved as a result of better project management, i.e., practical and proven PM methods for handling international projects and consistent consideration of cultural aspects and employee motivation, but meeting deadline commitments remains a serious problem. This is also true for European aerospace.

To win the competition in the procurement of high-tech products, it is important to pay attention to the following points:

- The product offered must fully meet the customer's requirements. This means that when analyzing the invitation to tender, great care must be taken to ensure that the required performance targets are met as fully as possible. However, over fulfillment, a nice-to-have solution, should be avoided because the customer is usually not willing to pay a higher price; see Sect. 14.3.
- The customer must be offered an efficient management structure based on short decision-making paths. It is best if a competent and financially strong company takes on the task of general contractor (GC) and assigns partial tasks to subcontractors. If a consortium is offered, appropriate structures and rules similar to the general contractor concept of a sole proprietorship should be proposed; see Sect. 5.3.
- Schedules are similar to the price offer; if all other factors are right, the shortest duration has the greatest chance of winning, provided that credibility is high. The customer is usually interested in the proposed schedule standing up to a credibil-

ity check. For example, are there past comparison situations, is the scheduling logic consistent, etc.? These factors can be included in the evaluation of the offer; see Sect. 14.5.

- The offered price and payment schedule must be competitive and cover possible risks as well. Deciding on the price is probably one of the most difficult points in bidding because, on the one hand, you want to maximize profits, and on the other hand, you do not want to jeopardize your chances of winning. Personal experience, cost databases, and risk appetite play a significant role in this context; see Chap. 19.
- The creation of offerings that are outstanding in terms of quality and content is a fundamental prerequisite for international competition; see Chap. 14. This applies in particular to the following points:
  (a) A thorough analysis of the tender documents; precise knowledge of what the customer requires and/or wants.
  (b) A thorough elaboration of the offer, highlighting all the competitive technology and management advantages.
  (c) The presentation of experience and skills.
  (d) A competitive and credible schedule offer.
  (e) A competitive price quote includes a payment schedule.
  (f) Presentation of a convincing management concept.
  (g) Description of the intended key personnel and their experience.
  (h) A qualitatively captivating visual representation, because "*ugliness sells badly*" [Raymond Loewi, 1953].

20.4 Project Management in Europe

20.4.1 PM Implementation Errors

In European authorities and industry, the topic of project management is becoming increasingly important, and many companies have it already firmly established and can no longer imagine doing without it. This is the case, for example, in power plant projects, plant engineering, the automotive industry, architecture projects, software houses, and aerospace. In addition, medium-sized companies also benefit from the introduction of PM.

But "*all that glitters is not gold*" says a well-known proverb. PM implementation occasionally encounters obstacles from European authorities and industry. Often, serious mistakes are still made, which can be typified as follows:

- PM implementation is only seemingly accomplished, but no effective methods are implemented.
- The project organization is complicated and does not assign responsibility precisely.
- Project managers have insufficient authority.
- Targets (specifications) and tasks (functional specifications) are insufficiently formulated.

- Interfaces are not defined and specified systematically enough.
- Configuration and change control are inadequately performed.
- Project planning degenerates into a stepchild, and systematic project meetings and progress reviews are poorly conducted.
- Equally damaging can be overly detailed and confusing planning that is not controllable.
- The cost estimate made is not sufficiently substantiated and does not sufficiently incorporate empirical values from the past.
- There is a lack of systematic cost monitoring and comparison to project progress, e.g., Earned Value Management *(EVM)*; see subsection 9.4.6.

20.4.2 Management Requirements for the European Industry

The demands on European industry, which like many other international industrial companies has to compete globally, are very high today. Therefore, it is necessary that the common management procedures and the associated software concepts be applied efficiently. Therefore, no project office should be without the following PM systems:

- Phasing concepts: life cycle, project objectives, impact studies, life cycle costs, economic feasibility studies, decision making; see Chap. 4.
- Systems engineering procedures: specification, interface definition, reviews, verification, acceptances; see Chap. 7.
- Quality assurance procedures: quality, reliability, maintainability, safety; see Chap. 8.
- Project planning and monitoring procedures: project structure (*WBS*), schedule and milestones, terms of reference, project planning, status monitoring, action control, cost planning and monitoring, schedule/cost analysis; see Chap. 9.
- Risk analysis procedures: risk identification, risk assessment, risk hedging, and risk mitigation measures; see Chap. 19.

European companies have gained good experience with the aforementioned management procedures, which have been tested over the last four decades. It has also been learned that the depth of implementation and the particular situation are very important, because both overly detailed and too superficial planning and monitoring are problematic for project execution.

20.4.3 The Right Application Is What Counts

The correct use of the known PM procedures requires a certain level of experience because ignorance often leads to the incorrect application of the existing PM methods that have been tested in practice.

For this reason, however, managers must also have the corresponding experience. The wealth of experience of PM employees and managers is the most important asset of the companies concerned. Often, planning is not thorough and systematic enough, and monitoring is not consistent enough. Both are fundamental mistakes in project management. It is important to hold regular internal status meetings, e.g., weekly, fortnightly, or monthly. The status meeting should not exceed 1 hour if possible; see Sect. 11.3. Enforcing this is not always easy in a global environment. Of course, this requires good meeting preparation and management. Personal contact plays a particularly important role because, even in today's so-called information age, it is easy to have first-class electronic media equipment, but lack interpersonal contacts. Successful project management, however, depends on good interpersonal contacts among project staff, which is particularly important when global partners are working together.

But there is also another almost opposite problem, namely the excessive use of PM software and procedures. It must not happen that inexperienced PM employees are confronted with the full force of the PM method arsenal when executing a small project in order to shoot with cannons at sparrows. Often, inexperienced managers make such mistakes out of insecurity. But also customers, and here especially government institutions with a bureaucratic preponderance, tend to require "*too much of a good thing*" for a variety of reasons. This results in highly skewed PM requirements (*high level of detail*), which leads to cost-intensive requirements and compulsory exercises for suppliers without increasing the success of the project.

At this point, reference is made to the responsibility of a central project office (*PMO*), which has the task of training and supervising newcomers to the project business; see Chaps. 15 and 25.

20.5 Importance of Project Management for European Industry

20.5.1 Increasing Competitiveness Through Project Management

The competitiveness of European industry is largely responsible for the standard of living. Today's tasks in industry are usually project tasks, which involve the development of new, better, and more cost-effective products. Not infrequently, this is accomplished in the context of European or global joint tasks similar to the aerospace industry projects described above. The ability to successfully manage global projects therefore plays a significant role for European industry.

20.5.2 Implement Project Management Consciously

It is important and necessary not only to put on the project management mantle for appearances sake, but also to effectively implement and professionally apply the methods and procedures of project management. Of course, it is also necessary to ensure that PM staff receive regular and thorough training; see Sect. 15.2.

Today's industrial projects are mostly complex and organizationally highly interconnected tasks. This is true for most project undertakings. However, technical and organizational complexity are the major problems in managing project work, and they are often the cause of technical problems, schedule delays, and cost overruns. The great challenge to the competing European industry is to be able to complete complex system tasks better, faster, and more cost-effectively in the future through efficient PM methods and procedures.

20.6 Project Management Performance

20.6.1 Increasing the Performance of Project Management

The performance of project management (*PM quality*) is largely responsible for the success of project execution, because better overall results can only be achieved through efficient PM. At the same time, however, the costs of project management must also be considered, which must be in a certain ratio to the overall project costs. But simply increasing PM costs will not automatically bring greater success. The goal is to achieve the greatest possible impact at the lowest possible cost.

The performance, or in other words the technical ability of the project manager and his team, is put to the test. PM is a specialist activity that requires the following qualities of the project staff in a global environment:

- In-depth knowledge of known PM methods in the areas of:
  - Systems engineering
  - Quality assurance
  - Project monitoring
  - Cost management
  - Contract Management
  - Risk Management
- Leadership skills and experience
- Sense of responsibility and decisiveness
- Prudent risk appetite
- Assertiveness
- Good leadership skills
- Inclination to work on management tasks
- Willingness to communicate and openness
- Knowledge of foreign cultures

Compliance with project targets can be used as a measure of PM quality.

20.6.2 Competitiveness of European Industry

The competitiveness of European companies can be measured, among other things, by PM quality, because competitive advantages result from an attractive technical project offer and deadlines, as well as favorable pricing, and from keeping promises because that is how the supplier is evaluated.

At the same time, however, companies also strive to maximize profits. The high PM requirements mentioned here describe the conflict that many project managers face, but are unable to solve the problem and therefore often even resign. Likewise, the usually existing cost pressure naturally also affects the project managers, and this then leads to the fact that not enough resources are available for the PM tasks and the PMs are usually hopelessly overloaded.

20.6.3 Mastering Globalization Through Project Management

But what needs to be done to decisively improve the situation? Admittedly, the ways to achieve these lofty targets for project management are not always easily attainable. But just like the international competitors in our environment, we must also seriously address the problem. The following is a summary of the decisive measures that will enable us to face globalization in the PM business better than before:

- Consistent and courageous implementation of a tailored project organization: *The project managers and their teams must have a home port,* e.g., *the PMO, and arrangements must be made for temporarily seconded employees; see subsection 5.2.5.*
- Professional use of familiar PM procedures and software: *companies must have appropriately concise manuals that do not only sit in the closet, but are also used in day-to-day operations.*
- Comprehensive training of PM staff, training to become PM professionals: *PM staff must be trained as comprehensively as possible, especially in the areas of economic and contractual processing; see Sect. 15.2.*
- Equipping PMs with the necessary powers to handle projects: *PMs must have an adequate level of authority to perform their duties: Decisions, personnel, approvals,* etc., *see Sect. 15.3.*
- Improving early customer relations: *Whenever reasonable, close and preferably collegial customer contact must be established as early as possible; see Sect. 14.3.*
- Critical evaluation of project tenders and avoidance of the nice-to-have *effect*: *It is not an isolated case that additional services,* e.g., *a higher quality that was not demanded, are offered by the supplier. Reason: people are reluctant to change their habits. On the other hand, it is of course not economical to offer a product quality which is not demanded (nice-to-have) but which is not paid for; see Sect. 14.5.*
- Strict adherence to project targets once agreed upon: *Once targets have been set, they should be strictly adhered to if possible. This is particularly important for*

follow-up business (reference project). Incidentally, this also applies to internal company agreements, because the project managers in the companies are, after all, the customers of the specialist departments. Unfortunately, this process often works particularly badly. On the other hand, profit potentials lie fallow here; see Sect. 9.4.

- Improving configuration and change monitoring: *Changes to the previously defined configuration must be recorded and evaluated in good time. The question of cost absorption must be clarified in connection with the contractual conditions. But without a systematic recording of the changes and the determination of the causes, evidence must be provided; see Sect. 12.2.*
- Early risk analysis and initiation of mitigation measures: *Early identification of risks and their impact on the project are of utmost importance for project success. As a measure, risk mitigation measures are to be implemented,* e.g., *through technical alternatives, planning for reserves, and/or other appropriate safeguards,* e.g., *through insurance; see Sect. 19.3.*

20.7 Cultural Influences

20.7.1 The Cultural Environment

Doing business in a global environment requires that responsible managers deal with the culturally different behaviors of their partners. What is considered normal in a country like Germany is not the same in other cultures. Harris aptly writes, "*Culture influences human behavior, moral beliefs, work productivity, value systems, and creates role models with influence on management*" [11]. Cultural and language differences complicate global collaboration because "*Cultural differences often lead to misunderstandings*" [12].

This means that employees who work in global environment should be prepared for their work through appropriate training. To strengthen efficiency across cultures, cross-cultural training should be a high priority [13]. Culture gives people an understanding of who they are, where they belong, how they should behave, and what they should do [12].

Harris continues, " … *people and countries with similar cultures come together easily, while people and countries with different cultures tend to go their separate ways*" and describes the cultural characteristics as follows: [14]

- Communication and language
- Clothing and appearance
- Food and eating rules
- Time and understanding of time (*reliability and punctuality*)
- Relationship to each other (*age, gender, status, degree of kinship, property, power, influence, wisdom,* etc.)
- Values and norms (*behaviors*)
- Belief and appearance

- Mental behavior and willingness to learn
- Working methods and practices.

Of course, the cultural difference also affects leadership behavior. The leadership style that is common in one's own country (*culture*) frequently does not work in other countries (*cultures*). It is not enough to say that the others just have to adapt or try to force certain things. Harris mentions 10 important concepts of global leadership [15]:

1. Global leadership: *Being able to work effectively in the global environment requires respect for different cultures.*
2. Cross-cultural communication: *learning foreign languages is important, but to avoid misunderstandings, you also need to know the meaning of language symbols and their meanings.*
3. Cultural sensitivities: *Recognizing culturally different behaviors and differences in order to work together effectively with understanding.*
4. Cultural adaptation: *Efficient adaptation and integration of specific cultural behaviors, including existing sub-cultures.*
5. Cultural management influence: *It is important to understand that the management philosophies and practices of different countries are deeply rooted in their respective cultures.*
6. Cross-cultural performance efficiency: *one should recognize and, if possible, apply the working practices of foreign cultures.*
7. Changing international business practices: *The manager with experience in global cooperation accepts cultural differences and business practices as well as loyalty and integrates them into his projects.*
8. Cultural synergy: *With reference to the very different systems, a successful cooperation is to be aimed at by skillfully combining the best cultural experiences in each case in order to obtain a greater whole.*
9. Work culture: *application, integration and merging of characteristics of different cultures through exchange of experience and intensive training measures.*
10. Global cultures: *With the understanding that the different characteristics of human cultures have always been universal, it should be possible to develop a global culture with the same characteristics.*

In order to get a picture of cultural differences and to establish measurable comparability, Meyer defined eight measurement points in her book, as well as a scale (*from—to*) that she uses to make global cultural differences measurable [16]:

1. Communication: *little content* versus *a lot of content*
2. Evaluation: *direct negative feedback* versus *indirect negative feedback*
3. Conviction: *principles first* versus *applications first*
4. Management: *equal* versus *hierarchical*
5. Decisions: *Consensus* versus *top-down*
6. Trust: *task-based* versus *relationship-based*

7. Disagreement: *confrontational* versus *confrontation avoidance*
8. Time keeping: *linear time* versus *flexible time*

In Fig. 20.7, the behavioral practices of two countries are reproduced as an example and one sees an almost complete sameness in some areas and a completely contrasting situation in other areas. "*When you compare cultures, there can be similarities in some areas and besides that, there can be very contrasting areas*" [17].

20.7.2 Languages and Their Meaning

The language used can be another problem, as word content and cultural background can play a major role. Meyer describes this vividly using various examples. In some cultures, information is expressed in direct form, while in other cultures, people communicate in indirect form. Meyer distinguishes between groups of countries with linguistically different complex languages and communication behaviors [18].

1. *Low-context: the group with good communication in precise but simple and direct language* (e.g. *USA)*
2. *High context: the group with good communication in cultivated, nuanced, complex, and indirect language* (e.g., *Japan).*

For the different cultures, there is a wide range between the two poles Low Context (*LC*) and High Context (*HC*), as shown by the example "*Communication*" in Fig. 20.8. According to this, Germany belongs to the LC group together with Australia, Canada, the Netherlands and the USA [18]. This means that according to Meyer, the communication of this group of countries is precise, simple, and clear. On the right side of the scale, we find countries of the HC group such as China, Japan, Indonesia, and Korea. The second group of countries communicates in a highly developed and nuanced way. That means you must read a lot between the lines.

Country group II (*HC*) can look back on a long historically grown cultural background and a relationship-oriented society that can often be traced back over many generations. Therefore, when meeting with partners in this country group, one should listen very carefully in order to better understand even the most codified information, i.e., to read the "*no*" between the lines. Meyer calls this "*reading the air.*" [18] She continues: New cultures in Country Group I (*LC*), only a few centuries old, such as the United States, for example, under the influence of heavy immigration, including different languages and cultures, depended on developing a simpler language to avoid misunderstandings.

Now, when representatives of both groups work together on a project, the different behavior easily leads to problems because both groups communicate in their usual thought patterns. In this context, a joint workshop is recommended in which the cultural differences of the team members are addressed in advance to better

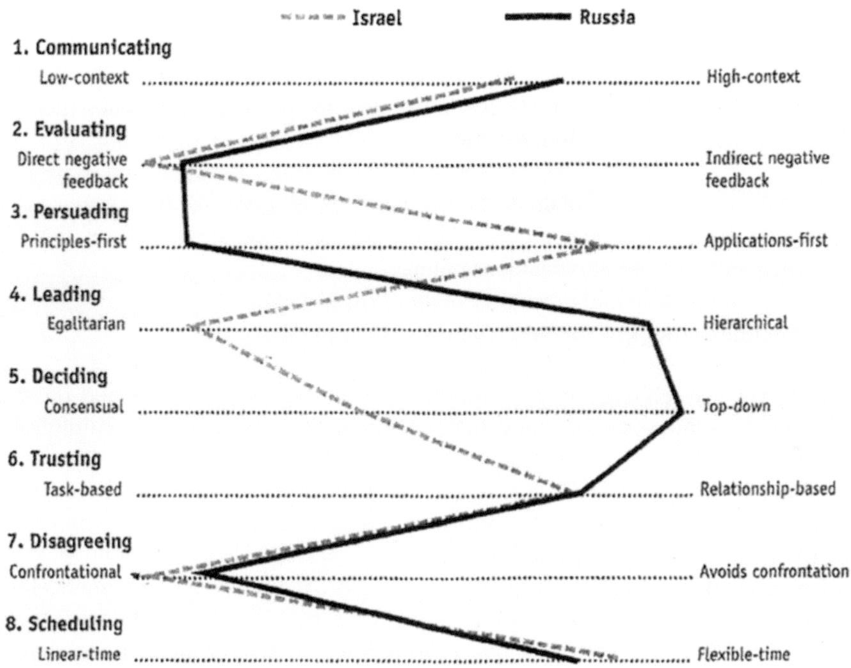

Fig. 20.7 Example of the behavioral practices of two countries (Source: Meyer)

Fig. 20.8 Communication behavior (Source: Meyer)

understand them. In particular, this involves behavioral patterns and linguistic subtleties.

With multicultural teams, the best way to ensure understanding is to use simple, i.e., LC, communication practices. Also, the following practical verification method is recommended to ensure understanding [18]:

- Verbal repetition of the main discussion points by all participants on a rotational basis to avoid misunderstandings. This is time-consuming but ensures that all participants have the same understanding
- Each participant should present what they will do next
- A meeting participant should prepare a written summary.

However, the tendency to put everything in writing that is considered professional in LC cultures is often met with suspicion by colleagues from HC cultures, as it is suspected that the verbal statement is not trusted.

Figure 20.9 shows a bird's-eye view of how colleagues in different cultures react to criticism. In countries such as Germany, France, and Russia, negative criticism is usually expressed honestly and directly, while negative criticism in countries such as Japan, Thailand, and Indonesia is expressed more indirectly and diplomatically.

The topic of leadership is also an important aspect for managers, and Meyer distinguishes between liberal (*egalitarian*) and hierarchical leadership structures; see Fig. 20.10 [19]. As can be seen, some European cultural groups tend toward the liberal leadership structure, while Germany and France, for example, are placed in the middle range. The following rules are common when working with colleagues from European cultural groups:

- Direct contact with the business partner without the supervisor is allowable
- Inform the supervisor only in case of higher-level matters
- Passing hierarchy levels is possible, but not the rule
- In some groups of countries, such as Scandinavian countries, Canada, and the USA, it is quite common to use the first name in correspondence.

Asian cultures are mostly found in the hierarchical or patriarchal realm. They tend to respect hierarchies and status much more than Western cultures. The following rules should be observed when working with colleagues from this culture:

- Communication should only be with colleagues at the same level.
- When communicating with lower-level employees, the supervisor should receive a copy.
- Contacting higher-level supervisors should be approved by the direct supervisor.
- In correspondence, the last name should be used unless the first name has been agreed upon.

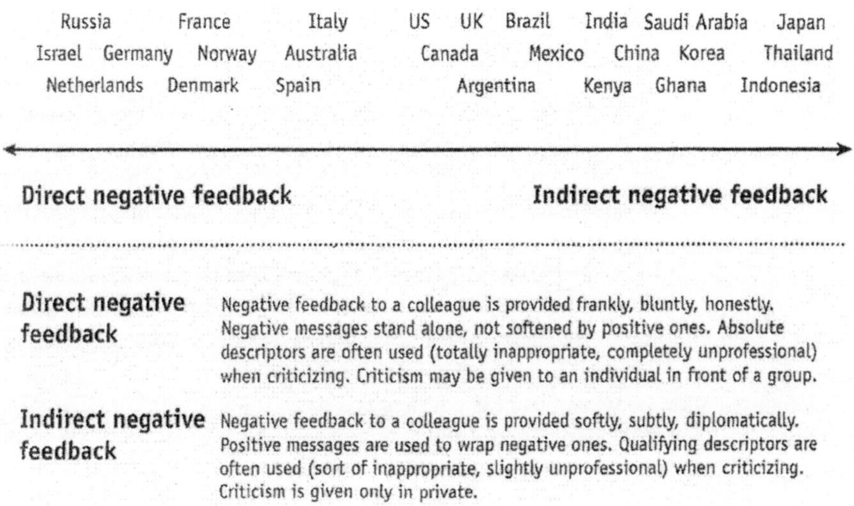

Fig. 20.9 Reaction to negative criticism (Source: Meyer)

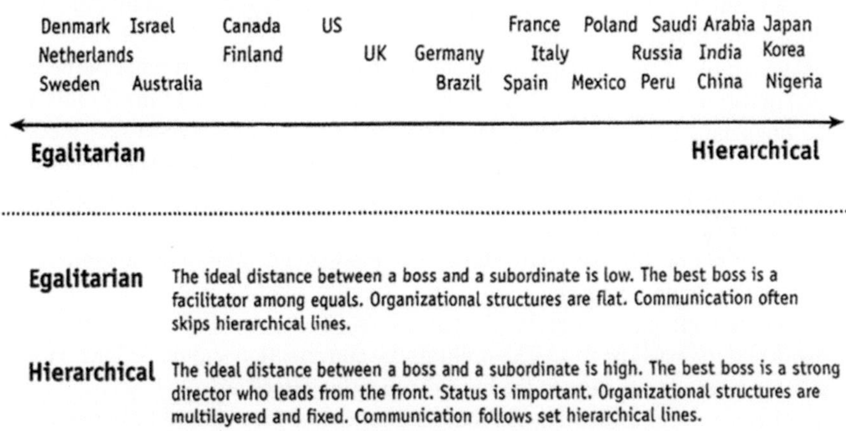

Fig. 20.10 Conduction behavior of LC and HC cultures (Source: Meyer)

Volker Mennel, a former work colleague of the author, reports from a business trip to China: "*My negotiating partner invited me to his home after difficult negotiations and repeatedly expressed his appreciation to me. I asked our delegation leader what gift I could/should bring when visiting my negotiating partner. He just laughed and said this invitation is not to be taken seriously but only a mere courtesy to me. He still said, should I really go there, the family with grandparents and children would faint*".

20.7.3 Creation of a Basis of Trust

Trust is an important prerequisite in business life and is of great importance in addition to contractual commitment since there is no such thing as an absolutely watertight contract. A loss of trust always makes constructive cooperation between partners difficult because in most cases they depend on close cooperation (*give and take*). Trust often develops in the course of close cooperation. On the other hand, the cultural background plays a decisive role. Meyer distinguishes between two different groups of countries whose trust *is* based on a task-oriented business relationship, and in this context names the USA and Switzerland ("*business is business*") and names a second group of countries whose trust *is* based on a relationship-oriented business relationship, and this is China and Brazil ("*business is personnel*") [20].

Communication between business partners is a basic prerequisite for successful cooperation. In global cooperation, linguistic and cultural hurdles must be overcome. Communication is essentially based on verbal and written media, such as one-on-one conversations, meetings, telephone calls, e-mails, letters, and reports that can be used in specific situations. In view of possible cultural differences, such as those to be expected in global projects, it may be appropriate to prefer a personal telephone call to the rather dry e-mail communication. This is especially true when dealing with complicated topics.

Meyer emphasizes the importance of informal face-to-face meetings and recommends that employees of global collaborations schedule sufficient time for this and, if the financial situation permits, plan appropriate travel for a face-to-face meeting (*one-on-one discussions and meetings*) if necessary [21]. The development of a trustworthy and resilient interpersonal relationship facilitates the interaction of business partners in resolving complex transactions.

In many countries, especially those of the so-called high-context group, it is customary to begin communication with a friendly greeting and a question about how you are feeling and your health. This establishes or maintains a personal relationship and the basis of trust. The preliminary conversation should be long enough to allow a relaxed transition to the business meeting afterwards.

In discussions, personal attacks that lead to loss of face should be avoided as a matter of principle. This applies in particular to multicultural teams. In Asian countries, such as China, Korea, and Japan, there is an emphasis on team harmony by avoiding loss of face by all participants. In China, it is more important to avoid a team member's loss of face than to express one's own opinion [22].

20.7.4 How to Deal with Disputes?

In disputes between employees, an openly expressed opinion can have a shocking effect. How employees from different cultures react to differences of opinion is shown in Fig. 20.11. To avoid unnecessary tension in disputes among globally

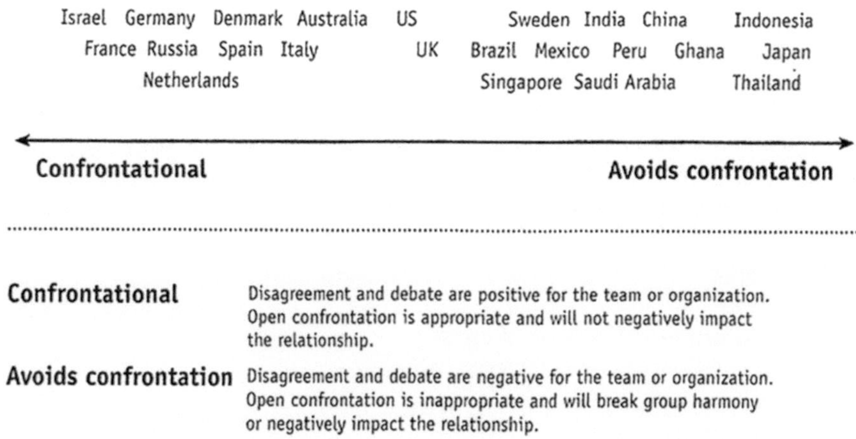

Fig. 20.11 Reaction to disagreement (Source: Meyer)

composed teams with different cultural backgrounds, the use of a neutral moderator is recommended. Another way to find a solution is the French method of *thesis, antithesis and synthesis* mentioned by Meyer to jointly consider the arguments of both sides and work out a solution. The Anglo-American *brainstorming system* is also a helpful method.

It is important that managers working in a global environment follow the recommendations described earlier. Especially when dealing with HC cultures, it is important to be able to read between the lines. Kilian aptly writes: "*Those who speak directly do not take up the dance with language, …* " [23]. Thus, one should engage with the other foreign culture and its language in the interest of successful business. Kilian therefore adds the following to the saying "*speech is silver, silence is gold*" "*skillful speech is platinum*" [24]. Since every culture has its own unique face [25], you win over more people if you empathize with the mentality of other countries and adapt your language behavior [26].

In this context, Kilian also deliberately addresses the topic of "*small talk*" because it is the shortest path to the emotional brain, the key to togetherness [27]. In a global business initiation involving many different cultures, small talk is often the bridge to the actual business.

20.7.5 Management of Global Projects

The management methods described in the previous sections are in principle also applicable to international projects; see Chap. 16. Therefore, project managers must first examine and consolidate their knowledge before taking on an international management task. They should then ask themselves the question, "*What is different about global projects?*" Is it the cultural difference described earlier, or do other aspects need to be considered as well?

Here is a summary of the specifics of managing global projects [28]:

- Cultural aspects (*see subsections 20.7.1 and 20.7.2*)
- Determination of an official project language
- Adoption of uniform management standards
- *Manuals*
- *Procedures*
- *Definitions*
- Project structure and division of tasks
- *Work Breakdown Structure* (*WBS*)
- *Specification*
- *Statement of Work* (*SOW*)
- *Work packages*
- Project financing (*national and international*)
- *Financing models*
- *Base currency and exchange rates*
- *Inflation rates*
- Legal requirements (*national and international*)
- *Contract law* (*country*)
- *Jurisdiction* (*country*)
- *Arbitration court* (*country*)
- *Premium regulation*
- Legal agreements
- *LOI and contracts*
- Overarching policy requirements and limitations
- *Military specifications*
- *Limiting the transfer of know-how*
- *ITAR Restrictions* (*USA*)
- *Regulatory requirements* (*national and international*).

The definition of binding rules is particularly important in international project management in order to avoid misunderstandings. In addition, it is important to consider the aspects described by Meyer in subsections 20.7.1 to 20.7.4 when managing global projects [16].

As in politics, misunderstandings often lead to problems. Because if one does not know the other culture, it is suspicious and sometimes even scary [29]. The only way out of the dilemma is to deal with the situation and acquire the necessary knowledge through appropriate books or courses, because there is usually no malicious intent behind not understanding the behavior of one's counterpart.

John Mole points out in his book "*Mind Your Manners*" that the different behavior of people in certain situations can be explained by their cultural background and writes: "*When people from different nationalities or cultures come together in teams, meetings, negotiations, or as employees of the same company, they bring with them different expectations and beliefs of how they should work together. The way others do things is not different out of stupidity or carelessness or incompetence or malice, although it may appear so. The judgement of what is right is rooted in habit, tradition, beliefs, values, attitudes, and accepted norms; in other words, the culture to which that person belongs*" [30].

Mole conducted surveys and asked several specific questions to people from different countries (*cultures*) and recorded the evaluation of the answers on a scale. Table 20.1 shows an example of ten corresponding statements to specific questions defined by Mole [31].

The question now arises of how to work efficiently with partners from a foreign culture in different constellations, as a supplier, partner, or customer. One can always learn from each other and should treat each other respectfully and try to understand the other culture. On vacation trips to foreign countries, one often brings home interesting objects that one was impressed by, in addition to souvenirs.

However, business relationships are about being successful, and it would be best if there was a sense of achievement ("*win-win*") for both sides. This should be the long-term goal of globalization, and project managers have a special leadership role in this context. Müller-Kästner writes: "*Globalization means realignment of supply and demand, new linkages between labor and capital. But beyond the economic aspects, it also leads to fundamental sociopolitical, cultural ethical reassessments.*" [32] Efforts must be made to really get to know other cultures. Jack Welch, CEO of General Electric until 2002, puts it this way:

- It's a Fact that Social and Educational Differences Exist between different Cultures.
- Solution: "*Global Training and Learning must become a Top Priority at Globally Operating Companies*".

Appropriately, a comparison of different leadership styles from three geographically far-flung nations, namely Japan, USA, and Germany; see Fig. 20.12 [33].

Hofstede et al., point out that cultures are learned and not innate [34]. This can be understood because people who immigrate from one culture to another and are forced to adapt often learn new behaviors very quickly. The authors mentioned above describe three mental levels of human beings:

1. the innate universal nature
2. they learned the culture of a specific cultural area and
3. the specifically developed and acquired personality.

Table 20.1 Comparison of questions and explanations (according to Mole)

	Question	Country-specific answer
A	Do the team members from different cultural groups say what they truly think, or do they speak about things indirectly? Mayer: Workers from Northern Europe speak clearly about what they think.	**"Do all speak all clearly what they think?"** **Unclear: < – > Clear**
B	Do team members express humor during business development? In some cultures, such as GB and Ireland, this is often the case and it relaxes the business atmosphere.	**"Does humor exist in business life?"** **Seldom: Sometimes: Often**
C	"He who writes, stays," is a common idiom in Germany and other German-influenced countries, while in Romance cultures, oral communications are preferred.	**"Information Orally or Written?"** **Oral: < – > Written**
D	Good personal relationships are often a requirement for successful cooperation. The prerequisite for good cooperation is that "the chemistry must exist."	**Are good personal relationships important?** **Yes: < – > No**
E	Are workers punctual, or is the academic 15-minute rule? Punctuality is the courtesy of Kings! Workers in Northern Europe, and also Japanese, are normally punctual.	**"How Punctual are you?"** **>Too Late: 15 m Late: Punctual**
F	Are team members flexible or do they follow strict instructions? This is a characteristic of Germans and Japanese, while Greeks and Chinese tend to be flexible.	**"Do all speak all clearly what they think?"** **Unclear: < – > Clear**
G	Teamwork is a "MUST" for complex international projects. Northern Europeans and Japanese work well in a team, while in Turkey and China.	**"Does humor exist in business life?"** **Seldom: Sometimes: Often**
H	"He who writes, stays", is a common idiom in Germany and other German-influenced countries, while in Romance cultures, oral communications are preferred.	**"Information Orally or Written?"** **Oral: < – > Written**
I	Good personal relationships are often a requirement for successful cooperation. The prerequisite for good cooperation is that "the chemistry must exist."	**Are good personal relationships important?** **Yes: < – > No**
J	Are workers punctual, or is the academic 15-minute rule? Punctuality is the courtesy of Kings! Workers in Northern Europe, and also Japanese, are normally punctual.	**"How Punctual are you?"** **>Too Late: 15 m Late: Punctual**

Japanese Management	USA Management	German Management
Draft visions	Specifying visions that motivate and inspire, and clear goals	Negotiation of clear goals
Encouragement of employees	Briefing of employees	Assignment of employee-specific tasks
Creating challenges	Taking advantage of opportunities	Develop the maximum technical possibilities
Inclusion of all employees in the identification of problems and opportunities	Appointment of specialists to observe the environment and diagnose opportunities and risks	Managers and employees are experts with detailed knowledge
Central Management instrument: mixture of motivating emphasis on strengths and indirect addressing of weaknesses (saving face is important!)	Feedback as a permanent, motivating strengths analysis	Feedback only as corrective error analysis

Fig. 20.12 Comparison of leadership styles (Source: Hoffmann, Schoper, Fitzsimons)

People from different cultures who work closely together as a team in a multinational company and are open to each other learn from and benefit from each other. The author worked for many years as a project manager in a globally oriented Luxembourg-based satellite operating company, which employed people from almost 30 nations. The cultural mix led to the development of the company's own individual corporate culture, which had a positive and motivating effect on the employees and, through the formation of creative project teams, also had enormous benefits for the company.

Boutellier, Gassmann, and von Zedtwitz write, "*Multinational companies (MNCs) have a tremendous impact on global innovation and the structure of the world economy. They:*

- *determine the international division of labor with production, R&D, marketing, and sourcing strategies*
- *transfer technologies, management capabilities, and financial capital*
- *control over 80% of the privately owned technological resources (Dunning 1993)*
- *influence regional restructuring with the establishment of local offices*
- *have a high negotiation potential in light of foreign direct investments, particularly in smaller economies*" [35].

International project management in a global environment is conducive to connecting cultures and building bridges to foreign cultures through acquired knowledge: "*Building Bridges between Islands of Knowledge*" [36].

The author has for many years been actively involved in building intercultural global bridges in close cooperation with experienced space colleagues from Germany, through training, consulting, and support work.

Literature

1. Müller, *Klaus*: Globalisierung, Campus Verlag, Frankfurt am Main *2002*
2. Preyer, Gerhard: The Global Challenge, FAZ Gableer Edition, 1998
3. USAF: System Program Management (AFSCM 375), May 1966.
4. Tathjen, Walter: Special issue of the journal Luft & Raumfahrt, p. 11, 2016.
5. INTELSAT was the first commercial communications satellite in geostationary orbit
6. SES is a satellite operating company based in Luxembourg
7. DLR-SART, Bremen: SpaceLiner Mission Requirements Document, SL-MR-SART-00001-1/2, Issue 1, Rev. 2, 08.07.2016
8. Orwing, Jessica: Business Insider Germany, February 2016
9. Sabagh, Karim, SES: Taking Control of O3b Networks.
10. Sabagh, Karim, SES: Interview with Mark Holmes, Via Satellites Digital, June 2016.
11. Harris, Philip R., Moran, Robert T., Moran Sarah V.: Managing Cultural Differ, 6th Ed., Elsevier Butterworth-Heinemann, Oxford, 2004, p. 4.
12. Lewis, James P.: The Project Managers Desk Reference, McGraw Hill, Bos ton, 1999, p. 1.
13. Source 11, p. 22
14. Source 11, p. 5
15. Source 11, p. 25
16. Meyer, Erin: The Culture Map, Public affairs, N.Y., 2014, p. 15ff.
17. Source 16, p. 17ff
18. Source 16, p. 39ff and p. 69
19. Source 16, p. 125ff
20. Source 16, p. 170f
21. Source 16, p. 190f
22. Source 16, p. 198ff
23. Kilian, Susanne: Don't let me be misunderstood! ARISTON; 2015, p. 53
24. Source 23, p. 73
25. Source 23, p. 76
26. Source 23, p. 93
27. Source 23, p. 105
28. Madauss, Bernd: Lecture "*Management of International Space Projects*", International Space University (ISU), Strasbourg, 2016.
29. Lientz, Bennet P.: Kathryn P. Rea, Project Management for the 21st Century, 2nd Ed., 1989, p. 5.
30. Mole, John: Mind Your Manners - Managing Business Cultures in the new Global Europe, Nicholas Brealey Publishing, London, 3rd Edition, 2003
31. Source 30, p. 14ff
32. Müller-Kästner Burkhard: Das Projekt Euro und die Globalisierung, Deut sches Projektmanagement Forum, Dresden 1998, GPM, Munich, p. 147
33. Hoffmann, Hans-Erland, Schoper, Yvonne-Gabriele, Fitzsimons, Conor, John: Internationales Projektmanagement, Deutscher Taschenbuch Verlag, 2004, p. 133

34. Hofstede, Geert, Hofstede Gert, Jab, Minkov, Michael: Cultures and Organiza tions, Mc Graw Hill, 2010, p. 8.
35. Boutellier Roman, Gassmann Oliver, von Zedtwitz Maximilian, Managing Glo bal Innovation, Springer, Berlin Heidelberg New York, 2000; p. 3.
36. Source 33, p. 203

Competition in a Global Environment 21

Contents

Abstract

Competition is a mainstay of the economy because, as in sports, it spurs those involved to achieve their best performance. Technological excellence has led to improvements in living conditions in many countries, but unfortunately it has also had negative effects, such as armed conflicts. The industrial competition of individual countries has developed into global competition in recent decades. German companies compete not only with their European neighbors but also with companies from other continents.

B. -J. Madauss, *Project Management*,
https://doi.org/10.1007/978-3-662-69057-4_21

Space projects are inherently global undertakings because satellites cross countries and continents at high speed. That's why cooperation between national agencies and companies is usually a given. The International Space Station (ISS) space station is a good example of this; agencies and companies from Canada, Europe (*ESA*), Japan, Russia, and the USA are successfully pulling together here, see also Table 16.2. You have to talk to each other if you want to place objects in space, because otherwise, collisions will inevitably occur. Efficient management is needed for successful multinational cooperation, not only in space but also in other industries.

In parallel with the development of space systems, project management was also successively developed and constantly improved in line with product enhancement. This was necessary in order to be able to successfully solve complex tasks, especially manned space flight. The foundation of project management as we know it today was laid in the 1950s in the context of complex projects by the U.S. Department of Defense and NASA. We still benefit from it today.

This chapter, which deals with competition in the global environment, contains inputs prepared by Professor Victor Billig from Florida, USA. He is a very experienced space engineer and has extensive knowledge of space management. I have successfully worked closely with Victor Billig for many years on several global projects in Asia, Europe, and the Middle East. Both of us have also shared our management knowledge through university lectures and seminars. Before coming to Germany, Victor Billig worked, among other things, on NASA's Apollo project. Therefore, I asked him to write down the history of spaceflight and management from his point of view, starting with Sputnik.

21.1 Visions and Planning for Global Customers

This chapter provides a historical link between the 50 years of space developments and the importance of adequate project management disciplines, representing access to opportunities in a world of competition by accessing Global Opportunities.

The highest goal of customers whose objectives are everlastingly intertwined in an eternal struggle to be on top, whether they are companies, civil institutions, or nations acting as servants to their public, is their vision of supremacy. The purpose is success in all forms of power, and the motivation is nature's inspiring desire to be first. In global/international competition for power, developments involving space, energy, and communication lead the way. Certainly, the closed-door intrigues, secret agreements, and plans of industrial leaders working co-dependently with their political counterparts set forth spirited dreams of enterprise, prestige, achievements, and fortune. However, the most remarkable fact of the matter is that without trained management, neither companies nor national customers alone are able to achieve technological breakthroughs. History has repeatedly demonstrated that the space industry, for example, moves forward only when effective cooperation exists between private and government forces.

21.1.1 Orderly Transition from One Management Leadership to Another

The only apparent commonality between the historically successful projects with highly intertwined technological content is their great complexity. The list of examples well known to the authors includes Sputnik, Apollo, the Manhattan Project, and many in the fields of medicine. Less known are the unsuccessful projects such as Iridium, Hermes Spaceplane, and TV SAT, obscured by failures in project management to master complex interfaces and by the corresponding extinction of whole companies devoured by their own projects. What are the lessons learned? Jim Collins, in his book entitled "*How the Mighty Fall,*" identifies five stages of decline symptomatic of U.S. companies studied after a 4-year research study of 60 companies. He concludes with the statement, "*Whether you prevail or fail, endure or die, depends more on what you do to yourself than on what the world does to you*" [1]. The untold losses for humanity are inestimable. Certain, however, is that the return on experience belongs to those who study project management, taking heed of the lessons learned. The most startling thing about the final stage of decline is that for large institutions/companies, it typically means the invocation of a legal process imposing an orderly transition from one management leadership to another. To protect the innocent, no one really dies, only the owners or leaders change. The example taken for illustration below is Sputnik.

On October 4, 1957, the Soviet Union launched the first artificial Earth satellite into orbit. Weighing only 184 lbs., this ball of electronics enclosed in a metallic sheath through which protruded several dipole antennas was broadcasting "*Sputnik's beeps*" of recognition to the world on two radio frequencies of 20 MHz and 40 MHz alternately. With this elegantly simple proof of Konstantin Eduardovich Tsiolkovsky's equation formulated in 1903 that given sufficient thrust, gravitational and centrifugal forces acting on a body will reach equilibrium to sustain orbital flight without further thrust, humankind was propelled into the space era. Never mind that the frequencies were not those agreed to by the Soviet scientists of the International Geophysical Year (*IGY*) organization. Never mind that the IGY forum was used to publicize the event at an IGE progress meeting in Washington, D.C., a perfect place to gain world attention through the media. All this was a stunning act of competitive maneuvering that successfully placed the USSR ahead and opened the "*space race*" between the USA and the Soviet Union [2]. Never before during the Cold War had so much fear been aroused in America since the Russians began atomic bomb testing. The American media began speculating about the "*what-ifs*" stemming from Russia's newly acquired capability to... orbit a nuclear warhead. Yes, a new era has arrived.

Sputnik encircled the globe every 96 minutes, or 20 times per day. It followed an elliptical path with a perigee of 141.7 miles and an apogee of 588 miles, which gradually declined due to atmospheric drag, and in some months it plummeted back to Earth. Despite the modest debut, the significance of this great event should not be diluted with extraneous references to trivia. But, of course, there were many earlier failures. And indeed, there is nothing unique in the above-mentioned specifications. Any one of many could have succeeded. The challenge is not the thrust or the choice of orbit. The real challenge is to manage the reliability engineering trade-off process leading to a fast convergence to any low-risk solution within budget; see Fig. 21.1.

In the case of Sputnik, autocratic leadership was exercised under the direct cognizance of Joseph Stalin. It was quite unlike the situation in the United States, where severe competition between the military services and their respective industrial rivals, retarded the convergence needed to launch the Earth's first artificial satellite. Only when President Eisenhower ended the competition was it possible to launch an equivalent mission named "*SCORE.*" Figure 21.2 shows Atlas B with SCORE on the launch pad on December 18, 1958, a rocket with a much heavier mass delivered into orbit compared to Sputnik 1 and with a more sophisticated communication payload. It broadcast the voice of Eisenhower sending his greetings around the world, and, like Sputnik, it then plummeted back to Earth. Before leaving office, Eisenhower gave a farewell speech delivered in 1961, warning the public to guard against the "*military-industrial complex*"; one of the most famous speeches in American history [3].

Looking closer at the global environment of that time, we see two nation-customers locked in a "*cold war*" and entwined with their industrial entities struggling to be on top for prestige, power, or any other perception of success then prevailing. Both sides had access to German experts who developed the V2 Rocket during World War II. The USA had integrated the V2 into the Redstone Rocket, and Wernher von Braun was in charge of the program for the US Army. The trade-off options for America's first orbital mission, "*Explorer 1,*" were sufficiently flexible to allow a wide range of specifications to be met, leading to the discovery of the Van Allen Belt; the only shortcomings were those of Explorer 1 (see Fig. 21.3). A modified Redstone missile, Juno 1, was launched 4 months after Sputnik 1 and had only one-sixth of the mass of Sputnik. Although the first race (into orbit) was lost, the more spectacular race, "*to put a man on the moon and bring him back safely,*" was President J. F. Kennedy's gift to the world. A key to the overall comeback of the USA was the highly effective project management discipline employed by NASA.

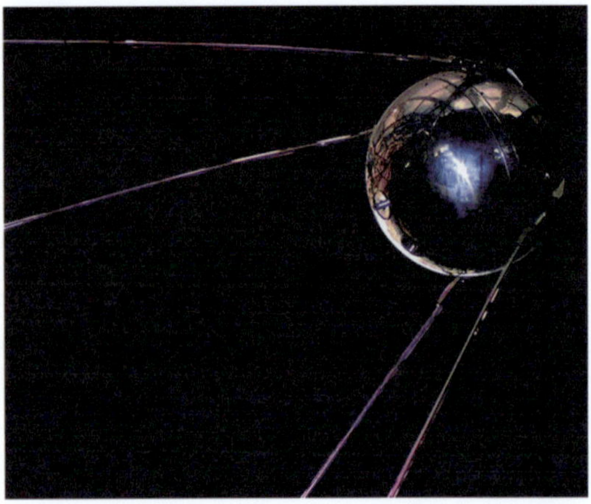

Fig. 21.1 Sputnik satellite. Source: Creative Commons

Fig. 21.2 Atlas B Launcher (Source: Creative Commons)

Fig. 21.3 Explorer I Satellite (Source: Creative Commons)

The sections below will illustrate the ways a customer or government agency can use and harmonize the strengths and weaknesses of a competitive industry. In the global acquisition process, Request for Information (*RFI*), Request for Proposal (*RFP*), and Negotiation (*NEGO*) are among the most vital of all project management disciplines. Students should learn that by applying these disciplines, they confer lasting value upon system management as a fundamental logic that transcends and survives the random forces of military-industrial ambition.

21.1.2 Inside Today's Industries

Inside today's industries, there are endless organizational changes taking place. Many of these changes, often accompanied by new partnerships, alliances, or mergers, are common in space. They frequently occur at times of economic crises or when political decisions have an impact on the marketplace of a country, a region, or the entire world. An important national election or a natural or man-made disaster could be the immediate trigger. We have seen costly space projects postponed until the last vote was counted. The moving about of leading officers of key organizations within the industrial landscape is another cause of change inside private as well as civil institutions. Sect. 21.4 will discuss how these organizations reconfigure themselves to enhance their business prospects and give their shareholders a better return on investment. China, for example, has made significant sales of communication satellites to numerous countries in Asia, Africa, and South and Central America, which China launched with its very successful Long March III B booster. Russian companies have cemented strong alliances with aerospace counterparts in the West and are finding new and interesting market outlets for global business of the powerful "*Proton*" rocket, for example, used a couple of times by SES. The great Yuzhnoye manufacturing center in Ukraine, which produced Russia's first intercontinental ballistic missiles in the early 1950s, is now teamed with Boeing for the "*Sea Launch*" *of* communication satellites and with Orbital Sciences for building the Antares Launch Vehicle used by NASA to transport supplies to the International Space Station. These are illustrations of how successful business transforms earlier achievements into prosperity for the future.

As world events gradually improve the economic standards of developing countries, it becomes increasingly important for business partners to periodically re-examine new market potential from all viewpoints in order to critically assess the political stability of emerging regions. Transient conditions in one market can easily mask cyclic behavior in another market, promoting false strategies in both private and government affairs and alliances. Too often, partners or licensing agencies assume the existence of "*common understandings*" that simply do not exist between widely differing cultures, thus making it necessary for potential agreements to ensure a coherent legal foundation across the four corners of any contract. While such idealistic agreements are typically not achieved in today's global space system projects, the resulting failures have served to signal new guidelines for future project managers striving for ultimate success. Meanwhile, students of the new management science contained herein should anticipate the inherent ups and downs in the affected markets and counter them with the alleviating best practices assembled in this textbook. Keep in mind that you are an emissary of progress. Good engineering and good management are like a hand in glove. In recent years, China and India have been expanding their economies at record growth rates. Other nations, like South Korea and Brazil, are impressively producing highly competitive products on a global scale. However impressive such gains may appear, whether in Asia, South America, Africa, or the Middle East, they must be considered with critical regard to political and economic stability in their respective regions. Project managers working in the commercial space industry should be mindful that application satellites, including observation, communication, remote sensing, relay, surveillance, tracking, navigation, etc., are all inherently suited to global services and therefore present us with a powerful new counterforce to recover from undesirable instabilities that confront us from time to time. We must use this new force wisely. We owe future generations the promise of continued prosperity that space can offer.

21.1.3 Fast Forward: 50 Years Later

At this point in time, an alert project management student might ask, "*What is the connection between Sputnik and the visions of global customers?*" Let us flash a few perspectives to illustrate the point in history. The horrors of Pearl Harbor, the soup and bread lines that existed from the great depressions preceding and following World War II, the Marshal Plan, and care packages sent to Europe were all fresh images in the minds of the American public. Sputnik's appearance 16 years after Pearl Harbor spread fear of an impending nuclear threat—a bomb that at any time could be dropped from space. It mobilized an enormous countereffort involving many talented people to design, build, operate, and manage military aerospace resources. Let us examine the situation evolving between the end of the war and 50 years later, as it will expose many of the phenomena illustrating the overlays of the system management process and how the efforts of certain individuals caused political transients to occur, which reset the economic playing field. Enemies became friends and potential business partners. Friends became competitors and potential enemies. It will also explain how the birth of new customers in global markets can help replace hostility and war. This kind of "*politically constructive marketing*" was practiced decades later by President Ronald Reagan, opening a new era of trade with China.

The end of the war in the Pacific was already in sight when at the Department of the Air Force, General Curtis LeMay [4] introduced his ideas for securing the peace. He became the Commanding General of the widely known Strategic Air Command (*SAC*) and played a key role in navigating through the so-called "*Cold War.*" Between the presidential terms of Eisenhower and J.F. Kennedy, he sought to shape U.S. defense policy, capitalizing on all known military advantages, including nuclear strength. At the same time, Robert McNamara [5], Secretary of Defense, was actively setting new policies to stop what was commonly regarded as counterproductive spending. He insisted on incorporating analytic discipline into military procurement. After all, the war was over, and with the national budget deficit standing at the highest in history, the government needed to manage projects wisely. Taking advantage of the longstanding tradition of industrial management expertise in the USA, McNamara began to extend it into the acquisition methodology of large-scale, complex weapon systems. What followed was the preparation of directives, manuals, plans, specifications, reports, procedures, and other tools for efficient project management beyond this; McNamara is credited with the highly publicized "*limited warfare*" strategy of 1961 to implement President J. F. Kennedy's position that the nation's defense had to be "*designed to reduce the danger of irrational or unpremeditated general war.*" Kennedy wanted choices in an emergency other than "*inglorious retreat or unlimited retaliation.*" Although McNamara's analytical approach to financial control was praiseworthy, he lacked political consensus in the synthesis of his recommended strategy, which failed to be approved by President Lyndon B. Johnson. Certainly, the fact that some of his contemporaries still favored nuclear alternatives in Viet Nam might have precipitated his early resignation in 1968, and his move to help developing countries as President of the World Bank and the limited, flexible warfare policy he initiated currently remain a cornerstone of American policy. Project management students should take note, however, that at no time during this historic conflict was politically constructive marketing visible; see the paragraph immediately above. Had government leaders initiated such business innovations earlier, it would most likely have improved the outcome.

Looking more closely at the talented individuals dominating the aerospace scene, perhaps the most important figure was General Bernard A. Schriever [6], Commanding General of the USAF Systems Command. He was regarded as the architect of the U.S. ballistic missile and military space programs. He set forth the formal regulatory documentation defining procurement disciplines, e.g., AFM 66-1 and AFM67-1 [7]. His organizational rigor further established standard operating procedures for research and development; for the design of highly reliable equipment and quality assurance; for manufacturing and product assurance; for testing and deployment of new space systems applications; as well as for life-cycle ground maintenance to support operations in space. General Schriever's work during the "*Space Race*" encouraged private investment in new technology urgently needed to meet the extreme performance demands on material strength, miniaturization, thermal hazard, and control, power generation, and, of course, human endurance.

In parallel with the Air Force, the U.S. Navy was actively extending its own technological boundaries. Admiral Hyman D. Rickover (*USN*), considered the "*Father of Nuclear Navy*" was making headlines while developing the nuclear-powered submarine personally managing this complex program. Figure 21.4 shows H. Rickover on the cover of Time magazine. Some years later in 1982, he delivered a speech at Columbia University entitled "*Doing a Job,*" which outlined his management philosophy of remarkable determination and clarity of purpose [8]. This speech is a precious rarity to be read and remembered by all project managers as a treasured classic in leadership; many of his ideas can be traced to system project

Fig. 21.4 Admiral Hyman Rickover

management practices (*SPMP*) currently followed inside some of the largest aerospace companies in the world.

Norman R. Augustine was another outstanding contributor whose work will be long remembered. In the mid-1970s, Augustine served as Undersecretary of the Army, following an impressive early career at DOD Defense Research and as VP of Advanced Programs and Marketing LTV Missiles and Space. Returning to the industry in his later career, Augustine was instrumental in the Lockheed Martin merger and served as the first CEO, controller, and corporate chairman until his retirement in 1997. He was renowned for his "*Augustine's Laws,*" widely published and quoted in numerous technical and commercial works of literature. All 52 laws are significant statements of management experience and philosophical irony. Among the most commonly studied and deliberated are:

- Law 5: "*... Increasing the number of participants (in a project) merely reduces the average output.*" This statement has been extensively analyzed from the standpoints of communication latency, logistical costs and time delays, contract complexity, and performance risk.
- Law 16: "*In the year 2054, the entire U.S. defense budget will purchase just one aircraft....*" This prediction was based on an extrapolation of budgetary and cost estimates. Recent accuracy checks show good agreement without mid-term course corrections.

What the world has inherited 50 years later, after the shocking experience of Pearl Harbor, is in a nutshell, a set of rules to wield over belligerent nations, adventuring demagogues, and clamoring factions, to show that war, if inevitable, can be mitigated when political leaders are adequately advised by enlightened system project management experts who, in the course of their missions, exert the following precepts in acquiring global customers:

1. Ascertain the competitiveness of incumbent sources worldwide.
2. Determine the potential customer's ability to procure your products and his legitimate need.
3. Synthesize a sale scenario, including government sanctions.
4. Prepare a legacy plan to ensure that your product is compatible with the customer's existing interfaces. This might include training, upgrading, or modifications to your product or its legacy.
5. Ease the financing and insurance conditions of the prospective customer.
6. Encourage offset procurements/turnkey support to the extent consistent with ITAR restrictions.

The above list illustrates the kinds of concerns that should be addressed by sincere SPM offers in order to establish trust with emerging nations and stimulate competitiveness and overall productivity.

21.1.4 Perspectives for the Next Years: 1991–2041

Since the time of the Soviet dissolution in 1991, there has been a noticeable upswing in global competition induced, on one side by technological breakthroughs and, on the other, by political concessions to incentivize local business. More governments are getting into the act of promoting their country's resources as a means to procure space and energy infrastructure, hoping to exploit new high-tech opportunities. And indeed, in many countries, industry has flourished top-down and bottom-up. India was among the first to benefit from the space race, creating one of the world's most powerful software giants.

Space communications is by far the peak sector of system project opportunities, and it can challenge a nation's industry at every supplier level. In the German state of Saxony, the so-called "*Verdrängungsstrategie*" was used in government (post-reunification) grants as a means to supercharge competition among suppliers at subsystem and equipment levels of space hardware, awarding contracts to start-up firms for bypassing an incumbent supplier. Although the practice penalizes the incumbent, it rewards the winning team and benefits the entire industry as a whole. The energy sector, while receiving frequent government intervention for R&D in the nuclear field, is less apt to receive government R&D money in other fields. Due to the high cost of launching into orbit, space projects frequently need to have government support at the system level, or even lower, if the risk is high. Later, however, these projects generally spawn new opportunities in every branch and sector of

Fig. 21.5 Major global space customers

Major Global Space Customers

• Intelsat	Global
• SES (*Astra & Americom*)	Global
• Eutelsat	Europe
• Intersputnik / RSCC	Russia
• Inmarsat	Global
• JSAT	Japan
• Echostar	US
• Arabsat	MENA
• Asiasat	China
• YahSat	UAE & GCC

industry and the government itself. Figure 21.5 shows the main global space customers and the propagation effect of commercial projects on GDP and employment/competitiveness based on progressive surveys. As missions become repetitive, private industry and public-private partnerships will become more common. Currently, space transport systems and tourism have been moving in that direction. Nevertheless, for countries getting started in space, the resources of their governments are still an absolute necessity, both financially and legally.

Moving ahead into future missions such as Mars or exploration of other planets in our solar system will surely bring about changes in the competitive environment. China endeavors to place a man on the moon in mid-2020. Russia has been involved in disputes with Ukraine, the birthplace of the launcher that carried the first man into space, Yuri Gagarin. Whatever the changes in the visions and planning of global customers may be, the system project management discipline associated with these visions will continue to be the backbone of successful conduct in competitive environments. The following sections of this chapter will elaborate on how this discipline is carried out in actual practice.

21.2 Understanding of Customer Requirements

The most challenging part of winning new contracts is meeting customer requirements. A full understanding of customer requirements is mandatory for writing a proposal; see Chap. 14. Forecasting how these requirements will evolve in the future is even more daunting as the customer himself is generally not certain about the future and often expects the bidders to state the performance of their offered system in response to a known requirement. In space projects, the stakes are high and markets are evolving as swiftly as customer requirements are being defined literally. ASTRA 1 K, a communication satellite of SES, was successfully negotiated by Alcatel, France, in 1999, but due to its failed launch in 2002, the business plan of SES for that satellite was changed to using another satellite instead. Thus, customer requirements evolved so quickly in 3 years that Alcatel was no longer requested to provide the replacement satellite. The old requirements were modified, and a new RFP was released to the industry. And this time, Lockheed Martin was the winner.

Predicting the evolution of markets is only one of the many factors that influence the needs of customers. Technology upgrading within the field of qualified competitors has proven to be another vexing factor, meaning that strategic products can and do drive the market. Companies that dare to introduce new technology into the market must accept high risks as mistakes in the high-tech industry can be unforgiving. Motorola's Iridium was such a case, the failure of which cost a whopping $5 billion in R&D money. On the receiving end of such flawed attempts are ICT consumers, such as the internet. In 2012, the Eutelsat Company launched KaSat, bearing a massive payload aimed at connecting local and long-distance networks for providing high-speed internet service across Europe, North Africa, and the Middle East. Although successfully implanted, the internet industry as a whole has suffered from systemic dysfunction, characteristic of an impending widespread collapse. The main stage of the overall life-cycle scenario is depicted in the following table:

1. Proof-of-Concept for U.S. DOD and early exercises (*in the 1970s*)
2. Commissioning and preliminary service for military units
3. Commercial definition for public operator service
4. Initial commercialization for private service
5. Innovative growth in user hardware and programs
6. Application phase for social networking
7. Security dilemma phase
8. Reliance decline—Friend or Foe
9. Divestiture capture of business assets
10. Reconciliation and transfer of assets to closed systems.

As this book is being written, we are somewhere between stages six and seven. Despite well-known progressive signs of internet collapse, few users are aware of how or when a final death will come. The turning point of life-cycle growth is at stage 7, which can be considered the start of end paralysis and reconfiguration inflicting 50% of the companies, taking into account certain effects not attributed to industry; see Collins [1]. We will see later in this chapter how competition in a global environment can trigger new solutions that are only possible on a global scale. Thinking globally is important to time phasing projects competing for the same customers in all markets, fluctuating by either natural or man-made causes. The origin of the fluctuation matters less than the dynamic and statistical consequences. We will examine how a system's life-cycle quality progressively fluctuates according to the life-cycle costs of its component subsystems, ultimately dooming the market development of the entire system. There are methods to achieve certain defenses against how rapidly the destruction proceeds, such as:

- Capitalizing on consumer loyalty and regional co-dependent financing
- Improved system development, e.g., upgrades, safety features, productivity
- Avoiding self-fulfilling prophecies and single-source suppliers

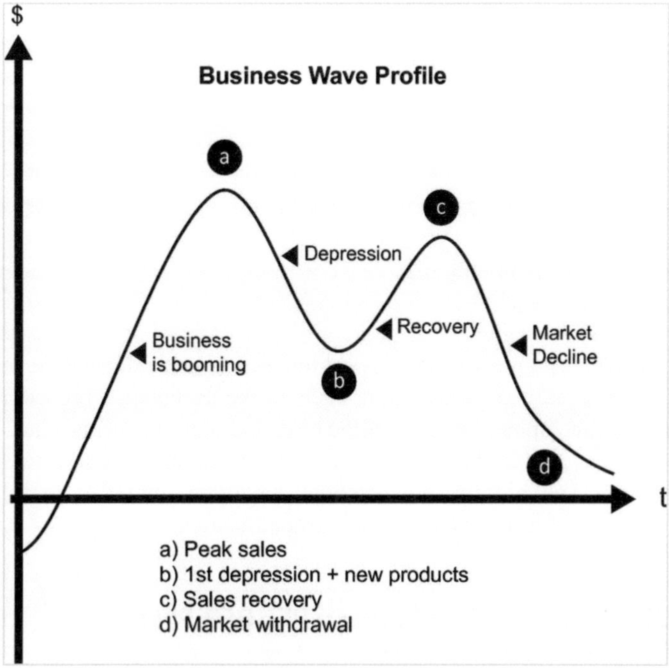

Fig. 21.6 Business development, prediction, and tracking

- System refinement for specific market penetration, e.g., by region
- Providing appropriate insurance coverage.

Depending on the defense methods used, the cost estimation validity and sales policy applied, as well as the general progress of other competing systems, system life-cycle business development can be predicted and tracked, as illustrated in Fig. 21.6. Take note that in major future programs, the idea of co-dependency via an international development team is a fundamental requirement. It not only supports good financing, but it also discourages over-ambitious political leaders from obtaining total system technology independence and thereby violating inter-regional treaties for aggressive purposes.

The consequences of not understanding the requirements of a single customer may result in the barring of suppliers from future bidding. The scale of components or services unsuited for expected applications or business environments may also cause restrictions on sales opportunities. In communication satellite procurement, for example, a moratorium is sometimes placed on the use of equipment that does not meet customer quality standards. It is prudent, therefore, for suppliers to exercise caution when interpreting what the customer asks for in formal RFP documents. Usually, the contents of the RFP include:

1. an introductory page giving background on the intended procurement

2. description of the required system
3. technical specifications
4. test specifications
5. statement of work
6. quality assurance plan
7. management plan
8. commercial requirements
9. draft contract
10. compliance statement and
11. evaluation criteria.

Special schedules/annexes may be additionally inserted depending on the particularities of the intended acquisition, such as the acceptance procedure, implementation and development plan, qualification and manufacturing plan, test plan, inspection, and audit stipulations. All such provisions are essential to successful contract awards, and bidding organizations are often reminded to follow RFP instructions to the letter in order to avoid disqualification.

21.3 Global Acquisition Process: RFI, RFP, Proposal, Evaluation, and NEGO

During the course of maturing and customizing space system projects for both developed and emerging regions of the globe, the methodology has come about to accelerate the acquisition process and responsiveness thereof, especially to latent deficiencies of the customer or the prime contractor. The use of this methodology delineates look-ahead programming to wrap up a workable contract. At the same time, it promotes effective integration of the customer team: that they are prepared for the detailed execution of the contract that they have helped to generate.

Let us first look at an example of an acquisition project, the procurement of a national communication satellite system in an emerging region. Typically, there is a pre-existing business plan for an assumed set of services agreed to by the leader of the national government and blessed by the participating ministries. What, at the outset, usually not known by that government, are the technical requirements to be established and the detailed description by which the project will be implemented to shape the desired system? At the end of negotiations with the down-selected candidate supplier, these details will become fully defined and the selected prime contractor will become committed to complying with the commercial service requirements, fully accepting the RFP model contract, or otherwise prevailing to secure customer acceptance of his non-compliant proposed changes. Only through such a process is it possible for the customer to acquire hardware and software previously not specifiable and for the supplier to maintain the necessary confidentiality imposed by law and industrial agreements or restrictions. The protection of restricted information must always be observed and appropriately managed by means prescribed for each phase of the project.

21.3.1 The Request for Information

The Request for Information (RFI) is one of the least appreciated customer tools, and, considering its great value, relatively little has been written about it in project management literature; see Sect. 14.2. Compared to proposal preparation, for which millions of dollars are allocated yearly to educate some of the largest companies in the world, not much attention is paid to the RFI, and when so, only at the last minute. In reality, a well-developed RFI policy can save a great deal of money when properly applied early in the acquisition process. Some governments, especially those below the federal level, like cities and counties, are starting to become aware of this. Of course, whenever a large number of consenting parties are involved in a potential project, consensus problems are more likely to occur; despite that, the psychological effects of the RFI will often overcome any conflict.

The principal documents to include in an RFI are ideally the same as those one would find in an RFP, as outlined below. The main problem is when the customer is doing it for the first time. Usually, he cannot rely on analogies from other branches, and then he must seek help from consultants. In so doing, it will be prudent for him to make certain that the consultants selected are able to give qualified and impartial advice. Here, it is not only a matter of CV or resume credentials, but also of personal integrity and character, as each consultant will bear the responsibility for the evaluation of information furnished in the disclosures made by candidate suppliers and in the acceptance of such information into the final RFP. The choices made at the transition point between RFI and RFP are crucial to the success of the project. A preference given to a particular technology might at some later point add unacceptable risk. The initial cost or schedule might be influenced. The budgets might be impartially slanted to more favorably agree with one particular bidder, or to a configuration difficult for other bidders to comply with. Nevertheless, in the final analysis, capable consulting must be trusted to bring about the necessary balanced approach, and experience has validated the wisdom of following an RFI step in the project roadmap.

The time allowed in the project schedule could be anywhere from 3 months to 1 year, depending on the expected size and type of contract, as well as the circle of bidders, each of which demands individual scrutiny. It will be essential for the evaluation to be organized in such a way that the same consultant reads and scores identical chapters for all bidders. Therefore, it is not workable to assign different consultants to the same chapters with the intention of reducing the RFI time allocation.

21.3.2 The Request for Proposal

Specific documents that constitute an RFP issued for a turnkey project recently carried out by a Middle Eastern country are representative of a large-scale communication satellite system; see Sect. 14.2. The subject titles are consistent with certain particularities of the customer, which typically vary according to the interfaces of the system with legacy infrastructure. However, the size and scope of the RFP seen below, are rather comprehensive, and the relationship of the documents to each other is quite standard within the space industry. This relationship will determine which document takes precedence in the event of a conflict between two documents. At the beginning of the RFP period, which may extend 3–4 months, a question-answer exchange is customary to allow bidders an opportunity to clarify their understanding of the text, according to bidder guidelines. The procedure is rigorously maintained to ensure fairness during the entire proposal cycle.

Commercial proposals are generally based upon a fixed price with incentives or penalties for lateness, in-orbit performance, and other factors impacting the agreed service quality, including lifetime. Bidders are always prone to stuff their budgets with excess margins in time, mass, reliability, etc., in order to avoid penalties. On the other hand, loss of contract award is the result of exaggerated margins, and therefore, the customer seeks to call for complete transparency through templates for presenting budgets and budget margins, which all bidders must follow identically. Link budgets are standardized by ESA, NASA, and other procurement agencies. The RFP will prescribe the applicable standard authority for parts procurement and materials, as well as the processes allowed in the construction of the system as well as in the associated testing.

21.3.3 Proposal Preparation

Although the preparation of proposals officially starts right after the RFP is received from the prospective customer, the actual start may be as much as several years earlier. Customer replacement satellites are predictable far in advance. Suppliers are frequently preparing long-term strategic plans for disruptive technical innovations or for entering into admission status as qualified bidders to operators such as SES, Intelsat, Eutelsat, and many others. Large aerospace companies like Lockheed Martin, Astrium, Space Systems Loral, and Thales Alenia are regularly setting the stage long in advance of the official RFP release, "*leaking*" the news of their latest designs and seeking early-bird offers to hopefully influence a forthcoming RFP at least as an option. Thus, the jockeying scene is for pre-RFP attention and enhancements to gain customer recognition through special conferences, workshops, and training sessions.

As a rule, most company finance departments impose a Go/No-go review of a major tender prior to kick-off. An organizational announcement will confirm the appointment of the proposal manager and the key personnel responsible for the preparation of each volume of the proposal and each professional discipline, such as system engineering. Instructions will be distributed concerning the proposal preparation schedule and storyboard headlines to capture in words and illustrations the company perceptions and scenarios of a winning proposal. Progress reviews take place almost daily, and a Blue-Team/Red-Team Review will defensively/offensively assess and modify the final proposal with top management officers of the company attending. Except for the above-mentioned Q&A, no communication should take place between customers and bidders while proposals are being written; see Sect. 14.3.

Competition between global aerospace companies is extremely fierce, and many companies resort to improving their chances by teaming with one another for particular proposals. While this occasionally works well, the practice seen from the customer standpoint is not desirable as it reduces competition in a sector where there are few qualified bidders. Admission to this select group is so restricted that even large countries like China are forced to sell in the non-competitive arena. What does this mean? Side-stepping the question for a moment, let us turn attention to Ukraine, a smaller country but one with a powerful history in space launch vehicles, as described earlier in the present chapter. When it was part of the Soviet Union, Ukraine performed in a non-competitive environment, meaning that external market forces had little effect on the evaluation of their merited achievements. Today, however, they are obligated to seek work in free markets and have experienced different challenges in managing cooperation with Boeing in the Sea Launch venture. A competitive venture in space implies self-reliance to maintain control of one's product. It should not be confused with "*vertical integration*" or doing everything yourself. It means securing long-term second-source supplier agreements for your system, which your customer's end user will score as qualified for space. Anything less implies that some part of your system may not be flightworthy.

As the evaluation by the customer moves forward, the plans and criteria for the evaluation of proposals by the customer, as generally announced already in the RFP, are put into action. The scheme followed is similar to the assessment of the RFI responses, where now each proposal is separately rated with respect to the given criteria. A down-selection of usually the two highest rated technical proposals is then announced, signaling the start of negotiations. In the event a clear winner is evident, the negotiations might proceed with only one down-selected candidate and all commercial proposals are then opened. This procedure supports an impartial judgment of the offered technical quality without consideration of the price tag. During NEGO, however, the customer may exercise the prerogative of reverting back to the second place candidate in the event this becomes necessary.

21.3.4 Proposal Negotiation (*NEGO*)

Let us assume that NEGO can be successfully carried out with one candidate bidder. The compliance statements of the selected bidder have been evaluated and show no evidence of a single "*show stopper,*" but indeed one could emerge in the course of NEGO. The course of our sample project is depicted in Fig. 21.7. It should be noted here again that we are dealing with a turnkey program, and the scope of responsibility of the contractor will be enlarged accordingly; see Sect. 14.6.

Preparation for NEGO is strongly dependent upon thorough logistic support and clearly defined organizational responsibilities on both sides of the table. Book captains are appointed for each volume, and separate rooms are provided to ensure an undisturbed flow of progress toward reaching a common text between the

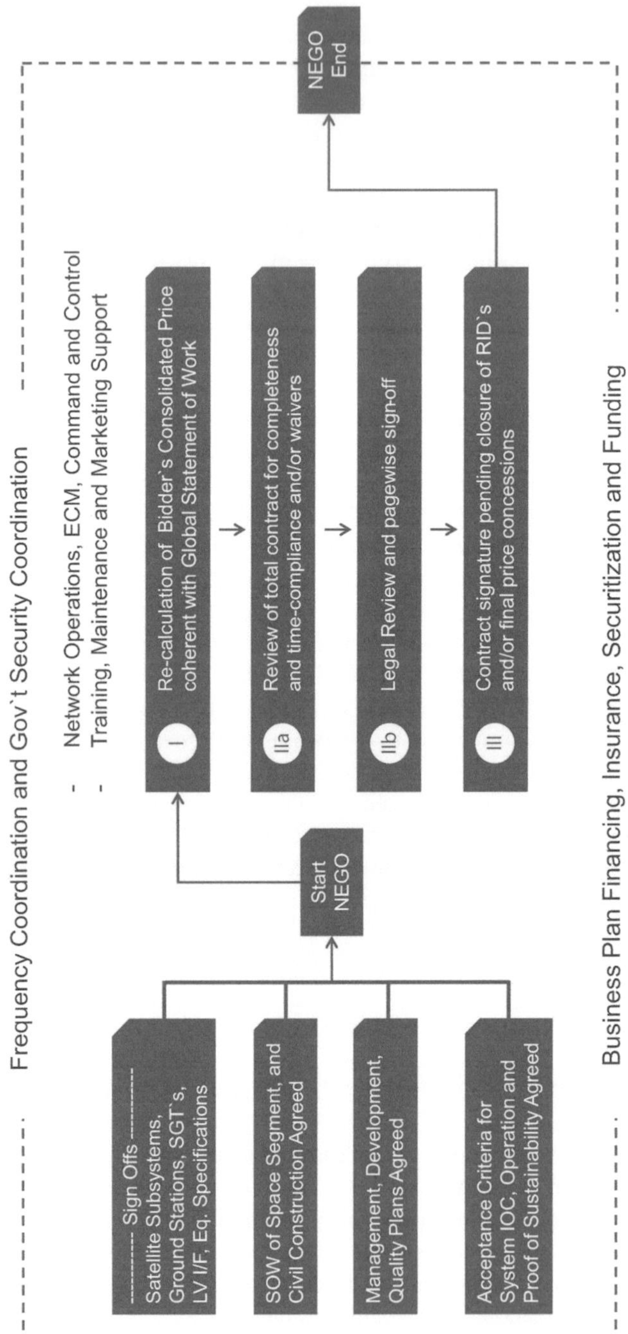

Fig. 21.7 NEGO Process—Course of Events for a Sample Turnkey Program

as-delivered proposal of the NEGO contractor and the as-written text of the customer's RFP. For example, the Statement of Work provided in the RFP will be compared with the NEGO contractor's annotated SOW template in his proposal. Wherever differences appear, the book captain of the proposal version shall provide justification for not complying or how compliance will affect his commercial offer. The objective of the first round of NEGO (*NEGO I*) is to agree upon a single SOW, which then becomes the first annex to the contract. Any and all such deviations agreed upon during NEGO shall be considered the formal contract version at the time that the customer approves the bidder's Best-and-Final offer, which opens the second round of the NEGO process.

The objective of NEGO II is to make a comprehensive review of all changes in the bidder's proposal emanating from NEGO I and to determine whether the Best-and-Final offer is complete and in compliance with the principal requirements of the customer with regard to technical, financial, schedule, and quality performance. NEGO II enables optimization in NEGO III.

The objective of NEGO III is to arrive at a finalization of all remaining open discrepancies and open actions that will be either waived by the customer or brought into compliance by the bidder. At the time that NEGO III is completed, a final round of contract review will take place, ensuring that all attachments and annexes to the contract have been signed off page by page by the respective officers of both parties, enabling the signatures of the contractor's and government's representatives to be affixed to the negotiated contract.

21.4 Industrial Partnership and Alliance in a Global Environment

In this section, we will explore in depth, newly recognized relationships and surprisingly revealing associations between industrial alliances and system project management, using breakthrough evidence to better understand success/failure histories of real projects performed by international alliances, both regional and global. Since the beginning of the 21st century, an avalanche of studies has been performed, helping to illuminate strange and sometimes counter-intuitive events that move us progressively into the human sphere of industrial psychology, culture, and behavioral science.

Already in the early 1960s, as the power of the mainframe computers astonishingly simplified such tasks as weather forecasting (UAC Research Labs, E. Hartford, Conn.), ballistic missile early warning [9] going to the moon [10] [detecting and treating human disease, the challenges began to outpace the concomitant solutions. While technical achievements were sensations headlined in the daily media, unsolved problems in political, social, and fiscal affairs far outstripped potential solutions, and remain so today. The connection between industrial partnerships and system project management was just becoming appreciated at the time, as "*think tanks*" appeared in companies like General Electric, TRW, Rand Corp., Mitre, and others [11]. One outstanding target was the ability to construct market models to steer the economy of the USA. Alan Greenspan, later to be the chairman of the U.S. Federal Reserve Board, was prominently active in using analytic software to help harness the new computer power for complex economic tasks [12]. In view of the enormous attention given earlier to simulation and prediction issues, it is now ironic that we still face a growing enigma. One factor that has become more clear, however, is the importance of good alliances in future undertakings, especially those having vast consequences, such as television broadcasting via satellite, where global investment could leverage a 1000 to 1 added-value industry (i.e., hundreds of trillions in U.S. dollars) [13]. These numbers do not include the added value created in collateral industries outside the telecommunications and media sector, and neither can the numbers account for the strategic impact on world opinion diffused across scattered markets around the world. In a recent example, the international alliance between Astrium and Thales Alenia Space was awarded the turnkey contract to provide UAE multiple spacecraft, "*YahSat,*" for dual (commercial/civil) communication services over a 15 year service lifetime [14]. While traditional considerations continue to motivate partnerships in all forms of industrial enterprise, awareness of new considerations as described in the text below, is vital for implementing major projects of the space age for economic and political world stabilization. Note: In 2007, the author of this book became a corporate advisor and acting CTO and I've served as the management consultant.

21.4.1 "The Only Independent Variable in the Formula is Technology versus Time"

When Alan Greenspan took office at the Fed in 1987, analytical economic tools were spawning widely, despite their obvious deficiencies. Algorithms were generated, and the DOD "*Whiz Kids*" optimized the defense budget but not the military outcome (see Sect. 21.1). In a hearing before Congress, Greenspan formally admitted an error of trust in the self-stabilizing forces of the U.S. economy [15]. Whether it was confidence based on analytical conclusions, his almost 20 year FRB empirical policy, or some combination of the two that led him astray, he seemed to feel betrayed by the events that finally precipitated the great mortgage crisis of 2008.] Currently, there is much controversy in the United States and in European countries, which also felt the shock, as to who should share the blame, and how the fallout can be minimized. We will have to let history be the judge. In the meantime, it will be useful for students to identify alleviating factors that might play a role when the chips are down and such large sums of money are at stake. Test questions appearing at the end of this chapter will stimulate your answers and offer possible opportunities to integrate the ideas learned in this section of the book regarding partnership and alliance in a global environment.

Among the most rewarding aspects of international cooperation is that it offers the chance for people of different cultural viewpoints to work together [16]. In space business, this is so common that we even take it for granted. Also, in space projects, professionals in economics, finance, law, and space policy have rather different viewpoints compared to professionals in technical disciplines. Engineers are particularly fortunate. They are frequently selected as project managers, a role in which they will work with the very same model they have designed and developed earlier. That means they are intimately familiar with the performance of the system they will deliver to their customers, unlike economists, who rarely can define a complex model with good accuracy. For that reason, executive engineers are found to be well suited to lead positions inside alliances and they generally appreciate the time-related critical issues on which project partnerships depend. The success of projects fosters strong alliances, great companies, and great nations. Coordinated efforts and timing are expressed not only in the project schedule; they are intrinsic periodicities of the delivered product, the daily reports, the operation of the product, the annual reviews, and the added value received by the customer.

As systems of the future become more complex, their interfaces in an ever-shrinking world are becoming progressively challenging to manage. This trend is expected to continue during the next decade. Cultural distinctions will intensify the cost of interfacing legacy hardware and the related need to replace failed or obsolete systems. Consequently, cost reductions are seen as essential. During these next critical years, key projects must create the platform for political stability and a recursive economic plan is one of the cornerstones of the recovery time concept. As economic changes become increasingly unpredictable, the only independent variable in the formula is technology versus time, i.e., technological advancement; see Chap. 7. Time constants of high-tech product market life are monotonically shorter, and so are the operating cycles. Due to the need for higher frequencies and bit rates, transaction speeds are now much faster than in the past, introducing new and unexpected problems in traditional communication. One such problem, sometimes called high-speed trading or "*flash crash*" of the stock market, has also been identified in the media, traffic, and other systemic operations. Does this mean that all operations utilizing high-tech communication systems will be subject to flukes and glitches? No, but it does mean that precautions must be taken to preclude them from the outset, and this generally, is neither done nor regulated by any civil or governmental authority. Thus, the advancement of technology presents new challenges that require system projects to build in precautionary measures for a host of reasons. In the case of a flash crash, as noted above, a filtering or data collection node could be used to delay the high speed trading and increase security authentication time. In more demanding cases like running a national economy, stabilizing operations is not quite that simple. However, when contemplating a global alliance, some of the dreams of Joseph Stiglitz might be achieved if judicious planning simultaneously integrates new regulatory devices to enhance long-term co-dependencies between the partners [17]. Alliances are therefore to be regarded as a natural opportunity to eliminate desegregation [18] and should be committed to the following objectives:

- Strategic timing of periodic disbursements, e.g., progress payments and dividends.
- Retention of alliance experts in start-to-finish system project activities.
- Supremacy in high-frequency ICT and secure data methodology.

The above objectives might seem to include goals that are extraneous to traditional joint ventures, acquisitions, mergers, etc. In fact, they are not on the conventional list of reasons that private companies seek out other companies as partners. However, if these goals are accomplished, they will significantly promote the avoidance of a time based dilemma as described above and contribute to the economic stabilization of the involved countries. Therefore, the governments of the participating companies should incentivize the associated undertakings and if appropriate, enter into public-private partnerships with their respective industries. Before elaborating further on the theory behind these new regulatory devices and how this theory can be effectively implanted into present day culture, let us take a look at traditional forms of partnership and alliance in the global environment and how they have evolved in the aerospace industry.

21.4.2 Industrial Cooperation and Alliances for Global Business

Turning attention to existing practices, the background, advantages, disadvantages, and evolution of various forms of partnership commonly practiced in aerospace business are discussed below. It is important for students to bear in mind that the space industry is still relatively young and has emerged principally out of government military organizations (Army, Navy, Air Force); research institutes like Jet Propulsion Laboratories, NASA, and ESA; satellite operators, including national and international, e.g., Intelsat, SES Astra, Inmarsat, and EUTELSAT; and private industry builders of spacecraft, launch vehicles, ground stations, and mobile infrastructure. The companies from which the space industry emerged were typically aeronautic, electronic, and automobile firms. It was not long before the various advantages of one or the other disciplines became apparent, and then mergers or acquisitions became the order of the day by top management to make them more attractive to government agencies. An example of how large aerospace companies emerged in the USA is shown in Fig. 21.8. The process in Europe, shown in Fig. 21.9, was rather similar, whereby the companies sought to reinforce their deficiencies across the borders of Italy, Germany, Great Britain, etc., to better attract customers from two or more different governments. Ultimately, only two major competitors remain: Thales Alenia Space and Airbus Space and Defense, a continuation of EADS.

Partnership motivation is usually sparked by a constant desire to improve long- or short-term business, either through sharing know-how, markets, or access to capital. The common organizational alternatives are: mergers, acquisitions, and strategic partnerships. A strategic partnership is a joint venture or long-term business relationship.

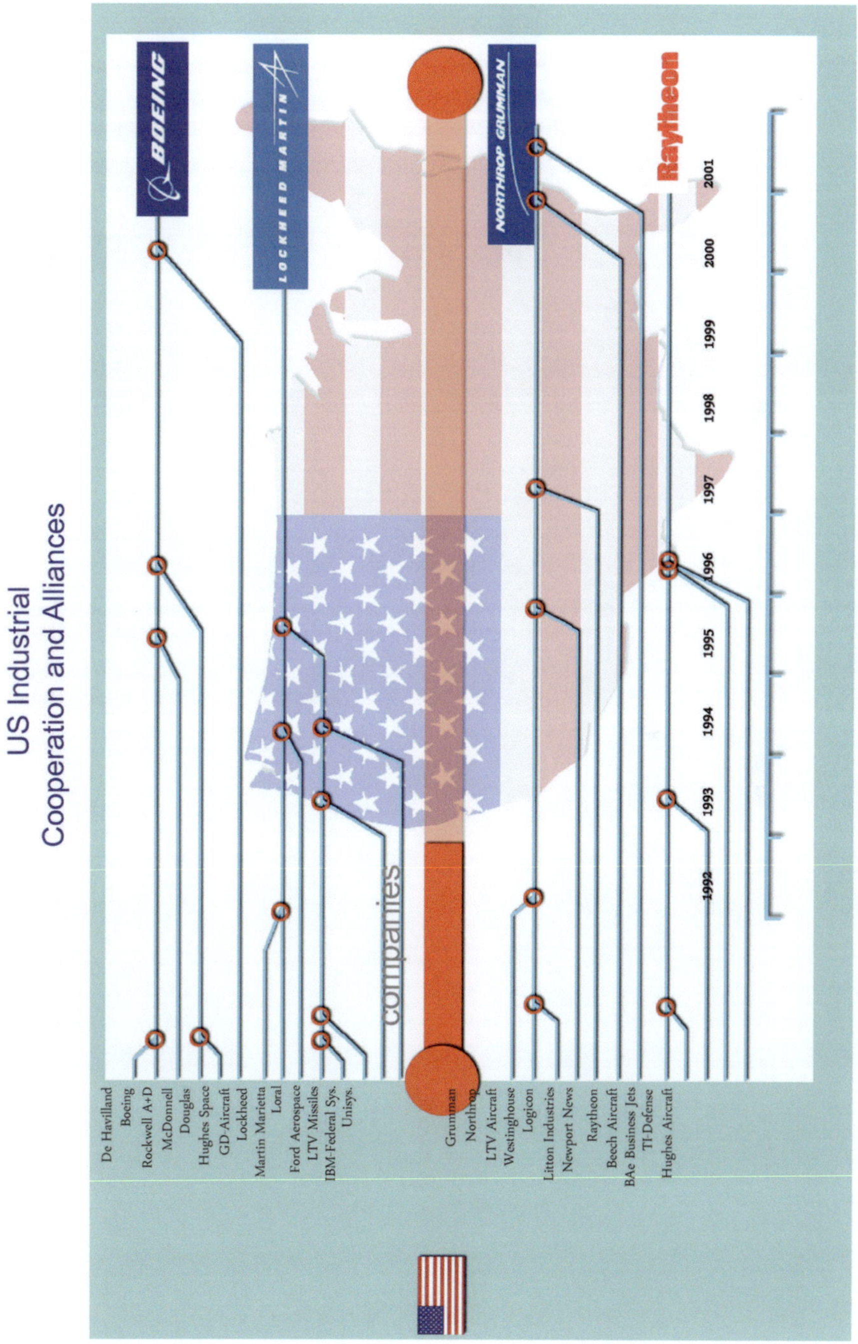

(Source: U. Beck, SSP05)

Fig. 21.8 Industrial Cooperation and Alliances of US Space Companies

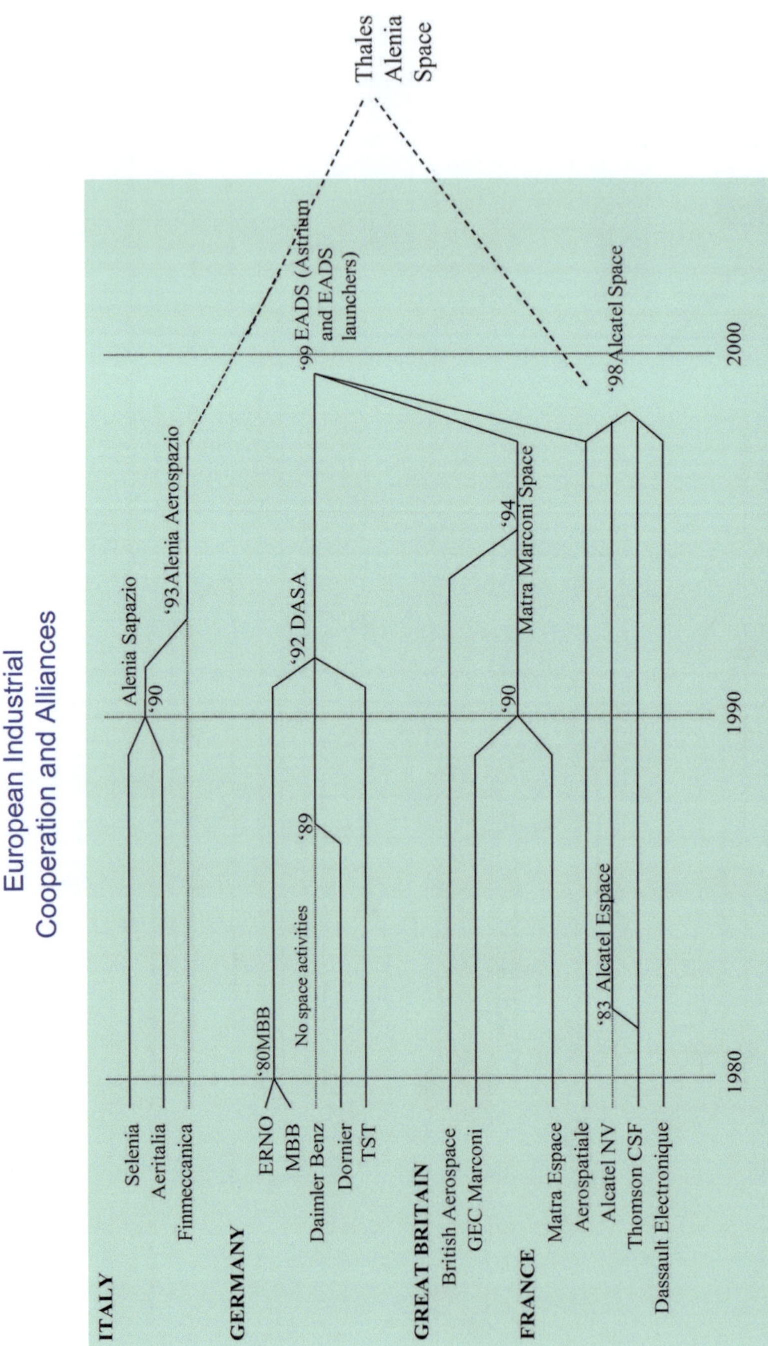

Note: EADS has merged with AIRBUS D&S (status as of 2017)

Fig. 21.9 Industrial Cooperation and Alliances of European Space Companies. Source: ISU EMBA MODULE D Boulder, Col. 2012

21.4.3 Space Alliances to Help Stabilize Global Economy

To recap the above, throughout the world, the space industry has seen a gradual, steady, and voluntary grouping among companies converging into new partnerships and alliances. The process can be viewed as a traditional success course, likely to continue on a global scale for many years to come, The author of this chapter has taken part in numerous international strategic alliances and mergers, including the due diligence of some of the largest space companies. Accumulated experience also demonstrates a surprisingly robust potential for using the space industry to help significantly stabilize national economies. This can be accomplished in three steps:

1. controlling periodic disbursements of leveraging funds
2. blocking and countermeasures against high-frequency interventions
3. cementing cultural synergies to perform space projects more efficiently and with greater co-dependency.

The next important space projects with a typically 360 degree inter-sectoral spread include the U.S. NASA mission to Mars [19], ESA's mining of the Asteroid Belt, DLR's supersonic orbital passenger transportation vehicle "*SpaceLiner*" (*see subsection 7.5.5*), and numerous space tourism and STEMS projects. Here, the acronym STEMS refers to "*Satellite Telemedia Services,*" introduced in Dresden by the author of this chapter to prepare the newly reunified German state of Saxony to weigh in for world competition. At the end of the 1990s, the main issues were:

1. Recovery time to sink unemployment from 23% to the general population rate.
2. Training time, learning curve, and market entry time/project considerations.
3. Partnership organization structure and "*best use*" strategy.
4. Technology of choice in order to win new contract opportunities.

The employer of the Regional Office for Aircraft and Space (ROAS) was the State Ministry of Economics and Labor, and the customers were ESA, DLR, European space companies, and the Federal Ministry of Research and Technology. The goal of the assignment was to evaluate aerospace companies and research institutions statewide and to facilitate their participation in space projects through training, business development workshops, or conferences. For example, in the year 2000, experienced trainers collaborated with ROAS at the Congress Center of the Leipzig Fair to organize lectures representing the products and services of satellite builders/operators from around the world, including SES, Alcatel, Deutsche Aerospace (now Airbus), Globalstar, Lockheed Martin, Motorola, and Space Systems Loral. These lectures provided an opportunity for dozens of Small- and Medium-Sized (SME) firms to test their competence against the most demanding commercial requirements and to set their sights on upgrades or adaptations needed for qualifying their hardware.

The lessons learned in space projects have yielded a rich foundation for the Executive MBA program of the International Space University (*ISU*) at Strasbourg.

Relevant lectures given in that program have been taken into account to provide further insights for this section on "*Partnership and Alliances in a Global Environment.*" Thanks to these insights, carried over with appropriate updating, we are able to present a most complete picture of theoretical aspects and compare them with results obtained from actual projects insofar as they impact the national economy. Lessons learned from space system projects deeply penetrated design concepts of emergency operation and carry over to future design of all systems, whether they are rockets, satellites, aircraft, automobiles, ships, ground systems, amphibious or hybrid space vehicles that can travel in air and in space, like the SpaceLiner mentioned above. Especially when dealing with legacy, however, the human interface has bio-memory constraints that are often overlooked by engineers' jubilant innovative automaticities. That was the case when automatic car window lifters were first introduced, choking passengers who somehow got their heads caught above a closing window. The human brain cannot erase its own legacy of impulsive reactions, which can cause destructive commands in new systems, either by omission or commission of acts learned in prior operating routines. It is mandatory that operators of new systems are regularly tested to ensure that their reactions and reaction-times are not impaired as designers forge ahead into so-called secure emergency modes or failsafe man-machine systems. This will be a huge challenge for car designers in the motorized vehicle industry when entering the era of hands-off driving. In the course of dealing with space projects, a wide array of proven disciplines, is coming of age for the benefit of projects in all sectors and at all size scales from nano- and micro-organisms to systems operating across astronomical boundaries.

Schedule planning for typical procurement of large-scale commercial satellite systems prescribes approximately 3 years for the design and development of the first model and about 2 years for each recurrent satellite. The implications are that it could take 5 years to get into orbit if the first launch fails; the risk is 10%. Based on these numbers, a 15-year, one trillion dollar timetable is set into motion starting at EDC (effective date of contract) of the procurement contract. And with stakes of such magnitude, controlling the periodic disbursement of contract funds has serious economic consequences. Payment plans, it should be noted, are typically designed to penalize the prime contractor for lateness or reward him for flawless performance. This tends to smooth periodic payments within the industrial team and later, during operations, within collateral sectors of the user industries; for communication satellites, this means virtually all sectors of the economy.

There are several reasons why high-speed telecommunications are known to destabilize the economy, notably high-frequency trading (also known in the stock market as "*flash crash*") and the effects of hacker interventions with which we are routinely confronted. However, when there are two (or more) potentially resonating sources in the transfer function of the user's cash flow path, reinforcement occurs according to the closeness of one natural frequency (fn) to another, and economic stability is endangered. Stability is affected by "*maneuvering*" one of the two frequencies, e.g., an annual project dividend is changed to a monthly dividend. The second annual frequency associated with source "*b*" will be no longer in close proximity to the first annual frequency associated with source "*a,*" and the system

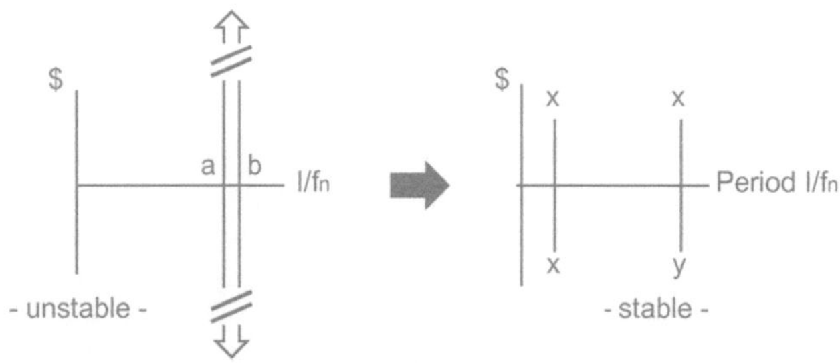

Fig. 21.10 Corrective measures for re-stabilization

intrinsically re-stabilizes. This corrective measure is illustrated in Fig. 21.10, depicting the reinforcement and non-reinforcement conditions as seen in "*functions of a complex variable.*" For those students interested in further reading, a standard text on Operational Calculus will be helpful.

Depending on the particularities of the intervention needed, system projects can be designed that provide wide maneuverability to adjust the stabilization effect on the economy.

21.4.4 Technology, Economy and Humanity: One Frontier

Back in the 1980s, as the mission of the European SpaceLab was first being designed with NASA, DLR, and the industrial partners of the projects associated with this ambitious mission, one could sense the excitement of exploring not only new technology but also new dimensions of financial partnership. Little did we know that the frontiers of "*social and cultural*" limits were being approached with the same success formula of speed and maneuverability. And, as noted earlier, the only independent variable was technology.

The space era is destined to bring people together. Although it does so at high speed when it concerns market investment, there are rather different time manifestations observed in international space projects where it also concerns pride. One important difference observed inside multi-cultural teams even before SpaceLab is the phenomenon of human embarrassment as an expression of pride. Nowhere in the animal world is embarrassment so quickly evidenced as in people, and it also influences decision-making, geo-politics, and industry—some authorities regard it as the last frontier in predicting behavior. Space project managers, however, are accustomed to dealing with the varied cultural idiosyncratic behavior of team members. There may be a "*prima donna*" among them, but they are all aware of the common dangers, up to and including total failure.

The USA, for example, is the "*land of the second chance,*" where mistakes are tolerated, but in some countries, the culture affords no second chance. Nevertheless, educators in teaching System Project Management at NASA and ESA, following empirically proven training techniques, have converted the space project environment into a laboratory for partnership and alliance in the global environment. The future of world progress and human rights are simultaneously in action. Focusing on the lessons learned in hundreds of projects, the reference level as provided in this text book represents state-of-the-art as currently practiced in space. Jointly, Bernd Madauss and I trained project management students of about 25 nationalities from around the world. Planning a turnkey project for an emerging nation is an ideal setting to exercise the experience gained from the lessons learned. Students working on such a project should feel especially fortunate to experience the transformation in behavior of employees of different socio-cultural background after project exposure with coaching and on-the-job training (*OJT*). Time constants and transfer functions are also essentially different from those observed in economic transitions. Despite the presence of non-repeatable responses from individual trainees and stochastic discontinuities, it usually takes 4–6 months of continuous immersion to achieve the level of confidence required for solo operations. This is to be followed by yearly re-evaluation sessions with full-team interaction simulating various emergency situations. These multi-objective sessions provide an environmental assessment of:

(a) employee qualification
(b) speed and quality of response, including psychological conditioning
(c) team stability profile
(d) single point failures.

This evaluation is comparable to Kirkpatrick's evaluation model [20] but modified for space management. Results of periodic assessments and health checks are elaborated together with HR for carrier development purposes.

In summarizing the social and cultural aspects of international alliances in the global environment, it is imperative to emphasize the importance of space system management in setting forth a new standard of best practices for industry and governments around the world. The ethical and commercial value of international partnerships to plan and carry out future space projects has been proven to be unquestionably inspired. But how does this translate to the national and then to the global environment? How a relatively small and young space industry permeates the general population. And will the political authorities of different countries/cultures be against working together for the common good? First of all, the sequence of permeation is likely to spread from global to national faster than the spread from national to global, owing to traditional resistances and obstructions inside the political machinery of national governments. People today are communicating over the boundaries of national boundaries creating a kind of global consensus. Top-down, crowd sourcing, and cross-border media are natural spin-offs of the Internet's social networking, Twitter, Facebook, and other networking innovations that bring new business, new work, and new opportunities.

Although it brings new headaches for lawmakers, it still creates an unstoppable force. Take, for example, the so-called "*uber movement*" in the USA. Here, modern communications have invaded the traditional cab industry nationwide. California and Massachusetts lawmakers are having fits, while the courts are penalizing "*uber*" companies for engaging freelance workers without providing social employment benefits [21]. But moving ahead are other, less invasive forms of progress, and for better or for worse, this process is here to stay. The criteria for the transition time frame is, of course, not the same for political/cultural regimes as they are for worker/family groups. Although likely to be longer, it will surely rise or fall depending on the incentives that accompany the transition. As technology advances, the erratic episodes of unconnected people will become more frequent and grave. System science teaches us that logarithmic decay accompanies every limited linear process. Therefore, the space partnership will ultimately diffuse Earth's retributive obsessions.

In April of 2015, a Falcon 9 launch vehicle lifted the TurkmenAlemMonaco-Sat52E into geostationary orbit, symbolizing a partnership between Turkmenistan, Monaco, and Luxembourg, where Luxembourg will provide the global marketing. Luxembourg has also expressed deep interest in Asteroid Mining, an ESA project gaining considerable attention [22]. Turkmenistan is a country of some five million people, a mineral rich emerging economy on a territory of 491,210 square kilometers with a predominantly Muslim population. Monaco is a predominantly Catholic principality with about 40,000 citizens on about 2 sqkm of land. It is well known for its tourism, festivals, auto racing, and, most of all, its gambling. What will be the economic distillates emerging from this partnership and with SES, Luxembourg marketing, and the media broadcast output for the next 15 years? The GDP growth potential from media is perhaps better than the potential from the Asteroid Belt, based on initial estimates. But returns on intangible benefits could be priceless. Industrial development is often incentivized through intangible benefits, and well-conceived telemedia capitalizes on the co-dependencies between dissimilar national strengths and customs. These questions will be left to the future 15 years of imaginative marketing of SES, one of the best global satellite operators, and to the test answers of student readers of this present text book.

One must wonder: will space create a new gold rush like the one in California during the 1800s? Perhaps, but in the opinion of the authors, such a bonanza is rather unlikely for the foreseeable future, aside from the comparison of hauling costs. While the price of gold per kg might be almost twice as much as the launch cost per kg of payload (*about $20 thousand*), the actual project cost, including mission execution, would far exceed the price of a conjectured payload. One thing appears to be certain, however: despite all the spontaneous diversions that space offers to reward our earthly fantasy, the pull of the sky will continue to be a perennial attraction for mankind. And whether our efforts succeed or not, individually or collectively, we can never erase our flirt with space. It is incumbent upon system project managers to populate their teams with key personnel having the right "*stuff*" *to* prepare as-built configurations for adaptability into future missions—not an easy order, especially within partnerships.

Engineers today, accustomed to fast prototyping, must extend their design tools in order to explore and rapidly synthesize virtual "*what-if*" scenarios before impending environments overtake us, causing unexpected failures to be suffered in situ.

Outstanding people are needed for such demanding responsibilities. Foresighted scientists, astro-physicists, materials engineers, chemists, biologists, procurements, legal/political and financial experts are essential to predict new markets and charting new human territories. The future system project is a train moving forward. The front of the train is breaking up new ground, and at the back, the ground is breaking up the end of the train. In the middle of the train, useful cargo is stored for long-term system productivity. Project teams of experts have to be suitably selected for the different qualifications needed across the time zones of each part of the train. The journey of the PM will become progressively rapid, and the life-cycle durations in the space era will become progressively shorter. Let's not get stuck along the way with useless prototypes. In the following.

During the 1980s, all eyes were on the RCA company as it introduced the first economy-of-scale communications satellite production series, RCA 4000 [23] with sufficient capacity to accommodate up to 45 medium power transponders. In 1985, SES Astra bought the fourth satellite of the 4000 series, which became ASTRA 1A.

The new design not only incorporated all the payload advances (e.g., *HDTV*) of their earlier systems, but also the 3-axis stabilized and propulsion systems successfully advanced in Europe. Soon after, other U.S. companies like Hughes and Lockheed also offered competitive concepts for buyers to procure the same class of satellite, including the already-accepted electric propulsion embedded in a highly flexible bus and an in-orbit reconfigurable payload to match the elusive communication demands throughout the mission's lifetime. More important for gaining high productivity, advanced manufacturing technology applied was becoming more cost-attractive to commercial as well as government users [Aviation Week, Sept. 18, 2014, Amy Svitak, "*Lockheed unveils upgrades to A2100 satellite bus*"]. The great managers of the above mentioned companies were instinctively aware that innovative satellite technical performance would attract customer attention, but only innovations in productivity could retain customers over the long haul and thereby achieve stable business for their respective companies. They were aware of the correlation between learning curve cycles and operating lifetimes, of development time versus production time, and of investment cost versus amortization. But it takes rare talent to control the project participants, who are responsible for delivering all these resources according to plan. Continuous reference to the "*train process*" is imperative. Are you at the front of the train or somewhere between the front and the middle? This is a crucial question because long-term stability in the middle of the train brings the capital necessary for breaking new ground at the front of the train. Also, technical attractiveness at the front of the train brings the capital necessary for long-term stability in the middle of the train. Therefore, keeping the correct level of effort among participating industrial partners in view of time planning, life cycles, and time constants is the key to mastery of transitioning system projects. A compilation of satellites ordered from Hughes and Lockheed in 1995–2015 shows that about 150 high powered spacecraft were launched worldwide.

With all that power and bandwidth blossoming forth in orbit, the next natural step was to go after the biggest prize of all internet mobile services. Motorola, with its longstanding reputation for open roads and car radios, seemed preordained to capture such a market. Borrowing from their extensive experience with satellite constellation technology as applied to the Iridium program (*see Sect. 21.2*); Motorola moved ahead, for better or for worse, with a globally distributed RFP. We were called to Phoenix, Arizona, to present one of the most ambitious proposals "*ever conceived by man.*" By the time we arrived, the headquarters building was swarming with contender teams from all over the world and we had gotten word that we were allocated exactly 1 hour, including Q&A, to deliver our presentation. Our slides, depicting the offer from Contrives Switzerland, bristled with illustrations of the most powerful high speed ISL system known anywhere. The silence in the auditorium at the end of the review was a tribute to the outstanding European research and development, led by Dr. Reinhard Czichy with an eminent international bidder team of experts, most of whom contributed to laser R&D projects promoted by the European Space Agency. Having been denied earlier, an FCC pioneer preference for Iridium earlier, Motorola declined to try again and looked warmly at the impressive link budgets flashed on the screen. But at this point, there were still a few open questions to be answered before the winning team would be down-selected for contract negotiations. Turning to events taking place in Europe, a brief detour follows, describing the landscape at that time.

21.4.5 Worldwide Success Stories in Space over the Past 50 Years

During the post-Apollo era, a host of new programs entered the space arena. On the U.S. side of the Atlantic, NASA introduced the Space Shuttle, an effort to replace costly non-reusable manned launchers, and took leadership of the Space Lab project, a newly conceived multinational collaboration that set the stage for the International Space Station (ISS). Despite some political accommodation, the ISS proved to be a huge success in spearheading world cooperation. On the European side of the Atlantic, ESA was active, together with its member states, in the highly successful development of the Ariane heavy launchers. Germany, within the frame of a cooperative BMFT/GfW-NASA treaty, awarded MBB a contract to design, construct, test, and space qualify two HELIOS scientific spacecraft, both launched by the United States: HELIOS A in 1974 and HELIOS B in 1976. MBB, supported by General Electric consultants, set new standards for conducting project management in European space programs. In fact, the accomplishment of the management goals of the HELIOS project was the second objective of the program after the success of the technical mission. A Final Report summarizing the lessons learned was published by MBB in December 1976 [24].

Bernd Madauss and Victor Billig, who at the time of the HELIOS mission were in charge of MBB's Project Control and Subcontract Management, respectively, are credited with bridging Europe into the PM methodology needed to carry out commercial space programs under the industrial policy of the European Space Agency. Thus, through HELIOS, Bernd Madauss imported PM know-how into Luxembourg for procuring and operating SES's global Astra commercial satellite fleet. He exercised the same intensive disciplines in his contracts with Ariane Space, and other launch agencies to provide the best launchers and launch services available worldwide, contributing directly to the SES success story. The ILS, founded in 1995, is a prime example of Russia's highly competitive spirit for cooperative marketing of the Russian Proton launcher for commercial projects. Despite the joint venture founded in 1995, Russian know-how in competitive space business continues to suffer from the same isolation practices we encountered in other countries behind the "*Iron Curtain.*"

After our retirement, Bernd and I, conducted several PM training/coaching programs to help launcher/telecommunications companies and other agencies or companies organize their own policies for space system procurement or upgrading of legacy infrastructure. These programs, typically given in 1–4 weeks depending on the extent of topic content in marketing, quality, and engineering, have been delivered to many countries, including Germany, Oman, the UAE, China, Bolivia, Ukraine, Japan, Indonesia, Austria, Poland, and, as mentioned above, Luxembourg. In 2007, Bernd joined the YahSat satellite company in the UAE for 1 year as a Corporate Advisor and also became acting CTO.

Sponsoring jointly with the International Space University, the consultancy company ACT/PMT, founded by Bernd Madauss and located in southern Bavaria, has enhanced the character of many space programs and broadened the opportunities for executive cross-exchanges. Since the start of the new millennium, a growing demand has been evident for satellite communication and ACT/PMT has expanded its scope of activity to encompass team consultancy projects wherein groups of ACT/PMT consultants performed well defined tasks for specific time periods or milestones, such as providing guidance to a customer for proposal preparation or at a Critical Design Review. Examples of consultant activity provided are:

- The United Arab Emirates received ACT/PMT support services for the preparation of RFI/RFP, tender evaluation, negotiations, and contract preparation for the turnkey project YahSat. Executive officer staff was provided by three consultants over a period of 2 years, including field office services.
- ACT/PMT conducted in 2010 an intense 25-day training program about Satellite Technology and Management Training at the Royal Army of the Sultanate of Oman.
- The Bolivian Space Agency ABE (*Agencia Boliviana Espacial*) received for several months field office support from ACT/PMT for the TKSAT-1 communication satellite manufactured and launched by Chinese companies China Academy of Space Technology (CAST) and CALC. Five highly experienced consultants were provided on a rotating schedule for ABE continuous advisory and field office support at CGWIC (*China Great Wall Inc.*) and CAST (*China Academy of Space Technology*); see Fig. 21.11.

It should be noted, that, with the genius spirit of Ludwig Bölkow in mind, we, the former employees of MBB, continued to carry the flag of system engineering and project management through consulting and training. At some point in time, it might not matter who the flagbearer was, as long as the mission was accomplished.

Fig. 21.11 ACT/PMT TKSAT-1 Field Office Team Members. From left to right: Arnold Jablonsky, Victor Billig, Klaus Polzer, Bernd Madauss, and Hans-Jürgen Tscheulin

21.4.6 When Working on the Frontiers of Globalization Think Globally

Owing to the pervasive influence of communication systems in general, the new trends witnessed in space began to spread to many fields of application. Successful industrialists like Elon Musk (Space X), Sir Richard Branson (*Virgin Galactic*), Robert Bigelow (*Genesis Space Habitat*), Burt Rutan and Paul Allen (*Space Ship One*), and Jeff Bezos (Blue Horizon Space Tourism) gravitate towards the USA but operate XYZ business from anywhere in the world and headline regularly in the daily media. Space medicine in Europe has also followed a similar scenario. In the 1980s, major advancements in cardiology centered on the progress and technological developments of the United States.

This implies, of course, that human, capital, and other physical resources are available, but moreover, "*Making Globalization Work*" as Nobel Prize laureate Joseph Stiglitz sees it in his well-known book, also means that the timing of project efforts must be globally maneuvered: that costly and time-consuming programs are scheduled, managed, and implemented with respect to globally significant time periods. Keep in mind that mastery of world health and prosperity will depend heavily on our working intelligently with our solar environment, exploiting the cyclic interfaces of its interactive systems, and counting on great people to move us forward from land to sea, to air, to space.

System engineering and project management, originally developed in the aerospace domain, have now penetrated other high-tech fields of technology, medicine, biology, and other domains. May the seventh edition of this book about project management continue to distribute practical knowledge and wisdom to our worldwide younger generations of engineers and business people. They will be obligated to efficiently manage and execute, to the benefit of today's society, the envisaged future projects of growing technical and social complexity in a challenging and competitive global environment.

Literature

1. Collins, Jim: How the Mighty Fall, Publisher Harper Collins, 2009
2. Dickson, Paul: SPUTNIK: The Shock of the Century, Walker Publishing Company, New York, 2001.
3. Eisenhower, Dwight D.; Farewell Speech, Jan. 17,1961, Yale Law School
4. Hickman,Kennedy: General Curtis E. Lamay: Father of the Strategic Air Command https://militaryhistory.about.com
5. Editors of Encyclopedia Britannica "*Robert S. McNamara*" https://www.britannica.com
6. US Air Force: "*General Bernard Adolph Schriever*" www.af.mil
7. US Air Force: Manual 66-1 Maintenance Management Policy April 1998
8. Admiral Hyman, Rickover G.: "*Doing a Job*" GovLeaders.org 1982.
9. Stars & Stripes Media, Patrick Dickson, 2014 "*BMEWS, Thule.*"
10. NASA Johnson Space Center, Houston, updated May 26, 2016 (JSC from Gemini and Apollo Projects onward).
11. Washington Post Feb. 23, 1996 "*The Think Tank That Went Out for a Spin...*" (explains why companies split in two for separation from government work)
12. An example of U.S. economy model consulted by FRB: "*Tool for Macroeconomic Policy Analysis*" FRB Feds Notes, Apr. 2014
13. List of 419 active satellites in geostationary orbit as of Nov. 20, 2015. Satsignals.net
14. YahSat: How to locate@52.5East, complete tutorial https://Worldsatellitehouse.blogspot.com
15. Associated Press, NBC Lisa Myers report: "*Greenspan admits 10/23/2008 the economic crisis was his fault but revealed a flaw in a lifetime of economic thinking and left him in a state of shocked disbelief*"
16. Billig, Victor: "*Cultural and Social Impact on Project Management*" ISU Seminar Boulder,Colorado March 2012.
17. Stieglitz, Joseph: "*Making Globalisation Work*" W. W. Norton and Co., 2006
18. Billig, Victor, Dodel, Hans, Madauss, Bernd: "*Broadband via Satellite - The Great Desegregation SAGETTE*" (Initiative for China) Presentation ESA 2nd Conference Space and Society, Noordwijk Feb. 27, 2007.
19. Aldrin, Buzz: Mission to Mars, National Geographic Society, Washington, D.C., 2013.
20. Kirkpatrick, Donald: https://en.m.wikipedia.org
21. Isaac, M., Scheiber, N.: "*Uber offers concessions but drivers stay freelancers*" International New York Times (San Francisco) Apr. 23, 2016.
22. MIT Technology Review, Michael Reilly, Feb. 14, 2016
23. The Digital Transformation Agenda 2016 Encyclopaedia Astronautics www.astronautics.com
24. HELIOS Final Report, Design and Development Contract, QC-90-00 VP 6/70, DFVLR, Cologne December 1976.

Standardized Project Management Plan (PMP) 22

Contents

B. -J. Madauss, *Project Management*,
https://doi.org/10.1007/978-3-662-69057-4_22

Abstract

In most cases, carrying out projects is a complex and extremely varied task dependent not only on the use of the correct methodological approach, but also on dealings with customers, superiors, subcontractors, suppliers and co-workers.

The author has complied with the frequent requests from project engineers who are entrusted with the execution of international projects to publish a "Standardized PMProject Management Plan (PMP)" which was developed in the course of his international work as a free-standing element of this book. This PMP is based on the author's experience and is recommended as a guideline for use in the performance of international projects.

22.1 Introduction

22.1.1 Scope and Applicability

The PMP project provides detailed descriptions of the management approach and defines the organizational concept and the rules, methods, and procedures to be applied by the responsible Organization (*i.e. Customer or Contractor*) for the execution of the project.

The respective Organization understands the importance of project management for the successful execution of the project and has chosen a professional approach to project management, which is reflected in the selected and well-proven project management procedures contained in this PMP. The project management procedures fully comply with the requirements stipulated in the relevant documentation.

22.1.2 Role of the Customer and Contractor

The Customer role is the enabler of project development, serving the Customer objectives. The Contractor [*name*] for the project will be fully responsible vis-à-vis the Customer [*name*] for the implementation and execution of the project. The Customer understands, that the Contractor and its subcontractors that they are business partners.

The contractor's role is to manage the work to be performed by the Contractor and its subcontractors, and to ensure the timely completion of the project activities. The Contractor has the leadership for all technical, administrative, and contractual tasks to be performed by the Contractor and the selected subcontractors, to achieve the contractually defined project objectives in accordance with the agreed system specification, the Statement of Work (SOW) and the required quality assurance standards within the frame of the agreed schedule and cost targets.

For all contractual and/or technical directions between the Customer and the Contractor, as well as between the Contractor and the subcontractors, the designated Project Manager of the Contractor will act as the single focal point.

22.1.3 Subcontractor Control

The Contractor has the responsibility to lead and control the assigned subcontractors. As a pre-requisite for successful subcontractor control, the Contractor and its subcontractors have jointly established detailed project plans and procedures. All subcontractors agree with the proposed management principles laid down in this PMP.

The Subcontractor Manager (SCM) of the Contractor continuously monitors and controls the progress and the status of the project work performed by subcontractors. The SCM acts as the formal point of contact between the Contractor and the associated subcontractors. The Contract Officer (CO), Project Control Manager PCM), System Engineering Manager (SEM), and Quality Assurance Manager (QAM) will support the SCM in his duties.

22.2 Project Management Approach

22.2.1 General Project Requirements

The key project management requirements can be summarized as follows:

- Implementation of a project management organization.
- Appointment of the Project Manager.
- Appointment of the project managers at subcontractors.
- Assignment of authority to Contractor and subcontractor project managers.
- Identification of key personnel at the Contractor and subcontractors.
- Implementation of effective communication means, including reporting to the Customer and the company management, and.
- Implementation of efficient planning and control methodologies.

22.2.2 Customer Reference Documents

As a general rule Customer requirements shall principally govern the PMP application. For space or high-tech-projects following ESA management requirements and procedures may be applied as a guideline for performing the project management functions (ECSS reference: ecss-secretariat@esa.int):

• ECSS-M-00A	Policy and principles
• ECSS-M-10A	Project breakdown structures
• ECSS-M-20A	Project AM organization
• ECSS-M-30A	Project AM phasing and planning
• ECSS-M-40A	Configuration management
• ECSS-M-50A	Information/documentation management
• ECSS-M-60A	Cost & schedule management
• ECSS-M-70A	Integrated logistics support

22.2.3 Industrial Organization and Management Structure

22.2.3.1 Project Structure

The project structure proposed by the Contractor follows the established Work Breakdown Structure (WBS), which is further detailed in Sect. 22.3 of this PMP.

22.2.3.2 Commitment of the Project Team

The members of the project team, the Contractor, and the associated subcontractors, have agreed to apply the management ground rules defined in this PMP. A joint team agreement will be stated in the Memorandum of Understanding (MOU) and the individual subcontracts will define the working relationship, the obligations, and the rights of the subcontractors. Immediately after the contract award, the industrial organization of the team will be implemented. It is important, that the contractual relationship between the Customer, Contractor, and subcontractors is transparent and clearly structured.

The management proposals submitted by the subcontractors to the Prime Contractor form an integral part of this PMP. Said proposals contain a description of the subcontractor management approach, the extension of the WBS as applicable to each subcontractor, the associated Work Package Descriptions (WPDs), and the subcontractor schedules manpower & cost plans.

22.2.3.3 Management Approach

In line with the Customers management guidelines, the Contractor proposes the following management approach for the project:

- Project management guidelines implemented by the Contractor: The project will be organized along proven management guidelines and principles. The project manager will be the only single point of contact for all communications between the Customer the Contractor and the subcontractors.
- Project Management Office (PMO): At contract award, a centralized PMO will be implemented. The head of the PMO will be the appointed Project Manager. The assigned Project Manager (PM), Mr. [*name*] will lead the management team of the Prime Contractor.
- The assigned PMO team members will support him in his management function, see Fig. 5.16.

- Project Structure and Work Sharing: The work-sharing concept between the Contractor and subcontractors shall be agreed upon by the parties. It also defines the responsibility of the subcontractors.
- Work Package Responsibility: The management responsibility of individual work packages resides with the appointed WP managers at the Contractor and at its subcontractors. The appointed project manager has the overall responsibility for the performance of the project tasks, including those of the subcontractors.
- Change of task assignment: Any change of WP assignment must be immediately reported to the Contractor's PM.
- Planning and Control Software: For planning and control purposes, dedicated project management and control software will be used to assist in planning, scheduling, and reporting of the project status.
- Regular Status Monitoring: Progress will be regularly monitored and measured for tasks performed by the Contractor and subcontractors through means of reports and meetings between all team members concerned.
- Change Control: All change requests (engineering change proposal–ECP) shall be assessed by the Contractor for potential impacts on the design, performance, schedule, and costs of the project prior to implementation.

22.3 Project Breakdown Structure

22.3.1 General

This part of the PMP provides details of the proposed project organization and an introduction to the Product Tree (PT), the WBS, and the related WPDs.

22.3.2 Program Organization Breakdown Structure

22.3.2.1 Responsibility and Authority of the Program Manager

If possible all PM staff members shall perform their duties within the close vicinity of the central PM organization of the Contractor at *[address]*.

The PM is fully authorized to plan, monitor, and control the work of the project and he is responsible for meeting the contractual requirements and maintaining cognizance of the progress of the activities being conducted for the project. The project management staff will support the PM in his duties.

The PM of the project [*name*] will have the authority to give directions to all PMO members. Similar to the Contractor, the subcontractor PMs will also give directions to their teams involved in the project. Authority to control the work in progress at the Contractor as well as at subcontractors will be assigned to the assigned PMs. The PM of the Contractor maintains and controls the managerial interface between the Customer, the functional departments of the Contractor, and all subcontractors involved in the project. A typical project organization of a satellite operationg company is shown in Fig. 22.1.

See the sample in Fig. 5.16

22.3.2.2 PM Support Functions

The PM support functions, including a detailed definition of tasks and responsibilities, are defined as follows:

Contract Officer (COF)

1. Contract negotiation with Customer
2. Contract negotiations with subcontractors
3. Contract implementation at Contractor level
4. Negotiation and implementation of CCNs
5. Control of milestone payments
6. General COF tasks

Project Control Management (PCM)

1. Implementation and control of SOW, WBS and WPDs
2. Schedule monitoring and control
3. In-house manpower and cost control
4. Overall project cost control
5. Contract change control
6. Inventory control
7. Configuration and documentation control
8. Management information system

Technical Management (TEM)

1. System specification
2. Interface specification
3. Subsystem Specifications
4. Standardization
5. Demonstration
6. Product Management
7. Manufacturing supervision
8. Process analysis and control
9. AIT planning and organization
10. Supervision of AIT activities

Subsystem Control Management (SCM)

1. Subsystem engineering supervision
2. Subsystem analysis and specification control
3. Subsystem design analysis and review
4. Technical monitoring and control
5. Progress monitoring (in close cooperation with the PCM)
6. Subcontractor change control
7. Action item control

Quality Assurance Management (QAM)

1. QA supervision of the total Project
2. QA planning and control

3. Quality assurance
4. Safety analysis
5. Reliability analysis
6. Minutes of meeting

22.3.2.3 Assigned Key Personnel

The successful execution of a project largely depends on the qualifications of the managers and engineers recruited for the project. The staff of the Contractor and its subcontractors will be selected based on the relevant technical and management experience and practical know-how. The PM key personnel is responsible for completing the project within the technical and financial targets and the given schedule limitations. Résumés of the selected key personnel must be provided.

22.3.2.4 Company Organization of the Contractor

The PM will draw the required resources from the functional departments of the Contractor and/or the assigned subcontractors. Work Packages identified in the WBS will be assigned to the appointed Work Package (WP) managers at Contractor and/or subcontractors. In doing so, the Contractor will apply the principles of matrix organization. Therefore, each WP manager has to report to his line manager at Contractor and/or at subcontractors but must also report to the PMO concerning the technical planning and the progress made of the particular WP. Further details are described in paragraph 22.4 of this PMP; schedule planning and Control.

22.3.2.5 Reporting to Company Management

Because of the priority assigned to the project, the PM reports directly to the Chief Technology Officer (CTO). He will report regularly about the progress of the project and will also highlight any critical problems that he has encountered including proposed solutions to solve the problems.

22.3.2.6 Project Advisory Board[1]

A Project Advisory Board (PAB) consisting of highly experienced staff members of the Contractor has been set up to monitor and control the activities of the project team. The members of the PAB will be:

Based on the review results the board will make recommendations to the PM and will ask for corrective actions if felt necessary.

22.3.3 Product Tree and Work Breakdown Structure

See the example at Fig. 9.8

The WBS shall reflect the natural breakdown of the system into subsystems, components, and WPs. The WBS will be the basic reference for schedule planning and control (paragraph 22.4), for manpower and cost planning and control

[1] The PAB members must be identified and indicated by their names.

(paragraph 22.5), for configuration and documentation control (paragraph 22.6 and 22.7), for risk management (paragraph 22.8), for contract change control (paragraph 22.9) and for information management (paragraph 22.10).

22.3.4 Work Package Description

WPDs have been prepared with reference to the established WBS and in conjunction with the technical definition and the established project schedules. The WPDs are contained in [WPD document] of the project.

22.3.5 Program Directory

A table with names, companies, and coordinates (telephone, fax, and e-mail) must be indicated by the user of this PMP.

22.4 Management and Control

22.4.1 Introduction

The Contractor must prepare a control procedure with the aim of establishing a systematic process for schedule planning and control. This will be the responsibility of the assigned PCM.

The Master Project Bar chart (MPB) and the associated detailed project schedules will form the baseline for schedule monitoring and control. All schedules are established based on detailed analyses of the work to be performed by the Contractor and their subcontractors in accordance with the established WBS, in order to achieve the established project goals. The PCM coordinates, monitors, and controls all schedules established by the departments of the Contractor and subcontractors involved in the project.

22.4.2 Master Project Bar-Chart

The MPB is subdivided into the major project phases. The first phase of the MPB covers the concept phase (phase A) followed by definition (phase B) and development (phase D). In addition to the MPB, detailed plans must be prepared. The MPB will be used for overall project schedule control and for status reporting to the Customer. The established detailed schedules include margins for critical activities in order to cope with small problems, which may occur during the design, development, integration, and testing of the particular subsystems or WP. The Prime Contractor and its subcontractors will make sure that adequate resources are available to complete the project task in accordance with the schedule targets. Consequently, the following points, which are not exclusive, are considered to be included:

- Sufficient lead times for procurement processes
- Margins for rearrangements of test and integration sequences
- Margins for changes in the availability of planned facilities
- Margins due to normal downtime of facilities for maintenance and
- Allowance of time to solve any interference and incompatibility of system/subsystem/units detected during integration and test.

The Contractor shall apply appropriate software for project schedule control.

22.4.3 Definition of Program Milestones

The Contractor and the subcontractors must select a series of major milestones, which shall be integrated into the MPB. Each milestone denotes a significant progress checkpoint in the work to be performed. As a guideline, about four major Program events per year have been defined. Typical project milestones are "*Deliverables,*" "*SRR,*" "*PDR,*" "*CDR,*" "*QR,*" "*AR,*" etc.

22.4.4 Schedule Monitoring and Control

The Contractor must implement a schedule monitoring and control system, which ensures the timely provision of all goods and services specified in the contractual arrangement of the project. The schedule monitoring and control system will in particular ensure the following:

- Effective communication of the approved schedules to subcontractors responsible for the execution of the work;
- Adherence to the requirements of these plans and schedules;
- Use of these plans to control actively the execution of work;
- Timely reporting of any actual or foreseeable deviation from the plan to the PMO at Prime Contractor;
- Effective processing of such reported deviations including the definition and implementation of any required corrective action;
- Maintenance of an up-to-date schedule database, which permits effective progress reporting to the management of the Prime Contractor and to the Customer.

Schedule monitoring and control will be a prime responsibility of the PCM. The PCM shall coordinate, monitor, and control all schedules established in accordance with the WBS and as applicable to the functional departments of the Prime Contractor and the relevant subcontractors. He also shall prepare schedule status summaries.

22.4.5 Biweekly Schedule Reviews

On a weekly or biweekly routine, the PCM of the Contractor will review the planning status, taking into account the actual accomplishments made in the various work areas. He will call for status meetings with the Contractor and teleconferences with subcontractors.

If a series of changes have been made to the planning logic, duration, or interfaces, the PCM shall forward to the PMO of the Contractor and the subcontractors concerned an updated and legible version of the revised MPB. The changed planning documents shall be provided as soon as possible but no later than 1 week after such changes have been introduced. The actual dates for completed activities will be clearly shown. As part of the documentation submitted prior to each subsystem progress meeting the PCM will update and issue the appropriate plans on the basis of which a review of the schedule status will be conducted.

Subcontractors are required to inform the SCM of the Contractor as soon as possible of any changes in the project which has an effect on the overall planning. The SCM will in close cooperation with the PCM review the overall planning and if appropriate, notify the PM of the Contractor, the Customer and the subcontractors concerned of the actual status situation.

22.4.6 Schedule Reporting

If appropriate the Customer may be invited to attend the Monthly Progress Meetings (MPMs) between the Contractor and all subcontractors. Progress information will be made available during the MPM to all parties involved in the project. On a quarterly basis the Prime Contractor will provide the Customer 2 weeks after completion of each quarter, with a schedule status overview. The overview will be generated on the basis of the monthly project reviews. This information will be transmitted in a brief report and will be related to the MPB. A complete analysis of the plans will be given, such that the completion of all major activities and milestone achievements can be seen on the MPB. If necessary the critical area, derived from the analysis, will be identified on the MPB. The schedule reports will contain the following information:

- Major activities started or completed, giving actual dates;
- Major activities in progress if the expected completion date has shifted, giving new expected date;
- Major activities involved in any other change: e.g. added, deleted, change of duration, start or end dates, network logic including interfaces;
- Explicit confirmation that all activities not included in the list have not been re-planned or modified in any other way.

22.5 Manpower and Cost Control

22.5.1 Introduction

The manpower and cost control procedure for the project has been prepared with the aim to establish guideline for manpower and cost planning and control at the Contractor and its subcontractors. The assigned PCM has the prime responsibility for manpower and cost control. The Contractor aims to apply a disciplined manpower and cost control system. The basic elements of this system are:

- The agreed WBS
- The finalized WPDs
- The agreed project schedules (see paragraph 22.4).

The project Manpower and Cost Plan (MCP) provided contains the total estimated manpower and costs of the entire project. The MCP summarizes the manpower and costs of all WPs at the Contractor and subcontractors as identified in the WBS. The control process of the estimated resources and the associated costs has to be performed on the basis of the established MCP. In line with the project schedules the PM will issue the Work Authorizations (WAs). For tasks to be performed by the associated subcontractors, WAs will be issued in line with the terms and conditions of the respective subcontract.

22.5.2 Preparation of Manpower and Cost Plan

Jointly with the project team the PCM of the Contractor shall prepare in conjunction with the established WBS, the WPDs and the project schedules a detailed MCP, which contains all efforts, goods and services, which the industrial team of the project intends to provide. The cost breakdown is strictly related to the established WBS, which provides details of the hierarchical structure of the Program and which shows at its lowest level for all non-purchase items the individual WPs. The objectives, contents and outputs (deliverables) of the WPs are defined in the relevant WPDs and form the basis for the manpower and cost estimating process.

The anticipated manpower efforts and costs for the entire project, as well as for summary levels defined in conjunction with the WBS, have been presented against time, using generic formats of the MCP. Manpower and costs will be subdivided into cost categories and into time elements. Usually the timely manpower and cost allocation is provided by monthly intervals for the first year and into quarters for the remaining part of the project duration.

22.5.2.1 Labor Efforts

Proposed labor efforts have been listed according to the following labor categories:

- Management and administration
- Engineering and scientists
- Drafting and documentation
- Manufacturing
- Others, e.g. technicians (to be specified)

In cases the breakdown of labor categories as specified above was not sufficient, other corresponding categories of the project are shown.

The number of man-hours from start of work has been estimated by the project on a normal working time basis. It has also been taken into account, that neither overtime, nor any unproductive time such as holiday, vacation, sick leave, etc. is included in the proposed labor effort, (e.g. no hours will be shown in if the factory is closed for holiday and no work is performed during that period). The project will not take into account any labor efforts for supporting services already covered by an overhead function or claimed elsewhere. Exceptions may be made for special cases, such as the use of internal special facilities, which are used longer than a normal working day.

For the preparation of the MCP for the project the Prime Contractor has applied standard hourly labor rates, as defined by the Finance Department of Prime Contractor and subcontractors. These labor rates include all overheads, but exclude any fee, profit or any other supplementary charges.

22.5.2.2 Manpower Deployment

Manpower deployment histograms will be submitted by the Contractor showing the total efforts in man-hours and costs subdivided into the periods used in the MCP. In addition one histogram with similar information has been submitted for each firm involved in the project.

22.5.2.3 Non-Labor Costs

The non-labor costs will be given for the appropriate periods of the WP duration in the following categories:

- Materials
- Internal Special Facilities
- External Major Products
- External Special Equipment
- External Services
- Travel and Subsistence
- Packing and Shipping
- Others, e.g. Insurance (to be specified)

Where Internal special facility time is quoted, the charging method and the type of facility will be specified. Where the purchase of external major products or external special equipment is indicated, the name of the product or equipment and expected supplier has been specified in the WP description.

Travel cost has been presented in the relevant WPDs. For the subsistence cost, each participating firm (Contractor or subcontractor) will indicate the per diem rate where applicable and the total number of travel days utilized in its cost figure. Preparation of the travel and subsistence information is based on the company regulations regarding travel and subsistence expenses.

The cost figures for non-labor involving outside procurement (purchase of materials and/or supplies, rental/purchase of special equipment and/or tools, external services) will be indicated on an expenditure basis; that is, the estimated cost will be shown in the period when the costs are expected to be incurred.

22.5.2.4 Total Cost per Work Package

The total cost per work package is the grand total of the total labor cost and the total non-labor cost (as quoted in accordance with the above requirements). Each firm will submit a separate MCP for its own part of that work, quoting the costs in the local currency. For all cost estimates the Euro will be applied as currency.

22.5.2.5 Cost References

The detailed MCP prepared by the Contractor and subcontractors involved in the project, and as subsequently negotiated, resubmitted and agreed to by the Contractor management and/or the Customer, will serve as the basis for manpower and cost control. In addition it will serve as a reference for reporting and monitoring for all cost reimbursement areas, and as a reference for assessing any changes throughout the contract.

In cases of renegotiating the total cost of the project or parts thereof, or in cases of converting one type of cost to another, the project will upon request by the Prime Contractor management and/or he Customer refine the appropriate detailed cost documents compatible with the renegotiated cost conditions.

22.5.2.6 Manpower and Cost Summaries

The Company MCP will be a summary of all manpower and cost data contained in the relevant WPs. Therefore, for each company involved in the project, i.e. for the Contractor and the associated subcontractors, a summarized Company Manpower and Cost Plan will be prepared.

22.5.2.7 Resubmission of Cost Documents

If during subsequent negotiations between the project team and the Contractor management or between the Customer and the project team, the cost of the total project or of individual WPs or other areas of work is revised, the project will resubmit the affected cost documents. Such documents will be in conformity to the provisions set forth above. All rates and costs will be shown in Euro.

22.5.3 Manpower and Cost Control at Prime Contractor

22.5.3.1 Work Authorization

The WA documents serve as a formal agreement between the PMO and the team leaders at the functional departments of the Contractor and/or the subcontractors, which are responsible for the performance of the tasks. The WA may contain a complete WP or part of a WP. The total work content of a WP has been defined in the appropriate WPD. Any negotiated contract modification that may cause changes to the system will be implemented through an ECP. Upon approval of the ECP, a Contract Change Notice (CCN) will be issued; see Sect. 22.9.

22.5.3.2 Company-Wide Accounting

A company-wide computerized accounting system must be used to collect and control project related cost information and compare the actual costs with the planned budgets. The cost data will be organized by the cost account number, which is directly related to the WBS code. Actual cost data are generated from personnel time allocation and invoices for items and services been procured externally.

22.5.3.3 Cost Reporting at the Contractor

Reporting of manpower used and costs incurred for work packages in progress will be performed by using the same cost categories and accounting periods as the negotiated and agreed cost reference. The project will supply Contractor management with the documents and information specified in the following sections. Every month a manpower report will be submitted to the Contractor management as follows:

- 5 business days after end of reporting period for WPs at Prime Contractor
- 10 business days after end of reporting period for WPs at subcontractors.

This report will show for each active WP, the planned and actual man-hours for the reporting period.

22.5.3.4 Manpower and Cost Control at Subcontractors

The Manpower and Cost control procedures applied at subcontractors are described in subcontractor commitments to the Contractor.

22.5.4 Manpower and Cost Reporting to Customer

As part of the quarterly progress report to the Customer, the Contractor will provide the Customer with a report of manpower and cost been occurred at the Contractor and at subcontractors;. The report will include actual spending of manpower, material, travel, etc. of each WP and summaries per company and at project level.

22.6 Configuration Management and Control

22.6.1 Introduction

The Configuration Management and Control function is related to identification, control and reporting, of the following:

- Identification and documentation of functional and physical characteristics of systems, subsystems and products;
- Control of changes to the characteristics of the identified items, with verification of compliance to agreed requirements;

- Detection of implication of changes applied to one configuration items on others and on systems;
- Record of changes, documentation and reporting of change processing and implementation status;
- Enable all participants to a program to view and use identical data during the life cycle of the program, in a consistent and controlled manner.

22.6.2 Organization of the Configuration Management and Control

The aim is to ensure that the product(s) of the project are conforming to an agreed system configuration, based on the approved Technical Specifications (TS) related to the elements of the final product(s) for which they accept design responsibility.

The PM of the Contractor will delegate the authority for configuration and documentation control to the PCM. He will be responsible for establishing and monitoring of the standard configuration and documentation control practices and procedures available at the Contractor in order to ensure configuration and documentation control compliant with the contractual requirements, including specifications, WPDs and CDRL. The configuration control system covers the following topics:

- Identify the Configuration Items (CIs)
- Identify, maintain and update the configuration baseline
- Identify and control all physical and functional interfaces
- Organize Configuration Control Boards (CCBs) in order to control and release baseline documents and their updates.
- Process engineering and contractual changes (ECPs, CCNs)
- Control the implementation of approved changes
- Provide configuration reporting
- As design configuration status
- As-built configuration status
- Change status list
- Waiver status list

The documentation control system consists of a centralized project library, which covers following responsibilities:

- Record of all documentation issued
- Prepare and dispatch document/data lists
- Dispatch and distribute the documentation
- Administer a formal review flow for accepting/approval of documents issued.

During the project set-up, the product(s) of the project, including its components will be described. The Contractor will produce and maintain a list of all configuration items identified, based on inputs from all parties (Customer, Contractor and subcontractors). At project go ahead, the Prime Contractor will produce a Configuration Item Data List (CIDL). A version of the CIDL showing the "*as-designed*" status is produced for each review. The version submitted for the acceptance reviews contains the "*as-built*" status (ABCL).

Technical notes and handouts are listed if the content has an important relation to the product without being described in another document. Applicable documents issued by the Customer, subcontractors or Contractor will be listed as well. Before formal configuration control, the documents for the CIDL are baselined. Change control is then set up and implemented, as well as the control of interfaces.

Verification is made that configuration has been made with respect to the requirements specified in the project requirements description document established during project organization.

22.6.3 Configuration Baselines

After project go ahead, the configuration baselines (functional, development and production configuration baselines) will be established as references to the approved definition of the project, taking into account the requirements identified during the earlier stages of the project.

The documents related to the configuration baselines describe characteristics of the product(s) and serve as a reference for verifying at key milestones of the project the conformance to the initial specifications. In case of required changes of the configuration these will be negotiated taking into account the configuration baselines.

22.6.4 Configuration Items

A configuration item can be treated as a black box in the configuration management process, with or without interfaces to other configuration items. All configuration items are identified and listed in the (CIDL).

The generation and issuance of specifications, standards and drawings that define the CI will maintain configuration identification. The PCM will assign unique identification numbers to be issued to the project line item specifications, procedures and drawings. Each item will be assigned a unique identification number. Previously released drawings of proven designs, standard specifications and common hardware will retain previously assigned numbers. Subcontractor items will be identified by the manufacturer's part number or by the assigned Prime Contractor part number. New drawings and associated lists will conform to engineering standards and procedures implemented at Prime Contractor.

Interfaces are controlled using Interface specifications and/or Interface Control Documents/Drawings (ICDs). This includes interfaces to components external to the system, interfaces between system components and interfaces among configuration items.

22.6.5 Configuration Control Procedures

The configuration control procedures list is established during the preliminary project definition. This will be confirmed at the preliminary design review. The formal configuration control based on the specification approved by the Contractor is started at the end of the System Requirement Review (SRR). The documents required to control a configuration item are identified during the first project phase and confirmed latest at the Preliminary Design Review (PDR) of the corresponding product.

A Change Control Document (CCD) is established to document any evolution of a configuration baseline. This document includes a section for description, justification and classification of the change request, and for review and approval by the Change Control Board (CCB).

Any subsequent request for change to a document, drawing or parts list will initiate a change request that will be processed by the CCB. The class I change requests will be transformed to a formal ECP, for which the Customer's approval is required prior to its implementation. Class II changes will be processed by project CCB and the Customer will be regularly informed of these changes and their status.

Change control will begin with the establishment of an allocated baseline. Changes to the baseline configuration will be processed through project control. This process will consist of a uniform and systematic origination, analysis, preparation, review, approval, authorization and release of engineering changes.

The change control system will include a submission of an ECP and a firm definition of change classification; i.e. class I or II. Change classification will be the primary determinant of the manner in which a change shall be processed. The configuration control process defined in succeeding paragraphs will apply to software and hardware.

The CCB will be an organization that approves or disapproves the ECP or document change requests. Permanent members of the CCB will be the key people of the PMO. Specialists from Prime Contractor functional departments and/or subcontractors may be invited to participate to the CCB. All CCB members, whether permanently assigned or specially invited, must be fully authorized to take decisions on behalf of their respective organization, and to accept actions placed upon the organization they represent.

All chosen contractors have demonstrated to the Contractor that their Configuration Management and Control system is matching the requirements stipulated by Prime Contractor. All subcontractors will be audited by the PM of the Contractor to ensure that the methods employed by the participating contractors are compatible with the Contractor requirements.

22.6.6 Configuration Management Reporting to Customer

The PCM will maintain configuration status and will issue a report as part of the quarterly progress report to the Customer; see also Sect. 11.2.5. Records of the hardware and software configuration (as-built) will be compared to the released (as-designed) configuration. The as-designed configuration will be derived from released engineering records. An "*as-designed*" report will be maintained and updated as changes are approved by CCB. PCM will prepare an as-built configuration status report for each deliverable system at the control item level. Each control item will be identified in terms of location, part number and serial number. The report will be initially prepared at the start of control item integration and maintained through completion of integration and test activities.

22.7 Documentation Management and Control

22.7.1 Introduction

The Contractor documentation control system consists of a centralized Program library, which covers following aspects:

- Record of all documentation issued
- Prepare and dispatch document/data lists
- Dispatch and distribute the documentation
- Administer a formal review flow for accepting/approval of documents issued.

Documentation control falls under the responsibility of the appointed PCM. The PCM will be obligated to control the actual status of all project documents prepared by Prime Contractor and the subcontractors. The PCM has the responsibility to implement and maintain the Program related documentation control system (DCS). A Document Management System (DMS) in place at Prime Contractor is available for automated handling and control of document management principles.

Each document required for the project will be filed in accordance with the Contractor documentation coding system. The next step will be the immediate distribution of documents to the assigned recipients at Customer, Contractor or at the subcontractors in accordance with the implemented project organization. Status monitoring of all documents presently available or under preparation will be performed with the DMS. All parties involved in a particular project, shall receive regular document status reports.

22.7.2 Documents Format and Presentation

Prime Contractor's DCS has been designed to ensure proper handling of documentation for projects, which are to be executed at Prime Contractor. The DCS function covers identification, preparation, implementation, distribution, and status control of all relevant project documents. This includes the following tasks:

- **Identification**. All documents required for completion of a program must be identified prior to start of the project. This includes also the documents to be delivered to the Contractor management and/or Customer. It is mandatory that all document requirements be shown in the Deliverable Requirements List (DRL) of the project. All contractually deliverable documents shall be contained in the Contractual Deliverable Requirements List (CDRL). During the course of the project the Contractor may generate additional project documents, and/or the associated contractors or any other organization may produce additional documents, which must also be included in the DCS.

- **Preparation**. Preparation of project documents must follow the basic rules (structure and formats), that the Contractor has commonly implemented for programs to be executed at the Contractor.
- **Approval and Release**. All official documents stated in the CDRL, require approval by the author and the responsible PM prior to release of the document. The assigned PM shall approve all documents, which require approval by company management or Customer, prior to release of the document.
- **Distribution**. Distribution of documents and keeping track of such distribution is the responsibility of the appointed PM or the assigned PCM. Timely updates of documents have to be provided accordingly.
- **Status Control**. The PCM has to perform status control of all project documents.

22.7.3 Document Preparation

All formal documents of the project will be prepared in a generic format and structure. Therefore each formal document generated for the project must be structured to contain following elements:

- Cover Page
- Distribution Record
- Change/Revision Record
- Table of Content
- Text and Figures
- Abbreviations (if applicable)

22.7.4 Generic Documentation Numbering System

The generic documentation numbering system to be applied for projects performed at the Contractor is composed of following elements:

Program	Code
• Document Type	
• Originator Code	
• Running Number	
• Issue Number	
• Revision Code	

The composition of the entire documentation numbering system for the contractors projects is as shown hereafter:

PPP-DDD—OOO—RRRRR—I—R	
PPP	Project code
DDD	Document type
OOO	Originator code
RRRRR	Running number
I	Issue number
R	Revision code

Note: It must be noted, that the drawing numbering system at the Contractor may slightly deviate from the document numbering system

22.7.5 Document Type

A two or three digit alphanumerical code is foreseen for the definition of the specific document types. A list with typical codes for document types is provided hereafter. If required the code may be extended to more digits.

Document Type	Code
Handbook (any kind of)	HB
Instruction (any kind of)	IN
Plan (any kind of)	PL
Procedure (any kind of)	PR
Reports	RE
Review documents	RD
Specification (any kind of)	SP
Minutes of meeting	MM
Fax	FX
Letter	LT

22.7.6 Originator Code

Use of an agreed short originator code or the initials of the person who has issued the particular document may indicate the document originators, companies, departments and/or authors. Usually such code should not exceed three digits.

- Running Number: A Running Number (RN) shall be assigned to each document, so as to allow for unique identification of all documents prepared for a particular program. Pending on the size of the program the RN may be a four, five or six digit code. The first document prepared and registered shall receive the RN "*0001,*" the second document shall be numbered "*0002*" and so forth.
- Issue Code: All FDs shall be issued together with an Issue Code, starting with "*I*" for the first issue, "*II*" for the second issue and so on.

- Revision Code: All officially released documents are initially published with Issue Code "*I*" but at the first issue have no revision code assigned. Any revision in data or text to a particular issue of a document, for example to issue "*I*" of a specification, shall be identified through a revision code, i.e. "*a*," "*b*," "*c*," etc. for the first, second and third revision. Revisions are to be clearly marked in the relevant page of the particular document and shall be identified in the Change and Revision sheet of the document.

Updating of documents takes place in a controlled manner, following the rules established under paragraph 22.7.6 The available DMS at the Contractor ensures upon registration that no previous document with identical issue/revision code exists and requires a different one.

22.7.7 Document Categories

The Program documents may be grouped into formal and working documents as defined hereafter:

- Formal Documents (FDs): FDs are all documents, which are required for the completion of a program and therefore the PM must approve them. FDs carry official signature of the appointed PMs of Prime Contractor. FDs prepared by the associated contractors, may also require approval by the Prime Contractor PM. All FDs must be formally released and are to be listed in the Data or Document Requirements List (DRL) respectively in the CDRL; for example specifications, plans, handbooks, etc. FDs are also documents under strict documentation and/or configuration control.
- Working Documents (WDs): WDs are documents not required by a DRL but which may carry important project information and which will be used as back-up data. For example minutes, memos, and any other documents prepared during the course of program execution.
- Documents for "*Read Only*" and Documents with "*Restricted Access*": Because of the nature of commercial confidentiality, certain documents may not be made available by the prime and/or subcontractors involved in the Program. The definition for such documents are defined below:
  - "*Read-only*" documents are documents that can only be consulted at the subcontractor's premises in a so-called "*data room.*" In principle, if no derogations are given, no notes can be taken in this data room.
  - "*Restricted Access*" documents are documents made available to a limited group of people at Customer and/or Prime Contractor, which have signed respective None Disclosure Agreements (NDAs). The people having signed an NDA with the respective subcontractor(s) will receive a numbered identified paper copy of these documents from the subcontractor. These people will personally be responsible for the confidentiality of these documents.

22.7.8 Contractual Data Requirements List

The CDRL has to be prepared and agreed upon prior to start of the Program. The CDRL will become part of the contractual documentation. A typical CDRL contains following information (see also following sample):

(a) Document Identification (number and title)
(b) Originator (name of author/company/etc.)
(c) Planned completion date (calendar date or months after receipt of order)
(d) Submittal requirements (number of copies and destination)

Sample CDRL

a. Document identification		b. Originator	c. Completion	d. Submittal requirements	
Number	Title	Name	Date	Copies	Destination

22.7.9 Document Approval and Release

It is common praxis that for any prepared document the author will sign the document prior to distribution. In some cases a document may be distributed as draft only or a preliminary version. However also in case the author should be clearly identified.

All FDs must be formally approved and officially released prior to distribution. At project start the PM must assign authorities for approval and release of documents. In line with the established project tree, the PM, who has the overall approval authority, may delegate his power for approval to other members of the project team. Release of documents requires the particular document has been approved by the PM or one of the project team members. The PM has the authority to release any project document and he may delegate his power for release to other members of the project team. In some cases Customer and/or Prime Contractor management may require to approve any high-level project document prior to its release. Details hereof have to be agreed upon prior to start of the project.

The Contractor has implemented a central filing system for projects to be executed at Prime Contractor, in order to identify all project documents.

The appointed PCM for the project will issue and update a Documentation Status List (DSL), which provides the project management teams at Prime Contractor, CUSTOMER and/or of the subcontractors with a current status overview. The project documents listed in the CDRL shall be made available at the times indicated in the CDRL. Since the completion dates for the particular documents are related to Program schedules, the PM or the PCM must have at all times of project execution full visibility of the documentation status. He must be in the position to identify the status of the project documents as follows:

- Not yet available: Documents which has been requested by the CDRL but are not yet available, and which are to be prepared at a later stage of the project; see CDRL.

- Under Preparation: Documents which are not yet available but which are required in the CDRL and are under preparation.
- Been Approved: Documents, which are completed and approved and have been made available (released and distributed) to the project teams at Prime Contractor, customer and/or the associated contractors.
- Under Review: Documents, which are in use but presently under revision.

22.7.10 Document Distribution

The PCM is responsible for the distribution of documents to the project team members at Prime Contractor, and if applicable also project team members at Customer and/or the subcontractors. It is a general rule at Prime Contractor, that all project documents (i.e. drafts, approved issues, changes, etc.) shall be distributed as soon as possible and without any delay. In this respect another important rule is to be followed is the distribution of documents according to need. Therefore the PCM will decide on the distribution lists. The PCM must keep track of the distribution in order to provide the team members with updates of the documents and he must make sure that older issues are not in use anymore. Distribution of documents may be done by hard copies or by means of electronic data processing or by e-mail. In order to reduce the amount of paperwork involved, and to increase the speed of document submittal, electronic distribution of documents shall be applied as much as possible.

22.7.11 Documentation Management Reporting to Customer

The PCM will maintain documentation status and will issue a report as part of the quarterly progress report to the Customer.

22.7.12 Document Review

A Review Data Package is to be submitted to the Customer 10 working days prior to each major review (SSR, PDR, CDR). Customer and/or partners are to examine the documents and summarize their comments and recommendations/questions in the form of Review Item Discrepancy (RID).

The RIDs will be made available to the Contractor and partners five working days before the review meeting, in order to allow them to analyze them and organize the review according to the RIDs content. Written answers will be presented and discussed during the Review Meeting. At the Review Meeting, the Customer and Prime Contractor shall evaluate the answers to the RIDs and define jointly a decision per RID (closed or AI). The minutes shall be written with the following information:

- RID forms, including question, answer and decision
- action items

– remaining discrepancies/findings
– review-board conclusions (outcome of the review, with possible conditions for closeout).

22.8 Risk Management

22.8.1 Introduction

The Contractor will perform a risk management function in order to determine risks of the project. Risk management is aimed to minimize or mitigate any risk that might result from failure to complete the project in compliance with the relevant project requirements. Prime Contractor will perform continuously risk assessment functions in order to identify, evaluate, monitor and control any potential risk. The primary objective of risk management for the project includes:

(a) Keeping project management informed on any potential and actual risk which becomes visible.
(b) Identification of risks for each Work Package in order to inventory potential causes and assess consequences on project objectives.
(c) Classify all identified risks according to the ranking of their criticality.
(d) Implement on a quarterly basis a risk abatement plan.
(e) Establish and implement a quarterly Risk Status Report.

22.8.2 Definitions and Policy

Risk management and mitigation is the task of identifying and analyzing all of the potential risk areas and to establish a risk mitigation/retirement procedure. Risk mitigation steps are initiated for reduction of the probability of occurrence and/or the elimination and/or total reduction of the consequences of the identified risk.

Risk is defined as an uncertainty of successfully achieving the commercial, technical and/or programmatic requirements of the project; i.e.:

- Can the goals of the project business case be achieved?
- Can the performance stated within the specification be achieved?
- Can the planned tasks stated in the WPDs be achieved within the agreed financial budgets and within the frame of the project milestones?
- Are all of the required resources fully available for the project?

The provisions of risk management will be implemented on all WBS levels of the project. Identification of risks, the immediate introduction of risk assessment actions and elimination of risks shall be the primary formal means for risk management of the project. Prime Contractor and all subcontractors will strictly follow the risk management requirements as specified in this PMP.

The risk management activity is an integral part of the PMO performance. The PM of the Contractor will assign the responsibility for the routine tasks of risk management to the PCM, but contributions are required from all members of the

PMO. In all cases when a major risk has been identified, this event will immediately be brought to the attention of the PM, and he will get directly involved in the risk retirement action.

22.8.3 Risk Assessments and Management

Risk is defined as an uncertainty of successfully achieving the technical and/or programmatic requirements of the project. The following are the steps to be taken for implementation of the risk management processes to be applied for the project.

See also example at Table 19.2

Prime Contractor will perform for each major element within the agreed WBS a risk assessment in order to recognize as early as possible any potential risks. Initially all potential causes for risks will be identified and the consequences for the project objectives will be analyzed. The Prime contractors PM will assign a risk owner responsible for conducting mitigation actions.

22.8.4 Risk Control and Reporting

The main part of risk management is the risk control process. Any risk has two dimensions:

(a) the probability of the event and
(b) the impact of risk to the project.

In other words, the risk control process consists of risk assessment, reduction or elimination and management. In cooperation with all specialists involved, the technical specifications (performance, operational, environmental requirements) are to be checked with respect to the adequacy of information and the precise and identical interpretation by both Customer and the Contractor supported by subcontractors.

The Contractor will produce a Risk Status Report (RSR) as part of the Quarterly Progress Report to the Customer. The RSR will contain for each major risk identified the following information:

The RSR will contain following information:

- Identified risks
- Date of creation
- Expected date of releasing the risk
- Cause and type of risk
- Consequences (evaluation on performance, schedule and cost)
- Risk criticality ranking (Low, Medium, High)
- WBS element or Work package concerned
- Designated risk owner
- Initiated mitigation action and their current status.

The Prime Contractor will include in the Quarterly Progress Report information about the risks status applicable to project.

22.9 Contract Change Control

22.9.1 Introduction

This contract change control procedure has been prepared with the aim to establish a procedure for control of contract changes to the project to be executed by Prime Contractor. The contract officer at Prime Contractor will perform the contract administration task of the project, applying the implemented standard procedures of Prime Contractor. The basic tasks, which he will perform, will be as follows:

- Negotiation and implementation of the contract and its attachments to the total project, including subcontractors and suppliers.
- Implementation of changes; determination and assessment of change impact and approval.
- If approved, the change will be incorporated into the applicable documents.

22.9.2 Definition of Changes

A Change request is needed if there are either:

- Changes to the product/system requirements/specifications
- Changes to the general conditions of the agreement or the procedures
- Changes to the deliverables
- Changes requested to the interfaces.

22.9.3 Classification of Changes

Change request will be differentiated as follows:

- Class I: Changes affecting system level which may be stipulated by the Contractor
- Class II: Changes affecting at subsystem level which may be stipulated by the Contractor or a subcontractor.

Procedure to Introduce an Engineering Change: The introduction of an ECP will be issued on a change request form. Each form must be numbered as defined in the document management control procedure (see Sect. 12.2) applying the ECP standards of the Customer.

22.9.4 Contract Change Notice (CCN)

For all alterations/changes to the contractual conditions of the Agreement the Contractor shall submit to the CCB a CCN for approval. The approval of a change submitted to/by the Contractor with a CCN is given when the members of the CCB have signed the CCN; see Chap. 13. The change that impact the overall project cost and/or schedule will have to be accepted by a chosen representative from Customer before the Contractor shall implement the change.

The quarterly report to Customer shall contain status information of all changes requested and approved.

22.10 Information Management System

22.10.1 Introduction

Information exchange is an important pre-requisite for the management of complex systems. Therefore the Prime Contractor must implement a project Management Information System (MIS), which takes care of information management and control between the Contractor and the Customer on the one side and between subcontractors on the other side.

An important part of the proposed MIS for the is the submission of reports and general information exchange between the parties involved. The appointed PCM will be in charge of the MIS.

The provisions of the information management will be implemented at all project levels. This system shall be the primary formal means for official recurring status reports about technical matters, schedule, financial and legal information of the project efforts. All project team members shall strictly follow the reporting requirements as specified in this plan.

The MIS is the primary mechanism for providing early warning of potential changes to approved schedules and resource requirements for the project. The system will also be used to document reasons for changes. Customer decisions and approvals for changes based on information received through the MIS will be promptly transmitted by the Prime Contractor to the Customer, subcontractors and WP managers concerned.

The level concept is a formalized and disciplined approach for identifying levels of information within the WBS. The MIS requires compatibility, consistency and accountability for all data from the top level for which Contractor is held accountable, down to the lowest level data provided by subcontract management. Levels of reporting in line with the WBS are:

- Level 1: Total system
- Level 2: Subsystems
- Level 3: Work Packages.

22.10.2 Responsibility and Accountability

Inherent in the MIS, is the concept that responsibility for accomplishment of an element of work includes responsibility for reporting the progress of the work and the status of resources utilization. The name and signature of this individual are clearly identified on all associated schedules and resources plans. The duties of the accomplishment individual include:

- Scheduling and subsequent rescheduling
- Accountability for resources employed
- Coordination of project effort in accordance with the project objectives
- Insuring consistency between his reports and those on related activities
- Making comparative time/cost evaluations
- Schedule cost and manpower programming
- Identification of potential problems
- Determination of alternative corrective actions
- Initiations of authorized corrective actions
- Early warning of Prime Contractor management when self-initiated corrective action is inadequate to deal with the problem.

Also included within the system is the concept that the Contractor management one level above the accomplishment responsible individual will review and approve all reports. Provision is made on all schedules and resource plans for inserting the names and signatures of the official making this review.

22.10.3 Meetings and Reviews

The planned meetings and reviews and the anticipated meeting timelines are summarized and reviews are defined as follows:

1. **Program review meetings**

The required system-level review meetings are SRR, PDR, CDR, MTR, QR, and AR.

2. **Biweekly Progress Meetings with WP managers at Prime Contractor and subcontractors**

The Prime Contractor Program Manager will conduct biweekly progress meetings with WP managers at Prime Contractor and teleconferences with subcontractors, at which progress will be measured, problems be identified, and workaround solutions be initiated.

3. **Monthly Progress Meetings with subcontractors**

Each month a progress meeting has been scheduled with subcontractors, to perform progress reviews based on monthly status information, and to identify potential problems and appropriate corrective actions. As an option, the Customer is invited to attend, as an observer and/or to avoid the need to duplicate these meetings with the Customer.

4. **Technical Coordination and QA Meetings**

On an as-required basis, special technical and QA meetings between the Customer and the Contractor, or Prime Contractor and its subcontractors will take place. To the extent possible technical meetings may be scheduled next to the monthly Progress Meeting, in order to keep the travel times and costs low.

With regard to travel times and costs, the Contractor may suggest the implementation of telephone- and/or videoconference procedures for the routine meetings. The Prime Contractor shall establish and submit to the Customer an overall project meeting plan indicating the main meetings foreseen at the project.

The project will be monitored at regular intervals by means of progress meetings, technical meetings, and regular communication between the project members. MPMs between the Customer and the Contractor will be held to address technical and programmatic issues. The MPM shall consist of two parts:

(a) Technical meetings:
   - Involvement of PM and engineering project staff
   - Attendees shall provide a detailed presentation on the activities performed by WPs already activated according
   - Draft/updated versions of documents and presentations must be delivered in advance; at least five working days prior to the meeting
   - Prepare and distribute the meeting agenda 2 weeks prior to the meeting

(b) Programmatic meetings:
   - Involvement of PM, commercial, and administrative project personnel
   - Attendees shall provide detailed schedule and cost information
   - Prepare and distribute the meeting agenda 2 weeks prior to the meeting.

The Contractor shall produce the monthly progress report templates prior to the monthly progress meeting.

22.10.4 Company Internal Management Review Meetings

The Contractor will conduct regular company internal management review meetings, in order to keep upper management informed concerning the progress of work, coordinate project activities with line managers, and in case of problems decide on actions to overcome any problems.

Project reviews will be performed as required by the Contractor. At each review the decision as to whether the review criteria have been met will be determined by a review board consisting of the Contractor appointed members. Achieved status of the review objectives shall be reported to the Customer at the end of each review. In case the objectives of a review are not met, the review board shall identify the reasons and remedial actions. The following reviews are planned for the project:

- SRR
- PDR
- Critical Design Review (CDR)
- Mid-Term Review Phase 2 (MTR-P2)
- Qualification Review (QR)
- Mid-Term Review Phase 3 (MTR-P3)
- Acceptance Review (AR).

The Contractor will invite experts to form the review board for any specific review described above. This will consist of the Contractor's PM technical staff of the Contractor and representatives of the subcontractor and the Customer. Project Reviews shall:

- Last TBD days
- Include all partners and optional Customer representatives

- Include presentations from all partners presenting their own deliverables documents/products according to the contractual data package
- Act as a technical meeting and all the partners shall provide a detailed technical presentation on the activities completed per each WP
- Include draft/updated versions of documents and presentations to be delivered in advance; at least 10 working days
- Have a proposed Agenda well in advance (2 weeks) to be agreed upon with the Customer, and once agreed the subs shall be informed.

22.10.5 Action Items

The purpose of action item management is to maintain a record of all the action items (AIs) decided during meetings—such as questions to be answered, documents to be provided, and investigations to be made—to monitor their implementation, and to keep their history throughout the project.

The action items must be formally recorded by the PM. The AI management tool shall allow monitoring and provision of an AT status list. The AIs are controlled formally via the list of action items attached to each minute of the meeting and the action item status list. The action item list shall be reviewed during each meeting. Action items are raised and generated by mutual agreement between the participants during meetings or teleconferences. They are summarized and listed on a dedicated page of the minutes of the meeting. An updated action item status list shall be issued on a regular basis and attached to the monthly progress report when useful.

When the action items are fulfilled, generally by fax or e-mail, they are registered as "*proposed for closure*" in the action items database. This list shall be then examined during the next progress meeting and the decision shall be taken to close the action item if both parties agree that the document referenced fulfills the action item satisfactorily. The minutes of the meeting record this decision and the status of these action items is changed to "*closed item.*" Action items shall never be closed without a common agreement between the originator and the actioner.

Originators or actioners at any meeting can be the Contractor, the subcontractors, and, at the full progress meeting and Reviews, also the Customer.

The required reporting and exchange of information can be summarized as follows:

- Quarterly Progress Report (QPR): A QPR shall be submitted by the Prime Contractor to the Customer. The QPR shall be concise and shall include the following details:
  - Highlights and trends (narrative description)
  - Tasks achieved during the past month (including an updated MPB); see Sect. 22.4
  - Manpower and cost status; see Sect. 22.5
  - Configuration status; see Sect. 22.6
  - Documentation status; see Sect. 22.7

  - Risk status; see Sect. 22.8
  - Contract change status; see Sect. 22.9
  - Quality assurance (as may be required).
- Monthly Progress Reports (MPR): Prior to the monthly progress meetings, progress reports will be distributed to all the partners summarizing the programmatic status of the project. This report is the summary of progress reports that each subcontractor needs to deliver to the Prime Contractor on a monthly basis, with the following content:
  1. Narrative Progress Report: This report is a summary of the activities performed during the reporting period, a status on deliverables, a summary of problems encountered, and the corrective actions to mitigate the problems followed by planned activities for the next period.
  2. Schedule progress: This report is meant to communicate the status of each work package. The report includes the actual start date of activities, the variance between planned and actual finish dates, and the steps planned to recover eventual delay.
- Cost report: Each subcontractor shall provide the Prime Contractor with the cost encountered and the manpower spent per work package. The reporting will be in an Excel template. It is imperative to maintain documentation for manpower spending and procurements to allow audits. The milestone payments will be on the basis of the cost report.
- Action Item Reporting by the Contractor: AIs will be contained in an Action Item Control Document on a Database accessible by the team.

22.10.6 Risk Management Reporting

Risk assessment and mitigation will be part of the project management responsibility. The Contractor will produce a quarterly RSR as part of the QPR. The RSR will contain for each major risk identified the following information:

- Risk identification and designation
- Cause, source, and type
- Consequences evaluation on perforce, schedule, safety, and cost
- Risk criticality ranking
- Work Package concerned
- Designated risk owner
- Corresponding mitigation actions and their current status.

22.10.7 Report Requirements

Reports to the Customer shall be updated on a quarterly basis. A list of specific dates indicating when each monthly report is due will be published and maintained by the PMO of the Prime Contractor in conjunction with the updated meeting schedule.

Distribution of Reports: If required, the number of hard copies and distribution of individual project reports must be specified. All relevant project information between the Customer, the Prime Contractor, and the subcontractors shall principally be communicated via e-mail.

Minutes of Meetings: At each meeting, Minutes of Meeting (MOM) shall be issued by the Prime Contractor. The minutes shall clearly identify all decisions and agreements made and actions accepted during the meeting. An updated Action Item List (AIL) and the Document Requirements List (DRL) shall be attached to the MOM).

As a guideline, the MOM shall contain the following:

- a standard Project front sheet, indicating the date, place, and subject of the meeting
- the agenda and the list of participants
- the conclusions of the meeting, with reasons for disagreement, if any
- a brief point of the overdue actions
- a list of new actions, precisely formulated, coded in accordance with an overall coding system and indicating actionee's and due dates assigned
- a brief summary of the discussion on major problem areas
- copies of technical notes viewgraphs, drawings, etc. as presented at the meeting.

A draft of the full progress and review meeting minutes shall be signed at the end of every meeting by the authorized senior representative of each party, to indicate that the wording is correct and properly reflects the agreements reached. Any matters having contractual impacts shall be subject to the agreed change procedures for implementation. The PM is responsible for making sure that the MOM is signed by the representatives of the parties and distributed in due time to all concerned.

For major reviews, the minutes shall as a minimum contain:

- RIDs, including question, answer, and review-board decision
- Action items
- Review-board conclusions (outcome of the review, with possible conditions for closeout).

Management Concept of a Plant Project 23

Contents

B. -J. Madauss, *Project Management*,
https://doi.org/10.1007/978-3-662-69057-4_23

Abstract

The realization of a complex plant project requires the use of effective and proven management procedures. This chapter describes a management concept for the execution of a major plant project, for example, a power plant, which has been drawn up on the basis of management methods and procedures for major projects which have been tried and tested in practice. Emphasis is provided to particularly important project management methods applicable for plant projects.

23.1 Project Management and Organization

The efficient project execution of a plant project requires as an important prerequisite the setup of a dedicated project management office (PMO) and the implementation of an effective project management system. These basic requirements are particularly true for projects of plant engineering and for research and development (R&D) projects. A plant Project Manager (PM) must be appointed, who has the overall responsibility to manage the plant project. The PM must ensure that at least the following major functions are performed (see Figs. 5.3 and 5.4):

- Project planning and monitoring
- System Engineering
- Product Assurance
- Safety and Permits
- Information Management
- Documentation and Configuration Management
- Risk Management
- Contract Management
- Implementation Management
- Operations Management

23.1.1 Responsibility of the Project Management

Project management (PM) responsibility for the plant project rests with the lead firm (*prime contractor*) and should fully cover the following areas:

- Technical responsibility
- Administrative responsibility
- Technical/administrative interface responsibility

Integrated technical–administrative project responsibility, including interface control, is absolutely essential in the interest of effective project management and has also proven necessary in practice. In contrast, a separation of responsibilities often leads to wrong decisions, because technical measures inevitably entail administrative (*deadlines, costs,* etc.) measures. It is also essential that the interface responsibility lies fully with the project management. In this context, the European space projects Helios and Europa III (now *Ariane*) deserve special mention as good examples; see Subsection 7.2.3. In both cases, a clear and workable management concept was applied, by assigning full project responsibility to the PM. A detailed responsibility overview for a typical plant project is given below:

- Technical responsibility:
  - Definition of project requirements
  - Trade-off analyses
  - System design
  - System specification
  - Technical project planning and monitoring
  - Technical change control
- Administrative Responsibility:
  - Management and organization
  - Staffing of the project office

  - Company selections
  - Industrial contracting
  - Financial/cost planning and monitoring
  - Scheduling and monitoring
  - Change control
- Technical/administrative interface responsibility:
  - Technical interface definition
  - Administrative interfaces
  - Interface monitoring
  - Interface change control

23.1.2 Authority of the Project Management

It is imperative that the project manager (PM) for the plant project be provided with appropriate authority, which is necessary to manage a project of this magnitude. In order to meet the requirements outlined in Subsection 23.1.1, the following authority shall be granted to the project manager and his team:

- Technical authority:
  - Selection and implementation of the system concept and system specification
  - System-relevant decisions, especially in the case of changes
  - Determination of the main milestones of the plant project
  - Approval of all system level technical documentation
- Administrative authority:
  - Selection and implementation of the management and organizational concept
  - Selection of key personnel
  - Evaluation and acceptance/rejection of key personnel
  - Implementation of a company selection process
  - Conducting contract negotiations
  - Implementation of a status and financial control system
  - Approval authority over all system level administrative documents (i.e., contracts, organizational and management plans, etc.).
- Interface authority:
  - Implementation of an interface control procedure
  - Approval of all system level interface documents

23.1.3 Organizational Concept of the Project Management

The organization of the project management (PM) is to be carried out according to the aspects of an optimal project flow. It is important to structure the necessary project tasks (*functions*) of the PM in such a way that the overall responsibility of the PM can be broken down into sensible subareas of responsibility. Since the PM responsibility, as described in Subsection 23.1.1, covers both extensive technical and administrative tasks, it makes sense to structure them according to project disciplines. This means that specialists in the various fields of knowledge are assigned to the PM. The range of tasks of the PM is subdivided into the following subareas:

- Project planning and monitoring:
  - Work breakdown structure and work packages
  - Schedule and sequence planning
  - Financial and cost planning
  - Action item monitoring
  - Deadline and process monitoring
  - Financial and cost monitoring
  - Performance evaluation
- Product Safety:
  - Reliability
  - Quality
  - Operational safety
  - Availability
  - Maintenance
- Safety and Permits:
  - Safety concept
  - Radiation protection concept
  - Contacts with authorities
  - Safety report
  - Approval procedure
- Systems Engineering (SE):
  - System optimization (trade-offs) and design
  - System specification
  - Subsystem specification
  - System and subsystem planning
  - Technical interface control
- Information Management:
  - Data management
  - Meeting coordination
  - Information and reporting
  - Documentation and configuration control
  - Review coordination
- Contract Management:
  - Contract (*legal terms*)
  - Specifications
  - Contract amendments
  - Subcontractor control
- Implementation Management:
  - Site planning
  - Test planning

  - Production and test facility
  - Site management and inspection
  - Acceptances
- Customer Management:
  - Customer contacts
  - User requirements
  - Utilization plan
  - Operating concept
  - Implementation, training, and education

The PM is the superior of the subarea managers and is responsible for the overall management. The subarea managers of the plant project report to the project manager of the overall plant. In addition to the PM, they are part of the key personnel of the plant project. Recruitment of the key personnel is to be carried out by the PM in cooperation with HR.

23.1.4 Integration of the Plant PM into the Company Organization

The plant PM is assigned to the Projects/Programs Board of Management area. The project manager reports to the member of the Executive Board responsible for the Projects and Programs area. It is imperative that the employees of the PM, even if they belong to other companies/institutes, are seconded to the PM team for the duration of their task assignment. In addition, appropriate specialized personnel from industrial and consulting firms may need to be involved to handle project management tasks. However, care should be taken to ensure that the key personnel required to complete the tasks described in Subsection 23.1.3 are, if possible, employees of the lead company. For larger projects, the PM team should preferably be housed in a closed building complex (*PMO office)* in order to improve communication within the project team. In addition, appropriate information systems (*email, teleconferencing, videoconferencing,* etc.) are an integral part of the PMO (see Sect. 11.4). With reference to modern communication technologies, virtual teams are possible under certain circumstances.

To support the PM (e.g., *to carry out special studies, investigations,* etc.), the resources of the company's specialist departments and, if necessary, appropriate positions of an industrial architect (IA) may also be involved; see Subsection 23.2.1. The involvement of specialist departments is preferably carried out within the framework of the matrix organization (*allocation of work packages to the specialist departments*). The assignment of the IA is to be regulated within the framework of a support contract. In both cases, the planned activities must be planned in a timely manner.

23.1.5 Management Interfaces with Ministries, Offices, and Authorities

The central point of contact for the plant project in the company is the PM. The PM of the project has the responsibility to perform the management interface with the ministries, offices (*state offices*), and authorities (*licensing authorities, TÜV,* etc.). For this reason, the PM shall appoint appropriate staff for the interface task and provide them with appropriate responsibility and authority. In Sect. 5.1, the possible management interfaces to ministries, offices, and authorities are shown, as they are common in public sector plant projects.

23.1.6 PM's Right to Issue Instructions to Industry Participants

Effective project management can only be ensured if the PM has an appropriate authority to issue instructions to the industrial contractors, i.e., the subcontractors and the IA, see Sect. 23.2. For this reason, the following industrial policy measures are required:

1. The selection of industry partners must be made according to the rules of free competition.
2. In industrial contracts, appropriate provisions guaranteeing the right of the lead company to issue instructions must be included.

23.1.7 Implementation of the Plant PM

The implementation of the PM has to be accomplished step by step and in accordance with the respective project phase; see Sect. 23.4. In other words, starting with the development phase, the PM shall be officially implemented and then gradually built up to such an extent that the lead company has an appropriate project team at the start of the deployment phase. For this purpose, a corresponding personnel recruitment plan shall be prepared showing the personnel buildup in connection with the main milestones of the project.

The qualification requirements for the PM personnel, i.e., for the key personnel (*project managers and the subarea specialists*) and the technical experts shall be precisely defined by job descriptions.

23.1.8 Project Commission

For monitoring and accountability purposes, a project commission is often established in larger community projects to monitor the activities of the PMs at regular intervals and to assist the PMs in solving serious problems.

23.2 Industrial Organization and Key Personnel

Large-scale industrial and governmental projects require the simultaneous deployment of many governmental and industrial companies, as the special experience and capacities of a single company are often exceeded. The joint cooperation of institutions and companies requires a workable organizational concept and the provision of experienced key personnel. In the interest of optimal handling of such projects, a procedure for industrial organization and staffing of key positions, which has been tested in practice, is therefore proposed, the details of which are described below; see Chap. 5.

23.2.1 Interindustry Project Organization

A typical project organization for a major project is shown in Fig. 5.18; it shows the Apollo program organization with different levels for example. NASA is the client for the Apollo project and the only central point of contact for all project matters. Within the lead company, full responsibility for the project has been assigned to the project manager; see Sect. 23.1.

In the interest of efficient project execution, it must be prevented that the customer (*authorities, offices,* etc.) bypasses the lead company and establishes direct contacts with the subcontractors. The PM of the lead company shall be informed of any communications in this regard and shall have the right to reverse any unauthorized instructions at any time, unless agreed to by the PM.

It makes sense that an IA is called in to support the PM for the handling of a plant project. The IA should primarily take on tasks from the areas of systems engineering, interface control, training and education, and special tasks in the area of project management.

The lead company should engage external freelance consultants (*experts*) or consulting firms on a short-, medium-, or long-term basis whenever necessary to accomplish its tasks. The main purpose of this is to achieve the following:

- Gaining know-how from other projects (*lessons-learned*)
- Bridging personnel bottlenecks, e.g., to handle temporary work (*review, coordination,* etc.)
- Obtaining external expert opinions

23.2.2 Contractor: Project Management

It is of crucial importance for plant projects that the contractors (*industrial main contractor and subcontractors and the industrial architect and any research institutions to be included*) also set up a corresponding project management structure mirroring the client. For reasons of better communication between the individual partners, it is specified in this context that the internal project organization of the participating companies is to be set up as far as possible identically to the internal project organization of the PM; see Fig. 5.19.

23.2.3 PM Right of Instruction

The PM's right to give instructions to the lead company is as defined in Subsection 23.1.6. The binding contact person for the various project managers (*prime and subcontractors*) is the respective project manager. The authority of all project managers shall be defined in such a way that the right to issue instructions covers all technical (*specification, design,* etc.), administrative (*costs, contract,* etc.), and interface-relevant aspects.

23.2.4 PM Monitoring Function

The PM is responsible for the supervision and the related control of the overall project. Monitoring of the work at the contractors (*prime and subcontractors*) shall be ensured by the following measures:

1. Reporting by subcontractors to the PM; see Sects. 23.4 and 23.5
2. Conducting audits by the PM (e.g., *quality control audits,* etc.)
3. Conducting system and subsystem reviews at their subcontractors; see Sect. 23.7. System reviews are conducted by the PM and subsystem reviews by the respective subcontractors. The PM shall be involved in the subsystem reviews as an observer.

23.2.5 Qualifications of the Industrial Companies

The industrial companies (*IA*, *prime- and subcontractors*) to be used for the plant project must have appropriate qualifications and sufficient industrial know-how and must provide appropriate proof of this in their offers. In both cases, the respective company and the key personnel intended for the project (*see subsection 23.1.3*) must be described in detail. This point must be specifically addressed in the tender documents.

23.2.6 Industrial Tendering Procedure

The lead company should, whenever possible, implement a tendering process that allows the client (*customer*) to make a contractor selection (*IA*, *prime- and subcontractors*) according to objective criteria. This is also linked to the intention that the contractors in turn select the subcontractors according to similar criteria.

23.3 Project Structuring

The systematic structuring of projects (especially, very large projects) into smaller subareas that can be clearly controlled and understood by management is a basic prerequisite for successful project management. Projects should be structured in such a way that the entire project can be divided into manageable subareas, which in turn can be further subdivided; see Subsection 9.3.3.

23.3.1 Work Breakdown Structure as a Central Management Tool

"The work breakdown structure (WBS) of a project is an important central management tool, since all technical and administrative project data can be brought together via the WBS." For many projects, the WBS is intended to be the main structuring tool of the following management systems:

- Financial and cost structure (*numbering key*)
- Schedule assignment
- Specification tree
- Documentation tree

The individual management systems are assigned by including the WBS numerical code.

23.3.2 Determining the WBS Levels for a Plant Project

The upper levels of the WBS are defined in the technical area according to standard aspects. The individual levels are defined as follows:

- Level 1: Project areas
- Level 2: Main systems
- Level 3: Subsystems and individual functions
- Level 4: Hardware/software elements
- Level 5: Work package level

It is an important lesson learned from practice that too much detail in the WBS, i.e., definition of too many levels, can lead to considerable difficulties in project execution. The first outline block of the WBS defines the system and functional areas of the project. The second code block then defines the hardware elements of the project. The work package code, which is made up of the marking structure for the standard work packages and the company code, is to be regarded as the third marking block, which is directly assigned to the higher level WBS structure in the non-hardware area and to the hardware outline in the hardware area.

23.3.3 WBS Development and Change Monitoring

The WBS for large projects can only be developed step by step and in accordance with the phase progress to the deepest level; see Fig. 9.6. In detail, the following procedure is envisaged:

1. Gradual freezing of the first three WBS levels at the start of the development phase
2. Final elaboration of the hardware/software elements
3. Early definition of standardized work packages

The WBS for plant projects shall be frozen in its base configuration during the development phase, and changes to the WBS configuration shall be monitored continuously from that point forward.

23.3.4 WBS and Organizational Structure

When creating a WBS, the following principle applies: "*The WBS should not be used as an organizational chart.*" This means that the WBS is to be regarded as a task breakdown structure, while at the same time an organizational chart is created from which the responsibility for the individual WBS elements and/or work packages can be derived; see Fig. 9.11. In this context, another principle should be mentioned: "*It must be ensured that for each WBS element or work package a responsible body, but only one body, can be identified in the organization chart. However, it is possible for an organizational unit to be responsible for several WBS elements or work packages at the same time.*"

23.3.5 Developments of a WBS Procedure

A WBS procedure shall be developed by the PM to ensure that all project personnel from the lead firm, industrial, and regulatory partners work according to consistent guidelines. The WBS procedure should include at least the following points:

- Numbering system explanations
- Integration of different number blocks
- Creation *of WBS content dictionaries* and work package descriptions
- WBS implementation and change control

23.4 Project Planning and Monitoring/Project Control

The planning and monitoring process is of extraordinary importance for any project. For plant projects, it is advantageous to have a planning concept that is based on the WBS (see Sect. 23.3), which includes the following planning tools:

- Definition of the project phase
- Schedule and sequence planning
- Cost planning

The monitoring concepts and resulting measures for the project can be summarized as follows:

- Status control (deadline status)
- Action item (AI) control
- Cost control
- Performance measurement

At this point refer to Sect. 9.2.

23.4.1 Definition of Project Phases

Project execution according to sequential phases separates the chronologically different activities of a project (*concept studies, definition work, development work, construction activities, operations, and disposal)* by firmly defined completion points (*major milestones*) of the individual phases and creates the basis for determining meaningful decision points. The phasing concept provides that, after completion of the individual phases, very specific, previously defined, and also measurable final results are available before the initiation of the subsequent phase. This means, for example, that with the exception of a few particularly deadline-critical activities, the development phase cannot be started until the definition phase has been fully completed.

The sequential approach ensures consistent procedures and largely excludes the carryover of system errors into the expensive realization phases (*construction and operation*) by eliminating errors in the early phases.

23.4.2 Scheduling and Process Planning

Scheduling and process planning is a very important management tool for any project. For this reason, a thorough and well-founded plan showing the individual activities and milestones must be drawn up at a very early stage, i.e., before the start of each respective project phase. In the case of large-scale projects, it is necessary that the planning documents are based on the WBS (see Subsection 23.3.2).

Bar charts and network plans are preferred for scheduling and process planning. All defined milestones must be included in the bar charts. Project milestones are determined and defined according to the following levels:

- Level 1: Milestones to be released and monitored by the customer
- Level 2: Milestones to be released and monitored by the prime company
- Level 3: Milestones to be released and monitored by subcontractors

23.4.3 Cost Planning

Cost planning is closely linked to the WBS, work package descriptions, and time schedules. In other words, the cost plans must be aligned with the WBS coding and the planning documents.

Cost planning is to be carried out in several steps. During the planning phase (*project definition*), cost analyses are to be carried out by the PM in order to develop corresponding target costs for the main areas of the project (e.g., *Levels 1 and 2 and Partly Level 3*). Different estimation methods shall be used to determine the target costs, namely:

- Parametric and computational cost estimation models
- Ratio calculations (*analogies to other projects*) and/or
- Detailed level of effort estimates

In this context, reference is made to Sect. 10.1. Accompanying these estimates, investigations of the estimation accuracy and sensitivity analyses are to be carried out to clearly and unambiguously highlight the factors influencing costs (*cost drivers*). Existing cost databases (*empirical values*) are to be used wherever available. Wherever possible, cost estimation models should be used for independent estimation to increase the validity of the cost estimate, see Subsection 23.10.3. In this context, it should be mentioned that the cost analysis must also include the operating phase.

In connection with the industrial offers, the industrial partners shall submit binding and detailed cost offers, which shall then, after a negotiation is conducted, be converted into a final cost or price offer; see Sect. 23.11. It is useful to apply cost estimation models for independent comparative estimates. The detailed and binding financial offers of the industry are to be prepared on the basis of work package descriptions and in connection with the time planning. The prime contractor must base his cost proposal on binding subcontractor and supplier bids.

It is assumed that both fixed and cost-reimbursement prices may be applied. Where cost-reimbursement price bids occur, the cost bids shall be broken down by work package according to the appropriate cost categories and according to the time allocation. Typical cost categories are (see Fig. 10.3):

- Personnel costs
- Cost of materials

- Plant costs
- Rental costs
- Travel expenses
- etc.

Cost summarization to the project level shall be performed in accordance with the WBS. In addition to the previously mentioned offers, each contractor shall submit a detailed cash flow plan.

23.4.4 Status Control

An essential component of project monitoring is the regular status control of the project. On the basis of the information provided in Subsection 23.2.4, the project status is to be recorded monthly and compared with the specified planned values (*target/actual analysis*). Deviations from the planned values are to be analyzed by the respective responsible companies/organizations and, if necessary, converted into suitable corrective measures to eliminate delays. Figure 9.25 shows the monitoring control loop.

If any identified schedule delays cannot be avoided, they must be reported to the next higher management level. A planning analysis must then be carried out at the next higher management level.

Unavoidable schedule delays of the project are to be reported to the PM according to the rules described in Subsection 9.4.3. The PM will then perform a higher level project analysis for the overall project based on the individual management reports. It goes without saying that the information flow between the addressed management levels must be kept going and that project reports and meetings are the main link between the partners. The procedure for status control is to be worked out in detail in the project management manual.

23.4.5 Action Item Control

In addition to routine project status control, AI control (also known as action control) has proven to be an effective management tool. When problems arise on an ad hoc basis, the AI process provides an ideal management system. The AI system is easy to install and does not require any project staff to have any particular prior knowledge of special planning techniques. Therefore, an institutionalized AI control system should not be missing from any project. Especially, in large projects, it is an absolute necessity. For this reason, the PM should introduce an AI control system at the initiation of the project. However, the definition of actions must follow certain rules; see Subsection 9.4.4 (*Action item control*).

23.4.6 Cost Control

In addition to status control (*including complementary AI control*), cost control is a key project monitoring tool. The following procedures are in the foreground:

- Control of cash outflow and/or payment schedules
- Control of hours and costs incurred (*cost-reimbursement contracts only*)—commitment control (*committed funds*)
- Target/actual comparison (*variance analysis*)
- Cost-to-completion (*CTC*) projections

The control of the outflow of funds and the payment transactions is to be carried out on the basis of previously prepared planning documents. The release procedure for project areas or sections and the status control are closely related to this. The payment plans and the outflow of funds plans are directly related to the project planning (*sequence and deadlines*) and the work progress via the corresponding milestones.

In areas for which cost-reimbursement agreements are made, the PM is obliged under the rules of German pricing law to monitor the hours and costs incurred by the respective companies. Here, too, target/actual deviations must be analyzed in connection with project planning.

Commitment control is an important control tool of the industrial company to create an overview of the committed funds for orders made. The PM, who is interested in a constantly updated commitment overview, will work toward ensuring that subcontractors implement a corresponding control system; see Subsection 23.4.6 (*Cost monitoring*).

23.5 Information Management

Project management tasks are particularly dependent on a well-functioning information system. Above all, information must be exchanged quickly and in a targeted manner between those involved in the project. In view of the size and complexity of plant projects, the timely implementation of an information system is of the utmost importance; see Chap. 11. The following media are available for information transfer:

- Meetings
- Reporting
- Phone
- Email
- Teleconference
- Video conferencing

23.5.1 Information System in the Project

The design and installation of project information systems must be adapted to the specific needs of each project. First of all, in accordance with the implemented project organization (including the industrial organization), the information needs must be defined. Therefore, it is necessary to investigate who (*organizational area, group of people,* etc.) has to transmit which information (*content*), when (*time*), and to whom (*recipient*).

For the present project, an information system is proposed that meets the specific needs of this project. It is envisaged that the project management information system will not be implemented solely at the lead company site but will include the selected industrial partners. The proposed procedure builds on the experience of the aerospace industry.

23.5.2 Reporting

The efficiency of project management can be significantly increased by a well-functioning reporting system. The reporting system, an important part of the information system, should be set up in such a way that at regular and predetermined times, e.g., on the tenth of each month, the industrial partners send an integrated status report to the PM reporting on the most important project parameters. This report should provide information on the following areas; see Sect. 11.2.

1. Project organization
   - Changes and their justifications (*key personnel,* etc.)
   - Approval status and implementation status
2. Configuration description (*system*)
   - Technical changes and their justifications (*change request*)
   - Approval and implementation status
3. Summary of the main events
4. Critical events
5. Action overview (*list*)
6. Work breakdown structure
   - Changes
   - Justifications
7. Timetable and schedule
   - Status report (*target/actual*)
   - Corrective actions
   - Changes
   - Justifications
8. Cost/financial planning
   - Cost report (*target/actual*)
   - Change report
   - Justifications

The reporting cycle should be based on a monthly cycle.

23.5.3 Information Distribution

The passing on of project information (*distribution*) must be clearly regulated because it must be prevented that a flood of information occurs due to indiscriminate distribution and project employees are not sufficiently informed. For this reason, the distribution of information for the project is based on the defined project organization. Regular status reports, for example, almost always have a fixed distribution list.

23.5.4 Meetings and Minutes

Regular and ad hoc meetings are an important information tool in the project. The main advantage lies in the rapid exchange of information, the possibility of immediate reaction, and the personal contact between all meeting participants. Identifiable problems can be analyzed immediately, and corrective action can be taken without delay.

In this context, reference is made to the importance of status meetings, as they represent a particularly important and efficient form of project information. It is important that status meetings are held regularly and prepared thoroughly. The following measures should be mentioned in particular:

- Preparation of an agenda
- Holding of pre- and post-meetings
- Preparation of a protocol of results

Project meetings do not serve an end in themselves but have a very specific task to fulfill, a task that must be clearly and unambiguously defined in the context of the agenda. Consequently, project meetings should not be finished without a record of the results. What is meant, however, is not a verbatim protocol, but a protocol that summarizes the meeting results and measures and actions to be initiated. It is of great advantage if the meeting minutes (*possibly handwritten*) are prepared during the meeting and read out to the participants immediately so that corrections can be made on the spot and the minutes can be approved immediately. Particularly, in the case of meetings with other companies, clients, etc., the results of the meeting can thus be established immediately and become binding through the signing of the minutes by the respective meeting leaders. This procedure will be introduced by the PM for the present project.

23.5.5 Conducting Tele- and Videoconferences

Travel time and costs can be saved by introducing tele- and videoconferencing. Teleconferencing therefore represents an interesting alternative for the project. Teleconferences are particularly recommended when conducting status meetings (*status inquiries*) in the interest of saving time and costs. It is advisable for the participants to send each other their reports or presentation documents a few days in advance (e.g., *by email*) and to display these simultaneously in both conference rooms during the meeting; see Sect. 11.4.

23.6 Configuration and Documentation Management

The quality of the implemented configuration and documentation management is of the utmost importance for successful project execution. The most important thing here is that the required project documentation is clearly and unambiguously identified at a very early stage. The plans, specifications, and procedures (*document types*) required for project execution must be clearly identified for each WBS level in the form of a documentation tree and their content defined by a requirements description (*documentation requirement description*); see Sect. 12.1 (*Documentation requirements*).

However, regular recording and monitoring of the documentation status is also important; that is, the PM must be informed at regular intervals about the completion, change, and release status of all identified documents. Changes to the technical configuration must be recorded immediately and completely and checked for their financial impact. Only then can a technical change be implemented; see Sect. 12.2.

23.6.1 Documentation Identification

The identification of the necessary documentation requirements is very important for the implementation of projects. Documentation requirements (*data requirements*) must be defined at an early stage in the project. The required documents (*e.g., specifications, plans, procedures, manuals,* etc.), their purpose (*e.g., system specifications, test plans, management manual,* etc.), the time of completion (*according to the project schedule*), and the documentation contents (*brief description*) must be determined.

To create a documentation overview, it is common to display the entire documentation of a project in a documentation tree; see Fig. 12.1. In this context, it is important to make the following documentation principles particularly clear:

1. In addition to the mandatory documentation, which must be planned as far in advance as possible, documents are also created in every project that do not belong to the mandatory documentation. These are, for example, additional analyses and technical reports whose necessity only becomes apparent later.

2. Project correspondence (*letter, telex, and fax*) is not part of the project documentation but must be handled as part of the normal correspondence arrangements.

23.6.2 Documentation Creation Procedure

It is envisaged that the project documentation will be prepared according to uniform guidelines. This requirement is not intended for the PM area of the lead company alone but includes the documentation systems of the industrial architect and the subcontractors. Specifically, the following measures are envisaged:

- Application of a uniform numbering system and use of uniform cover sheet formats
- Establishment of standard distribution lists

The unified numbering system to be introduced must meet the following conditions:

1. Identification of the document type
2. Assignment of the respective document to the *WBS*
3. Identification of the documentation creator
4. Definition of a sequential number (*1–n*) for the unique identification of a document
5. Marking of the issue and revision index

23.6.3 Documentation Monitoring

Documentation monitoring is an important project management function, as the PM and his team must be aware at all times about which documents still need to be created (*requirements*):

- When
- By whom
- Content
- etc.

What is the status of the documents already created?

- Released
- Withdrawn
- Changed

The central instrument of documentation control is the documentation control list. The documentation control list must be kept up to date at regular intervals, at least monthly.

23.6.4 Configuration Control

Closely interlinked with documentation is configuration management. The construction configuration of a technological project is subject to considerable change from the very beginning, i.e., already during the concept and definition phases. The purpose of configuration control is to monitor the timely freezing and the changes of the technical system in terms of cost-efficient project control. The freezing of the technical system must take place gradually and in accordance with the phase planning.

Changes to the respective frozen configurations, which are to be reviewed with regard to their scheduling and financial impact prior to implementation, are subject to a monitoring process by the Change Control Board (CCB), which meets not only at regular intervals (e.g., *weekly*) but also on an ad hoc basis, if necessary, to review submitted configuration change requests for approval or rejection.

An important measure for configuration management is the definition of change classes; see Sect. 12.2 (*Configuration monitoring*). The following change classes are introduced for the present project:

- Class A: Concept changes (*project level*)
- Class B: Functional changes (*main system*)
- Class C: Design changes (*changes that do not affect the abovementioned Class A and Class B changes*)

Class A changes are to be approved by the prime company management, Class B changes by the contractor, and Class C changes by the respective subcontractors. However, if it is found that a requested change to Class C will impact the system, Class C shall be changed to B. An analogous action shall be taken in cases where Class B changes which impact system level project areas.

Construction deviations requested, e.g., due to material shortages, etc., or identified by quality control based on an inspection, shall always be approved by a fast-track release (*regulation by deviation and waivers*) if they do not cause any detectable product deterioration (i.e., no impairment of *form, fit, and function).*

23.7 Review Coordination

In practice, it has proven useful to ensure the project status at specific milestones by conducting detailed project reviews; see Fig. 4.2. Typical reviews are:

1. Requirements reviews (RRs) at the end of the concept phase
2. Specification reviews (SPRs) at the end of the definition phase

3. Preliminary design reviews (PDRs) at the start of the development phase
4. Critical design reviews (CDRs) to begin prototype development
5. Acceptance reviews (ARs), or *First Article Configuration Inspection (FACI)*, at the conclusion of prototype development
6. Production review (PR) at the start of the production or deployment phase

The purpose of each review is to critically examine the content of the project results (*documents, hardware, and/or software*) created up to that point. The coordination of the reviews, which can take place at different project levels (*system level, subsystem level,* etc.), should be performed by the respective project management. For this purpose, appropriate flowcharts, checklists, and review procedures should be prepared; see Sect. 7.3.

23.7.1 Review Plan

For the power plant project, the previously mentioned review milestones shall be performed. The review milestones are to be integrated into the respective schedules and flowcharts of the prime- and subcontractors in such a way that the dependencies of the project activities with the reviews are clearly and unmistakably recognizable.

The review milestones represent the completion of specific development steps. For example, the system concept review (SCR) should be set at the end of the concept phase and the CDR at the end of the design work for the prototype. The reviews conclude specific project phases and freeze their configuration. The respective date of the individual reviews can only be determined in connection with the overall schedule for the project at hand. Therefore, postponements of the overall schedule naturally led to postponements of the review dates.

23.7.2 Review Coordination

The individual reviews must be planned and managed in detail and the results evaluated. For this reason, it is necessary to designate a review coordinator as part of the information management process who is responsible for coordinating the review activities. The main activities of the review coordinator are:

1. Composition of the review teams
2. Determination of review responsibilities including review chairs
3. Creation of the review schedules
4. Definition of the review input data
5. Definition of the review output data
6. Oversee review preparation and follow-up actions

The composition of the review teams shall generally be based on the following considerations:

- Permanent Review Team Members:
  - PM who usually also takes the chair or appoints a representative
  - Heads of the organizational units subordinate to the PM or the deputies appointed by them
  - Review coordinator
- Supporting Review Team Members
  - Experts in specific disciplines, e.g., design, maintenance, reliability, etc.
  - Observers, e.g., industrial or regulatory assessors

Review plans shall be communicated to project staff as early as possible. One to two months prior to conducting a review (*SPR, PDR, CDR,* etc.), a detailed review plan showing the intended review input data, review team composition, and exact schedule information shall be provided.

The input data must be available to the review team approximately 2 weeks before the start of the review in order to be able to carry out an appropriate pre-evaluation before the review; see Subsection 23.7.2. Wherever possible, the review process should be supplemented by *splinter meetings* to clarify detailed issues. At the beginning of the review, the input data provided shall be checked for completeness and then in detail for technical correctness. Each review is to be concluded with the preparation and adoption of a catalog of measures (*action items*). The actions published in the action item catalog are to be completed as soon as possible by the individuals and/or organizations named in the action item catalog to fully complete the corresponding review. The rework period should not exceed the planned period of 1 month; otherwise, schedule delays will inevitably occur due to the correlation of the review milestones with the project schedules; see Subsection 23.4.4.

23.8 Systems Engineering

SE is a central topic in the project management of technical projects. Considering all aspects, the SE team is responsible for overall system design, system and subsystem specifications, and the coordination of all subject related parameters in the interest of an optimal system design. The design of modern systems, in which many specific individual disciplines are closely interlinked, requires system technical holistic considerations in the interest of an optimal system solution.

With reference to comparable projects, the SE team will be at the center of the project tasks and will be the team with the largest number of personnel. On the one hand, the SE leader must have experts representing exclusively the system aspects, but at the same time, he or she must also have subject-oriented experts to be able to maintain the dialog with the industrial partners.

23.8.1 System Design

The system engineering department of the project is responsible for the overall design of the system. This task refers to the definition of the main physical and technical parameters of the overall system. In doing so, the system requirements (*performance requirements for the system*), the boundary conditions (*location, environment, safety,* etc.), and the technological feasibility limits must be coordinated in such a way that a physically/technically and economically optimal system design is achieved.

The preliminary system design for the project must be largely completed during the definition phase (*Phase B*). This means that the main parameters of the system must be determined at this time and defined in a system specification (*see Subsection 23.8.2*). However, the optimization process of the overall system is not yet fully completed, and the system specification can only be frozen at the end of the definition phase. During the development phase (*Phase C*), further system analyses must be carried out, primarily relating to the following tasks:

- Technical compatibility analyses of the system and subsystems:
  - Mechanical configuration
  - Electrical interfaces
  - Mounting conditions
  - etc.
- Reliability analyses
  - Meantime between Failure (MTBF) requirements
  - Redundancy concepts
  - Component selection
  - etc.
- Profitability analyses:
  - Life cycle costs (LCC)
  - Cost optimizations
  - etc.
- Safety analyses:
  - Environmental requirements
  - Location issues
  - Safety concepts
  - etc.

To reconcile the above factors in terms of a technically/economically optimal system design, a series of system and subsystem comparison analyses (*trade-offs*) must be carried out during the definition and development phases involving the various experts, e.g., physicists, technicians, economists, etc.

23.8.2 Specification Processes

The system specification is the main technical reference document of any system. As described in Subsection 23.8.1, the first, but still preliminary, edition of the system specification shall be released during the definition phase (*Phase B*). In accordance with MIL-STD-490, the specification shall be organized as follows:

1. Overview
2. Reference documents
3. Requirements
4. General description
5. Quality controls
6. Establishment regulations
7. Notes
8. Enclose attachments if necessary

The system specification is regarded as a higher level specification affecting the entire system, while second-level specifications are subsystem specifications intended to support the system specification.

23.8.3 System Management

System management of plant projects is at the heart of project management. The task of SE can be divided into several complexes. On the one hand, the main system aspects (*design, specification,* etc.) are to be established and then the closely related system aspects. The design and specification of the system, subsystems, and equipment are the responsibility of the contractors, but they must be checked for compatibility with the overall system specification.

Prior to the development phase (*Phase C*), all specifications must be frozen. Change control is the responsibility of configuration management; see Subsection 23.6.4. The monitoring of the technical performance data (*technical performance control*) is carried out in connection with the performance measurement described in Sect. 9.4.

23.9 Product Assurance

Reliability, safety, and availability are important criteria of modern products. They are directly related to the quality of the system and its components and parts.

The reliability of a system is a decisive factor in system design, both economically and in terms of producer liability (*safety*), because achieving high reliability values, whether through redundancies or special material selection, costs a lot of money. It is important that the relevant factors influencing product safety are given full consideration at the early stages of a project. In a sense, the product assurance experts must be involved in the system optimization process. In this context, timely consideration of official requirements is of course an important matter; see Chap. 8.

23.9.1 Reliability of the System

High reliability of the system is important in two respects. First, in connection with system safety (see Sect. 8.3), because the requirements proscribed by the authorities must be included in the safety concept. In the second case, the requirements resulting from an economically justifiable availability must also be considered.

The achievement of high system reliability for the project at hand, taking into account the environmental conditions (*temperature, vibrations,* etc.), can be achieved mainly through two measures:

- Use of durable components
- Redundancy

However, both measures must be seen in the context of the costs required. Increasing the individual reliability of components (*increasing the service life*) is associated with considerable costs.

In many cases, redundant replacement devices (*passive redundancy*) will be provided, or redundant units (*active redundancy*) will be installed in the respective devices. Derived from the reliability requirements (*required total reliability of the system*), the product assurance team must develop the optimum reliability concept, considering economic constraints.

23.9.2 System Safety

The requirements for system safety in the construction and operations phases can in principle be derived from the valid safety regulations of the federal government and/or the states, but in the case of large-scale projects, they are usually overlaid by additional special requirements.

The task of the product assurance team is to summarize the various official safety requirements in a safety specification and, based on this, develop a safety concept. The safety concept created, as described in security requirements (see Subsection 23.9.1) among other things, is also an important input parameter for defining the reliability requirements. The regulations formulated in the safety concept refer equally to the development, manufacturing, and operation phases. However, special importance is attached to the consideration of safety aspects during the development phase since the influence on the project is greatest at this time.

23.9.3 Quality Control Program

The achievement of high reliability values depends primarily on the quality of the components used. For this reason, an appropriate quality control program must be established in the early stages of the project. The main points of the quality control program are:

- Formulation of the quality requirements
- Definition of appropriate design measures
- Description of the inspection and sampling procedures
- Establishment of test programs

For the practical implementation of quality controls (inspections, sampling, etc.), which must be carried out in accordance with the quality control requirements established for the present project, reference shall be made to the departments established at the industrial companies.

23.10 Handling Interfaces

Interface management plays a particularly important role in large and technically complex projects, since it must be always ensured that the defined interfaces really do fit together. If a change occurs in one part of a project without the other part (i.e., *the opposite part with the matching interface*) being informed, interface incompatibilities can easily occur; see Subsection 23.8.2 (*Importance of the interface specification*).

For the project at hand, for which many technical interfaces are to be expected, effective interface management is required to ensure that the congruence of the interfaces is guaranteed at all times. The interface specification is a special specification supporting the system or subsystem specification(s) for the exact definition of interfaces. It is used to record and control the interfaces between the individual specifications. The interfaces can be divided according to the following areas:

- Mechanical
- Electric
- Electronic
- Software

23.11 Contract Management

For the present plant project, a project-specific contract management is proposed to be able to attend to the project-relevant issues in direct dialog with the project management; see Chap. 13 for details.

23.11.1 Types of Contracts and Prices

Wherever possible, the lead company intends to conclude fixed-price contracts. In this context, it is assumed that fixed-price contracts can only be concluded in connection with a price escalation clause when contracts are concluded over a longer period of time. For particularly critical project areas, cost-reimbursement prices may also be applied. The final determination of the type of price to be applied depends on several conditions, for example, the state of the technology or the accuracy of cost estimates. Section 13.1 summarizes the common contract or price types and their application criteria.

The decision on which price types to apply to the project at hand must be made in the context of the particular project phase and an assessment of the technical risk of the project area under discussion. *Cost Plus Incentive Fee* (*CPIF*) pricing is a viable alternative to fixed pricing for R&D projects with significant risk areas. The CPIF is designed to encourage the contractor to improve performance based on a suspension of the premium. The premium scheme can be based on the following three parameters:

- Cost reduction (see Fig. 13.3)
- Schedule improvement
- Technical performance improvement

That is, the client includes the contractor in cost savings by increasing the normal profit and proposes further profit improvements due to early schedule fulfillment (project completion, delivery dates, development milestones, etc.). The award of technical performance bonuses (increase in nominal profit) refers to parameters such as:

- Weight (reduction)
- Speed (increase)
- Volume (reduction)
- Reliability (increase)
- Quality (improvement)
- etc.

The application of a premium scheme can also provide for a reduction in profit (penalty) for poor performance.

23.11.2 Standard Contracts

Standard contracts shall be used for the project. This concerns both the main part of the contract and the attachments. The contract shall be structured as follows (see Sect. 13.2):

- Main part of the contract:
  1. General (rules to be applied, language, etc.)
  2. Special guarantees
  3. Legal provisions (applicable law, right of withdrawal, etc.)
  4. General financial arrangements (price, taxes, etc.)
  5. Project implementation (general conditions, task transfer, subcontractors, etc.)
  6. Project acceptance
  7. Other conditions (patents, licenses, etc.)
  8. Other additions, if necessary
- Annexes to the contract:
  1. Price and cost information; see Subsection 23.4.3
  2. System specification; see Subsection 23.8.2
  3. Specifications; see Subsection 23.11.3
  4. Other special conditions

The modular and uniform contract structure of all contracts leads to a uniform approach to contract preparation and a better overview of the individual processes. The modular structure also allows the respective specialists, lawyers, technicians, financial experts, etc., not to be constantly involved in contract preparation work at the same time, but to be occupied initially only with the module that concerns them. Only when the contract documents are summarized are joint discussions and negotiations are then necessary.

23.11.3 Service Specifications (Specifications)

The statement of work (SOW) for a project should also be structured in modules wherever possible to decouple the specific areas of responsibility of the individual specialists (technicians, schedulers, documentation specialists, etc.) from one another for a certain period of time. For the present project, the following uniform structure of the statement of work is proposed:

- Main part
  1. Foreword
  2. Project goals
  3. Requirements
  4. Tasks
     - Project management
     - Development
     - Production
     - Test
     - Product assurance
     - etc.
  5. Documentation

6. Deliveries
   - Documentation
   - Hardware
   - Software

- Attachments
  1. Applicable documents (*list*)
  2. Documentation requirements (*list*)
  3. Documentation request descriptions
  4. Hardware deliveries (*list—item, recipient, and delivery date*)
  5. Software deliveries (*list—program, recipient, and delivery date*)
  6. Provisions by the customer (*list—articles/program and delivery date*)
  7. Work breakdown structure; see Subsection 9.3.3.
  8. Schedule; see Subsection 9.3.4.

When drawing up the bills of quantities, the criteria specified in Subsection 9.3.2 should be strictly followed.

23.11.4 Contract Management

The contract management is to be fully integrated into the PM organization. This means that the contract manager is to be seconded to the PM for the duration of the execution of the project. He shall report to the PM. It is envisaged that the subcontractors will take an analogous organizational measure to integrate contract management as part of the general management guidelines.

23.12 Manufacturing and Testing Equipment

Any realistic project planning must include the availability of production and test facilities. Often, the availability or nonavailability of a plant is a decisive criterion for the feasibility and/or schedule compliance of a project.

It is therefore important that a list of existing facilities (production and test facilities) and their specification (*capacity*) is drawn up very early in the project process (i.e., *during the concept and definition phase*). The construction of new facilities or the expansion of existing facilities is usually very costly, so existing facilities should be used wherever possible. Often, this leads to corresponding system optimizations in the early project phases, where plant availability is partly responsible for the system design.

23.12.1 Need for the Present Project

At a very early stage, the required plant needs for the project in question must be defined to include this information in the overall project planning (*see Subsection*

23.4.2). The system requirements must be defined in detail, i.e., for each system area and the individual phases to be defined.

- Development (C)
- Production/setup (D)
- Operation/use (E)

The intended installations shall be specified in detail. In doing so, the location issue, the time and duration of use, and the technical requirement for the facilities in question must be specified in detail.

23.12.2 Plant Availability

As part of the handling of the early project phases (*Phases A and B*), it must be determined whether the required facilities are fully available with regard to the technical specifications of the site and the schedule (*including corresponding schedule reserves*) or whether some of the required facilities must first be constructed. It must also be considered that appropriate alternative concepts must be identified and construction of new facilities shall be included in the overall project schedule (see Subsection 9.3.4).

23.13 Operations Management

The production/test phase (*Phase D*) of the present project cannot be started until the development phase has been completed and the approval process has been finalized. The head of engineering management, who also reports to the PM, is responsible for site management during the implementation phase.

Its main tasks are defined as follows:

- Site planning
- Implementation planning
- Production and test facilities
- Site management and inspection
- Acceptance tests

In carrying out its tasks, it has at its disposal the results of the project functions set up in the preliminary phases (*Phases A–C*). The engineering team has to be built up slowly with the start of Phase B work in order to include the concerns of the deployment phase already in the system design to be carried out in the preliminary phases. The engineering team shall be significantly strengthened in terms of personnel before the start of the deployment phase.

23.14 Utilization Management

In order to be able to take the interests of the future users of the present project into account, the utilization management will be firmly incorporated within the PM from the beginning. It is planned that at the beginning of the project (*Phases A and B*), a relatively small utilization team will identify the user interests and incorporate them in the system design. As part of the work in Phase C (*development*), the utilization team is to be expanded to be able to adequately prepare for the utilization phase (*Phase E*), which begins after completion of the deployment phase (*Phase D*). The primary responsibilities of the team are as follows:

- Maintaining contact with potential users
- Input of user requirements
- Development of a utilization plan (*operational concept*)
- Elaboration of an operating concept
- Implementation of the use phase including training and education

Practical Example Cost Estimation Model 24

Contents

Abstract

Right at the beginning of a project, the initiator (customer, sponsor, or stakeholder) usually wants to know what financial risk he is getting into. Is it possible to produce a serious cost estimate at an early stage of a project? This is a difficult task and one that is difficult to solve without tools. That is why those responsible often fall back on rather unreliable data. This is not surprising, because at the beginning of the project, no project structure or planning has been created. Therefore, one is often groping in the dark, not only at the beginning of the project but also during the course of the project.

B. -J. Madauss, *Project Management*,
https://doi.org/10.1007/978-3-662-69057-4_24

The question is, how can the situation be improved? Experience has shown that the problem can best be solved with a cost estimation model that has been tested in practice. The "*4cost-aces*" cost estimation model is used as a practical example to demonstrate the procedure of parametric cost accounting. The model is a software tool for parametric cost estimation that has been tested in practice and is continuously adapted to the advancing technology. The author has had the opportunity to test the model in the past and is convinced that it is a top-of-the-line product in the industry.

24.1 Operation of the Model

24.1.1 Consistent Application of Cost Estimation Models

This Chapter contains contributions prepared by Herbert T. Spix, Glashütten, Taunus, Germany. Herbert Spix is member of the 4Cost management team. The importance of parametric costing (parametric) is demonstrated by the successful application of the method to complex and large aerospace or other high-tech products, such as automotive. The model is applicable at the system level and for subsystems, e.g., a car engine, and at the component level.

A cost estimation model can be used to determine relatively reliable costs already in the early stages of a project, e.g., in Phase A. The prerequisite is, of course, that the input data are realistic. However, it must be taken into account that the range of estimation accuracy is still relatively large in Phase A and only decreases in the subsequent phases. However, the life cycle costs determined during Phase A (see Subsection 10.5.2) should be accurate enough to make important project decisions. Mr. Seldon mentioned in this context that an accuracy of ±20% in Phase A is still acceptable (Seldon 1979). A greater accuracy is then already expected for the subsequent phases. In summary, it can be said that the model can be used not only in the early phases but also in accordance with the phase model described in Sect. 24.4, throughout all phases, i.e., from conception (Phase A) to disposal (Phase F). This means that the costs of all project phases (life cycle costs—LCCs) can be presented as an integrated solution with the *4cost-aces* model; see Fig. 24.1. For companies, this has the advantage that an initial end-to-end cost determination of the overall project (life cycle cost—LCC) can be carried out at the start of the project and then used as the basis for a profitability analysis (ROI analysis). The result is a structured and comprehensible cost representation, as required for negotiations with the customer and at the same time with subcontractors and/or suppliers.

As new information continuously emerges during the course of the project, it is necessary to regularly repeat the model calculations performed at the beginning of the project to accompany the respective project status in order to further refine the estimation accuracy, as shown in Fig. 24.1. The aim is that the degree of accuracy is adjusted to a small range of only a few percent. In the early stages of a project, cost optimizations (*trade-offs*) also occur in conjunction with the technical optimization processes, for which model calculations are very helpful, because often only a few

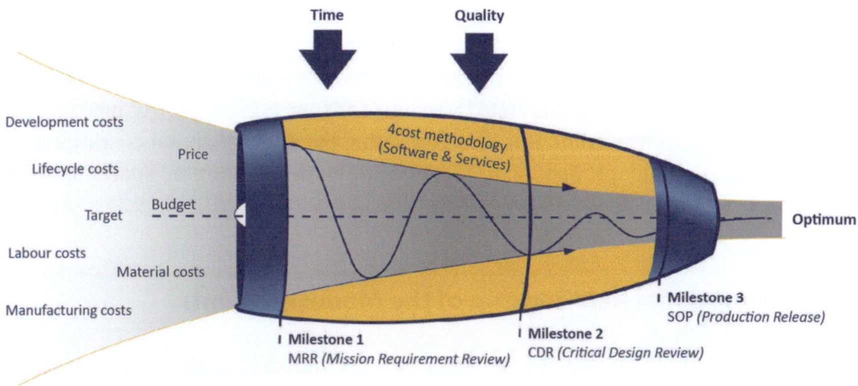

Fig. 24.1 Continuous model application

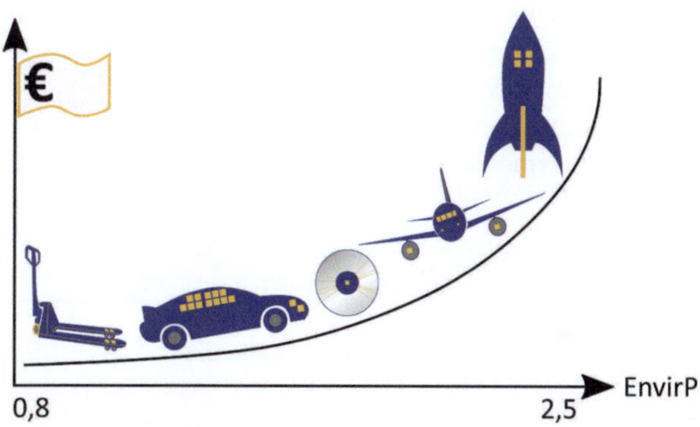

Fig. 24.2 Symbols of the insertion areas of the cost model 4cost-aces

factors need to be changed to represent a new LCC version. Model calculations also offer a great advantage in the dialog with customers, subcontractors, and suppliers, because one can react very quickly to new requirements. It is important to mention that model calculations should not be seen as competition to traditional *bottom-up costing,* but as a supplement and confirmation.

24.1.2 Areas of Application of 4Cost-Aces

The cost estimation model *4cost-aces* is universally applicable from the simple stationary product to the complex space device; see Fig. 24.2. The figure illustrates that the relative product costs are, among other things, very strongly dependent on the environmental conditions or the environment (EnvirP). The complexity hidden in this is described in practice by the product specification. As you can see, the values 0.8–2.5 are not a linear progression but an exponential curve. It is easy to explain from the figure that the extreme environmental conditions of a rocket are higher in relation to other products, such as a car or a passenger aircraft, which also leads to higher costs.

The environmental conditions (EnvirP) are shown symbolically in Fig. 24.2 and reproduced in Table 24.1 (abbreviated version) for four typical environmental conditions (quality requirements). It is easy to see that the environmental conditions for development (D) and production (P) are many times higher in space than on the ground.

24.1.3 Functional Relationship of the Model Elements

Figure 24.3 symbolically shows the functional relationship of the model elements using the example of a cost calculation for the development costs of a passenger car. The cost estimation model can be applied at all levels. Using the work breakdown structure (WBS), a product, in this case a passenger car, can be divided into different levels, e.g., into mechanical and electronic components. Granularity is critical to the accuracy of the cost estimate. The development costs are one-off costs, the so-called *none recurring costs (NRCs).*

The production costs, the so-called *recurring costs (RCs), are* closely related to the NRC. Figure 24.4 symbolically illustrates the functional relationship of the

Table 24.1 Different environmental conditions

Quality requirements/to deploy	ENVIR D/P
Stationary	
Own use/commercial	0.3–0.6
Mass production, limited support	0.6–0.7
Commercial/telecommunication	0.8–1.2
Typical software quality requirements	1.0–1.4
Military/good industrial quality	1.0–1.4
Mobile	
Commercial/telecommunication	0.9–1.2
Car manufacturing/good industrial quality	0.8–1.3
Ships/submarine	1.4–1.6
Military systems	1.3–1.7
Typical software requirements	1.4–1.7
Aircraft	
Commercial (FAA)	1.5/1.6–1.8
Military planes	1.8/1.9–2.0
Software requirements	1.9–2.1
Space	
Unmanned (typical), e.g., satellites	1.9–2.2
Manned space	2.5
Manned deep space	2.5–2.8
High software quality requirements	2.5–2.8

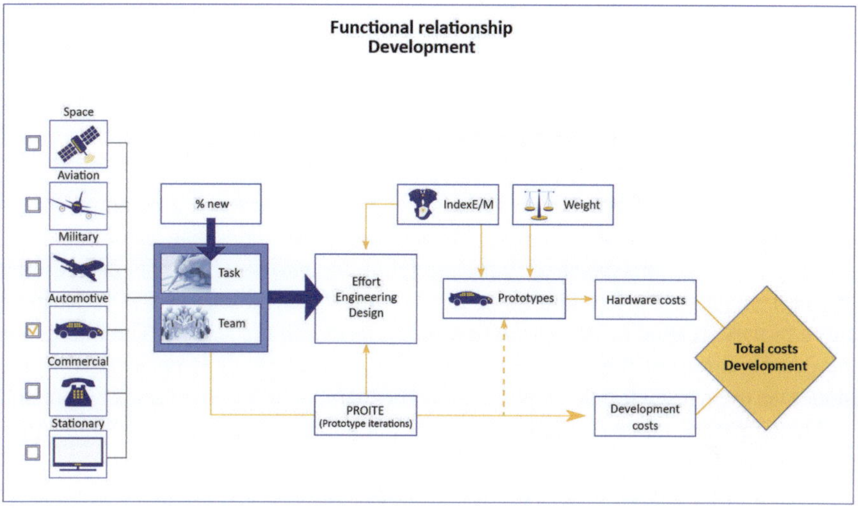

Fig. 24.3 Calculation logic of development costs including prototype construction

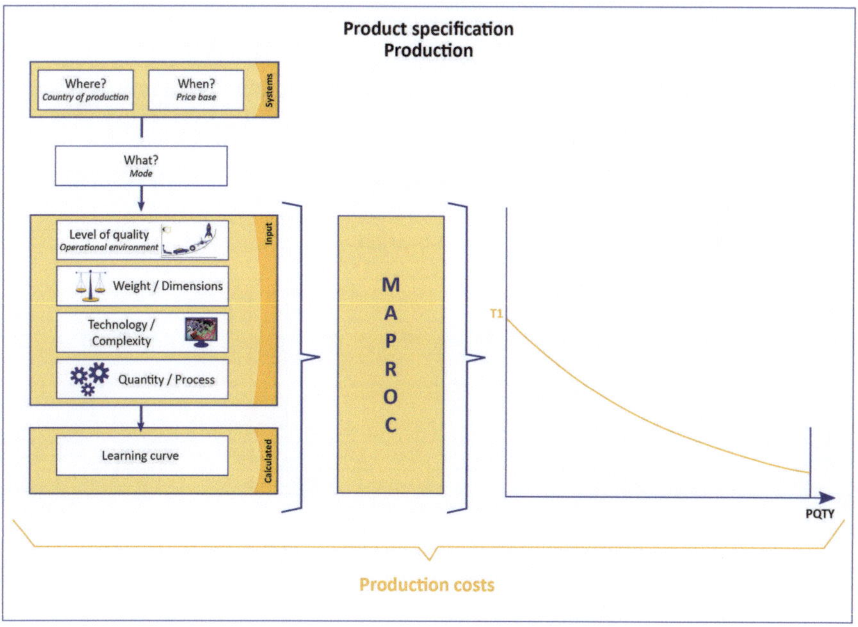

Fig. 24.4 Calculation logic of production costs including learning curve

model elements using the example of a cost calculation for the production costs of a passenger car, including an associated learning curve for larger production units.

24.1.4 Production Costs Considering Learning Curves

The determination of production costs using learning curves is a common procedure; see Fig. 24.5. The philosophy of the learning curve states that, according to the learning curve factor (*specified in the model in decimal %*), the costs decrease by this percentage when the number of units doubles. In the case of a 90% learning curve by 10% related to 100% (*the equation is stored in the model*). The linchpin is the determination of the theoretical "*first unit cost*" (T1). T1 is determined in the model via the parametric description of the mechanical or the electronic component.

The Crawford method is a recognized way to calculate cost degression for production. The Crawford equations are defined as follows:

The learning curve (ULC) can have a value between 0.6 and 1.00. This factor determines the cost degression. The appropriate equivalent of 0.95 ULC is a 95% learning curve. Component costs (*unit costs*) are the costs of an individual production unit, identified by the number of units (i). Total production costs are the sum of all production units.

A 100% ULC is a straight line and indicates that no learning is taking place. The 100% ULC would require an automatic process. A reduction in cost is only possible

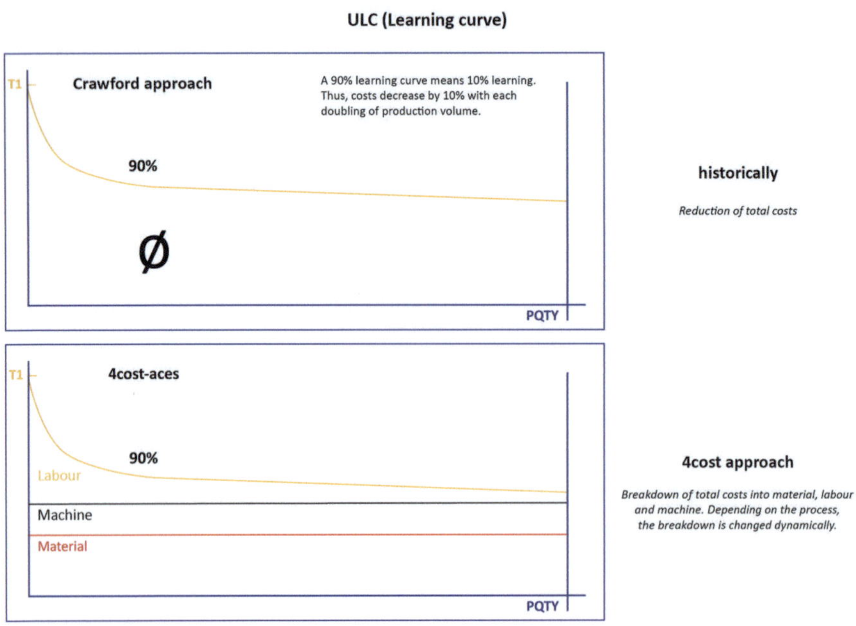

Fig. 24.5 Breakdown of costs into material, machine, and labor according to 4cost-aces

by reducing T1 (*first piece cost*). An 80% curve represents a large learning effect that specifically affects the first production units. The more units are produced, the smaller the "*learning effect*" becomes. The area under a curve up to the respective number of units corresponds to the "*sum of the series.*"

The graph shown in Fig. 24.6 illustrates the fundamental relationship between the manufacturing process (MAPROC), the ULC (learning curve) associated with this process, and the corresponding number of units. In principle, the following approach is anchored in the model:

1. Manual production high learning rate (75%), low quantity 1–10 pieces per year
2. Automated manufacturing low learning (98%), high volume 10,000–1,000,000 units

All other combinations are suggested by the model and automatically used in the model typical costing mode. The innovative extension of the Crawford algorithms to a cost distribution in three (3) typical cost categories is a special feature of the *4cost-aces* model. Early and decisive switch points are possible and allow the application of the parametric in the early phase of a production and accompanying until the end of the production. More advanced analyses, such as *LCC*, can easily be performed with this data set.

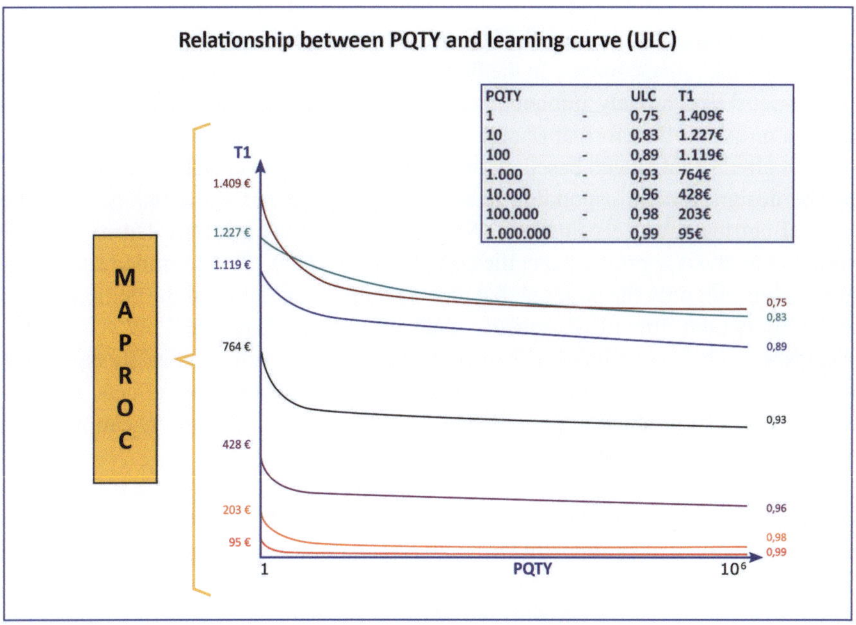

Fig. 24.6 Relationship between the manufacturing process (MAPROC) and the associated ULC and number of pieces

24.2 Parametric Costing

24.2.1 Determination of Hardware Costs

The parametric model "*4cost-aces*" (the parametric) calculates costs based on "*technical*" parameters, such as mass, and not on hourly rates. This would result in results not being comparable within the industry. It is a fundamental feature of parametric costing to derive the development and manufacturing costs for hardware and software elements of a system from historical data (parameters). This derivation is also called calibration. Modern models calculate costs by means of wizards, which make it possible to describe technological conditions. The result can then be the technology index (IndexE/M).

The quantity structure is defined in the *4cost-aces* cost estimation model by the following basic elements, for example:

WE =	Electronic weight
WM =	Mechanical weight
EnvirP/D =	Field of application, quality standard
Qty =	Project quantity, component quantity
PQTY =	Theoretical annual production quantity, optional
IndexE/M =	Complexity and technology metric
Start date =	Development and/or production start date
EcBase =	Economic base, year and month

The processing and import of 3D CAD models into the parametric model *4cost-aces* is an innovative concept. In the flowchart shown in Fig. 24.7, it can be seen that "*4cost-aces*" has already implemented this. First, the 3D information is analyzed using a universal data format (*.stp). During this analysis, a weight of the components is already calculated based on the geometries. If special materials are required by the design, the corresponding material can be selected via a dialog. Once the general entries have been made, a WBS is automatically created. During the 3D structure analysis, a product-specific complexity (INDEX) is determined and calculated. Thus, not only the structure but also a computable model of the component or assembly is built directly. Parametric analyses can then be carried out very easily. Analyses, such as quantity scenarios or quality optimizations, can easily be carried out in relation to the application area (ENVIRP/D) and their cost effects. In addition, changes to production periods and their cost impact (*penalties*) can be made visibly, quickly, and easily.

The calculation model for hardware elements is shown in Fig. 24.8 (mechanical components) and Fig. 24.9 (electronic components). Data input can be done in two ways, manually or electronically via a universal data format (*.stp) that analyzes the 3D information of a model. 3D models are in that case the basis for the automatic creation of the PSP/WBS and thus the basis for the parametric cost calculation.

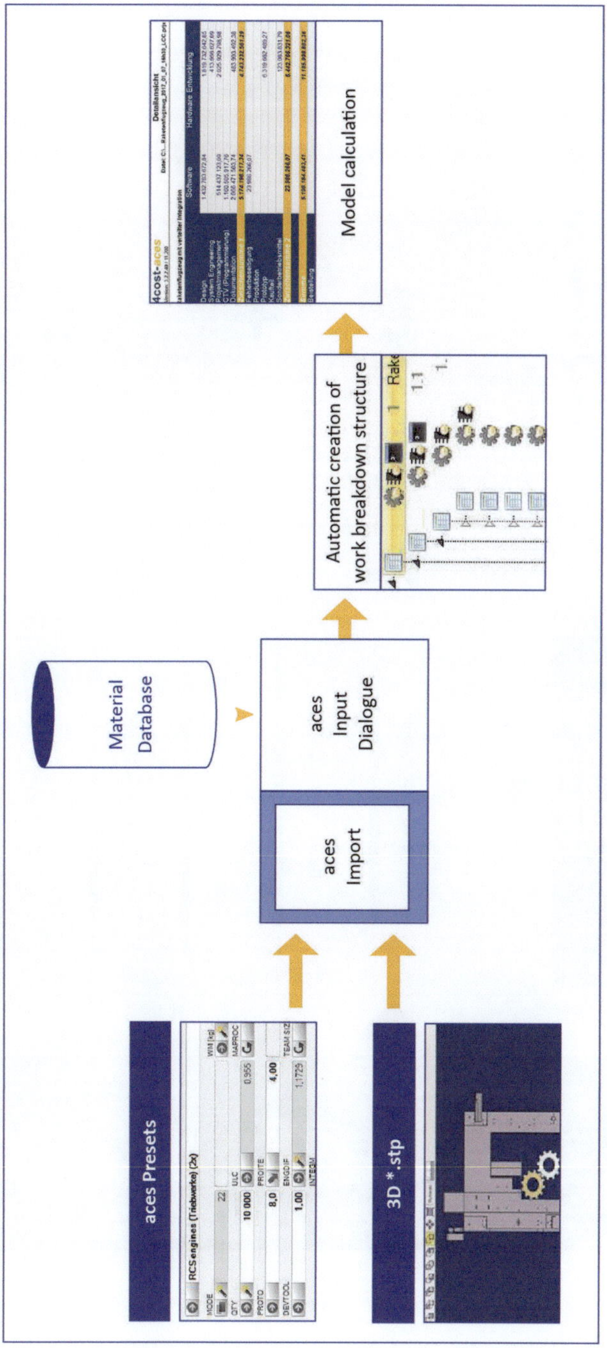

Fig. 24.7 3D import flowchart

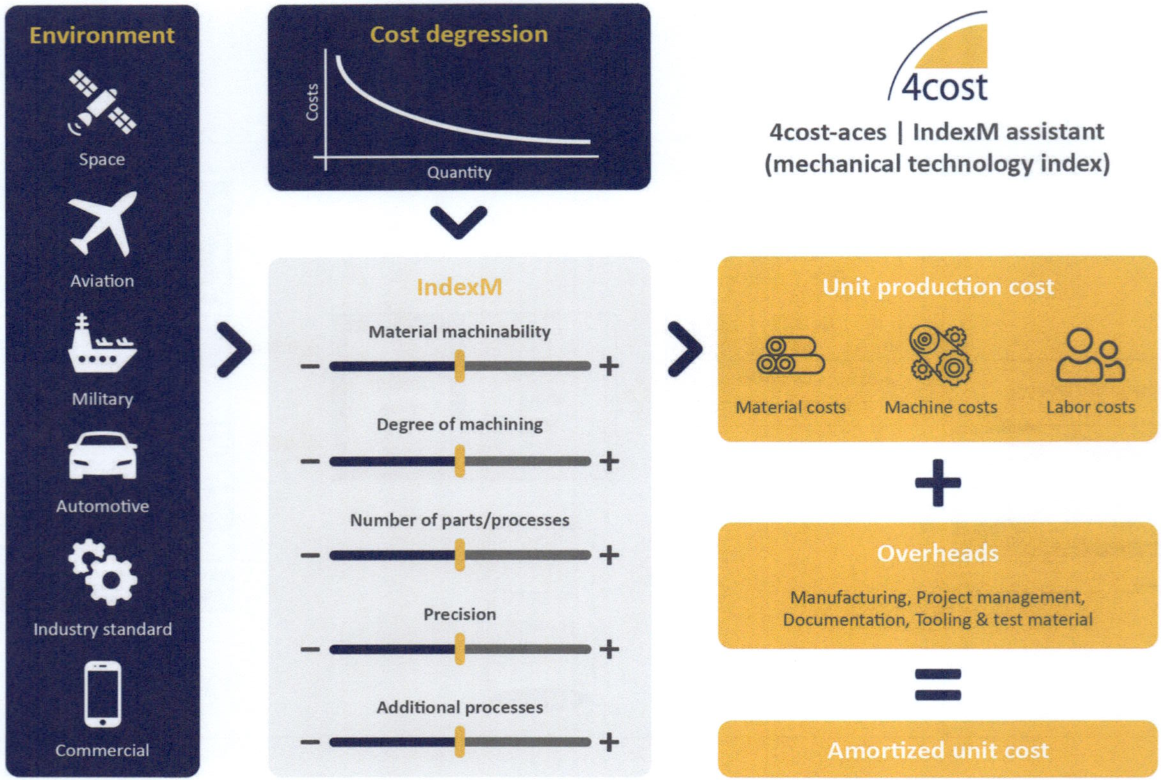

Fig. 24.8 Calculation model for mechanical hardware costs

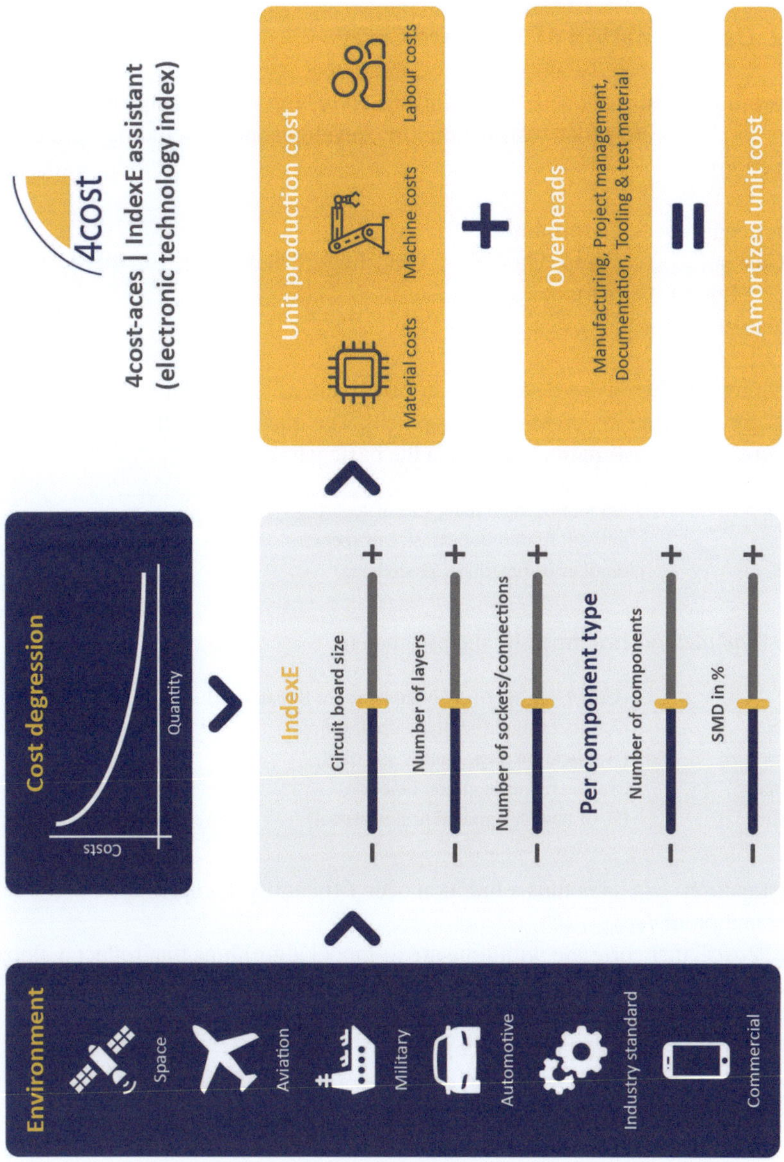

Fig. 24.9 Calculation model for electronic hardware costs

Depending on the question, an inflation factor can be applied to the calculated costs. This is done automatically in the model with the latest statistically determined multiplication factors; see Fig. 24.10.

24.2.2 Determination of Software Costs

In software cost estimation in particular, the following relationships apply analogously to hardware cost estimation in development (excluding prototype construction):

- Functional processes simplified:
  - Cost = $_{f\text{expense-Time}}$ (Inp, Outp, Obj, Uint, Engdif, Smodfrac, Envird) * *f*Effort $_{\text{costs}}$ (Engdif, Envird).
- Explanation of $_{f\text{Expenditure-time}}$:

In contrast to hardware development, no "*physical*" basic parameters (such as weight) are available in software development. The quantity structure is defined in the *4cost-aces* cost estimation model via the basic parameters:

Inputs =	(amount of input data, size of records)
Outputs =	(amount of output data, size of records)
Objects =	(number of functions, procedures)

The "*complexity*" is defined by the parameters:

Uint =	(User/Unit Interface; complexity of SW design)
EngdifS =	(Engineering difficulty/team descriptor)
Envird =	(Application area, quality standard)
Smodfrac =	(Software modification function)
Newdsft=	(% of new design and design repeat)

The function $_{f\text{Effort-Zeit}}$ is thus a function over ***f*** (quantity, complexity)

Explanation of $_{f\text{ Expense-Cost}}$:

The "*costs*" that correlate with time are in-model algorithms that reflect a global industry standard historically and regressively derived. Independent of:

- Inflation (Federal Statistical Office)
- Currency (purchasing power, market basket ratio, and value added)
- Economic basis (German Federal Statistical Office)

This so-called base set is modified by again technical parameters and is not dependent on inputs. Thus, the definition (evaluation) of the team factor and the evaluation of the working environment within the parameter "*EngdifS*" (*Engineering Difficulty for Software*) have a direct influence on the abovementioned basic set.

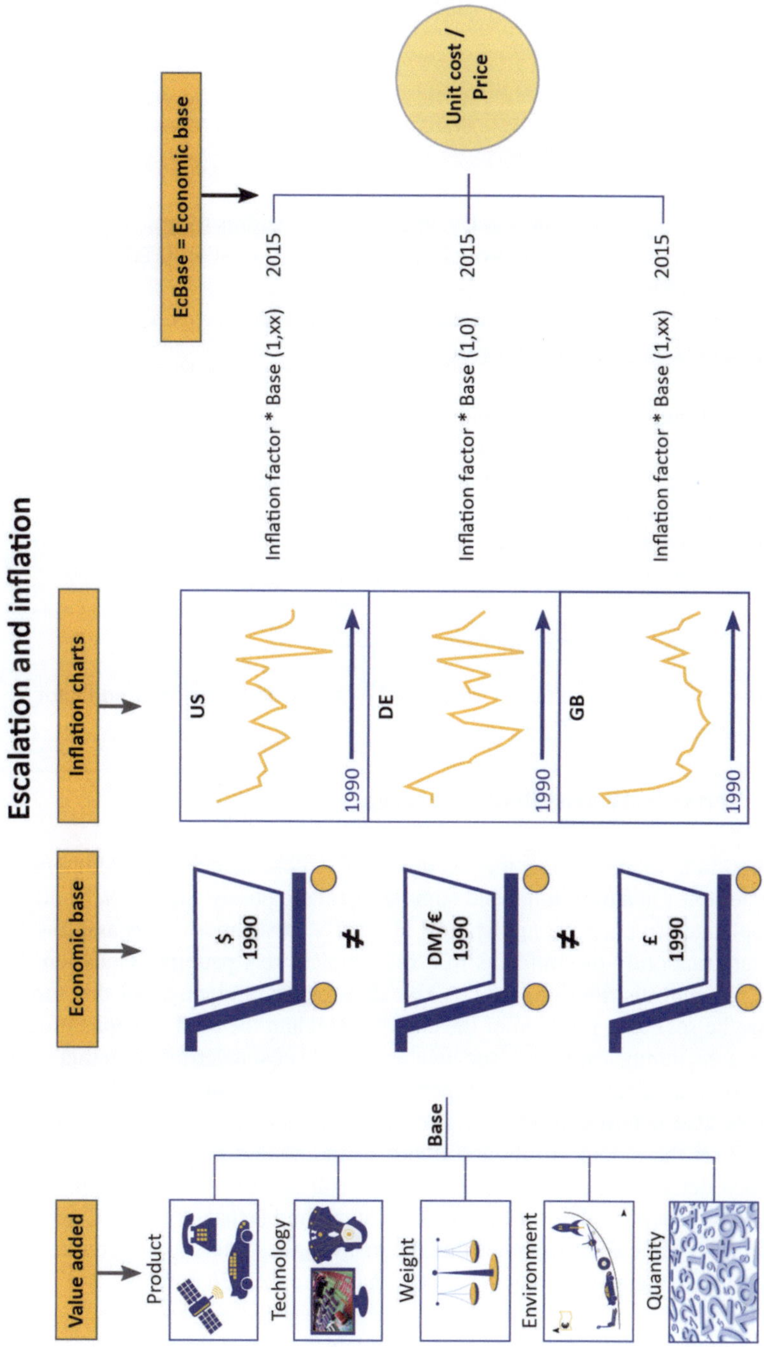

Fig. 24.10 Relationship between technology/product and economic conditions

The functional outcome: $_{f\ \text{Expense-Cost'}}$ is thus parametrically derived and in no way subject to an internal or external (price) hourly rate. The previously mentioned factors:

- Inflation
- Currency
- Economic basis

can and must be applied, if necessary, to the base costs thus technically calculated; in particular, the base costs are brought into a currency relation (Federal Statistical Office, Value Added).

To summarize, the *4cost-aces* model always calculates costs, i.e., the expenses incurred in the creation of the project in the areas:

- Concept design (preliminary concept)
- Full design (detailed, fine concept)
- CTV (Coding Test and Validation), (Spiral Development, Realization).
- Project Management
- Data (formal designs, documentation, etc.)
- Enhancement/maintenance (troubleshooting)

Prices and offers are usually structured like this:

$$\text{Price} = \text{Cost}\left(\text{parametrically calculated}\right)^{*}\ \text{Overhead factor}^{*}\ \text{margin factor}$$

24.3 Sensitivity and Risk Analyses

It is important to determine the accuracy of a cost calculation. Mathematical approaches exist in a multitude, and some of them are purely "*academic*" and often too complicated in practical application. Basically, these approaches assume that a risk occurs randomly or conforms to a rule (typical risk patterns: Gaussian, beta, triangle, Weibull distribution, etc.). A risk does occur randomly, but this randomness must be seen as a function of the environment and the field of application.

There are a number of parameters in a parametric cost determination that contain uncertainty in the determination. All parameters therefore contain a so-called *residual risk* or also determination risk. The "*residual risk*" is analyzed in *4cost-aces* depending on the model profile and is not a randomness within a chosen profile, which then determines a possible cost distribution and risk with the help of a random number generator. The most conventional method is that of Monte Carlo determination. The name already shows that the mathematics refers to pure randomness. In fact, it is like a game of roulette, except that more than 36 numbers and 2 colors are considered.

The "*runoff profiles*" included in these models determine the area where the randomness occurs, i.e., it forces the randomness into a profile determined by the user.

Determining a profile that seems right for an estimate is as much a risk as the estimate itself.

The "*Risk Analysis*" in *4cost-aces* takes into account all cost-relevant parameters. This integrated sensitivity analysis makes it possible to show a calculated range for the result (*chances and risk*), to determine a "*Probability*" and to look at the cost distribution. This simulation of a wide variety of cost items can be iterated up to 10,000 times per simulation. Figure 24.11 shows an example of a risk analysis. Each element in the work breakdown structure is considered and becomes part of the overall calculation.

24.4 Life Cycle Costs

The cost estimation is an important management task that should be tackled as early as possible, preferably already in Phase A, see Sect. 10.1. However, usually only few technical documents are available at this stage. When making investment decisions, however, it is necessary to consider LCC, i.e., the costs for all phases from the conceptual to the operational phase and sometimes including the disposal phase, already in the early stages of a project; see Sect. 24.4. Cost estimation models can be successfully used already in the early phases to determine LCC (*LCC*) even the *total cost of ownership* (*TCO*) can be considered, since development cost and production cost are estimated at the same time with an acceptable accuracy using the parametric approaches.

24.4.1 The Challenge

In the case of utilization costs, there is often little information available at the start of the project, but this must be verified in the course of the project by means of appropriate analyses. Procurement costs, on the other hand, are generally in the similar cost range for different suppliers, since similar technological requirements also generate comparable costs. This could only change if products were procured from low-wage countries, but this would give rise to questions of quality. This is precisely where the key to evaluating the cost of usage (*LCC*) through parametric simulation lies, because it is not the cost that is decisive, but the technical definition of the product, expressed in hours of the MTBF factor (*Mean Time Between Failure*). The MTBF factor determines the failure rate and is thus the main key to the utilization cost calculation; see subsection 8.2.2.

24.4.2 The Method

4cost-aces has an approach to implement this complex topic in an LCC simulation. The fundamental feature of the parametric is that the detailed information missing in the early phase is substituted. The important key figure MTBF is derived and

4cost-aces
Autor: OFFICE12-PC (HS)

Sensitivität

Seite 1 / 1
17.11.2023
4cost

€ (2001) - nicht inflationiert

Grafik der Trefferverteilung und Kosten

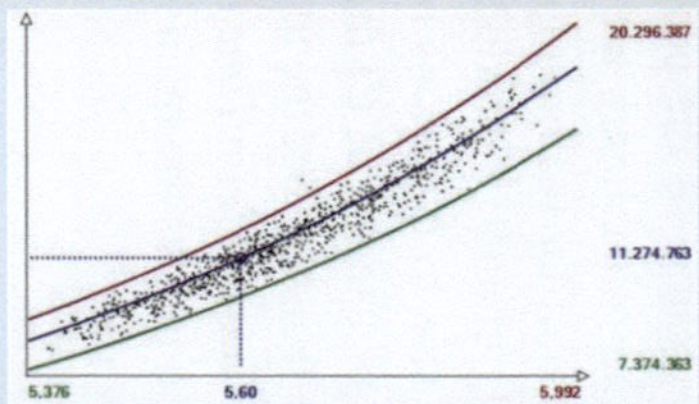

Kosten (Y) über X-Selektiertem Parameter (X)

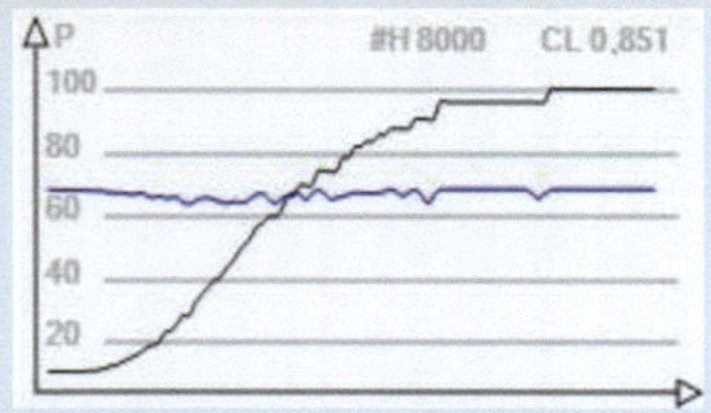

Treffer (Y) über X-Selektiertem Parameter (X)

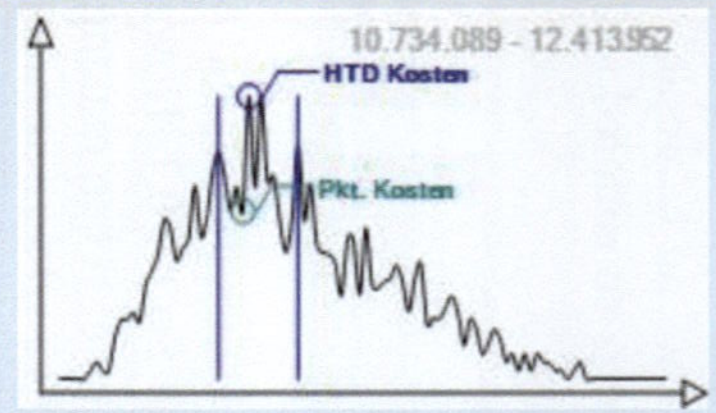

Treffer (Y) über Kostenspektrum (X)

Iterierte Parameter und deren Bereich

Parameter Name	- %	Min Wert	Punkt Wert	Max Wert	+ %	X	Parameter Name	- %	Min Wert	Punkt Wert	Max Wert	+ %	X
WE	3	1,940	2,000	2,060	3	Nein	WM	14	31 832	37 014	40 716	10	Nein
IndexE	14	4,542	5,282	5,652	7	Nein	IndexM	4	5,376	5,600	5,992	7	Ja

Kosten Statistik

Anzahl der Iterationen	7 683	Zuverlässigkeits Index		0,851								
Minimum (Absolut)	7 374 363	Punkt Kosten		11 274 763	Maximum (Absolut)		20 296 387					
Unterer Bereich %	3,55	Punkt Bereich(+/- Toleranz) %		70,32	Oberer Bereich %		26,12					
Minimum (Punkt Toleranz)	9 019 811	Punkt Kosten		11 274 763	Maximum (Punkt Toleranz)		13 529 716	Punkt Toleranz %		20,00	PT Wahrscheinlichkeit %	71,00
Minimum (Most likely)	10 734 089	Zentrum (Most likely)		11 638 631	Maximum (Most likely)		12 413 952	MLC Threshold %		25,00	MLC Wahrscheinlichkeit %	34,65
Kosten bei Wahrscheinlichkeit	> 99 %	20 296 387	95 %	15 902 899	90 %	15 127 577	85 %	14 481 476	80 %	13 964 595	75 %	13 447 714
											70 %	13 060 053
	65 %	12 672 393	60 %	12 284 732	55 %	12 026 292	50 %	11 767 851	45 %	11 509 411	40 %	11 250 970
											35 %	10 992 530
	30 %	10 734 089	25 %	10 475 649	20 %	10 217 208	15 %	9 829 547	10 %	9 441 887	5 %	9 054 226
											< 1 %	7 374 363

Version Anwendung: 4.3.3.128 - Version Report: 4.3.3.128
C:\....Bernd _beispielStruktur_komplett_2006_02_15_hs_neu_spiel.prjx

Fig. 24.11 Result representation of the risk analysis

calculated from early available parameters. It should be emphasized here that this can be done independently of information from the manufacturer. However, the manufacturer's MTBF values can of course also be used as a basis or as a benchmark.

The structure of the product is necessary for the LCC simulation. This structure is supplemented by technical parameters such as weights and technology key figures, areas of application, and products manufactured and in use. Complex maintenance concepts worldwide use and the primary transport methods can be evaluated easily and quickly.

A time-based cost representation, as well as the calculation of the required spare parts demand depending on the desired availability, is also given. Many additional aspects, such as spare parts reporting, stock quantities, spare parts availability, and the required general availability, are calculated. The MTTR (*Mean Time to Repair*), i.e., the average time of a repair, is also determined. The MTTR values are included in the cost-of-use analysis. The ability to quickly examine a wide variety of scenarios makes it possible to illustrate cost influences and offers a decisive advantage in product selection and project decision-making.

It is thus possible to carry out optimization processes and corresponding *trade-offs in* early Phases A and B (see Sect. 24.4) before the product is developed and manufactured as a hardware or software component. The early consideration of the LCC is a valuable basis for decision-making for the customer or operator of a system.

24.5 *4cost* Method Mix

The problem is often to combine the different cost estimation methods. As a rule, the so-called *Excel method* is the most used method. With this, everyone can create a cost estimate or better a cost calculation according to his/her own ideas. The big disadvantage, however, is that there is then no standardized procedure.

4cost GmbH has created a consistency and cost transparency that is unique in the world, because there are the following modules and models that communicate with each other, which complement each other and can show company-specific costs via reporting (data display, outputs, and also prints). This has the advantage that cost calculations can be controlled centrally. The procedure ensures that protected cost data can be changed only by an administrator password and not by unauthorized persons. The modules (software models) include the following options. The modules developed by 4cost GmbH and their integration are shown in Fig. 24.12.

The modules are described in detail below:

1. **Parametric** *(4cost-aces)*: *4cost-aces* can be used for all areas in the early project phases:
   - Hardware development costs (electronic/mechanical)
   - Software development according to 4costs own developed method
   - Hardware production costs (electronic/mechanical)

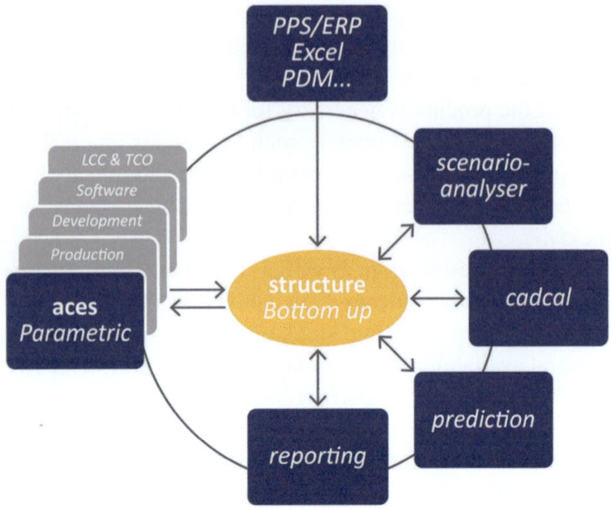

Fig. 24.12 4cost modules and integration

- LCC or also LCC/TCO (total cost of ownership)
- Sustainability calculations are also possible, considering means of transport, distance, and country of origin.
- As explained earlier, a combination of all cost areas of the LCC analysis can be considered here.

2. **Bottom Up**: The *4cost-structure* module is a "*bottom-up tool*" that can be used independently, even in the early phase, because it can be fed with the data from the *4cost-aces* parametric estimate, thus saving a lot of time. There is a bidirectional connection here, i.e., a parametric data set can be generated from a "*bottom up*" calculation, which then provides the best basis for further cost calculations. All data are checked in themselves and do not have to be reentered but, if necessary, adapted to the new project to be evaluated.
3. **Scenario Analyzer***:* It is a module for displaying the cost structure within a company or even a provider. Extensive analyses are possible. The result can be used in all modules. The calculation surcharges (sales, administration, etc.) are added to the parametric model to determine prices. This module also includes a parametric markup calculation, which makes it possible to calculate the required markups without much effort.
4. **Prediction***:* Here, there is a possibility to perform multidimensional regression analyses. Provided that sufficient historical data are available, a cost forecast can be carried out with this tool. All common mathematical procedures are available. A special feature is that the regression can be performed with independent 4cost mathematics. The results confirm the new, innovative method. However, sufficient historical data must be available to be meaningful.
5. **Cadcal**: As previously described, it is possible to have a work plan created during the design of a component and a statement of the production costs. This

Fig. 24.13 Military aircraft A400M (Creative Commons)

module allows the designer to identify cost influences during his/her design, e.g., simple considerations; is a "*counterbore*" mandatory, can it be drilled through, and what are the cost savings associated?

24.6 A Practical Example of an Independent Cost Estimation with 4cost-aces

At the early project phase of the military aircraft A400M (see Fig. 24.13), the German Military Procurement Office (Customer) approached the author of this book with a request to perform in very short time (1 week) an independent model-based cost estimation. The results were urgently needed as reference for a top-level negotiation meeting between the French and German government officials. The author of this book had been known as one of Germany's first cost model user with an outstanding reputation in the field of model-based cost estimations.

In order to perform a credible model-based cost estimation within a very short time, the author of this book joined forces with Herbert Spix, the inventor and partner of the 4cost company, Berlin. In his role as managing director, Herbert was leading the model-based cost estimating process taking into account the expertise of the author in aircraft design (*Transall C-160*) and project cost estimation.

Immediately after a kickoff meeting with the customer, the 4cost estimating team started the process with the following immediate actions:

1. Initial tasks and settings
   - Preparation of a Product Tree and WBS
   - Implementation of mass budget schedule
   - Identification of manufacturing data
   - Typical settings like:
     - Economic base
     - Technical improvements
     - Development scope (how much is really new)
     - Time frame: starting dates of development and or production
     - Software: since no detailed software information was available, a built-in software assistant was used to set the required inputs for software development and estimating SW Team size and cost.
2. Collection of available project information and structure of the project:
   - Overview of the cost calculation approach (customer)
   - Provision of project documents (customer)
   - Submission of a cost results to customer (4cost)
     - Aircraft summary
     - Subsystems (body, fuselage, power, etc.)
     - Furnished items (material chosen by the manufacturing company)
     - Standard configuration items (development only and no production cost)
     - Software development and implementation
3. Procedure of the estimating exercise:

   The parametric analysis was carried out on the basis of the technical documents and oral explanations provided to 4cost by the customer costing team:
   - Structuring of the WBS
   - Mass distribution of the components according to the WBS
   - Determination of the technical/economic/quantitative parameters
   - Determination of the product-specific indices (complexity of the individual aircraft items)
   - Determine the development parameters (e.g., new design, team, task, and estimated duration (schedule))
4. Important information

   The cost analysis was not based on an aircraft that had already been produced, but the technical and material related parameters were determined and applied by the model.

   The approach applied was not based on an approach of using already existing products and technologies. It confirmed the applicability of the model by relaying on the model's cost relations (CERs—cost estimating relations) built into the model and guarantied good results.
5. Results of the cost model example
   1. The independent cost model exercise provided an acceptable cost range for negotiations and confirmed the previously prepared industrial quotations.
   2. The exercise has shown that already in an early project phase a reliable cost estimation based on technical parameters can be generated with a parametric cost model in a very short time.

III. All costs were determined taking into account the following economic factors:
   - Currency: Million € (M€)
   - Price basis: July 1998
   - Costs: Total amortized production cost including profit, markup
   - Learning curve (computed by the model): ULC 85%
IV. For reason of confidentiality, 4cost did not publish detailed financial results of the calculations.

Project Management: Training Concept 25

Contents

Abstract

The author has developed a training concept for project management of which the structure is reproduced here. The training concept includes several weeks of theoretical instruction and practical exercises and can be performed in an individually arranged sequence (Madauss, Training plan "project management," 1986).

The theoretical part covers all project management topics described in this book. The courses will be conducted with reference to this book, and, in addition, seminar handouts will be provided [1]. For the practical part, the completion of an end-to-end project entitled "*Light Rail Transit*" is planned, a case study developed by the author [2]. This project example contains a number of very different project aspects from the fields of mechanics, electronics, structural and civil engineering, systems engineering, SW, etc., which can easily be adapted to the needs of the participants.

Field trips to companies with typical project management tasks should be carried out at regular intervals. In the following training plan, the temporal/content sequence

B. -J. Madauss, *Project Management*,
https://doi.org/10.1007/978-3-662-69057-4_25

and the logical linkage are reflected. The study times are to be individually adapted to the curriculum content. The training plan is divided into the following three main sections:

(a) Basic subjects
(b) Specialization subjects
(c) Industrial internship

Corresponding course documents (*handouts*) are to be prepared for the respective teaching blocks.

25.1 Seminar Elements and Their Titles

25.1.1 Theoretical Part (*Lectures*)

T. 01: Brief overview/teaching material/teaching body
T. 02: Basics of intercultural communication (*example international projects*)
T. 03: PM introduction/principles
T. 04: Life cycle of a project
T. 05: Project planning (*performance, schedule, and costs*)
T. 06: Network planning technique
T. 07: Project accounting
T. 08: Project monitoring
T. 09: Integrated project monitoring
T. 10: Conflict management
T. 11: Modern methods for cost estimation
T. 12: Project organization
T. 13: Systems engineering
T. 14: Quality assurance
T. 15: Management planning
T. 16: Information systems
T. 17: Documentation and configuration control
T. 18: PM software
T. 19: Contract management
T. 20: Competition/offers
T. 21: Project personnel
T. 22: Risk management

25.1.2 Practical Part (*Case Studies*)

P. 01: Techniques and teamwork
P. 02: Life cycle of a project
P. 03: Project planning

P. 04: Network planning techniques
P. 05: Project monitoring
P. 06: Integrated project monitoring
P. 07: Modern methods for cost estimation
P. 08: Project organization
P. 09: Systems engineering
P. 10: Quality assurance
P. 11: Management planning
P. 12: Information systems
P. 13: Documentation and configuration control
P. 14: Software training
P. 15: Contract management
P. 16: Competition/offers
P. 17: Rhetoric (persuasion)/negotiation techniques
P. 18: Industrial internship

25.1.3 Excursion Examples

E. 01: S/W house for PM systems
E. 02: S/W house for cost estimation models
E. 03: Large enterprises (*international projects*)
E. 04: PM consulting firm

25.2 Project Case Study *"Urban Rapid Transit System"*

Based on a rapid transit project, the author laid the foundation for a case study in 1978, which was tested during several seminars. The documents of this case study have since been used in various forms at practical seminars. With reference to the existing documentation, a comprehensive project was created by the author. In the initial phase, several simple small projects are to be carried out, which are then combined. In connection with the skills learned, the individual projects become larger and more complex, to be integrated into the urban rapid transit project at the end of the training course.

The project allows course participants a broad-based introduction to the practical characteristics of a project. Civil and structural engineering (*right-of-way*), infrastructure, rolling stock, electrification, control, etc. are covered. The following is a press release from the first practical seminar and the detailed report on the results of the seminar (excerpt from MBB/INTERN 7/8, 1978): *"Project Managers in the hardness test."*

"*Ottobrunn. Seventeen participants in a practical seminar organized by MBB's Space Division recently rehearsed tough competition for three days. The aim of this event was to impart project management experience in all management roles up to project manager.*" The participants had previously acquired the theoretical tools in a 20-hour seminar entitled "*Planning and monitoring of research and development projects,*" which was led by the author.

Project management exercises were completed by seminar participants in a role play scenario. The seminar leader in the role of Mr. Traffic, representative of the Ministry of Transport of an imaginary state, X-country, had asked the participants to bid for a new transport system for the capital of his country. The group was divided into three consortia of providers, with the sonorous names of

- INTERCONSULT
- EUROCONSULT
- MULTICONSULT

and appointed a project manager for each. With the tried and tested means of experienced industrial architects, the competition for the client's contract began.

For learning reasons, one project team presented its plans and financial framework in individual steps, while the other teams temporarily played the client together with Mr. Traffic. Thus, one by one, all the bids were heard collectively and evaluated according to a point system. In the end, the winner of the competition was the MULTICONSULT group.

The result of the three seminar days, which demanded a lot of work and concentration from all participants, can be described as excellent. Three complete proposals were developed on flip charts, and the professional knowledge of all participants was significantly deepened by the teamwork and led to a positive feeling about their own performance and their team colleagues. The seminar participants mentioned: "such seminars should definitely be continued."

25.3 Curriculum Drafts

With reference to Sect. 25.2, the draft curricula are reproduced here:

- Seminars: Blocks T. 01–T. 22
- Practice cases: Blocks P.01–P.18
- Excursions: Blocks E.01–E.04

The following tables contain the individual drafts. One lecturer should be responsible for each course block. The instructor in charge is responsible for creating a detailed lesson plan with the involvement of other accompanying subject matter experts. This schedule is then to be coordinated with the program's seminar supervisor.

The management of the course blocks P.02–P.16 is to be assigned to the training management on a superordinate level in order to be able to pursue the simulation project "*urban rapid transit*" as an independent, homogeneous project. Together with those responsible for the seminar blocks (P.03–P.22), the project processing will be coordinated and integrated. Especially, in this segment, the practical experience of projects already carried out should benefit the participants.

PART 1: Seminar Schedule of the Theoretical Part
(Learning content, learning objective, learning method, and success monitoring)

Block T. 01

- *Course content:* Course overview and introduction to the subject matter and faculty
- *Learning objective:* To provide participants with an overview of the training program progression and key milestones
- *Teaching methods:* Lecture/discussion
- *Performance review:* Not applicable

Block T. 02

- *Learning content:* Basics of intercultural communication
- The communication models
- Different values
- The technique of listening
- *Learning objective:* To introduce course participants to the basics of communication
- *Teaching methods:* Lecture, exercise, and group work/role play
- *Success control:* Evaluation of the exercise results

Block T. 03

- *Learning content:* PM introductory seminar. What is project management?
- Why PM?
- History of PM
- PM and systems engineering
- PM, a significant task for the future
- Definitions in PM
- PM use:
- In industry
- In construction
- With authorities, etc.
- Use of various practical examples to explain the learning content
- *Learning goal:* Participants should not only learn about the importance and benefits of PM for business but should also recognize the benefits for their own field of expertise.
- *Teaching methods:* Lecture/lecture/individual work
- *Success control:* Evaluation of the individual work; see Blocks T.02 and T. 03

Block T. 04

- *Learning content:* Life cycle of a project
- Project goals
- Project flow from A to Z
- Phase concepts
- Important milestones in the project
- Project decisions
- Costs of the individual phases

– *Learning goal:* Participants will learn about setting project goals, the typical flow of projects, and how to define project phases and key milestones.
– *Teaching methods:* Lecture/group work/individual work
– *Success control:* Evaluation of the individual work; see Block T. 04

Block T. 05

- *Learning content*: Project planning
- Work breakdown structure (WBS)
- Work packages
- Schedule
- Resource Plan
- Cost plan
- Capacity planning

– *Learning objective*: To familiarize participants with common methods and procedures for project planning (including quantities and costs)
– *Teaching methods*: Lecture/group work/individual work
– *Success control*: Evaluation of the individual work; see Sects. 9.1–9.3

Block T. 06

- *Learning content:* Network planning technique
- The different NP methods (*PERT, CPM,* etc.)
- Activity node networks
- Activity arrow networks
- Time estimate
- Calculation of the critical path
- Software application overview
- See also Blocks T. 05 and T. 03 (scheduling)

– *Learning objective:* The course participants should learn in detail the application of the network planning technique for project work.
– *Teaching methods:* Lecture/group work/individual work
– *Success control:* Evaluation of individual work; see *Groh/Gutsch: Netzplanchnik, 1982* and also Sect. 9.2

Block T. 07

- *Learning content*: Accounting
- Basics of cost accounting
- Cost of sales/overhead/profit margin
- Project/accounting interface
- Cost tracking
- *Learning objective:* The course participants should become familiar with the basic principles of accounting and the interfaces to the project.
- *Teaching methods:* Lecture/lecture/individual work
- *Performance review:* Evaluation of individual work; see Sect. 10.1

Block T. 08

- *Learning content:* Project monitoring
- Work release
- Progress control
- Actions–control
- Cost monitoring
- See Blocks P. 05–P. 07
- *Learning objective:* To familiarize course participants with common project monitoring procedures and methods
- *Teaching methods:* Lecture/lecture/individual work
- *Success control:* Evaluation of individual work; see Sect. 9.4

Block T. 09

- *Learning content:* Integrated project monitoring
- The power triangle
- Technical performance control
- Determination of the value of work
- Cost to completion (CTC)
- Trend analyses
- *Learning objective:* To enable course participants to analyze and monitor the project parameters of engineering/schedules/costs in an integrated manner
- *Teaching methods:* Lecture/lecture/individual work
- *Success control:* Evaluation of individual work; see Sect. 9.4

Block T. 10

- *Learning content:* Conflict management
- Conflict dimension
- Conflict analysis
- Conflict resolution
- *Learning objective:* The course participants should learn how to manage conflicts in the project quickly and effectively.
- *Teaching methods:* Lecture/exercise/group work

– *Success control:* Evaluation of the exercise results; see Chap. 15

Block T. 11

- *Learning content:* Modern methods for cost estimation
- Importance of cost analysis
- Cost estimation methods
- Cost estimation models
- Design to cost (DTC)
- Life cycle costs (LCC)
- Cost reduction method

– *Learning objective:* To familiarize the participants with the various estimation methods and procedures
– *Teaching methods:* Lecture/group work/individual work
– *Success control:* Evaluation of the individual work; see Chap. 10

Block T. 12

- *Learning content:* Project organization
- Management functions
- Team strength
- Key Personnel
- Matrix organization
- Industrial collaboration
- Working groups
- Consortia
- Management companies

– *Learning objective:* The course participants should learn the principles for organizing a project.
– *Teaching methods:* Lecture/group work/individual work
– *Success control:* Evaluation of the individual work, see Chap. 5

Block T. 13

- *Learning content*: Systems engineering
- System technical control loop—system specification
- Review concepts
- Systems Engineering Management

– *Learning objective:* The course participants should learn about and apply the principles of systems engineering (*which plays a role in the field of technology*).
– *Teaching methods:* Lecture/group work/individual work
– *Performance review:* Evaluation of individual work; see Chap. 7

Block T. 14

- *Learning content:* Quality assurance (QA)
- Task and function
- Quality and economy
- Availability
- Producer liability

- *Learning objective:* To familiarize course participants with the principles of quality assurance
- *Teaching methods:* Lecture/group work/individual work
- *Success control:* Evaluation of individual work, see Chap. 8

Block T. 15

- *Learning content:* Management planning
- Necessary management plans
- Implementation of the management plans
- Management plan content
- Management plan checklists

- *Learning objective:* To help participants learn the techniques for creating management plans
- *Teaching methods:* Lecture/group work/individual work
- *Success control:* Evaluation of individual work, see Chap. 6

Block T. 16

- *Learning content:* Information systems
- Information types
- Reporting
- Project meetings
- Modern information systems

- *Learning objective*: The participants should become acquainted with the common information systems in the project.
- *Teaching methods*: Lecture/group work/individual work
- *Success control*: Evaluation of the individual work, see Chap. 11

Block T. 17

- *Learning content:* Documentation and configuration control—documentation types
- Documentation requirements
- Numbering systems
- Release and status
- Configuration base
- Change control

- *Learning objective:* The course participants should become familiar with the common procedures of documentation and configuration control in the project.

- *Teaching methods:* Lecture/group work/individual work
- *Success control:* Evaluation of individual work, see Chap. 12

Block T. 18

- *Learning content*: PM software
- Project requirements
- Marketable programs
- Software interfaces project/company
- Software basic course
- PC software; see Dworatschek, S. and Hayek, A.: Marktspiegel Project management-Software, Verlag TÜV, Rheinland, 1993
- *Learning objective*: The course participants should learn necessary theoretical requirements as a basis for the practical course. See Block T. 14.
- *Teaching methods*: Lecture/group work/individual work
- *Success control*: Evaluation of individual work; see Chap. 18

Block T. 19

- *Learning content:* Contract management
- Contract types and pricing
- Expense reimbursement with premium regulation
- Performance bonuses
- Contract structure
- Service specifications
- *Learning objective:* To provide course participants with a basic knowledge of contract management
- *Teaching methods:* Lecture/group work/individual work
- *Success control:* Evaluation of individual work, see Chap. 13

Block T. 20

- *Learning content:* Competition and supply
- The role of the client
- The role of the contractor
- Tender documents
- Offer strategies
- Quotation
- Offer evaluation
- *Learning objective:* To provide course participants with an understanding of the basic principles of the different roles of client and contractor in the management of project assignments
- *Teaching methods:* Lecture/group work/individual work
- *Success control:* Evaluation of individual work, see Chap. 14

Block T. 21

- *Learning content*: Project personnel
- Motivation
- Project managers and their teams
- Key personnel
- PM training
- Leadership style in the project
- Increase and decrease of personnel in the project

– *Learning objective*: The course participants should learn the importance of personnel deployment and personnel management in the project.
– *Teaching methods*: Lecture/group work/individual work
– *Success control*: Evaluation of role-play results and individual work; see Chap. 15

Block T. 22

- *Learning content*: Risk management
- Risk philosophy
- Risk factors
- Classification of risks register
- Risk management

– *Learning objective*: The course participants should learn the methods of risk management.
– *Teaching methods*: Lecture/group work/individual work
– *Success control*: Evaluation of the role play results and individual work; see Chap. 19

PART 2: Seminar Schedule of the Internship Part Learning Content, Learning Objective, Learning Methods, and Success Control

Block P. 01

- *Learning content:* Techniques of teamwork
- Target descriptions
- Task definition
- Visualization techniques
- Role plays/role changes

– *Learning objective:* The course participants should become familiar with the techniques of teamwork based on practical examples. This is an important prerequisite for the course participants, especially with regard to the practical case "*urban rapid transit*" to be carried out later.
– *Teaching methods:* Lecture/group work/individual work
– *Success control:* Evaluation of the exercise results

Block P. 02

- *Learning content:* Life cycle of a project
- *Learning objective:* The course participants are to apply the knowledge they have learned from Block T.04 based on the practical case of the "*urban rapid transit system.*"
- *Teaching methods:* Lecture (presentation of the project "*Stadtschnellbahn*")/case study
- *Success control:* Evaluation of the created working documents and survey of the course participants

Block P. 03

- *Learning content*: Project planning
- *Learning objective*: The course participants are to put into practice the knowledge they have learned from Block T. 05 based on the practical case of the "urban rapid transit system."
- *Teaching methods*: Case study
- *Success control*: Evaluation of the created working documents and survey of the course participants

Block P. 04

- *Learning content:* Network planning technique
- *Learning objective:* The course participants are to put into practice the knowledge they have learned from Block T. 06 based on the practical case of the "*urban rapid transit system.*"
- *Teaching methods*: Case study
- *Success control:* Evaluation of the created working documents and survey of the course participants

Block P. 05

- *Learning content:* Project monitoring
- *Learning objective:* The course participants are to put into practice the knowledge they have learned from Blocks T. 07 and T. 08 based on the practical case "*urban rapid transit system.*
- *Teaching methods*: Case study
- *Success control:* Evaluation of the created working documents and survey of the course participants

Block P. 06

- *Learning content:* Integrated project monitoring

- *Learning objective:* The course participants are to put into practice the knowledge they have learned from Block T. 09 based on the practical case "*urban rapid transit.*"
- *Teaching methods*: Case study
- *Success control:* Evaluation of the prepared working documents and survey of the course participants

Block P. 07

- *Learning content:* Modern methods for cost estimation.
- *Learning objective:* The participants are to put into practice the knowledge they have learned from Block T. 11 based on the practical case "*urban rapid transit system.*"
- *Teaching methods*: Case study
- *Success control:* Evaluation of the created working documents and survey of the course participants

Block P. 08

- *Learning content:* Project organization
- *Learning objective:* The participants are to put into practice the knowledge they have learned from Block T. 12 based on the "*urban rapid transit*" practical case.
- *Teaching methods*: Case study
- *Success control:* Evaluation of the created working documents and survey of the course participants

Block P. 09

- *Learning content*: Systems engineering
- *Learning objective:* The course participants are to put into practice the knowledge they have learned from Block T. 13 based on practical case of the "*urban rapid transit system.*"
- *Teaching methods*: Case study
- *Success control:* Evaluation of the created working documents and survey of the participants

Block P. 10

- *Learning content:* Quality assurance (QA)
- *Learning objective:* The course participants are to put into practice the knowledge they have learned from Block T. 14 based on the practical case of an *urban rapid transit system.*
- *Teaching methods*: Case study
- *Success control:* Evaluation of the created working documents and survey of the participants

Block P. 11

- *Learning content:* Management planning
- *Learning objective:* The course participants should put into practice the knowledge they have learned from Block T. 15 based on the practical case of the "*urban rapid transit system.*"
- *Teaching methods*: Case study
- *Success control:* Evaluation of the created working documents and survey of the course participants

Block P. 12

- *Learning content:* Information systems
- *Learning objective:* The course participants are to put into practice what they have learned from Block T. 16 based on the practical case of the "*urban rapid transit system.*"
- *Teaching methods*: Case study
- *Success control:* Evaluation of the created working documents and survey of the course participants

Block P. 13

- *Learning content:* Documentation and configuration control
- *Learning objective:* The course participants are to put into practice the knowledge they have learned in Block T. 17 based on the practical case "*urban rapid transit.*"
- *Teaching methods*: Case study
- *Success control:* Evaluation of the created working documents and survey of the course participants

Block P. 14

- *Learning content*: Software training
- Instruction in operating programs
- PC application programs: Planning, control, documentation, graphics, etc.
- *Learning objective*: The course participants should learn to master the use of project management software programs (see Block T. 18).
- *Teaching methods*: Group work and case studies
- *Success control*: Evaluation of the case studies

Block P. 15

- *Learning content:* Contract management
- *Learning objective:* The course participants are to put into practice what they have learned from Block T. 19 based on the practical case of the "*urban rapid transit system.*"

– *Teaching methods*: Case study
– *Success control:* Evaluation of the created working documents and survey of the course participants

Block P. 16

- *Learning content:* Competition and offers
– *Learning objective:* The course participants are to put into practice the knowledge they have learned from Block T. 20 based on the practical case "*urban rapid transit system*."
– *Teaching methods*: Case study
– *Success control:* Evaluation of the prepared work documents and survey of the course participants

Block P. 17

- *Learning content:* Rhetoric/negotiation techniques
- Talk freely
- Presentations
- Appearance in front of an audience
- Negotiation
- Employee training
– *Learning goal:* To introduce course participants to the common practices of rhetoric and negotiation techniques and the principles of employee training
– *Teaching methods:* Lecture/role play
– *Success control:* Evaluation of the role play results

Block P. 18

- *Learning content:* Industrial internship
- Completion of an internship in an industry appropriate for the individual course participant
– *Learning objective:* Practical application of the knowledge learned in theory and practical case studies (*Blocks T. 01–T. 22 and P. 01–P. 17*) in industry
– *Teaching methods:* Teamwork
– *Success control:* Assessment of the course participants by the industrial supervisor with the involvement of the responsible manager of the corresponding industrial company

PART 3: Seminar Schedule of the Excursion Part (Examples) Company, Institute, and Program

Block E. 01

- S/W house (*PM systems*)

– *Learning objective:* Getting to know common software programs for large and small projects (*see also Block P.14*)

Block E. 02

- S/W house (*cost models*)

– *Learning objective:* To become familiar with and test a computer-based cost estimation model in practice (*see also Block P.07*)

Block E. 03

- Large company (*international PM*)

– *Learning objective:* To become familiar with the systems engineering aspects of project management using practical examples

Block E. 04

- Management consulting firm

– *Learning objective:* To become familiar with management procedures in consulting practice

Literature

1. Madauss, Bernd-J.: Training Plan "Project Management", 1986
2. Madauss Bernd-J.: Project Management Case Study, 1978/1980 (unpublished).

Project Management Training Proposal

26

Contents

B. -J. Madauss, *Project Management*,
https://doi.org/10.1007/978-3-662-69057-4_26

Abstract

The efficiency of project management is mainly dependent on two aspects: the availability of existing methods and tools and the qualifications of project personnel. The available methods of Project Management and tools are generally fully developed and have been described in the previous chapters of this book. The required qualifications can be gained through both training and/or on the job experience. The best approach, however, is when future project team members (Project Manager and Team) are fully trained before making their way into the project world.

In Chap. 25, a Project Management (PM) training concept is described, of which the elements have been subjected to practical testing in the past. This chapter contains excerpts from the author's training program developed in 2012 at the International Space University (ISU), Strasbourg.

The author was supported by Prof. Victor Billig, Florida, USA on this chapter. They taught at the ISU within the framework of the Master Program MSS with focus on Space Flight Management. Furthermore, they were essential participants in the planning and performance of the Project Management Module "Space Executive MBA (EMBA) Program" of ISU. Based on their extensive PM experience, the ISU entrusted the authors with leadership of the development of a Project Management Training Program for the European Spaceflight Agency ESA, Paris. The completed "Project Management Training Program" (PMTP) consisted of the following segments:

1. Introduction and Overview
2. Technical Process Segment
3. Programmatic Segment

The training processes contained in the PMTP are not only applicable to Space Flight projects but also in adapted form equally applicable to high-tech projects in other branches because from the view of management, there is synergy between spaceflight and similar complex projects in other high-tech branches. For the future training organization, the designation **TO** (Training Organization) was chosen and for the Agency or Company receiving training, the designation **Client**.

26.1 Introduction and Overview

ISU has developed and teaches space PM at its postgraduate campus in Strasbourg, France. In particular, ISU's Master of Space Management (MSM) and the added EMBA programs for space executives represent an outstanding heritage of professional project management training. Much of ISU's success in professional project management training is due to its open academic model that provides for the inclusion of experienced project managers from international government and industry space-related organizations. In past decades ISU and its Extended Faculty have grown in the project management arena. Further details about ISU are shown in Sect. 26.4.

In 2012 ISU prepared in response to a request from the European Space Agency (ESA) a project management training program. The task for preparation of the PM training program was assigned to the authors of this section. Since much of it is generic and applicable also for high-tech projects of other branches, they developed a PMTP, tailored to the aforementioned original training program developed by ISU for the ESA.

The focus of this Section is on the design, development, implementation and evaluation of a PMTP applicable for future space and high-tech projects. As mentioned before, it takes the work previously prepared by ISU into account. The details of the major steps of the PMTP are provided in Sect. 26.2. The headings of the steps are shown below:

- Objectives of the PMTP
- Proposed Approach and Design and Development Logic
- Task Description of the Four Training Phases
- Training Schedule and Deliveries
- Training Locations and Facilities
- Identification of Training Professionals

The details of training management are shown in Sect. 26.3.

During preparation of the PMTP, a comprehensive review of the state-of-the-art PM requirements was prepared, to ensure that the developed training concept can be implemented cooperatively between a Training Organization (TO) and any future Clients. The authors are confident that this program can effectively meet the challenges of solidifying the Client's leadership cadre. The PMTP provides a generic model for future PM training applicable for governmental and/or commercial Clients.

26.2 Process Description

26.2.1 Objectives of the Project Manager Training Course

Design and development (D&D) of the Project Manager Training Course follows a document structure of training material to be generated according to task requirements and planned events. The logical interrelationships between these events are schematically shown in Fig. 26.1. This diagram depicts, in a simple way, the concept for the initial design of the baseline training material and shows how iterations are performed by the Client and ISU to update the course during development phase and post-development implementation.

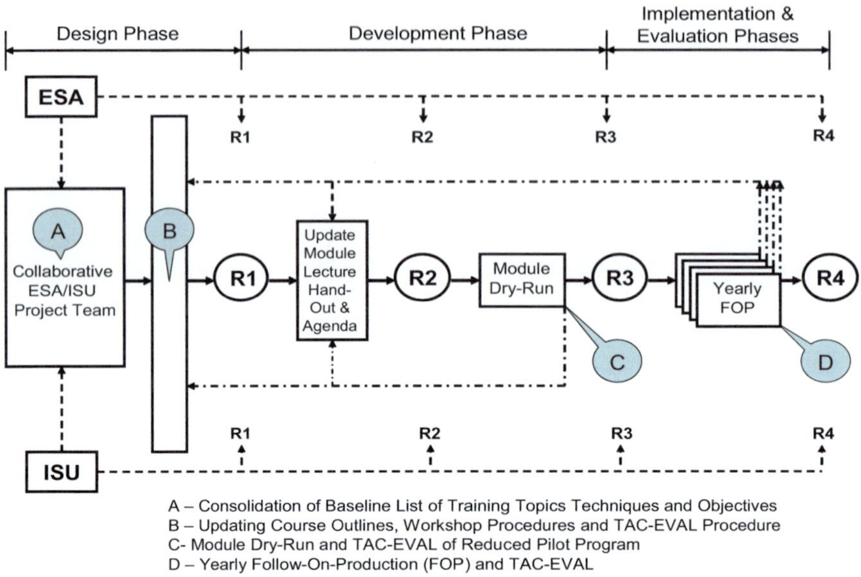

Fig. 26.1 Design and development logic of the PM training course concept. Note: ESA may be replaced by client and ISU by TO

The approach taken in configuring the logic of work flow during the program execution is to minimize interphase ripple effect while focusing strongly on the central objective of each phase. Priority is given to the following:

1. Flexibility for the TO to introduce early performance measurement feedback reducing the impact of modification after the conclusion of development.
2. Utilizing transitional means to upgrade a standard course with direct reference to the Client projects either completed or projected, e.g., a project manager's syllabus for a particular class of projects such as Public Private Partnership (PPP) programs.
3. The facilitation of Training Assessment and Course (TAC) Evaluation (EVAL); that is, TAC-EVAL after each training session is performed, including the yearly Follow-On-Production (FOP) training sessions.
4. The smooth integration of Client staff learning through in-house management training within the overall context of PM development and continuity of the Client culture.

26.2.1.1 Proposed Classroom Activities

Classroom activities consist of lectures, hands-on exercises, case studies, simulations, and other instructor supervised training in which students are working individually or in teams, pre-coordinated by an instructor. The time, the place, and content of the activity are prescribed by an agenda, participation roster, and procedure, which define the type and methods of instruction.

TO will apply a wide range of classroom activities those have proven to be effective in conveying project manager skills. Also addressed are leadership acumen, poise, and tactfulness in dealing with critical situations. Opportunities will be given to apply online techniques during classroom activities on a noninterference basis. The aim of this advanced approach is to develop a distinct managerial capability of the Client to confront crisis situations spontaneously and to display good judgment without undue reliance on outside sources.

Section 26.2.2.2 describes proven training methodology to be applied. These methods have evolved from classroom training with various situational characters, e.g., case studies, random participant mix, and workshop exercises in comparative negotiation, failure review simulations.

Contact time in the classroom will be fixed by the overall agenda of each module in the PM Training Course. Outside the classroom, students will be expected to perform collateral study in preparation for lectures and simulations and to take on additional research as may be required by the specific training objective. Most of the time commitment outside the classroom will be voluntary. However, instructors are encouraged to organize ad hoc discussions beyond the fixed time schedules—e.g., "fireside" chats.

26.2.1.2 Coordinated Lectures and Simulations

As noted in Sect. 26.2.1.1, course instructors are encouraged to organize informal contact time with students outside the classroom. Conversely, the Client is encouraged to take an active part in the tailoring of content introduced by the instructor into the training both during the initial course design (Phase 1) and during the actual implementation (Phase 3). While formal design scope of each agenda item will be frozen upon completion of course development (Phase 2), the ISU design and development logic as proposed in Sect. 26.2.3 will motivate the Client to participate in a gradual but transcendental refinement of the initial design baseline. How does this process work? The details of the refinement coordination process will be found in Sect. 26.2.2.4. At this point, however, it should be noted that the proposed training will be inherently responsive to the changes occurring by the Client, in order that project manager talent of the Client stays abreast of its own technological and industrial environment.

As a starting point, published lecture materials, lesson-learned, case studies, simulations, and curriculum of ISU's EMBA courses will be applied. In addition, extensive project management training and development experience programs with industry and governments will be taken into account.

Other departures could become necessary in case the course content is not sufficiently challenging for a particular class year. Thus, easy upgrading of the content is a key feature of the configuration control of the handout set distributed to the students. This will be further elaborated in the management concept; see Sect. 26.3.

Preparation of course material will begin immediately after approval is given in the Design Review, R1 (see Sect. 26.2.3). The work has to be performed by the TO with inputs and concurrence of the Client. As training proceeds from one instantiation to the next, it is likely that new material will be incorporated to keep up with evolving technology on project management. If required, the TO will update the Project Management Training Course.

26.2.1.3 Training Assessment Plan

The Training Assessment Plan (TAP) will be prepared to measure and assess the effectiveness of the Project Manager Training Course. "Assessment" in the context of benefit to the Client or to all stakeholders of the Project Manager Training program is treated as "Evaluation" of training at Level 4, see Sect. 26.2.1.6.

Detailed contents of the TAP will be prepared during Phase 1, according to the following outline:

1. Introduction
2. Definitions
3. Description of Measurement Process:
   Level I: Reaction of Student
   Level II: Impact on Learning
   Level III: Impact on Behavior at Work
4. Determination of Effectiveness at Each Level
5. Assessment Organization, Coordination, and Control
6. Measurement Templates for Survey/Feedback
7. Effectiveness Assessment Worksheet

26.2.1.4 Training Topics

The training has been structured into eight (8) major topic areas; see Table 26.1. It is intended that the Client and the TO work together in Phase 1 to determine the best list of training topics for different projects, i.e., large-scale and medium-size projects.

26.2.1.5 Client Participation Plan

The cornerstone of delivering effective training to Client project managers is to design, develop, and implement the program in close cooperation and with frequent concurrence of Client collaborators. Hence the TO proposes a Project Management Office (PMO) as the focal point for a team effort that coordinates and controls the required tasks prescribed under the contract.

26.2.1.6 Evaluation of Training Assessment and Course Effectiveness (TAC-EVAL)

The enhancement of Client project results will strongly depend on the proposed TAC-EVAL system, which contains methodology used widely in the space industry and government agencies. It should be noted that the key personnel chosen for executing the PMTP contract have in-depth experience in this subject. For Training Course Evaluation, the following Key Performance Indicators (KPIs) will be applied as a test follow-up:

1. Participant satisfaction (Level I)
2. Measurement of learning based on course examinations (Level II)

Table 26.1 Training topic areas and associated topics

Training Topics		
The below shown eight (8) training topic areas are subdivided into a number of topics described; the actual need for application to different projects has to be determined.		
A. Setting up the project 1. Stakeholder (customer) expectation and MGMT 2. Requirements generation and MGMT 3. System architecture and optimization 4. Feasibility studies and trade-offs 5. Life cycle cost considerations and DTC 6. Project plan development	**B. Procurement process** 1. Procurement strategy 2. Work sharing requirements 3. Invitation to tender (ITT) 4. SOW, specifications and plans 5. Delegations—role/relation 6. Proposal evaluation and IPC 7. Contract negotiation (NEGO) 8. Contract type (CPIF, FFP, etc.) 9. Kickoff and ATP	**C. Implementation process** 1. Organizational Structure and Internal Goals 2. PMO implementation 3. Importance of key personnel 4. Industrial organization 5. Subcontractor management
D. Project management skills 1. Leadership style 2. Personnel goal setting and motivation 3. Team management and delegation 4. Decision making 5. Cultural/social/ethical factors	**E. PM processes** 1. Project phasing concept 2. Product tree/WBS/WPDs 3. Time schedule management 4. Cost estimating methodologies 5. Manpower/cost estimating and budgeting	6. The new ESA financial system 7. Manpower/cost control 8. PC, CTC, CAC, and EVM 9. DOC/COFI management 10. Risk/conflict management 11. Project reviews 12. Reporting and Communication
F. Technical management 1. System engineering (SE) 2. Subsystem engineering (S/SE) 3. Design and development 4. I/F management (ICD) 5. AIV management 6. Product assurance (PA)	**G. External interfaces** 1. Industry understanding 2. User—involvement and MGMT 3. Partnerships, JV, and PPP 4. Other agencies—cooperation 5. ESA support (HR, etc.)	**H. MGMT procedures and tools** 1. Standard PM/SE procedures hand books and guidelines 2. Standard PM software tools

3. Measurement of participant improvement at workplace (Level III)
4. Assessment of course effectiveness as far as the overall Client performance is concerned (Level IV)

Prior to implementation, an Evaluation Worksheet will be applied during the FOP TAC-EVALs. Some of the recent failures in managing large projects have stimulated new literature adding to the wealth of knowledge available for the evaluation process. An example is the 4D approach stemming from the Hubble Telescope lessons learned at NASA. The "Four Dimensions" coined by Charles J. Pellerin, former NASA Director Astrophysics Division, in his book entitled *How NASA Builds Teams* depicts team success as being promoted by 1—Cultivating, 2—Visioning, 3—Directing, and 4—Including (people) [1].

26.2.2 Approach and Development Logic

26.2.2.1 Topics to Be Trained

Table 26.1 provides the list of required training topics that will be reviewed and modified in Phase I. The topics selected will be aligned in a matrix which shows a correlation between the topics, simulation, case studies, and lectures in which they will appear.

26.2.2.2 Proven Training Methodology

In order to develop the project manager capabilities desired by the Clients leadership, it is necessary to do as much hands-on experiential activities as possible, so participants can learn by doing. The instructors will combine pre-course activities and lectures, with in-class, competitive, team-based learning activities to make sure those participants are engaged and actively learning. Table 26.2 provides guidance for the development and implementation of the Project Management Training Program with the goal of enhancing trainees' knowledge, comprehension, application, analysis, synthesis, and evaluation of the practice of Client project management.

In the evolution of the PM Training Course, primary classroom-related methods include pre-course assessment, pre-reading, lectures by the TO and industry experts, hands-on in-class exercises demonstrating key concepts, team competitions, threaded simulation activities and associated coaching, and post-course assessment. The threaded simulation is used to allow trainees' to experience the full spectrum of learning (knowledge through evaluation) and have opportunity to reflect on the outcome of their decisions and make adjustments. The goal is to limit lectures to 50% or less of total classroom time; see Training and Development Methods Based on Bloom's Taxonomy [2].

In order to develop the project manager capabilities desired by then Clients leadership, it is necessary to do as much hands-on experiential activities as possible, so participants can learn by doing. Instructors will combine pre-course activities and limited lectures, with in-class, competitive, team-based learning activities to make sure those participants are engaged and actively learning.

Table 26.2 Methods based on bloom's taxonomy

Bloom's taxonomy level	Explanation	Example	ESA PM threaded simulation
Knowledge	Recall/recognize information	Identify an outline of a project plan	**Reinforce** prior learning from lectures
Comprehension	Understand meaning	Explain the meaning of each element of the project plan	**Review** current learning
Application	Use or apply knowledge	Determine key governance and guidance documents for the project plan	**Determine** what learning is appropriate
Analysis	Interpret knowledge	Identify issues in existing project plan	Use learning to find **needs**
Synthesis	Develop new structures	Suggest alternative approaches to solving a schedule, cost, or risk problems	Use learning to find **possible solutions**
Evaluation	Assess concepts	Recommend the BEST solution to a particular problem	**Make decisions** and **implement solution**
Reflection and modification	Understand impact of previous decisions	Implement solution and determine impacts on the system	**Play out** the implementation and **adjust** as needed

26.2.2.3 Simulations and Real-Life Case Studies

The simulations integrate project management, systems engineering, science, technical, and administrative issues into a life-like experience that is able to provide a rich context for learning. The proposed simulation to design and develop includes the following elements:

1. Create project management-related exhibits defining a complete small- to mid-sized commercial or science missions.
   (a) Needs, goals, and objectives
   (b) Potential stakeholders and associated expectations
   (c) Proposed concept of operations and associated mission architecture
   (d) Cost, schedule, and performance requirements and constraints
   (e) Realistic systems and engineering information
2. Assign 3 or 4 balanced competing teams that are challenged to create credible project plans that provide detailed requirements to conceive, design, develop, and operate the simulated mission design and development.
3. Once teams have created a complete and credible project plan, the coaches will select a certain phase of the project and introduce typical changes or issues into the mix, allowing teams to demonstrate their ability to modify the plan and adjust to new realities. Potential issues include but are not limited to reduced budget, descope, major assembly test failure, and launch delay.
4. Experienced space-related project and program managers will coach the activity and assess the performance of each team.

5. The simulation activity will span the entire course and culminate in final briefings by each team that will include major project management exhibits with focus on cost, budget, technical performance, logic networks, schedule, risk, team performance, leadership communications, and interpersonal skills. Coaches will assess each team and provide detailed feedback to each.

Case studies will also be used to engage participants in real-life missions and systems. The case studies provided by the TO span the topics listed in Sect. 26.2.3.1. Simulations and case studies provide powerful learning opportunities for participants. The simulations integrate project management, systems engineering, commercial, and science, technical, and administrative issues into a life-like experience that is able to provide the rich context needed for learning. Case studies are also very useful in engaging participants in real-life missions and systems. They will focus on collections of lessons learned from past failures that challenge trainee ability to synthesize preventive measures for the future.

26.2.2.4 Special Features for Advanced PM Methodologies

The advanced methodologies anticipated for the training will make extensive use of ICT developments currently reaching the market, along with existing Client tools and facilities, which are available on a noninterference basis. The list below also shows PM aspects most likely to benefit from the advanced features:

1. Concurrent management techniques: cost/schedule estimation and diagnostics
2. Crowdsourcing survey: R&D for commercial applications
3. Cam Studio over Skype: R&D for technical applications
4. Video collaboration: PM dispute resolution HR search
5. Simulation via "cloud" storage: team proposals/invitations to tender
6. Remote virtual (robotic) exchange: PM skill and dexterity
7. New software tools and security systems:
   (a) Security dwell and information backhaul vulnerability
   (b) Document sharing, versioning, and e-storage
   (c) Synchronized conferencing and work sessions (simultaneous management)
   (d) Web-collaborative publishing
   (e) Integrated Terrestrial Space PM information technology

It should be noted that although advanced IT tools have been around for many years, the market has experienced an 80% growth in the last 2 years. Therefore, continued contact with advanced systems for upgrading the Client project management training methodologies will have high priority.

26.2.3 Task Description of the Four Phases

The following chapter depicts the tasks that will be performed in the execution of the program. Reference is made to the training objectives presented in Sect. 26.2.1

and to the proposed approach and Design and Development logic given in Sect. 26.2.2. This paragraph covers:

- Phase 1: Course Design
- Phase 2: Course Development
- Phase 3: Course Implementation
- Phase 4: Course Evaluations

26.2.3.1 Phase 1: Training Course Program Design

Immediately after program kickoff, TO and the Client will collaborate in Phase 1, the Design of the Project Management Training Course (PMTC). Table 26.3 identifies key steps of the design process.

The keys to the success of the Project Management Training Course are the TAC-EVAL philosophy and procedure detailed in previous sections. During the design phase of the contract, the TO/Client team will prepare a first draft of the TAC-EVAL procedure to include the following:

1. Identification of survey sheets for course reaction
2. Mandatory pre- and post-testing configuration document
3. Interview sheet design for feedback of OJT performance change
4. Joint Client/TO iteration of the TAC-EVAL worksheet

Table 26.3 Design of the project manager training course

Phase 1: Activity design of the ESA PM training course
1. Review and validate overall course objectives, requirements, and required topics from AD-1
2. Working with ESA HR and PM leadership, establish selection criteria for the course configuration of training and architecture
3. Working with ESA HR and PM leadership, develop alternative configurations of training and course architectures. Possibilities include Theme-Based Structure with supporting Simulation and Simulation-Based Structure with Supporting Lectures, both with Case Studies
4. For at least two of the alternative configurations of training and course architectures, develop course modules supporting training topics, configurations of training, and course architectures
5. Using criteria from Step 2, working with ESA HR and PM leadership, select the best configuration of training and course architecture for the ESA PM Training Course
6. Using the selected required PM topics from AD 1, configuration of training, and course architecture from Step 5, identify course modules and associated learning objectives, and populate modules with appropriate training topics
7. Using the configuration of training, supporting architecture, and course modules and associated learning objectives, prepare a first draft of the TAC-EVAL procedure
8. Integrate all aspects of the ESA Project Manager Training Course to prepare for Milestone R1 to include required training topics; proposed training architectures and configuration of training, course modules and associated learning objectives (tied to required training topics), and overall course flow
9. Participate in Milestone R1 and obtain Design approval from ESA HR
10. Document results in the ESA PM Training Course Design

Table 26.4 Possible baseline for the project management training course

Item	Comments
Duration	16 days, separated into two 8-day sessions, each conducted T-F, sat/sun off, M-Th., sessions nominally separated by 7–8 weeks (depending on holiday schedule)
Number of participants (assumed)	12–15 participants, three competing teams
Location	Noordwijk area, off ESTEC campus
Residence options	In-residence (lodging and meals required) or daily commute (snacks, lunch, and dinner required)
Daily in-session	7 h per day of in-class activities, split 50/50 between lectures and hands-on or simulation activities. Short hourly breaks. Strict enforcement of no cell phones/texting
Presession and intersession activities	Pre-assessments (to support Kirkpatrick Level 2), pre-reading, intersession project
Instruction format	Balanced between on-site instructor-led, on-site ESA speakers, on-site guest speakers, and web-enabled lectures from ASI, DLR, JAXA, NASA, and other organizations
Communications with participants	All official communications with participants, including pre-reading, course materials, assessments and evaluations, homework, and projects, will be through proven web-enabled methods. Use of emails and phones allowed, but kept to a minimum
Computers and software required	Fully capable PC (preferred) or MAC with Microsoft Office (or compatible), MS Project loaded and operational (2007 or later), Task (2008 or later) for the MAC, @Risk (2012)
Books provided	Project Management Handbook, 6th Edition, Bernd Madauss, 2000 Applied Project Management for Space Systems, Chesley, Larson, Menrad, McGraw Hill, 2009 Shoot the Moon, a Project Manager Case Study, Dan Andrews (NASA/Ames), McGraw Hill, 2012
Work sharing	Participant teams need access to Google groups or similar capability for work/product sharing

At this point in time the TO team has demonstrated that they know the process for designing the Client Project Management Training Course to meet the requirements for the course design. The TO team has also provided examples of key exhibits that may be developed in the process. This design process culminates in a brief overview of the Project Management Training Course; see Table 26.4 that can be taken to Phase 2, Training Course Development.

26.2.3.2 Phase 2: Training Course Development

Contingent upon satisfactory completion of the R1: Design Review, Phase 2 tasks will begin. The work is directed toward preparation of a representative module dry run. For Design and Development of new courses, a top-down approach will be taken, applying systematic education principles; see "Training and Development Models based on Bloom's Taxonomy" and alsoTable 26.2.

Table 26.5 Phase 2: Course development process

Phase 2: Course development
1. Review and validate course requirements, configuration of training, and course architecture
2. Create learning objectives for each course module
3. Validate module objectives with ESA HR and PM leadership
4. Compare course modules and associated required topics with existing materials from ISU and its partners to identify existing course modules that could be used, as is, or modified to suit
5. Identify required topics that need modification or totally new development
6. Identify potential developers/providers from ISU, partners, ESA, or other contributors of new materials
7. Compile the list of required topics, status of development, and provider credentials to ESA HR
8. Complete development of all course materials
9. Create a complete agenda for the ESA PM Training Program and associated presentation materials, tools, exams, instructors, and speakers
10. Provide the exhibits from Step 9 to ESA HR and PM leadership
11. Prepare for and conduct a dry run of a selected module with ESA HR and PM leadership
12. Conduct the R3 review
13. Provide final agenda, presentation materials, tools, exams, proposed instructor, and speaker names to ESA HR and PM leadership

Comments are provided below to support the course development process described in Table 26.5.

- Step 1: Review and validate course requirements, configuration of training, and course architecture. The compliance matrix captures course requirements that will be validated.
- Step 2: Create learning objectives for each course module. The learning objectives for each module establish what the trainee should be able to do after completion of the module. The module learning objectives are the basis for the content of the course module and the basis for the TAC-EVAL.
- Step 3: Validate module objectives with Client HR and PM leadership. In this step, the configuration of training, course architecture, training topics, course modules, and associated module learning objectives will be reviewed, modified as necessary, and documented for future work.
- Step 4: Compare course modules and associated required topics with existing materials from TO and its partners to identify existing course modules that could be used, as is, or modified to suit. The primary purpose of Steps 4–6 is to save Client funds. Instead of creating a totally new set of *Space-Related* course content, examples, lessons-learned, case studies, and simulations, TO is able to select from a vault of up-to-date and timely professional course materials that have been generated to include:
    1. Executive MBA from ISU with a focus on project management
    2. Books and courses, titled *Applied Project Management for Space Systems and Project Management*

3. Various certificate programs for space project manager development with industry, ESA, and NASA

- Step 5: Identify required topics that need modification or totally new development. Because of the extensive experience of TO in creating and teaching space-related project management and its unsurpassed vault of course content, examples, lessons learned, case studies, and simulations, it is possible to identify and use materials that require little or no modification and, if necessary, create totally new materials to suit the necessary required topics.
- Step 6: Identify potential developers/providers from TO, partners, Client, or other contributors of new materials. TO is able to select from a wide range of international faculty, extended faculty, and partners. This provides the Client with unparalleled flexibility in developing the Project Management Training Course.
- Step 7: Compile the list of required topics, status of development, and provider credentials to the Client HR. This step culminates with an approval from the Client HR to proceed to develop all course materials.
- Step 8: Complete development of all course materials. Steps 1–7 are focused on determining the overall scope of the required development effort, and in this step the TO and its extended international faculty and partners will accomplish the following:
  (a) Create an overall time-phased agenda supporting the allotted time-periods.
  (b) Establish the format and style of the presentation slides to include, if possible, both the Client and TO logos.
  (c) Provide an approved slide template to all developers.
  (d) Work with the development and presentation team to create final drafts of all presentation materials.
  (e) Create the flow of the threaded simulation to include presentation materials, background information, key project management and systems engineering exhibits, and trainee guided assignments.
  (f) Conduct an end-to-end review of all materials with a focus on technical integration, logistics required, and quality of workmanship.
  (g) Develop a complete and final Client PM Training handout (possibly in printed and electronic form).
- At this point in the process, the Client and TO have created a configuration of training, course architecture, addressed all required training topics, created learning objectives for all course modules, created all lecture materials and associated items, and delivered a complete and final draft of the course materials.
- Step 9: Create a complete agenda for the Client PM Training Program and associated presentation materials, tools, exams, instructors, and speakers. In this step, final checks are made, and any nagging errors or inconsistencies are corrected and all materials finalized.
- Step 10: Provide the exhibits from Step 9 to Client HR and PM leadership. This will allow the Client stakeholders an opportunity to review materials prior to the module dry run.

- Step 11: Prepare for and conduct a dry run of a selected module with the Client HR and PM leadership. In this step, for a selected module, the following items will be provided: test sheets and scoring rules, HR observer requirements for each day, and configuration validation guideline for newly developed course material.
- Step 12: Conduct the R3 review. The review includes a summary of the R3 deliverables, an agenda, and procedures for conducting the review, as well as documented minutes of the R3 review to include RIDs and/or potential changes.
- Step 13: Provide final agenda, presentation materials, tools, exams, proposed instructor, and speaker names to the Client HR and PM leadership. At this point all presentation materials, tools, exams, exercises, case studies, simulation materials, and TAC-EVAL materials for Kirkpatrick Levels I, II, and III have been developed, reviewed, and approved for implementation. The R3 review has been completed successfully, and the team is ready to implement the Project Manager Training Course.

A "go-ahead" to perform the dry run requires a successful R2: Development Review. The development of the Project Manager Training Course is iterative and involves the Clients HR, the TO's Project Management leadership, and the TO project team. Table 26.5 provides the anticipated overall process for Phase 2 along with comments below the table.

26.2.3.3 Phase 3: Course Implementation

The proposed TO team has a wealth of highly successful experience in implementing professional training courses. During Phase 2 all aspects of the Client PM Training Course are developed, presented, evaluated, and improved to create a BASELINE-Client-PM Training Course. Table 26.6 provides an overview of the TO successful and efficient course implementation approach that will be applied to all instantiations of the Course.

Preparation for the Follow-On-Production (FOP) of courses to be carried out in yearly intervals will proceed after satisfactory completion of the R3: Pre-implementation Review. This Phase 3 activity will exercise the following process in close cooperation with the Client project office counterparts and submit a Client PM Training handout (possibly in printed and electronic form).

During the implementation phase there are several activities that inevitably cause issues for the sponsors and professional training organization, cause frustration and aggravation for the trainees, and cost the sponsoring organization unnecessary time, energy, and money. These activities include:

(a) Overall logistics associated with the setup of the venue and execution of the program
(b) Distribution of pre-course materials, textbooks and associated course materials, and pre-course knowledge surveys (and associated processing)
(c) Communications among sponsors, trainees, and instructors

Table 26.6 Process for training course implementation

Phase 3: Training course implementation
1. Review and validate all ESA PM Training Course requirements, course materials, course agenda, instructors, venue, room arrangements, and capabilities as documented at the end of Phase 2
2. Work with ESA representatives to determine desired schedule and number of participants for FOP 1
3. Work with ESA representatives to confirm ESA speakers' and ISU speakers' availability
4. Prepare all materials for course announcement and place all necessary materials on web-enabled support platform. The course announcement will include all logistics information, detailed agenda, instructors' names, ESA, and guest speaker names and backgrounds, as well as an overview of the entire ESA Project Manager Training Course
5. Complete venue arrangements to include contract, room arrangements, audiovisual support, and arrangements for food and lodging (if required)
6. Conduct pre-course telephone briefing with all participants, instructors, and ESA speakers to coordinate activities and set expectations
7. Enable learning management system so participants can complete online registration, pre-course and knowledge survey (Kirkpatrick Level II), and obtain pre-reading materials and electronic textbooks (where available)
8. Support participant arrival and registration and kickoff the ESA PM Training Course
9. Conduct ESA PM Training Course per the configuration of training and course agenda described in Subset. 26.2.3.3
10. Participants complete post-course knowledge survey and course evaluations (Kirkpatrick Levels I and II), ISU compile results of both and provide to ESA HR. ESA and ISU work together to conduct a full TAC-EVAL of FOP 1
11. Working with ESA HR and PM leadership, prepare for and conduct Review R4—Evaluation Review at completion of FOP 1
12. Repeat Steps 1–11 for each FOP of the ESA Project Manager Training Course

(d) Post-course evaluations and post-course knowledge surveys (TAC-EVAL Levels I and II)

The TO, including Extended Faculty, and partners have conducted literally hundreds of professional training courses over the last 5 years and propose to work with ESA to mitigate many of the nuisance items associated with items (a)–(d) above.

The TO has created exhaustive and fairly sophisticated checklists that can be applied to the Client Project Management Training Course implementation, which should minimize (not necessarily eliminate) logistics-related problems.

Pre- and post-course TAC-EVAL (Kirkpatrick Level II) surveys can be loaded by the TO, surveys completed by trainees, and results processed by using the web-enabled learning management system that has been in use for over 4 years in other professional training applications [3]. This capability should save the Client sponsors time, energy, and money. Post-course evaluations (Level 1) can be implemented in the same way. The Client Project Management Training Course communications among sponsors, trainees, and instructors can be enhanced by use of the learning management system as well. The TO and its professional staff has decades of experience in successfully implementing professional training and educational programs

Table 26.7 Process for training course evaluation

Phase 4: Training course Evaluation	Comments
1. Develop an overall strategy for course evaluation for Kirkpatrick Levels I–IV	Kirkpatrick Levels I and II can be implemented by ISU after working with ESA HR to define the Level I course evaluation and Level II pre- and post-course knowledge surveys. Levels III and IV will require very close integration between ESA and ISU to be successful
2. Gather and analyze Kirkpatrick Level I trainee course assessment	At the end of each year, trainee satisfaction will be measured by using the Level I course evaluation form to assess participant's reaction to the course. ISU/ESA will collect and track each trainee's reaction to key factors influencing how well the program is received and what can be done for improvement. Level I trends will be tracked 3 years
3. Conduct Kirkpatrick Level II pre- and post-course knowledge surveys and analyze the results	Trainee learning will be measured by comparing results of knowledge surveys before and after the training and by quality of participation in class. Knowledge enhancement and quality of trainee participation will be used to assess Level II progress. Results for individuals will be available to the individual trainees, and the cohort aggregate data will be provided to ESA
4. Gather information and assess the trainee's progress in Kirkpatrick Level III performance	Together with ESA HR, measurements of the Course impact on trainee career progress will be summarized. Based on interviews with supervisors and HR personnel, changes in trainees' on-the-job (OJT) performance will be project-correlated with goals, schedule, and cost achievement in managing their respective programs at ESA
5. Gather information and assess progress in Kirkpatrick Level IV performance	Kirkpatrick Level IV performance enhancement can only be assessed by ESA, working with ISU, to determine the value of the ESA Project Manager Training Course in improving the performance of ESA. Weighted evaluation of Level IV results in task Step IV will be performed. Yearly evaluations will further account for other activities or changes at ESA having a bearing on overall performance of the Agency. Results of Levels I–IV are summarized in TAC-EVAL worksheets
6. Prepare for and present results in Review R4	Report of findings will be summarized in the R IV—FOP reviews and will be presented and at the end of each yearly session. RIDs and/or proposals from the TAC-EVAL findings will be described in TAC-EVAL results

globally. The TO's efficient approach to implementation can help minimize many of the nagging issues that occur in the conduct of the Client Project Management Training Course.

26.2.3.4 Phase 4: Training Course Evaluation

The TAC-EVAL approach described in Sect. 26.2.1.6 provides a viable approach to evaluation by applying Kirkpatrick Levels I–IV to obtain essential feedback and metrics associated with the Client Project Management Training Course [4]. Table 26.7 details the process for Phase 4, Training Course Evaluation. A yearly top-level evaluation of trainee assessments and course effectiveness will be made on each successive FOP of the course. The worksheet serves as a collection point for analysis of the course impact on trainee career development and for synthesis of an overview to continuously readapt the program. Project Management Training at the Client can, in this way, lead rather than lag behind the industry. Adhering closely to the evaluation process defined in Sect. 26.2.1.6, the tasks required to achieve the desired results are described chronologically below.

Course evaluations that terminate the first 3 years of follow-on training will have a far reaching significance for the Agency and its role in shaping Europe's industrial policy. TAC-EVAL is consequently a springboard for keeping project manager training responsive to the evolving needs of the Client. It is with this viewpoint in mind that the TO has proposed a training configuration that offers a bottom-up quick-start heritage coupled to a top-down growth potential, deeply rooted in traditional proven methodology.

The TO design, development, implementation, and evaluation processes are fully integrated and set up to allow cost-effective, efficient, and accurate design and development of the Client Project Management Training Course. All processes are focused on meeting the Client's course and topic requirements. Working closely with the Client HR and PM leadership, TO has a solid approach to creating the baseline Project Management Training Course, and by applying the TAC-EVAL process, it will be possible to improve and enhance the Course after each offering. TO proposes to apply highly efficient approach and web-enabled tools to enhance the overall implementation of the program and alleviate many of the logistical problems that often occur in such programs. A complete set of course objectives will be created for each course module to support the TAC-EVAL process.

26.3 Training Management and Logistics

26.3.1 Training Schedule and Deliveries

Based on the design and development logic of the course provided in Fig. 26.1, a Master Project Bar-chart (MPB) has been prepared; see Fig. 26.2. The MPB that shows the key milestones and major activities of all project phases will become the reference for the training project status control and reporting to the Client; see Sect. 26.3. The MPB is backed-up by detailed schedule of all project phases.

26.3.2 Training Heritage of the Training Organization

The proposed TO will assign highly qualified teachers in all of the project management disciplines by taking some of the EMBA hand-out material into account. True to the ISU tradition, the EMBA provides excellence in training of space and high-tech PM leaders in an uncompromising manner. The modular nature of the program allows high-value professionals to remain dedicated to their relevant tasks, while investing in their careers and future potential by joining a like-minded class. At the same time, some residential weeks during the training period allow the absorption of the relevant material and the creation of a balanced curriculum that seeks to

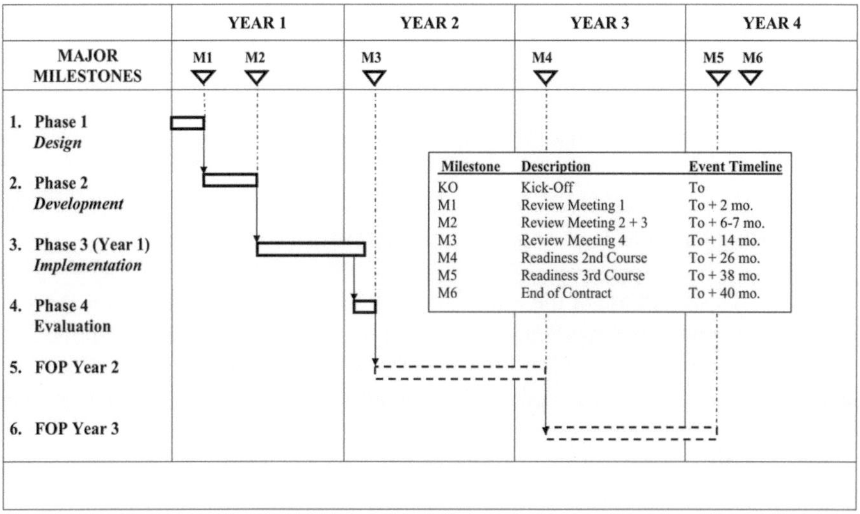

Fig. 26.2 Master project bar-chart. Note: the time line may be changed to periods instead of years

exceed expectations and provide all the necessary skills for doing business in the space and high-tech sector.

Standard reference will be the experience of the International Space University (ISU) who teaches Space Project Management at its postgraduate campus in Strasbourg, France. In particular, the degree programs Master of Space Management (MSM) and the added MBA education program to space executives (EMBA) represent an outstanding heritage of professional project management training at ISU; for more information see Annex A.

26.3.3 Proposed Training Locations and Facilities

TO and the Client representatives will identify the proper course location. The preferred location would be in a conference center at the Client or in a hotel that provides conference facilities. There are a number of venues available and the appropriate one can be identified in due time.

Alternatively, the facilities at ISU campus Strasbourg may be considered. This location has the advantage of using the facilities of ISU, which include:

1. Two rooms having a capacity for 30 people fully equipped with audiovisual equipment to aid presentations (white boards, projectors, and microphones)
2. Access to the ISU Concurrent Design Facility (CDF), which is an excellent presentation room for topics that require enhanced visualization elements. The ISU CDF features three large screens with independent input that can aid significantly in the presentation of complex material.
3. Access to two large amphitheatres

4. Access to the ISU library
5. The possibility of having lunch prepared or delivered in house
6. A Holiday Inn hotel is located just next to ISU, facilitating the housing of trainees and trainers.
7. Constant support from the ISU academic support stuff

26.3.4 Identification of Training Professionals

During the Design Phase of the PM course, the TO will identify professional training experts who have the relevant training experience of topics identified in Table 26.1 and in addition related management experience with space and/or high-tech projects. They must have proven knowledge of a large amount of experience in lecturing as well in the conduction of simulation tasks.

All generic training topics such as schedule and cost control but also for special topics like modern communication tools such as modern information methodologies will be provided. During Phase I professional trainers will be recruited which includes training sources currently available. In addition experts with teaching experiences will be invited for special lectures and/or simulation and fire side discussions coming from organizations like:

- ESA
- NASA
- National Space Agencies (i.e., CNES, DLR, and others)
- Satellite Operators (i.e., Eutelsat, SES, and others)
- Industry (e.g.: Airbus, Boeing, Lockheed Martin, Thales Alenia, and others)
- Management consulting companies
- Individuals with a specialized background

In addition, the TO proposes to invite senior project management specialists to provide special lessons learned information during actual class work and at fire side discussions. These guest speakers will be identified, in coordination with the Client, during Phase 1 of the project.

26.3.5 Training Management

The training management concept is tailored to the required needs of an efficient control of the development and execution of the Project Management Training Course.

The MPB provided in Fig. 26.2 will become the basis for schedule control and progress reporting. A key PM responsibility is to make sure that all deliverable training items specified in the related work packages are submitted on time to the planned destination.

TO confirms its obligation to the following:

1. Allocation of most suitable training resources to meet the training objectives as specified agreed with the Client
2. All applicable training tools to be used by the participants during the events
3. To define jointly with the Client the specific delivery dates for all planned events
4. Ensure that any relevant information is properly cascaded and distributed within the TO/Client team.
5. All documents and training material will be provided in the English language.
6. To the extent possible the TO will maintain the same trainers throughout the entire life cycle of the PM training courses until the year 3.

The TO training manager will perform monthly schedule control on the basis of the training schedule, which will also be used for status reporting to the Client; see Fig. 26.2. In case of schedule delays the cause for delay will be analyzed, and if possible work-around solutions will be initiated. Any delay that affects the training schedule will be reported to the Clint management.

Meeting and review coordination and progress reporting are important responsibilities of the training management. This includes:

1. Minutes: Preparation, signing, and distribution of minutes of meetings held in connection with contractually specified review events (milestones). The minutes will clearly state all agreements made and actions assigned at such meeting (action item list).
2. Regular Meetings: A series of regular review and ad hoc meetings will be held during the course of the training. Regular meetings are as shown in Table 26.8.
3. Monthly Progress Report: Preparation and submission of a monthly progress report (MPR) to the Client's representative. The monthly report shall contain the following:
   - A clear description of work accomplished by making reference to the relevant milestone
   - In case of slippages and/or problems, the TO will also provide appropriate corrective or work-around solutions with the aim of meeting the scheduled milestone dates of the project.
   - The MPR shall also include an action item register indicating following three status situations: actions done, completed, or open.
   - A forecast of the anticipated tasks during the next reporting period

Table 26.8 Proposed meeting schedule

Meeting	*Location*	*Date*
KO-meeting	ESTEC	To
Review meeting #1 at completion of Phase 1	ESTEC	To + 2 mo.
Review meeting #2 during Phase 2 prior to the dry run	ESTEC	To + 7 mo.
Review meeting #3 at completion of Phase 2 including evaluation and dry run discussion	ESTEC	To + 14 mo.
Review meeting #4 at completion of Phase 4	ESTEC	To + 26 mo.

Table 26.9 Deliverable documents

Doc ID	*Title*	*Review*	*I*	*A*	*D*
RP-1	ESA Project Managers Training: High Level Design including Baseline Topics, Schedule, Speakers	R1		X	
RP-2.1	ESA Project Managers Training: Training Material including Lesson Outlines, Handouts Grading, Simulations	R2	X		
RP-2.2	ESA Project Managers Training: Reduced pilot program proposal including Handouts, Agenda, instructor's CV	R2		X	
RP-3	ESA Project Managers Training: Candidate/participant assessment report including Worksheet sets Levels I, II, III	R4 + after each training	X		
RP-4	ESA Project Managers Training: Program Evaluation report including worksheet sets Level IV	R4 + after each training	X		
RP-4.1	ESA Project Managers Training: Report of Findings—Data Collection	R4	X		
RP-4.2	ESA Project Managers Training: Report of Findings—Data Analysis	R4	X		
RP-5	ESA Project Managers Training: Training Assessment Plan	R1		X	
RP-6	ESA Project Managers Training: ESA Participation Plan	R1		X	

Notes: I = for Information, A = for Acceptance, D = for Decisions

- Note: At the request by either party, some review meetings may be organized as tele-/videoconference.

4. Ad hoc Meetings: If necessary ad hoc meetings will be organized as tele-/videoconference.
5. Acceptance: Completion of tasks requires concurrence by the Client prior to continuation of the follow-on task as indicated in the proposed tasks.
6. Collaborative Project Team: TO proposes a CPT consisting of the Client and TO experts (the learning and development team) to establish during all four phases close interaction and communication between the parties encompassing the co-design, co-development, and co-delivery of the PM training course program.

The TO will provide the Client with electronic versions (searchable PDF and native format) of the documents as indicated in Table 26.9 and confirms that the Client has the extended right to use, reproduce, and modify all associated training material designed and developed under the agreed contract, including but not limited to session plans, trainer, and learner guides. The enclosed deliverable items list (DIL) includes all documents required as supplemented and/or amplified according to the TO proposed documentation of work. Delivery will be made for either information (I) or approval (A) as shown in Table 26.9.

ESA may be replaced by Client.

26.4 Information of International Space University

The ISU is a nonprofit private interdisciplinary university founded in 1987 in Cambridge, Massachusetts, USA that offers a Master of Science in Space Management, a Master of Science in Space Studies, and a Space Executive Master of Business Administration. These are in addition to the flagship Space Studies Program, which has convened annually during the Northern Hemisphere summer since 1988 at locations around the world. ISU was founded on the "3-Is" philosophy—to provide an Interdisciplinary, Intercultural, and International environment for educating and training space professionals and postgraduate students. There are now over 3300 ISU alumni from more than 100 countries. The French Government recognized ISU as an institution of higher education in 2004 (decree MENS0400386A of February 27, 2004). In line with its increasing international recognition, ISU has had permanent observer status with the United Nations Committee on the Peaceful Uses of Outer Space (COPUOS) of the United Nations Office for Outer Space Affairs since 1998. ISU was also granted full membership of the Space Agency Forum (SAF) in 1995.

The ISU Central Campus and global headquarters, established in 1994, are located in Illkirch-Graffenstaden, near Strasbourg, France. Since 2002, ISU has had its own building thanks to the support of the local authorities. The campus hosts the students in both Masters' programs. The ISU faculty members include astronauts, space engineers, space scientists, managers, and experts in space law and policy. The faculty comprises experts in technical and nontechnical space-related fields. Its members come from countries around the world.

The programs offered by ISU are dedicated to the future career development of graduate students and professionals from all nations seeking advancement in space-related fields. Tailored to the needs of postgraduates and professionals in the space sector or those who wish to work in this arena, ISU offers two kinds of programs:

(a) Programs delivered each year on a regular basis
    - A 12-month Master of Science in Space Studies (MSS)
    - A 12-month Master of Science in Space Management (MSS)
    - An 18-month Space Executive MBA (EMBA)
    - A 9-week Space Studies Program (SSP)
    - A 1-week Executive Space Course (ESC)
(b) Short programs delivered on demand
    - Professional Development Programs
    - Workshops
    - Short Courses
    - Forums

Participation in these programs is open to individuals and institutions of all nationalities. As an open academic forum, ISU welcomes open and free discussions.

Literature

1. Pelerin Charles J.: How NASA Builds Teams, John Wily & Sons, Inc., New Jersey, 2009
2. Bloom's Taxonomie von Lernzielen im kognitiven Bereich. ("Taxonomy of educational objectives", 1974). 5. Auflage. Beltz Verlag, Weinheim 1976, ISBN 3-407-18296-1.
3. Kirkpatrick Donald L. and Kirkpatrick James: Implementing the Four Levels, pg. 47, Berret-Koehler Publication, Inc., San Francisco, 2007
4. Kirkpatrick Donald L. and Kirkpatrick James: Evaluating Training Programs, 3rd Edition, Berret-Koehler Publication, Inc., San Francisco, 2005

Abbreviations

ACWP	Actual Cost Work Performed (*current cost including commitment*)
ADM	Architecture Development Method
AU	Astronomical Unit (*mean Sun-Earth distance*)
AFSC	US Air Force Systems Command
AI	Action Item
AI	Artificial Intelligence
AIT	Assembly, Integration and Test
AIV	Assembly, Integration and Verification
AIRBUS	European aerospace company
AKM	German management working group (*Arbeitskreis Management*)
APPEL	Project and engineering leadership
ARGE	Working group of companies; see subsection 5.3.3
ARIANE	European satellite launch system of Arianespace
ASTRA	Direct-to-home satellite television systems of SES, Luxembourg
ATC	After-training career
ATP	Authority to Proceed (*work release*)
AU	Astronomical Unit (*mean Sun-Earth distance*)
BAC	Budgeted Actual Costs
BAFO	Best And Final Offer
BCE	Baseline Cost Estimate
BCWP	Budgeted Cost Work Performed
BCWS	Budgeted Cost Work Scheduled
BER	Berlin capital airport
BMBW	German federal ministry of education and science (*now BMFT*)
BMFT	German federal ministry for research and technology
BMVg	German federal ministry of defense
BOM	Board of Management
BS	Bachelor of Science
BU	Business Unit
CAC	Cost at Completion
COFI	Configuration
CAST	China Association for Science and Technology

B. -J. Madauss, *Project Management*, https://doi.org/10.1007/978-3-662-69057-4

CALC	China Aircraft Leasing Group
CCB	Change Control Board
CCN	Change Notice
CDRL	Contract Data Requirement List
CDR	Critical Design Review
CER	Cost Estimation Relationship
CEO	Chief Executive Officer
CFE	Customer Furnished Equipment
CGWIC	China Great Wall Industry Cooperation
CIF	Cost Insurance Freight
CPO	Collaborative project office
CPT	Collaborative project team
CNES	Centre National d'Dtudes Spatiales (*French space agency*)
COS-B	ESA research satellite
CM	Configuration Management
COTS	Components Of-The-Shelf
CPFF	Cost Plus Fixed Fee
CPIF	Cost Plus Incentive Fee
CPM	Critical Path Method
CST	Center Spatial de Toulouse (*French test center*)
C/SCSC	Cost/Schedule Control System Criteria
CTC	Cost to Completion
CWBS	Contractor WBS
DASA	Deutsche Aerospace AG (*German Airbus Industries*)
DCAA	Defense Contract Audit Agency
D&D	Design and development
DGLR	Deutsche Gesellschaft für Luft und Raumfahrt (*German Aerospace Society*)
DFVLR	German Aerospace Agency e. V.
DIL	Deliverable Item List
DIN	German Industrial Standard
DLR	German Aerospace Agency (*deutsches Zentrum für Luft- und Raumfahrt, Cologne*)
DOC	Documentation
DOD	Department of Defense
DOE	Department of Energy
DRD	Data/Document Requirements Description
DRL	Data/Document Requirement List
DSARC	Defense System Acquisition Review Council
DTC	Design to Cost
ECA	Extracurricular activity
ECP	Engineering Change Proposal
ECSS	European Cooperation for Space Standardization of ESA (*www.ecss.nl*)
EDC	Effective Date of Contract

EDP	Electronic Data Processing
ELB	Elbphilharmonie Hamburg
ELDO	European Space Vehicle Launcher Development Organization (*now ESA*)
EM	Engineering Model or Engineering Manager
EMBA	Executive master of business administration
EPP	ESA participation plan
EQ	Equipment
ERNO	ERNO-Raumfahrttechnik GmbH (*now Airbus D&S*)
ESA	European Space Agency, Paris (*successor organization of ELDO and ESRO*)
ESOC	ESA Space Operations Centre, Darmstadt, Germany (*sub-organization of ESA*)
ESRIN	ESA Research and Information Centre, Rome (*sub-organization of ESA*)
ESTEC	European Space Technology Centre Noordwyk (*sub-organization of ESA*)
EUROPA	I/II/III, ELDO launch vehicles
EUTELSAT	European satellite operator, Paris
EVA	Earned Value Analysis
EVM	Earned Value Management
FACI	First Article Configuration Inspection
FAME	Forecasts and Appraisals for Management Evaluation
FFP	Firm-Fixed Price
FM	Flight Model
FMEA	Failure Modes and Effects Analysis
FMECA	Failure Modes Effects and Criticality Analysis
FOB	Free on Board
FOP	Follow on production
FP	Fixed Price
GE	General Electric
GEO	Geostationary Satellite Orbit (*altitude 36,000 km*)
GFW	Gesellschaft für Welraumforschung (German society for space research)
GTZ	Ass. of technical Cooperation (*Gesellschaft für Technische Zusammenarbeit*)
GPM	German PM-society (*Gesellschaft für Projektmanagement, Nuremberg*)
GSFC	Goddard Space Flight Center (*a sub-organization of NASA*)
HELIOS	US-German Solar Probe
HEOS	ESA research satellite
HIREL	High Reliability
HR	Human Resource
HW	Hardware
IA	Industrial Architect

I&E	Implementation and evaluation
IABG	Industrieanlagen-Betriebsgesellschaft, Ottobrunn near Munich
ICD	Interface Control Document
ICT	Information and communication technology
IGA	Intergovernmental Agreement
ILS	International Launch Service (*U.S.-Russian JV for Proton and Angara Launchers*)
INTELSAT	International Telecommunications Satellite Organization, Washington DC
IPC	Industrial policy committee
IPCE	Independent Parametric Cost Estimate
IPMA	International Project Management Association founded in 1995
IPMI	Institute for Project Management and Innovation
IPT	Integrated Project Team
IRIDIUM	Global network of Motorolas LEO satellites orbiting at low altitude
ISPA	International Society of Parametric Analysts
ISS	International Space Station (*see also Table 16.2*)
ISU	International Space University
ITAR	International Traffic in Arms Regulations of the USA
ITT	Invitation to Tender (*s.a. RFP*)
ITU	International Telecommunication Union of the UN headquartered in Geneva
JFM	Junkers Flugzeug- und Motorenwerke AG (*now Airbus DS*)
JV	Joint Venture
KPI	Key Performance Indicator
LCC	Life Cycle Cost
LCROSS	Lunar crater observation and sensing satellite
LEO	Satellites flying in low earth orbit (*approx. 300 to 1000 km*)
LKE	Lockheed-Kunichev-Energia-International
LLI	Long Lead Item (*components of long delivery times*)
LMSC	Lockheed Missiles & Space Corporation
LoI	Letter of Intent
LoL	Limit of Liability
LCC	Life Cycle Cost
JV	Joint venture
MBA	Master of Business Administration
MBB	Messerschmitt-Bölkow-Blohm GmbH (*part of Airbus D&S*)
MDT	Mean-Down-Time
MENA	Middle East and North Africa
MGR	Manager
MGMT	Management
MICS	Management Information and Control System (*NASA*)
MIS	Information System
MO	Matrix Organization
MOU	Memorandum of Understanding

MPB	Master project bar-chart
MPP	Master Project Plan
MPR	Monthly progress report
MRCA	Multi Role Combat Aircraft (*see also Tornado*)
MRD	Mission Requirement Definition
MRB	Material Review Board
MSM	Master of space management
MTBF	Mean-Time-Between-Failure
NASA	National Aeronautics and Space Administration
NAMMA	NATO-MRCA Development and Production Management Agency
NATO	North Atlantic Treaty Organization
NDA	Nondisclosure agreement
NEGO	Negotiation
OJT	On-The-Job-Training
PA	Product Assurance
PACCT	PERT & Cost Correlation Technique
PANAVIA	Trilateral management company for the Tornado project (*see also Tornado*)
PAR	Program Appraisal Review
PC	Personal Computer
PC	Project Control
PDR	Preliminary Design Review
PERT	Program Evaluation and Review Technique
PFM	Proto/Flight Model
PL	Project Leader (Projektleiter, Project Manager)
PM	Project Management (Manager)
PM	Prototype Model
PMI	Project Management Institute
PO	Project Operation
PMO	Project Management Office
PMP	Project Management Plan
PMTC	PM training course
PPP	Phased Project Planning
PPP	Public private partnership
PROJACS	Planning program of IBM
PRICE	Programmed Review of Information for Costing and Evaluation
PPS	Project Planning System (*developed by Dornier System*)
PSP	Projektstrukturplan (WBS)
Q&A	Question & Anywhere
QA	Quality Assurance
QM	Qualification Model
RAMS	Reliability, Availability, Maintainability and Safety
RFD	Request for Deviation
RFI	Request for Information
R&D	Research and Development
RFP	Request for Proposal (*see also ITT*)

RFW	Request for Waiver
RID	Review Item Discrepancy
RM	Risk Management
RMS	Resource Management System
ROI	Return on Investment
ROM	Rough-Order-Of-Magnitude
SCR	System Concept Review
SE	Systems Engineering
SES	Société Européenne des Satellites, Luxembourg (Operator of the ASTRA satellite fleet)
SETIS	Société Européenne d'Etude et d'Intégration des Systemes Spatiaux, Paris
SINET	Planning program developed by Siemens AG
SME	Small-Medium Enterprise
SOW	Statement of Work (*Work Statement*)
SPACELAB	Manned space laboratory of ESA
SPR	System Production Review
SR	System Requirement
SS	Subsystem
S/SE	Subsystem engineering
SSCAG	Space Systems Cost Analysis Group
SSR	System Specification Review
STM	Structure Model
STOP	Sequential Thematic Organization of Publication
SPM	System Project Management
SW	Software
SWOT	Strengths-Weaknesses-Opportunities-Threats (*Stärken-Schwächen-Chancen- Risiken*)
TAC-EVAL	Training assessment and course evaluation
TAP	Training assessment plan
TBD	To be defined/determined
TBM	Technical Business Management
TKC	Turn-Key-Contractor
TORNADO	Trinational combat aircraft of Germany, England and Italy (*see also MRCA*)
TRB	Test Review Board
TRL	Technology Readiness Level
TRANSALL	German-French military carrier
TPC	Technical Performance Control or Task Performance Control
TÜV	German association for technical inspection
TV-SAT/TDFl	Franco/German television satellite system for direct reception
UAE	United Arab Emirates
VDI	Association of German Engineers (*Verein Deutscher Ingenieure*)
VHF	Very high frequency
WBS	Work Breakdown Structure (Projektstrukturplan (PSP))
WP	Work Package (Arbeitspaket (AP))
WPD	WP Description (Arbeitspaketbeschreibung (APB))

MIX
Papier aus verantwortungsvollen Quellen
Paper from responsible sources
FSC® C105338

If you have any concerns about our products,
you can contact us on
ProductSafety@springernature.com

In case Publisher is established outside the EU,
the EU authorized representative is:
Springer Nature Customer Service Center GmbH
Europaplatz 3, 69115 Heidelberg, Germany

Printed by Libri Plureos GmbH
in Hamburg, Germany